Boron and Refractory Borides

Edited by V. I. Matkovich

With 271 Figures

Springer-Verlag Berlin Heidelberg New York 1977

Dr. Vlado I. Matkovich
The Carborundum Company
Niagara Falls, NY 14302/USA

Presently at:
Pall Corporation, 30 Sea Cliff Avenue,
Glen Cove, NY 11542/USA

For explanation of the cover motive see legend to Fig. 2, page 80.

ISBN 3-540-08181-X Springer-Verlag Berlin Heidelberg New York
ISBN 0-387-08181-X Springer-Verlag New York Heidelberg Berlin

Library of Congress Cataloging in Publication Data. Main entry under title. Boron and refractory borides. 1. Boron compounds. 2. Borides. 3. Refractory materials. I. Matkovich, V. I., 1924-. QD181.B1B 66.546 '671'2. 77-5056.

Offsetprinting: Beltz Offsetdruck, Hemsbach/Bergstr. Bookbinding: Brühlsche Universitätsdruckerei, Lahn-Gießen.
2132/3130-543210

Contents

D. Special Applications

List of Contributors

BECHER, H.J., Anorganisch-Chemisches Institut der Universität Münster, Gievenbecker Weg, 4400 Münster/Federal Republic of Germany

BEREZIN, A.A., Physico-Technical Institute im. Ioffe, Academy of Sciences, Polytechnichescaya ul. 26, Leningrad/USSR

BUSCHOW, K.H.J., Bell Laboratories, 600 Mountain Avenue, Murray Hill, N.J. 07974/USA

CASTAING, J., Office National d'Etudes et des Recherches Aerospaciales, Paris/France

COSTA, P., Office National d'Etudes et des Recherches Aerospaciales, Paris/France

CUEILLERON, J., Université de Lyon 1, 43, boulevard du 11 novembre 1918, 69621 Villeurbanne/France

DAHL, R.E., Pacific Northwest Laboratory, Richland, WA/USA

ECONOMY, J., IBM Corporation, 5600 Cottle Rd., St. Jose, CA 95193/USA

EPIK, A.P., Institute of Material Problems, Academy of Sciences, Kiev 142, Krzhizhanovski St. 3/USSR

ETOURNEAU, J., Université de Bordeaux I, Laboratoire de Chimie Minerale Structurale, 351, cours de la Liberation, 33405 Talence/France

FELDMAN, C., The Johns Hopkins University, Applied Physics Laboratories, 8621 Georgia Avenue, Silver Spring, MD 20910/USA

GAL'CHENKO, G.L., Lomonosov State University of Moscow, Chemical Faculty, Moscow/USSR

GEIST, D.†, Institut für Angewandte Physik der Techn. Hochschule Clausthal, Leibnizstraße, 3392 Clausthal-Zellerfeld/Federal Republic of Germany

GOLIKOVA, O.A., Physico-Technical Institute im. Ioffe, Academy of Sciences, Polytechnichescaya ul. 26, Leningrad/USSR

GURIN, V.N., Physico-Chemical Institute, Academy of Sciences, Leningrad/USSR

HAGENMULLER, P., Université de Bordeaux I, Laboratoire de Chimie Minerale Structurale, 351, cours de la Liberation, 33405 Talence/France

KAZANIN, M.M., Physico-Technical Institute im. Ioffe, Academy of Sciences, Polytechnichescaya ul. 26, Leningrad/USSR

KORSUKOVA, M.M., Physico-Chemical Institute, Academy of Sciences, Leningrad/USSR

KOVENSKAYA, B.A., Institute of Material Problems, 252142 Kiev 142, Krzhizhanovski St. 3/USSR

KRUKONIS, V., Arthur D. Little, Inc. , Acorn Park, Cambridge, MA 02140/USA

LAVUT, E.A., Lomonsov State University of Moscow, Chemical Faculty, Moscow/USSR

LAVUT, E.G., Lomonsov State University of Moscow, Chemical Faculty, Moscow/USSR

LIN, R., The Carborundum Company, Niagara Falls, NY 14302/USA

LUNDSTRÖM, T., Uppsala Universität, Kemiska Institutionen, Box 531 751 21, Uppsala 1/Sweden

MAHAGIN, D.E., Pacific Northwest Laboratory, Richland, WA/USA

MAKARENKO, G.N., Institute of Material Problems, Academy of Sciences, 252142 Kiev 142, Krzhizhanovski St. 3/USSR

MATKOVICH, V.I., The Carborundum Company, Niagara Falls, NY 14302/USA. Presently at: Pall Corporation, 30 Sea Cliff Avenue, Glen Cove, NY 11542/USA

MATTES, R., Institut für Angewandte Physik der Techn. Hochschule Clausthal, Leibnizstraße, 3392 Clausthal-Zellerfeld/Federal Republic of Germany

MERCURIO, J.-P., Universitê de Bordeaux I, Laboratoire de Chimie Minerale Structurale, 351, cours de la Liberation, 33405 Talence/France

MOORJANI, K., The Johns Hopkins University, Applied Physics Laboratories, 8621 Georgia Avenue, Silver Spring, MD 20910/USA

NASLAIN, R., Universitê de Bordeaux I, Laboratoire de Chimie Minerale Structurale, 351, cours de la Liberation, 33405 Talence/France

NOWOTNY, H., Institut für Physikalische Chemie der Universität Wien, Wien/Austria

ORLOV, V.M., Physico-Technical Institute im. Ioffe, Academy of Sciences, Polytechnichescaya ul. 26, Leningrad/USSR

PASTOR, H., Central Laboratories, Ugine-Carbone, Grenoble/France

PERKINS, P.G., Department of Pure and Applied Chemistry, University of Strathclyde, Glasgow, CL 1 XL/Scotland

PRUDENZIATI, M., Laboratorio di Electtronica, Via Vivaldi 70, Modena/ Italy

ROGL, P., Institut für Physikalische Chemie der Universität Wien, Wien/Austria

SAMSONOV, G.V.†, Institute of Material Problems, 252142 Kiev 142, Krzhizhanovski St. 3/USSR

SINELNIKOVA, V.S., Institute of Material Problems, Academy of Sciences, 252142 Kiev 142, Krzhizhanovski St. 3/USSR

SMITH, W.D., Airco Spear, Carbon Graphite division, 4861 Packard Rd., Niagara Falls, NY 14302/USA

SOLOVIEV, N.E., Voronezh, University pl. 1, Voronezh State University/ USSR

SPEAR, K.E., Associate Professor of Ceramic Science, Material Sciences Department, The Pennsylvania State University, University Park, PA 16802/USA

STASINEVICH, D.C., Ministry of Chemical Industries, Kirov St. 20, Moscow/USSR

THEVENOT, F., Ecole Nationale Supérieure de Mines, 158, cours Fauriel, 42023 Saint-Etienne Cédex/France

TKALENKO, E.N., Physico-Technical Institute im. Ioffe, Academy of Sciences, Polytechnichescaya ul. 26, Leningrad/USSR

UGAI, Ja.A., Voronezh, University pl. 1, Voronezh State University/ USSR

VIDAVSKY, L.M., Lomonsov State University of Moscow, Chemical Faculty, Moscow/USSR

WILKINS, M.L., Lawrence Livermore Laboratory, University of California, Livermore, California/USA

WOHRER, L.C., The Carborundum Company, Niagara Falls, NY 14302/USA

WOSILAIT, A.A., The Carborundum Company, Niagara Falls, NY 14302/USA

ZAITSEV, V.R., Physico-Technical Institute im. Ioffe, Academy of Sciences, Polytechnichescaya ul. 26, Leningrad/USSR

ZHIGACH, A.F., Ministry of Chemical Industries, Kirov St. 20, Moscow/ USSR

A. Introduction

V. I. Matkovich

During the meeting of the International Symposium on Boron held in October, 1972 in Tbilisi, U.S.S.R., the idea was proposed to assemble a review of boron and refractory borides by the specialists present. The advantages of such a work were immediately apparent. Such diverse applications of borides as in protective armor, nuclear reactors, coatings, reinforcement, etc. can hardly all be presented in sufficient detail by a single author. On the other hand it was also recognized that with so much specialization, some areas of interest may not be covered.

Within the last decade or two a number of areas have been developed in which the use of refractory borides is growing and improvements are being actively explored. Thus, a number of borides have considerable potential as reinforcing material for plastics or light metals, though only boron fibers have been firmly established up to the present. Application of flakes and films for two-dimensional reinforcement appears attractive, although the high cost of materials and development represents a considerable barrier.

A number of borides have been used to manufacture lightweight protective armor. In this area relatively fast changes seem to be taking place as improvements in performance and weight are made. Boron carbide has found considerable use in this application and new developments exploit the light weight of beryllium borides.

It should also be mentioned that refractory boron compounds have been used extensively in nuclear applications. Boron is one of the very few elements that can be used as neutron-absorber material in fission reactions. With the advent of fast breeder reactors and world energy shortages, the use of boron for these applications should increase substantially in the future.

It is hoped that further investigations of properties of various refractory borides may lead to new interesting applications. One of such areas, where a considerable amount of work is being done, is in electronic properties related to purification of boron, crystal growth, stability of various modifications, etc.

Difficulty in contamination control has traditionally been the biggest single obstacle in the understanding of boron chemistry, particularly of the higher borides. This problem is further complicated by the problems in boron analysis, many of which, concerning composition and stoichiometry of otherwise common borides always seem to recur. A few examples are mentioned below.

When boron carbide was first discovered at the turn of the century, Moissan assigned it the formula of B_6C. Somewhat later a homogeneity range was established with the composition of B_4C representing its high carbon-content limit. Furthermore, structure determinations all but proved that this composition was stoichiometric. The unit cell contains 15 atoms, 12 grouped in the form of an icosahedron and three forming a linear chain. As boron and carbon could not be distinguished (at that time) by X-ray methods, the assignment of 12 boron atoms to icosahedron and three carbon atoms to the three-atom chain was the only reasonable choice.

At the present time it is believed that $B_{13}C_2$ ist the only compound in this composition range. The central atom of the three-atom linear chain appears to be boron, and carbon atoms may substitute for boron in their icosahedral positions. Consequently, B_4C composition apparently represents a solid solution between $B_{13}C_2$ and carbon.

Another interesting example is the so-called "diamond-like boron", reported more than a century ago. Even as early as 1876 Hampe reported this compound to be $C_2Al_3B_{48}$. Later, this phase was reported as a mixture of AlB_{12} and B_4C and still later as just AlB_{12}. It appears now that the old composition of $C_2Al_3B_{48}$ was correct, but that the crystals consist of two inter-twinned orthorhombic phases. Similarly a composition reported as AlB_{12} was corrected to rhombohedral boron but is now believed to have been an aluminum boride derivable from beta boron structure.

Finally the existence of pure boron in simple tetragonal modification is being questioned and a series of often quoted compositions such as B_7O, $B_{13}O_2$, $B_{13}P_2$, $B_{13}As_2$, AlB_{10}, etc. apparently do not represent compounds.

The above problems are also the consequence of the fact that boron is a somewhat unique element, not resembling the elements that surround it in the periodic table, and many of its properties are still not completely understood. As the presence of impurities is uncertain, and analysis is difficult, the conclusions based on analogies and similarities are repeatedly being proven wrong as improved purities and methods of determination are developed.

Thus, under the circumstances, it is not surprising that theoretical considerations regarding electronic structure, nature of bonding, atomic coordination, and general nature of compound formation constantly run into inconsistencies. In a simple way, a B_{12} icosahedron may be understood as having twelve bonds pointed directly outward from its center, providing a pentagonal pyramidal coordination for each boron atom and requiring two electrons to close its electronic shell. The composition and structure of boron carbide of the B_4C formula fits this picture almost perfectly, and includes a tetrahedral coordination of the two carbon atoms. Yet a close examination indicates that no electrons are transferred from the three-atom chain to the icosahedra and presumably the electron-deficient composition of $B_{13}C_2$ actually has a higher stability. In more complex situations, such as in alpha rhombohedral boron, the electron deficiency may be resolved by assumption of three-electron bonds. However, no known scheme can satisfy the electron requirements of either tetragonal or beta-rhombohedral boron, or for that matter many borides discovered later.

It could be argued, with some justification, that the theories are not incorrect but that insufficient understanding of the complexities of the electronic and bonding nature of boron results in apparent discrepancies. It may also be hoped that treatment of polyhedral units as separate entities may reduce the present complexities to a simple electronic understanding involving boron polyhedra. After all, it is the tendency of boron to form polyatomic groups which stand out as a distinct boron characteristic in these complex structures.

Along these lines it is interesting to note that a classification of borides according to the periodic table does not result in the usual change in stoichiometry of compounds from one group to another. Instead, a classification in terms of boron groups and size of other atoms often appears to be more meaningful. Thus, a division into diborides, tetraborides, hexaborides, dodecaborides, etc. actually arises from classification into compounds based on the presence of single boron atoms, connected pairs, chains, multiple chains, planar networks and finally cage-type polyhedral grouping of many boron atoms.

In lower metal borides the structures can be described in terms of a continuous metal network where boron atoms are found in openings between the metal atoms. In higher borides there is, of course, a continuous three-dimensional network of boron atoms. However, while the network of metal atoms in predominantly metal structures can be described in terms of repeated arrangements of a few atoms, the basic boron network may require describing an arrangement of over 150 atoms before the nature of the systematic repetition itself can be described. The problem of complexity can, however, be simplified. Thus, for example, one may describe a group of 12 atoms as a polyhedron (such as an icosahedron) followed by a description of 13 icosahedra forming a unit of 156 atoms, and finally describing the repetition of these units as approximately consistent with a face-centered cubic arrangement. As entities, polyatomic boron groups may thus be formally compared to arrangements of large metal atoms while as polyatomic groups, one may be inclined to compare them with molecules. Clearly, they are neither. Comparisons of this type may be helpful but also indicate a need for better understanding of the nature of these entities.

The description of structures in terms of arrangements of polyatomic entities leads further to the concept of isotypic series and structural derivatives. Such derivatives are based on a fixed three-dimensional framework where variations arise from accommodation of additional atoms. The fact that such accommodations may involve a variety of atoms in different positions and number and that the basic framework may exist also in the absence of such atoms cannot be satisfactorily explained at present. This should not be surprising as the concept of derivatives is relatively new and most of the compounds are not clearly understood in terms of variation of their compositions, role of impurities, stabilities, etc. A better understanding may thus require a broader basis of more reliable data.

This review is divided into major chapters, i.e., theoretical considerations, preparation and properties, and special applications. Thus, some sections are concerned with elements of one or two specific groups of the periodic table or a larger group, such as transition metals. Other groups are, however, selected on the basis of similarity of compounds such as diborides or hexaborides which involve elements regardless of their position in the periodic table. In this way some emphasis is given to the nature of borides to form structural derivatives. Inevitably, however, some subjects are covered only lightly while others are covered in considerable detail. It is hoped that the latter subjects are generally of greater significance.

B. Considerations of Theoretical Interest

I. The Electronic Structure of Boron Compounds

G. V. Samsonov and B. A. Kovenskaya

The problem of elucidating the electronic structure of compounds ranks high in their theoretical development. Knowledge of the electronic structure of various compounds clears the way for further progress not only in theory but also in material manufacture. It is thus natural that simultaneously with the investigation of physico-chemical properties of borides, scientists should also investigate their electronic structure models.

Boron compounds are characterized by a variety of compositions and structures, arising from the specific electronic nature of the boron atom. The existence of an unpaired electron, possessing three-bonding functions, provides the possibility of forming a series of electronic configurations of differing stability.

Theoretical and model concepts of the electronic structure of boron compounds published in the literature are contradictory. The principal questions discussed are those of direction of electron transfer participation of boron electrons in the formation of electron bonds and metal-boron bonding.

The models suggested by authors vary from those in which the boron atoms are the donors of electrons to the metallic energy bands, to those in which the boron actively participates in the formation of its own tight bonds. Realization of the latter requires additional electron participation of a partner.

JURETSCHKE and STEINITZ (1958) postulate a metallic model of the MeB_2 band structure. They propose that an adequate number of boron electrons occupy the metal d-band, so that the boron atoms simply stabilize the hexagonal structure with no strongly pronounced boron-boron bonds. The three available boron electrons occupy the d-band, overlapping the s-band, in such a way that the Fermi level lies on its boundary. From this assumption it follows that upon the formation of a diboride, the boron electrons occupy higher energy levels. This would contradict the high stability of the diborides in at least Groups IV and V, and thus the model does not seem very convincing. The formation of similar boride phases by metals of different valencies, and insignificant variations of the interatomic B-B distances in diborides (as compared to the standard length of the covalent B-B bond) indicate that the stability of these compounds is conditioned to a great extent by the bonds between the boron atoms. KIESSLING (1950, 1959) and ROBINS (1958), the authors of the above theory, proceed from the assumption that the donor ability of the boron atoms plays an important role in electron distribution, and that boron electrons, donated to the system, stabilize the metallic lattice.

DEMPSEY (1963) also suggested a metallic model for the MeB_2 band structure for Group IV. Unlike JURETSCHKE and STEINITZ (1958), he believes that the boron atom contributes only one electron to the d-band of the metal. In accordance with this model the Fermi level is in the middle of the d-band. Because the analysis of the physical properties of Group IV diborides indicates that their carrier concentration is somewhat insignificant, this model would be reasonable only with the assumption of splitting or deep minimum of the d-band.

Information on the character of the metal d-band is highly contradictory (CALLAWAY, 1961; NEMNONOV, 1965). As has been shown on TiC and TiN compounds by BILTZ (1958) and ERN and SWITEDICK (1965), in cases of marked interaction in the Me-X state, density curve of compounds has a deep minimum.

PIPER (1966) assumes that the electronic structure of diborides is analogous to the electronic structure of the carbides and nitrides and that the Me-B bonds may be responsible for the analogous minimum and low-state density on the Fermi level.

CADEVILLE (1966) and DANIEL (1966) are among those who advocate the model in which the boron atoms contribute electrons to the d-bond of the metal. Assuming a primary role of the metal d-band in the electronic structure of borides, they interpret their experimental results on electrical and magnetic properties of sub- and monocarbides of transition metals at the end of the period (from chromium to nickel). They presume that the number of electrons in the band increases on the average by 1.7. - 1.8. electrons per atom of boron. Further, these authors believe that the energy d-band comprises a completely occupied subband and a partially occupied subband of the opposite spin overlapping a wide conduction band. The authors indicate that this scheme is applicable only to boride phases of the Me_2B and MeB type of transition metals at the end of periods and may not apply to "middle" transition metals (Ti, V). Furthermore, the electronic structure of borides must greatly depend on the Me/B ratio. Nevertheless, CADEVILLE (1966) applied this model to subborides, monoborides and diboride of chromium. Proceeding from the magnetic-moment value for CrB_2, he concluded that each boron atom donates 1.7 electrons to the 3d-band of the metal. This procedure involves the improbable assumption that electrons contributed by boron are uniformly distributed between the two opposite spin subbands and determine a certain concentration in the subband. The author points out that this hypothesis requires confirmation by additional experimental evidence (other than magnetic).

ZHURAKOVSKY et al. (1969) also interpret the results of X-ray spectra analysis for sub- and monoboride of iron as evidence of the partial transfer of valence electrons from boron atoms to the metal atoms. They extend this conclusion to the electronic structures of other types of borides.

In the work of TYAN et al. (1969) the results of the investigation of low-temperature specific-heat capacities are interpreted on the basis of borides retaining the initial band structure of pure metals. In this case each boron atom would contribute 1 - 1.5 electrons to the d-band of the transition metal. The inadequacy of such an approach was apparent to the authors who indicated the difference between the crystal structure of metals and their borides. The authors also noted the lack of theoretical computations of the electronic structure of transition metal borides.

All the above models are based on the assumption of transfer of boron atom valence electrons (all or a part of them) to the d-band of the metal.

In contrast, another group of workers suggest that boron atoms use their valence electrons to form their own tight bonds.

SILVER and BRAY (1960) and SILVER and KUSHIDA (1963) interpret the experimental results of nuclear magnetic resonance of titanium and zirconium diborides as a distinct indication of the strong probability of a graphite-like electronic structure of boron sublattice in diborides. Each boron atom is assigned four electrons in the plain net for B-B bonds, and diborides of Group IV metals have two free electrons per metal atom.

JOHNSON and DAANE (1963), assuming that B-B bonding is tight and that the metal electrons are responsible for the metallic bond, carried out a study aimed at obtaining the experimental data needed to evaluate two features of the electronic structure of borides. First, they intended to find out if the boron atom possesses an adequate number of electrons to ensure the B-B bonds. Then they planned to determine whether all the metal valence electrons retained after transfer of some to the B-B bond are localized. If not, they would determine the free carrier concentration using the Hall effect.

The selection of yttrium diboride as the object of investigation for the MeB_2 phase was natural. Since metals of different valencies can form isotypic boride phases, the assumption that metal atoms donate all their valence electrons to the boron was to be applied only to metals of lowest valencies; that is, those that are trivalent. The assumption that an yttrium atom donates three electrons for the boron-boron bonds should be verified since the existence of isomorphous metal borides of lower valency is open to question. Furthermore, because the temperature dependence of the electrical resistance of yttrium diboride shows a metallic type of conduction, some of the valence electrons of the metal pass into the conduction band and the number of electrons donated to the boron should be less than three.

Two models have been analyzed:
1. A metal atom donates two electrons to the boron atoms to form graphite-like B-B bonds. In this assumption, the magnitude of the HALL coefficient was calculated. Calculations agreed well with the experiment.

2. The boron atoms possess a sufficient number of valence electrons to ensure covalent boron-boron bonds. In this case two electrons are added to the conduction band. However, this alternative is rejected upon comparison with the experimental data of the Hall coefficient.

The authors concluded that in diborides, four electrons per boron atom are available for the B-B bonds. Each metal atom donates two electrons to the boron atoms. KREBS (1956) also supports this opinion.

The authors of a number of studies, such as NEMNONOV and KOLOBOVA (1966), FISCHER and BAUN (1968), NEMNONOV et al. (1969) and LYAKHOVSKAYA et al. (1970) interpret the results of the X-ray spectral analysis of diborides as an indication of the transfer of valence electrons from metal to boron.

Using measurements carried out by electron spectroscopy methods, RAMQUIST et al. (1969a b) concluded that in the transition metal diborides, the metal atoms give a part of their valence electrons to boron.

On examination of electrophysical properties, the same conclusions have been arrived at by ETOURNEAU et al. (1970) for hexaborides of the CaB_6 structure type.

MAR and STOUT (1972), in the course of investigations of the thermodynamic properties of erbium and uranium dodecaborides, also deduced that the metal atoms give two electrons to the boron sublattice.

The primary reason for the existence of a number of contradictory models derives from the fact that data on certain experiments used by researchers do not provide sufficient information from which to draw clear conclusions as to the electronic structure. Moreover, an attempt to extend a particular model of electronic structure to all types of borides should be regarded as misguided, because of the great variety of crystal structures and possible bond combinations. A real electronic structure should define a particular sublattice of boron and metal for each type of boride which could thus be considered separately.

A model close to reality might result from a generalization of available data on crystal-chemical characteristics, physico-chemical properties and semi-empirical regularities that can be deduced for various boride phases of the same metal and from boron compounds of the same type with different partners. In addition, it seems necessary to generalize the experimental and theoretical investigations and compare them in order to analyze the available experimental data on the physico-chemical properties of borides from the standpoint of the peculiarities of their chemical bond and energy spectra of electrons.

The lack of accurate solutions to Schrödinger's equation leads to the use of methods based on simplified concepts that result in approximated solutions. Quantum-mechanical computations of the electronic states of solids carried out in one-electron approximation have been widely used lately. GELD (1974) emphasizes the great importance of such theoretical investigations of the bond structure of refractory compounds in facilitating an understanding of their physical, physico-chemical and mechanical characteristics.

At present the basic methods used for calculation of the electronic character of compounds are joint plane waves (JPW), orthogonalized plane waves (OPW), tight bond, and molecular orbitals. Calculations carried out by the OPW method for TiC (NEKKEL and NOVOTNY, 1971) agreed well with the results obtained by the PPW method (ERN and SWITENDICK, 1965).

Recently, several studies on electronic energy spectra of metals and compounds were performed through the use of the tight bond method, LCAO (LAFON and LIN, 1966; TYLER and FRY, 1970; NORWOOD and FRY, 1970, and others). This method, though inferior to the PPW and OPW methods in accuracy and suitability of results, provides a possibility of obtaining information relatively easily. The LCAO method provides data on the primary features of the electronic structure of complex composites and the effect of the nature of the component atoms on the band structure of the crystal.

In a number of papers, (ALSTRUP and JOHANSON (1968) and LYE (1967)), it has been shown that the accuracy of the LCAO method is highly satisfactory if the effect of the three nearest neighbors is taken into account.

A variation of this method is the GO-LCAO method (group orbitals-linear combination of atomic orbitals) suggested by GORYACHEV et al. (1974b). This method is particularly convenient for computing the electronic structures of compounds of the homologic series. It should also be noted that this method results in electronic structure parameters which are most clearly comparable with the experimental results.

The number of papers devoted to the computation of electronic spectra of borides is still insignificant.

Quantum-mechanical considerations of the electronic structure of MeB_6 borides are given in papers by FISCHER and FLODMARK (1954), LOUGUET-HIGGINS and ROBERTS (1954), JAMAZAKI (1957) and FLODMARK (1959). In these calculations, carried out in approximation of tight bond, only s- and p-electron states of boron atoms have been considered. The calculations concluded that a transfer of two electrons from metal atoms to the B_6 complex take place. Calculations made by GARF et al. (1970) for CaB_6, through the use of the MO-LCAO method accounting for s- and p-states of boron and metal atoms, introduced another step in determining the electronic structure of hexaborides. The authors of these calculations also conclude that there is a transfer of electrons from the metal to the boron. Although these results permit analysis of some similarities of the properties of hexaborides, more reliable conclusions could be drawn if the d-states of metals are taken into account.

LIPSCOMB and BRITTON (1960), considering the electronic structure of the di-, hexa- and dodecaborides, presume that in all cases the transfer of at least two electrons occurs from the metal atoms to the boron complex.

The data on calculation of the energy spectrum of the diborides and tetraborides is presented in papers by GORYACHEV and KOVENSKAYA (1974). These calculations, carried out by the GO-LCAO method, confirmed the transfer of electrons from metal atoms to boron atoms. It should be noted, however, that in all instances the number of transferred electrons appeared to be less than two.

GORYACHEV et al. (1971) computed the energy spectrum of some dodecaborides (YB_{12}, YbB_{12}, LuB_{12}), carried out by the MO-LCAO method and concluded that the direction of transfer of electrons in the dodecaborides is uncertain.

From the foregoing it is clear that the electronic theory of boron compounds is far from complete. Comparing the data of theoretical studies with the experimental characteristics permits the formulation of models of electronic structure of various types of borides that are reasonably consistent with the available information.

Although a large number of compounds are formed by the interaction of various simple substances with boron, the available data give a sufficiently reliable picture of the electronic structure of the d- and f-transition metal borides only. These borides form isotypic series of similar compounds. They are, therefore, very convenient for studies of the basic regularities of changes in physical properties depending on the main quantum number and the number of electrons of the metal atoms (that is, for explaining the role of the metal atom in the boride lattice). The diborides are an example of this general type since their physical properties have been well studied.

Calculation of the electronic energy spectrum was carried out for the diborides of Ti, Zr, Hf, V, Nb, Ta, Cr and Mo (SAMSONOV et al., 1972; GORYACHEV et al., 1974b).

The calculations were performed by the GO-LCAO method for a fragment of the crystal lattice of diborides of the AlB_2 structure type, allowing for interactions of atoms of not only the nearest neighbors but also of those in the 4th and 5th coordination spheres. The s- and p-states of boron and s-, p- and d-states of metals were used as initial orbitals.

The basic elements of the electronic structure of compounds are the redistribution of electrons from level of isolated atoms into the bands

Table 1. Calculated parameters of electronic structure of diborides

Calculated: property:	AlB_2	TiB_2	ZrB_2	HfB_2
Charge of atoms	$Al^{+0.38}B_2^{-0.165}$	$Ti^{+1.9}B_2^{-0.95}$	$Zr^{+1.6}B_2^{-0.8}$	$Hf^{+1.5}B_2^{-0.75}$
Contribution into the conduction band, %	-	Ti - 82.6 B - 17.4	Zr - 81.5 B - 18.5	Hf - 81.0 B - 19.0
Occupation of d-states of metals	0.80	1.06	1.20	1.33
SWASC d^5, % [a]	16.0	21.2/43/	24.0/52/	26.6/55/
Occupation of sp-states of boron	3.165	3.95	3.80	3.75
SWASC sp^2+p^x, %	79.1	98.7	95.0	93.7
Higher level activity, eV	-11.3943	-10.5309	-10.1168	-8.9361
Dissociation energy into the plasma state, eV [b]	546.19	561.8	492.6	451.4
Dissociation energy into atomic state, eV	391.86	318.9	160.6	120.9
Dissociation energy in relation to the components	234.17	256.3	93.7	54.1
Energy bond of boron in a cell, eV	=	443.5	420.3	370.3

[a] In brackets are given the values of SWASC of pure metal
[b] Energetical parameters are related to the calculated fragment

of compounds, the participation of electrons of the partner's atoms in the conduction and the formation of the Me-Me, Me-X, and X-X bonds.

The evaluated parameters of the electronic structure of diborides are given in Table 1.

The calculations confirmed that the transfer of electrons from metal to boron occurs and that the occupancy of electronic states of boron always exceeds three electrons per atom ($p^2 + p^x$). However, in contrast to the conclusions drawn by JOHNSON and DAANE (1963) and LIPSCOMB and BRITTON (1960), the number of transferred electrons is close to two only for titanium diboride. The donor ability of metal atoms decreases with the increase in the number of d-electrons and their main quantum number. In this direction, the contribution of d-states of metal atoms to the common energy states is increased and the contribution of s- and p-states of boron atoms is decreased. These relationships agree to a large extent with the configuration model. It can be seen from the results listed in Table 1 that the SWASC of the d^5-electrons of metals in the diborides is always less than that of pure metal, but there is a marked tendency of these values to approach each other as they increase.

VB_2	NbB_2	TaB_2	CrB_2	MoB_2
$V^{+1.0}B_2^{-0.5}$	$Nb^{+0.9}B_2^{-0.45}$	$Ta^{+0.7}B_2^{-0.35}$	$Cr^{+0.7}B_2^{-0.35}$	$Mo^{+0.26}B_2^{-0.13}$
V - 81.0 B - 19.0	Nb - 74.0 B - 26.0	Ta - 71.5 B - 28.5	Cr - 75.3 B - 24.7	Mo - 64.4 B - 35
2.80	3.18	3.40	3.60	4.3
56.0/63/	63.6/76/	68.0/81/	72.0/73/	86.0/90/
3.50	3.45	3.35	3.35	3.13
87.5	86.2	83.7	83.7	78.2
-10.3029	-8.7680	-8.1353	-9.0262	-8.2082
561.8	529.0	489.4	487.8	-441.18
179.7	185.2	134.1	118.1	72.5
115.8	114.9	62.0	57.9	4.0
412.5	395.3	350.8	358.3	332.2

The transition of atoms from the free state into the diboride occasions a rearrangement of their configurations in accordance with the following:

$$B : s^2p \rightarrow sp^2 + p^x$$

$$Me : s^nd^m \rightarrow s^{n-y}p^ed^{m-z}$$

Computation of the electronic spectrum of aluminium diboride showed that its semi-conductive properties are formed due to the significant remoteness of the last occupied state from the upper free one ($\Delta E \approx I$ ev).

In transition metal diborides, an incomplete occupation of the upper energy level overlapping the free states determines the metallic type of conduction.

From the results of calculations it also follows that the conduction band in these compounds is formed by a group of electronic states of metal and boron atoms. The degree to which the electronic states of boron participate in the conduction band is insignificant.

These findings agree satisfactorily with the experimental data concerning metallic type of conduction, insignificant participation of electronic states of boron in the conduction band (LYAKHOVSKAYA et al., 1970), and transfer of electrons from the metal to the boron (SILVER and BRAY, 1960; SILVER and KUSHIDA, 1963; JOHNSON and DAANE, 1963).

VINNITSKY (1972) also draws the conclusion from the measurements of boride properties, that the metal atoms in diborides donate a part of their valence electrons to the boron atoms.

Based on calculated data for the electronic structure model of diborides, it is natural to suppose that the formation of diborides appears to be possible only when a certain minimum number of electrons are donated to the boron atoms by its partner in the compound (≈ 0.3). Starting from this concept, it may be explained (by insignificant donor ability) that under normal conditions it is not possible to prepare stable molybdenum and tungsten diborides (see Table 1). It is also natural to assume that for metals with more than five electrons in the isolated state, the possibility of the formation of diborides depends on the relationship of the competitive SWASC of d^5 and d^{10} states.

Metals with SWASC d^5 > SWASC d^{10} possess a donor ability and, therefore, the existence of their diborides (RuB_2, OsB_2, IrB_2) is natural. Consequently, it is also natural that the metal acceptors (Pt, Pd, Ni) have no diborides.

The boride phases of the MeB_4 type have been studied to a lesser degree than the diborides. Data on their physical properties are given in papers by PADERNO et al. (1974), SEVERYANINA et al. (1973) and SEVERYANINA (1974).

Detailed results of the calculation of the electronic structures of zirconium, neodymium, gadolinium, terbium and holmium tetraborides by the GO-LCAO method are given by SAMSONOV et al. (1974b) and ARABEY et al. (1974). Some of these parameters are listed in Table 2. Comparison of experimental and calculated data permit some correlations. Thus, the donor ability of atoms of all rare-earth metals (REM) for which the calculations were carried out, agrees with the results of the X-ray spectral analysis (SEVERYANINA, 1974), electromagnetic measurements (JOHNSON and DAANE, 1963) and confirms the conclusions of LIPSCOMB and BRITTON (1960).

The donor ability of metal atoms is determined by the tendency of REM atoms to form stable f^0, f^7, f^{14}-configurations and, therefore, the character of changing atomic charges in the series CeB_4 - NdB_4 - GdB_4 - TbB_4 - HoB_4 is expected.

The data shows the primarily tetravalent state of cerium and terbium atoms in their tetraborides. This is also in agreement with experimental results (FISK et al., 1972; BUSCHOW and GREYGHTON, 1972, SEVERYANINA, 1974).

The computation indicated a metallic character of the above tetraborides (incomplete occupation of the upper levels overlapping with the free states) which is also substantiated by experiment. As in the case of the diborides, the conduction bond consists, to a great extent, of electronic states of the metal.

Computation of the electronic spectrum of MeB_6 borides was made by use of s- and p-states of metal alone (GARF et al., 1970). This is, apparently, insufficient. Nevertheless, a comparison of experimental and calculated data enabled the authors to draw some conclusions with re-

Table 2. Calculated parameters of electronic structure of tetraborides

Calculated property	CeB_4	NdB_4	GdB_4	TbB_4	HoB_4
Charge of atoms	$Ce^{+2.125}B_4^{-2.125}$	$Nd^{+1.14}B_4^{-1.14}$	$Gd^{+1.14}B_4^{-1.14}$	$Tb^{+1.58}B_4^{-1.58}$	$Ho^{+1.07}B_4^{-1.07}$
Occupation of sp-states of boron atoms	3.53	3.29	3.28	3.50	3.26
Statistical weight of sp-states of boron atoms, %	88.2	32.2	82.0	87.5	81.5
Contribution of d-states of metal atoms	0.703	0.783	0.18	0.45	0.12
Dissociation energy into the atomic states, eV	378	362.8	369.57	379.68	291.98
Contribution of ionic bond component, %	5.2	8.0	10.7	---	10.8
Metal atoms configuration	$4f^{0}5d^{0.7}$ $6s^{0.5}6p^{0.6}$	$4f^{3}5d^{0.78}$ $6s^{0.36}6p^{0.71}$	$4f^{7}5d^{0.18}$ $6s^{0.52}6p^{1.16}$	$4f^{7}5d^{1.22}$ $6s^{0.52}6p^{1.22}$	$4f^{10}5d^{0.12}$ $6s^{0.74}6p^{1.07}$
Boron atoms configuration	$2s^{1.62}2p^{1.91}$	$2s^{1.58}2p^{1.70}$	$2s^{1.58}2p^{1.71}$	$2s^{1.06}2p^{2.25}$	$2s^{1.56}2p^{1.70}$

gard to the electronic structure of hexaborides. As in the case of di- and tetraborides, the calculations confirmed the transfer of electrons from metal to boron.

This work also indicates the occupation of the upper energy levels, mainly by metallic states.

The authors affirm the role of the electronic states of boron in the formation of hexaboride properties and of the significant role of f-d-transitions in REM atoms in their donor activity.

Of the REM dodecaborides only YbB_{12} and LuB_{12} were calculated (GORYACHEV et al., 1971). The findings revealed a slight transfer of electrons from boron to metal in YbB_{12} (0.58) and from metal to boron in LuB_{12} (-0.12). To obtain more definitive results on the direction of electron transfer in borides of the MeB_{12} type it is necessary to take into account a greater number of interacting atoms. At any rate, considerable localization of electronic states of boron atoms in the MeB_{12} lattice (compared with the MeB_6 lattice) indicates that during the formation of MeB_{12} type borides, a transfer of electrons from metal to boron occurs. This transfer becomes more intense as the difference between the ground state of the 4f- shell of the REM atom and stable f^7 or f^{14} configurations increases.

Calculations show that the conduction band in borides of the MeB_{12} type, as in MeB_2, MeB_4 and MeB_6, consists mainly of metallic states.

The results of theoretical and experimental studies of diborides and higher borides leads to the following generalizations.

It is characteristic of these compounds to have additional electronic states of boron atoms in all boride phases beginning with the diborides. The extent of localization of the electronic states of the boron atoms exceeds by 1.5 - 2 times that of metal atoms and increases in the direction from diborides to dodecaborides.

For borides of d-transition metals, the dependence of metal donor ability on the number of their d-electrons has been established by calculation. An analogous dependence on the number of f-electrons is defined for the REM borides.

The role of metallic states in borides decreases from the di- to the tetra-hexa- and dodecaborides and the localization of electronic states on the sublattice of the boron atom is particularly necessary for the formation of stable dodecaborides.

The investigation of borides by various methods, including X-ray, electron spectroscopy and nuclear magnetic resonance plus data on their electro-physical and electro-magnetic properties show that in borides such as MeB_2 and higher which are rich in boron, the transfer of electrons is directed towards the boron sublattice and the sp-state of boron plays the main role in its electronic structure.

Calculations of electronic structure of various lower boride phases have not so far been reported. Therefore, a conclusion on the participation of electronic states of metal and boron atoms in the formation of their properties may be based only on analysis of the whole complex of information on their crystal-chemical and various physical properties.

The lower borides are formed primarily from d-transition metals. The latter have high heats of atomization and, consequently, a strong tendency to retain their lattices. Therefore, variations in electronic structure caused by the introduction of small amounts of boron cannot result in a significant change in the lattice. In lower borides of the Me_xB type, the boron atoms are "dissolved" in the interstices of the crystal lattice of the metal. The electronic states of boron are delocalized and the boron-boron bonds are weak (their formation does not require participation of all valence electrons of boron). Therefore, the assumption that boron valence electrons participate in metal-boron bond is reasonable.

This leads to an understanding of the basic role of metallic states in the electronic structure of lower borides which is consistent with the data on their physical properties.

The varying character of properties which defines the strength of interatomic bonds, such as melting points, thermal expansion coefficients, and values of mean square shifts of atoms, indicates that metals with higher energetical stability (those with higher values of SWASC of stable d^5 and d^{10} configurations) form lower borides with stronger interatomic bonds (KOVENSKAYA and SEREBRYAKOVA, 1966, 1970). This observation points to a basic role of metallic states in the formation and properties of lower borides.

Measurements of the magnetic and electrical properties of sub- and monoborides of transition metals performed by CADEVILLE (1966) and DANIEL (1966) indicate that boron atoms donate 1.7 to 1.8 electrons (per boron atom) to the d-band of the metal. This may also be an indication of the basic role of d-states in the electronic structure of lower borides. The same inference follows from studies of X-ray spectra of sub- and monoborides (ZHURAKOVSKY et al., 1969); from investigations of low-temperature specific heat capacities (TYAN et al., 1969); from magnetic measurements for niobium, chromium, manganese and cobalt monoborides reported by LUNDQUIST and MYERS (1961) and LUNDQUIST et al. (1962) and for the sub-boride of cobalt reported by FRUCHART (1963). Based on these measurements LUNDQUIST et al. (1962) postulate the following scheme of distribution of energy bands for lower borides.

Interaction of metal atom orbitals leads to the formation of two bands: a close, localized, rather narrow 3-d band, in which electrons are localized on metal atoms, and a wide low-density 4s-conduction band. An additional band is provided by bonds between the boron atoms (type 2s2p). The distribution of electrons among the bands depends on the relationship of the bands to each other and to the Fermi level. Variation in the number of boron magnetons is linearly dependent on the total number of electrons. Assuming one occupied spin subband, the number of boron magnetons is associated with the number of vacancies in the second subband. In accordance with this, the number of 3d-electrons per metal atom was evaluated as: 8.1 in MnB; 8.9 in FeB; 9.7 in CoB; 10.0 in NiB; 8.2 in Fe_2B and 9.2 in Co_2B. In all instances (except nickel) the number of 3d-electrons obtained is greater than the number of valence electrons in free atoms. Hence, we are led to a direct indication of the transmission of a part of the boron valence electrons to the metal band. To explain metallic conduction in these compounds, it should be assumed that the Fermi level intersects the overlapping 3d-, 4s- and 2sp bands.

COOPER et al. (1964) interpret the results of measurements of the Mössbauer effect for the sub- and monoborides of iron in the same manner.

Results of an evaluation of electronic state densities made by KOVENSKAYA (1971) on the basis of a whole complex of experimental characteristics of the electro-physical properties of different boride phases are also interpreted by the author, as evidence of the transfer of electrons from boron to metal in lower boride phases.

In summary, it is convenient to divide the multiformity of the structural types of electronic states of borides into two groups. The first group is characterized by high metal content (from Me_3B up to MeB_2), and the second by tight boron sublattices (from MeB_2 up to MeB_{12}). Diborides, the most extensive class of borides, are the transmission link between both groups.

In borides of the first group, rich in metal (from Me_xB up to MeB_2), the transfer of electrons is directed toward the metal. Metallic states play the main role in the electronic structure of these compounds. In borides rich in boron (such as MeB_2 and higher) the transfer of electrons is directed toward the boron sublattice and the main role in the electronic structure belongs to the sp-states of boron. Therefore, we observe the formation of more stable borides of the second group by metals which are active electron donors and lower borides by metals capable of accepting electrons.

References

ALSTRUP, J., JOHANSON, K.: Energy bands of silicon obtained by empirical LCAO method for the third nearest neighbors. Phys. Stat. Sol. 28, 555-561 (1968)

ARABEY, B.G., GORYACHEV, Yu.M., KOVENSKAYA, B.A.: Ob electronnom strojenii tetraboridov RZM yttrijevoj grouppy. J.Exp. Teoret. Chimii 10, 1012-1016 (1974)

BILTZ, H.: Über Elektronenzustände von Hartstoffen mit Natrium-Chlorid-Struktur. Z. f. Phys. 153, 338-342 (1958)

BUSCHOW, K.H.J., GREYGHTON, J.H.N.: Magnetic properties of rare earth tetraborides, J. Chem. Phys. 57, 3910-3916 (1972)

CADEVILLE, M.G.: Magnetic properties of Mn and Cr diborides. J. Phys. Chem. Sol. 27, 667-676 (1966)

CALLAWAY, J.: d-Bands in cubic lattices. Phys. Rev. 121, 1351-1354 (1961)

COOPER, J.D., GIBB, T.C., GREENWOOD, N.N., PARISH, R.V.: Chemical applications of the Mössbauer effect. Trans. Faraday Soc. 60, 2097-2102 (1964)

COSTA, P., CONTE, R.: Properties of the carbides of the transition metals. Nucl. Metal. (AIME) 10, 3-8 (1964)

DANIEL, L.E.: Sur la structure electronique de certaines borures des métaux de transition, J. Phys. 27, 449-456 (1966)

DEMPSEY, E.: Bonding in the refractory hard metals. Phil. Mag. 8, 285-291 (1963)

ERN, V., SWITENDICK, A.C.: Electronic band structure of TiC, TiN and TiO. Phys. Rev. A 137, 1927-1931 (1965)

ETOURNEAU, J., MERCURIO, J.P., NASLAIN, R., HAGENMÜLLER, P.: Electronic structure of certain hexaborides of CaB_6 type. J. Sol. State Chem. 2, 332-338 (1970)

FISCHER, J., BAUN, A.: Band structure and X-ray spectra of pure metals and of titanium oxides, nitrides, carbides and borides. J. Appl. Phys. 39, 4757-4766 (1968)

FISCHER, J., BAUN, A., FLODMARK, S.: Covalent B-B bonds in certain metal boride crystals. J. Chem. Phys. 22, 11-16 (1954)

FISK, Z., COOPER, A.S., SCHMIDT, P.H., CASTELLANO, R.N.: Preparation and lattice parameters of the rare earth tetraborides. Mat. Res. Bull. 7, 285-291 (1972)

FLODMARK, S.: Electrons distributions and energy bands in the metal borides. Arkiv. for Fysik 14, 513-516 (1959)

FRUCHART, R.: Moments magnetiques des borures ferro-magnetiques de cobalt Co_3B, Co_2B et du borure de fer FeB, Compt. Rend. 256, 3304-3306 (1963)

GARF, E.E., GORYACHEV, Yu.M., PADERNO, Yu.B.: Redkozemelnyje metally i ich sojedinjeniya. pp. 101-108, Kiev: Izdatelstvo Naukova dumka

GELD, P.V.: Dopolnjenje k monographii TOT L. Karbidy i nitridy perechodnych metallov. pp. 264-291, Moskow: Izdatelstvo Mir 1974

GORYACHEV, Yu.M., KOVENSKAYA, B.A., ARABEY, B.G.: Electronnyj energeticheskij spectr tetraborida neodima. J. Exp. Teoret. Chimii 10, 862-868 (1974a)
GORYACHEV, Yu.M., KOVENSKAYA, B.A., SAMSONOV, G.V.: Raschjet electronnogo spectra sojedinjenij perechodnych metallov metodom GOLKAO. Izvestiya Vysschych uchebnych zavjedenij, Phisika 8, 35-41 (1974b)
GORYACHEV, Yu.M., ODINTSOV, V.V., PADERNO, Yu.B.: Structura energeticheskich polos v kubicheskikh dodecaboridach metallov. J. Prikladnoj Chimii 12, 344-349 (1971)
JAMAZAKI, M.: Calculation of energy bands in the metal borides MeB_6 by the group theory method, J. Phys. Soc. Japan 12, 1-6 (1957)
JOHNSON, R.W., DAANE, A.H.: Electron requirements of bands in metal diborides. J. Chem. Phys. 38, 425-432 (1963)
JURETSCHKE, H., STEINITZ, P.I.: Hall effect and electrical conductivity of transition metal diborides, J. Phys. Chem. Sol. 4, 118-124 (1958)
KIESSLING, R.: The borides of some transition elements. Acta Chem. Scand. 4, 209-216 (1950)
KIESSLING, R.: Interstitial structures, Powd. Metallurg. Bull. 3, 177-184 (1959)
KOVENSKAYA, B.A.: Issledovaniya nekotorykh physicheskikh svojstv i electronnogo strojeniya boridov perekhodnykh metallov IY-YI grupp periodicheskoj sistemy. Kiev: Avtoreferat Kandidatskoj Dissertacii (1971)
KOVENSKAYA, B.A., SEREBRYAKOVA, T.I.: Phisicheskije svoistva boridnykh phaz khroma. Izvestiya AN SSSR. Neorganicheskije materialy 12, 2134-2138 (1966)
KOVENSKAYA, B.A., SEREBRYAKOVA, T.I.: Phisicheskije svoistva boridnykh phaz niobiya. Poroshkovaya Metall. 4, 42-48 (1970)
KREBS, H.: The effect of the homopolar bond character on the structure of inorganic salts II, semi-conductors and alloylike phases. Acta Crystallogr. 9, 95-99 (1956)
LAFON, E.E., LIN, C.C.: Energy band structure of lithium by tight-binding method. Phys. Rev. 152, 579-581 (1966)
LIPSCOMB, W., BRITTON, D.: Valence structure of the higher borides, J. Chem. Phys. 33, 275-279 (1960)
LONGUET-HIGGINS, ROBERTS, M. de V.: The electronic structure of the borides MeB_6. Proc. Roy. Soc. A 224, 336-341 (1954)
LUNDQUIST, N., MYERS, H.: The saturation magnetization of the monoborides of Mn, Fe and Co. Arkiv for Fysik 20, 463-468 (1961)
LUNDQUIST, N., MYERS, H., WESTLIN, R.: Paramagnetic properties of the V, Cr, Mn, Fe, Co, Ni monoborides. Phil. Mag. 7, 1187-1196 (1962)
LYAKHOVSKAYA, I.I., ZIMKINA, T.M., PHOMICHEV, V.A.: K-spectry bora v diboridakh perekhodnykh metallov i v sojedinjeniyakh LaB_6, BaB_6 i AsB: Physica tverdogo tjela 12, 174-181 (1970)
LYE, R.G.: Band structure of titanium carbide. Colloq. Intern. Centre Nat. Rech. Sci. No. 157, 207-210 (1967)
MAR, R.W., STOUT, N.D.: High temperature enthalpies of binary dodecaboride. J. Chem. Phys. 57, 5342-5346 (1972)
NEKKEL, A., NOVOTNYJ, G.: Novyje tugoplavkije materialy, pp. 9-16, Moscow: Izdatjelstvo Mir 1971
NEMNONOV, S.A.: Electronnaya structura i nekotoryje svoistva perekhodnykh metallov i splavov I, II i III bolschich periodov. Physika metallov i metallovedeniye 19, 4-11 (1965)
NEMNONOV, S.A., KOLOBOVA, K.M.: Rentgenovskije spectry, electronnaya structura i svojstva metallicheskikh sojedinjenij titana. Physika metallpv i metallovjedjeniye 22, 680-686 (1966)
NEMNONOV, S.A., MENSCHIKOV, A.Z., KOLOBOVA, K.M., KURMAEV, E.Z., TRAPEZNIKOV, V.A.: A study of electronic structure and interatomic bonds in certain compounds and binary alloys by X-ray spectroscopy method, Trans. Met. Soc. AIME 245, 1191-1199 (1969)
NORWOOD, T.E., FRY, J.H.: Energy bands of VO. Phys. Rev. B. 2. 472-474 (1970)
PADERNO, Yu.B., DUDNIK, E.M., SEVERYANINA, E.M.: Trudy nauchnogo seminara po konfiguratsionnoj modeli kondensirovannogo sostoyaniya veshchestva pp. 45-51. Kiev: Izdatel: Institut problem materialovedeniya AN USSR 1974
PIPER, J.: Galvanomagnetic effects in ZrBr single crystals. J. Phys. Chem. Sol. 27, 1907-1910 (1966)
RAMQUIST, L., EKSTIG, B., KALLNE, E., NORELAND, E., MANNE, R.: X-ray study of inner level shifts and bond structure of titanium carbide and related compounds. J. Phys. Chem. Sol. 30, 1849-1860 (1969a)

RAMQUIST, L., HAMRIN, K., JOHANSSON, G., FAHLMAN, A., NORDLING, C.: Charge transfer in transition metal carbides and related compounds studied by electron spectroscopy for chemical analysis. J. Phys. Chem. Sol. 30, 1835-1847 (1969b)

ROBINS, D.A.: Bonding in carbides, silicides and borides. Powd. Metallurg. Bull. 2, 172-176 (1958)

SAMSONOV, G.V., GORYACHEV, Yu.M., KOVENSKAYA, B.A., TELNIKOV, E.Ya.: Electronnyj spectr i physicheskije svoistva diboridov titana, vanadiya i khroma. Izvestiya Vysshikh uchebnykh zavjedjenij. Physica N 6, 37-42, (1972)

SAMSONOV, G.V., GORYACHEV, Yu.M., KOVENSKAYA, B.A., ARABEY, B.G.: Raschjet energeticheskogo spectra tetraborida tzeriya. J. Exp. Teoret. Chimii 10, 716-720 (1974)

SEVERYANINA, E.N.: Poluchenije i issledovanije physicheskikh svoistv tetraboridov redkozemelnikh metallov. Kiev; Avtoreferat kandidatskoj dissertacii 1974

SEVERYANINA, E.N., DUDNIK, E.M., PADERNO, Yu.B.: Teplovoje rasschirenije tetraboridov nekotorykh redkozemelnych metallov. Poroschkovaya metall. 12, 72-76 (1973)

SILVER, A.H., BRAY, P.J.: NMR study of bonding in some solid boron compounds. J. Chem. Phys. 32, 288-292 (1960)

SILVER, A.H., KUSHIDA, N.: Nuclear magnetic resonance in the transition metals diborides. J. Chem. Phys. 38, 865-868 (1963)

TYAN, Y.S., TOTH, L.E., CHANG, Y.A.: A study of electrons transition theory in the hard-melting metals borides by the investigation of the low-temperature heat capacity. J. Phys. Chem. Sol. 30, 785-792 (1969)

TYLER, J.M., FRY, J.H.: Energy bands of hexagonal NiS. Phys. Rev. B. 1, 4604-4606 (1970)

VINNITSKY, I.M.: Galvanomagntitnyje svojstva monokarbidov i diboridov perekhodnykh metallov. Kiev: Avtoreferat kandidatskoj dissertacii 1972

ZHURAKOVSKY, E.A., KOTLYAR, V.I., SCHASCHKINA, T.B.: Ob electronnom strojenii boridov zheljeza. Doklady AN USSR N 7, 12-16 (1969)

II. The Nature of the Chemical Bond in Borides

G. V. SAMSONOV and B. A. KOVENSKAYA

The character of the chemical bond between the atoms of a compound depends on its electronic structure.

The peculiar electronic structure of boron atoms (the existence of an unpaired electron possessing three-bond functions) defines the realization of numerous combinations of all types of bonds during interaction with various elements. The nature of the chemical bond in borides has not so far been sufficiently studied. There are many hypotheses on the relative contribution of covalent, metallic and ionic components to the bond energy and the physical properties of borides.

Thus, JURETSKE and STEINITZ (1958) postulated a metallic model for the MeB_2 phase, presuming that metal-metal bonds (Me-Me) define the basic properties of diborides and that boron atoms simply stabilize the hexagonal structure without pronounced boron-boron (B-B) bonds. The idea of the basic role of Me-Me bonds in diboride lattices is advocated by ROBINS (1958) and KIESSLING (1959).

In a scheme of interatomic bonds in diborides, suggested by DEMPSEY (1963) the Me-Me bonds are also considered to be essential, however, a significant role is attributed also to the Me-B interactions.

PIPER (1966) considered the scheme of interatomic bonds (for carbides and nitrides) suggested in the papers by BILTZ (1958), ERN and SWITENDICK (1965) to be acceptable for diborides if the B-B bonds, which have considerable significance for borides, are added to the Me-Me and Me-X bonds.

The primary role of Me-Me bonds in the lattice of borides is emphasized in a model by CADVILLE and DANIEL (1966). They describe an interatomic bond scheme in sub- and monoborides as a strong covalent bond Me-X and a less essential B-B bond. TYAN et al., (1969), interpreting the results of investigation of low-temperature specific heat capacities, assigned a primary role to the Me-Me bonds and an insignificant role to other interactions in the boride lattices.

According to a model suggested by JOHNSON and DAANE (1963) for higher borides, beginning from MeB_2, the primary role in their structure belongs to tight covalent B-B bonds. The metal electrons form a metallic type bond.

Describing the scheme of interatomic bonds in borides, GOLDSCHMIDT (1971) starts by observing that the boron atom size defines the formation of direct B-B bonds in boride lattices, and Me-Me and Me-X bonds are not of such exceptional significance as in carbides and nitrides. The author considers the Me-B bonds to be basic for the lower borides of the type Me_2B, while for higher borides beginning with MeB_2, the main role is played by the B-B bonds. However, the author points to an important role of Me-Me bonds in all these structures. Because of their metallic character they influence high melting points, and hardness.

In this author's view, the main factor determining the structure of borides is considered to be the atomic sizes of their components rather than their electronic structure.

A contradiction in interatomic bond schemes suggested by various workers arises from the fact that there are no well-elaborated methods for reliable determinations of the type of chemical bond. The experimental data used in these papers proved to be insufficient for a construction of a final model. This particularly applies to compounds like borides, with strongly pronounced diversity of bonds. Therefore, consideration of the bond character from the standpoint of electronegativity of atoms or their dimension factors only, useful in the case of simple types of bond, cannot give reliable results for various types of borides. More reliable information on the character of interatomic bonds might be obtained from the available data on the crystal-chemical features of various borides and the results of investigation of their physico-chemical properties and theoretical computation of their electronic structure.

Consideration of crystal-chemical properties of borides results in a conclusion that structures of lower (metal-rich) borides, (Me_3B, Me_2B), are mainly defined by the metal lattice. Boron atoms are isolated from each other and the effective distance between them (2.1 Å) exceeds that observed in other boride phases by a considerable amount. Lower borides may be related to "interstitial" phases. The Me-Me coordination remains the same as in the initial metal, and though a marked distortion of lattice occurs because of the insertion of boron atoms, the main role in interatomic bonding belongs to the metal-metal bonds.

Data on various physical properties of lower borides characterizing interatomic bonding are very few, and theoretical calculations of their electronic structure do not exist.

A general characteristic of cohesive forces can be obtained by analyzing such data, as melting points, sublimation heat, thermal expansion coefficients, and elastic moduli.

Reliable data for lower borides of different metals are available on melting points only. A comparison of experimental values shows that the character of their change is determined by the electronic structure of the metal. Thus, rise in melting point values in the boride series, Cr_2B-Ta_2B-Mo_2B-W_2B, corresponds to an increase in the energy stability of the metals themselves [i.e., increase in SWASC d^5 characterizing the covalent component of the Me-Me bond, (SAMSONOV et al., 1971 a.b.)].

Estimates of electronic state density values for different boride phases (KOVENSKAYA, 1971) showed their dependence upon the SWASC of the d-metal present, indicating the main role of the Me-Me bond in the structure of lower borides. Therefore, the assertion by GOLDSCHMIDT (1971) on the basic role of the Me-B bond in borides does not appear justified.

The character of rearrangement of boron atom configurations and the change of relations of the Me-Me and B-B bond lengths, (the latter being reduced upon the increase of boron content), testify to the increasing role of the boron-boron bonds.

During the formation of lower borides (M_2B, M_3B), before direct B-B bonds are formed, the Me-B bonds are added to the basic covalent-metallic Me-Me bonds upon the insertion of boron atoms into the metal lattice. As the increase of boron content results in formation of direct B-B bonds, their significance increases with the increase in boron content. The increase of boron content without the formation of B-B bonds would lead to its occupation by delocalized electrons. The formation of its own covalent bonds is more advantageous from the energy standpoint. Analysis of available experimental results allows a deeper understanding of the relationship of different types of bonds and how this relationship changes from lower to higher borides. The change of physical properties from lower borides to diborides has been studied in a series of phases formed by niobium and chromium metals, giving a great number of phases (KOVENSKAYA and SEREBRYAKOVA, 1966, 1970). Conclusions made on the basis of experimental data and estimates of a number of parameters characterizing the physical nature of the investigated phases (SAMSONOV et al., 1971 a, b) may be formulated in the following way.

All boride phases studies are compounds of the metallic type. An increase of the B/Me ratio leads to a decrease in the specific electrical resistance, its thermal coefficient and thermal expansion coefficient. The same tendency is displayed for the calculated quantities of the mean-square shift of atoms, effective carrier mass and Fermi energy. Carriers concentration values decrease and their mobility increases from lower to higher borides.

Analysis of physical properties related to the strength of interatomic bonds, such as melting points, thermal expansion coefficients, values of mean-square shift of atoms, etc., indicates that all types of lower borides (up to MeB_2) formed by metals with high energy stability (high values of SWASC d^5) have stronger interatomic bonds. The authors interpret these results to mean that the basic bonds defining the structures of lower borides (Me_4B, Me_2B) are covalent Me-Me bonds. The Me-X bonds are metallic and formed by sp-states of boron and sd-states of metal.

With the increase of boron content the covalent component of the bond in boride lattices increases due to the appearance of direct B-B bonds. However, in borides of the MeB type, the basic role of Me-Me bonds is, apparently, retained. For these phases, as in the case of Me_2B, a clear dependence of all properties characterizing the lattice strength on the electronic structure of the metal can be traced. Further increase in boron content leads to a greater role of B-B bonds and a decrease in the metallic bond character.

As mentioned earlier CADEVILLE and DANIEL (1966) interpret their results on measurements of magnetic and electrical properties of sub- and monoborides of transition metals as indicative of a major role of the Me-Me bonds in lattices of these compounds. The same conclusions are drawn from the results of investigations of low-temperature specific heat capacities carried out by TYAN et al. (1969).

KUENTZLER (1971) too, analyzing the results of measurements of low-temperature specific heat capacities of lower borides of transition metals, inferred an important role for Me-Me bonds in these borides.

The density of electronic states on the Fermi level can be regarded as characteristic of the role of direct Me-Me bonds. When estimated according to the results of calculated values of current carrier effective masses and Fermi energies (Table 1), it confirms the significant role of the Me-Me bonds in lower borides and their weakening on transition to phases with a greater boron content. With the increase of the boron content the localization of electronic states of boron is also increased accompanied by a strengthening of boron-boron bonds (formation of pairs, chains and, finally rigid sub-lattices of boron). An increase in the extent of localization leads to a decrease in the collective

Table 1. Computed values of electronic state densities of borides

Boride	TiB_2	ZrB_2	HfB_2	NbB_2	CrB_2	Nb_3B_2	NbB	Nb_3B_4	"Cr_4B"	"Cr_3B_2"	CrB	Cr_3B_4
State density 10^{-22} (ev^{-1}/cm^{-3})	0.45	0.48	0.55	4.28	5.43	17.75	13.76	6.65	21.93	20.76	14.04	7.92

stock of electrons (weakening of Me-X bonds, mainly metallic). This is confirmed by a decrease in carriers concentration obtained by computation and values of Fermi energy in the series Me_4B-Me_2B-MeB-Me_3B_4-MeB_2. Decrease in the thermal expansion coefficient values in this succession and increase in melting points, the phonon heat-conduction component, micro-hardness and other "strength" parameters are natural. These indicate the strengthening of interatomic bonds in the lattice.

There is a similar pattern of property changes observed in the series of lower boride phases of beryllium, Be_5B-Be_2B-BeB_2, (the only non-d-transition metal forming lower borides) (MARKEVICH et al., 1960; MARKEVICH and MARKOVSKI, 1960). Transition from metallic conduction of lower borides to semi-conduction of higher borides is the consequence of an increase in the covalent bond component (increase in the extent of localization of electronic states of boron) and a decrease in the metallic component (NESPHOR and SAMSONOV, 1971).

Therefore, a conclusion may be reached as to the correlation between the increasing role of B-B bonds in boride properties with the increase in boron content. The existence of stable boride phase of the platinum group metals may be explained by the ability of boron atoms to form strong bonds. In comparison, carbides and nitrides of these metals are metastable or do not exist at all.

A scheme of interatomic bonds in di- and higher borides can be proposed by comparing their physico-chemical properties with computed data. For diborides of transition metals the results of investigations of physical properties and some estimated parameters (KOVENSKAYA and SEREBRYAKOVA, 1970; SAMSONOV et al., 1971 a, b) may be summarized as follows.

All transition metal diborides display a metallic type of conduction. Diborides of group IV transition metals have great mobility with insignificant (for compounds of metallic type) carriers concentration (1/4 electron per cell). Diborides of groups V and VI show some occupancy of the conduction band (1 electron per cell) and considerably smaller carrier mobility. Diborides of group IV metals differ from groups V and VI diborides by having lower electrical resistance, thermal expansion coefficients and mean square atom shift, and higher thermal coefficient of resistance and heat conduction.

Examination of electronic structure parameters of diborides (SAMSONOV et al., 1971 a, b; Table 1) shows that from group IV metal diborides to group V and VI metal diborides, the donor activity of metal is reduced, the occupancy of their d-states is increased (SWASC d^5 is increased), the extent of localization of sp-states of boron atoms is decreased (there is a decrease SWASC $sp^2 + p_\chi$), the activity of the upper occupied level is increased and dissociation energies are decreased.

Estimated values of SWASC of d^5-states of metals and $sp^2 + p\chi$ states of boron atoms express the extent of localization of these states and can serve as an estimate of the covalent bond in these compounds (SAMSONOV et al., 1971a, b).

Comparison of the computed results with the experimental characteristics (Table 2) reveals correlations between them. Thus, moving from group IV metals diborides to groups V and VI, the succession of change of properties characterizing the strength of interatomic bonds, such as decrease in melting points, phonon conduction component, microhardness, elastic modulus, characteristic temperatures, increase in the thermal expansion coefficients, values of mean-square shift of atoms etc., agree well with the reduction of the extent of occupancy by electrons of sp-states of boron atoms. A smaller contribution to the $p\chi$-states of boron atoms in the above series causes the weakening of interatomic bonds both in the net planes and between the layers. This influence, derived by calculation, completely agrees with experimental data on stresses in the lattices of transition metal diborides (GILLES and LEWIS, 1968). They conclude that strong bonding exists in all directions in titanium diboride and reduction of interatomic bonds (especially that of interlayer) takes place from group IV to group V diborides. Information on strong interatomic bonding in diborides of group IV metals in all directions, including those of interlayers, has been reported in works by VAHLDIEK (1967) and KEIHN and KEPLIN (1967). They found an almost total lack of anisotropy in a number of elastic properties (elastic modulus, thermal expansion coefficient) in diborides of group IV transition metals.

Table 2. Some physical properties of diborides

Property	TiB_2	ZrB_2	HfB_2	VB_2	NbB_2	TaB_2	CrB_2
Melting point, ^oC	2980	3040	3250	2400	2900	3200	2200
Thermal expansion coefficients $x10^6$/deg.	4.6	5.9	6.3	7.6	8.0	8.2	10.5
Phonon component of heat conduction wt/m deg.	20.6	18.9	16.6	13.1	6.9	6.0	10.4
Elastic modulus $Ex10^{-6}$, kg/cm^2	5.6	4.3	-	3.5	-	2.6	2.5
Characteristic temperature, ^oK	1100	765	550	880	701	545	726
Mean-square shift of atoms, Å	0.075	0.081	0.083	0.088	0.088	0.090	0.092
Density of electronic states							
$gx\ 10^{-21}$,	4.50	4.76	5.51	26.8	42.8	32.6	54.6
$eV^{-1}\ cm^{-1}$	7.79	-	-	30.6	-	-	88.8
Work function, eV	4.19	3.94	3.83	-	-	-	3.18
Heat of formation KCal/mole	71.4	76.0	80.0	-	41.9	50.2	31.0

Examination of data listed in Tables 1 and 2 leads to the conclusion that the lattice properties of diborides are mainly determined by hybride sp-states of boron atoms ($sp^2 + p\chi$). The extent of their hybridization is, in turn, determined by the donor ability of metal atoms. Since group IV metals posses higher donor ability than group V and VI metals, their diborides realize a greater extent of localization of states of boron atoms. Consequently, the rigidity of the lattice of these diborides should be higher. This is displayed in lower values of thermal expansion coefficients, mean-square shift of atoms and higher values of melting points, microhardness, phonon component of heat conduction, elastic modulus and characteristic temperatures.

Reduction of the lattice rigidity of diborides moving from group IV to groups V and VI correlates with a systematic decrease in the mobility of current carriers. This correlation indicates the advantageous electron-phonon mechanism of carriers scattering in the compounds in question.

Systematic decrease in computed values of dissociation energies in the series of diboride groups IV-V-VI corresponds to the successive change of experimentally determined values of their enthalpies of formation.

Correlation of calculated and experimental data is distinctly displayed for the group of surface properties. Change in "surface activity" in the series of groups IV-V-VI metals diborides (TiB_2-Vb_2-CrB_2) such as decrease in the work function and surface energy, and increase in weting and rate of interaction with acids, agrees with the increase in the computed parameter of activity of the upper level in this series.

Finally, a correlation exists between the calculated parameters of contribution of d-states in the series TiB_2-VB_2-CrB_2 and an increase in the magnetic moment and the extent of electronic states localization (Table 2). The latter was estimated after the experimental results obtained by CASTAING et al. (1969), CASTAING (1971) and KOVENSKAYA (1971).

An analogous correlation of experimental and computed characteristics may be traced in diborides of the same group. In this case the situation becomes complicated by the increasing role of d-states of metal in interatomic bonds with the increase of the main quantum number of d-electrons.

Correlation of computed results with the experimentally determined properties and also with the results of other investigations of the electronic structure of transition metals diborides may lead to the conclusion that the results of computation reflect qualitatively the true physical picture and, therefore, may be employed to expain the physico-chemical properties and character of their change in the diborides series.

The calculations indicate that the main role in interatomic bonding in diborides belongs to the covalent B-B bonds (Table 1), formed by sp-boron valence electrons and partly non-localized valence electrons of metal.

The extent of localization of tight B-B bonds is higher, the higher the donor ability of metal atoms. On the other hand, the higher the energy stability of d-states of metal, the more significant are the Me-Me bonds. These are mainly of covalent character and formed by use of a localized part of valence electrons of metal. Metallic bond Me-B is formed by hybrid Me- and B-states. The participation of boron electrons in the formation of this electron stock is considerably smaller than that of metal electrons.

This scheme of interatomic bonds in the above phases is a generalization of calculated results of electronic energy spectra and wide-range investigations of physico-chemical properties of diborides. It can apparently be extended to borides of the Me_2B_5 type. In the latters, only the extent of interatomic Me-Me interactions is reduced (in comparison with MeB_2 compounds for the same metal), and Me-B bonds are increased. This conclusion is derived from considerations of crystal-chemical features of these borides and estimates of a number of carrier transition parameters (SAMSONOV et al., 1972).

Phases of MeB_4 composition alone represent a wide class of higher borides with the clearly expressed network sublattice of boron atoms. In respect to their composition and structure, they assume an intermediate position between the di- and hexaborides. Like di-borides, they form isostructural series of compounds and are therefore very convenient to study the character of the interatomic bond. Recently, data were reported on a variety of physical properties of tetraborides (SEVERYANINA, 1974). The author found a metallic type conduction for all phases studied (YB_4, PrB_4, NdB_4, GdB_4, TbB_4, DyB_4, HoB_4, ErB_4, TmB_4). The values of electrical resistance for these phases were significantly lower than those of pure metals. Analysis of "lattice" characteristics of tetraborides (melting points, thermal expansion coefficients, estimated values of mean-square shifts of atoms and Debye temperatures) indicated that the strength of interatomic bonds in tetraboride lattice exceeds that of pure metal considerably. In combination with the crystal-chemical characteristics of tetraborides (close values of B-B distances for all tetraborides except CeB_4 and YbB_4), these data lead to the assumption that the B-B bonds are responsible for the "lattice" properties of tetraborides. A slight difference in experimental results in the REM series (correlating with the electronic structure of the latters), should be assigned to the effect of the Me-Me bonds. The subgroups comparison of experimental data on tetraborides of cerium and yttrium shows a higher energy stability of the former in comparison with that of the latter.

By computing the electronic energy spectrum (SAMSONOV et al., 1974; GORYACHEV et al., 1974; ARABEY et al., 1974; Table 2) is obtained the following information on the character of interatomic bonds in tetraborides.

1. The superposition of the three classical types of bonds-covalent, metallic and ionic is characteristic for the tetraborides.

2. The main role in the lattice of tetraborides belongs to the covalent bonds between the boron atoms. The formation of covalent bonds occurs by using their own valence electrons and some non-localized electrons donated to the boron by the partner's atoms.

3. The Me-Me bonds have covalent-metallic character and their formation does not require participation of all valence electrons of metal. Part of the covalent component the Me-Me bond is defined by its ability to form f^0-, f^7-, f^{14}-electronic configurations. Me-B bonds are mainly metallic with an insignificant share of ionic component. They are formed largely by use of electrons of the metal and to a lesser degree by use of boron electrons.

4. The high SWASC of f^0-electronic states of Ce atoms in the first subgroup and f^7 of Tb atoms in the second, determines mainly tetravalent states displayed in tetraborides. Their electronic configurations rearrange in the following way:

Ce: $4\,f^2\;5\,d^0\;6\,s^2$ $\quad$ $4\,f^0 5\,d^{0.7} 6\,s^{0.5} 6\,p^{0.6}$

Tb: $4\,f^8\;5\,d^1\;6\,s^2$ $\quad$ $4\,f^7 5\,d^{1.2}\;s^{0.5} 6\,p^{1.2}$

This defines the extreme values of electronic structure parameters for their tetraborides in each subgroup.

5. Tetraborides of the cerium subgroup show higher values of parameters characterizing the strength of interatomic bonds.

A comparison of calculated data with the experimental results permits a correlation. Thus, the high extent of localization of electronic states of boron atoms obtained by calculation agrees with the high values of all properties studied in the experiments which characterize lattice rigidity (Table 3). Their change in the REM series corresponds to the change of calculated parameters defining the strength of interatomic bonds (Table 2). A high extent of localization of boron states and their weak participation in the electron bond (providing high electrical conductivity) are also confirmed by results of the X-ray spectra investigations of REM tetraborides (SEVERYANINA, 1974). The assumption of the tetravalent state of cerium atoms in its tetraboride is consistent with the conclusions derived by FISK et al. (1972) and BUSCHOW and CREYGHTON (1972) on the basis of measurements of lattice parameters and the magnetic properties of tetraborides.

Table 3. Physical properties of tetraborides

Property[a]	NdB_4	GdB_4	TbB_4	HoB_4
Thermal expansion coefficient, $\chi 10^{-6}$/deg.	5.84	7.0	6.55	7.6
Mean-square shift of atoms, Å	0.022	0.024	0.023	0.029
Characteristic temperature, °K	733	632	661	517
Melting temperature, °K	3926	3300	3515	2965
Microhardness, kg/mm^2	1950	1830	1900	1680

[a] Data from the papers by SEVERYANINA (1974) are used

Neither the calculations of the role of ionic bond in tetraborides (Table 3) nor the data on physical properties of tetraborides (SEVERYANINA, 1974) agree with the statement of GOLDSCHMIDT, (1971) on the essentially ionic character of the chemical bond in these compounds.

Satisfactory agreement of computed and experimental data permits a proposal of a scheme of interatomic bonding in tetraborides similar to that suggested for diborides.

In tetraborides, as in diborides, the prevailing role in the formation of lattice properties belongs to the covalent B-B bonds. The influence of localized states of metal atoms is smaller, and that of their delocalized states greater than in the diborides.

The existence of a rigid space network of boron atoms in higher borides and of a similarity in lattice parameters for the isomorphous higher borides shows that the Me-Me bonds in these compounds have no significant importance. The diborides are a transition link between the groups of lower and higher borides, and the shortest B-B distance depends on the metal. In MeB_4 this dependence becomes insignificant. In MeB_6 phases the dependence of the shortest B-B distance on the metal is even smaller. For instance, BLUM and BERTAUT (1954) showed that replacing of thorium atoms in ThB_6 by sodium atoms has no influence on the lattice size up to $Th_{0.23}Na_{0.77}B_6$. The authors attributed this to the high rigidity of the boron sublattice and the main role of the B-B bond in the hexaborides.

The MeB_6-type borides belong to a group of higher borides that has been studied most extensively (SAMSONOV and PADERNO, 1961; PADERNO et al., 1969).

All the REM hexaborides (except SmB_6) have metallic type conductivity. Other parameters, such as high melting points and hardness, also testify to a significant role of a covalent bond component in these compounds. The phonon component of heat conduction is also connected with high lattice rigidity, defined by strong covalent B-B bonds (L'VOV et al., 1965).

Theoretical estimates of interatomic bonds made by LONGUET-HIGGINS and ROBERTS (1954) and JAMAKAZI (1957) affirm the basic role of covalent B-B bonds in the structure of borides of the MeB_6 type. Calculations with a greater degree of approximation, which take into account the s- and p-states of metal atoms were carried out by GARF et al. (1970). These data indicate that significant localization of electronic states of boron atoms determine the covalent boron-boron bonds. The extent of localization of electronic Me-B states is significantly smaller. The Me-B bonds are metallic in character since the conduction band is mainly formed by metallic states. The Me-Me bonds are covalent-metallic in character. The extent of localization of electronic states of metal atoms is smaller than in tetraborides. Compared to the latter the ratio of the covalent and metallic Me-Me bonds is shifted in the direction of the metallic state. Thus, the more strongly pronounced metallic character of hexaborides and the smaller values of electrical resistivity in comparison with tetraborides may be explained by a larger contribution of the metallic component of the Me-Me bond, as well as by a smaller electron-phonon scattering in a more resilient lattice.

Higher borides of the MeB_{12} type have been studied less than hexaborides. However, the available data permit the possibility of a scheme of interatomic bonds in these compounds. From the crystal-chemical point of view one may expect a greater extent of localization of electronic states of boron in these borides than in hexaborides. The role of metallic states in their formation should be smaller. Indeed, the values of the lattice periods of dodecarborides depend very little on the atomic number of the metal, and the Me-Me distance is approximately 25% larger than in hexaborides.

In addition, while the unit cell size curve of hexaborides has a typical maximum for Eu and Yb atoms and depends upon the atomic number of the elements, a similar curve for the dodecaborides is comparatively smooth. The results of investigations of physical properties of dodecaborides carried out by ODINSOV (1971) have shown that all of the REM dodecaborides (except YbB_{12}) have a metallic type of conduction with values of electrical conductivity exceeding those of pure metals. They have high values of thermal expansion coefficients, microhardness, phonon component of heat conduction, characteristic temperatures, melting points and mean-square shifts of atoms.

A comparison of these parameters with those of the hexaborides indicates a higher strength of interatomic bonds in the MeB_{12} lattice in comparison with those of MeB_6. On the other hand, the Me-Me distance in dodecaborides is approximately 25% greater than in hexaborides. This indicates that the change in the "lattice" characteristics in dodecaborides is caused by a further strengthening of B-B bonds (as compared to hexaborides). Calculation of the electronic energy spectrum of dodecaborides (LIPSCOMB and BRITTON, 1960; GORYACHEV et al., 1971) also shows that localization of electronic states of boron, defining rigid covalent B-B bonding takes place to a greater extent than in di-, tetra- and hexaborides. The extent of localization of metallic states is decreased and, consequently, the covalent-metallic Me-Me bond becomes even less significant than in hexaborides. Correspondingly, there is an increase in the metallic component of the Me-Me bond and the contribution of metal states to the mainly metallic Me-B bond.

Agreement between this scheme of interatomic bonds and the experimental data is apparent. While all the parameters characterizing the strength of dodecaboride lattices have higher values than those of the metals forming them, the variation of electrical conductivity of dodecaborides in the REM series depends on the metal and the corresponding curve has a strongly expressed minimum for YbB_{12}.

The character of variation of interatomic bonds in the series MeB_2-MeB_4-MeB_6-Meb_{12} can be further illustrated by examples for borides of transition and non-transition metals. Thus, the thermophysical and thermochemical properties of magnesium borides (GERMANIDZE et al., 1966; MOROSOVA and RYBAKOVA, 1970; ARIA et al., 1971; RYBAKOVA, 1972) indicate the strengthening of interatomic bonds in the following order: MgB_2-MgB_4-MgB_6-MgB_{12}. In the series YB_2-YB_4-YB_6 and LaB_4-LaB_6- the values of thermal expansion coefficients steadily decrease while melting points and chemical stabilities increase (KUDINTSEVA et al., 1958; MEERSON et al., 1968; ZHURAVLEV et al., 1970).

Therefore, the rigidity of the boride lattice is a function of two factors: the quantitative ratio of B/Me components and the number of d- (or f-) electrons of the corresponding metals. In the borides, with the small B/Me ratio, lattice rigidity is determined mainly by the Me-Me bonds. The role of the Me-Me bonds increases with the number of metal d-electrons from 0 to 5 and from 5 to 10 and f-electrons from 0 to 7 and from 7 to 14. The role of B-B bonds in the formation of borides increases with the B/Me ratio, and in MeB_χ (χ > 2) borides it becomes predominant.

In the general scheme of interatomic bonds in borides there is a clear tendency to a shift of the dominating role from Me-Me bonds in the lower borides to the B-B bonds in the higher borides.

References

ARABEY, B.G., GORYACHEV, Yu.M., KOVENSKAYA, B.A.: Ob electronnom strojenii tetraboridov RZM yttrijevcj grouppy. J. Exp. Teoret. Chim. 10, 1012-1016 (1974)

ARIA, R.M., MOROSOVA, M.P., SEMENOV, G.A., RYBAKOVA, G.A.: Mg-B sistema i opredelenije entalpii obrazovaniya Mg-B_χ. J. Phys. Chim. 45, 181-182 (1971)

BILTZ, H.: Über Elektronenzuständen von Hartstoffen mit Natrium-Chlorid-Struktur. Z.f. Physik 153, 338-358 (1958)

BLUM, P., BERTAUT, F.: To the investigation of borides with augmented contents of boron. Acta Cryst. 7, 81-86 (1954)

BUSCHOW, K.H.J., CREYGHTON, J.H.N.: Magnetic properties of rare earth tetraborides. J. Chem. Phys. 57, 3910-3916 (1972)

CADEVILLE, M.G., DANIEL, E.: Sur la structure électronique de certaines borures des métaux de transition. J. Phys. 27, 449-456 (1966)

CASTAING, J.: Structure électronique et propriétés magnetiques des diborures des métaux de transition. Office national d'etudes et de recherches aerospatiales publ. N 140, Paris (1971)

CASTAING, J., CANDRON, R., TOMPANCE, G., COSTA, P.: Electronic structure of the transition metals diborides. Solid State Commun. 7, 1453-1456 (1969)

DEMPSEY, E.: Bonding in the refractory hard metals. Phil. Mag. 8, 285-291 (1963)
ERN, V., SWITENDICK, A.C.: Electronic band structure of TiC, TiN and TiO. Phys. Rev. A137, 1927-1931 (1965)
FISCHER, J., FLODMARK, S.: Covalent B-B bonds in certain metal borides crystals. J. Chem. Phys. 22, 11-16 (1954)
FISK, Z., COOPER, A.S., SCHMIDT, P.H., CASTELLANO, R.N.: Preparation and lattice parameters of the rare earth tetraborides. Mat. Res. Bull. 7, 285-291 (1972)
FLODMARK, S.: Electrons distribution and energy bands in the metal borides. Arkiv Fysik 14, 513-516 (1959)
GARF, E.S., GORYACHEV, Yu.M., PADERNO, Yu.B.: Redkozemelnyje metally i ikh sojedinjeniya, pp. 101-108. Kiev: Izdatjelstvo Naukova Dumka 1970
GERMANIDZE, M.S., GELD, P.V., LJETUN, S.M.: Teplojemkost, entropiya i davlenije dissociacii dodecaborida magniya. J. Prik. Chim. 39, 1941-1944 (1966)
GILLES, D.C., LEWIS, D.: Bond strength in certain group IV and group V metals diborides. J. Less-Common Metals 16, 162-171 (1968)
GOLDSCHMIDT, H.J.: Splavy vnedreniya, pp. 364-378. Moskwa: Izdatelstvo Mir 1971
GORYACHEV, Yu.M., KOVENSKAYA, B.A., ARABEY, B.G.: Electronnyj energeticheskij spectr tetraborida neodima. J. Exp. Teoret. Chim. 10, 862-868 (1974)
GORYACHEV, Yu.M., ODINTSOV, V.V., PADERNO, Yu.B.: Structura energeticheskikh polos v kubicheskikh dodecaboridakh metallov. J. Prik. Chim. 12, 344-349 (1971)
JAMAZAKI, M.: Calculation of energy bands in the metal borides MeB_6 by the group theory method. J. Phys. Soc. Japan 12, 1-6 (1957)
JOHNSON, R.W., DAANE, A.H.: Electron requirements of bands in metal diborides. J. Chem. Phys. 38, 425-432 (1963)
JURETSCHKE, H., STEINITZ, P.I.: Hall effect and electrical conductivity of transition metals diborides. J. Phys. Chem. Sol. 4, 118-124 (1958)
KEIHN, F.G., KEPLIN, E.I.: High-temperature thermal expansion of certain group IV and group V diborides. J. Am. Ceram. Soc. 50, 81-86 (1967)
KIESSLING, R.: Interstitial structures. Powd. Metallurg. Bull. 3, 177-184 (1959)
KOVENSKAYA, B.A.: Issledovanije nekotorykh phisicheskikh svoistv electronnogo strojeniya boridov perekhodnykh metallov IV-VI groupp periodicheskoj sistemy. Avtoreferat Kandidats. Dissertacii, Kiev 1971
KOVENSKAYA, B.A.: SEREBRYAKOVA, T.I.: Phisicheskije svoistva boridnykh phaz chroma. Izvestiya AN SSSR. Neorganicheskije Materialy N 12, 2134-2138 (1966)
KOVENSKAYA, B.A., SEREBRYAKOVA, T.I.: Physicheskije svoistva boridnykh phaz niobia. Poroshkovaya Metal. N 5, 42-48 (1970)
KUDINTSEVA, G.A., POLYAKOVA, M.D., SAMSONOV, G.B.: Prigotovljenije i nekotoryje svoistva geksaborida yttriya. Phys. Metal. Metal. 6, 272-275 (1958)
KUENTZLER, R.: Structure électronique de borures des métaux de transition par mésures de chaleur spécifique. J. Phys. 32, 634-672 (1971)
LIPSCOMB, W.N., BRITTON, D.: Valence structure of the higher borides. J. Chem. Phys. 33, 275-280 (1960)
LONGUET-HIGGINS, H.C., ROBERTS, M. de V.: The electronic structure of the MeB_6 borides. Proc. Roy Soc. A224, 336-341 (1954)
L'VOV, S.N., NEMCHENKO, V.F., PADERNO, Yu.B.: Vysokotemperaturnyje neorganicheskije sojedinjeniya. pp. 445-451, Kiev: Izdatjelstvo Naukova Dumka 1965
MARKEVICH, G.S., KONDRASHOV, Yu.D., MARKOVSKII, L.Ya.: O novoj boridnoj phazje v sistjemje berillij-bor. J. Neorganicheskoj chim. 5, 1783-1787 (1960)
MARKEVICH, G.S., MARKOVSKII, L.Ya.: O chimicheskoj stojkosti boridov berilliya po otnosheniju k kislorodu, azotu i ugljerodu pri vysokikh temperaturakh. J. Prik. Chim. 33, 1008-1012 (1960)
MEERSON, G.A., ZHURAVLEV, N.N., MANELIS, R.M., RUNOV, A.D., STEPANOVA, A.A., GRISHINA, L.L., GRAMM, N.V.: Nekotoryje svojstva boridov yttriya. Izvestiya AN SSSR. Neorganicheskije Materialy N 2, 608-616 (1968)
MOROSOVA, M.P., RYBAKOVA, G.A.: Entalpii obrazovanija di-i tetraboridov magniya. Trudy Gosudarstvennogo Inst. Prik. Chim. N 66, 35-41 (1970)
NESHPOR, V.S., SAMSONOV, G.V.: Electrophysicheskije svoistva i kharacter svyazi v sojedinjeniyakh berilliya s borom i ugljerodom. Izvestiya AN SSSR. Neorganicheskije Materialy N 1, 50-54 (1971)
ODINTSOV, V.V.: Poluchenije i nekotoryje physicheskije svoistva dodecaboridov so structuroj tipa UB_{12}. Kiev: Avtoreferat kandidatskoj dissertacii 1971

PADERNO, Yu.B., GARF, E.S., NEMYSKY, T.: Electrophysicheskije svoistva plavlennykh geksaboridov shchjelochno- i redkozemelnykh metallov. Poroshkovaya Metal. N 10, 55-58 (1969)

PIPER, J.: Galvanomagnetic effects in ZrB_2 single crystals. J. Phys. Chem. Sol. 27, 1907-1910 (1966)

ROBINS, D.: Bonding in carbides, silicides and borides. Powd. Metal. Bull. 2, 172-176 (1958)

RYBAKOVA, G.A.: Termokhimicheskoje izuchenije boridov berilliya i magniya. Leningrad: Avtoreferat Kandidat. Dissertacii 1972

SAMSONOV, G.V., GORYACHEV, Yu.M., KOVENSKAYA, B.A.: O nekotorykh osobjennostyakh temperaturnoj zavisimosti electrophisicheskikh svoistv Mo_2B_5 i W_2B_5. Doklady AN SSSR A10, 943-946 (1972)

SAMSONOV, G.V., GORYACHEV, Yu.M., KOVENSKAYA, D.A., ARABEY, B.G.: Raschjet energeticheskogo spektra tetraboridov tzeriya. J. Exp. Teoret. Chim. 10, 716-720 (1974)

SAMSONOV, G.V., KOVENSKAYA, B.A., SEREBRYAKOVA, T.I.: O chimicheskoj svyazi v boridakh perekhodnykh metallov. Doklady AN SSSR A12, 976-980 (1971a)

SAMSONOV, G.V., PADERNO, Yu.B.: Boridy redkozemelhykh metallov. Kiev: Izdatjelstvo AN USSR 1961

SAMSONOV, G.V., PRYADKO, I.F., PRYADKO, L.F.: Konfiguratsionnaya model veshchestva. Kiev: Izdatelstvo Naukova Dumka 1971b

SEVERYANINA, E.N.: Poluchenije i issljedovanije physicheskikh svoistv tetraboridov redkozemelnych metallov. Kiev: Avtoreferat Kandidat. Dissertacii 1974

TYAN, Y.S., TOTH, L.E., CHANG, Y.A.: A study of electrons transition theory in the hard-melting metals borides by the investigation of the low-temperature heat capacity. J. Phys. Chem. Sol. 30, 785-792 (1969)

VAHLDIEK, F.W.: Electrical resistivity, elastic modulus and Debye temperature of titanium diboride. J. Less-Common Metals 12, 209-212 (1967)

ZHURAVLEV, N.N., BJELOUSOVA, I.A., MANELIS, R.M., BJELOUSOVA, N.A.: Rentgenographicheskoje opredjeljenije koefficientov termicheskogo rasshireniya boridov lantana i yttriya. Kristallographiya 15, 836-838 (1970)

III. The Electronic Structures of the Hexaborides and the Diborides

P. G. PERKINS

Abstract. The current state of knowledge of the electronic structures of the binary di- and hexaborides is reviewed. New work is presented which allows rationalization of the stability of these materials as well as the metal-deficient and mixed-metal phases. Band structure and density-of-states calculations lead to values of electrical conductivities and Hall coefficients for these materials which agree well with experimental data.

1. Introduction

There exists a large number of binary metal borides of formula AB_n in which n can take a considerable range of values. The complexity of the borides is further increased by the widespread occurrence of "non-stoichiometry". Many of these materials exhibit physical properties of great actual and potential technical importance, and it is clearly of some interest to advance a convincing general rationale of their electronic structure so as to explain present observations and to indicate and underpin future work. However, very little *quantitative* theoretical work has been carried out and it is only for hexa- and diborides that any band-structure calculations have been performed (LONGUET-HIGGINS and ROBERTS, 1954; YAMAZAKI, 1957). Hence, in the present review we shall restrict coverage to these systems, and a substantial amount of hitherto unpublished work will be discussed (ARMSTRONG et al., in press). There also exists a number of excellent reviews of the structural chemistry and physical properties of the borides (ARONSSON, 1965; THOMPSON, 1970; MUETTERTIES, 1967) and it is not our purpose to repeat this. It is, however, of some worth first to abstract from these reviews, and briefly consider relevant data which will be of some importance in evaluating and interpreting the theoretical work.

2. Physical Properties

a) Hexaborides

There are known approximately 20 hexaborides of metals which crystallize in the simple cubic CsCl structure (Fig. 1) and belong to space group O_h^1 (m3m) Some of the hexaborides are metal-deficient and this phenomenon produces some variability in the lattice translation vectors. Overall, with metals as disparate in the Periodic Table as calcium and plutonium, the lattice constant varies only between 0.4096 (HoB_6) and 0.4269 nm (BaB_6). It seems that only large-sized metals are able to form stable hexaborides; this should be rationalizable by a suitable theory.

A number of these hexaborides, e.g. LaB_6, are excellent electrical conductors, being comparable in this respect with metals. Indeed, the conductivity of the hexaboride frequently exceeds that of the parent metal. Moreover, the metallic nature of the conductivity is evidenced by the temperature coefficient which, for some of the substances, has the same sign as that of the metal. However, it should be pointed out that

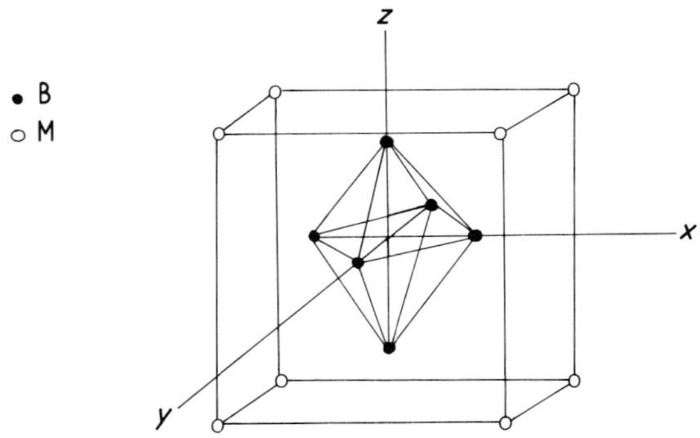

Fig. 1. Crystal structure of hexaborides, MB_6

the experimental determination of such physical properties can be a matter of considerable difficulty, as they are frequently strongly affected by chemical impurities, method of treatment, and homogeneity. When the total number of electrons per unit cell indicates a closed-shell structure, e.g. the alkaline earth hexaborides, then the property of metallic conductivity is lost. These materials appear to possess small band gaps, lower absolute conductivity, and negative temperature coefficients. This is indicative of intrinsic semi-conducting properties.

Measurements of Hall coefficients have been carried out on these materials and reveal different classes of behavior: (i) the rare-earth hexaborides have small Hall coefficients (-4.2 to -5.0) $\times 10^{-4}$ cm^3 $coulomb^{-1}$; these values correspond to approximately one free conduction electron per unit cell; (ii) the Hall coefficients for alkaline earth hexaborides are large and correlate with a small free-electron population.

The work functions of some hexaborides are remarkably low, that of YB_6 being 2.22 eV (DECKER and STEBBINS, 1955; SAMSONOV and GRODSTEIN, 1956) and of LaB_6 2.74 eV (DECKER and STEBBINS, 1955; LENK, 1956; MIKHAILOVS'KII, 1962). This property correlates also with the reported high thermionic emissivity of the latter material (WESTBROOK, 1960).

b) Diborides

The structures of diborides are typified by that of AlB_2 (Fig. 2). The space group is D^1_{6h} and the unit cell has a basis of one aluminum and two boron atoms. Overall, layers of metal atoms alternate with layers of boron atoms. In the boron layer, graphite-like structure is manifest, with B-B distances varying between 0.191 nm (GdB_2) and 0.1714 nm (CrB_2). However, it should be noted that departure from MB_2 stoichiometry is common and this may affect the lattice constants. It is interesting that the AlB_2 structure can accomodate a variety of metal atoms, in contrast to the hexaboride case, where only large atoms form stable systems.

As for many hexaborides, the diborides are good electrical conductors although, again, this property is somewhat sensitive to the physical characteristics of the samples. Coupled with the measured resistivities, the Hall coefficients can generally be interpreted in terms of high concentrations of conduction electrons. For the metals Ti, Zr, and Hf the Hall coefficients are large and it is necessary to suggest high mobility for the conduction electrons in order to explain the observed experimental data.

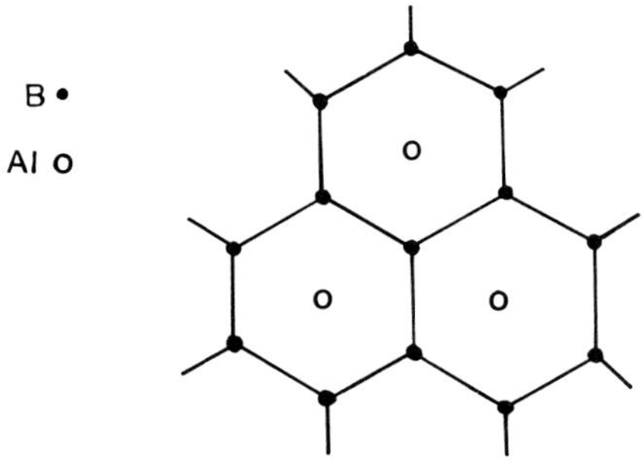

Fig. 2. Crystal structure of diborides, MB_2

3. Theories of Electronic Structure

a) Hexaborides

Since the hexaboride sub-lattice is built from "molecular" B_6 cages, it is natural to attempt to rationalize the electronic properties of the system from the molecular viewpoint. LONGUET-HIGGINS and ROBERTS (1954) first built up delocalized molecular orbitals for the B_6 cage unit. They then used these molecular levels to generate an approximate band structure for the systems. They concluded that, for a closed-shell system, two further electrons were needed, giving the cage anion $B_6{}^{2-}$. Since these can be provided by a metal with at least two valence electrons, their result afforded a simple and elegant explanation of most of the observed facts. Indeed, although doctrinaire applications of the theory leave unexplained some features, e.g. the existence of the fairly stable hexaboride KB_6 (HAGENMULLER and NASLAIN, 1963) it has turned out to be the most fruitful approach and has been quoted by almost all workers in the area. This is because it offered an insight into the bonding pattern for the "molecular" unit, while at the same time showing how crystal energy bands may be produced. The bonding levels for the isolated cage were arranged in the order $a_{1g} < t_{1u} < t'_{1u} < t_{2g}$ and valence bands formed from these were shown to just accommodate 20 electrons. This is exactly the right number for the formulation of the unit cell as $M^{2+}B_6^{2-}$. Furthermore, a gap appeared between the valence and conduction bands and the authors concluded that this would operate throughout the Brillouin Zone. Later calculations have underpinned the validity of this result. ARMSTRONG et al. (1973) calculated that, for the related $B_6H_6{}^{2-}$ ion, the orbital energies fall in the order $a_{1g} < t_{1u} < e_g < a_{1g} < t_{2g} < t_{1u}$ (the last triply degenerate set is required for the six extra hydrogen electrons) and presently *ab initio* calculations on neutral $B_6{}^{2-}$ using Gaussian basis sets (ARMSTRONG et al., 1973) have afforded the order of filled cage orbitals as $a_{1g} < t_{1u} < a_{1g} < t_{2g} < 2e_g$, with a bonding-antibonding gap of 7.4 eV. Moreover, in both the latter investigations electron density plots were derived which show that the B_6 cage incorporates a high level of electron density and the whole appears to be held together both by this and by electron concentration on the trigonal faces. This will be referred to again later.

A band structure for CaB_6 was constructed by YAMAZAKI (1957) by setting up and diagonalizing an energy matrix. Nearest-neighbor interactions only were included. Space-group theory was used in order to construct symmetry-projected eigenfunctions and so to factorize the secular determinant. The diagonal elements were formed from the one-electron orbital energies of the valence orbitals of B (2s, 2p) and Ca (4s) while overlap integrals were used for scaling the off-diagonal elements. In YAMAZAKI'S procedure the rather drastic assumption of orthonormality between Bloch functions was made and the three high-symmetry directions ΓR, ΓM and ΓX were studied. The treatment concurs with that of LONGUET-HIGGINS and ROBERTS in that the valence bands are just filled by the 20 valence electrons of CaB_6. A small direct band gap (1.5 eV) occurs at the point R and this is rather greater than that which was measured by JOHNSON and DAANE (1963) (0.40 eV). No estimate of the polarity of the system or density of states was obtained from YAMAZAKI'S calculation and hence no light was shed on the question of whether such materials are $M^{++}B_6{}^{2-}$.

More recently, band-structure calculations on the hexaborides of Ca, Sr, Ba, and La have been carried out by ARMSTRONG et al. (in press). Essentially here a crystal wavefunction was set up

$$\chi_i(K) = \sum_r C_{ir}(K)\psi_r(K)$$

where r ranges over all unit cell orbitals and ψ_r is a Bloch function

$$\psi_r(K) = \sum_{R_j} e^{iKR_j}\,\phi_r(R - R_j).$$

K is the wave vector and R is built from the three primitive translation vectors of the lattice; $\phi_r(R_j)$ is an atomic orbital situated at vector distance R_j from the origin. The Hamiltonian and overlap matrices, respectively, are set up as follows

$$H_{ij}(K) = \sum_{R_j} e^{iK\cdot R_j}\,H_{ij}{}^{(O,R_j)}$$

$$S_{ij}(K) = \sum_{R_j} e^{iK\cdot R_j}\,S_{ik}{}^{(O,R_j)}$$

where the subscript (O,R_j) indicates an element between orbital i in the original unit cell and orbital j in the cell displaced by the lattice vector R_j. In general, matrix elements converge when the summation over R includes all overlap integrals between orbitals on centers separated by less than 1.0 nm. In evaluating the matrix elements the integrals $H_{ii}{}^{(O,O)}$ are set equal to the negatives of the relevant ionization potentials I_m for the orbitals of the free atoms. The basis set in all cases comprised the ns orbitals of the metal and the 2s,2p set for each boron atom. Inclusion of metal 3d orbitals is desirable, but computer limitations precluded this in the investigations. However, neglect of these bases should not seriously affect the conclusions of the work as a whole. Indeed in molecules, it has been found that the empty d orbitals of the elements at the left-hand side of the Periodic Table play little part in bonding (LEVISON and PERKINS, unpublished). The off-diagonal terms in the Hamiltonian matrix are calculated from the expression

$$H_{ij}{}^{(O,R_j)} = -A\,S_{ij}{}^{(O,R_j)}(I_i \times I_j)^{\frac{1}{2}}$$

i.e. the "geometric" version of the Mulliken-Wolfsberg-Helmholtz expression (BALLHAUSEN and GRAY, 1962). From the above matrices real, double matrices ${}^{D}H$ and ${}^{D}S$ are formed and the Fock matrix is reduced to diagonal form by inversion of ${}^{D}S$ (to remove overlap terms in the matrix) followed by a Cholesky decomposition into lower and upper triangular forms, i.e.

$$D_S^{-1} = L \cdot U.$$

Application of $L^D H U$ yields the diagonal matrix H^O. This affords complete sets of eigenvalues and eigenvectors and the latter enable definition of the distribution of a particular atomic orbital throughout the states. Furthermore, addition of each contribution affords an atomic orbital population. The polarity of the system under investigation is thus revealed and, simultaneously, partial and total densities of states are produced. In order to get satisfactory coverage of the representation domain of the central Brillouin Zone, the calculation is performed for a large number of k values covering the main symmetry directions. In such calculations, one aims to obtain information on (i) the disposition of bands with respect to energy - this throws light on problems of bonding, stability, absorption and emission spectra, and electron emissivity; (ii) the polarity of the system; (iii) the shapes of bands and the topology of Fermi surfaces which help with rationalizing the electrical properties of the solid.

α) *Ca, Sr, and Ba Hexaborides*

Figure 3 shows the representation domain for the simple cubic Brillouin Zone and Figure 4 gives the band structure for CaB_6 in a number of specified symmetry directions. Since the energy pattern for SrB_6 and BaB_6 is so similar, details are omitted here. Figure 5 shows the corresponding band structure for the metal-free B_6 lattice. For Ca, Sr, and Ba hexaborides, the 20 valence electrons just fill the first 10 bands. The smallest band gaps in each case are direct and occur at the X point, i.e. on the center face of the Brillouin Zone. The band gaps are CaB_6, 2.11 eV; SrB_6, 3.68 eV; and BaB_6, 2.64 eV. These are all somewhat higher than those reported by JOHNSON and DAANE (1963) but the calculation would undoubtedly be improved in this respect by the inclusion of the empty higher d orbitals of the metal, since in general one finds that these bases interact with and lower the energy of the lower conduction band (McALOON and PERKINS, 1972). One would, moreover, expect d orbital participation to be more effective in the order Ca < Sr < Ba. In this context, it is interesting to observe that (Fig. 5) the lower conduction bands of the B_6 cage *alone* are quite broad and themselves would give rise to appreciable electronic conductance. Hence although, as stated above, metal d orbitals will mix in here, there is no *a priori* requirement for their presence. There are

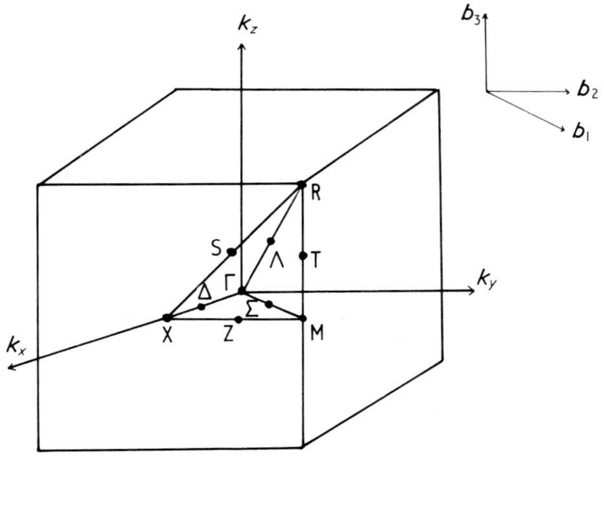

Fig. 3. Representation domain for cubic reciprocal lattice

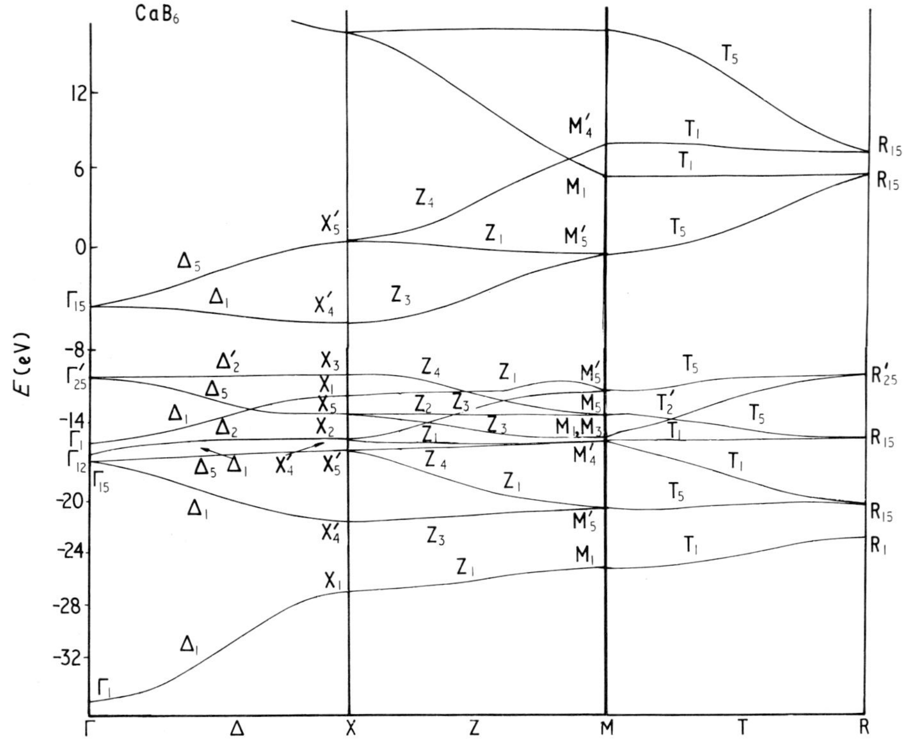

(a)

Fig. 4a and b. Band structure for CaB_6

some differences from YAMAZAKI'S calculation which no doubt stem from his non-inclusion of overlap between Bloch functions of the same symmetry. In general, YAMAZAKI'S bands are flatter in the higher valence-lower conduction region and he also reported the narrowest band gap to occur at R. The present calculations concur rather more closely both with the *ab initio* and with those of LONGUET-HIGGINS and ROBERTS (1954), differing at the Γ point in one feature only.

In all the metallic systems the metal s states fall low in the energy scale and broaden out the lowest valence band considerably along Σ, Λ and Δ. This is clearly shown by reference to Figures 4 and 5. This broadening is strongly indicative and stabilizing metal-metal bonding and is consistent with the observation that only large elements form very stable hexaborides. Other than this, the presence of metal does not appear to perturb the band structure of B_6 dramatically. Essentially, incorporation of the metal in the unit cell adds a level in the conduction bands, since the metal-free B_6 exhibits the same number of valence bands as MB_6. From the band structure, by interpolation between the K points, the density of states can be calculated, both *in toto* and broken down into its orbital components.

For calculations of this nature, it is of great value to make comparisons with experimental information wherever possible. For BaB_6, the boron X-ray K emission spectrum has been measured (LYAKHOVSKAYA et al., 1970). Boron X-ray K spectra from solid borides exhibit narrow maxima,

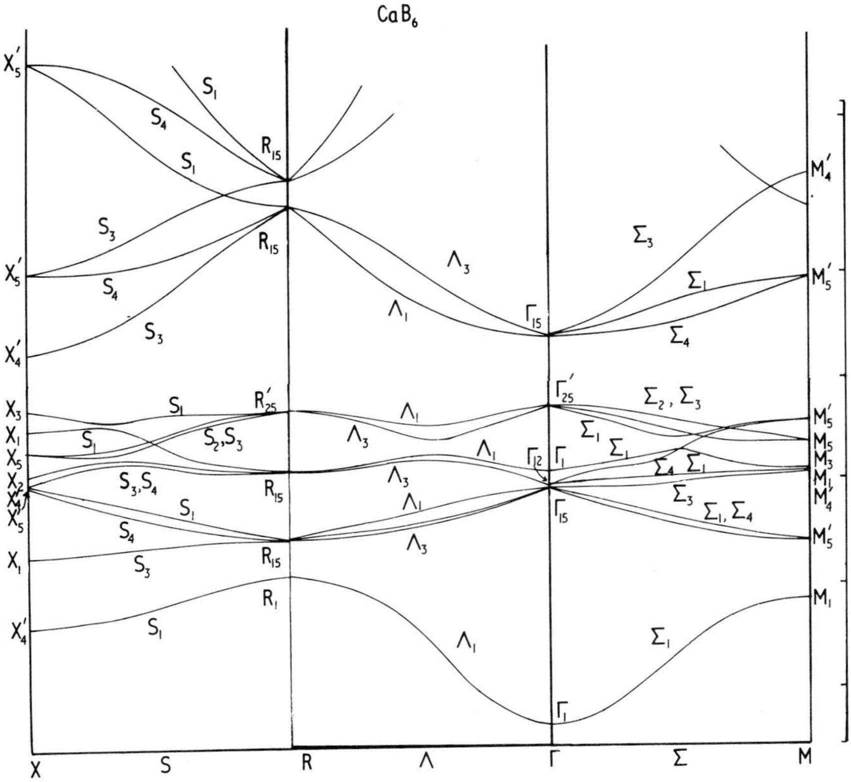

(b)

a feature consistent with excited states, created by the ejection of an electron from the boron K shell, being localized around the boron atoms. This K-shell electron is replaced by one from the higher boron 2p level with emission of radiation, and hence we are interested in the availability of boron 2p electrons at various energy levels. Although the K emission spectrum does not, in principle, give the density-of-p-states directly, since all transition probabilities are not necessarily equal, we can reasonably assume this and comment on any situation where the assumption appears to break down. Hence we can compare directly the calculated density of 2p boron states for BaB_6 with the X-ray emission spectrum for that solid.

Figure 6 shows the partial density of 2p states and the total density of states, together with the emission spectrum for BaB_6 and Table 1 lists the separations between the peaks of the calculated function compared with those in the experimental spectrum. The agreement is very satisfactory, particularly when one considers the approximate nature of the calculation. It is interesting that the density-of-states function shows, in addition, a prominent peak A' and this corresponds to a weak shoulder in the experimental X-ray spectrum. The latter was not explicitly listed by the authors of the experimental paper. With regard to the intensities of the experimental peaks, in general these also correlate well with the density of boron 2p states and indicate that our assumption of constant transition probabilities is not too severe an approximation.

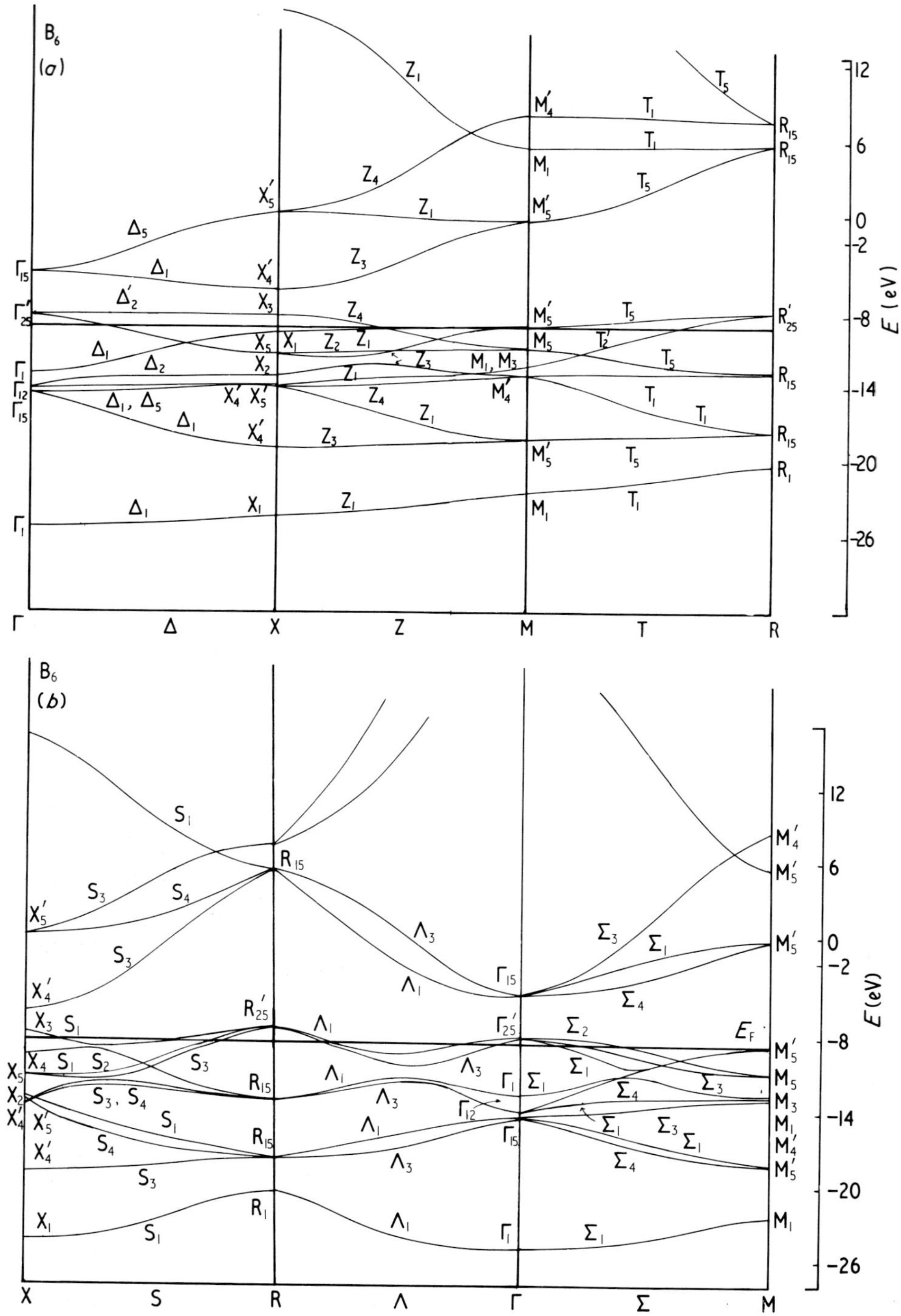

Fig. 5a and b. Band structure for metal-free B_6

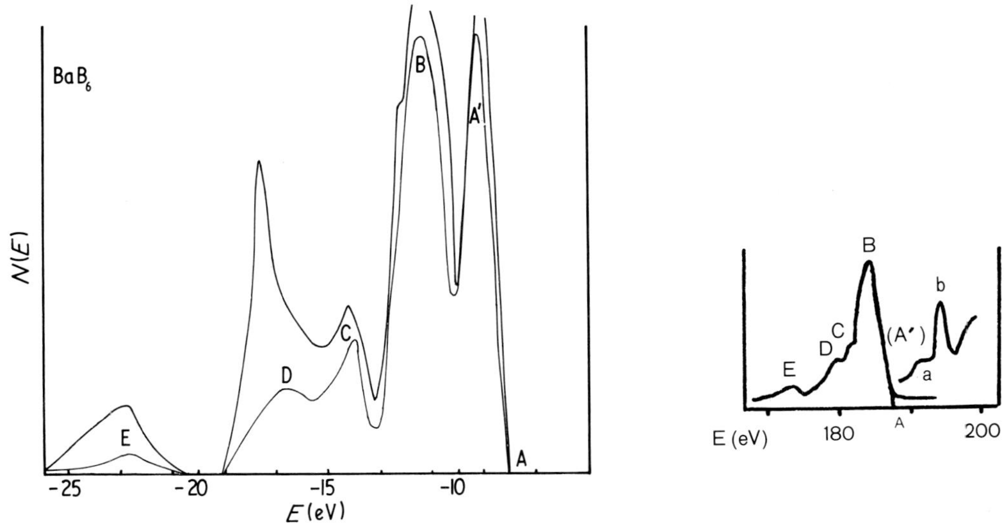

(a) (b)

Fig. 6a and b. (a) Total density of states (outer line) and density of boron 2p states (inner line) for BaB_6; (b) Experimental boron K emission spectrum for $Ba3_6$. (Fig. reproduced by permission of the authors and publishers)

In all three solids the metal atoms transfer 0.9-1.0 electrons to the B_6 cage in each unit cell. Thus, whereas it has hitherto been supposed that each B_6 cage had to acquire two electrons for stability (as in the molecular situation), it can now be seen that there is no absolute requirement for this in the solid state. In order to get maximum stability, all that is necessary is that the first ten valence bands of the overall system are filled. The highest valence level for the systems is, of course, that of the highest point in the band system and it is seen that this quantity does not vary much over the Ca, Sr and Ba hexaborides. Essentially the same value is obtained as was calculated by YAMAZAKI (1957).

A study of the band structures of $M^{2+}B_6{}^{2-}$ and neutral B_6 is revealing with regard to the stability of hexaborides in general. The density-of-states calculation for B_6 shows that the Fermi level lies at -8.8 eV and many valence states are unoccupied. The eigenvectors for the band show that the two missing electrons in each unit cell weaken both the intercage p_σ-p_σ, p_π-p_π bonding and that which permeates the eight trigonal faces of the cage unit. The boron s orbitals are little affected. This effect, coupled with the formation of a Fermi surface having a number of flat areas, causes drastic destabilization of the cubic lattice leading to collapse. The lack of electrons in the intercage bonds is particularly serious for the structure, since the associated bond distances are shorter than those in the cage (LAPLACA, 1965). Conversely, if electrons are fed into the B_6 system, it gains stability rapidly and the inter- and intra-cage bonding is enhanced. Thus the metal ex-

Table 1. BaB_6 boron K emission spectrum

Point on exp. spectrum	Energy differences (eV) exp.	calculated
E		
	5.9	6.1
D		
	1.6	2.6
C		
	3.2	2.6
B		
		2.2
A'[a]	2.0	
		1.1
A		

[a] prominent peak calculated which coincides with a very weak shoulder in experimental spectrum. Corresponds to point A' in LaB_6 case (Table 2) but not explicitly mentioned in reference

erts a direct effect on the stability of the B_6 sub-lattice and this explains the observation that certain alkali metals form hexaborides (HAGENMULLER and NASLAIN, 1963), since here only one electron per unit cell is missing. The alkali metals have low s-orbital ionization potentials and hence the electron would be strongly delocalized to the boron cage, while the lowest energy band would be less broadened. Similarly, removal of metal from a hexaboride can be regarded simply in terms of removal of a proportion of electrons from the structure. If the metal contributes two electrons, then a 20% metal deficiency is equivalent to a loss of 0.4 electrons from the valence band. This merely empties a proportion of states at the top of the valence band. The very large number of states in this region acts as a buffer towards metal loss and so we would expect the overall lattice to remain fairly stable. Metal loss, of course, does also reduce the Madelung energy and hence one would not expect the lattice to be stable to metal loss equivalent to one electron per unit cell. It has also been observed that there are narrow limits to the ranges of first and second ionization potentials of elements which form hexaborides. The band structure for neutral B_6 provides a natural explanation of this point. High first and second ionization potentials cause the metal orbitals to appear in the valence area of the band structure rather than the conduction bands. These extra valence bands would then absorb electrons, leaving the higher valence bands, of cage character, unfilled. Essentially, this means that the metal electrons remain localized on these atoms. The LCAO band-structure calculation thus gives a guide to polarity and the properties of two-electron bonds. However, it yields no information about the distribution of electrons in space.

The *ab initio* calculation of the electronic structure of molecular B_6^{2-}, carried out using a basis set of Gaussian orbitals, shows an interesting electron distribution. Figure 7a shows that there is a build-up of electrons inside the cage - the section is across the fourfold axis of the octahedron - and Fig. 7b shows a similar concentration on the eight trigonal faces of the cage. Simply looked at, the stability of the system can be understood in terms of a central mass of electron density to which each positive nucleus is coulombically attracted. In the solid this situation still obtains, supplemented by the additional inter-cage bonds.

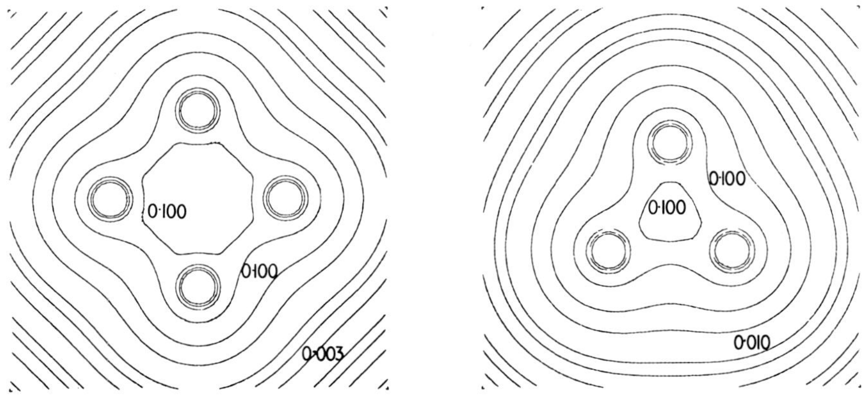

(a) (b)

Fig. 7a and b. (a) Electron density contours for the B_6^{2-} cage across a C axis; (b) Electron density contours for the B_6^{2-} cage across a C axis

β) *La Hexaboride*

The band structure for this system is shown in Figure 8. As expected, the general levels of the bands in the absolute scale are rather similar to those of BaB_6. The difference is that not only are the ten valence levels filled but the third electron of La occupies the conduction levels. Using the density-of-states calculation for the system, we can assign the Fermi level, which lies at -3.2 eV. This value is in satisfactory agreement with the experimental work function which, at 2.74 eV (DECKER and STEBBINS, 1955; LENK, 1956; KU, 1962), is exceptionally low and in part accounts for the high thermionic emissivity of the material (WESTBROOK, 1960).

The calculated total density of states and partial density of boron 2p states are shown in Figure 9, and Table 2 lists the energy separations between peaks on Figure 9 and corresponding peaks on the measured boron K emission and absorption spectrum. The agreement is pleasing and encourages the belief that the calculational method is capable of affording reliable semi-quantitative results.

As in the alkaline earth hexaborides, there is considerable intermetallic bonding, as witnessed by the wide lowest band, which is partly of lanthanum s character. This intermetallic binding persists even though the La-La distance is increased from 0.374 nm in the pure metal (PEARSON, 1958) to 0.4154 nm in the hexaboride. The lower conduction states exhibit little La s-orbital participation, since the bands are degenerate at all symmetry points, so that mixing here is symmetry-forbidden. There can, of course, be s participation along line S,(XR) along Λ(ΓR) and along Z(XM), since there are non-degenerate respresentations in these directions. Of these, possibly the second is most important. This does not mean that electronic conduction does not involve the metal - if La d orbitals were included in the basis set, then they would undoubtedly turn up in this region of the band structure and would transform in the correct way to be involved in the conduction states.

The calculations described here enables a rough analysis of the electrical properties of LaB_6. In order to effect this, the Fermi surface is required and can be derived. Fermi surface occurs in three bands,

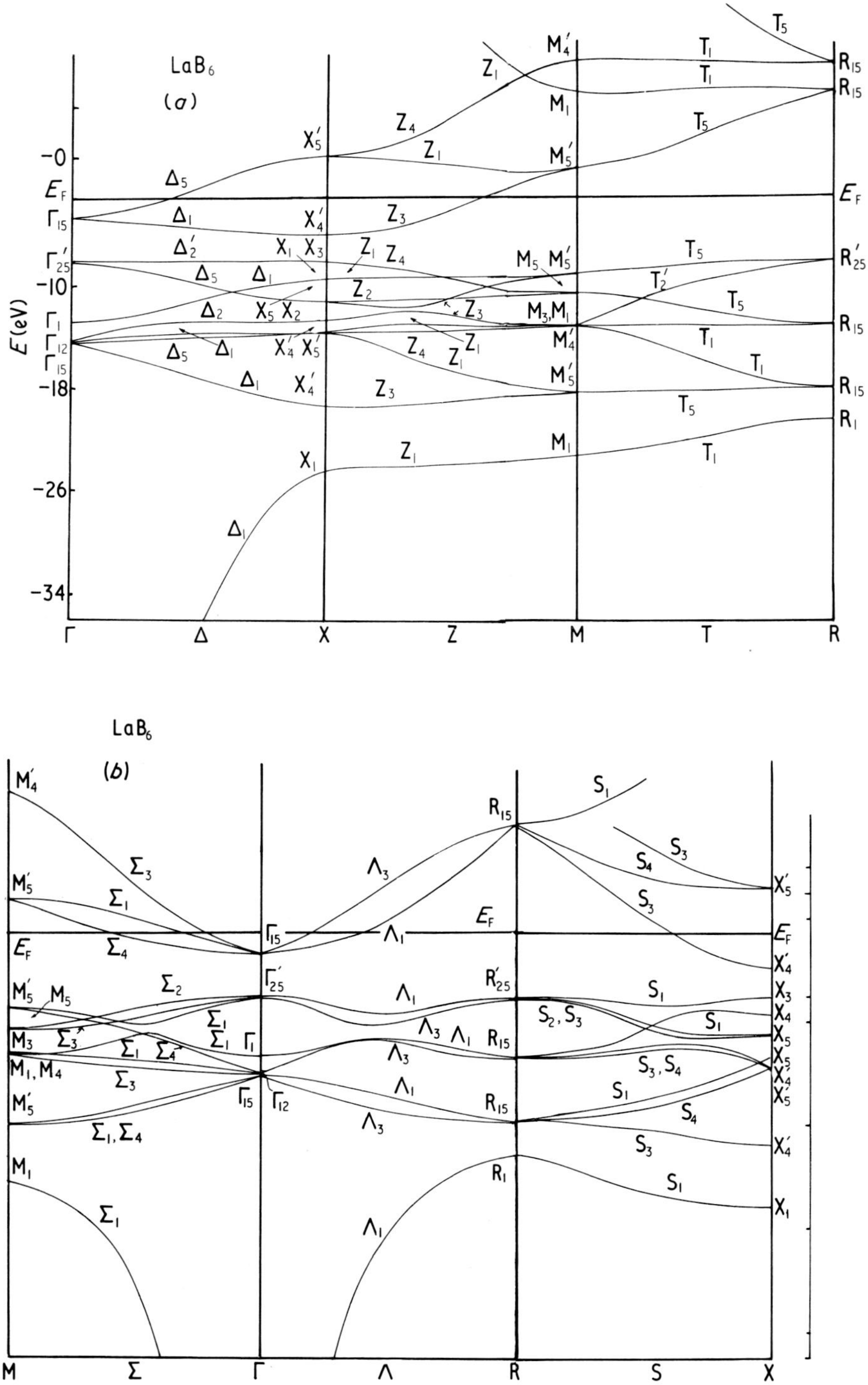

Fig. 8a and b. Band structure for LaB_6

Table 2. LaB_6 boron K emission and absorption spectrum

Point on exp. spectrum	Energy differences (eV) exp.	calculated
E		
	5.9	5.8
D		
	4.4	3.2
C		
	1.2	2.2
B		
	2.9	2.6
A'		
	1.1	1.2
A		
	3.0	2.8
a[b]		
	2.7	2.5
b[b]		
	3.9	>2.5[a]
c[b]		
	2.4	-
d[b]		

[a] the calculation does not extend far enough to identify the position of the relevant absorption peak

[b] absorption peaks

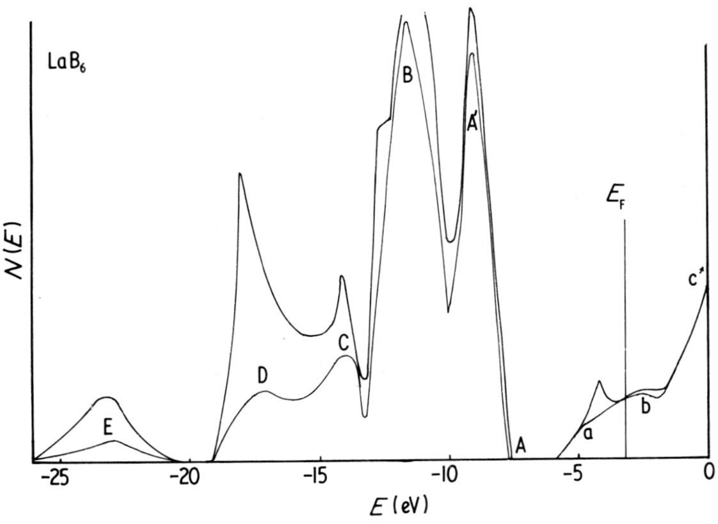

Fig. 9. Total density of states (outer line) and density of boron 2p states (inner line) for LaB_6

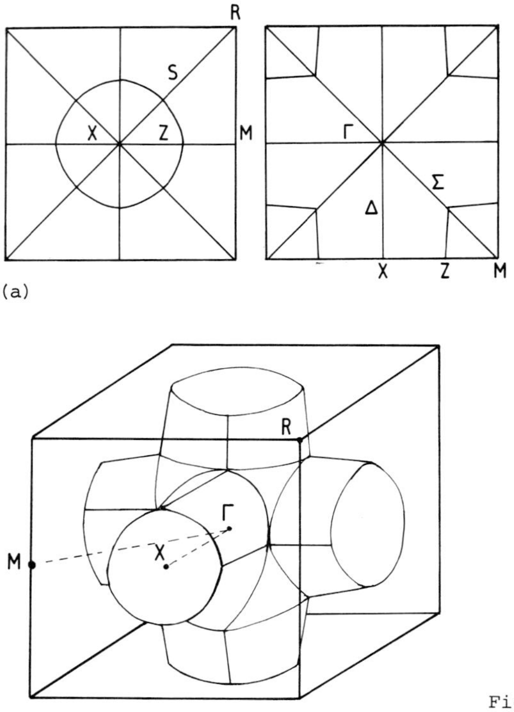

(b)

Fig. 10a and b. Fermi surface in 11th band for LaB_6

the contribution of the eleventh band (counting up from the lowest valence band) being most important. It is essentially based on the free-electron sphere but impinges strongly on the XMR face, giving a large, circular filled region around X. The Fermi surface of this band is probably fairly reliable, since spin-orbit coupling is only likely to modify the band energies at Γ_{15} in the conduction bands. This point has neither La d orbital nor s orbital involvement and so spin-orbit effects will be small. Figure 10 illustrates the situation. This suggests a Fermi surface for this band which is multiply connected through K space, and the accuracy of this prediction could be checked by a study of the magneto-resistance of the crystal at high magnetic fields. In the twelfth and thirteenth bands the Fermi surfaces are essentially distorted free electron spheres around Γ (Fig. 11a and b).

Neither meets the zone boundary. It is somewhat difficult to make direct comparisons between this picture and the Fermi surface of the normal form of La metal itself since the latter has the double hcp. structure, with four atoms per unit cell. However, we can note that dhcp. La metal also has a multiply-connected region of electrons in one band which comprises a large proportion of the overall Fermi surface (FLEMING, LIV and LOUCKS, 1968). However, the band structure of the high-temperature fcc. phase of La calculated by MYRON and LIU (1970) shows a simpler picture, i.e. the Fermi surface appears to be restricted to one band only. This is more comparable with the present result for LaB_6. The Fermi surface of La (fcc) in the (100) plane resembles that of the 11th band of LaB_6 quite markedly. The widest conduction band in La metal is ~2.4 eV from MYRON and LIU's and FLEMING et al. calculations and from one performed by the same method as here for LaB_6. In the hexaboride a number of conduction bands in the ΓM,

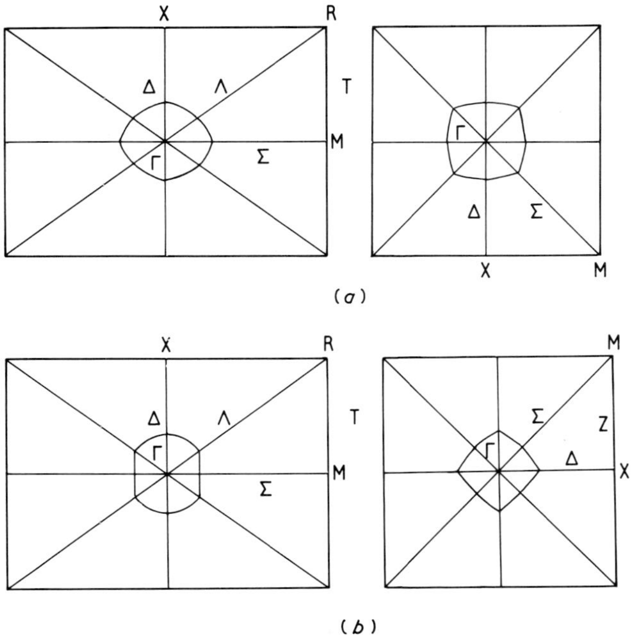

Fig. 11a and b. (a) Fermi surface in 12th band for LaB_6. (b) Fermi surface in 13th band for LaB_6

ΓR, and ΓX directions are considerably wider than this, and we would therefore expect to find rather lower effective masses in the hexaboride and possible higher electrical conductances. All the bands in the conduction region are free-electron-like and are associated with low principal effective masses for the conduction electrons. The areas in reciprocal space of the Fermi surfaces of LaB_6 and La itself have not been compared, but it does not seem that this quantity will be significantly less for the hexaboride than for the metal. Moreover, the phonons associated with the B_6 sub-lattice are rather less easily thermally excited than those of a heavy metal lattice, and hence the increase of electrical resistance with increased temperature, though still "metallic", should be rather less marked than it is with a metal. In this context, the vibration frequencies of the B_6 cages correlate with a high melting point for LaB_6 (> 2500°C) (JOHNSON and DAANE, 1961). Use of these data affords a very high mean free path (ZIMAN, 1963) for conduction electrons, although no doubt the B_6 cages are efficient scattering sites.

Measurement of the Fermi surface of each band allows rough computation of the individual contributions to the conductance (taken for convenience to be a scalar). The value which emerges is $6 \times 10^5\ \Omega^{-1}\ cm^{-1}$. In a recent paper MERCURIO et al. (1974) report this quantity as $1.16 \times 10^5\ \Omega^{-1}\ cm^{-1}$ at 300 K. The agreement between the two figures is very satisfactory, and hence overall one can rationalize well the observed electrical conductivity. The Hall coefficient is inversely proportional to the Fermi surface area and to the Fermi velocity of the electrons. Because of the multiple-band composition of the Fermi surface, the conduction electrons have a range of Fermi velocities, but if we again restrict ourselves to the 11th band the separate relevant contributions to the Hall effect can be computed. The quantity

$$\left(\frac{\partial^2 E}{\partial K^2}\right)$$

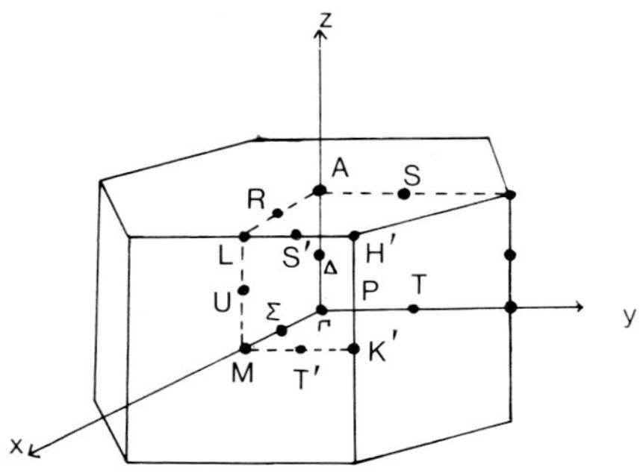

Fig. 12. Representation domain for hexagonal reciprocal lattice

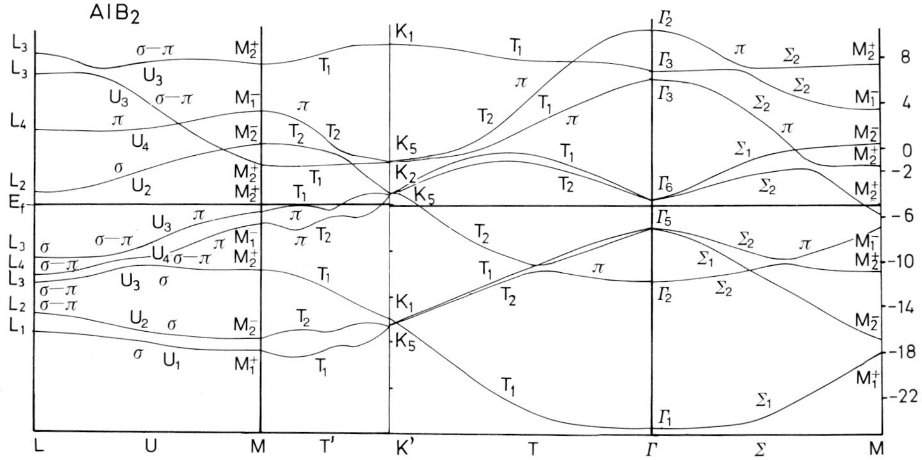

Fig. 13 a

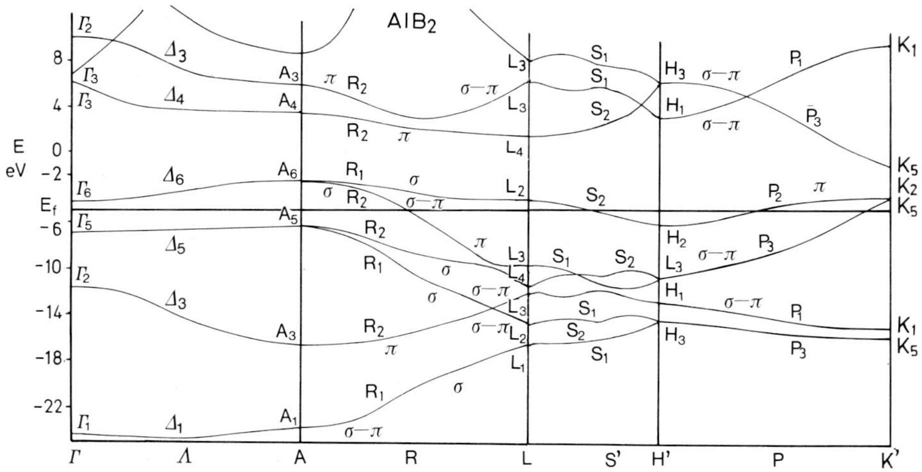

Fig. 13 b

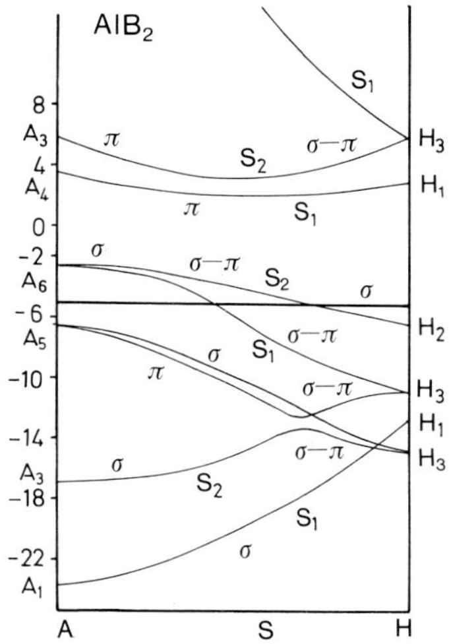

Fig. 13 c

Fig. 13a-c. Band structure for AlB_6

is positive, and the Hall coefficient is in satisfactory agreement with the experimental value (~4.0 × 10^{-4} cm^3 $coulomb^{-1}$ (PADERNO and SAMSONOV, 1960)).

The density-of-states and partial-density-of-states calculations lead to prediction of the polarity of the overall system as $La^{+1.75}B^{-1.75}$. Thus one sees that the third electron on the central metal atom is almost completely transferred to the B_6 cage. This should "saturate" the cage and hence it is expected that central atoms with four valence electrons available (eg. Th) will retain two and only increase the negative charge on the cage marginally.

b) Diborides

α) *Aluminum Diboride*

Early considerations by LIPSCOMB and BRITTON (1960) suggested that, in a diboride such as AlB_2, two electrons were transferred to the B_2 unit by the metal, so giving rise to a graphite-like two-dimensional net. The remaining free electrons on the metal would then account for the metallic conductivity exhibited by many of these materials. JURETSCHKE and STEINITZ (1958), however, postulated that electrons were transferred from the borons into the empty metallic d orbitals forming the metal bands. Of these alternatives, that of LIPSCOMB and BRITTON appears to be more consistent with the known structural data on diborides.

The only band-structure calculation for a diboride is that by ARMSTRONG et al. Here the same method as for hexaborides was employed. The Brillouin Zone is shown in Figure 12 and the band structure around the symmetry lines is illustrated in Figure 13. The main changes are seen to be in the radial directions: the bands from Γ-K, Γ-M broaden considerably. The main point of interest lies in the fact that the odd electron of aluminum lies in unfilled bands. If one follows around the K space circuit ΓKMΓ then pronounced graphite structure of the

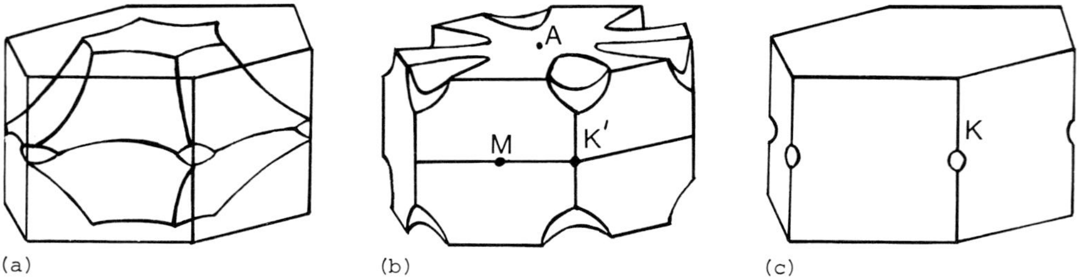

Fig. 14a-c. (a) Fermi surface in 5th band for AlB_2; (b) Fermi surface in 6th band for AlB_2; (c) Hole surface in 4th band for AlB_2

bands is evident. These bands separate effectively into σ and π types, and these are strongly associated with the boron net. In fact, it is quite clear from the filling of the band that electrons are delocalized from aluminum into the boron orbitals, which are π-bonding in the two-dimensional symmetry of the layer. Indeed, the analysis of the density of states leads to polarity $Al^{+1.884}$ $B^{-0.942}$. This is just the description earlier proposed by LIPSCOMB and BRITTON, and indeed by symmetry; since the diboride net has almost one extra electron per atom, the π-bond order between each boron pair must approach 2/3. It is difficult to see what experimental evidence of e.g. spectroscopic nature one would seek to clarify the nature of the boron-boron π bonding, since there is little in boron chemistry with which to compare the diborides. The nearest relatives are members of the B_2X_4 series of molecules e.g. B_2Cl_4 and $B_2NMe_2)_4$ and here there is only a minor π component in the B-B bond (ARMSTRONG et al., 1971; PERKINS and WALL, 1966). However, it is interesting that the latter bonds are associated with a B-B interatomic distance of 0.170 nm, i.e. almost the same as that in the solid diboride (~ 0.172 nm).

The polarity of the system means that ~1.1 electrons are retained on each aluminum atom and can contribute to the electrical and galvanomagnetic effects. The Fermi surfaces for electrons and holes are shown in Figure 14. The main surface (Fig. 14a) stems from the 5th band, while in Figure 14b there is only a spilling over of electrons into the 6th band in the vicinity of the points H. There is also a hole surface (Fig. 14c) in the 4th band and hence a negative component of the effective mass tensor. This latter band resembles the situation in graphite somewhat (HERMAN, 1958, STEWART, pers. comm.), although in AlB_2 the states stem from the Al 3p orbitals. In the conduction bands as a whole, the states are mixtures of boron π and the aluminum $3p_{x},_{y}3p$ degenerate pair of levels.

The conductivity is obviously fairly anisotropic because of the different effective masses in each direction in k-space. However, unlike graphite, where the longitudinal/transverse mass ratio is 10^2, there should be an appreciable contribution to the conductivity along the c-axis. This will come from electron waves along the KH' line, where there is fairly free-electron-like behavior.

If, for simplicity, the conductance is assumed scalar, then it is calculated to be $3.5 \times 10^5\ \Omega^{-1}\ cm^{-1}$. This falls in the region of the experimental value for the closely-related YB_2 (JOHNSON and DAANE, 1963). The Hall coefficient also depends on the Fermi surface and is obtained from the additive effects of electrons and holes. Overall this has a negative sign, which is consistent with the main bulk of charge carriers being electrons.

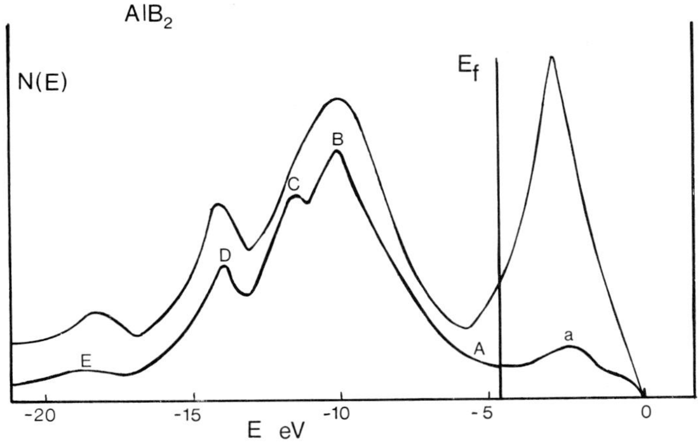

Fig. 15. Total density of states (outer line) and density of boron 2p states (inner line) for AlB_2

Table 3. AlB_2, TiB_2 boron K emission and absorption spectrum

Point on exp. spectrum	Energy differences (eV) exp. (TiB_2)	calculated (AlB_2)
E		
E–D	5.4	4.7
D–C	2.4	2.4
C–B	4.3	1.6
B–A	4.2	4.3
A–a[a]	3.0	3.0

[a]absorption peak

The total density of states and that of boron 2p states are plotted in Figure 15, and in Table 3 the calculated energy separations between peaks of the latter are compared with those computed from the boron K emission and absorption spectrum of TiB_2 (LYAKHOVSKAYA et al., 1970). The main differences arise in the high-intensity band, where the titanium d orbitals play their major role.

The band structure of AlB_2 allows reasonable speculation on the electronic properties of MgB_2. From a simple electron count, one would naively expect that in MgB_2 there are four filled valence bands and a band gap. Reference to Figure 13 shows that this is not the case. Because of band overlap and degeneracy of the 4th and 5th bands along P, substantial sections of a number of bands remain exposed as Fermi surface, since the Fermi level decreases only to -5.4 eV. Because of the lower s ionization potential of Mg, the intermetallic bonding (in the lowest band) will be substantially raised in the energy scale and weakened. Similarly, the 3p orbitals of magnesium will lie higher in the absolute energy scale and so the strongly interacting "π type" orbitals on boron and aluminum which give rise to the 4th band in AlB_2

will be shifted upwards in MgB_2. This means that the "extra" electron of Al not only stabilizes the diboride system by improving the energy of the metal-metal bond but, perhaps more cogently, it stabilizes the Al-B bonds. The relative instability of MgB_2 is consistent with this pattern.

One can get a very rough idea of the band structure of TiB_2 by adding one electron to the AlB_2 picture. This makes a new Fermi level at ~ 3 eV and cuts down the Fermi surface area considerably. This is consistent with the large value of the experimental Hall coefficient (JURETSCHKE and STEINITZ, 1958). The high electrical conductivity (JURETSCHKE and STEINITZ, 1958) of this and other group IVA diborides will be due to: (i) the long mean free paths for the electrons (the melting point for this material (THOMPSON, 1970) is in the region of 3000°C); (ii) the low effective electron mass in the basal plane.

References

ARMSTRONG, D.R., BREEZE, A., PERKINS, P.G.: On the electronic structure of some metal hexaborides. J. Phys. (C) 8, 3558-3570 (1975)

ARMSTRONG, D.R., PERKINS, P.G., STEWART, J.J.P.: Calculation of the electronic structure of boranes by the self-consistent molecular orbital method. Part I: $B_{10}H_{14}$ and some neutral chloroboranes. J. Chem. Soc. Dalton Trans., 3674-3679 (1971)

ARMSTRONG, D.R., PERKINS, P.G., STEWART, J.J.P.: Calculation of the electronic structure of boranes by the self-consistent molecular orbital method. Part II: Highly symmetrical cage anions. J. Chem. Soc., Dalton Trans., 627-632 (1973)

ARONSSON, B.: In: Borides, Silicides and Phosphides. B. Aronsson, R. Lundström and S. Lundquist (eds.). New York: Wiley and Sons, 1965, p. 32

BALLHAUSEN, C.J., GRAY, H.B.: The electronic structure of the vanadyl ion. Inorg. Chem. 1, 111-122 (1962)

BLUM, P., BERTAUT, F.: Contribution à l'étude des borures à teneur élevée en bore. Acta Cryst. 7, 81-86 (1954)

DECKER, R.W., STEBBINS, D.W.: Photoelectric work functions of borides of La, Pr and Nd. J. Appl. Phys. 26, 1004-1006 (1955)

FLEMING, G.S., LIU, S.H., LOUCKS, T.L.: Fermi surfaces for DHCP La, Nd and Pr. Relation to magnetic ordering and crystal structure. Phys. Rev. Lett. 21, 1524-1526 (1968)

HAGENMULLER, P., NASLAIN, R.: An alkali boride: the hexaboride, NaB_6. C. R. Acad. Sci. (Paris) 257, 1294-1296 (1963)

HERMAN, F.: Theoretical investigation of the electronic energy band structure of solids. Rev. Mod. Phys. 30, 102-121 (1958)

JOHNSON, R.W., DAANE, A.H.: The lanthanum-boron system. J. Phys. Chem. 65, 909-915 (1961)

JOHNSON, R.W., DAANE, A.H.: Electron requirements of bonds in metal borides. J. Chem. Phys. 38, 425-432 (1963)

JURETSCHKE, H.J., STEINITZ, R.: Hall effect and electrical conductivity of transition metal diborides. J. Phys. Chem. Solids 4, 118-127 (1958)

KUDINT'SEVA, G.A., TSAREV, B.M.: Radistkh. Elektron 3, 428 (1958)

LAPLACA, S.: Meet. Amer. Cryst. Assoc. Suffern (N.Y.) Feb. 1965

LENK, R.: Czech. J. Phys. 6, 625 (1956)

LEVISON, K.A., PERKINS, P.G.: unpublished

LIPSCOMB, W.N., BRITTON, D.: Valence structure of the higher borides. J. Chem. Phys. 33, 275-280 (1960)

LONGUET-HIGGINS, H.C., ROBERTS, M.D.: The electronic structure of the borides MB_6. Proc. Roy. Soc. 224, 336-347 (1954)

LYAKOVSKAYA, I.I., ZIMKINA, T.M., FOMICHEV, V.A.: Boron K spectra from transition metal diborides and LaB_6, BaB_6 and AsB compounds. Sov. Phys. Solid State 12, 138-143 (1970)

McALOON, B.J., PERKINS, P.G.: Semiempirical LCAOMO theory for infinite systems. J. Chem. Soc. Faraday Trans. II 68, 1121-1126 (1972)
MERCURIO, J.P., ETOURNEAU, J., NASLAIN, R., HAGENMULLER, P., GOODENOUGH, J.B.: J. Solid State Chem. 9, 37 (1974)
MIKHAILOVS'KII, B.I.: Ukr. Fiz. Zhur. 7, 75 (1962)
MUETTERTIES, E.L.: The chemistry of boron and its compounds. New York: Wiley and Sons, 1967
MYRON, H.W., LIU, S.H.: Energy bands for FCC La and Pr. Phys. Rev. B1, 2414-2417 (1970)
PADERNO, Y.B., SAMSONOV, G.V.: Dokl. Akad. Nauk. SSSR 137, 646 (1960)
PEARSON, W.B.: Handbook of lattice spacings and structures for metals and alloys. New York: Pergamon Press, 1958
PERKINS, P.G., WALL, D.H.: Self-consistent molecular orbital calculations on aminoboranes. J. Chem. Soc., 1207-1211 (1966)
SAMSONOV, G.V., GRODSTEIN, A.E.: Zhur. Fiz. Khim. 30, 379 (1956)
STEWART, J.J.P.: personal communication
THOMPSON, R.: The chemistry of metal borides and related compounds. In: Progress in Boron Chemistry. New York: Pergamon Press 2, 173-230 (1970)
WESTBROOK, J.H.: Proc. Int. Symp. High Temp. Technol. New York: McGraw Hill, 1960
YAMAZAKI, M.: Group theoretical treatment of the energy bands in metal borides. J. Phys. Soc. Japan 12, 1-6 (1957)

IV. Boron and Aluminum Dodecaboride as the Specific Type of Hopping Conduction Materials

A. A. BEREZIN, O. A. GOLIKOVA, V. R. ZAITSEV, M. M. KAZANIN, V. M. ORLOV, and E. N. TKALENKO

1. Introduction

It is known that hopping conductivity is one of the main physical features of many low mobility materials. Hopping conduction, in contrast to the more common zonic type, is the movement of the charge carriers (electron or hole) by a series of successive hops from one local level to another.

At the present time we may tentatively divide hopping conduction materials into three groups according to the method by which they generate local (or quasilocal) levels:

1. Substances with polaron conduction (e.g. NiO, MnO, TiO_2 etc.).
2. Doped (partly compensated) semiconductors.
3. Amorphous semiconductors.

Early reports on electrical properties of boron and some borides presented strong arguments for the existence of hopping conductivity in these materials. This article contains a review of our investigations of crystalline β-rhombohedral boron, amorphous boron and aluminum dodecaboride (α - AlB_{12}). The results of these investigations provide a definite basis for grouping the above-mentioned materials with the more commonly accepted group of the hopping conduction type.

First, we shall briefly summarize some of the principal experimental arguments in favor of the nonstandard (nonzonic) character of the electroconductivity of the materials. Next, we will consider in more detail the conductivity of β-boron at low temperatures and in strong electrical fields. Finally, we shall discuss the conductivity properties of amorphous boron.

2. Nonstandard Character of the Conductivity of Boron and Aluminum Dodecaboride

In recent years, many workers on boron and borides have reported data on conductivity (σ), thermeolectric power (α), Hall coefficient (R), magnetoresistance ($\Delta\rho/\rho$) and frequency dependences of σ at various temperatures and different carrier concentrations. (GOLIKOVA et al., 1969, 1970a, b, c, 1972; AVDEENKO et al., 1968) The results of these studies have established the following characteristics of the boron conductivity mechanism:

1. Low mobilities (from Hall effect); $u_H \lesssim 1$ $cm^2/V{\cdot}sec$).
2. Strong disagreement between the values of Hall mobility and the mobility from magnetoresistance ($u_{\Delta\rho/\rho} \gg u_H$).

3. In heavily doped β-boron conductivity increases exponentially with temperature (300°K - 1500°K) and thermoelectric power increases simultaneously.

4. The conductivity depends on the power-type frequency (where s = 0.4-0.7).

All of these boron properties cannot be explained by zonic-like movement of holes in the valence band and electrons in the conduction band. Thus, it is possible to arrive at some conclusions as to the existence of the hopping conduction mechanism in β-boron. On the basis of its properties β-boron cannot be definitely classifies into any known group of low-mobility crystalline materials (molecular crystals, semiconductors with impurity hopping conduction, or materials with current carriers in the form of small polarons). At the same time, large dimensions and a very complex structure of the unit cell of β-boron suggest behavior similar to that of amorphous substances in which the hopping character of the current movement is confirmed by experiments at low temperatures and in strong electric fields. This has been referred to as "amorphous conception" (BEREZIN et al., 1971).

It should be mentioned that β-boron and amorphous solids have been compared with respect to vibrational spectra and heat carrying ability. PETROV et al. (1969) have interpreted the relationship between temperature and the thermal conductivity of β-boron using hypothesis that long-wave-length ("crystalline") phonons play a dominant role in heat transfer at low temperatures while short-wave-length ("amorphous") phonons become more important at higher temperatures.

Investigation of aluminum dodecaboride (AlB_{12}), which may be considered a crystal-chemical analog of β-boron, has shown that the electrical properties of AlB_{12} and β-boron are similar in many respects (BEREZIN et al., 1973a). For example, the activation conductivity and thermoelectric power of AlB_{12} increase simultaneously. Also, the Hall coefficient for AlB_{12} remains practically constant for a wide range of temperatures (BEREZIN et al., 1973a). These findings are similar to those observed earlier for doped boron (GOLIKOVA et al., 1968).

3. Hopping Conduction of β-Boron and Amorphous Boron in the Strong Electric Field

The most characteristic feature of hopping conduction materials is in nature of dependence of their conductivity on temperature T and on the strength of electrical field E. When the conduction occurs as a result of thermally activated hops of carriers and the local trapping levels are randomly distributed, the conductivity obeys the relation known as Mott's law provided that the temperature is sufficiently low:

$$\sigma \sim \sigma_0 \exp\left[-\left(\frac{To}{T}\right)^{\frac{1}{4}}\right] \qquad (1)$$

When σ obeys Eq. (1), a plot of lg σ versus 1/T represents a sliding activation energy which decreases as the temperature decreases.

In the case of hopping conduction in a strong electric field at low temperatures, the conductivity is determined by the mechanism of activationless tunneling. MOTT (1970, 1971) was also the first to consider this type of hopping conductivity.

The appropriate form of the dependence of conductivity in a strong electrical field was obtained by SHKLOVSKY (1972a). This law (the fieldhopping law of MOTT and SHKLOVSKY) is represented as follows:

$$\sigma \sim \sigma_0 \exp\left[-\left(\frac{E_o}{E}\right)^{\frac{1}{4}}\right] \qquad (2)$$

It should be noted that E_o does not depend on temperature. The transition from Eq.(1) to Eq.(2) takes place when E increases or T decreases. This transition is caused by the predominance of activationless hops of the carriers over the temperature activated hops. The probability of such hops is determined by the tunnel factor $e^{-2\gamma R}$, where R is the distance between a pair of trapping centers and γ is the decay constant of the wave function of the localized state.

The temperature dependence of $\rho \equiv \sigma^{-1}$ for zone-melted β-rhombohedral boron in different electric fields is given in Figure 1(A). The plot of ρ (T) dependence for the lowest field used (1470 V/cm) is shown in a plot of lg ρ versus $T^{-1/4}$ [Fig. 1(B)]. The same temperature dependences for amorphous boron (low field case) are given in Figure 2(A/B), and for AlB_{12} in Figure 3(A/B). At low temperatures we can see the $T^{-1/4}$-dependence of lg ρ which indicates a hopping conduction mechanism (WERHEIT and LEIS, 1970; BEREZIN et al., 1973b). At T> 600K the region of intrinsic conductivity begins in both amorphous and β-boron. The depth of the thermal energy gap in amorphous boron (Eg = 1.4 eV) is somewhat less than that in β-boron.

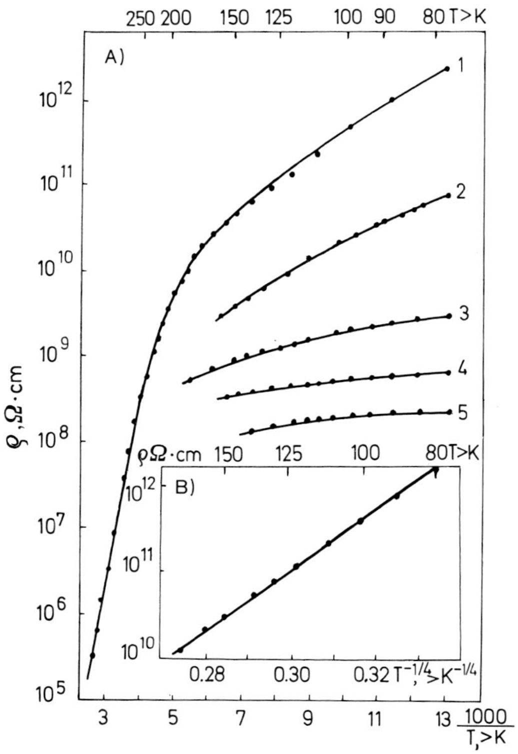

Fig. 1. (A) The temperature dependences of lg ρ versus 1/T for zone-melted β-boron in various electric fields (1: - 1470 V/cm, 2: - 14,700 V/cm, 3: - 59,000 V/cm, 4: - 10,000 V/cm, 5: - 150,000 V/cm). The thickness of the sample is 170 μ. (B) The dependence of lg ρ on $T^{-1/4}$ for β-boron in low field (1470 V/cm). The inclination of the curve corresponds to value T_o = 6.10^7 in Mott's law [Eq. (1)]

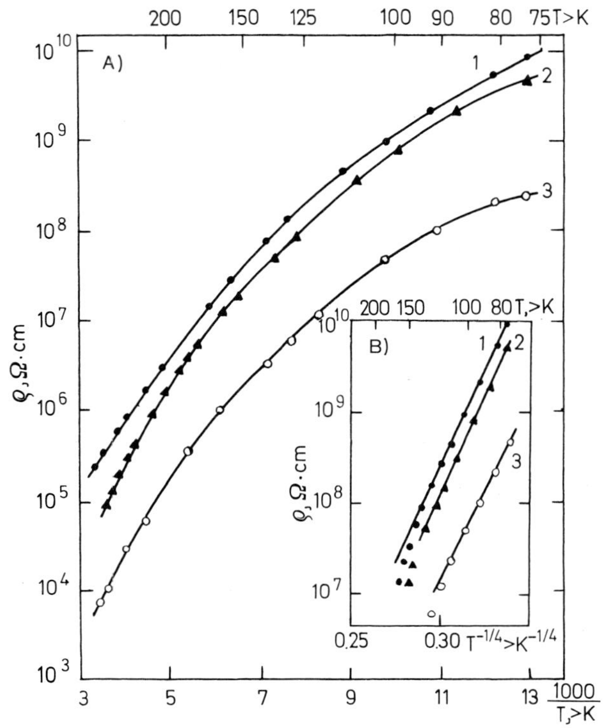

Fig. 2A and B. The dependence of lg ρ on $\frac{1}{T}$ (A) and on $T^{-1/4}$ (A) for three samples of amorphous boron (1 - 3) with different resistances. The value of T_o in Mott's law is near to 10^8 °K

The inclination of the curves in Figures 1(B) and 2(B) indicates that the values of T_o for both β-rhombohedral and amorphous boron are near 10^8K. The theory (AMBEGAOKAR et al., 1971) allows us to relate the characteristic temperature To in Eq.(1) to the value of G_o, which is the density of localized states at the FERMI level. In the case of properly disordered (i.e. impure or amorphous) semiconductors, the following relation holds:

$$KT_o \simeq \frac{\nu\gamma^3}{G_o} \qquad (3)$$

where γ is the decay constant of the wave function of the localized carrier ($\psi \sim \varepsilon^{-\gamma}$), and ν is the dimensionless constant (approx. 16). Let us use the formula:

$$G_o \sim \frac{\varepsilon_o N^{2/3}}{e^2} \qquad (4)$$

(SHKLOVSKY, 1972b) for the relationship between G_o and the density of localized levels N. Then we shall obtain the relationship:

$$N \sim \frac{(\nu\gamma^3 e^2)}{\varepsilon_o\ \kappa\ T_o} \qquad (5)$$

where ε_o is the static dielectric constant (for β-boron $\varepsilon_o \simeq 10$ and for amorphous boron $\varepsilon_o \simeq 9$).

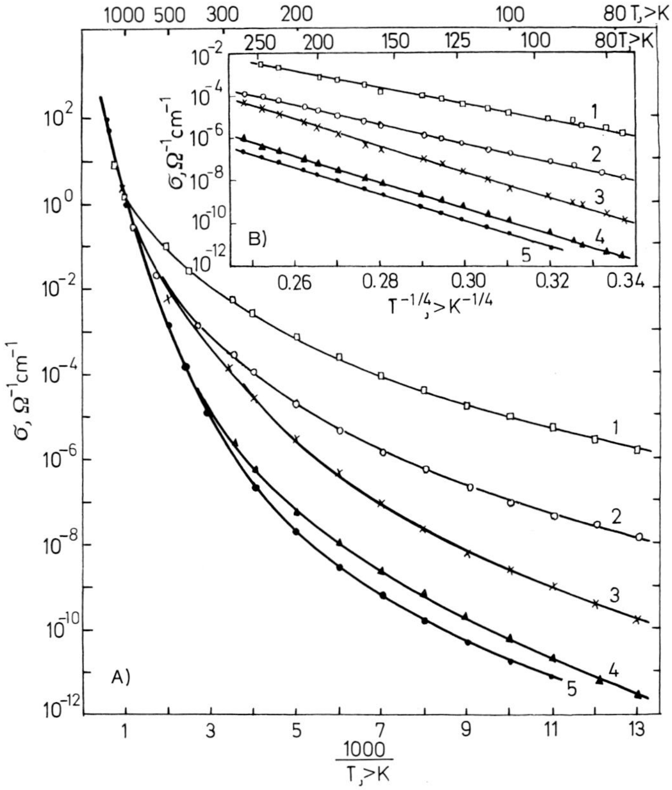

Fig. 3A and B. The dependence of lg σ on 1/T (A) and on $T^{-1/4}$ (B) for five samples of AlB_{12}. For these samples $T_o = 10^8 - 10^9$ °K

To estimate the value of N it is necessary to know the value of γ. For an appropriate estimation of N we shall take the value $\gamma^{-1} \sim 3$ Å which was previously calculated by BEREZIN (1973) in his work on the polarized point atom lattice model. (For the sake of simplicity the same value of γ is taken for the amorphous boron.) By use of Eq.(5) the estimation for N is given by:

$$N \sim 10^{18}\ \text{cm}^{-3} \tag{6}$$

Here we have ignored the difference between the values of N for β-rhombohedral and amorphous boron as the value represented is only an order-of-magnitude approximation.

Now let us consider the conductivity of β-boron and amorphous boron in a strong electrical field.

The current-voltage characteristics (IV-characteristics) of β-boron were measured at different temperatures (T = 77-143 K). The IV-characteristics of amorphous boron were measured at T = 77 K. Our IV-characteristics of β-boron and amorphous boron are given in Figures 4(A) and 5(A) respectively (see also Fig. 1(A), 4(B) and 5(B)).

As far as we know, the IV-characteristics of β-boron were accurately measured by PRUDENZIATI (1970) and PRUDENZIATI et al., (1973) and IV-characteristics of amorphous boron by MOORJANI and FELDMAN (1970). However, the IV-characteristics of β-boron were measured at comparatively low field levels. In the papers mentioned above, the IV-charac-

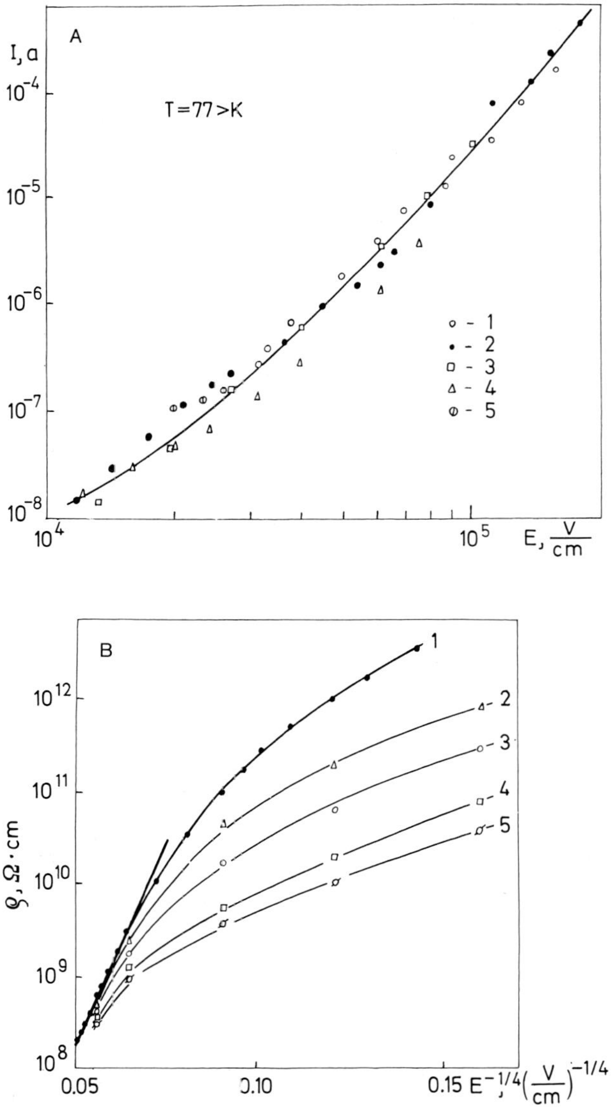

Fig. 4. (A) The dependences of current I on field E for the samples of β-boron with various thicknesses (1: 90 μ, 2: 170 μ, 3: 300 μ, 4: 500 μ, 5: 1500 μ), at T 300 = K.
(B) The dependence of lg ρ on $E^{-1/4}$ for β-boron for various temperatures 1: 77 K, 2: 83 K, 3: 100 K, 4: 125 K, 5: 143 K. The thickness of the sample is 170 μ

teristics were explained on the grounds of space-charge-limited currents. In order to check whether space-charge-limited currents have any appreciable influence on the formation of the IV-characteristics, we carried out a special investigation of the role of sample thickness on the specific resistance for thicknesses of 90 - 1500 μ for β-rhombohedral boron, and 90 - 780 μ for amorphous boron. In the case of our samples, space-charge-limited currents were not essential to the formation of IV-characteristics in β-boron and amorphous boron. The role of the contacts (injection of carriers from the contacts into the samples) was shown to be negligible. The IV-characteristics of amorphous boron for somewhat higher fields (E > 10 kV/cm) are consistent with the Poole-Frenkel law.

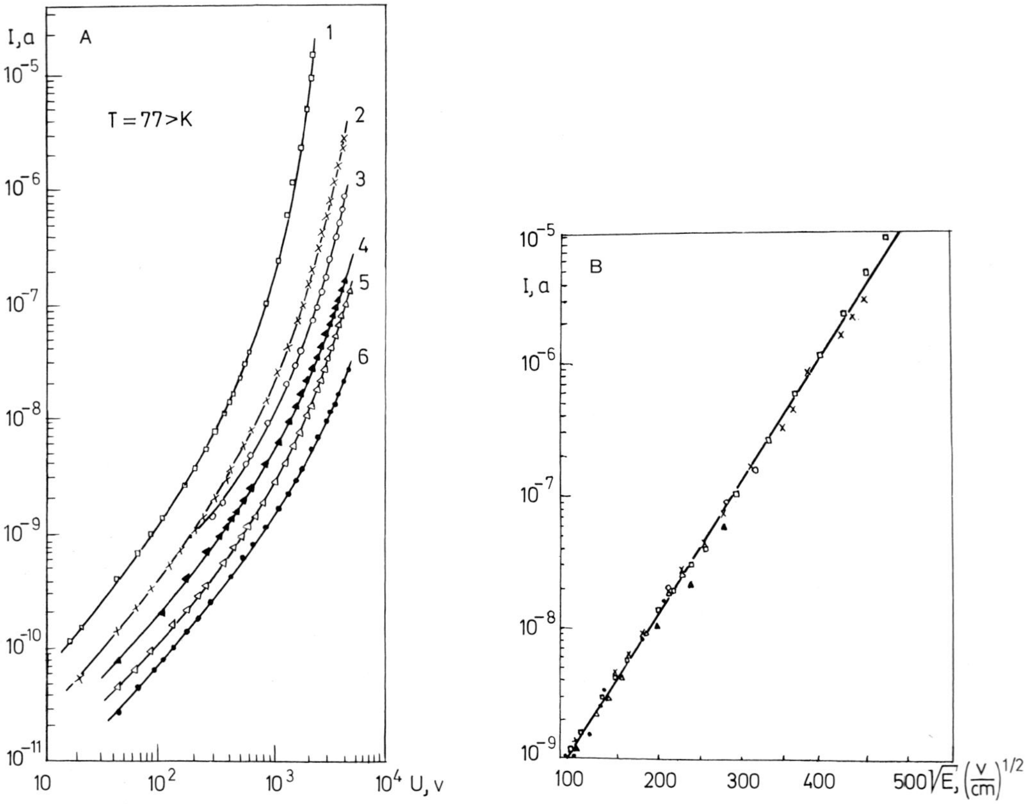

Fig. 5 (A, B). (A) The IV-characteristics of the amorphous boron at 77°K. The thickness of the samples are: 1: 90 μ, 2: 180 μ, 3: 200 μ, 4: 350 μ, 5: 450 μ and 6: 780 μ. (B) The dependence of I on $\sqrt{E}$ (Poole-Frankel coordinates) for the different samples of amorphous boron

$$\sigma \sim \sigma_0 \exp \frac{(\beta \sqrt{E})}{K\ T} \tag{7}$$

The straight part of lg σ (E) versus $\sqrt{E}$ plot in Figure 5(B) gives the following value for the Poole-Frankel constant:

$$\beta \text{ exper} \simeq 0.35 \text{ at. units} \tag{8}$$

This value agrees well with the theoretical calculated value:

$$\beta \text{ theor.} = \frac{\sqrt{e^3}}{\varepsilon_0} \simeq 0.33 \text{ at units} \tag{9}$$

The IV-characteristics of β-boron behave differently than those of amorphous boron. Figure 1(A) shows distinctly that when the applied electric field is increased, the conductivity becomes independent of temperature. This may be interpreted as the transition from temperature-dependent conduction ($T^{-1/4}$ Mott's law) to activationless field-stimulated hopping (MOTT, 1970, 1971; SHKLOVSKY, 1972a). The asymptotical inclination of all plots of Figure 4(B) in the strong field limit corresponds to the value $E_o \simeq 10^8$ V/cm in the field hopping law.

The applied electrical field E, the optimal hopping length R_{opt} and density of states at FERMI level g_o are related in this way: (SHKLOVSKY, 1972a; BEREZIN et al., 1973b).

$$eE\, R_{opt} = \frac{1}{4/3\ \pi\ R_{opt}\ G_o} \qquad (10)$$

Comparing Eq.(10) with Eqs.(4) and (2) one obtains

$$E_o = \frac{12\gamma^4}{\pi e G_o}; \qquad N \sim \left(\frac{12\gamma^4 e}{\pi \varepsilon_o E}\right)^{\frac{3}{2}} \qquad (11)$$

and consequently for N

$$N \sim 10^{22}\ cm^{-3}. \qquad (12)$$

In the region of validity of Eq.(2) the optimal hopping length is $R_{opt} \sim \left(\frac{1}{2\ \gamma}\right)\left(\frac{E_o}{E}\right)^{1/4}$. If we take $E = 3.10^4$ V/cm as an example, we obtain (γ^{-1} = 3 Å); i.e. $R_{opt} \simeq 11$ Å (the value of R_{opt} is not sensitive to E). This allows us to assume that the trapping centers responsible for such hops are generated by the presence of atoms, with coordination numbers of 8 and 9 which represent about 13% of all atoms in the complex lattice of β-rhombohedral boron.

By comparing the densities of local levels in β-boron resulting from the temperature-dependent Mott's law [Eq.(1)], and field-independent Mott-Shklovsky's law [Eq.(2)], we see that Equation (6) gives lower value than Equation (11). This difference may be the consequence of the existence of the temperature-activated polaronic contribution to T_o in the case of thermally activated hopping [Eq.(1)]. In the case of temperature-independent field assisted hopping [Eq.(2)] this contribution disappears.

It should be mentioned that one needs very low temperatures (several °K), for the observation of the specific-hopping dependences [Eqs.(1) and (2)] in the impure compensated semiconductors. However, in the case of β-boron the corresponding region shifts to higher temperatures. This shift is the result of the relatively broad distribution in energies of the β-boron traps.

This is related to the fact that atoms with different coordination numbers have different chemical bonds. Furthermore, the relatively low dielectric constant of β-boron leads to comparatively small Bohr radius.

4. Other Results

a) Optical Measurements

Since the optical absorption of β-rhombohedral boron has been investigated in many papers, here we will only report our results on the reflection and absorption spectra of amorphous boron (Figure 6) and AlB_{12} (Fig. 7). The difference in the spectral behavior of the absorption coefficient of two samples of AlB_{12} (Fig. 7) appears to have been conditioned by the growth of the absorption gap caused by some elec-

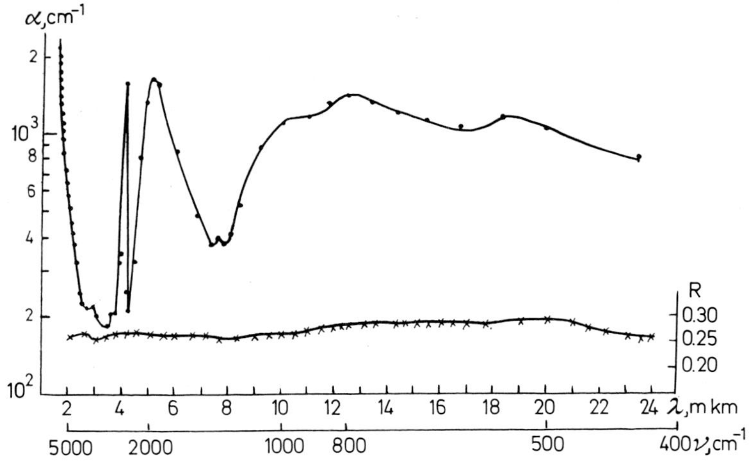

Fig. 6. The absorption and reflection spectra of amorphous boron

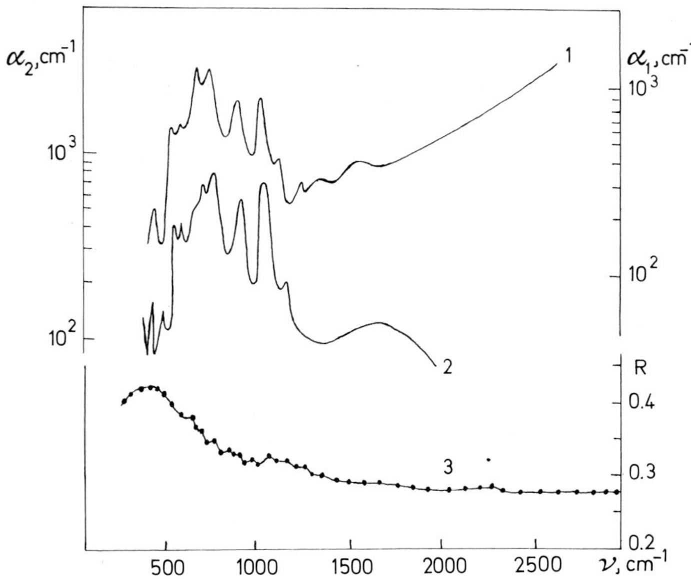

Fig. 7. The absorption (1,2) and reflection (3) spectra of AlB_{12}. The samples 1 and 2 have $\rho = 10^2$ Ω/cm and 10^6 Ω/cm at 300 K

tronic transitions. Thus one sample had values of $\rho = 10^2 \Omega$.cm at T = 300 K which can be compared with values of $\rho = 10^6$ Ω.cm at T = 300 K for the other.

It has been reported that AlB_{12} has an appreciable dispersion of dielectric constants ($\varepsilon_0 \simeq 19$, $\varepsilon_\infty \simeq 11$; BEREZIN et al., 1973a) and a noticeable dispersion ($\varepsilon_0 \simeq 10$, $\varepsilon_\infty \simeq 8$) has been reported for β-rhombohedral boron (GOLIKOVA et al., 1969). Yet in the case of amorphous boron there is no noticeable dispersion.

Consequently, the crystal lattices of β-boron and AlB_{12} should have some degree of ionization leading to a possible appearance of the polaronic effect. This would allow the current carrier to hop on the level of quasilocalization (see Sect. 3; BEREZIN et al., 1973). In the case of β-boron the appearance of an ionic component is obviously related to the presence of nonequivalent atoms with various coordination numbers (6, 8 and 9) in the unit cell. For AlB_{12}, the possibility of ionization is most clearly evident from the heteroatomic character of the crystal lattice. In amorphous boron the statistical equivalence of all the atoms and the absence of a specific lattice site leads to the absence of a noticeable ionic component (GOLIKOVA et al., 1973).

b) The Frequency Dependence of Conductivity

AVDEENKO et al. (1968) and GOLIKOVA et al. (1970c) reported data on conductivity of β-boron as a function of frequence. Their results are illustrated in Figure 8. These results may also be interpreted in favor of hopping mechanism. According to DAVIS and MOTT (1970):

$$\sigma(\omega) = \frac{\pi}{3}\{G_o(E_F)\}^2 \kappa T e^2 a^5 \omega \left[\ln \frac{\omega_{ph}}{\omega}\right]^4, \quad (13)$$

where Go (E_F) is the density of states at Fermi level. $a \equiv \gamma^{-1}$ is the radius of the wave function of the localized carrier and ω_{ph} - is the effective phonon frequency.

For approximation Eq.(13) may be illustrated by using the power function $\sigma \sim \omega^s$ $(s < 1)$. The results are presented in Figure 8 for s = 0.4 - 0.7.

For example, with values of $\omega = 10^6$ sec^{-1}, $\omega_{ph} = 485$ cm^{-1} (GOLIKOVA et al., 1969); T = 77 K and a = 3 Å one obtains

$$G_o\ (E_F) \sim 10^{20}\ eV^{-1}cm^{-3},\ N \sim \left(\frac{G_o e^2}{\varepsilon_o}\right)^{\frac{3}{2}} \sim 10^{18} cm^{-3} \quad (14)$$

which seems a reasonable estimation consistent with Eq.(6).

c) Magetoresistance

Up to some extent the magnetoresistance $(\Delta\rho/\rho)$ of doped β-boron (GOLIKOVA et al., 1968, 1970c) may also be interpreted in favor of the hopping model.

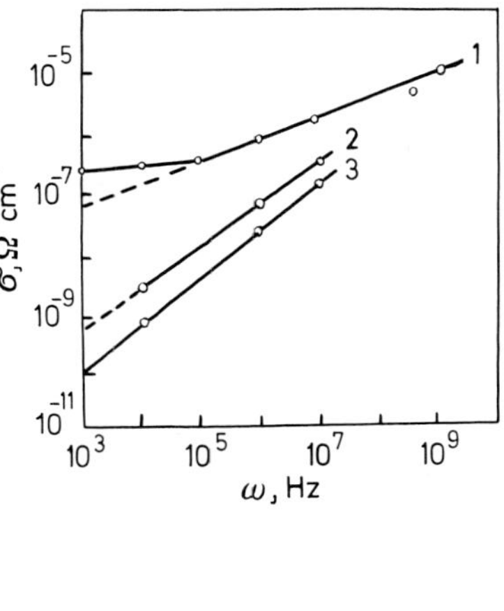

Fig. 8. The frequency dependence of conductivity of β-boron (1: 300^oK, 2: 166^oK, 3: 77^oK)

By producing the experimental results by the square dependence on the magnetic field

$$\frac{\Delta\rho}{\rho} \sim H^2 \tag{15}$$

and taking for $\Delta\rho/\rho$ relation obtained by SHKLOVSKY (1972b),

$$\frac{\Delta\rho}{\rho} \sim 0.01 \frac{e^2H^2}{\gamma^4 e^2 t^4} \left(\frac{To}{T}\right)^{\frac{3}{4}} \tag{16}$$

one may obtain an independent estimation for γ. If we take as an example $H = 1.4 \times 10^4$ Oe, T = 300 K, $\Delta\rho/\rho = 6.8 \times 10^{-3}$ and $T_o = 10^8$ K, we obtain

$$\gamma^{-1} \simeq 18\ \text{Å} \tag{17}$$

However, it should be mentioned that our measurements were carried out for doped, not pure boron. T_o in Eq.(16) was taken to have the value 10^8 K appropriate for pure zone melted β-boron. Due to this inconsistency, the value obtained by Equation (17) is doubtful.

d) Thermoelectric Power

The temperature dependence of thermoelectric power for β-boron, amorphous boron and AlB_{12} were reported recently (GOLIKOVA et al., 1968, 1973; BEREZIN et al., 1973a). Since there is no generally accepted theory of thermoelectric power in the region of hopping conductivity, it is suggested that the comparatively weak temperature dependence of thermoelectric power in these materials is a qualitative argument in favor of the hopping conductivity. The situation resembles the cases in which Morin's formula for thermoelectric power may be applied.

e) Temperature Dependence of Conductivity of AlB_{12}

The dependence of conductivity of AlB_{12} on temperature is illustrated by Figure 3. The low-temperature section (below 300 K) may be described satisfactorily by Mott's $T^{-1/4}$ - relation (1) (with $T_o = 10^8 - 10^9$ K). However, the σ (T) dependence for temperatures higher than 300 ^oK is of the activational type with the activation energy 0.2 - 0.4 eV (BEREZIN et al., 1973a). For the temperatures $T \gtrsim 900 - 1000$ ^oK the region of intrinsic conduction begins with $E_g = 2.0 - 2.5$ eV.

5. Conclusion

We will now summarize the results of our comparative study of β-boron, AlB_{12} and amorphous boron.

β-boron is somewhat similar to amorphous material because of the large dimensions and the complex structure of its unit cell. The unit cell of β-boron contains 105 atoms and is constructed of a complex arrangement of B_{12} icosahedra. There are three groups of atoms with respect to their coordination numbers. The coordination numbers (Z) for 91, 12 and 2 atoms are Z = 6, 8 and 9, respectively. This circumstance, along with the distortion of icosahedra, results in the existence of perfect equivalence only for atoms separated from each other by a space of about ten interatomic distances. This holds true at least for atoms with Z = 8 and 9. Moreover, if the degeneration between the energy levels in such complex lattices is not very strong, then this degeneration can be comparatively easily influences by either thermal vibrations or by impurities.

AlB_{12} is similar in complexity to β-boron. It has about 200 atoms in its unit cell. In addition to β-boron, AlB_{12} has B_{12} icosahedrons as the main structural units of its crystal lattice.

Thus, it seems reasonable that the properties of β-boron, as well as those of its analog AlB_{12}, may be determined partly by the short-range order as in amorphous semiconductors (for the phonons and electrons with short wave lengths), and partly by the long-range order as usually found in crystalline solids.

Indeed, the same specific dependences of conductivity on temperature and electric field appropriate for the amorphous semiconductors are observed for β-boron and AlB_{12}. The same situation exists for the thermal conductivity at high temperatures (PETROV et al., 1969).

At the same time some of the properties of these materials are the same as those of crystals. Thermal conductivity is temperature dependent at low temperatures. Also, structure of the lattice absorption bands is distinct, although there is a definite relation between the change of lattice absorption spectrum in the region 8 - 25 μ and the change of short-range order.

It should be mentioned that the numerical estimates obtained for concentration of hopping sites (10^{18} - $10^{19}/cm^3$) are smaller than the concentration of unit cells ($\sim 10^{21}/cm^3$). Thus, we may assume that some trapping centers produced by nonequivalent atoms do not act as hopping sites in the conduction mechanism of β-boron and aluminum dodecaboride.

References

AMBEGAOKAR, V., HALPERIN, B.I., LANGER, J.S.: Hopping conductivity in disordered systems. Phys. Rev. B4, 2612-2627 (1971)

AVDEENKO, B.K., GOLIKOVA, O.A., ZHUBANOV, M.Zh., KISKACHI, A.Yu., MATVEEV, V.N.: Frequency and temperature dependences of boron conductivity. Fizika tverdogo Tela 10, 2546-2547 (1968)

BEREZIN, A.A.: On the theory of polaron effect in boron. Fizika tverdogo Tela 15, 1937-1939 (1973)

BEREZIN, A.A., GOLIKOVA, O.A., MOIZHES, B.Ya., KHOMIDOV, T.: Density of states and mobility of current carriers in β-boron. Fizika Technika Poluprovodnikov 5, 2320-2323 (1971)

BEREZIN, A.A., GOLIKOVA, O.A., KAZANIN, M.M., TKALENKO, E.N., ZAITSEV, V.K.: Studies of a conductivity mechanism of β-rhombohedral boron in a strong electric field. Phys. Stat. Sol. 20, 447-458 (1973b)

BEREZIN, A.A., GOLIKOVA, O.A., ZAITSEV, V.K., KAZANIN, M.M., ORLOV, V.M., STIL'BANS, L.C., TKALENKO, E.N.: Electrical properties, reflection spectrum and conduction mechanism of AlB_{12}. Fizika Tverdogo Tela 15, 952-955 (1973a)

DAVIS, E.A., MOTT, N.F.: Conduction in non-crystalline systems V. Conductivity, optical absorption and photoconductivity in amorphous semiconductors. Phil. Mag. 22 (179), 903-933 (1970)

GOLIKOVA, O.A., KISKACHI, A.Yu., KHOMIDOV, T.: Electrical properties of β-boron at high temperatures. Fizika i Technika Poluprovodnikov 4, 804-806 (1970a)

GOLIKOVA, O.A., MIRLIN, D.N., UMAROV, A.S., KHOMIBOV, T.: Thermoelectrical and optical properties of amorphous boron. Fizika i Technika Poluprovodnikov 7, 1633-1635 (1973)

GOLIKOVA, O.A., NIKITIN, E.N., TKALENKO, E.N.: Electrical properties of doped boron. Fizika i Technika Poluprovodnikov 4, 1399-1400 (1970b)

GOLIKOVA, O.A., ZAITSEV, V.K., PETROV, A.V., STIL'BANS, L.S., TKALENKO, E.N.: Thermoelectric properties of semiconductors with non-standard conduction mechanism. Fizika i Technika Poluprovodnikov 6, 1724-1728 (1972)

GOLIKOVA, O.A., ZHUBANOV, M.Zh., KLIMASHIN, G.M.: Electrical properties of boron. Fizika i Technika Poluprovidnikov 2, 548-551 (1968)

GOLIKOVA, O.A., ZHUBANOV, M.Zh., KLIMASHIN, G.M.: Electrical properties of boron. Electron Technology Warsaw 3, 1/2, 269-273 (1970c)

GOLIKOVA, O.A., ZHUBANOV, M.Zh., MIRLIN, D.N.: Lattice absorption in boron crystals. Fizika tverdoga Tela 11, 1655-1659 (1969)

MOORJANI, K., FELDMAN, C.: Electrical conduction in amorphous boron and silicon. J. Non-Cryst. Solids 4, 248-255 (1970)

MOTT, N.F.: Charge transport in non-crystalline semiconductors. Festkorperprobleme 9, 22-45 (1958)

MOTT, N.F.: Conduction in non-crystalline systems. IV. Phil. Mag. 22, 7-29 (1970)

MOTT, N.F.: Conduction in non-crystalline systems. VII. Phil. Mag. 24, 911-958 (1971)

PETROV, A.V., GERMAIDZE, M.S., GOLIKOVA, O.A., KISKACHI, A.Yu., MATVEEV, V.N.: Thermoconductivity and crystal structure of β-boron. Fizika i Technika poluprovodnikov 11, 907-910 (1969)

PRUDENZIATI, M.: Space-charge-limited currents in β-boron. Phys. Stat. Sol. 3, K81-K85 (1970)

PRUDENZIATI, M., LANZI, A., MAJNI, G., MALAVASI, G.: Switching effect in β-boron. Phys. Stat. Sol. 18, 651-656 (1973)

SHKLOVSKY, B.I.: Hopping conduction in semiconductors in a high electric field. Fizika i Technika Poluprovodnikov 6, 2335-2347 (1972a)

SHKLOVSKY, B.I.: Hopping conductivity in low doped semiconductors. Fizika i Technika Poluprovodnikov 6, 1197-1226 (1972b)

WERHEIT, H., LEIS, H.G.: On the conductivity mechanism of β-rhombohedral boron. Phys. Stat. Sol. 41, 247-253 (1970)

V. Electron Paramagnetic Resonance (EPR) in Boron Nitride, Boron and Boron Carbide

D. GEIST

1. History and Introduction

EPR in semiconductors and insulators mostly results from foreign atoms or from lattice defects. Yet the first measurements on boron nitride (GEIST and RÖMELT, 1964) revealed resolved hyperfine splitting and a detailed picture of the relevant defects will be given later (Sect. 2). In contrast to this situation, in α-rhombohedral boron only a single line was found (GEIST, 1962); this is true for nearly all material investigated so far (Sect. 3). There seems to be a correlation between EPR intensity and electrical conductivity. The purest single crystals give the smallest EPR intensity in the dark. But such crystals also clearly exhibit photo-EPR (EPR under illumination (GEIST and KLEIN, 1967)). The changes in time of the photo-EPR (Sect. 4) allow conclusions on the distribution in energy of the states involved in the electron transitions. The line width varying with temperature or with ilumination gives some insight into the interaction between the spin-carrying centers (Sects. 3 and 4). EPR in boron carbide (GEIST, 1964a) is similar to that in boron (Sect. 5); combined measurements of electrical conductivity, EPR intensity and Hall coefficient show some correlation between these properties.

2. EPR in Boron Nitride

a) The Three-Boron Center

α) The Spectrum

In polycrystalline hexagonal boron nitride of good but not extreme purity a ten-fold hyperfine split EPR line has been found; see Fig.1a and Table 1 (GEIST and RÖMELT, 1964; RÖMELT, 1966). The well - known rules for the combined hyperfine interaction of several similar nuclei show, that three ($\underline{n}$ = 3) nuclei of spin $\underline{I}$ = 3/2 - corresponding to ^{11}B, the most abundant isotope - will give a spectrum with $\underline{s}$ = 2In + 1 = 10 components, if the electronic spin is $\underline{S}$ = 1/2. This consideration has stimulated the investigation of BN enriched in ^{10}B, the isotope with $\underline{I}$ = 3, expecting $\underline{s}$ = 19 components. But the individual line widths are large, and only one rather broad line can be observed (Fig. 1b). Most experiments have been performed at about 10 GHz. One investigation at 35 GHz (GEIST, unpublished, 1972) fully confirms the interpretaion given above, which is further supported by the intensity ratio 1:3:6:10:12:12:10:6:3:1 of the components (RÖMELT, 1966a; KHUSIDMAN and NESHPOR, 1967).

The observation that the spectrum is not isotropic in oriented material (RÖMELT, 1968) has proved correct. Further investigation (MOORE and SINGER, 1972) on carbon-doped, highly oriented compression - annealed pyrolytic BN of approx. 2° mosaic spread of the c-axis gave spectra (Table 1) well resolved for θ = 90° (θ angle between magnetic field and c-axis) and poorly resolved for θ = 0°.

a

b

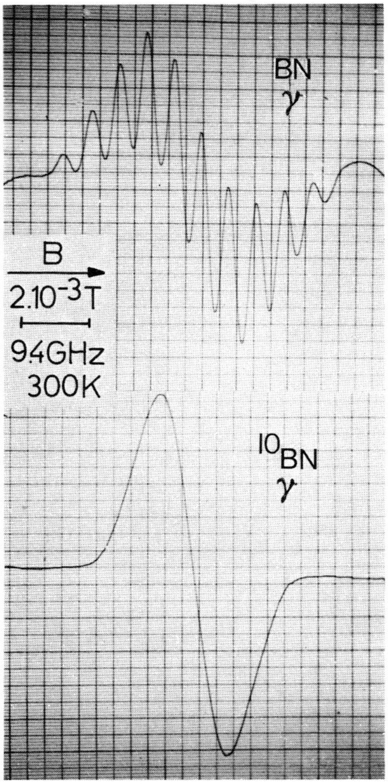

Fig. 1. (a) 10-line spectrum of BN (three ^{11}B nuclei). (b) Unresolved 19-line spectrum of ^{10}BN (three ^{10}B nuclei). In both cases gamma-irradiated material (RÖMELT, 1966b)

β) Description of the Center

Hexagonal boron nitride has a layer structure similar to (but not identical with) graphite. In the hexagonal network in each layer each nitrogen atom is surrounded by three boron atoms (distance B - N $1.45 \cdot 10^{-10}$ m). The boron electron states in this plane are σ states (sp^2 hybrides). The π states perpendicular to the planes come from the p_z states of the atoms. Estimates of the effective charges in BN give -1.4 for the σ-charge and 0.4 for the π-charge, resulting in an overall charge of -1.0 on the nitrogen and +1.0 on the boron (KHUSIDMAN and NESHPOR, 1967, 1970a). This result supports the hypothesis that the center responsible for EPR is a positively charged nitrogen vacancy incorporating an unpaired electron, similar to the F center in alkali halides.

The EPR spectrum will only appear in suitably prepared material. Heating above 1850°C and good crystallization are necessary (RÖMELT, 1966). A thorough investigation on many different materials, and especially with two different high-temperature furnaces, one containing graphite and the other completely free of graphite (constructed with BN and TiB_2 only) leads to the conclusion that carbon atoms are vital for the appearance of the EPR spectrum (MOORE and SINGER, 1972). But the carbon atoms do not participate in the EPR spectrum; they only stabilize the nitrogen vacancies acting as F centers, and facilitate the creation of nitrogen vacancies by forming volatile CN gas. Experiments with ^{13}C do not contradict this view.

The creation of the three-boron center can also be accomplished by irradiation with UV light (Hg lamp), X-rays (25 - 60 keV) or gamma-rays

Table 1. Constants of the EPR spectra in hexagonal[a] boron nitride

Three-boron center. BN of natural isotopic composition (81.17% ^{11}B)

Temp. [°K]	g-value	h.f. constant $A/g\mu_B [10^{-4}T]$	2 h[b] $[10^{-4}T]$	Reference Remarks
1.7	2.0023±0.0010	-	7	10 GHz
77	2.0052±0.0020[c]	-	>10	GEIST and
300	2.0027±0.0003	7.8	6±0.5	RÖMELT, 1964
500	-	-	6.5±0.5	RÖMELT, 1968
300	-	7.2	-	35 GHz; GEIST, unpubl.
300	2.0029±0.0002	7.1±0.2	≈ 5.7	KHUSIDMAN and
			$\Delta H_i = A\mu_B$	NESHPOR,1968,1970b
300	2.0030±0.0001	7.9±0.1	-	MOORE and
300	$g_\perp$ = 2.00321±0.00002[d]	a = 7.85±0.07[e]	-	SINGER, 1972
	$\Delta g = (-9.5\pm0.2)\cdot10^{-4}$	b = -1.27±0,07		

Three-boron center. BN enriched in ^{10}B (91% ^{10}B)

Temp. [°K]	g-value	$A/g\mu_B$ $[10^{-4}T]$	overall width $\Delta B[10^{-4}T]$	Reference Remarks
1.7	2.0035±0.0005	-	17.5±1	10 GHz
77	2.0026±0.0003	-	18 ±1	RÖMELT, 1968
300	2.0029±0.0003	2.6	19 ±1	

One-boron center. $hf = g\mu_B B + Am + D\{I(I + 1) - m^2\}$.

	Splitting (2I + 1)	g-value	$A/g\mu_B$	$D/g\mu_B$	Reference
^{11}BN	4	2.004±0.002	122±4	2.5±0.5	RÖMELT, 1968
^{10}BN	7	2.003±0.001	40±2	-	

[a] $D_{3h}^1$2; C $\bar{6}$m2. a = 2.504; c = $6,661\cdot10^{-10}$ m (MOORE and SINGER, 1972)

[b] Full width at half power of one conponent $2h\sqrt{2\ln 2}$

[c] Measurement disturbed by underlying line

[d] $g = g_\perp + \Delta g \cos^2\theta$; θ = angle between c-axis and magnetic field B (induction)

[e] $A/g\mu_B = a + b\cos^2\theta$

μ_B = Bohr-magneton, m = nuclear quantum number

(ca. 1 MeV; ^{60}Co) of not too pure material (RÖMELT, 1966). High temperature-annealed (1900°C) yellow BN may contain up to 10^{19} spins/g, and when irradiated may reach $8 \cdot 10^{17}$ spins/g; then of $3 \cdot 10^4$ B-atoms (B_3N_3 rings) one carries a spin (RÖMELT, 1966). Carbon - free BN contains less than 10^{16} spins/g (MOORE and SINGER, 1972; KHUSIDMAN and NESHPOR, 1968). Isochronal annealing in the range 300...800°K (KRASNOPEROV et al., 1969) will destroy the three-boron centers. The loss in EPR signal is accompanied by thermoluminescence and thermally stimulated current, indicating an activation energy for the center of 1.0 ± 0.1 eV (KATZIR et al., 1972).

b) The One-Boron Center (RÖMELT, 1968)

By UV light, X- or gamma-rays, another defect, a one-boron center, is produced in hexagonal boron nitride (Fig. 2). The fact that just one boron nucleus is involved follows clearly from the four-fold and seven-fold hyperfine splitting in ^{11}BN and ^{10}BN respectively (Table 1). The highest defect density amounts to one spin on $2 \cdot 10^5$ boron atoms (10^{17}/g). The EPR spectrum seems to be somewhat anisotropic; the EPR intensity is reversibly temperature-dependent in a particular way (RÖMELT, 1968). High temperature annealing (300 - 500°K) will destroy the center; from the accompanying thermoluminescence and thermally stimulated current an activation energy of 0.7 ± 0.1 eV has been deduced (KATZIR et al., 1972).

c) EPR in Cubic Boron Nitride

EPR has only been found in an impure specimen (presumably containing Mg), but not in a (very small) pure one (KHUSIDMAN and NESHPOR, 1970a).

a

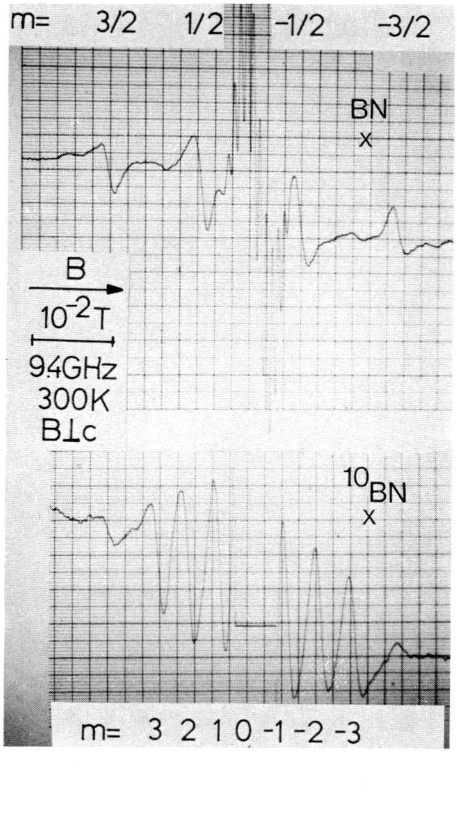

b

Fig. 2 a and b. EPR of X-irradiated boron nitride. (a) ^{nat}BN: 4 lines from a center with *one* nucleus ^{11}B ($I = 3/2$, $\mu/m = 1.8$). 10 lines from a center with *three* ^{11}B. (b) ^{10}BN: 7 lines from a center with *one* ^{10}B ($I = 3, \mu/I = 0.6$) 19 lines from a center with *three* ^{10}B, merged into one line, which was suppressed to avoid overloading of the recorder (see Fig. 1) (RÖMELT, 1966b)

3. EPR in Boron

a) The Single Symmetrical Line at *g* = 2.0029 (300 °K) in Beta-Rhombohedral Boron

α) *Line Intensity and Carbon Content*

In a previous publication it was proved that a bulk and not a surface effect is involved (GEIST, 1962). Carbon, known to be electrically active (WEINTRAUB, 1911), seems to be involved (GEIST, 1962, 1963, 1964b; GEIST and GLÄSER, 1965), but in the low concentration range ($< 10^{-3}$) there is no proportionality between spin concentration and chemically determined carbon content; furthermore ^{12}C (nuclear spin $I = 0$) and ^{13}C ($I = 1/2$) have the same EPR effect (Table 2), with no hyperfine splitting from ^{13}C (Fig. 3) (KLEIN and GEIST, 1967). Experiments with a large fraction of carbon (up to 10%) led to the suggestion that boron carbide (found by X-rays) may be responsible for the EPR (Table 2). But it remains to be explained why of 100 molecules $B_{12}C_3$ only one carries a spin (KOULMAN et al., 1968, 1970; KOULMAN and TAGLANG, 1968).

Table 2. *g*-values and line widths (between the extrema of the differentiated absorption line, in 10^{-4} T) of the single, isotropic, symmetrical EPR line in beta-rhombohedral[a] boron

Temp. [°K]	B: *g*-value	B: ΔB (±0.2-1)	B + ^{12}C (<10^{-4}): *g*-value	B + ^{13}C (< 10^{-4}): *g*-value	ΔB (±0.2-3)
1.5	2.005 ±0.001	11-17	2.005 ±0.001	2.0052±0.0008	14-20
77	2.0030±0.0001	2-6.7	2.0030±0.0001	2.0031±0.0001	2.6±0.2
300	2.0029±0.0001	4.4-10	2.0029±0.0001	2.0029±0.0001	2.2-6
600	2.0030±0.0002	7-13	2.0030±0.0002	-	10±1
900	2.002 ±0.001	13-27	2.002 ±0.001	-	19 (800K)
			(KLEIN and GEIST, 1967)		

Temp. [°K]	B: Δ B	B: *g* (± 0.0002)		B: Δ B	B + C(10^{-2}): Δ B	*g* (± 0.0001)		B + C(10^{-1}): Δ B
1.7	4.3	-	-	-	-	-	-	-
4.2	3.4	2.0033	-	-	-	-	-	-
77	1.3	2.0028	-	-	-	-	-	-
100	-	-	-	-	1.4	-	-	2.1
300	1.3	2.0028	2.0028	8	2.3	2.0032	2.0035	4.0
400	1.3	-	-	5	3.6	-	-	6.2
600	-	-	-	6	-	-	-	-
900	-	-	-	17	-	-	-	-
	(KOULMAN et al., 1972b)		(GEWINNER et al., 1973)		(KOULMAN and TAGLANG, 1968)			

[a] D^5_{3d}; $R\bar{3}m$; $a = 10.14 \cdot 10^{-10}$ m, $\alpha = 65^\circ 17'$ (HOARD et al., 1958, 1970; HUGHES et al., 1963; GEIST et al., 1970a)

β) *g-values, Line Widths and Line Shapes (Table 2)*

There is no correlation between the number of spins and the line width. Between about 77 and 300°K) the line shape is Lorentzian. Above 300°K, the line broadens in the wings and sharpens in the center (KLEIN and GEIST, 1967; KOULMAN et al., 1972b). For some samples the line width reaches a minimum near 500°K and broadens again; see Table 2. It is

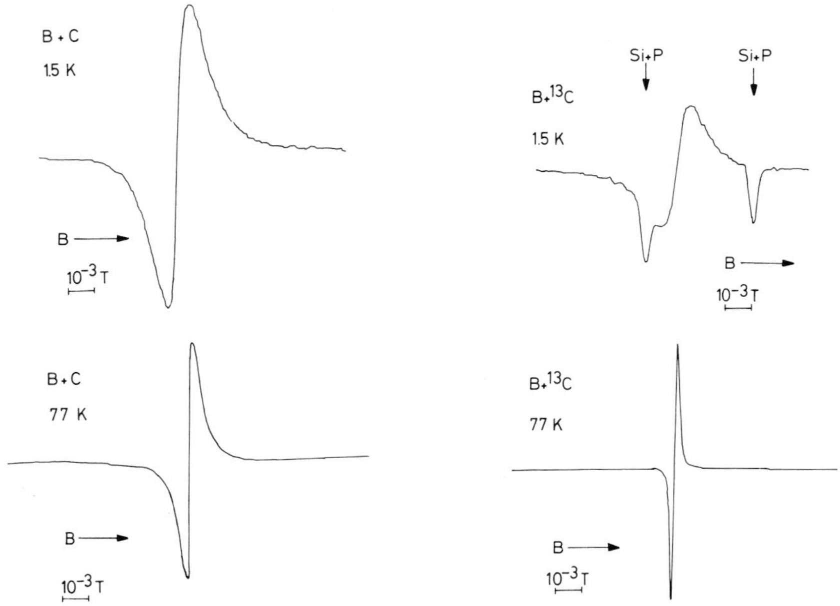

Fig. 3. Single isotropic EPR line in beta-rhombohedral boron at 1.5 and 77°K. Differentiated absorption line. *Left:* Boron with residual carbon of natural isotopic composition (98.9% ^{12}C with $\underline{I} = 0$). *Right:* Boron doped with carbon enriched in ^{13}C (50%; $\underline{I} = 1/2$). *Upper right:* $\underline{Si}$ + $\underline{P}$ calibration (KLEIN and GEIST, 1967)

assumed (KOULMANN et al., 1972a; GEWINNER et al., 1973), that there is a separate signal at high temperatures, with the same g-value but from a different origin: from special centers with localized carriers. The corresponding number of spins increases with the temperature with an activation energy of 0.4 eV. In other erarlier measurements about 0.12 eV had been found in the lower half of the temperature range. It is proposed (GEWINNER et al., 1973) that this high temperature line and the photo-EPR-line (see Sect. 4) come from the same center, especially because of some similarity in line width behavior.

The line width may be explained by motional narrowing (ADIROVICH et al., 1971). The delocalization frequency is

$$\underline{f}_d = \gamma A^2/\Delta B \approx 10^{10} \ldots 10^{11}\ s^{-1} \qquad (1)$$

(ΔB = line width, γ = magnetogyric ratio, A = local (hyperfine) fields from boron nuclei, estimated (100 - 200) · 76 A/m).

γ) Spin Concentration v e r s u s *Conductivity*

In the high temperature range (> 300 K) the intensity of the single line corresponds to a spin density of n_s increasing ($\propto \exp(\Delta\underline{E}/\underline{kT})$) with an activation energy of $\Delta\underline{E}$ = 0.4 eV (GEWINNER et al., 1973). The conductivity ($\sigma \propto \exp(\Delta\underline{E}/2\underline{kT})$) measured by other workers on other samples at high temperatures has an activation energy of $\Delta\underline{E}$ = 1.4 eV (BORCHERT et al., 1965; GEIST and GLÄSER, 1965). Nevertheless, guessing the mobility μ from the delocalization frequency $\underline{f}_d$ [Eq.(1)], the expression $\underline{e}_0 \mu \underline{n}_s = 8 \cdot 10^{-4} (\Omega\ cm)^{-1}$ just reproduces the conductivity otherwise measured at 500°K. Here the correlation between spin density $\underline{n}_s$ and conductivity in boron (see Fig. 4) is recalled. It is of course not neces-

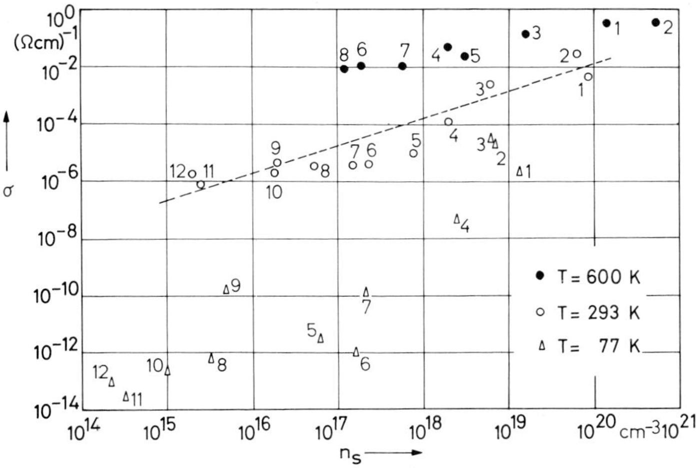

Fig. 4. Electrical conductivity σ versus spin density n_s in pure and carbon-doped beta-rhombohedral boron. Each sample has its individual number (GEIST and KLEIN, 1965; KLEIN and GEIST, 1967)

sary to conclude that the moving carriers (density $\underline{n}_c$) causing conductivity are also the spin-carrying species responsible for the EPR; one should realize however, that the assumption $\underline{n}_s \approx \underline{n}_c$ results in mobility values compatible with several estimates (WERHEIT, 1970).

It should also be remembered that levels in a semiconductor may change their charge state and spin dependence with temperature (GEIST, 1957). As an example a doubly ionizable donor ($\underline{g}$ = 2) will have a susceptibility

$$\underline{\chi} = \frac{(1/\underline{H})\,\mu_{\underline{B}}N\,\sinh\,(\mu_{\underline{B}}\underline{H}/\underline{KT})}{\frac{1}{2}\exp\left(\frac{E''-\zeta}{KT}\right) + \frac{1}{2}\exp\left(\frac{\zeta-E'}{KT}\right) + \cosh\left(\frac{\mu_B H}{KT}\right)}$$

if the singly ionized state is the spin-carrying one ($\underline{E}$': activation energy for the transition neutral-singly ionized, E": activation energy for the transition singly-doubly ionized, ζ: Fermi energy). These and other centers may give a temperature dependence not at all like 1/T. There may be a maximum susceptibility at a certain temperature.

δ) EPR Under Breakdown Conditions (KLEIN, 1970)

In an electric field above about 10^4 V/cm breakdown occurs with a modest (2- to 10-fold) increase in EPR intensity and a large (above 10^6-fold) increase in electrical conductivity. KLEIN assumes a corresponding increase in mobility.

b) Other EPR Spectra in Beta-Rhombohedral Boron

Hydrogen diffused into solid boron produces a single EPR line of nearly 10^{17} cm^{-3} centers and with an intensity maximum around 700°K. During the reduction of BCl_3 to boron by H_2 up to $3\cdot10^{16}$ cm^{-3} EPR-active centers may be introduced; mass spectrometer analysis confirms the correlation between EPR intensity and hydrogen content (APPENHEIMER et al., 1970). Beryllium and silicon introduce in boron a wider EPR line than carbon does, but without a clear-cut correspondence between line widths and magnetic moments of the isotopes involved. Substitution by foreign atoms does not seem to take place; instead the formation of beryllium boride e.g. may occur (KLEIN and GEIST, 1967; KHUSIDMAN and NESHPOR, 1970b). Boron doped with Eu, Gd, La, Ce, Nd or Ca exhibits one broad EPR line. A hexaboride phase formed may be EPR-active (KLEIN and GEIST, 1967; PRACKA et al., 1970). Mn, Ta or Fe seem to be responsible for EPR in boron. Mößbauer results with Fe are at variance (KLEIN and GEIST, 1967; PRACKA et al., 1970; STANKE and PARAK, 1972).

c) EPR in Other Boron Modifications

There have been observations of EPR in amorphous boron (KOULMAN and TAGLANG, 1966), in I-tetragonal (D_{4h}^{12}; $P4_2/nnm$; $a = 8.73 \dots 8.771$, $c = 5.030 \dots 5.10 \cdot 10^{-10}$ m) boron (APPENHEIMER et al. 1970), in II-tetragonal (D_4^2; $P4_122$; $a = 10.12$, $c = 14.14 \cdot 10^{-10}$ m) boron (KOULMAN et al., 1968) and in unspecified material (BRUNGS and JACOBSMEYER, 1964; KHUSIDMAN and NESHPOR, 1970b).

4. Photo-EPR in Beta-Rhombohedral Boron

a) Spectral Dependence

The initial experimental results on photo-EPR in beta-rhombohedral boron (GEIST and KLEIN, 1967; KLEIN and GEIST, 1967) were extended (GEIST and MEYER, 1970; GEIST 1973) and confirmed (ADIROVICH et al., 1971; NADOLNY, 1974) by later work. Pure boron, single crystals as well as polycrystalline material, can be (almost) free of an EPR signal. Under illumination with radiation in the neighborhood of the absorption edge, a single symmetrical EPR line appears with $g = 2.0029 \pm 0.0001$. The spectral dependences of the EPR intensity and the photoconductance are the same. Also, the temperature dependences of both the EPR and the photoconductance have similar and peculiar features.

b) Time Behavior

At 77°K, most changes are slow with time constants in the order of hours, so the effects may be called "persistent". At room temperature, on the other hand, time constants are in the order of seconds and stationary states are easily reached. It is necessary to discuss the low temperature behavior in more detail. Fig. 5 shows some EPR spectra in the dark and under illumination. The weak sharp line present in the dark is unaffected by the light. At the highest spin numbers (produced by long illumination with white light) the width of the photo-EPR line decreases to about 1/10 of its highest value of $35 \cdot 10^{-4}$ T. Spin number N and photoconductance G increase monotonously under illumination. After an interruption of the light, there is at first a slight, rapid decrease of N and G. Most of the photo-EPR and photoconductance is a *persistent* effect, i.e. N and G decrease with a time constant much greater than an hour (GEIST and MEYER, 1970; ADIROVICH et al., 1971; GEIST, 1973). The measurements yielded the spin number N; to convert this number to the spin concentration, the absorption behavior of boron must be considered.

The decrease of the photo spins n_s in the temperature range 150 - 260 K after a *homogeneous* excitation at 0.95 µm follows (NADOLNY, 1974)

$$n_s = n_o kT \left[Ei\left(-\frac{t}{\tau_\alpha}\right) - Ei\left(-\frac{t}{\tau_\beta}\right) \right] \approx -n_o kT \ln t + \text{const.} \tag{2}$$

($\tau_j = \exp(E_j/kT)$; $j = \alpha, \beta$; E_α, E_β constants; ln approximation if $\tau_\alpha \ll t \ll \tau_\beta$); in fact the same law as has been found for photoconductivity (NADOLNY, 1972). The concentration n_s reached $2 \cdot 10^{19}$ cm^{-3} after 4 to 5 h of illumination with about 10^{17} quanta/cm^{2}s.

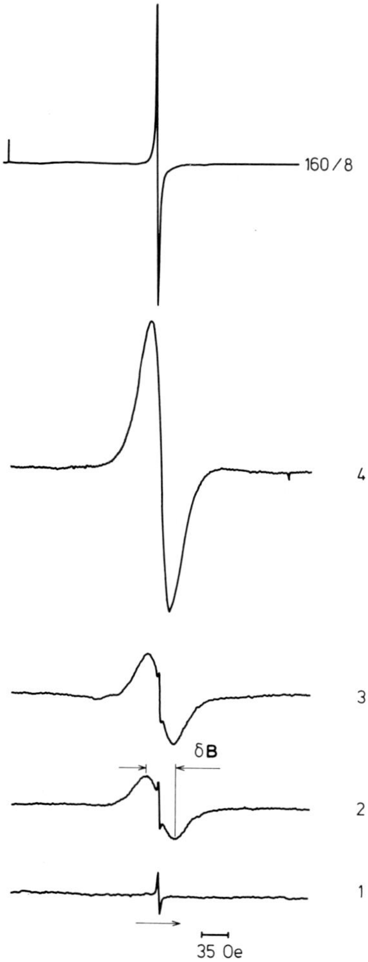

Fig. 5. Photo-EPR in beta-rhombohedral boron. 77°K. Differentiated absorption line [GEIST, 1973 (MEYER)]

Curve No.	Sample	Amplification	Wavelength range	Duration of illumination
1	P72-15.2 single crystal	x 200	dark	-
2			1000 - 630 nm	10 min
3				25 min
4				250 min
160/8	12 Me 1.1 polycrystal	x 8	2700 nm-UV	18.5 h

c) Model Considerations

Two models, both working essentially with four levels, will be considered: model G (GEIST, 1973) with valence band, S states (0.23 eV higher, few acceptors and many donors), traps T (0.97 eV higher) and conduction band (0.36 eV higher); model N (NADOLNY, 1974) with VL states distributed in energy ("valence localized"), SVL states, a single energy level ("special valence localized states"), traps T single level, and excited states distributed in energy (alternatively, traps could be distributed and excited states could be single level).

Unoccupied S donors (model G) and unoccupied VL states (model N) are assumed responsible for EPR. Because of slight transition probabilities, the traps cause the slow changes, especially the persistent effect. On the other hand, perturbation of the equilibrium between valence band and S states (G) leads to the rapid effect. The incoming light is assumed only to bring electrons from the valence to the conduction band (G) and from VL to excited states (N). The changes of the populations of the other states are then a consequence of thermal transitions.

The special time dependence Eq.(2) is a direct consequence of the distribution of excited states or traps in energy (N).

d) Line Width

At 77°K, the line width decreases with increasing spin concentration (GEIST, 1973). Working with only weak illumination, a temperature dependence of line width was measured; see Fig. 6. At higher temperatures, e.g. 187°K (ADIROVICH et al., 1971), the line width increases with increasing spin number.

The spin on a center [S state (G) or VL state (N)] is localized only in the sense that it can see a few neighboring boron nuclei often enough (ADIROVICH et al., 1971) to average out their magnetic moments. This effect will suppress the hyperfine splitting, which is in fact missing in the experimental EPR lines. At the same time, the line width may be explained with this model. The frequency Eq.(1) may be used (GEWINNER et al., 1973) to explain partly the line width at all temperatures. Above 200°K, transfer of the carriers from one center to another may further reduce the line width, again according to Eq.(1). The same effect is achieved with a higher excitation and corresponding higher number of spins. In this case, exchange or dipolar interaction may be involved.

Only speculations are possible at present about the nature of the centers. Their high concentration supports the view that the boron structure itself, and not some foreign atom, is responsible.

5. EPR in Boron Carbide

Table 3. g-values (± 0.0002 mostly) and line widths ΔB (10^{-4}T) of the single, isotropic, symmetrical EPR line in boron carbide

Temp.	Single crystal		Poly-crystalline		Powder powder + 5%C		Crystalline, also with Be	
[K]	g	ΔB	g	ΔB	g	ΔB	g	ΔB
1.7	2.0030	14	2.0035	10	-	-	-	-
77	-	5	2.001	8	-	4	-	6
300	-	8	2.000	cond.	2.0032	9	2.0030 (asym.)	-
	(GEIST, 1964a)		(GEIST, 1964a; GEIST et al., 1970b)		(KOULMAN et al., 1970)		(KHUSIDMAN and NESHPOR, 1970b)	

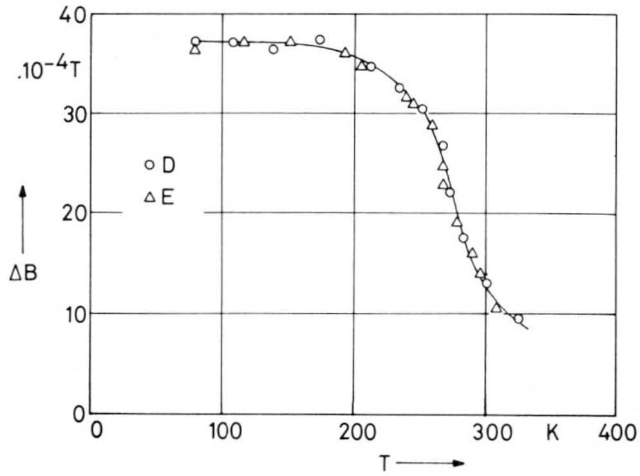

Fig. 6. Temperature dependence of the photo-EPR line width under weak illumination (low spin number). D: single crystal, E: polycrystalline sample (GEWINNER et al., 1973)

The EPR line in boron carbide (GEIST, 1964a) has been tentatively ascribed to boron atoms substituting a carbon atom (KOULMAN and TAGLANG, 1968) in the linear chain of carbon atoms in the unit cell ($B_{12}C_3$; D^5_{3d}; $R\bar{3}m$; $\underline{a} = 5.167 \cdot 10^{-10}$ m, $\alpha = 65°36'$ (CLARK and HOARD, 1943); $B_{13}C_2$; D^5_{3d}; $R\bar{3}m$; $\underline{a} = 5.18 \cdot 10^{-10}$ m, $\alpha = 65{,}7°$ (LARSON and CROMER, 1972; PLOOG, 1973)), a controversial assumption (KOULMAN et al., 1970). Beryllium greatly enhances the resonance (KHUSIDMAN and NESHPOR, 1970b).

At all temperatures, the line behavior in bulk material is compatible with localized EPR-active centers. At 300°K, in samples which are thick compared to the skin depth (> 0.15 mm), the sample conductivity results in a line shape (BLOEMBERGEN, 1952; DYSON, 1955; GEIST, 1967) proportional to $(\chi' + \chi'')$, χ' being the real, χ'' the imaginary part of the susceptibility (GEIST et al., 1970b). (Powdered samples may be misleading due to surface effects). From a double integration of the differentiated EPR absorption curve, the concentrations of the spins n_s can be determined by using copper sulphate (e.g.) as a standard; see Table 4. On the other hand, with sufficient care, Hall coefficients $\underline{R_H}$ (positive!) can be measured around room temperature (GEIST et al., 1970b). It seems remarkable that the concentrations $\underline{p} = (\underline{e_o R_H})^{-1}$ are nearly the same as the spin concentrations $\underline{n_s}$. Always taking these values as equal, carrier mobilities can be deduced within the range 10^{-2} to 1 cm^2/Vs. For a full discussion, the band structure and the energy levels of defects and foreign atoms would of course be necessary.

Table 4. Some representative values of electrical conductivity σ, Hall coefficient $\underline{R_H} = (\underline{e_o p})^{-1}$ and spin concentration $\underline{n_s}$ in boron carbide at room temperature (GEIST et al., 1970b)

Material	$\sigma[(\Omega\ cm)^{-1}]$	$p = (e_o R_H)^{-1} [cm^{-3}]$	$n_s\ [cm^{-3}]$
Coarse crystalline	0.154 - 0.2	-	$3 \cdot 10^{16}$ $<10^{17}$
Crystalline	10.0 - 15.6	$9 \cdot 10^{19}$	$7 \cdot 10^{19}$
Sintered	2.04 - 4.0	$4 \cdot 10^{19}$	$4 \cdot 10^{19}$ $6 \cdot 10^{19}$

6. Concluding Comment

A short comment may be allowed on the curious role of carbon in the substances considered. In hexagonal boron nitride, carbon seems necessary to stabilize the F-center-like three-boron center, but without immediate influence on the EPR spectrum. Again, in beta-rhombohedral boron, carbon is the impurity suspected of being involved in the EPR, but again without an immediate influence on the EPR line. Finally, the boron carbide structure itself is once again subject to discussion (whether $B_{12}C_3$ or $B_{13}C_2$), and the role of the central atom - C or B - with respect to the EPR is still unresolved.

References

ADIROVICH, E.I., BENDERSKII, V.A., BRIKENSHTEIN, V.Kh., KOROSTELEV, Y.A.: Investigation of the EPR signal observed during the illumination of boron single crystals. Sov. Phys, Semicond. 5, 13 (1971)

APPENHEIMER, S., NIEMYSKI, T., JABLONSKI, R.: Hydrogen sorption. In: Boron (T. NIEMYSKI, ed.) Volume. 3, p. 29 Warsaw: Pol. Sci. Publ. 1970

BLOEMBERGEN, N.: The magnetic resonance absorption in conductors. J. Appl. Phys. 23, 1383 (1952)

BORCHERT, W., DIETZ, W., HERRMANN, H.: Semiconducting properties of boron, Z. angew. Physik 19, 485 (1965)
BRUNGS, R.A., JACOBSMEYER, V.P.: Activation energies of single cryst. β-rhombohedral boron. J. Phys. Chem. Solids 25, 701-6 (1964)
CLARK, H.K., HOARD, J.L.: The crystal structure of boron carbide. J. Amer. Chem. Soc. 65, 2115 (1943)
DYSON, F.J.: Electron spin resonance in metals. Theory of electron diffusion and the skin effect. Phys. Rev. 98, 349 (1955)
GEIST, D.: Magnetic susceptibility of impurity atoms in semiconductors. Z. Naturforschg. 12a, 873 (1957)
GEIST, D.: Electron paramagnetic resonance in boron. Proc. Int. Conf. Physics Semiconductors Exeter. (A.C. STRICKLAND, ed.) p. 633. Inst. Phys. and Phys. Soc. London, (1962)
GEIST, D.: Electron Paramagnetic Resonance in Semiconductors: Festkörperprobleme (F. SAUTER, ed.) Vol. 2, p. 93. Braunschweig: Vieweg 1963
GEIST, D.: Int. Conf. Phys. Semicond. Paris. (M. HULIN, ed.) 767, Paris: Dunod 1964a
GEIST, D.: Cyclotron and electron spin resonance in diamagnetic semiconductors. Phys. stat. solidi 5, 217 (1964b)
GEIST, D.: Paramagnetic resonance of conduction electrons in solids. Z. angew. Phys. 23, 129 (1967)
GEIST, D.: Mechanism and quantitive description of the photoconductivity and photo EPR in β-rhombohedral boron. Z. Naturforschg. 28a, 953 (1973)
GEIST, D.: GLASER, H.J.: Electron paramagnetic resonance, electric conductivity and diffusion of impurities in boron. J. Phys. Chem. Solids 26, 57 (1965). Also In: Boron (A.K. GAULE, ed.) Vol. II, p. 203 New York: Plenum Press 1965
GEIST, D., KLEIN, W.: Electron paramagnetic resonance and electrical conductivity of boron. Phys. Lett. 17, 186 (1965)
GEIST, D., KLEIN, W.: Photoconductivity and photo-influenced electron paramagnetic resonance of zone-melted β-rhombohedral boron with low carbon content, Proc. Magnetic Resonance and Relaxation, 377 XIV Coll. Ampere Ljubljana, Amsterdam North Holland 1967
GEIST, D., KLOSS R., FOLLNER, H.: Structure refinement for β-rhombohedral boron. In: Boron (T. NIEMYSKI, ed.) Vol. III, p. 109. Warsaw: Pol. Sci. Publ. 1970
GEIST, D.: MEYER, J.: Carrier mobility in high-purity boron single crystals. Proc. 10. Int. Conf. Physics Semiconductors Cambridge, Mass. (S.P. KELLER, J.C. HENSEL, F. STERN, eds.) p. 597. Oak Ridge: U. S. Atom. Energy Comm. 1970
GEIST, D., MEYER, J., PEUSSNER, H.: Electrical conductivity and movable carrier electron paramagnetic resonance (EPR). In: Boron (T. NIEMYSKI, ed.) Vol. III, p. 207. Warsaw: Pol. Sci. Publ. 1970
GEIST, D., RÖMELT, G.: Paramagnetische Electronenresonanz in Boronitrid. Solid State Commun. 2, 149 (1964)
GEWINNER, G., KUBLER, J., KOULMANN, J.J., JAEGLE, A.: EPR investigation of paramagnetic centers in β-rhombohedral boron. Phys. Stat. Solidi (b) 59, 395 (1973)
HOARD, J.L., HUGHES, R.E., Sands, D.E.: The structure of tetragonal boron. J. Amer. Chem. Soc. 80, 4507 (1958)
HOARD, J.L., SULLENGER, R.D., KENNARD, C.H.L., HUGHES, R.E.: Structure analysis of β-rhombohedral boron. J. Solid State Chem. 1, 268 (1970)
HUGHES, R.E., KENNARD, C.H.L., SULLENGER, D.B., WEAKLIEM, H.A., SANDS, D.F., HOARD, J.L.: The structure of β-rhombohedral boron. J. Amer. Chem. Soc. 85, 361 (1963)
KATZIR, A., SUSS, J.T., HALPERIN, A.: Thermally stimulated currents and thermoluminescence of paramagnetic centers in hexagonal boron nitride. Phys.Lett.41A, 117 (1972). Bull. Israel Phys. Soc., 25 (1972)
KHUSIDMAN, M.B., NESHPOR, V.S.: Study of hexagonal boron nitride by the electron paramagnetic resonance method. Theor. Exper. Chem. (USA) 3, 152 (1967). (Orig. TEKh)
KHUSIDMAN, M.B., NESHPOR, V.S.: F-centers in hexagonal boron nitride enriched with the B^{10} isotope. Sov. Phys. Solid State 10, 975 (1968)
KHUSIDMAN, M.B., NESHPOR, V.S.: EPR study of hexagonal and cubic boron nitride. Porosh. Met. 10(8), 72 (1970a)
KHUSIDMAN, M.B., NESHPOR, V.S.: EPR study of some semiconductor compounds of boron. Porosh. Met. 9, 67 (1970b)

KLEIN, W.: Semiconductor properties of boron in the electrical breakdown range, J. Appl. Phys. 39, 5797 (1968). Also In: Boron (T. NIEMYSKI, ed.) Vol. 3, p. 259. Warsaw: Pol. Sci. Publ. 1970

KLEIN, W., GEIST, D.: Relations between electron spin resonance, electrical conductivity, and photoconductivity in pure and doped boron. Z. Physik 201, 411 (1967)

KOULMANN, J.J., GEWINNER, G., KUBLER, L., TAGLAND, P.: Magnetic properties of boron at high temperatures, C.R. Acad. Sci. (Paris) 273B, 900 (1972a)

KOULMANN, J.J., GEWINNER, G., KUBLER, L., TAGLAND, P.: Magnetic properties of boron at low temperatures. C.R. Acad. Sci. (Paris) 274B, 449 (1972b)

KOULMANN, J.J., KAPPEL, H., TAGLAND, P.: Boron carbide paramagnetic resonance. C.R. Acad. Sci. (Paris) 270B, 445 (1970)

KOULMANN, J.J., PEROL, N., TAGLAND, P.: Electronic properties of boron-carbon compounds. C.R. Acad. Sci. (Paris) 267B, 15 (1968)

KOULMANN, J.J., TAGLAND, P.: Electron paramagnetic resonance study of polycrystalline and amorphous boron. C.R. Acad. Sci. (Paris) 262C, 1830 (1966)

KOULMANN, J.J., TAGLAND, P.: Magnetic properties of boron-carbon compounds, C.R. Acad. Sci. (Paris) 266B, 759 (1968)

KRASNOPEROV, V.A., VEKSHINA, N.V., KHUSIDMAN, M.B., NESHPOR, V.S.: Luminescence and paramagnetic resonance of boron nitride. Zh. Priklad. Spektrosk. USSR 11, 299 (1969)

LARSON, A.C., CROMER, D.T.: Reexamination of the crystal structure of B_4C. Acta Cryst. Suppl. 28A, S53 (1972)

MOORE, A.W., DINGER, L.S.: Electron spin resonance in carbon-doped boron nitride. J. Phys. Chem. Solids 33, 343 (1972)

NADOLNY, A.J.: Verhandl. DPG (VI) 9, 619 (1974) Soc. Phys. Solid. State (b) 52, 69 (1972)

NADOLNY, A.J.: Capture of electrons in β-rhombohedral boron. In: Boron (F.N. TAVADZE, ed.) p. 88-91, Moscow: Nauka 1974

PLOOG, K.: Darstellung eines rhomboedreschen Borcarbides der Zusammensetzung $B_{13}C_2$. J. Less Common Metals 31, 177 (1973)

PRACKA, I., JUN, J., NIEMYSKI, T., JABLONSKI, R.: Doped boron: (T. NIEMYSKI, ed.) Vol. III, p. 57 Warsaw: Pol. Sci. Publ. 1970

RÖMELT, G.: Electron paramagnetic resonance of thermally and irradiation produced centers in boron nitride. Z. Naturforschg. 21a, 1970 (1966)

RÖMELT, G.: Electron paramagnetic resonance of unstable centers in photon-irradiated BN and ^{10}BN. In: Lattice Defects Semiconductors (R.R. HASIGUTI, ed.) p. 407. Tokyo: Univ. Tokyo Press 1968

STANKE, F., PARAK, F.: Moessbauer effect studies of cobalt 57 and iron 57 in β-rhombohedral boron. Soc. Phys. Solid State (b) 52, 69 (1972)

WEINTRAUB, E.: On the properties and preparation of the element boron. J. Ind. Eng. Chem. 5, 106, 299 (1911)

WERHEIT, H.: Semiconducting properties of boron; Festkörperprobleme (O. MADELUNG, ed.) Vol. X, p. 189, Braunschweig: Vieweg 1970

VI. Structural Determinants in the Higher Borides

V. I. MATKOVICH and J. ECONOMY

1. Introduction

One of the most interesting properties of boron is its ability to combine with itself to form polyhedral structures. Of the various polyhedral forms, those in which boron atoms occupy the twelve vertices of an icosahedron or cubo-octahedron are the most common, although an octahedral unit consisting of six boron atoms at the vertices is well known and is characteristic of the hexaborides and tetraborides. This review is concerned with establishing the structural determinants which are common to the borides and modifications of boron which contain B_{12} polyhedral units. In addition, the case of the tetraborides is considered in some detail, since it is in this system that the polyhedral boron units cease to act as the structural determinant.

The critical observation permitting a general categorization of the higher borides is based on the fact that the B_{12} polyhedra are much larger than single atoms. Hence, the unit cell of structures involving this group must be determined by the packing mode of these large units. The interstitial atoms, on the other hand, are presumed to be a factor in stabilizing the various structures. Thus, the unit cell dimensions should be derivable (with reasonable approximation) from the distance of closest approach (5.1A) of the B_{12} groups. A summary of the observed unit cell dimensions of practically all of the known B_{12} borides is shown in Table 1. These values are compared to the calculated dimensions obtained by multiplication by square roots of integral numbers as such relationships can be derived from various packing arrangements of equal size spheres. As may be abserved in Table 1, close agreement exists between observed and calculated values.

Table 1. Comparison between observed unit cell dimensions and those calculated by assuming various crystallographic relationships to the average B_{12} group diameter of 5.1A

Observed dimensions (A)	Calculated dimensions (A)
5.06 - 5.69	5.1×1 - 5.1
6.17	$5.1 \times \sqrt{3}/\sqrt{2}$ - 6.2
7.41 - 7.50	$5.1 \times \sqrt{2}$ - 7.2
8.82 - 8.89	$5.1 \times \sqrt{3}$ - 8.8
10.12 - 10.16	5.1×2 - 10.2
12.12 - 12.72	$5.1 \times \sqrt{2} \times \sqrt{3}$ - 12.5
14.14 - 14.28	$5.1 \times 2\sqrt{2}$ - 14.4
16.56	$5.1 \times 4\sqrt{2}/\sqrt{3}$ - 16.6
17.53	$5.1 \times 2\sqrt{3}$ - 17.6

In the case of the hexaborides the basic structural features are also determined by the B_6 octahedral groups which are invariably arranged in the form of a simple cube. The various hexaborides differ only by the nature of the interstitial metal atom. Again the octahedral boron unit appears to be significantly larger than any of the metal atoms and the structure is therefore determined by the arrays of the B_6 units.

The tetraborides also contain octahedral boron groups but are better characterized by the arrangement of the two-metal two-boron tetrahedral units. In this case the octahedral group is defective and does not provide the structural continuity of the hexaborides.

In this paper the following subjects are considered: (1) the nature of the polyhedral unit, (2) examination of the various stacking arrangements of equal sized spheres, (3) classification of all the B_{12} borides with respect to the arrangement of the B_{12} unit, (4) the role of the unique pentagonal pyramidal coordination in influencing the packing mode and (5) use of this methodology to explain the structural features of the tetraborides in terms of an M_2B_2 tetrahedral unit. The M_2B_2 model is also useful in interpreting other structures.

2. Nature of the Polyhedral Unit

The icosahedron (of boron atoms) is a geometrical form of cubic symmetry which has 20 equilaterial triangular faces with 12 boron atoms at the vertices (see Fig. 1). The cubo-octahedron also has 12 vertices each occupied by boron atoms but the polyhedron is made of eight equilateral triangles and six squares (see Fig. 1). The cubo-octahedron is about 5 percent larger in average diameter than the icosahedron. For purposes of this study both polyhedra are treated as large, nearly spherical entities.

The B_{12} units can also group together to form larger units which act like individual entities. One such group, reported recently is made up of 84 atoms where the central icosahedron is surrounded by twelve others. The individual unit in this case includes only one half of each of the surrounding twelve icosahedra. The unit can, therefore, be described as consisting of a central icosahedron and twelve pentagonal pyramids (see Fig. 2). This grouping with an average diameter of about 10.2A (twice that of a simple B_{12} icosahedron) has a total of 84 atoms and is designated as B_{84} or $(B_{12})_7$. Structures containing this unit are also amenable to analysis in terms of packing of B_{84} units. The B_{12} groups may thus appear as units of 12 atoms, referred to as B_{12} units or as units of 84 atoms, referred to as B_{84} units.

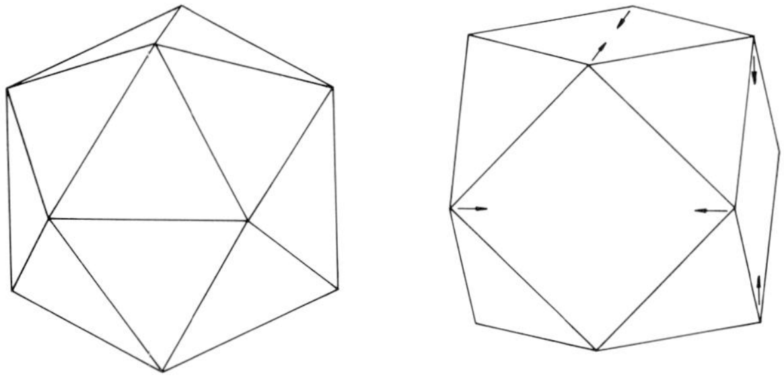

Fig. 1. Comparison between icosahedron *(left)* and cube-octahedron *(right)*

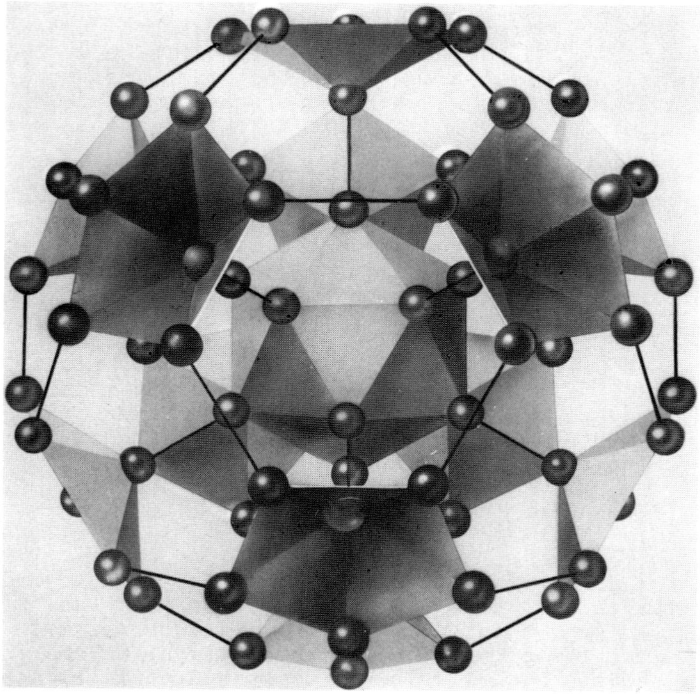

Fig. 2. Ideal arrangement of boron atoms in the B_{84} polyatomic unit. The arrangement consists of a central B_{12} icosahedron and 12 pentagonal pyramids centered radially at the vertices of the central icosahedron. The 60 peripheral atoms occupy vertices of a truncated icosahedron (12 pentagonal and 20 hexagonal faces)

In the case of the hexaborides and tetraborides the B_6 unit is octahedral with 4 boron atoms at the vertices of a square and two boron atoms directly above and below the center of the square. The unusual feature of the tetraboride is the fact that the B_6 unit can be defective and may not provide a continuous boron framework throughout the structure. The defective character of the B_6 unit in the tetraboride suggests that stability of these structures should be associated with metal-boron interaction. This led the authors (GIESE et al., 1965b) to interpret these structures in terms of a network of M_2B_2 tetrahedra. The valence structure of MB_4 compounds has been discussed by LIPSCOMB and BRITTON (1960) who calculated that the electron requirement of this structure is provided nicely by the 16 B atoms and the 4 M^{2+} ions. This led them to predict a closed-shell form a MgB_4.

3. Derivation of Possible Stacking Arrangements in Higher Borides

a) Packing Arrangements in Higher Borides

The concept of packing in higher borides is based on the observation that quasi-spherical polyhedral groups of 12 or 84 boron atoms form three-dimensional arrangements which can be described in terms of stacking arrangements of closes-packed layers of polyhedral units (MATKOVICH et al., 1965b). There are in principle six basic models that can be derived from closest-packed layers of equal size spheres. The units may have coordination of eight, ten or twelve (referred to

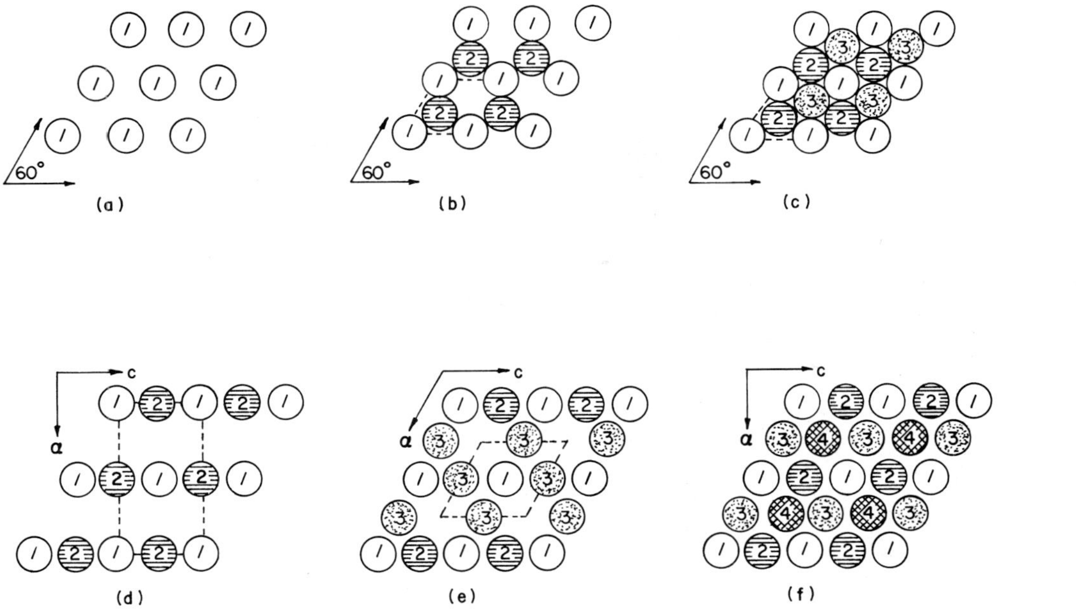

Fig. 3a-f. Packing models that can be derived by stacking of closest packed layers of spheres: (a) HP-hexagonal primitive arrangement, (b) HCP-hexagonal closest-packing arrangement, (c) CCP-cubic closest-packing arrangement, (d) TCP-tetragonal close-packing arrangement, (e) CPH-close-packed hexagonal arrangement, and (f) OCP-orthorhombic close-packing arrangement

as primitive, close-packed and closest-packed arrangements, respectively). A coordination of eight is obtained by stacking the closest-packed layers one above the other, so that each unit has only one close neighbor in the adjacent layers (see Fig. 3a). In the idealized case of equal size spheres, only one model can be classified in this group and is denoted as HP (hexagonal primitive arrangement). A coordination of ten is obtained by stacking the closest-packed hexagonal layers in such a way that each unit has two close neighbors in each of the adjacent layers. In the idealized case of equal size spheres, three different models can be derived in this manner. These are: tetragonal close packing (denoted as TCP), composed of two different layers (illustrated in Fig. 3d); close-packed hexagonal arrangement (denoted as CPH), composed of three different layers (Fig. 3e): an orthorhombic close-packed arrangement (denoted as OCP), composed of four different layers (Fig. 3f). A coordination of twelve is obtained by stacking of closest-packed hexagonal layers in such a way that each unit has three close neigbors in each of the adjacent layers. In the idealized case of equal size spheres, two different models can be derived in this manner and these are the well-known hexagonal closest packed (denoted as HCP) and cubic closest packed (denoted aus CCP) arrangements illustrated in Figures 3b and 3c, respectively.

The polyhedral atomic groups are not ideal spheres and hence the observed arrangements are as a rule somewhat distored and often result in a change of symmetry of the unit cell. Thus, a variety of models can be derived from the six basic ones listed above. In the proposed classification scheme (MATKOVICH et al., 1965b), the specific model is denoted simply by use of the capital letters describing the basic packing

Table 2. CCP model derivatives

Packing model designation	Calculated a(Å)	Calculated b(Å)	Calculated c(Å)	Observed a(Å)	Observed b(Å)	Observed c(Å)	Conventional formula	Structural formula	Reference
CCP (Cubic structures) (B_{12}-cubo-octahedra)	7.2	-	-	7.408			$B_{12}Zr$	$(B_{12})_4M_4$	La PLACA et al. (1960); MATKOVICH et al. (1965a)
				7.404			$B_{12}Lu$		
				7.473			$B_{11}U$		
				7.476			$B_{12}Tm$		
				7.484			$B_{12}Er$		
				7.492			$B_{12}Ho$		
				7.500			$B_{12}Y$		
				7.501			$B_{12}D_y$		
CCPt (Tetragonal structures) (cubo-octahedra)	5.1	-	7.2	5.22	7.3	7.35	$B_{12}Sc$	$(B_{12})_2M_2$	MATKOVICH et al. (1965a)
1/2CCPυt (Tetragonal structures) (B_{84}-polyhedra)	10.2		14.4	10.16		14.28	αAlB_{12}	-	KOHN et al. (1960)
				10.12		14.14	B	-	TALLEY et al. (1960)
				10.16		14.28	BeB_6	-	SANDS et al. (1961)
1/4CCPr (Rhombohedral structures) (B_{12}-icosahedra)	(hexagonal presentation) 5.1		12.50	4.908		12.567	B	B_{12}	DECKER et al. (1959)
				5.37		12.31	B_6O		
				5.984		11.850	B_6P	$B_{12}X_2$	La PLACA et al. (1961), MATKOVICH (1961)
				6.142		11.892	B_6As		
				5.67		12.19	$B_{13}C_2$	$B_{12}C_2B$	ALLEN (1953)
				5.63		12.29	$B_{12}C_2Al$	$B_{12}C_2Y$	LIPP et al. (1966)
				5.65		12.35	$B_{12}C_2Si$		
				6.330		12.736	$B_{3-4}Si$	$(B_{12-a}Si_a)Si_2$	MATKOVICH (1960)
				5.60		12.07	B_4C	$(B_{12-a}C_a)C_2B$	LIPP et al. (1966)
				5.642		12.367	$B_{40}C_4Al$	$(B_{12-a}C_a)(C_{2-x}B_x)(B,Al)_n$	NEIDHARD et al. (1970)
1/4CCPυr (Rhombohedral structures) (B_{84}-polyhedra)	10.2		25.0	10.944		23.811	B	$(B_{84})(B_{10})_2B$	HUGHES et al. (1963)
				10.98		23.82	$B_{32}Al$		
				11.13		23.83	$B_{14}Si$	$(B_{84})(B_{10-a}L_a)X$	GIESE et al. (1965a)

model. A numeral preceding these letters is used to describe the degree of multiplicity of the actual cell volume over that of the basic model and small case letters following the capital letters to indicate the actual cell symmetry (r, t, o, etc. to indicate rhombohedral, tetragonal and orthorhombic symmetry, respectively). Use of thos type of notation is illustrated in Tables 2 - 5 and the following discussion of B_{12} borides. Presence of B_{84} units rather than B_{12} is indicated by small case letter *v* immediately following capital letters, for example, 1/4CCP*v*r indicates a cubic closest-packing arrangement of B_{84} units with a rhombohedral unit cell which has 1/4 of the normal CCP volume.

The B_{12} polyhedral groups with dimensions of 5.1Å are considerably larger than single atoms, so that interstitial openings present in these structures may accommodate not only individual but even groups of atoms. The atoms present in these interstitial openings can play a role in satisfying the steric and electronic requirements of the icosahedral units. For example, such atoms may provide cross-linking between the icosahedral boron groups and contribute to the bonding coordination of boron atoms (HOARD and HUGHES, 1967) (MATKOVICH and ECONOMY, 1970). In addition, the interstitial atoms may act as electron donors to B_{12} icosahedra, which according to LONGUET-HIGGINS and ROBERTS (1955) require two additional electrons for stability. Since the interstitial openings are well defined, it would appear a simple matter to define the stoichiometric compositions of B_{12} borides in terms of full occupancies of the interstitial sites. Unfortunately, as discussed in a companion paper, derivatives of B_{12} borides display, in general, a variable composition, which seldom can be explained in a simple manner.

b) Classification of the Higher B_{12} Borides Based on Stacking Arrangement of Polyhedral Units

α) Cubic Closest-Packed Structures - CCP Derivatives

The cubic closest-packed structure which can be described as a face-centered cubic arrangement of B_{12} or B_{84} units is illustrated in Figure 4. Providing the units are ideally oriented, this packing results in a cubic cell with a = 7.2A for B_{12} units and a = 14.4A for B_{84} units. As indicated in Figure 4, various unit cells can results, depending upon orientation of the units in the structure. On this basis the following borides have been classified in this category.

Description	Subclassification
1. Cubic dodecaborides (series)	CCP
2. Scandium dodecaboride	1/2 CCPt
3. Alpha boron	1/4 CCPr
4. Rhombohedral borides (series)	1/4 CCPr
5. Beta boron	1/4 CCP*v*r
6. Alpha - AlB_{12}	1/2 CCP*v*t
7. Complex tetragonal boron	1/2 CCP*v*t
8. BeB_6	1/2 CCP*v*t

Various dimensions that can be derived from the cubic closet-packed arrangement of spheres of 5.1A diameter are illustrated in Fig. 4. Depending upon the nature of distortion from ideality, many combinations of unit cell dimensions are possible aside from the size difference of the B_{12} and B_{84} units. This includes orthogonal as well as non-orthogonal distortions. A comparison of the observed symmetries and dimensions for various structures derivable from the CCP model is discussed below.

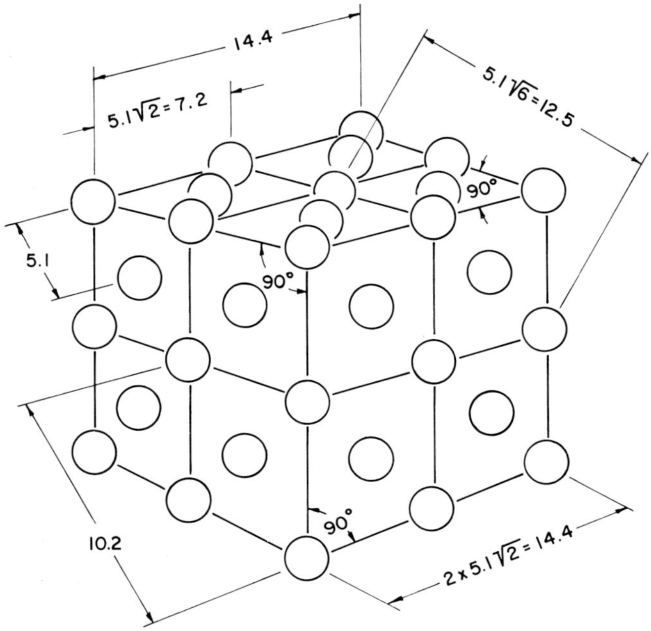

Fig. 4. Illustration of various dimensions that could be derived from the cubic closest-packed system. The simple face-centered cube of 7.2 A edge is denoted as CCP. The large cube of 14.4 A edge is denoted as 8CCP. (Each circle represents a B_{12} unit.) The same arrangement applies also to the B_{84} units providing all dimensions are multiplied by a factor of two

β) Cubic Structures - Dodecaborides

A case of simple cubic closest-packed arrangement of the B_{12} groups occurs in cubic dodecaborides. Here the unit cell contains four B_{12} groups. A face-centered cubic unit cell that would result by use of equal size spheres of 5.1A diameter would have the edge of 7.2A. The observed values ranging between 7.4A and 7.5A indicate that the effective diameter of the B_{12} group is somewhat larger. This is consistent with the fact that the B_{12} group in this structure assumes a cubo-octahedral form which is about 5% larger than the icosahedral groups from which the average diameter of 5.1A was derived.

The interstitial metallic atoms (U, Zr, Y, rare earths, etc.) are readily accommodated in the octahedral openings among cubo-octahedra and their size can have little effect on the unit cell dimensions. However, it is apparent that presence of metal atoms stabilizes this structure and particularly the cubo-octahedral form of the B_{12} group.

The unit cell dimensions of various metal dodecaborides are listed in Table 2.

γ) Tetragonal Distortions of the CCP Model - 1/2 CCPt and 1/2 CCPvt Derivatives

The smallest metal that forms a metal dodecaboride is scandium. Probably due to its small size a slight deviation from the cubic symmetry occurs and this structure is reported to be tetragonal (MATKOVICH et al., 1965a) (PESHEV et al., 1970). Detailed structure analysis has not been reported. However, all the known data are consistent with the assumption of a tetragonally distorted cubic closest packing of cubo-octahedral B_{12} groups.

In addition to the above there are a few structures with large unit cell dimensions which can clearly be derived from a CCP model. An example of this case are the tetragonal unit cells of a = 10.16Å and c = 14.14Å. The dimension of 10.16Å corresponds closely to the diagonal of the cube with edge of 7.2Å as shown in Fig. 4, while the dimension of 14.14Å, corresponds to twice the edge of the same cube. In a simple way these dimensional requirements can be interpreted in terms of closest-packed arrangements of sixteen B_{12} groups (with diameter of 5.1Å) or two B_{84} groups with diameter of 10.2Å. Probable presence of two B_{84} boron groups has been suggested. It is also probable that the polyhedral boron units in these structures have a different form than those observed so far and polyhedral groups lacking a center of symmetry should be considered.

It is of interest that apparently diverse compounds such as boron, BeB_6 and AlB_{12} have (as observed in Table 2) almost identical unit cell dimensions in spite of considerable differences in the unit cell content.

δ) Rhombohedral Distortions of the CCP Model - 1/4 CCPr and 1/4 CCPvr Derivatives

A rhombohedral unit cell is inherently present in any cubic closest-packed arrangement of equal size spheres as illustrated in Fig. 5. In many cases involving B_{12} icosahedral or complex B_{84} units, a distortion of the cubic closest-packed arrangement occurs in such a way that the three-fold symmetry of the system is preserved, resulting in a rhombohedral unit cell. Because the volume of the simple rhombohedral cell corresponds to one quarter of the cubic cell from which it is derived the isotypic series of this type are denoted as 1/4CCPr for B_{12} units and 1/4CCP*v*r for B_{84} units.

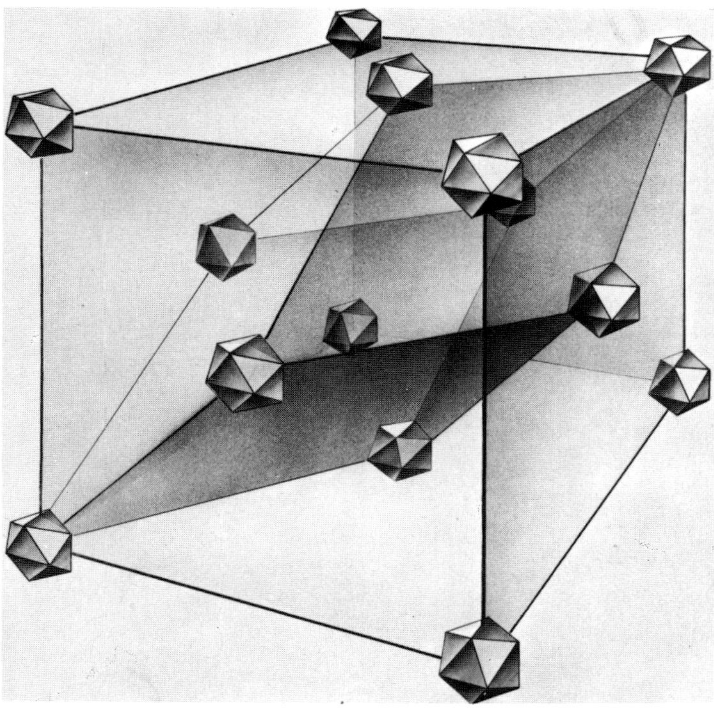

Fig. 5. An illustration of the relationship between the basic cubic model (CCP and the corresponding rhombohedral cell (1/4 CCPr)

The octahedral interstitial openings in this structure are not occupied in alpha boron but are clearly occupied by a number of atoms in various derivatives. In addition, substitution of other atoms for boron in icosahedra has been observed. These observations lead to interesting crystal chemistry of this series which is discussed in Chapter B.VII of this Volume.

Calculated and observed unit cell dimensions are compared in Table 2. It is of interest to observe that this particular configuration can accomodate a great variety of interstitial atoms and is stable in spite of considerable dimensional distortions. Furthermore the arrangement of B_{84} polyhedra bears at least a formal resemblance to the arrangement of the smaller B_{12}-icosahedra, although the interstitial openings are much larger and can accommodate two B_{10} groups joined by a single boron atom in a manner similar to a -C-B-C chain in boron carbide.

ε) Hexagonal Closest-packing - HCP Derivatives

A hexagonal close packing of B_{12} (or B_{84}) units results in a unit cell (HCP or HCPv) of a = 5.1Å (or 10.2Å) and c = 8.3Å (or 16.6Å). Figure 6 illustrates the HCP cell and the various cell dimensions that could be derived from this type of packing. As in the CCP model, various unit cells are derived by changes in orientation of the units.

There are no known B_{12} borides that display a hexagonal symmetry; however, the two orthorhombic borides listed below are classified on the basis of hexagonal closest packing of the polyatomic units (Table 3).

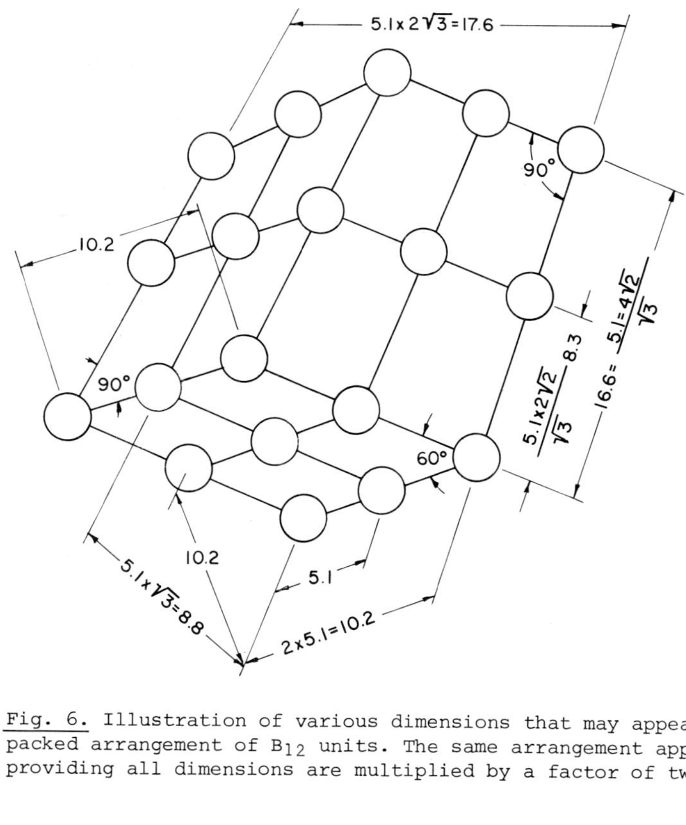

Fig. 6. Illustration of various dimensions that may appear in hexagonal closest-packed arrangement of B_{12} units. The same arrangement applies also to the B_{84} units providing all dimensions are multiplied by a factor of two

Table 3. HCP derivatives

2HCPo	a obs.	a (calc.)	b obs.	b (calc.)	c obs.	c (calc.)	Reference
C_4AlB_{26} or $(B_{12})_4C_8(AlB)_n$	8.89Å	(8.8Å)	9.10Å	(8.3Å)	5.69Å	(5.1Å)	PERROTTA et al. (1969)
2HCPvo / gamma-AlB_{12}	16.56	(16.6)	17.53	(17.6)	10.16	(10.2)	KOHN et al. (1961)

The structure of C_4AlB_{24-26} is based on a hexagonal closest packing of the B_{12} iscosahedral groups, and the positions of the interstitial atoms have been well established (PERROTTA et al., 1969). This compound may be regarded as an orthorhombic modification of boron carbide where aluminum is a stabilizing impurity. In fact when the compound is heated to high temperatures it undergoes a topotactic conversion into the well-known rhombohedral form of boron carbide.

The gamma AlB_{12} is related to the alpha AlB_{12} modification in terms of the hexagonal and cubic closest packing of quasi-spherical polyatomic units of about 10-16Å diameter (LONGUET-HIGGINS and ROBERTS, 1955). These two forms are usually intertwinned and there is evidence of a transition of one into the other. The often proposed B_{84} group for this polyatomic unit is consistent with dimensional requirements, however, units of similar dimensions are possible by different arrangements of atoms in the polyatomic unit.

ζ) Close-packed Structures - TCP Derivatives

As explained earlier, the close-packed structures are obtained by stacking of closest-packed layers in such a way that each unit of one layer is in contact with only two units in each of the neighboring layers. This results in coordination of 10 for each unit and the possible stacking sequences consistent with these requirements are illustrated in Fig. 3d, e and f. The dimensions of the unit cell of the three packing models mentioned above can be calculated by the following formulas (where R denotes the sphere diameter):

		value for R = 5.1Å
1. Tetragonal close packing		
TCP model (body-centered)	$a = R\sqrt{3/2}$;	6.2A
	$c = R$;	5.1A
2TCP model (all face-centered)	$a = R\sqrt{3}$;	8.8A
	$c = R$;	5.1A
2. Hexagonal close packing		
HCP model	$a = R$;	5.1A
	$c = 3/2R\sqrt{3}$;	13.3A
3. Orthorhombic close packing		
OCP model	$a = R\sqrt{3}$;	8.8A
	$b = 2R\sqrt{3}$;	17.6A
	$c = R$;	5.1A

Relationship between various dimensions of the TCP model is illustrated in Fig. 7.

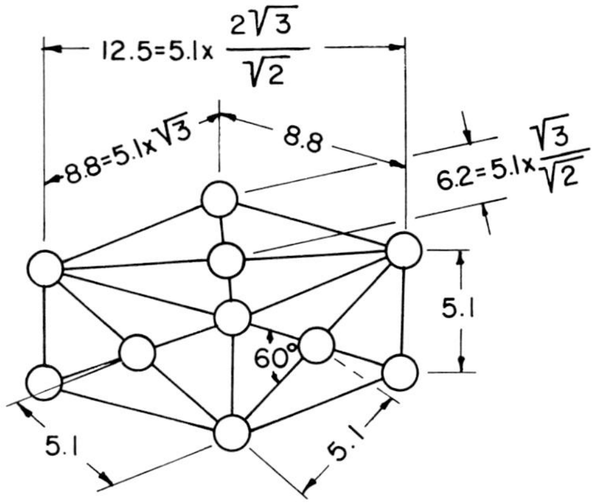

Fig. 7. Face-centered cubic arrangement of the tetragonal closest-packed model (2TCP)

Among the B_{12} borides only structures that can be derived from tetragonal close-packing arrangements of B_{12} icosahedra have been established. They are listed in Table 4, and are discussed in some detail in Chapter B.VII of this Volume. As in the case of the boron carbide-type series of compounds, the variation of composition and stoichiometry of the members of the TCP model series indicates considerable stability of this configuration of icosahedral groups. In fact, it has been shown that the interstitial atoms may be exchanged while retaining the same icosahedral framework (BECKER, 1960). It is thus possible that the structural changes leading to the 4TCP and 4TCPo models involve essentially placements of interstitial atoms, though changes in orientation of the B_{12} groups cannot be excluded at this time.

η) Hexagonal Primitive Structures - HP Derivatives

The hexagonal primitive structures of B_{12} units are obtained when the closest stacked layers of B_{12} groups are stacked directly one above the other as shown in Fig. 3a. Figure 8 illustrates the dimensional

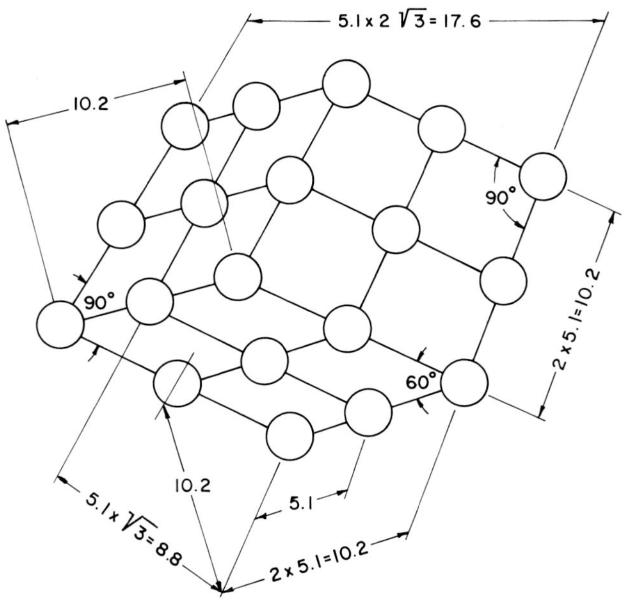

Fig. 8. Illustration of various dimensions that may appear in hexagonal primitive arrangement of B_{12} units (HP model)

Table 4. TCP derivatives

Packing model designation	Calculated a(Å)	Calculated b(Å)	Calculated c(Å)	Observed a(Å)	Observed b(Å)	Observed c(Å)	Conventional formula	Structural formula	Reference
				8.82		5.08	$B_{24}BeAl$	$(B_{12})_4Be_2Al_2$	KROGMAN et al. (1972)
				8.75		5.06	B	$(B_{12})_4B_2$	HOARD et al. (1958)
				8.82		5.09	$B_{48}C_2Al_3$ (800°C)	$(B_{12})_4C_2Al_4$	BLITZ (1908)1910), MATKOVICH et al. (1965c)
2TCP (B_{12}-icosahedra)	8.8	-	5.1	8.80		5.08	$B_{12}Be$	$(B_{12})_4Be_4$	BECKER (1960)
				8.986		5.078	$B_{25}Ni$	$(B_{12})_4B_2Ni_2$	DECKER et al. (1960)
				8.722		5.08	$B_{48}B_2C_2$	$(B_{12})_4B_2C_2$	PLOOG (1974)
				8.646		5.127	$B_{48}B_2N_2$	$(B_{12})_4B_2N_2$	PLOOG (1974)
4TCP (B_{12}-icosahedra)	8.80		10.2	8.75		10.15	B	$(B_{12})_8B_4$	GORSKI (1965)
4TCPo (B_{12}-icosahedra)	6.2 12.5	12.5 12.5	10.2 5.1	6.17 12.34	12.63 12.63	10.16 5.08	$B_{48}C_2Al_3$	-	MATKOVICH et al. (1965a)

relationships in this type of arrangement. The stacking of closest-packed layers in this manner may exist only when stabilized by presence of interstitial atoms. As a consequence the number of interstitial atoms is relatively large in these compounds. All of the known compounds with hexagonal primitive arrangement of icosahedra have an orthorhombic modification classified as 4HPo. These isotypic series are listed in Table 5. As in the other icosahedral isotypic boride series the polyhedral framework is retained even though the interstitial atoms do not occupy equivalent positions from one structure to another.

Table 5. 4HPO Derivatives

	a		b		c		
	obs.	(calc.)	obs.	(calc.)	obs.	(calc.)	Reference
$MgAlB_{14}$ or $(B_{12})_4Mg_4Al_4B_8$	10.313Å	(10.2Å)	8.115Å	(8.8Å)	5.848Å	(5.1Å)	MATKOVICH et al. (1970)
NaB_{15} or $(B_{12})_4Na_4B_4B_8$	10.298Å	(10.2Å)	8.415Å	(8.8Å)	5.847Å	(5.1Å)	NASLAIN et al. (1970)

4. Comparative Role of Bonding versus Packing on Structural Determinants in the Higher Borides

The concept of packing has brought a considerable degree of order among apparently unrelated complex structures of higher borides. Many borides with widely differing compositions can now be properly classified into isotypic series where the polyhedral framework is common to all the members (See Tables 2 - 5). The stoichiometry within a given isotypic series varies with the number and type of extrapolyhedral atoms which are accommodated in the interstices between polyhedra. A detailed analysis of crystal chemistry of such interstitial compounds is given in Chapter B.VII in this book by the same authors.

Structure determination of new compounds where the presence of B_{12} groups is indicated is greatly facilitated by use of the packing principle. This has been clearly demonstrated in the structure determination of $MgAlB_{14}$ (MATKOVICH and ECONOMY, 1970) where analysis of unit cell dimensions and space group requirements indicated that the orthorhombic unit cell was based on a primitive hexagonal arrangement of B_{12} icosahedra (See Fig. 9). Placement of the interstitial atoms was obtained by calculation of electron-density difference maps using only icosahedral boron atoms.

One of the clear advantages of the concept of packing is a simplified visualization of these very complex structures. Thus, a very large number of atoms in the unit cell may be described sequentially in terms of arrangements of a small number of entities (polyhedral groups) and then by the arrangement of the interstitial atoms between these entities.

This concept may also be utilized to describe complex transitions that occur in higher borides. For example, orthorhombic C_4AlB_{24} undergoes a topotactic transition into a rhombohedral structure of the same composition (usually written as $B_{12}C_2Al_{.5}$) retaining many of the symmetry elements of the precursor crystal (GIESE R.F.JR., et al., 1966). This conversion can be described as a transition from hexagonal

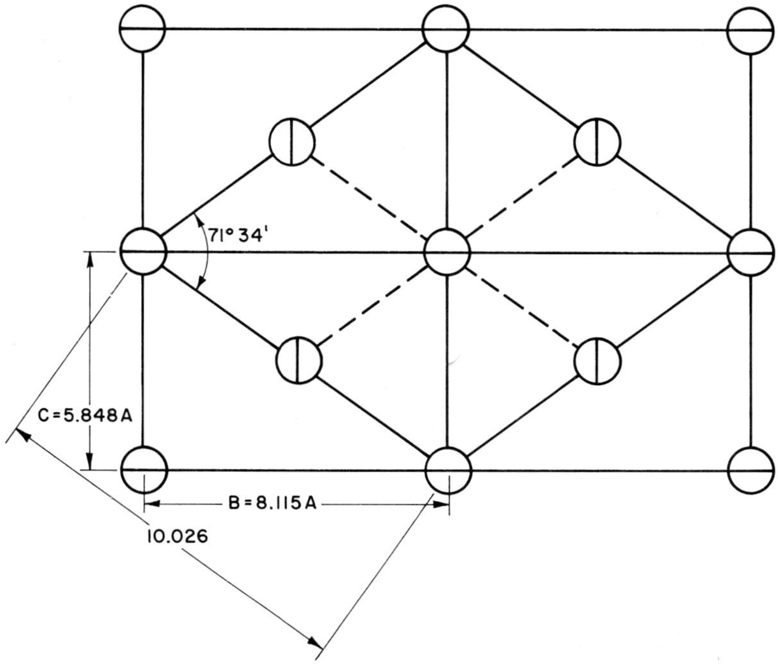

Fig. 9. Interpretation of unit cell dimensions of $MgAlB_{14}$ in terms of an icosahedral arrangement. The figure illustrates how a distorted closest-packed arrangement of B_{12} units can be obtained by placing a differently oriented unit at the center of the a face

closest packing to cubic closest packing of the B_{12} groups. The high temperature tetragonal form of $C_2Al_3B_{48}$ has been shown to disproportionate on cooling into two intertwined orthorhombic forms retaining a pseudotetragonal symmetry (MATKOVICH et al., 1965c). Similarly, AlB_{12} can crystallize, depending on the rate of cooling into an α or γ modification which may be described in terms of cubic or hexagonal closest-packing arrangement of a large quasi-spherical unit (probably B_{84}) (MATKOVICH et al., 1965b).

A theoretical treatment of the electronic structure of the icosahedral B_{12} group has been reported by LONGUET-HIGGINS and ROBERTS (1955). According to their calculations, the most stable form of the B_{12} icosahedra corresponds to a model where each boron atom has five intraicosahedral bonds and one external bond directed along the five-fold icosahedral axis. To permit this bonding the B_{12} group needs a total of 38 electrons i.e. two more than are available from the 12 boron atoms. This model has obvious merit when applied to molecular icosahedral boron structures and in fact LONGUET-HIGGINS and ROBERTS predicted correctly the existence of the $B_{12}H_{12}^{-2}$ ion. However, the applicability of this model to refractory borides has been less than satisfactory. As a rule, the refractory boride structures tend to be electron-deficient, and frequently the more electron-deficient structures appear to be more stable. Thus, tetragonal BeB_{12} on treatment with BCl_3 converts readily into the more electron-deficient structure of alpha tetragonal boron (BECHER, 1960), in another case $B_{13}C_2$ which lacks one electron from the required 38 has a higher melting point than the presumably ideal B_4C. Also, in contrast to the ideal bonding observed in $B_{12}H_{12}^{-2}$, with the refractory B_{12} borides the number of icosahedral boron atoms with a coordination of six may vary sharply. This is clearly illustrated in Table 6. The fact that this coordination can vary so widely among the higher borides suggests the need for a more sophisticated understanding of the electronic nature of the B_{12} unit in these structures.

Table 6. Pentagonal pyramidal coordination of icosahedral boron atoms (in structures based on simple B_{12} units)

Compound		Number of icosahedra per unit cell	Number of icosahedral atoms with coordination of six
Alpha boron (rhombohedral)	(1/4CCPr)	1	6
Alpha boron derivatives (B_4C, B_6P, etc.)	(1/4CCPr)	1	12
Tetragonal boron (simple)	(2TCP)	4	8
Tetragonal boron derivatives (BeB_{12}, $NiTaB_{24}$, etc.)	(2TCP)	4	4-6
$MgAlB_{14}$	(4HPo)	4	2
C_4AlB_{24-26}	(2HCPo)	4	12

Among the refractory borides there is of course one obvious limitation to the presence of 12 external icosahedral bonds directed along the pentagonal axis of the icosahedon.Namely , in an idealized boron structure made up of B_{12} icosahedra where all atoms display pentagonal pyramidal coordination, one is confronted with a polyhedral framework which preserves six five-fold symmetry axes and this is incompatible with lattice periodicity. When this type of icosahedral grouping occurs, it leads invariably to enlargement of the polyhedral unit to a B_{84} group observed in beta rhombohedral boron. The periodicity of the crystalline lattice of borides containing B_{84} or even B_{156} units is obtained by arrangements of these units into simple packing models. This does not mean that a high degree of directional external icosahedral bonds could not be obtained in packed arrangements of the B_{12} units. Thus the openings among the B_{12} groups may accommodate extra-icosahedral (or interstitial) atoms which often contribute to the required bonding directionality of the icosahedral boron atoms. This occurs in boron carbide where one-half of the external icosahedral bonds are directed toward interstitial carbon atoms.

Thus, the predictive utility of pentagonal pyramidal coordination of boron atoms and packing of the polyhedral boron groups can be summarized as follows:

1. The tendency for boron to desplay highly directed pentagonal pyramidal coordination can be used effectively to derive the polyhedral form of the B_{12}, B_{84} and B_{156} units.

2. The concept of packing can be used to derive various three-dimensional arrangements of polyhedra based on stacking of closest packed layers of quasispherical units.

3. The orientation of polyhedra and placement of extra-icosahedral atoms, tend to accommodate as many directional external polyhedral bonds as possible.

5. Structural Determinants in the Tetraborides

The boron arrangement in tetraborides was described by earlier workers (ZALKIN and TEMPLETON, 1953), (BERTAUT and BLUM. 1949) as a combination of structural types, that is, octahedral boron groups, similar to the hexaborides, and interstitial boron pairs accommodated in the manner observed in diborides. In addition it was pointed out that the octahedral boron groups, in conjunction with the boron pairs, provided a continuous three-dimensional framework.

Other spatial configurations which would provide a clearer picture of the MB_4-type structure have been studied. The MB_4 structure can be described very simply in terms of a continuous framework of M_2B_2 tetrahedra in which the ocatahedral boron groups are placed interstitially (see Fig. 10). The M_2B_2 units are stacked in a tetragonal close-packed array (ten nearest M_2B_2 neighbors) (MATKOVICH et al., 1965b). This type of representation in terms of tetrahedral units apparently can be applied to the V_3B_2 type borides indicating a possible structural similarity. Thus both YB_4 amd V_3B_2 structures would be based on continuous tetrahedral network of M_2B_2 units. At present it is not clear whether these tetrahedral groups possess any physical significance; they do, however, provide a clearer presentation of the spatial relationship between atoms. Figure 10 illustrates the structure of V_3B_2 and YB_4 represented in terms of tetragonal close packing of M_2B_2 tetrahedra. While the tetrahedral arrangement is the same in both structures, the differences in the accommodation of interstitial atoms are significant. The interstitial openings lie on the four-fold axis in the unit cell as indicated by the dotted squares in Figure 10. The

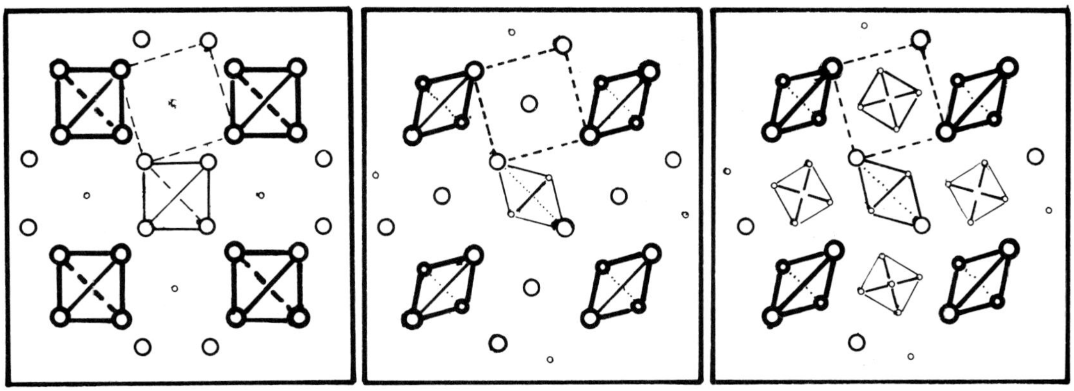

a b c

Fig. 10a-c. Comparison of the Fe_2B, V_3B_2 and YB_4 structural types. *Large circles:* metal atoms, *small circles:* boron atoms, *dotted squares:* interstitial openings. (a) Fe_2B structural type. A body-centered arrangement of M_4 tetrahedra. Boron atoms are placed interstitially. (b) V_3B_2 structural type. The same spatial arrangement of tetrahedra (M_2B_2 in this case) as in (a) but the central and corner tetrahedra are differently oriented. Interstitial openings accommodate metal atoms in this structure. (c) YB_4 structural type. The same arrangement of M_2B_2 tetrahedra as in (b). Interstitial openings accommodate up to 6 boron atoms in form of octahedra manner in which these openings are occupied in V_3B_2 and YB_4 is illustrated in Figure 11. In V_3B_2 (Fig. 11b) the interstitial opening accommodates one metal atom per unit-cell length, and in YB_4 (Fig.11c) the interstitial opening accommodates six octahedrally grouped boron atoms. Thus the YB_4 structure can be derived from the V_3B_2 structural type by substituting B_6 octahedra for the metal atoms on the four-fold axis.

In an actual crystal structure determination of YB_4 it was shown that the octahedral boron groups are defective. In the particular crystal that was examined, it was found that the B(1) octahedral site (see Figs. 10 and 11) is fully occupied; in the B(3) octahedral site, however, almost every fourth atomic position is vacant. As the contact between the boron octahedra and the boron pairs can be made only through the B(3) atomic sites (Fig. 10), partial occupancy of these positions argues against the suggestion that this structure is based on a continuous three-dimensional boron network. The B(2) sites which correspond to the boron pairs are apparently fully occupied.

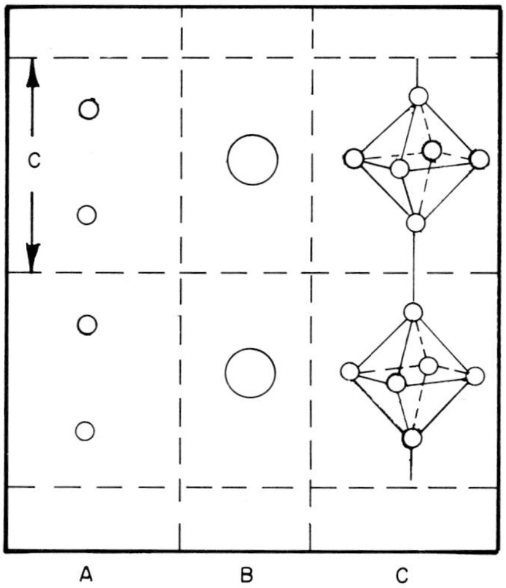

Fig. 11. Comparison of the interstitial sites in (A) Fe_2B, (B) V_3B_2 and (C) YB_4. In Fe_2B the interstitial sites accommodate two boron atoms along the four-fold axis (per unit-cell length). In V_3B_4 and YB_4, the M_2B_2 tetrahedra provide a larger opening which enables accommodation of four additional boron atoms in YB_4 or a single metal atom in V_3B_2

References

ALLEN, R.D.: The solid solution series, boron-boron carbide. J. Amer. Chem. Soc. 15, 3582 (1953)

BECHER, H.J.: Beryllium boride, BeB_{12}, with tetragonal boron structure. Z. anorg. allgem. Chem. 306, 266 (1960)

BERTAUT, F., BLUM, P.: Structure of uranium borides. C. R. Acad. Sci. (Paris) 229, 666-667 (1949)

BLITZ, H.: Über krystallisiertes Bor. Ber. deut. Chem. Ges. 41, 2634 (1908) and 43, 297 (1910)

DECKER, B.F., KASPER, J.S.: The crystal structure of a simple rhombohedral form of boron. Acta Cryst. 12, 503 (1959)

DECKER, B.F., KASPER, J.S.: The crystal structure of a nickel boride related to tetragonal boron. Acta Cryst. 13, 1030 (1960)

GEISE, R.F., JR., ECONOMY, J., MATKOVICH, V.I.: Interstitial derivatives of β-boron. Z. Kristallogr. 122, 144 (1965a)

GEISE, R.F., JR., ECONOMY, J., MATKOVICH, V.I.: Topotatic transition in C_4AlB_{24}. Acta Cryst. 20, 697 (1966)

GEISE, R.F., MATKOVICH, V.I., ECONOMY, J.: The crystal structure of YB_4. Z. Kristallogr. 122, 423 (1965b)
GORSKI, L.: Tetragonal modifications of boron. Sov. Phys. Solid. State, 9, K169 (1965)
HOARD, J.L., HUGHES, R.E.: In: The Chemistry of Boron and its Compounds. 25-154 (E.L. Muetterties, ed.), New York: Wiley and Sons 1967
HOARD, J.L., HUGHES, R.E., SANDS, D.E.: The structure of tetragonal boron. J. Amer. Chem. Soc. 80, 4507 (1958)
HUGHES, R.E., KENNARD, C.H.L. SULLINGER, D.B., WEAKLEIM, H.A., SANDS, D.E., HOARD, J.L.: The structure of β-rhombohedral boron. J. Amer. Chem. Soc. 85, 361, (1963)
KOHN, J.A., ECHART, D.W.: Aluminum boride, AlB_{12}, Analyt. Chem 32, 296 (1960)
KOHN, J.A., ECHART, D.W.: γ-AlB_{12}, a new structure and its relation to α-AlB_{12} as shown by twinned space groups, Z. Kristallogr. 116, 134 (1961)
KROGMANN, J., BECHER, H.J.: Structural determination of tetragonal boron containing aluminum and berylium with the formula $AlBe_{0.8}B_{24.2}$. Z. anorg. allgem. Chem. 392, 197 (1972)
LaPLACA, S., POST, B.: The boron carbide structure type. Plannseeber, Pulvermet. 9, 109 (1961)
LaPLACA, S., BINDER, I., POST, B.: Binary dodecaborides, J. Inorg. Nucl. Chem. 18, 113 (1960)
LIPP, A., ROEDER, M.: Boron carbide that contains aluminum. Z. anorg. allgem. Chem. 343, (1966); New B-C-Si; and B-Si Compounds, 344, 225 (1966)
LIPSCOMB, W.N., BRITTON, D.: Valence structures of higher borides. J. Chem. Phys. 33, 275 (1960)
LONGUET-HIGGINS, H.D., ROBERTS, M. deV: The electronic structure of an icosahedron of boron atoms. Proc. Roy. Soc. A230, 110 (1955)
MATKOVICH, V.I.: A new form of boron silicide, B_4Si. Acta Cryst. 13, 679 (1960)
MATKOVICH, V.I.: Interstitial compounds of boron. Amer. Chem. Soc. 83, 1804 (1961)
MATKOVICH, V.I., ECONOMY, J.: Structure of $MgAlB_{14}$ and a brief critique of structural relations in higher borides. Acta Cryst. B26, 616 (1970)
MATKOVICH, V.I., ECONOMY, J., GIESE Jr., R.F., BARRETT, R.: Structure of dodecarborides. Acta Cryst. 19, 1056 (1965a)
MATKOVICH, V.I., GIESE Jr., R.F., ECONOMY, J.: Packing of B_{12} groups in boron and borides structures. Z.Kristallogr. 122, 116 (1965b)
MATKOVICH, V.I., GIESE Jr., R.F., ECONOMY, J.: Phases and twinning in $C_2Al_3B_{48}$ (β-AlB_{12}). Z. Kristallogr. 122, 108 (1965c)
NASLAIN, R., KASPER, J.S.: Crystal structure of the Ø-phase in the boron-sodium system. J. Solid State Chem. 1, 150 (1970)
NEIDHARD, H., MATTES, R., BECHER, H.: Preparation and structure of an aluminum containing boron carbide. Acta Cryst. 26, 315 (1970)
PERROTTA, A.J., TOWNES, W.D., POTENZA, J.A.: Crystal structure of $C_8Al_{2.1}B_{51}$. Acta Cryst B25, 1223 (1969)
PESHEV, P., ETOURNEAU, J., NASLAIN, R.: Boron scandium system. Mater. Res. Bull. 5, 319 (1970)
PLOOG, K.: Crystal chemical relations between the nonmetal borides boron carbide ($B_{48}B_2C_2$). J. Electrochem. Soc. 6, 846 (1974)
RICHARDS, S.M., KASPER, J.S.: The crystal structure of YB_{66}. Acta Cryst. B25, 237 (1969)
SANDS, D.E., CLINE, C.F., ZALKIN, A., HOENING, C.L.: Rhombohedral elemental B. Acta Cryst. 14, 309 (1961)
TALLEY, C.P., LaPLACA, S., POST, B.: A new polymorph of boron. Acta Cryst, 13, 271 (1960)
ZALKIN, A., TEMPLETON, D.H.: The crystal structures of CeB_4, ThB_4 and UB_4. Acta Cryst. 6, 209-272 (1953)

VII. Crystal Chemistry of Higher Borides

V. I. MATKOVICH and J. ECONOMY

1. Introduction

The tendency of higher borides to form distinct stacking arrangement of closest-packed layers of B_{12} units also leads to the formation of an equally ordered array of interstitial openings. When any of these sites is fully accupied by interstitial atoms, stoichiometries are obtained with integral atomic ratios. With this in mind, an attempt has been made in earlier publications to group the higher borides into isotypic series (MATKOVICH, 1961; ECONOMY et al., 1965; GIESE et al., 1965). Two factors which frequently tend to comlicate clear-cut stoichiometric assignments include partial occupancy of interstitial sites, and the tendency of certain atoms such as carbon and silicon to substitute for boron in the B_{12} incosahedra. Both of these problems arise from the ease with which solid solutions of two distinct compounds can form within a given series. Existence of a number of distinct crystallographic sites leading to multiple stoichiometries within the same B_{12} framework may also confuse the stoichiometries of these structures. In the light of these potential complications it becomes even more urgent that the crystal chemistry of the higher borides be expressed in a uniform and consistent pattern in the published literature. This paper provides a continuation of our effort to achieve that goal.

It should be noted that relatively few structures of the higher borides have been determined. A number of less well identified compounds can be classified by analogy to known structures.

In this paper the crystal chemistry of five different series of interstitial derivatives of B_{12} borides is discussed. These are designated (MATKOVICH et al., 1965b) as 1/4CCPr (simple rhombohedral arrangement of icosahedral B_{12} units), 2TCP (tetragonal arrangement of icosahedral B_{12} units with nonidentical icosahedral groups centered at all faces), 1/4CCPvr (simple rhombohedral arragement of quasi-spherical B_{84} units), 2 HCP (hexagonal closest packing arrangement of icosahedral B_{12} units with orthorhombic symmetry) and 4 HPo (hexagonal primitive arrangement of icosahedral B_{12} groups with orthorhombic symmetry).

Before getting into the discussion of each of these group it is of value to consider in general terms the problems obtaining clear-cut stoichiometries. Other higher borides whose structures have not been determined can at least be classified as derivatives of various packing models as described in chapter B VI. of this Volume.

a) Defective Occupancy

A very common cause of variation in composition of B_{12} borides arises from partial occupancy of interstitial sites. Such systems should be interpreted as solid solutions where the two extremes, i.e., full occupancy and absence of interstitial atoms represent the two stoichiometric forms. A good example of this is boron subphosphide ($B_{12}P_2$) where variations in phosphorus content indicate a solid solubility range between α-rhombohedral boron (designated as B_{12}) and $B_{12}P_2$ (ECONOMY et al., 1965).

Frequently a stoichiometric assignment may be further complicated by the presence of interstitial elements spread over multiple interstitial sites. This situation is observed in orthorhombic and rhombohedral aluminum boron-carbides where the Al atoms substitute for boron in the C-B-C interstitial chain, but not exactly in the unique boron position.

b) Substitutional Solid Solutions

When substitutional solid solubility involves a well-defined interstitial site, the assignment of stoichiometry is relatively straight forward. However, the substitutional mechanism may also involve icosahedral boron positions. An example of this is the homogeneity range in the boron carbon system where the composition may vary from B_4C to B_7C. A substitutional mechanism was proposed by ALLEN (1953) where the interstitial chain could vary from C-C-C to C-B-C resulting in well-defined compounds of $B_{12}C_3$ and $B_{12}C_2B$. More recent determinations (MATKOVICH and ECONOMY, 1972) indicate, however, that $B_{12}C_3$ does not exist as a stoichiometric compound but results from partial substitution of born by carbon at six of the twelve icosahedral sites. This homogeneity range is best described as a solid solution between $B_{12}C_2B$ and a hypothetical $(B_6C_6)C_2B$. A similar problem has been encountered in the boron silicon system where the composition range can be described in terms of $B_{12}Si_2$ and a hypothetical $(B_6Si_6)Si_2$. Only boron-rich sides of the homogeneity range have actually been observed in both of the above cases.

c) Existence of Multiple Interstitial Sites

Variations in compositions can be further complicated when crystallographically different interstitial sites are occupied. In this case two mechanisms of solid solution may take place either selectively or simultaneously. In comparing various α-rhombohedral boron derivatives (1/4CCPr) one observes that the same interstitial opening may accommodate three small atoms (as in boron carbide) or two larger atoms (as in boron phosphide). Finally in the isotypic situation of $MgAlB_{14}$ and NaB_{15} some of the interstitial atoms occupy completely different interstitial positions (MATKOVICH and ECONOMY, 1970a; NASLAIN and KASPER, 1970).

2. Stoichiometric Assignments

a) 1/4CCPr (α-rhombohedral Boron) Derivatives

The 1/4 CCPr series is based on a cubic closest packing of icosahedral units (see Fig. 1) with a simple rhombohedral unit cell. The crystal chemistry of these derivatives has been reviewed earlier (ECONOMY et al., 1965). Since that time little new work has been carried out to clarify the nature of the interstitial atoms present in boron subarsenide, boron suboxide and boron subsulfide. The existence of a ternary system $B_{12}C_2Al$ has been reported independently and almost simultaneously by LIPP and ROEDER (1966) and by our laboratory (GIESE et al., 1966). LIPP and ROEDER (1966) have also reported a new composition $B_{12}C_2Si$ which appears to be closely related to $B_{12}C_2Al$. A detailed structure analysis of an aluminum borocarbide of composition AlC_4B_{40} has been reported (NEIDHARDT et al., 1970). In our program we have carried out a careful X-ray analysis of single crystals of rhombohedral boron silicide (SiB_{3-4}) and boron carbide ($B_{4-6.5}C$) (MATKOVICH and ECONOMY, 1972). The boron silicide was analyzed for a composition of SiB_3 and the structure was found to be in agreement with the findings of MAGNUSSON and BROSSET (1962) where two silicon atoms are accommodated interstitially and lie along the three-fold axis of the rhombohedral

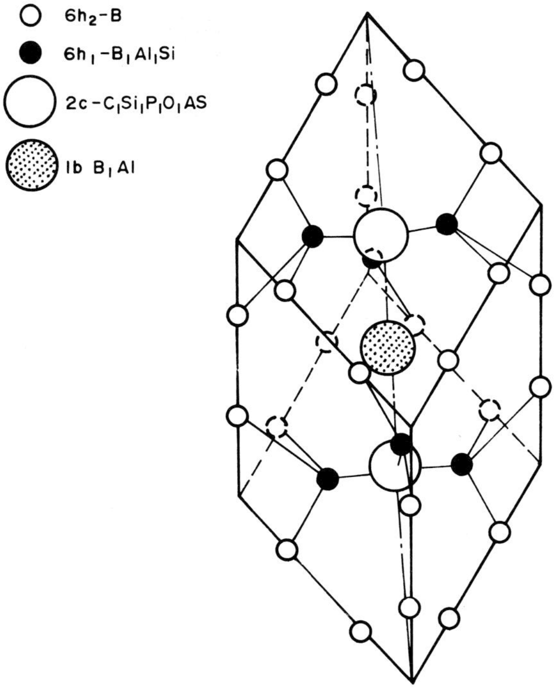

Fig. 1. Atomic positions in the structure of the 1/4 CCPr series of borides. In some cases the interstitial atoms do not occupy the 1b position exactly

cell. The rest of the silicon partially substitutes for the icosahedral boron atoms at the bottom and top of the icosahedral triangles. This composition is consistent with a $(B_{12-a}Si_a)Si_2$ stoichiometry where observed values of a vary between 1.0 and 1.5. Work on boron carbide crystals indicates that this compound resembles SiB_3 with carbon also substituting for icosahedral boron atoms but in a different position (see Figs. 2 and 3) (MATKOVICH and ECONOMY, 1972). The interstitial opening accommodates two carbon atoms and one boron atom in a C-B-C arrangement. These results lead to the conclusion that the commonly used formula of B_4C represents a solid solution of a hypothetical $(B_6C_6)C_2B$ and $(B_{12})C_2B$. Existence of linear C-B-C and C-B-B as well as non-linear C-Al-B chains has been reported by NEIDHARDT et al., (1970) for AlC_4B_{40} [or $(B_{12})(C_{2-x}B_x)B_{1-n}Al_n$]. Thus the composition of aluminum boron carbide may be written as $(B_{12-a}C_a)(C_{2-x}B_x)(B_{1-n}Al_n)$. The random distribution of the aluminum over a six-fold position centered on the 1b site suggests that the aluminum atom may be too large to be accommodated in a linear three-atom chain. Therefore, $B_{12}C_2B$ is the only stoichiometric compound in the rhombohedral phase of the boron-carbon system which has been observed (MATKOVICH and ECONOMY, 1972). $B_{12}C_2Al$ and $B_{12}C_2Si$ may also fall in this group and could be written as $(B_{11}Al)C_2B$ and $(B_{11}Si)C_2B$, respectively. Thus, the compounds of the 1/4CCPr series can be designated by a general formula of $(B_{12-2}L_2)$ $(X_{2-x}B_x)$ $(B_{1-n}Y_n)$. Interatomic distances in boron carbide and boron silicide are compared in Figs. 2 and 3. Various compounds of this isotypic series are listed in Table 1.

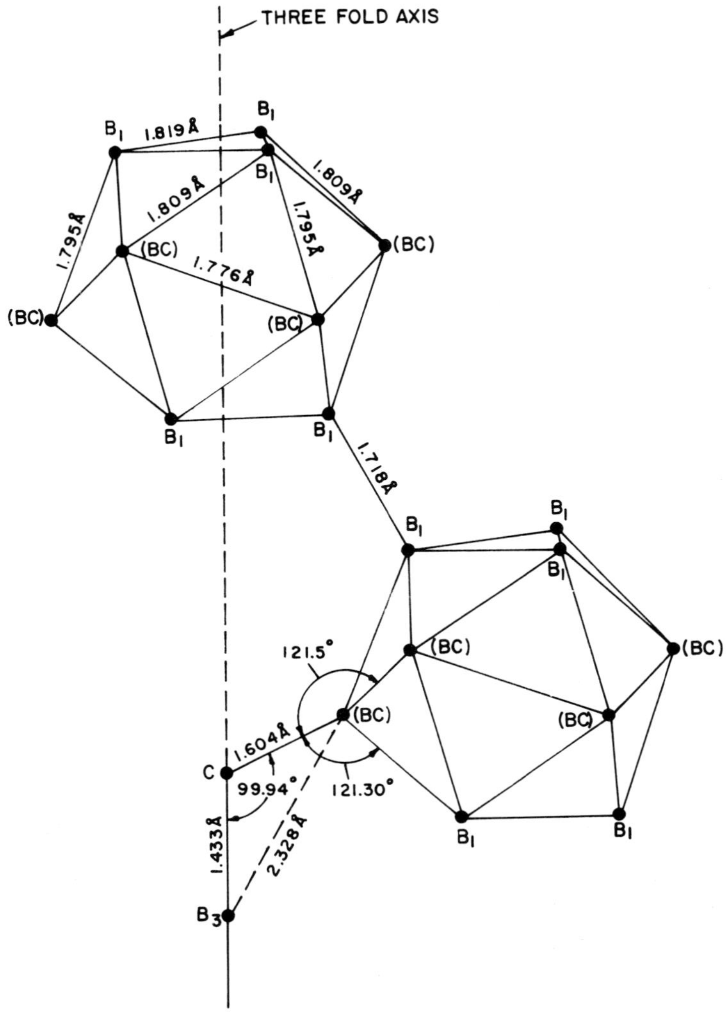

Fig. 2. Interatomic distances in $(B_{11.5}C_{0.5})C_2B$

b) 2TCP (Tetragonal Boron) Derivatives

The 2TCP derivatives are based on a tetragonal arrangement of B_{12} icosahedra and have available two two-fold and one four-fold interstitial sites (see Fig. 4). This isotypic series includes BeB_{12} (BECHER, 1960), NiB_{25} (DECKER and KSAPER, 1960), $C_2Al_3B_{48}$ (high temperature phase) (MATKOVICH et al., 1965a), $AlBeB_{24}$ (KROGMAN and BECHER, 1972), $B_{48}B_2C_2$ (PLOOG, 1974) and $B_{48}B_2N_2$ (PLOOG, 1974). In an earlier publication tetragonal boron was included in this series but it has now been suggested that the crystal contained carbon or other impurities. In the case of $AlBeB_{24}$, it has been shown (BECHER, 1960) that pairs of aluminum atoms occupy randomly a two-fold position and the same is the case with $B_{48}B_2C_2$ and $B_{48}B_2N_2$ (PLOOG, 1974). The interstitial atoms occupy two two-fold positions and groups of four atoms (randomly occupied) are centered around the four-fold position. With NiB_{25} it has been suggested (DECKER and KASPER, 1960) that the interstitial nickel and boron atoms occupy selectively two different sets of two-fold interstitial positions in the structure. In the case of

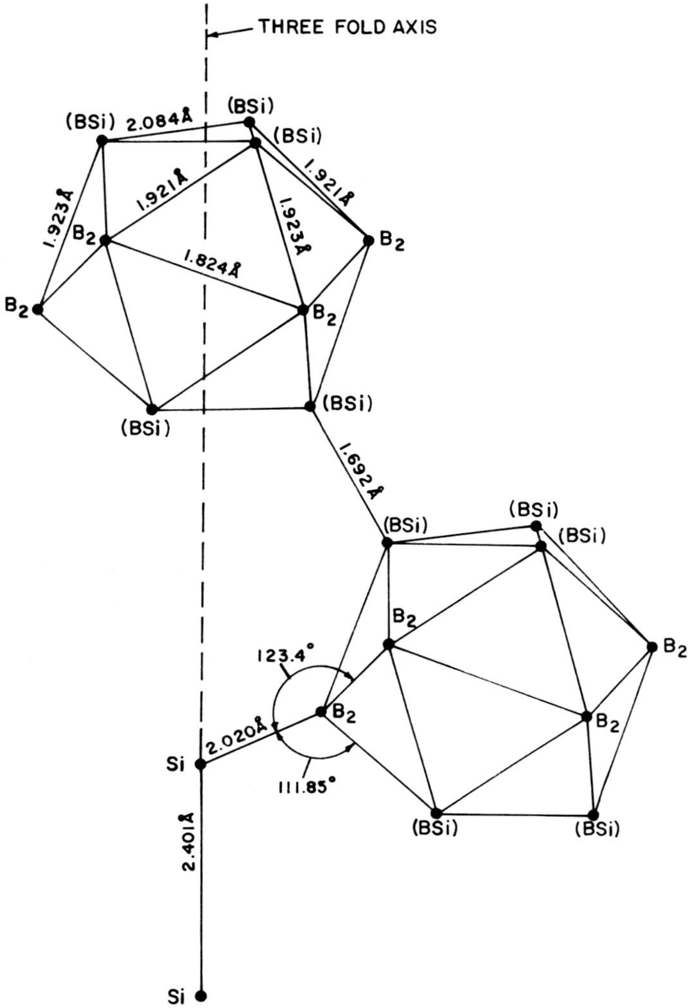

Fig. 3. Interatomic distances in $(B_{10.5}Si_{1.5})Si_2$

$C_2Al_3B_{48}$, the high temperature form is obviously a derivative of the 2TCP model. This was recently confirmed by MATTES et al. (1972). It is suggested that this phase correspond to a $(B_{12})_4C_2Al_4$ stoichiometry where the aluminum position is partially defective. In view of the above information a general formula for the 2TCP derivatives may be expressed as $(B_{12})_4X_2Y_2Z_4$. Various compounds of this series are listed in Table 2.

It is intersting to note the ease with these systems can convert from one to the other. Thus, Becher found that in reaction of BeB_{12} with BCl_3 the beryllium atoms are removed from the polyhedral framework while boron atoms are inserted, apparently in different positions (BECHER, 1960).

c) 1/4CCPυr (β-rhombohedral Boron) Derivatives

It has been suggested that $B_{14}Si$ and the so-called "monoclinic" AlB_{12} were isotypic with β-rhombohedral boron (MATKOVICH and ECONOMY, 1970).

Table 1. Compounds of the 1/4 CCPr series

	Formulas	
	Conventional	Structural
1. Boron carbides	a. $B_{13}C_2$	$B_{12}C_2B$
	b. B_4C	$(B_{11}C)C_2B$
	c. B_6C	$B_{12}C_2$
2. Boron silicide	a. B_3Si	$(B_{10.5}Si_{1.5})Si_2$
3. Boron subphosphides[a]	a. B_6P	$B_{12}P_2$
	b. $B_{13}P_2$	$B_{12}P_2B$
4. Boron subarsenide[a]	a. B_6As	$B_{12}As_2$
5. Boron suboxide[a]	a. B_6O	$B_{12}O_2$
6. Boron subsulphide[b]	a. $B_{12}S_2$	$B_{12}S_2$
7. Aluminum boro-carbide	a. $B_{12}C_2Al$	$B_{12}C_2Al$ or $(B_{11}Al)C_2B$
	b. $B_{40}C_4Al$	$B_{12}C_{1.33}B_{1.33}Al_{.33}$
8. Silicon boro-carbide	a. $B_{12}C_2Si$	$B_{12}C_2Si$ or $(B_{11}Si)C_2B$
9. α-rhombohedral boron[c]	a. B	B_{12}

[a] LaPLACA and POST, 1961

[b] MATKOVICH, 1961

[c] DECKER and KASPER, 1959

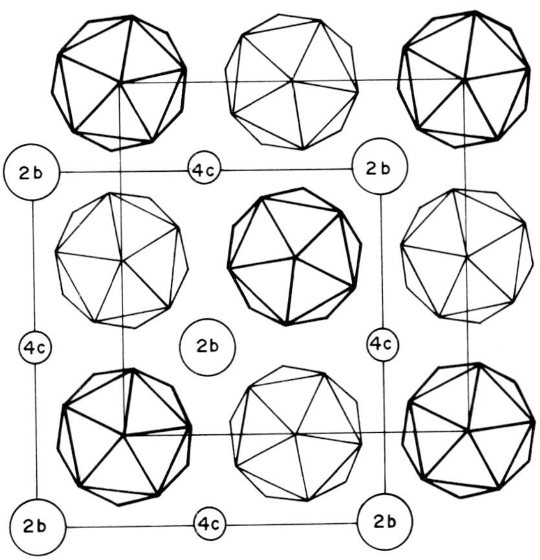

Fig. 4. The atomic arrangement in compounds of the 2TCP series of borides. Two-fold 2b sites and four-fold 4c sites are indicated. The exact position of the 4c site is not occupied, but a group of four atoms (random occupancy) is centered on that position

Table 2. Compounds of the 2TCP series

	Formulas	
	Conventional	Structural
1. Aluminum boro-carbide (high temperature form)	$C_2Al_3B_{48}$	$(B_{12})_4C_2Al_4$
2. Nickel boride	$B_{25}Ni$	$(B_{12})_4B_2Ni_2$
3. Beryllium boride	BeB_{12}	$(B_{12})_4B_2Be_2$
4. Aluminum beryllium boride	$AlBeB_{24}$	$(B_{12})_4Be_2Al_2$
5. Tetragonal boron carbide	$B_{25}C$	$(B_{12})_2B_2C_2$
6. Tetragonal boron nitride	$B_{25}N$	$(B_{12})_2B_2N_2$

This structure is based on a cubic closest-packed arrangement of B_{84} units (HUGHES et al., 1963). There is a similarity in the stacking arrangement of the B_{84} units and the B_{12} units in the rhombohedral boron series. In addition, the C-B-C interstitial chain in boron carbide bears a superficial resemblance to the interstitial B_{10}-B-B_{10} group in β-rhombohedral boron. In the case of $B_{14}Si$ and monoclinic AlB_{12} is has been suggested that the interstitial group consists of (B_7Si_3)-Si-(B_7Si_3) and (B_6Al_4)-Al-(B_6Al_4), respectively (GIESE et al., 1965). Recently an isotypic copper boride of composition CuB_{24} was reported by HIGASHI et al. (1974). A formulation for this isotypic series consistent with the above data would be $B_{84}(B_{10-a}N_a)_2X$. In this approach a group of atoms such as B_{10} is considered as an interstitial unit. Various compounds of this series are listed in Table 3.

Table 3. Compounds of the 1/4 CCPυr series

	Formulas	
	Conventional	Structural
1. β-rhombohedral boron	B	$(B_{84})(B_{10})_2B$
2. Boron silicide	$B_{14}Si$	$(B_{84})(B_7Si_3)_2Si$
3. Aluminum boride	$B_{32}Al_3$	$(B_{84})(B_6Al_4)_2Al$
4. Copper Boride	$B_{24}Cu$	- - -

d) 2HCPo Derivatives

The first compound of this series was reported as AlB_{10} (KOHN et al., 1958). Subsequent reports of isotypic aluminum borocarbide of composition C_4AlB_{24} (MATKOVICH et al., 1964) was believed to be a different compound. It appears now that significant amounts of carbon were present in all samples that were reported (PERROTTA et al., 1969). Chemical analysis (MATKOVICH et al., 1964) indicates some variations in composition which may be designed as C_4AlB_{24-25} while the X-ray analysis of the structure led PERROTTA et al (1969) to the composition $C_4Al_{1.1}B_{25.5}$.

This compound is based on a hexagonal closest packing of B_{12} units with an orthorhombic symmetry. The unit cell contains four B_{12} icosahedra arranged in the manner shown in Figure 5. The eight-fold carbon posi-

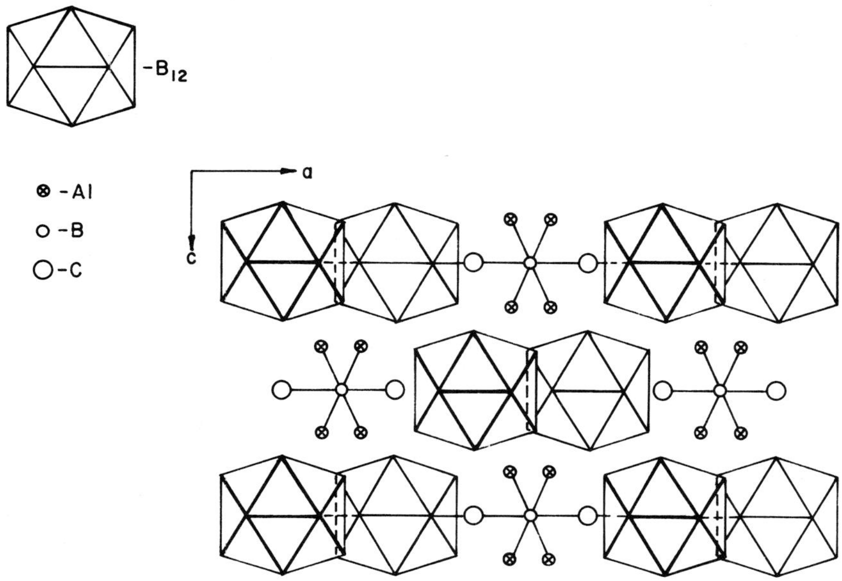

Fig. 5. Atomic arrangement in C_4AlB_{25}, the 2HCPo series of borides

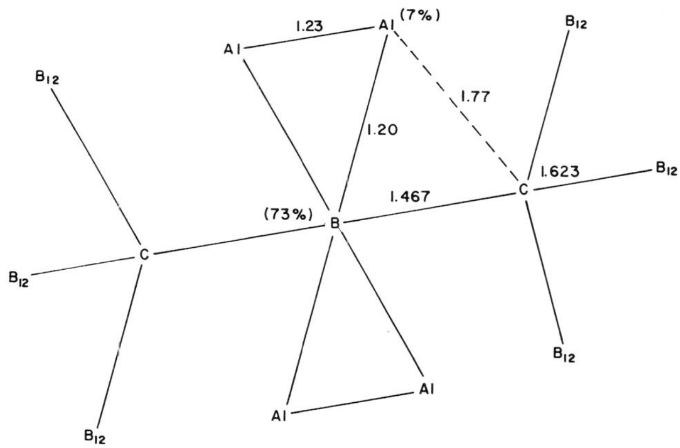

Fig. 6. Arrangement of interstitial atoms in C_4AlB_{25} as illustrated by PERROTTA et al. (1969). Short Al-Al distance indicates that all aluminum atoms cannot be present simultaneously

tions are fully occupied and the four-fold boron position (extra-icosahedral) is about 70% occupied. The aluminum atoms are however, distributed in small amounts along a number of sites (see Fig. 6).

It is of interest to note that C_4AlB_{24-26} converts on heating to the rhombohedral form of the 1/4CCPr series (GIESE et al., 1966). The difference in the arrangement of icosahedra can be described simply as corresponding to the cubic closest packing in the rhombohedral form and to the hexagonal closest packing in the orthorhombic form (see Fig. 7). There is, however, a close similarity in accommodation of the interstitial atoms. The C-B-C interstitial chain may exist in either of the forms. Furthermore, the boron atom may be replaced by aluminum atoms, but not in the same position resulting in a non-linear C-Al-C chain. This is illustrated in Figures 5 and 6. It is interesting that the aluminum position in the chain is multifold and cannot even

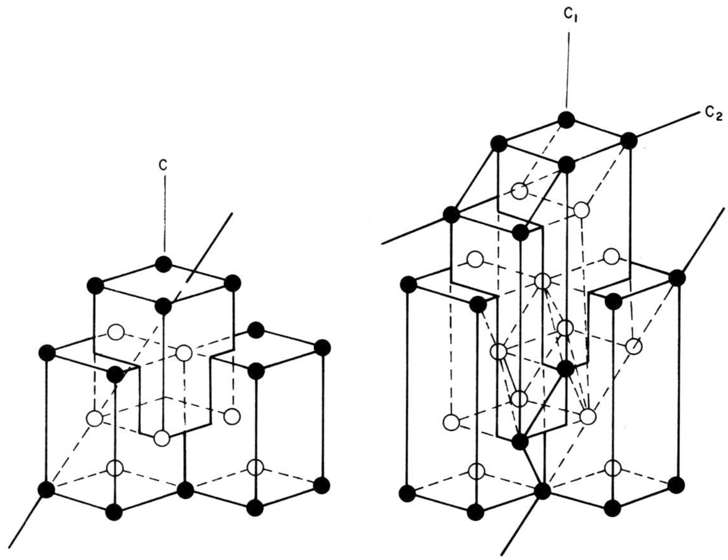

Fig. 7. Comparison of hexagonal closest packing *(left)* and cubic closest packing *(right)*. Assuming that the circles represent B_{12} units, the figures illustrate change in position of B_{12} groups during topotactic transition of C_4AlB_{25}, (GIESE et al., 1966)

theoretically be fully occupied. Thus in the rhombohedral form, aluminum position is six-fold but only one of the six positions may be occupied at any one time. Furthermore, no aluminum atom may be present when a boron atom occupies the central 1b position.

In summary, the composition C_4AlB_{26} with a unit cell containing two formula weights and a variable aluminum content may be formulated as $(B_{12})_4C_8B_4Al_n$ or by a more general formula of $(B_{12})_4X_8Y_4Z_n$.

e) 4HPo Derivatives

$MgAlB_{14}$ consists of a simple hexagonal arrangement of icosahedra with orthorhombic symmetry (MATKOVICH and ECONOMY, 1970a). The unit cell contains four B_{12} groups, eight additional boron atoms and up to eight metal atoms (see Fig. 8). The structural formula is best represented as $(B_{12})_4Mg_4Al_4B_8$. The interstitial boron atoms are accommodated in an eight-fold position while magnesium and aluminum atoms are distributed among a four-fold position. Both metal positions have been found to be partially occupied. The structure analysis indicated a tendency for some Al to be randomly distributed among the Mg positions. Another member of this series, NaB_{15} was reported by NASLAIN and KASPER (1970), with the structural formula $(B_{12})_4B_4Na_4B_8$. However, in comparison with the isotypic $MgAlB_{14}$ the interstitial aluminum position is not occupied in NaB_{15}. Instead the interstitial boron atoms occupy a new position between magnesium (sodium in the case of NaB_{15}) and the interstitial boron pair. Assuming full occupancy of interstitial sites the general stoichiometry of $(B_{12})_4X_4Y_4Z_8$ may be assigned to this series.

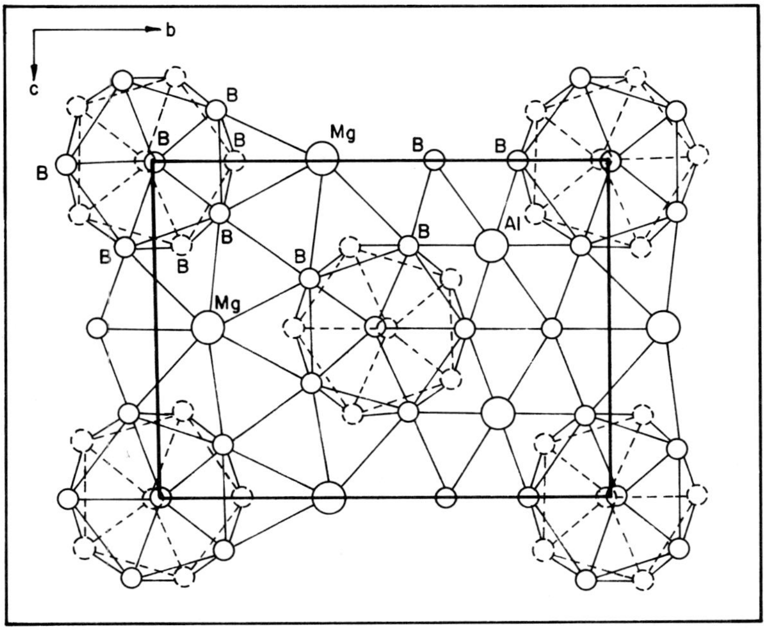

Fig. 8. A layer of icosahedra in $MgAlB_{14}$ with extra-icosahedral atoms. The atoms shown fall within x = ± 1/4

3. Summary

It has been found that all of the higher borides can be classified as derivatives of simple packing arrangements of B_{12} units. The stoichiometries within a given series can be expressed by a general formula denoting the interstitial atoms as X, Y and Z (each of which can assume values from zero to unity) and the substitutional icosahedral boron positions as L. A general description for stoichiometries of the five isotypic series may be listed as follows:

1/4CCPr	Derivatives $(B_{12-a}L_a)(X_{2-x}B_x)(B_{1-n}Y_n)$
2TCP	Derivatives $(B_{12})_4X_2Y_2Z_4$
1/4CCPvr	Derivatives $B_{84}(B_{10-a}L_a)_2X$
2HCPo	Derivatives $(B_{12})_4X_8Y_4Z_n$
4HPo	Derivatives $(B_{12})_4X_4Y_4Z_8$

References

ALLEN, R.D.: The solid solution series, boron-boron carbide. J. Amer. Chem. Soc. 75, 3582 (1953)

BECHER, H.J.: Über das Berylliumborid BeB_{12} mit β-Bor-Struktur. Z. anorg. allgem. Chem. 306, 266 (1960)

DECKER, B.F., KASPER, J.S.: The crystal structure of a simple rhombohedral form of boron. Acta Cryst. 12, 503 (1959)

DECKER, B.F., KASPER, J.S.: The crystal structure of a nickel boride related to tetragonal boron. Acta Cryst. 13, 1030 (1960)

ECONOMY, J., MATKOVICH, V.I., GIESE JR., R.F.: Crystal chemistry of alpha-boron derivatives. Z. Kristallogr. 122, 248 (1965)

GIESE JR., R.F., ECONOMY, J., MATKOVICH, V.I.: Interstitial derivatives of β-boron. Z. Kristallogr. 122, 144 (1965)

GIESE JR., R.F., ECONOMY, J., MATKOVICH, V.I.: Topotactic transition in C_4AlB_{24}. Acta Cryst. 20, 697 (1966)

HIGASHI, I., TAKAHASHI, Y., ATODA, T.: CuB_{24} single crystals. J. Less Common Metals 37, 199 (1974)

HUGHES, R.E., KENNARD, C.H.L., SULLINGER, D.E., WEAKLEIM, H.A., SANDS, D.E., HOARD, J.L.: The structure of β-rhombohedral boron. J. Amer. Chem. Soc. 85, 361 (1963)

KOHN, J.A., KATZ, G., GIARDINI, A.A.: AlB_{10}, A new phase, and a critique on the aluminum borides. Z. Kristallogr. 111, 1 (1958)

KROGMAN, J.K., BECHER, H.J.: Structural determination of tetragonal boron containing aluminum and beryllium with the formula $AlBe_{0.8}B_{24.2}$. Z. anorg. allgem. Chem. 392, 197 (1972)

LaPLACA, S., POST, B.: The boron carbide structure Type. Planseeber, Pulvermet. 9, 109 (1961)

LIPP, A., ROEDER, M.: Bòron carbide that contains aluminum. Z. anorg. allgem. Chem. 343, (1966); New B-C-Si; and B-Si Compounds, 344, 225 (1966)

MAGNUSSON, B., BROSSET, C.: The crystal structure of $B_{2.89}Si$. Acta Cryst. 16, 449 (1962)

MATKOVICH, V.I.: Interstitial compounds of boron. J. Amer. Chem. Soc. 83, 1804 (1961)

MATKOVICH, V.I., ECONOMY, J.: Structure of $MgAlB_{14}$ and a brief critique of structural relations in higher borides., Acta Cryst. B26, 616 (1970a)

MATKOVICH, V.I., ECONOMY, J.: Predictive nature of B_{12} polyhedral borides., Electron Technol. (Warsaw) 3, 167 (1970b)

MATKOVICH, V.I., ECONOMY, J., GIESE, JR., R.F.: Presence of carbon in aluminum borides, J. Amer. Chem. Soc. 86, 2337 (1964)

MATKOVICH, V.I., ECONOMY, J., SMITH, W.D.: Distribution of carbon atoms in boron carbide type structure. In: Boron (in Russian. R.N. TAVADZE, ed.) p. 183 - 195 Moscow: Nauka 1975

MATKOVICH, V.I., GIESE, JR., R.F., ECONOMY, J.: Phases and twinning in $C_2Al_3B_{48}$ (β-AlB_{12})., Z. Kristallogr. 122, 108 (1965a)

MATKOVICH, V.I., GIESE, JR., R.F., ECONOMY, J.: Packing of B_{12} groups in boron and boride structure., Z. Kristallogr. 122, 116 (1965b)

MATTES, R., BECHER, H.J., RETHFELD, H.: Relation of structure of $AlBe_eB_{24}$ and β-AlB_{12} to the tetragonal boron lattice. In: boron (in Russian F.N. TAVADZE, ed.) p. 202 Moscow: Nauka 1975

NASLAIN, R., KASPER, J.S.: The crystal structure of the Ø-phase in the boron sodium system. J.Solid State Chem. 1, 150 (1970)

NEIDHARD, V.H., MATES, R., BECHER, H.: Zur Darstellung und Struktur eines aluminum-haltigen Borcarbides. Acta Cryst. 326, 215 (1970)

PERROTTA, A.J., TOWNES, W.D., POTENZA, J.A.: Crystal structure of $C_8Al_{2.1}B_{51}$. Acta Cryst. 325, 1223 (1969)

PLOOG, J.: Crystal chemical relations between the non-metal borides $B_{48}B_2C_2$ and $B_{48}B_2N_2$ and the hypothetical "I-Tetragonal Boron". J. Electrochem. Soc. 121, 846 (1974)

WILL, G.: Crystal structure analysis of AlB_{10} by the convolution molecule method. Acta Cryst. 23, 1071 (1967)

VIII. Tetragonal Boron-I and Its Derivatives

H. J. BECHER and R. MATTES

1. Introduction

Elementary boron was obtained, in 1943, for the first time in the form of single crystals (LAUBENGAYER et al., 1943). This modification was named tetragonal boron-I. Later on, three additional modifications were found: two rhombohedral (α- and β-boron) (McCARTY et al., 1958); (SANDS and HOARD, 1957) and a second tetragonal one (TALLEY et al., 1960). In these modifications the crystal lattice is built up by more or less regular icosahedra which are connected by intericosahedral bonds or by single boron atoms. (The structure of tetragonal boron-II has not as yet been determined.) Similar structures exist also in some boron-rich compounds containing metals and non-metals. In this Section, therefore, tetragonal boron-I is discussed together with some borides of beryllium, aluminum and carbon which are built on the same structural principle.

2. Preparation of "Pure" Tetragonal Boron-I

On the basis of the published data, tetragonal boron-I can be prepared only by the pyrolytic reaction of BBr_3/H_2 - or BCl_3/H_2 - mixtures at the surface of a hot substrate (HOARD and NEWKIRK, 1960; OLEMSKA et al., 1966). It is deposited in the form of 0.1 - 1 mm size crystals, embedded in a dense microcrystalline or vitreous matrix which is considerably different from tetragonal boron-I. In spite of similar conditions the yield of tetragonal boron-I crystals varies in different investigations. Therefore in 1960 it was suggested that tetragonal boron-I is not a thermodynamically stable modification, but a monotropic one, produced by special nucleation and kinetics of deposition, its formation being induced eventually by impurity atoms so that it can be considered as a limiting case of a non-stoichiometric boride. Contrary to these statements, AMBERGER and PLOOG (1971) did not obtain crystals of tetragonal boron-I in the temperature region from 1100°C to 1250°C using zone molten tantalum wires as a substrate and very pure BBr_3/H_2 or BCl_3/H_2 mixtures. Below 1100°C amorphous boron was deposited, partially together with crystals of α-boron. At higher temperatures, microcrystalline β-boron was mainly formed. In a very small temperature range between 1150°C and 1170°C the authors obtained tetragonal boron-II. The deposition of crystalline tetragonal boron-I was induced, however, by addition of approximately 10 mole percent CH_4 to BBr_3 (PLOOG and AMBERGER, 1971). These crystals, however, contain carbon and correspond to the formula $B_{50}C_{1-2}$ (total number of atoms in one unit cell was found by elementary analysis and density measurements).

Crystals of the tetragonal boron-I type were also obtained (AMBERGER and SCHMIDT, 1971) when nitrogen was added to the reaction gas and were found to have a formula $B_{50}N_{1-2}$. Previously NIEMYSKI and coworkers reported a favorable influence of nitrogen on the growth of tetragonal boron-I (OLEMSKA et al., 1966) (MIERZEJEWSKA-APPENHEIMER and NIEMYSKI, 1967).

Yet the conclusion of AMBERGER, PLOOG and SCHMIDT (PLOOG and AMBERGER, 1971; AMBERGER and SCHMIDT, 1971) that carbon or nitrogen in the quantities stated by them were necessary to obtain tetragonal boron-I does not seem to be justified. Experiments with the deposition of boron from BCl_3/H_2 mixtures on inductively heated boron rods show that within the temperature region of 1300-1800°C, crystals of tetragonal boron are obtained in addition to a matrix of β-rhombohedral boron (OLEMSKA et al., 1966). The observations of LAUBENGAYER and coworkers (1943) about the growth of needle-like as well as of compact crystals have been confirmed. Contamination to the degree suggested by AMBERGER and coworkers (1971) can be excluded on the basis of the experimental conditions described. An important difference between the experimental conditions of the German (AMBERGER and PLOOG, 1971; PLOOG and AMBERGER, 1971; AMBERGER and SCHMIDT, 1971) and Polish (OLEMSKA et al., 1966) scientists consisted mainly in the use of the ratios $BCl_3:H_2$ or $BBr_3:H_2$ and the rate of deposition in relation to the surface of the hot substrate. The description of the experimental conditions suggests that the growth of boron on the substrate occurred faster with the German than with the Polish authors. According to the latter, the rate of deposition to obtain crystals of tetragonal boron-I must be less than 4 mg boron per cm^2 surface an hour at 1200°C. However, an increase of temperature up to 1800° allows a six-fold rate of deposit (OLEMSKA et al., 1966).

The influence of the rate of deposition on the crystallization is consistent with the opinion that crystalline tetragonal boron-I is formed only under suitable kinetic conditions. Very small impurities in the substrate zone may have an influence on the nucleation. However, it is not necessary to assume that impurity atoms of more than 1% are required.

In 1964 H. J. BECHER discovered a completely different way of synthesizing tetragonal boron-I but only in a microcrystalline powder form having 98-99 percent purity. Starting material was a powder of α-rhombohedral boron which was obtained by pyrolysis at 750-800°C using BBr_3/H_2-mixtures in quartz tubes (HOARD and NEWKIRK, 1960; BECHER, 1960). Formation of β-rhombohedral boron was carefully avoided. Pressed pellets of α-boron and powdered beryllium mixed in the atomic ratio 8:1 were heated at 900°C for several hours. After treatment of the reaction product with HCl, a powder was obtained with the composition BeB_{12} and 98-99 percent purity. Its X-ray diagram showed the same reflections as tetragonal boron-I (BECHER, 1960). A further reaction of BeB_{12} with BCl_3 in a closed tube at 375°C gives $BeCl_2$ and a boron powder with approximately 99 percent purity and a content of less than 0.5 percent Be. Its powder diagram agrees perfectly with tetragonal boron-I and on the basis of its density value the unit cell contains 50 boron atoms.

3. Structure

The crystal structure of tetragonal boron-I has been described extensively by HOARD and coworkers (HOARD et al., 1958). Along the [001] - direction, B_{12} - icosahedra are packed in such a way that the respective intericosahedral bonds almost exactly continue the vectors pointing from the center to vertices of the icosahedra. Each icosahedron is situated in the center of a tetragonal distorted cube with a basis of 6.24 x 6.24 Å and a height of 5.06 Å, formed by 8 icosahedra. The central B_{12}-unit is connected to each of these 8 icosahedra by one intericosahedral bond. One half of the bonds deviate significantly from the corresponding radial vectors (HOARD and HUGHES, 1967). The remaining two external bonds of each icosahedron lead to two single boron atoms, each of them combining four external bonds of four different icosahedra. The geometrical configuration corresponds to a bisphenoidally distorted tetrahedron. The bond angles, 96° and 136°, deviate considerably from tetrahedral symmetry. PAULING and KAMP (1959) have pointed out that in tetragonal boron-I the distortion of the icosahedra corresponds partially to the sterical strain caused by the tetragonal symmetry of the inter-icosahedral bonds. However, their discussion of boron-boron bond lengths as related to the bond strength and to the total number of filled orbitals seems to be beyond the present experimental facts. Nevertheless, nearly all intericosahedral distances are distinctly shorter than the intra-icosahedral ones (1.70 Å compared to 1.80 Å (HOARD et al., 1958). Exceptions are bonds which deviate most from the ideal radial direction (1.80 Å). The shortest bonds in the whole lattice originate from the bisphenoidally coordinated single boron atoms.

Accurate structural investigation is complicated by the fact that crystals of different shape and growing conditions have different unit cell dimensions and, therefore, atomic parameters differ beyond the limit of error. This is caused very probably by the large holes in the crystal lattice and the electron deficiency of the unit cell which contains only 50 boron atoms (LONGUET-HIGGINS and ROBERTS, 1955). HOARD and coworkers (1958) have found the presence of some electron densities in vacancies which may not be neglected. A perfect agreement between observed and calculated densities (with 50 boron atoms per unit cell) does not allow filling of lattice holes to a larger extent without simultaneously introducing unoccupied sites. Therefore, one must consider the existence of lattice defects consisting of unoccupied sites at vertices of icosahedra which cause displacements of atoms in the surroundings. Perhaps this could explain the observed variations in the unit cell dimensions. The following range of unit cell dimensions has been reported (HOARD et al., 1958; BECHER, 1964b; OLEMSKA et al., 1966):

	a	c
Needle-like crystals	8.74 ± 0.01 Å	5.07 ± 0.01 Å
Compact crystals	8.77 ± 0.01 Å	5.09 ± 0.01 Å
Powder	8.80 ± 0.02 Å	5.05 ± 0.02 Å
Needles and compact crystals	8.62 ± 0.02 Å	5.10 ± 0.02 Å

The low value for the a-axis as found by Polish authors (OLEMSKA et al., 1966) is surprising. Similar values have been published by AMBERGER and SCHMIDT for the tetragonal boron-I modification containing nitrogen (AMBERGER and SCHMIDT, 1971).

4. Further Investigations on Tetragonal Boron-I

Tetragonal boron-I resembles β-rhombohedral boron by its luster, hardness and semi-conductive properties. But no exact measurements are known because neither larger single crystals nor larger quantities of crystallographically equivalent bulk material have been obtained so far. Infrared spectroscopy of highly dispersed tetragonal boron-I may become informative with regard to substitution of the two single boron atoms by carbon or nitrogen atoms. Such investigations have been performed so far only on tetragonal boron-I of 99 percent purity and on the related BeB_{12} (BECHER, 1964b). Later results with infrared spectroscopy of very finely powdered $B_{12}C_3$ and its structural derivatives may prove promising also with respect to tetragonal boron-I and its structural variants (BECHER and THEVENOT, 1974).

5. Filled Variants of the Tetragonal Boron-I Structure

a) BeB_{12}

This compound was already mentioned in connection with the preparation of tetragonal boron-I from finely divided α-boron (BECHER, 1964) (BECHER, 1960). Attempts to get single crystals were not successful. However, on the basis of powder diagrams and by infrared spectroscopic examinations (BECHER, 1964b) the structural relation to tetragonal boron was quite obvious. The unit cell of BeB_{12} has the axes a = 8.80 Å and c = 5.08 Å in agreement with the upper limits of the corresponding tetragonal boron-I values.

The density D = 2.36 g/ccm corresponds to the presence of 48 boron and 4 beryllium atoms within the unit cell, which may be compared with 50 boron atoms of the tetragonal boron-I cell. It seems reasonable to put the 48 boron atoms of the BeB_{12}-cell at the vertices of 4 icosahedra, 2 beryllium atoms at the places of the bisphenoidally coordinated boron atoms (position 2b) and the remaining 2 beryllium atoms into the big holes of position 2a (BECHER, 1960). However, the short distances from the position 2b to the 4 nearest boron atoms are in better agreement with boron-boron than beryllium-boron distances. Therefore, the atoms in the position 2b should be boron atoms. This is also substantiated by the structural examination of aluminum substituted BeB_{12} described below.

b) $AlBeB_{24}$ or tetragonal and orthorhombic β-$(Al,Be)B_{12}$

α) Preparation

$AlBeB_{24}$, which has the same unit cell as tetragonal boron-I, has been obtained for the first time by reaction of sintered BeB_6 in aluminum melts containing silicon, followed by dissolving the remaining metal (BECHER and NEIDHARD, 1968). The phase is non-stoichiometric and the Al:Be ratio may vary. Similarly, the ratio (Al+Be):B varies but within a smaller range (BECHER et al., 1975). $BeAlB_{24}$ may also contain approximately 3% carbon if present at the preparation. In the pure ternary system Be-Al-B, this phase does not exist. In the boron-rich region of this system one obtains mixed crystals of the tetragonal compounds α-AlB_{12} and BeB_6 (BECHER, 1964a). Hence two different tetragonal boron-rich aluminum beryllium borides exist with a more or less extended homogeneity range: (i) mixed crystals of the α-AlB_{12} type and (ii) crystals of the tetragonal boron-I type to which β-AlB_{12} may also be assigned (BECHER, 1965). This second aluminum beryllium boride originates from aluminum melts only in the presence of carbon, up to 3 percent of which is accommodated in the lattice, or in the presence of silicon which is inserted in the lattice up to 0.2 percent. The same is observed at the formation of β-AlB_{12}. Because of this analogy we suggest the name "tetragonal β-(Al,Be) B_{12}" for the aluminum beryllium boride of the tetragonal boron-I type.

The compound can be prepared very easily by heating boron and beryllium to 1400 °C in Argon atmosphere in the ratio 10:1 together with a fifteen fold excess of aluminum. Carbon (approximately 1/20 of the quantity of boron) or silicon (approximately one half of the quantity of boron) must be added to obtain the desired phase. The regulus is extracted with HCl, the microcrystalline particles are sieved off if more than one phase is present and the remaining crystals can eventually be separated under a microscope, (BECHER et al., 1975). β-$(Al,Be)B_{12}$ can also be prepared aluminothermically (like β-AlB_{12}) in the presence of BeO and MnO_2 by use of the following mixture: 130g Al, 75g S, 60g B_2O_3, 20g BeO, 20g MnO_2, 2g C. This mixture yields β-$(Al,Be)B_{12}$ containing carbon. Without the addition of carbon we were not able to synthesize β-$(Al,Be)B_{12}$ nor β-AlB_{12} by aluminothermic reaction.

The structural relationship between tetragonal $AlBeB_{24}$ and β-AlB_{12} (BECHER, 1965), derived from the size of the unit cells, has been confirmed by the isolation of an orthorhombic compound $(Al,Be)B_{12}$, the unit cell of which has the same dimensions as the so-called A-phase of β-AlB_{12} (MATKOVICH et al., 1965). Crystals of the orthorhombic phase have been obtained by reducing the Be:B ratio to 1:20 during the reaction of beryllium and boron in aluminum melts containing carbon or silicon. This phase can also be prepared aluminothermically if one uses less BeO and no MnO_2 in the mixture mentioned above. The orthorhombic crystals analyzed so far had the Al:Be ratio of 0.67:0.33, but a greater range of homogeneity should be considered probable. Contrary to observations (MATKOVICH et al., 1965) with the beryllium-free β-AlB_{12} phase, in the presence of beryllium there are no indications of the presence of the so-called B-phase of β-AlB_{12}. In the former A- and B-components are always twinned intrinsically.

β) Structure

According to density and analysis the unit cell of tetragonal β-$(Al,Be)B_{12}$ contains 52 atoms like BeB_{12} (BECHER and NEIDHARD, 1968). The orthorhombic variant, the unit cell of which is twice as large, has 104 atoms per unit cell (BECHER et al., 1975). The structural investigation of the tetragonal form has been carried out with crystals containing no carbon. Their composition, $AlBe_{0,8}B_{24,2}$, was close to the ideal formula $AlBeB_{24}$ (KROGMAN and BECHER, 1972). As in tetragonal boron-I, the unit cell contains 4 B_{12} - icosahedra and two single boron atoms in the position 2b of $P4_2/nnm$ space group. No beryllium atoms are located in this position as electron density at 2b position is about 10% greater than at the icosahedral sites. The aluminum atoms are placed close to the position 2b, but are shifted statistically up and down the c-axes. Figure 1 shows the projection of the structure along this axis. The boron atoms in 2b and the aluminum atoms in 2a position are forming linear chains parallel to the c-axes. If the aluminum atoms were located exactly in the 2a portion, the Al-B-distance would be 2.54 Å. Due to the displacement of the aluminum atoms the distances are: 2.27 Å and 2.81 Å, respectively, in a random sequence. Furthermore, an aluminum atom in 2a would have four groups of three boron atoms as nearest neighbors, each forming an icosahedral face of one of the four adjacent icosahedra. Two thirds are at a distance of 2.22 Å, the remaining at 2.46 Å apart. Because of the random displacement, these distances increase or decrease. Altogether the asymmetrical coordination seems to be more favorable because with it aluminum has eleven boron atoms in the region of 2.19 - 2.29 Å as nearest neighbors. The positions of the beryllium atoms could not be determined so far. In the holes of the tetragonal boron lattice not filled with aluminum the electron density is zero within the limit of error at the present state of investigation. Also the filling of such holes with beryllium requires simultaneous vacancies in other positions due to the observed density. Therefore, it is more likely that beryllium is substituting for boron on the icosahedral positions. In that case the beryllium atoms should be shifted towards the positions 4c and 4d of the framework as boron-beryllium distances are usually larger than boron-boron distances (BECHER et al., 1975). These distortions must occur in a random manner because they could not be derived definitely from the calculated electron density map.

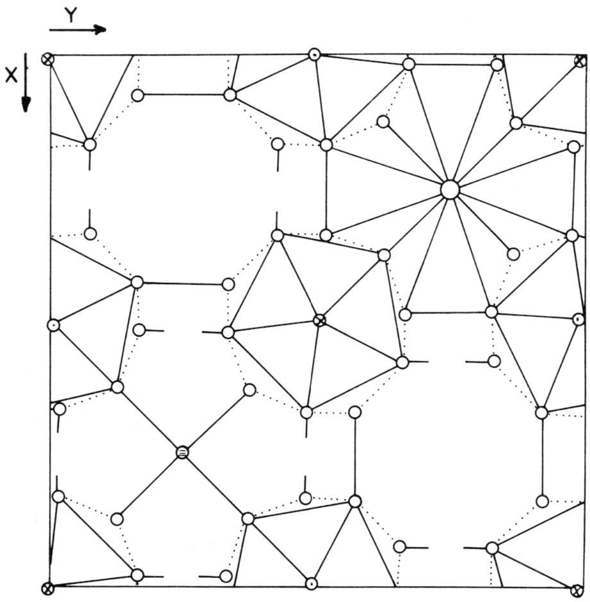

Fig. 1. View along the c-axis of the tetragonal $AlBeB_{24}$ or β-$(Al,Be)B_{12}$ unit cell
⊗ = atoms at the upper corners of the icosahedra centred at $00\frac{1}{2}$ and $\frac{1}{2}\frac{1}{2}\frac{1}{2}$
⊙ = atoms at the upper corners of the icosahedra centred at $0\frac{1}{2}0$ and $\frac{1}{2}00$
⊜ = boron at $\frac{3}{4}\frac{1}{4}\frac{1}{4}$
○ = aluminum at $\frac{1}{4}\frac{3}{4}\frac{1}{4}$
—— = edges of the upper pentagonal pyramids of the icosahedra and connecting bonds
······· = edges between the upper and lower pentagonal pyramids of the icosahedra
The following atoms within a unit cell are not shown:
(a) atoms at the lower corners of the icosahedra (b) boron at $\frac{1}{4}\frac{3}{4}\frac{3}{4}$ and aluminum at $\frac{3}{4}\frac{1}{4}\frac{3}{4}$

The crystal structure determination of orthorhombic β-$(Al,Be)B_{12}$ has been complicated by twinning. Therefore, three-dimensional data could not be utilized. The available results show unequivocally that the structure can be described by a pseudotetragonal subunit closely related to the unit cell of tetragonal β-$(Al,Be)B_{12}$. This subunit is obtained by equalizing the a- and b-axes of the orthorhombic lattice (12.34 Å and 12.62 Å, respectively) and choosing one half of their diagonal as the pseudotetragonal a-axis (8.82 Å). Within this subunit the arrangement of four icosahedra, their connection by bisphenoidally coordinated single boron atoms and the filling of the holes near the position 2a agrees with the network of tetragonal β-$(Al,Be)B_{12}$. In the orthorhombic variant, however, the aluminum atoms are not shifted statistically but ordered in the direction of increased z-parameter. This requires the directions [110] and [1$\bar{1}$0] of the tetragonal lattice to become unequal, leading to the lower, orthorhombic symmetry. Therefore, the difference between tetragonal and orthorhombic β-$(Al,Be)B_{12}$ can be referred to a random versus an ordered displacement of the aluminum atoms in the 2a positions. The observation of general twinning of the orthorhombic crystals can be explained as an occasional change of the ordered displacement in the submicroscopic region.

The transition from random to ordered displacement of the aluminum atoms obviously depends on the Be:Al ratio. When more aluminum atoms are present than necessary for filling the holes at the position 2a, the additional atoms may substitute for atoms of the icosahedral B_{12} groups as was proposed for the beryllium atoms in tetragonal β-$(Al,Be)B_{12}$. In those positions, the larger aluminum atoms will cause still greater distortions than the beryllium atoms, and may preclude the random displacement of aluminum atoms at 2a position for steric reasons.

The existence of the so-called B-component of beryllium-free β-AlB_{12} with the c-axis twice as large as that of the A-component demonstrates that an ordered, but alternating displacement of the aluminum atoms may also occur. A complete elucidation of the β-AlB_{12}-structure has not been possible because of multiple twinning. But it could be shown that in the tetragonal subunit (8.82 x 5.08 Å) of the A-component all boron atoms and one half of the aluminum atoms occupy the same positions as in the corresponding subunit of orthorhombic β-(Al,Be)B_{12}, (BECHER et al., 1975). Therefore, β-AlB_{12}, the carbon containing variant of which was formulated as $Al_3C_2B_{48}$ (MATKOVICH et al., 1964) is similarly to be regarded as a representative of the tetragonal boron-I structure type with filled holes in the 2a position.

c) $B_{50}C_{1-2}$ and $B_{50}N_{1-2}$

Amberger, Ploog and Schmidt have observed the deposition of tetragonal boron-I on tantalum filament or graphite plates coated with boron nitride only when they used BBr_3/H_2 mixtures containing carbon compounds or nitrogen (PLOOG and AMBERGER, 1971) (AMBERGER and SCHMIDT, 1971). The molar ratio of boron and carbon compound was about 5.5:1. However, the ratio of boron halide to nitrogen was about 1:10. The temperature of the hot surfaces was between 1100-1200 °C (B/C-system) and 1200-1600 °C (B/N-system). The crystals prepared by this method were shown to be filled variants of the tetragonal boron-I type. The carbon or nitrogen content varies within the range of the given formula depending on the experimental conditions. The results of an X-ray diffraction analysis are given for crystals with the following properties (PLOOG et al., 1972).

Formula	Density	Tetragonal Unit Cell
$B_{50}C_{1.9}$	2.43 g/ccm	a = 8.722 Å c = 5.08 Å
$B_{50}N_{1.8}$	2.46 g/ccm	a = 8.646 Å c = 5.127 Å

Fifty atoms within the cell were found at the same positions as the boron atoms in the tetragonal boron-I cell, i.e. 48 atoms in icosahedral positions and two single atoms in 2b positions. In this case the latter positions are clearly occupied by either carbon or nitrogen (PLOOG et al., 1972). No electron density was observed at the 2a positions, contrary to the aluminum-containing variants of tetragonal boron-I. However, there seems to be some electron density at the 4c positions (just above the limit of detection). Consequently, some atoms in excess of the above-mentioned fifty are assumed to be randomly distributed at these holes. A slight shift of the x- or z-parameter has been observed resulting in eight-fold positions, (designated as 8i and 8h near the original 4c position) (PLOOG et al., 1972).

6. Conclusions

The structural arrangement of tetragonal boron-I is obviously capable of undergoing changes leading to distortions, filling of holes and substitution by other atoms. These lattice defects are statistically distributed, and can, therefore, be observed only with difficulty. The known variants of the basic structure show the substitution of the bisphenoidally coordinated boron atoms at 2b position by carbon or nitrogen and the filling of holes at 2a position by aluminum atoms. Many other questions are not sufficiently clear at this time. It would be desirable to define the conditions of preparation still better and to complete the characterization of the products further by spectroscopic examinations.

References

AMBERGER, E., PLOOG, K.: Bildung der Gitter des reinen Bors. J. Less Common Metals 23, 21 (1971)

AMBERGER, E., SCHMIDT, H.: Stickstoffinduziertes Gitter beim Bor: I-tetragonales Bornitrid $(B_{12})_4B_2N_{1-2}$. Z. Naturforschg. 26b, 641 (1971)

BECHER, H.J.: Über das Berylliumborid BeB_{12} mit β-Bor-Struktur. Z. anorg. allgem. Chem. 306, 266 (1960)

BECHER, H.J.: Über die Kristallisation borreicher Berylliumboride und Aluminium-Berylliumboride aus Aluminiumschmelzen. Acta Cryst. 17, 617 (1964a)

BECHER, H.J.: Über die Bildung der tetragonalen Bormodifikation durch Substitution der Berylliumatome im Gitter des BeB_{12}. Z. anorg. allgem. Chem. 321, 217 (1964b)

BECHER, H.J.: Tetragonal boron and borides with similar structures. Boron, Vol. II, 89 ff., (G.K. Gaulé, ed.) New York: Plenum Press 1965

BECHER, H.J., NEIDHARD, H.: Siliziumhaltiges β-AlB_{12} und $AlBeB_{24}$ vom Typ des BeB_{12}. Acta Cryst. B 24, 280 (1968)

BECHER, H.J., RETHFELD, H., MATTES, R.: Zur Darstellung und Struktur von tetragonalem und rhombischen (Al,Be) B_{12} vom Typ des β-AlB_{12}. Z. anorg. allgem. Chem. 414, 203 (1975)

BECHER, H.J., THEVENOT, F.: Infrarotspektroskopische Untersuchung des Borcarbids und seiner isotypen Derivate $B_{12}O_2$, $B_{12}P_2$ und $B_{12}As_2$. Z. anorg. allgem. Chem. 410, 274 (1974)

HOARD, J.L., HUGHES, R.E.: Elementary boron and compounds of high boron content in: The chemistry of boron and its compounds. (E.L. Muetterties, ed.) New York: Wiley and Sons 1967

HOARD, J.L., HUGHES, R.E., SANDS, D.E.: The structure of tetragonal boron. J. Amer. Chem. Soc., 80, 4507 (1958)

HOARD, J.L., NEWKIRK, A.E.: An analysis of polymorphism in boron based upon X-ray diffraction results. J. Amer. Chem. Soc. 82, 70 (1960)

KROGMANN, K., BECHER, H.J.: Strukturbestimmung eines aluminium- und berylliumhaltigen tetragonalen Bors der Formel $AlBe_{0,8}B_{24,2}$. Z. anorg. allgem. Chem. 392, 197 (1972)

LAUBENGAYER, A.W., HURD, D.T., NEWKIRK, A.E., HOARD, J.L.: Preparation and properties of pure crystalline boron. J. Amer. Chem. Soc. 65, 1924 (1943)

LONGUET-HIGGINS, H.C., ROBERTS, M. de V.: The electronic structure of an icosahedron of boron atoms. Proc. Ray. Soc. (London) Ser. A., 230, 110 (1955)

MATKOVICH, V.I., ECONOMY, J., GIESE, R.R.: Presence of carbon in aluminium borides. J. Amer. Chem. Soc. 86, 2337 (1964)

MATKOVICH, V.I., GIESE, R.R., ECONOMY, J.: Phases and twinning in $C_2Al_3B_{48}$(β-AlB_{12}) Z. Kristallogr. 122, 108 (1965)

McCARTY, L.V., KASPER, I.S., HORN, N.F., DECKER, B.F., NEWKIRK, A.E.: A new crystalline modification of boron. J. Amer. Chem. Soc. 80, 2592 (1958)

MIERZEJEWSKA-APPENHEIMER, S., NIEMYSKI, T.: Vapor-phase crystallization of boron and boron carbide. J. Phys. Chem. Solids Suppl. 1, 229 (1967)

OLEMSKA, Z., BADZIAN, A., PIETRZAK, K., NIEMYSKI, T.: Crystalline modifications of boron deposited on boron substrates. J. Less Common Metals 11, 351 (1966)

PAULING, L., KAMP, B.: The discussion of tetragonal boron by the resonating-valence-bond theory of electron-deficient molecules. Z. Kristallk. 112, 472 (1959)

PLOOG, K., AMBERGER, E.: Kohlenstoff-induzierte Gitter beim Bor: I-tetragonales $(B_{12})_4B_2C$ und $(B_{12})_4B_2C_2$. J. Less Common Metals 23, 33 (1971)

PLOOG, K., SCHMIDT, H., AMBERGER, E., WILL, G., KOSSOBUTZKI, K.A.: $B_{48}B_2C_2$ und $B_{48}B_2N_2$, zwei Nichtmetallboride mit der Struktur des sog. I-tetragonalen Bors. J. Less Common Metals 29, 161 (1972)

SANDS, D.E., HOARD, J.L.: Rhombohedral elemental boron. J. Amer. Chem. Soc. 79, 5582 (1957)

TALLEY, C.P., LA PLACA, S., POST, B.: A new polymorph of boron. Acta Cryst. 13, 271 (1960)

IX. Compounds Based on Octahedral B_6 Units: Hexaborides and Tetraborides

J. ETOURNEAU, J.-P. MERCURIO, and P. HAGENMULLER

Atoms in a three-dimensional boron lattice have a five-fold coordination when situated in the vertex positions of a B_6 octahedron: they share four bonds with atoms of the same octahedron and the fifth one points outwards along one of the A_4 axes of the octahedron. The presence of this fifth bond is responsible for the occurrence of the three-dimensional structure. Only two families of boron-rich borides have structures based on the B_6 octahedron: the CaB_6-type hexaborides and the ThB_4-type tetraborides.

MOISSAN and WILLIAMS (1897) reported the discovery of CaB_6-type cubic hexaborides. Later ANDRIEUX (1929) proposed a general synthesis method for these compounds, by fused-salt electrolysis. Then Von STACKELBERG and NEUMANN (1932) determined the approximate structure, which was later refined (ALLARD,1932; PAULING and WEINBAUM, 1934; BLUM and BERTAUT, 1954).

Fused salt electrolysis had also allowed ANDRIEUX (1929) to prepare UB_4 uranium tetraboride. Later BREWER and SAWYER (1950) prepared CeB_4, ThB_4 and UB_4 isotypes; BERTAUT and BLUM (1949) fixed the positions of the metal atoms in the UB_4 lattice. ZALKIN and TEMPLETON (1950) established a complete model of the crystal structure using a ThB_4 single crystal.

1. Crystal Structure

a) CaB_6-Type Hexaborides

Elements that form hexaborides of a CaB_6-type are potassium, alkaline earths, yttrium and the larger lanthanides, as well as thorium and some actinides (NASLAIN and ETOURNEAU, 1966; EICK and MULFORD, 1969; McDONALD and STUART, 1960). The crystal structure of these compounds with cubic symmetry (Pm3m, O_h^1) is represented in Figure 1. It is characterized by a three-dimensional skeleton of B_6 boron octahedra, the interstices of which are filled by metal atoms.

Theoretical studies have shown that in the absence of an inserted metal the covalent boron sublattice is electron-deficient, and that electron transfer from the metal sublattice to that of boron is required (EBERHARDT et al., 1954; LIPSCOMB and BRITTON, 1960). The existence of thermodynamically stable KB_6, $Na_xTh_{I-x}B_6$ and $Na_xBa_{I-x}B_6$ phase indicates that the number of tranferred electrons is a least 1 (BLUM and BERTAUT, 1954; NASLAIN and ETOURNEAU, 1966).

According to different investigators, the structural parameters differ quite noticeably, which suggests the existence of homogeneity ranges. Many studies have been undertaken concerning the non-stoichiometry of MB_6 hexaborides: all of them lead to deviations from stoichiometry through cation defects. The reported results are summed up in Table 1.

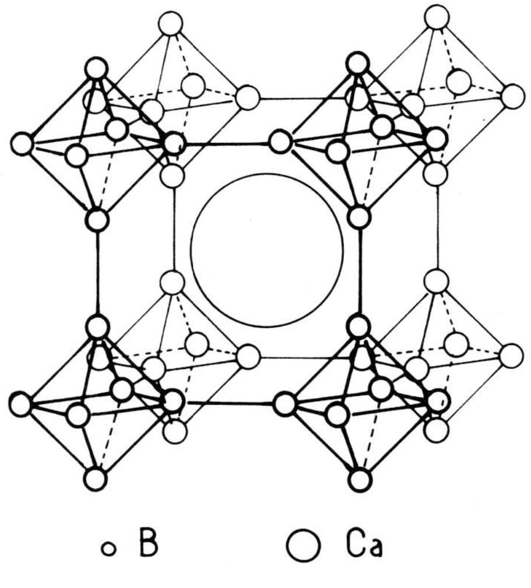

Fig. 1. Structure of CaB_6

Table 1.

M	x	Parameters (Å)	Reference
K	1.00	4.232	NASLAIN and ETOURNEAU (1966)
Ca	1.00	4.1520	ETOURNEAU (1970)
Sr	1.00	4.1981	ETOURNEAU (1970)
Ba	1.00	4.2706	JOHNSON and DAANE (1961)
Y	1.00	4.103	JOHNSON and DAANE (1961)
La	1.00	4.1561	JOHNSON and DAANE (1961)
	0.77	4.1561	JOHNSON and DAANE (1961)
Ce	1.00	4.1396	YAJIMA and NIIHARA (1971)
	0.70	4.1414	
Pr	1.00	4.1329	YAJIMA and NIIHARA (1971)
	0.69	4.1355	
Nd	1.00	4.1265	MERCURIO (1974)
Sm	1.00	4.1304	NIIHARA (1971)
	0.68	4.1278	
Eu	1.00	4.1780	SCHWETZ and LIPP (1973)
	0.90	4.1843	
Gd	1.00	4.1065	YAJIMA and NIIHARA (1971)
	0.70	4.1113	
Tb	1.00	4.1008	YAJIMA and NIIHARA (1971)
	0.75	4.1052	
Dy	1.00	4.0969	SPEAR (1974)
		4.1008	
Yb	1.00	4.1478	ETOURNEAU et al. (1970)
Th	1.00	4.1104	ETOURNEAU et al. (1971)
	0.78	4.1125	
Np	-	4.1129	EICK and MULFORD (1969)
Pu	-	4.115	McDONALD et al. (1960)
		4.140	
Am	-	4.1154	EICK and MULFORD (1969)

b) ThB_4-Type Tetraborides

Elements likely to form tetraborides of a ThB_4 type are yttrium, lanthanides except europium, as well as some actinides: thorium, uranium, neptunium, plutonium and americium (BERTAUT and BLUM, 1949; ZALKIN and TEMPLETON, 1953).

The crystal structure of the phases with tetragonal symmetry (P4/mbm, D_{4h}^5) can be described as follows: the boron covalent sublattice is formed by chains of B_6 octahedra, developing along the c-axis, and by pairs of B_2 atoms, bonding the octahedra in the xOy plane (Fig. 2). The resulting three-dimensional skeleton contains tunnels parallel to the c-axis that are filled by metal atoms (ZALKIN and TEMPLETON, 1953).

Because of the many common features of CaB_6 and ThB_4-type structures (covalent skeleton, B_6 octahedra, existence of boron "cages" ...), deviations from stoichiometry can be expected for tetraborides. The few studies made in this field concern only parameter values (EICK and MULFORD, 1969; JOHNSON and DAANE, 1961). Their results are too contradictory to allow a precise evaluation of the limits of possible homogeneity ranges. However it can reasonably be assumed that deviations from stoichiometry must take place on the deficit side of the metal.

2. Preparation

Since boron readily combines with most metals and reduces refractory oxides at high temperatures, different methods can be utilized to prepare borides. Since fused salt electrolysis is only of historical interest for hexaboride and tetraboride synthesis, we shall describe only general methods such as direct synthesis from elements and reduction of oxides by boron at high temperatures. Special attention will be given to methods of crystal growth.

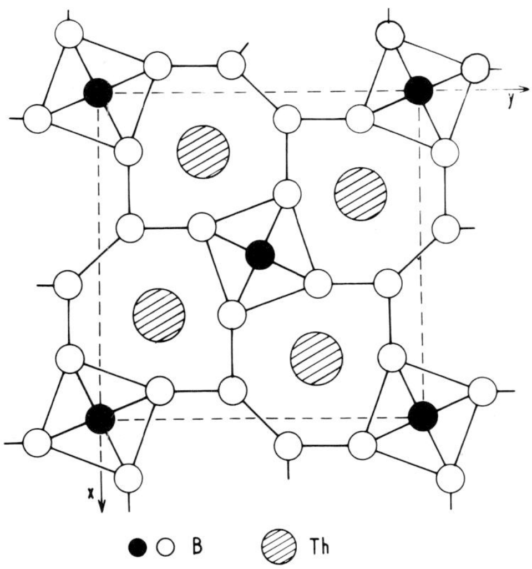

Fig. 2. Structure of ThB_4

a) Direct Synthesis from Elements

$M + yB \rightarrow MB_y$ (y = 4 or 6).

The choice of the specific experimental technique is determined by the degree of volatility of the metal to be used.

α) Direct Synthesis in the Presence of a Vapor Phase

This technique is utilized when the metals involved have high vapor pressures at the temperatures required for synthesis. This is the case for potassium, alkaline earths, samarium, europium, thulium and ytterbium. The utilization of these relatively volatile metals requires sealed containers, that resist the corrosive action of boron and of metal gases under pressure (ETOURNEAU, 1970b).

β) Direct Synthesis Involving Metals with Low Volatility

The reactions are generally carried out in vacuum or an argon atmosphere. To facilitate solid-state diffusion, the boron and the metal (often as a finely divided hydride) are shaped into tablets under high pressure (about 15 kb).

b) Indirect Synthesis

The principle of this method is the reduction, at high temperatures, of metal oxides by boron. As a general rule, the reduction reaction can be described as follows:

$$\text{metal oxide} + \text{boron} \xrightarrow{t\ >\ 1500^{\circ}C} \text{boride} + B_2O_2 \nearrow ,$$

where the nature of the volatile boron oxide is not well defined.

Oxide-boron mixtures pressed into cylindrical tablets under high pressure, are treated either by arc-melting, or by high-frequency induction heating in tantalum, molybdenum or tungsten crucibles (ETOURNEAU, 1970b). The elimination of oxygen in boron oxide generally requires that the reactions be carried out in vacuum.

Boron-carbon mixtures or B_4C boron carbide make it possible to reduce metal oxides at lower temperatures. However their utilization presents one drawback, that of a risk of contamination by carbon. Consequently the preferred method of synthesis involves only boron itself as the reducing agent, since it is the only element that can provide materials that are well characterized and of high purity.

c) Single Crystal Growth

Several crystal growing methods have been utilized in the preparation of tetraboride and hexaboride single crystals.

α) Chemical Transport

RABENAU, KAUER and KLOTZ obtained lanthanum hexaboride crystals (0.15 × 0.15 × 2 mm^3) by chemical transport in the presence of bromine. The reaction was carried out in a sealed ceramic (SiO_2 and Al_2O_3) flask with a thermal gradient between 1105 and 1095°C (KLOTZ, 1965). NIEMYSKI and KIERZEK-PECOLD used a $BCl_3 + H_2$ mixture as a transport agent that reacted with lanthanum oxide. Here the temperatures of both the cold and the hot sources were much higher (1000 and 1450°C). The edges of the crystals obtained were about one millimeter in length (NIEMYSKI and KIERZEK-PECOLD, 1968).

β) Melting with a Metal Excess

These reactions are carried out in open or sealed containers, according to the volatility of the metals involved. JOHNSON and DAANE, and later DEACON and HISCOCKS obtained prismatic LaB_4 needles 5 to 8 mm long as well as small SrB_6 single crystals (0.2 mm approximately) by slow cooling (16°C/h) from 1700°C (JOHNSON and DAANE, 1961; DEACON and HISCOCKS, 1971).

FISK et al. (1972) prepared single crystals of tetraborides of the first members of the rare earths (La to Nd) by this method.

γ) Zone-Melting

Zone-melting is an appropriate method for crystallization of congruent-melting compounds.

JOHNSON and DAANE (1963) obtained large yttrium tetraboride crystals (6 mm diameter and 10 cm long, approximately).

CURTIS and GRAFFENBERGER (1966), followed by NIEMYSKI and KIERZEK-PECOLD (1968) prepared, with a floating zone technique, lanthanum hexaboride single crystals from 4 to 10 mm long and with 3 mm^2 in cross section, approximately.

δ) Utilization of a Metal Flux

Aluminum is generally used as the metal flux. The method consists in heating up to 1550°C in an argon atmosphere a boron + metal mixture (in stoichiometric proportions) in the presence of a large excess of aluminum, then to let the molten mass cool down to 1000°C within 10 min, and finally to stop abruptly the heating.

FISK et al. (1972) thus obtained single crystals of tetraborides of the heavy rare-earths (from Sm to Lu, except Eu). With the same technique, STURGEON et al. (1974) prepared samarium hexaboride single crystals as well as (Eu, Sm) B_6-crystals a few millimeters in length.

d) Conclusion

A review of the methods of hexaboride and tetraboride preparation shows that indirect synthesis by borothermal reduction of oxides appears to be the most satisfactory and easiest method to apply when large quantities of borides are needed. It also makes it possible to obtain $M_xM'_{1-x}B_6$ (or $M_xM'_{1-x}B_4$) solid solutions with perfectly defined compositions.

However a detailed study of the physical properties of these materials - already quite developed concerning polycrystalline samples - more and more often requires single crystals. Consequently the previously described crystal growth techniques will have to be refined in order to fulfill the more and more demanding requirements of single crystal studies.

3. Thermal Stability

A study of the thermal stability of materials over a large temperature range provides a most helpful guide in the determination of ideal preparation conditions, for polycristalline samples as well as for single crystals.

a) Experimental Results

The thermal behavior of hexaborides and tetraborides can be described with the help of reaction mechanisms involving either congruent melting or dissociation through preferential evaporation of metal or boron (ETOURNEAU, et al., 1972a, b).

α) Hexaborides

There are four classes of compounds from the point of view of stability:

a) M = La, Ce, Pr, Nd, Sm or Am:

$MB_6 \rightarrow$ congruent melting.

b) M = K, Ca, Sr, Ba or Eu:

$MB_6 \rightarrow M^{\nearrow} + 6B$ (boron residue)

c) M = Y, Gd, Tb, Dy or Th:

$MB_6 \rightarrow MB_4 + 2B^{\nearrow}$

d) M = Yb:

$2MB_6 \rightarrow MB_{12} + M^{\nearrow}$

β) Tetraborides

There are two classes of compounds:

a) M = Y, Gd, Tb, Dy, Ho, Er, Tm, Lu, U or Pu:

$MB_4 \rightarrow$ congruent melting

b) M = La, Ce, Pr, Nd, Sm or Yb:

$MB_4 \rightarrow MB_6 + M$

b) Discussion

α) Relative Stability of Hexaborides

The interpretation of thermal behavior of hexaborides is based on two factors: the saturation vapor pressure of the metal, an increase of which increases the dissociation, and the stability of the B-B bonds within the boron sublattice, the strength of the B-B bonds decreasing as the size of the cubic lattice parameter increases.

The great stability of hexaborides from lanthanum to neodymium can be explained by the progressive contraction of the lattice, which compensates for the increasing volatility of the metal. Samarium hexaboride, thermally very stable, is the limiting case with regard to volatility of metal which is high, but the lattice parameter is still relatively small, since it is intermediate between those of CeB_6 and PrB_6 (Figs. 3 and 4).

The thermal behavior of EuB_6 and YbB_6 is a result of the high volatility of the metals, and also of the great dilatation of the unit cell and consequently of the fragility of the B-B bonds. The exceptional volatility of ytterbium explains the fact that YbB_6 is less stable than LaB_6, although their unit cells have very nearly the same size.

The dissociation of YB_6, GdB_6, TbB_6 and DyB_6 hexaborides into tetraborides with loss of boron at 1550°C, is a consequence of the volatility of boron at this temperature, and also of the very high stability of the corresponding tetraborides.

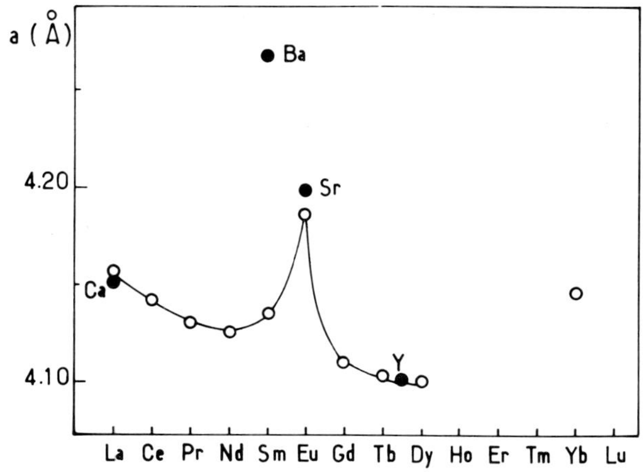

Fig. 3. Lattice parameters of rare-earth hexaborides

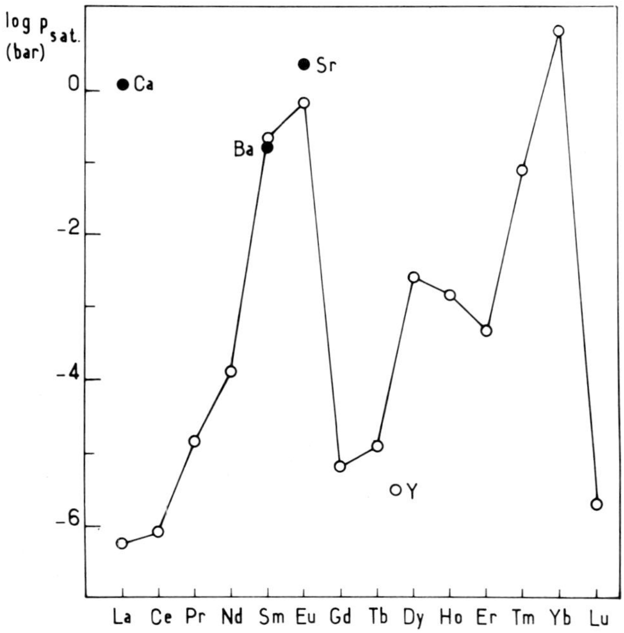

Fig. 4. Saturation vapor pressures of rare-earths at 1800 K

The interpretation proposed for the LnB_6 phases can be extended to alkaline-earth and potassium hexaborides. These dissociate through metal evaporation, yielding a β-rhomboedral boron residue, at the melting-point in the case of CaB_6, SrB_6 and BaB_6, and at 750°C in the case of KB_6 (ETOURNEAU, 1970b; NASLAIN and ETOURNEAU, 1966).

For a given volume of the unit cell, the most stable hexaborides will be that which corresponds to the least volatile metal: LaB_6 is then more stable than CaB_6, in spite of their identical parameters.

On the other hand, for comparable metal vapor pressure, the most stable hexaboride will be that which has the smallest unit cell: SmB_6 is indeed moore stable than BaB_6, which has a larger unit cell.

The fact that europium and strontium have comparable volatilities and also B-B bonds of identical strength accounts for the similar thermal behavior of EuB_6 and SrB_6.

In the same way one can explain the easy dissociation of KB_6 (a_{KB_6} = 4.232 Å, $p_K^{1700\ K}$ = 27 bar).

β) Relative Stability of Tetraborides

The temperatures at which dissociation of tetraborides (under 10^{-6} torr) begins are the following:

LaB_4: 1850°C; CeB_4: 2200°C; PrB_4: 1950°C; NdB_4: 1850°C;

SmB_4: 1650°C.

Thus, while the thermal stability increases from LaB_4 to CeB_4, there is a monotonic decrease with increasing atomic number from CeB_4 to SmB_4.

PADERNO and POKRZYWNICKI (1967a) showed, by studies of magnetic properties, that the inserted rare earth ion for GdB_4, DyB_4, HoB_4, ErB_4 and TmB_4 was in +III oxidation state. Figure 5 shows that this is also the case for the other tetraborides except of CeB_4 and YbB_4. The abnormal volume contraction for CeB_4 unit cell observed by many authors, can certainly be explained by the presence (at least partially) of Ce^{4+} ions (GSCHNEIDNER, 1961). On the other hand, the dilatation of YbB_4 could be the result of the presence of Yb^{2+} ions (FISK et al., 1972).

The trend in thermal stability observed for LaB_4, CeB_4, PrB_4, NdB_4 and SmB_4 is explainable by the counteracting effects of increasing metal volatility and decreasing unit cell volume. The first effect accounts for the lesser stability of LaB_4 relative to SmB_4 and the second for the maximum at CeB_4.

For comparable metal volatilities, the tendency to dissociate decreases with shortening of B-B distances. The high thermal stability of YB_4, GdB_4, TbB_4, DyB_4, HoB_4 and ErB_4 is the result of the combination of both a relatively low metal volatility and a small unit cell volume.

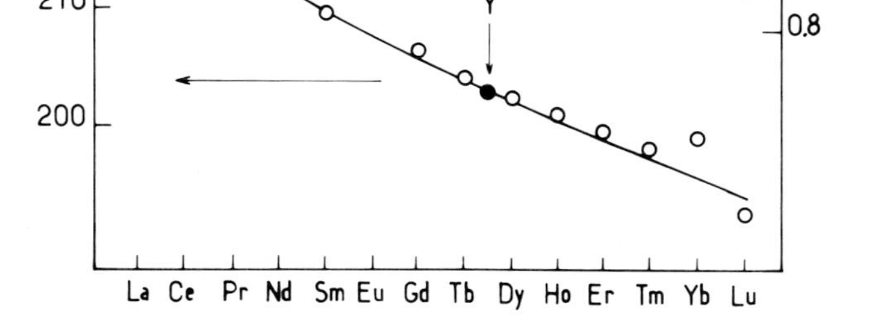

Fig. 5. Elementary volume of rare-earth tetraborides and ionic radii of rare-earth tetraborides in the +II oxidation state

EuB_4 and YbB_4 are difficult to obtain because of the high vapor pressure of europium and ytterbium. However this factor is not sufficient to explain why EuB_4 has never been prepared to date: an equally important factor could be the large unit cell that would result for a +II oxidation state, by analogy with YbB_4.

c) Structural Observations

A basis for comparison of the CaB_6 and ThB_4 structure types is the dimension of the site available for the metal atoms within the boron sublattice. Table 2 gives the extrema of the radii of these sites, calculated from average B-M distances, assuming the atomic radius of boron (r_B = 0.88 Å) does not significantly vary from one structure to another (NASLAIN, et al., 1973). It also shows the extrema for metal and ion radii of the elements that might form hexaborides and tetraborides (SHANNON and PREWITT, 1969). The site for metal occupancy for the hexaborides is noticeably larger than for tetraborides, as can be predicted from their respective coordinations (24 in MB_6 phases, 16+2 in MB_4 phases).

Table 2.

Structure type	Radii of metal sites (Å)	r_M (Å)	$r_{M^{3+}}$ (Å)
MB_6	2.26 - 2.14	2.23 - 1.75	1.28 - 0.92
MB_4	1.99 - 1.87	1.95 - 1.68	1.14 - 0.85

The available sites in hexaborides will then accommodate atoms larger than those in tetraborides. This is one of the reasons why large atoms such as Ca, Sr, Ba, Eu, and Yb can form hexaborides, whereas the small rare earths (Ho, Er, Tm, Lu) cannot.

The metal sites in tetraborides by contrast accomodate smaller atoms: this explains the great thermal stability of tetraborides of the heaviest trivalent rare earths (Gd to Lu). As in the rare earth series only europium - whose radius is large for the divalent state ($r_{Eu^{2+}}$ = 1.12 Å) - does not form any tetraborides, and because of the questionable existence of CaB_4, we consider the size of the calcium ion ($r_{Ca^{2+}}$ = 0.99 Å) as the upper limit for insertion of a metal ion into the boron sublattice. The fact that cerium is, at least partially, in a +IV oxidation state in CeB_4 and in a +III oxidation state in CeB_6, is in agreement with this conclusion.

4. Physical Properties

This section is concerned mainly with the study of electrical and magnetic properties of the ThB_4-type tetraborides and of the hexaborides.

a) Hexaborides

Since the publication in 1950 of the first physical data concerning CaB_6-type hexaborides, many band structure models have been proposed based on theoretical considerations, generally involving a LCAO-type approximation for atomic orbitals of boron (LONGUET-HIGGINS and ROBERTS, 1954; YAMAZAKI, 1957; FLODMARK, 1960; PADERNO et al., 1970).

The calculations made by LONGUET-HIGGINS and ROBERTS (1954) led to a model consisting of four bonding orbitals, which could contain 20 electrons per MB_6 group, separated from four anti-bonding orbitals by a forbidden band. For a $M^{II}B_6$ divalent element hexaboride, the 18 electrons donated by 6 boron atoms and the two valence electrons of the metal atom completely fill the bonding orbitals: thus $M^{II}B_6$ is an insulator or a semiconductor. For CaB_6 the calculated width of the forbidden band was 4.4 eV, but YAMAZAKI (1957) obtained a value of 1.5 eV. If the metal atom yields more than two electrons, the extra electrons partially fill an anti-bonding band and the hexaboride becomes metallic.

In FLODMARK's Calculations the wave functions relative to the 3s, 3p and even 3d boron orbitals were included (FLODMARK, 1960). The band occurring from the new orbital combinations and those obtained from solely 2s and 2p orbitals are different: in the new energy diagram, totaly occupied bonding bands can accommodate 12 electrons and partially occupied conduction bands the remaining electrons. CaB_6-type hexaborides would then be, according to this model, compounds of a metallic character, whatever the number of electrons provided by the metal.

Both methods of calculation lead to the same conclusion that for a metal in a +III or +IV oxidation state, the corresponding boride is metallic. They differ markedly, however, in their predictions for hexaborides containing divalent elements.

It is in order, however, to question the significance of either model, since they both lead to contradictions with certain experimental facts, for example CaB_6 and SrB_6 are all semiconductors contrary to the predictions of FLODMARK'S model.

The LONGUET-HIGGINS and ROBERTS model has shortcomings regarding its predictions for hexaborides containing less than two electrons per metal site. The model requires that the metal donates at least two electrons and that the number of electrons per MB_6 group be 20. But BERTAUT and BLUM have shown the existence of sodium substitutions in BaB_6 and ThB_6 corresponding to the formulas $N_{0.43}B_{0.57}B_6$ and $Na_{0.77}Th_{0.23}B_6$, where the number of electrons per group was respectively 19.57 and 19.69 (BLUM and BERTAUT, 1954). This difference is still greater in KB_6 prepared by NASLAIN and ETOURNEAU (1966), where only 19 electrons are available. The existence and the stability of KB_6 of the former ternary hexaborides show that their lattices are stable in the case of an electron concentration below two per inserted atom.

It was not possible to test the validity of the proposed models and possibly to make necessary modifications until significant progress had been made in the preparation of high purity poly- or single crystalline materials that were thoroughly characterized.

α) Transport Properties

$M^{III}B_6$ and $M^{IV}B_6$ hexaborides (M = Y, La, Ce, Pr, Nd, Gd, Tb, Dy and Th)

These hexaborides have a metal-type conductivity, the value of which, at room temperature, is in the range of 10^4 to 10^5 $\Omega^{-1}cm^{-1}$ (ETOURNEAU et al., 1970a; LAFFERTY, 1951; SAMSONOV et al., 1963; AUSKERN and ARONSON, 1968). The Hall effect measurements show that the number of conduction electrons is close to one per metal atom for trivalent hexaborides and to two per thorium atom for ThB_6 (JOHNSON and DAANE, 1963; SAMSONOV et al., 1963; AUSKERN and ARONSON, 1968). The thermoelectric power has a value of a few μVK^{-1} at room temperature, confirming the metallic character of these compounds (SAMSONOV et al., 1963; ETOURNEAU et al., 1970).

We must also note that studies on electronic specific heat at low temperatures have led ETOURNEAU et al. (1970) to find that the density of states at the Fermi level is not zero, but of the same order of magnitude as for the metallic rare earths.

$M^{II}B_6$ Hexaborides (M = Ca, Sr, Ba, Eu and Yb)

The high value of the lattice parameter of EuB_6 and YbB_6 compared to those of hexaborides of the other rare earths, as well as magnetic studies and Mössbauer resonance studies have shown that, among these compounds, europium and ytterbium were in a +II oxidation state (POST et al., 1956; PADERNO et al., 1967; COHEN, 1968).

By measuring electrical conductivity on single crystals, JOHNSON and DAANE (1963) established that CaB_6, SrB_6 and BaB_6 were semiconductors with activation energies of 0.40, 0.38 and 0.12 eV respectively. SAMSONOV et al. (1963) reported, from Hall effects measurements, conduction electron concentrations below 0.1 per metal atom for CaB_6, SrB_6, BaB_6, EuB_6 and YbB_6. Recently FISK (1969) confirmed the semiconductor behavior of EuB_6 and YbB_6 (ΔE = 0.38 eV for EuB_6), as well as MERCURIO et al. (1973) (ΔE = 0.30 for EuB_6 and 0.14 eV for YbB_6), by measurements on sintered bars (Fig. 6) (GOODENOUGH et al., 1973).

For all hexaborides containing a metal in the +II oxidation state, the thermoelectric power is high in absolute value (MERCURIO et al., 1973; GOODENOUGH et al., 1973). Measurements of specific heat with electron state densities at the Fermi level close to zero for YbB_6, CaB_6 and SrB_6 have confirmed the semiconductor nature of the $M^{II}B_6$ hexaborides (ETOURNEAU et al., 1970a).

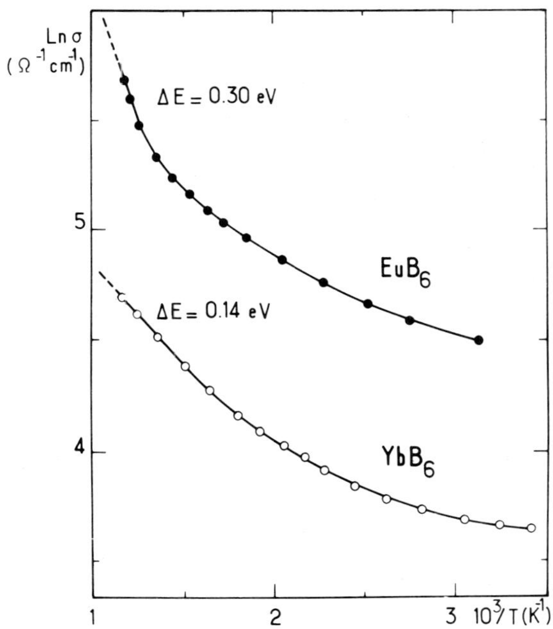

Fig. 6. Electrical conductivity of EuB_6 and YbB_6, plotted vs reciprocal temperature

$M^{II}_{1-x}M^{III}_{x}B_6$ Ternary Hexaborides

These compounds are interesting because of the simultaneous presence of elements in +II and +III oxidation states. By measuring electrical conductivity and thermoelectric power as a funtion of x, one can follow the development of their proerties from semiconductor ($M^{II}B_6$) to metallic ($M^{III}B_6$) behavior. In the case of $La_xEu_{1-x}B_6$ solid solutions, a sudden transition from a semiconductor to a metal has been observed (MERCURIO et al., 1974). The semiconductor character of EuB_6 disappears as soon as x = 0.01, while the metallic character grows as $x \rightarrow 1$ (Fig. 7). The Seebeck coefficient variation is high for small values of x, but quickly reaches a value near to that obtained for LaB_6 when x increases (Fig. 8).

Thus, the substitution of europium by La^{3+} ions - in small quantities - annihilates the semiconductor character of EuB_6, since the occurrence of electrons donated by lanthanum to the conduction band creates a metallic-type conductivity.

Samarium Hexaboride

For SmB_6 although the unit cell dilation is much less than for EuB_6 and YbB_6, it is still sufficient for samarium hexaboride to be considered as an intermediate compound between the $M^{II}B_6$ and $M^{III}B_6$ types (Fig. 3).

Electrical conductivity measurements carried out on polycrystalline samples as well as on single crystals, show that SmB_6 behaves as a semiconductor with a very low activation energy at low temperatures: $\Delta E \simeq 5.10^{-3}$ eV (Fig. 11) (STURGEON et al., 1974; MENTH et al., 1969; FALICOV and KIMBALL, 1969; COHEN et al., 1970; GEBALLE et al., 1970; NICKERSON et al., 1971). The thermal variations of the Hall coefficient

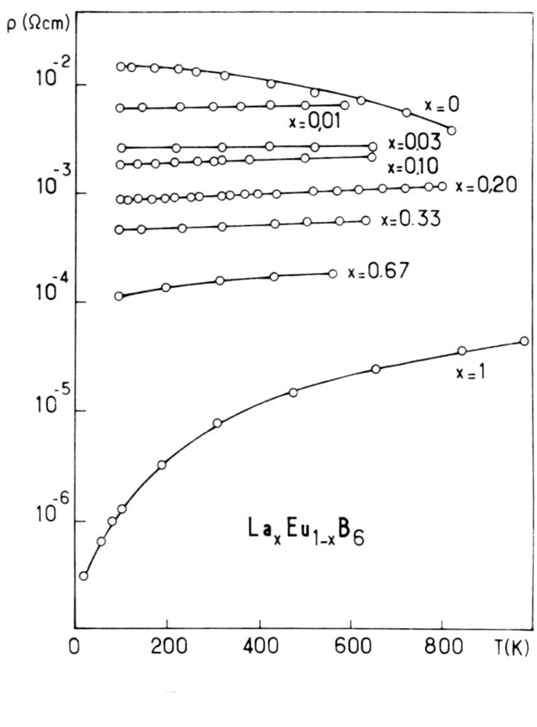

Fig. 7. Resistivity of $La_xEu_{1-x}B_6$ solid solutions as a function of temperature

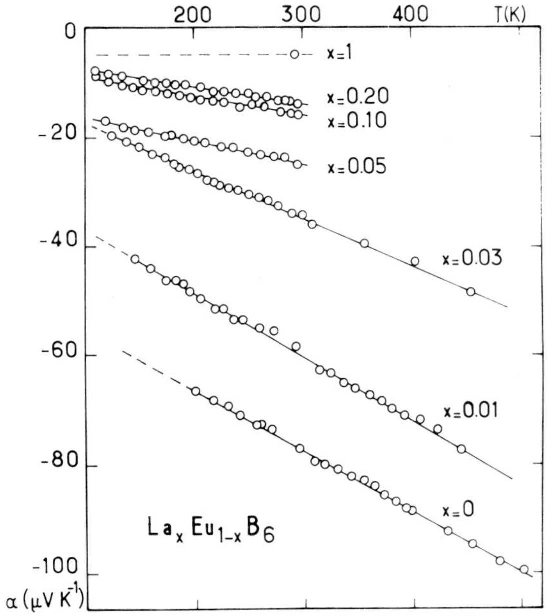

Fig. 8. Thermoelectric power of $La_xEu_{1-x}B_6$ solid solutions as a function of temperature

and of the thermoelectric power seem to correspond first to localized electron conduction and then to highly mobile carriers with rising temperature. It appears that $4f^6 \rightarrow 4f^5 5d^1$ transition takes place. The d-electrons are initially localized by disordered potentials (Anderson localization) at the bottom of the conduction band, they become progressively delocalized as higher energy levels are occupied.

Energy Diagrams

The models described above are appropriate only to CaB_6, taking simply into account the atomic orbitals of boron. To extend these models to the whole family of CaB_6-type hexaborides and particularly to rare-earth hexaborides, one must consider not only the $B_6\sigma$ bonding band and $B_6\sigma^*$ anti-bonding bands, but also the participation of the rare-earth 5d and 6s orbitals.

$M^{III}B_6$ Hexaborides. Two of the three valence electrons of the metal are donated to the boron skeleton and fill the $B_6\sigma$ bonding bands, the third one occupies a conduction band. It is reasonable to consider that the conduction electrons occupy a 5d-6s band originating from the rare-earths and not an anti-bonding $B_6\sigma^*$ band which would disturb the stability of the boron framework. This pattern can be extended to actinide hexaborides, involving 6d and 7s orbitals.

$M^{II}B_6$ Hexaborides. This concerns mostly europium and ytterbium hexaborides. For EuB_6 it can be assumed that the Eu^{2+}: $4f^7$ level is situated in the forbidden band (as for EuO). Europium vacancies would introduce holes in the $4f^7$ configuration. These holes are small polarons, i.e. Eu^{3+} ions trapped by electrostatic forces on a site near-neighbor to the vacancy. The density of these acceptor sites is N_A; the negative value of the Seebeck coefficient indicates that electron donor sites are also present actually with a $N_D > N_A$ density. These are rare earths in a state of substitutional impurity, represented in Figure 9 by the

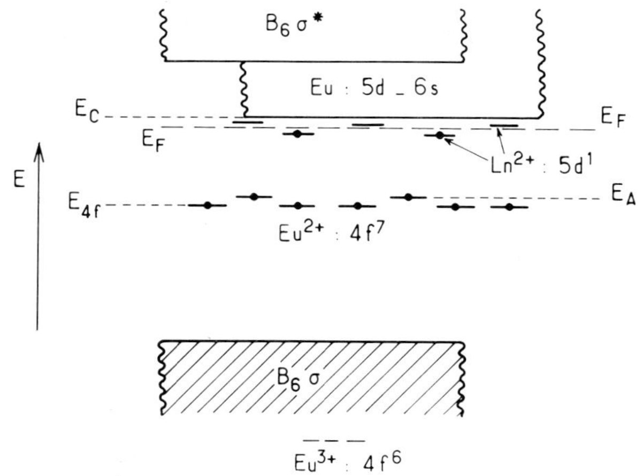

Fig. 9. Energy diagram for EuB_6

$5d^1$ donor levels of Ln^{2+}: $4f^{n-1}5d^1$ centers. When the density of impurity atoms is sufficient, a narrow $5d^1$ impurity band is formed: since $N_D - N_A < N_D$, the Fermi level drops down to this narrow $5d^1$ band.

Figure 6 shows that in the case of EuB_6 and YbB_6, the electrical conductivity, in a temperature range of 77 to 300 K approximately, occurs only because of electrons moving within the impurity band excited in the conduction band. Only at very high temperatures the excitations of electrons situated in $4f^7$ (EuB_6) or in $4f^{14}$ (YbB_6) levels have to be taken into consideration. The linear variation with temperature of the Seebeck coefficient shows, however, that the Fermi level remains located in the $5d^1$ band, even at the highest temperatures. Accordingly, the 0.30 and 0.14 eV values do not represent an intrinsic forbidden gap for $4f^7$ or $4f^{14}$ electrons.

In the case of $La_xEu_{1-x}B_6$ solid solutions, the mobility - and consequently the conductivity - increases with x, because the Eu^{2+} ions, whose number is decreasing, behave as high electron-scattering centers. For low values of x ($x \leq 0.03$) the electron mobility increases very quickly with x: indeed, the conduction takes place inside a narrow lanthanum impurity band which widens with x and progressively overlaps the bottom of the 5d-6s conduction band of europium as the number of lanthanum atoms increases.

A transition from semiconductor to metal appears between x = 0 and x = 0.01. For higher values of x, the Fermi level is higher than E_C, the energy at which the electrons are no longer correlated. This transition has a less noticeable influence on the thermoelectric power. The curves of Figure 8 can be interpreted by the presence of chemical composition inhomogeneities which would give rise to microphases, one corresponding to a polaron conduction, the other to an impurity band conduction. As long as the impurity band ranges are isolated, the material is a semiconductor, but when they become continuous, the conduction is metallic (MERCURIO et al., 1974).

Samarium Hexaboride. Since the rare-earth hexaborides in a +III oxidation state are metallic, the semiconductor character of SmB_6 seems to exclude the presence of Sm^{3+}: $4f^5$ ions. However, when samarium is in +II oxidation state, the Sm^{2+} ions can have both configurations $4f^65d^0$ and $4f^55d^1$, the former being obviously more stable than the latter because of the strong tendency of the rare earths to complete their 4f subshell. COHEN et al. (1970) showed by Mössbauer absorption measurements that both configurations coexist in SmB_6.

The very weak measured activation energy (5.10^{-3} eV) makes it plausible that the energy level diagram is similar to that of EuB_6 and that localized $5d^1$ levels could be situated in the lower part of the 5d6s conduction band (MERCURIO, 1974). The maximal energy E_C of this zone would constitute a separation limit between localized and non-localized states, according to Anderson localization (Fig. 10).

At low temperatures (<50 K) strong electron-electron and electron-lattice correlations take place within the $5d^1$ localized levels. With rising temperature the thermal excitation allows electrons to pass progressively from these $5d^1$ localized states to more and more delocalized states within a real conduction band Sm^{2+} : 5d6s : the change in the slope of the $\mathrm{Ln}\,\sigma = f(1/T)$ curve and the rapid decrease of $|\alpha|$ between 100 and 200 K can be explained by a progressive increase in conduction by highly mobile carriers, i.e. by the transition to a metallic state (Figs. 11, 12).

The presence of electrons in the hybrid 5d6s band reinforces their overlapping, which leads to a decrease in Sm-Sm distances. The variation of the lattice parameter of SmB_6 with temperature could be the result of the competition between this overlapping phenomenon and the normal lattice dilatation with temperature, which becomes predominant above 70 - 80 K (Fig. 13).

β) Magnetic Properties

Hexaborides of Non Magnetic Ions (Ca^{2+}, Sr^{2+}, Ba^{2+}, La^{3+}, Th^{4+})

These are diamagnetic, since their magnetic susceptibility is mostly caused by boron atoms. In the case of YbB_6: (Yb^{2+} : $4f^{14}5d^0$) the observed weak paramagnetism may be due to presence of impurities in very low concentrations (ETOURNEAU et al., 1970; ETOURNEAU, 1970; RABENAU et al., 1967).

Hexaborides of Magnetic Rare-Earths ($M^{III}B_6$)

These hexaborides become antiferromagnetic at low temperatures (MERCURIO, 1974; LEE and BELL, 1972; LEE et al., 1970; COLES and GRIFFITHS, 1961; BENOIT, 1955; PADERNO et al., 1969). While the effective moments calculated from experimental data in the paramagnetic range are close to the theoretical values obtained for free trivalent ions, there exists,

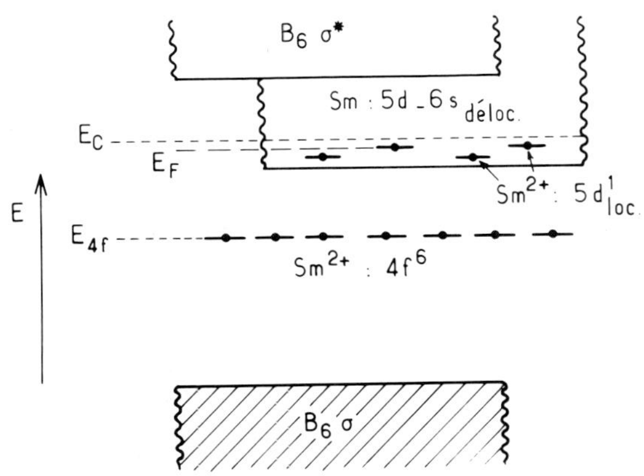

Fig. 10. Energy diagram for SmB_6

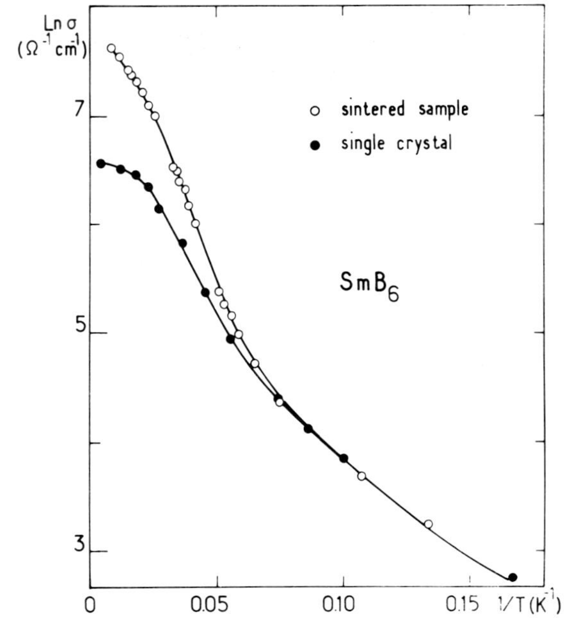

Fig. 11. Electrical conductivity of SmB_6

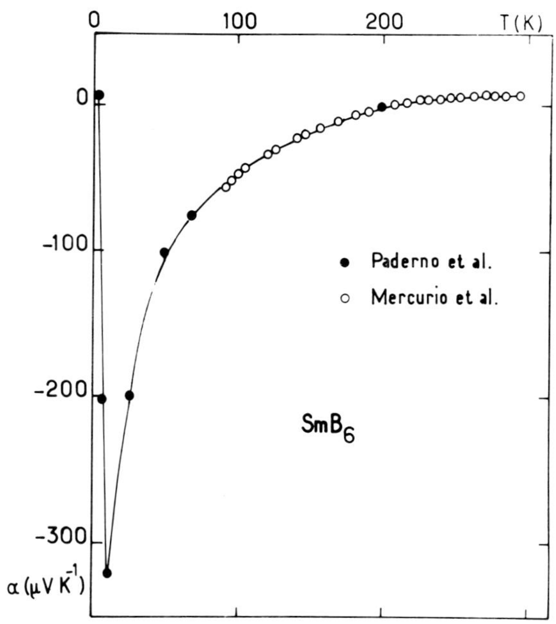

Fig. 12. Thermoelectric power of SmB_6

however, a large difference between the values of T_N and $|\theta_p|$, except for DyB_6. It seems that this difference is caused by deviations with respect to the straight lines that represent the behavior of the free ions of the $\chi^{-1} = f(T)$ curves, when T < 100 K (Fig. 14). Since $M^{III}B_6$-type hexaborides are metals, it is reasonable to expect interaction of the local moments with the conduction electrons. This interaction

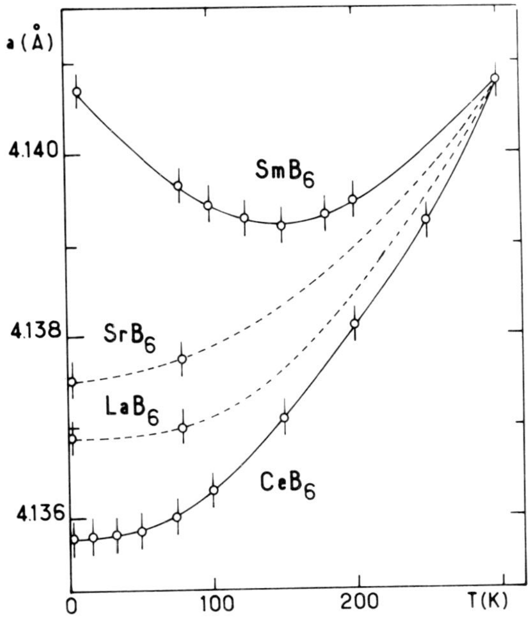

Fig. 13. Lattice parameters of some hexaborides as a function of temperature. The parameter values have been related to the CeB_6 parameter at room temperature

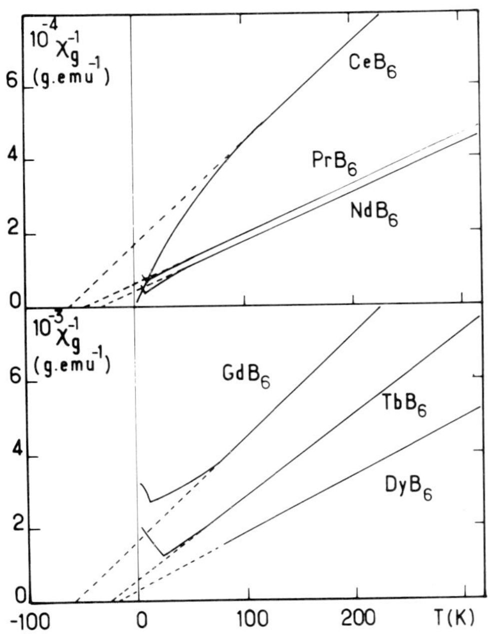

Fig. 14. Reciprocal magnetic susceptibility of rare-earth hexaborides

is of the form $H_{int.} = 2J_{sf}\vec{S}\vec{s}$, where J_{sf} is the exchange integral between a magnetic ion characterized by a $\vec{S}$ spin caused by 4f electrons and the conduction electrons carrying the $\vec{s}$ spin. J_{sf} polarizes the s electrons locally around the paramagnetic ion. Through the conduction electrons this local polarization gives rise to a large-distance interaction between localized moments. This phenomenon called Ruderman-

Kittel-Kasuya-Yosida interaction provides at least a qualitative understanding of the antiferromagnetism of the rare-earth hexaborides in the +III oxidation state (RUDERMAN and KITTEL, 1954; KASUYA, 1956; YOSIDA, 1957).

Europium Hexaboride and $La_xEu_{1-x}B_6$ Solid Solutions

EuB_6 orders ferromagnetically for $T < 13$ K, the Curie temperature varying from 8 to 13 K according to different investigators. The Curie constant determined in the paramagnetic range corresponds to the $4f^7$ ($^8S_{7/2}$) configuration of the Eu^{2+} free ion in its ground state (MERCURIO et al., 1974; GEBALLE et al., 1968; BACHMANN et al., 1970; HACKER et al., 1971).

FISK (1969) explained the ferromagnetism of EuB_6 by a modification of the RKKY interaction studied by BLOEMBERGEN and ROWLAND (1955) for semiconductors. According to these authors, the moments could be coupled through the valence band electrons. Their calculations, established with the aid of some simplifying assumptions lead to an exchange function that varies with the distance between inserted atoms; in the case of EuB_6 the interaction would be ferromagnetic.

In the case of $La_xEu_{1-x}B_6$ solid solutions, the paramagnetic Curie temperature increases at the beginning, up to 20 K for $x = 0.03$, and then decreases to -10 K for $x = 0.50$, after having passed through a zero value for $x \approx 0.33$. It starts increasing again for $0.50 < x < 0.75$ and vanishes for $x \geq 0.75$ (Fig.15).

For small values of x ($x \leq 0.05$) we saw that supplementary electrons were localized on $5d^1$ impurity levels, and then formed a narrow impurity band when x increased. These electrons interact with the localized spins of Eu^{2+} ions, giving a RKKY-type interaction. As long as x remains small, and consequently as long as the dilution of moment carriers is negligible, this interaction, which has a ferromagnetic nature, tends to reinforce the coupling between Eu^{2+} ions. A consequence is the remarkable increase of the paramagnetic Curie temperature (Fig. 15).

For compositions corresponding to $0.05 < x < 0.33$, there seems to be a noticeable difference between T_C and $|\theta_p|$, indicating that long-range antiferromagnetic interactions compete with ferromagnetic interactions.

Another increase of x ($0.33 \leq x \leq 0.75$) involves an increase of the average distance between Eu^{2+} ions; the RKKY interaction then becomes highly antiferromagnetic, which gives a negative paramagnetic Curie temperature; the long-range antiferromagnetic interactions become dominant.

For $x > 0.75$, the dilution of the magnetic moments in the lanthanum hexaborides lattice is such that the material is paramagnetic at all temperatures.

The previous analysis is confirmed by the study of $Sr_xEu_{1-x}B_6$ phases for $x \leq 0.2$: the substitution of Eu^{2+} ions by Sr^{2+} ions essentially creates a dilution of moment carriers. Even for quite a small dilution, the paramagnetic Curie temperature slightly decreases when x increases. This result is consistent with the diminution of the paramagnetic Curie temperature with x, observed for $Eu_xYb_{1-x}B_6$ phases (KRAUSE and SIENKO, 1973).

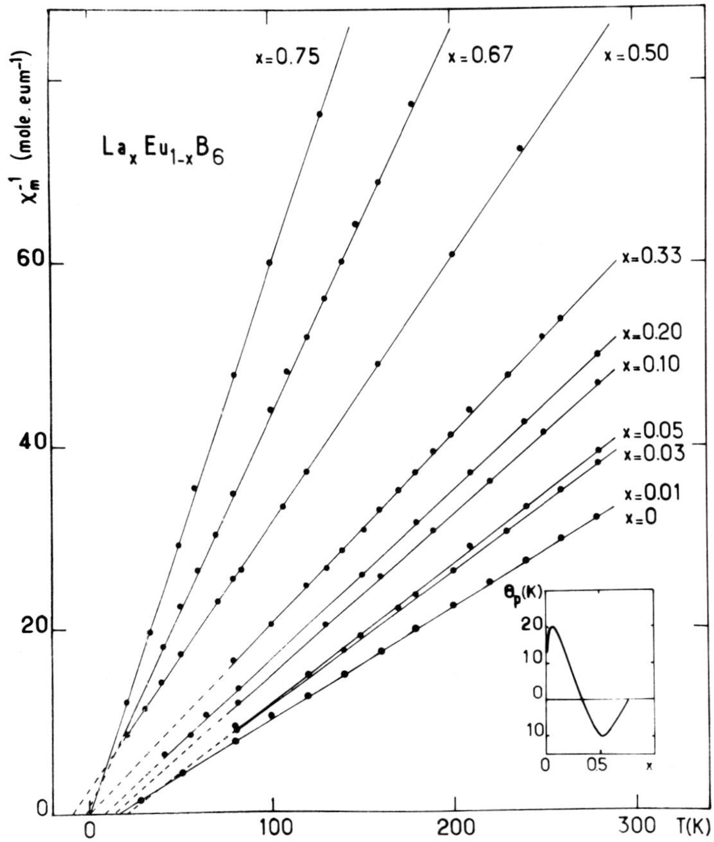

Fig. 15. Reciprocal magnetic susceptibility of $Eu_{1-x}La_xB_6$ phases as a function of T and paramagnetic Curie temperature as a function of x

Samarium Hexaboride

All the magnetic data show that at high temperatures the magnetic susceptibility is intermediate to those corresponding to $4f^6$ and $4f^5 5d^1$ configurations (MENTH et al., 1969; FALICOV and KIMBALL, 1969; MERCURIO, 1974).

The absence of magnetic order is in agreement with the simultaneous presence of both Sm^{2+} configurations: $4f^6 5d^0$ and $4f^5 5d^1$: the shape of the $\chi^{-1} = f(T)$ curve shows that at low temperatures strong correlations occur in the localized $5d^1$ levels (Fig. 16).

b) Tetraborides

The tetraboride unit cell contains 4 metal atoms and 16 boron atoms: 12 of them belong to two octahedra and the other 4 are bound by sp^2-type bonds. For stability each octahedron requires 20 electrons per $(MB_4)_4$ group (14 for internal bonding, 6 for external bonding) and each sp^2-type bond requires 4 electrons, i.e. a total of 56 electrons. Since 16 boron atoms provide only 48 electrons, an electron transfer from the metal sublattice to the boron one is necessary to ensure the stability of the structure (LIPSCOMB and BRITTON, 1960).

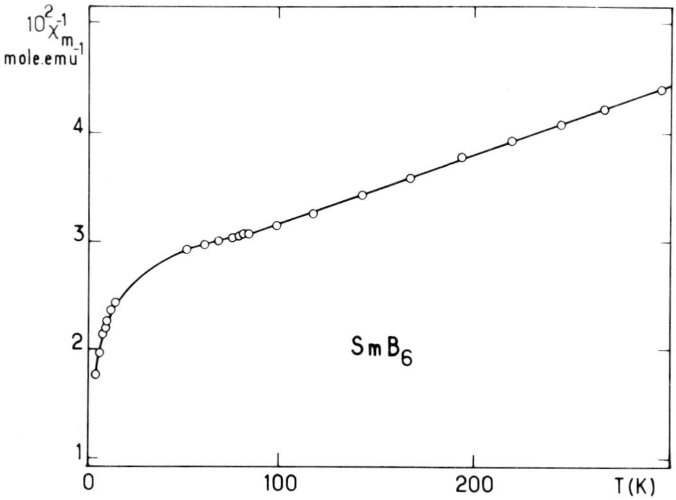

Fig. 16. Reciprocal magnetic susceptibility of SmB_6 as a function of temperature

Unlike the situation for hexaborides, there have been hardly any publications to date on the physical properties of tetraborides.

α) Transport Properties

Only the electrical properties of YB_4, ThB_4 and SmB_4 have been studied (JOHNSON and DAANE, 1963; AUSKERN and ARONSON, 1968; STURGEON, 1973): these are compounds with a metallic character. The number of conduction electrons per metal atom, deduced from Hall effect measurements, is 2 for ThB_4, but only 0.2 for YB_4. This result seems, in the case of yttrium, to require an oxidation state close to +II a localization at the bottom of the conduction band of the same type as that observed for SmB_6; a magnetic investigation could corroborate this supposition.

β) Magnetic Properties

Tetraborides of Non-Magnetic Ions (Y^{3+}, La^{3+}, Th^{4+})

YB_4 and LaB_4 are diamagnetic, whereas ThB_4 is Pauli paramagnetic (JOHNSON and DAANE, 1963; AUSKERN and ARONSON, 1964).

Tetraborides of Magnetic Rare Earths

The NdB_4, SmB_4, GdB_4, TbB_4, DyB_4, HoB_4 and TmB_4 phases order antiferromagnetically at low temperatures; the probable presence of two oxidation states for cerium and ytterbium in CeB_4 and YbB_4 is probably the reason for their particular magnetic behavior (no Curie-Weiss-type behavior). Only PrB_4 in the series of the rare-earth tetraborides is ferromagnetic (BUSCHOW and CREYGHTON, 1972). Just as for hexaborides, the magnetic properties of tetraborides may be interpretable on the basis of indirect RKKY interactions through conduction electrons.

However it is noteworthy that a shifting of the ordering temperature with composition has been observed in the UB_4-DyB_4 and UB_4-ErB_4 systems (ALI-ZADE et al., 1972). GIORGI et al. (1974) seem to have recently found a ferromagnetic ordering around 14 K in the $Y_xV_{1-x}B_4$ system for $0.62 < x < 0.70$.

5. Fields of Application

The CaB_6-type hexaborides are well known for their mechanical and refractory properties, such as high congruent melting points and hardnesses close to that of boron (3000 kg/mm^2). They are also characterized by a great resistance to stress deformation (5.10^4 kg/mm^2). Hexaborides have a high modulus of elasticity and a high tensile strength, which suggests their use as reinforcing elements in composite materials, in the way that alumina or carbides are actually used.

Most hexaborides and especially LaB_6 are very good electrical conductors at high temperatures; furthermore, they have excellent thermionic properties: not only is the electronic emission potential quite low (2.7 eV for LaB_6), but cathodes made of lanthanum hexaboride make it possible to obtain high current densities, better than those of standard cathodes such as W, ThO_2 or activated BaO (BROERS, 1973).

This property, together with the high chemical inertia and their thermal stability, causes lanthanum hexaboride to be more and more often substituted for materials generally utilized in demontable electronic tubes with high current densities, e.g. in scanning electron microscopes, electronic masks, etc.

References

ALI-ZADE, N.Kh., RAMAZANZADE, M.G., CHECHERNIKOV, V.I., GUSEINOV, N.G., SLOVIANSKII, V.K.: Magnetic properties of UB_4 - DyB_4 and UB_4 - ErB_4, Uch. Zap. Azerb. Inst. Nefti Khim., 9(8), 54-56 (1972)

ALLARD, G.: Etude roentgenographique de quelques borures. Bull. Soc. Chim. 51, 1213-1215 (1932)

ANDRIEUX, L.: Electrolyse d'oxydes métalliques dissous dans l'anhydride borique ou les borates fondus. Nouvelles méthodes de préparation du bore amorphe, des borures et de quelques métaux. Ann. Chim. 12, 423-507 (1929)

AUSKERN, A.B., ARONSON, S.: Electrical properties of thorium borides. J. Chem. Phys. 49(1), 172-176 (1968)

BACHMAN, R., LEE, K.N., GEBALLE, T.H., MENTH, A.: Spin scattering and magnetic ordering in EuB_6. J. Appl. Phys. 41(3), 1431-1432 (1970)

BENOIT, M.R.: Etude paramagnétique des composés binaires. J. Chem. Phys. 52(2), 119-132 (1955)

BERTAUT, F., BLUM, P.: La structure des borures d'uranium. C.R. Acad. Sci. (Paris) 229, 666-667 (1949)

BLOEMBERGEN, N., ROWLAND, T.J.: Nuclear spin exchange in solids: Tl^{203} and Tl^{205} magnetic resonance in thallium and thallic oxide. Phys. Rev. 97, 1679-1698 (1955)

BLUM, P., BERTAUT, F.: Contribution à l'étude des borures à teneur élevée en bore. Acta Cryst. 7, 81-86 (1954)

BREWER, L., SAWYER, D.L. quoted by ZALKIN, A., TEMPLETON, D.H.: The crystal structure of CeB_4, ThB_4 and UB_4. J. Chem. Phys. 18(3), 391 (1950)

BROERS, A.N.: High-solution thermionic cathode scanning transmission electron microscope. Appl. Phys. Lett. 22(11), 610-612 (1973)

BUSCHOW, K.H., CREYGHTON, J.H.N.: Magnetic properties of rare earth tetraborides. J. Chem. Phys., 57(9), 3910-3914 (1972)

COHEN, R.L.: Hyperfine structure in EuB_6 and EuTe studied with the Mössbauer effect in Eu^{151}. Bull. Amer. Phys. Soc. 13, 667 (1968)

COHEN, R.L., EIBSCHÜTZ, M., WEST, K.W., BUEHLER, E.: Electronic configuration of SmB_6. J. Appl. Phys. 41(3), 898-899 (1970)

COLES, B.R., GRIFFITHS, D.: Antiferromagnetic behavior of GdB_6. Proc. Phys. Soc. (London) 77(1), 213-215 (1961)
CURTIS, B.J., GRAFFENBERGER, H.: The floating zone crystal growth of lanthanum hexaboride. Mat. Res. Bull. 1, 27-31 (1966)
DEACON, J.A., HISCOCKS, S.E.R.: On the growth and properties of single crystal LaB_4. J. Mat. Sci. 6, 309-312 (1971)
EBERHARDT, W.H., CRAWFORD, J.B., LIPSCOMB, W.N.: The valence structure of boron hydrides. J. Chem. Phys. 22, 989-1001 (1954)
EICK, H.A., GILLES, P.W.: Precise lattice parameters of selected rare earth tetra- and hexaborides. J. Amer. Chem. Soc. 81, 5030-5032 (1959)
EICK, H.A., MULFORD, R.N.R.: Americium and neptunium borides. J. Inorg. Nucl. Chem. 31, 371-375 (1969)
ETOURNEAU, J.: Les hexaborures de type CaB_6. Le système bore-thorium. Thèse de Doct. Sci. Phys. Univ. BORDEAUX I (1970)
ETOURNEAU, J., MERCURIO, J.P., NASLAIN, R.: Etude comparative de la stabilité thermique des hexaborures de terres rares. C.R. Acad. Sci. (Paris) 275, 273-276 (1972a)
ETOURNEAU, J., MERCURIO, J.P., NASLAIN, R., HAGENMULLER, P.: Structure électronique de quelques hexaborures de type CaB_6. J. Solid State Chem. 2, 332-342 (1970)
ETOURNEAU, J., MERCURIO, J.P., NASLAIN, R., HAGENMULLER, P.: Etude comparative de la stabilité thermique des tétraborures de terres rares. C.R. Acad. Sci. (Paris) 274, 1688-1691 (1972b)
ETOURNEAU, J., NASLAIN, R., LaPLACA, S.: L'hexaborure de thorium non-stoechiométrique $Th_{1-x}B_6$. J. Less Common Metals, 24, 183-194 (1971)
FALICOV, L.M., KIMBALL, J.C.: Simple model for semiconductor-metal transitions: SmB_6 and transition-metal oxides. Phys. Rev. Lett. 22, 997-999 (1969)
FISK, Z.: Magnetic interactions in rare earth borides. Ph. D. thesis, Univ. California (1969)
FISK, Z., COOPER, A.S., SCHMIDT, P.H., CASTELLANO, R.N.: Preparation and lattice parameters of the rare earth tetraborides. Mat. Res. Bull. 7, 285-288 (1972)
FLODMARK, S.: A solid state spd-MOLCAO treatment of cubic-octahedral CaB_6 with variation of the orbital exponent. Ark. Fys. 18(4), 49-64 (1960)
GEBALLE, T.H., MATTHIAS, B.T., ANDRES, K., MAITA, J.P., COOPER, A.S., CORENZWITT, E.: Magnetic ordering in rare-earth hexaborides. Science 160, 1443-1444 (1968)
GEBALLE, T.H., MENTH, A., BUEHLER, E., HULL, G.W.: Properties of SmB_6 doped with Eu and Gd. J. Appl. Phys. 41(3), 904-905 (1970)
GIORGI, A.L., SZKLARZ, E.G., WHITE, R.W., HILL, H.H.: Evidence for ferromagnetism in alloys formed between UB_4 and YB_4. J. Less Common Metals 34, 348-351 (1974)
GOODENOUGH, J.B., MERCURIO, J.P., ETOURNEAU, J., NASLAIN, R., HAGENMULLER, P.: Mécanisme de conduction électrique dans les hexaborures d'europium et d'ytterbium. C.R. Acad. Sci. (Paris) 277, 1239-1242 (1973)
GSCHNEIDNER, K.A., Jr.: Rare Earth Alloys pp. 116-130. Princeton: D. Van Nostrand Co, Inc. 1961
HACKER, H., Jr., SHIMADA, Y., CHUNG, K.S.: Magnetic properties of CeB_6, PrB_6, EuB_6 and GdB_6. Phys. Stat. Sol. 4, 459-465 (1971)
JOHNSON, R.W., DAANE, A.H.: The lanthanum - boron system. J. Phys. Chem. 65, 909-915 (1961)
JOHNSON, R.W., DAANE, A.H.: Electron requirements of bonds in metal borides. J. Chem. Phys. 38, 425-432 (1963)
KASUYA, T.: A theroy of metallic ferro- antiferromagnetism on Zener's model. Prog. Theoret. Phys. 16(1), 45-57 (1956)
KLOTZ, H.: Chemischer transport von Lanthanhexaborid. Naturwiss. 52, 451-452 (1965)
KRAUSE, J.L., SIENKO, M.J.: Long-range ordering in magnetically diluted europium hexaboride. J. Solid State Chem. 6, 590-594 (1973)
LAFFERTY, J.M.: Boride cathodes J. Appl. Phys. 22(3) 299-309 (1951)
LEE, K.N., BACHMANN, R., GEBALLE, T.H., MAITA, J.P.: Magnetic ordering in PrB_6. Phys. Rev. B2(11), 4580-4585 (1970)
LEE, K.N., BELL, B.: Exchange interactions and fluctuations in CeB_6. Phys. Rev. B 6(3), 1032-1040 (1972)
LIPSCOMB, W.N., BRITTON, D.: Valence structure of the higher borides. J. Chem. Phys. 33, 275-280 (1960)

LONGUET-HIGGINS, H.C., ROBERTS, M. de V.: The electronic structure of the higher borides MB_6: Proc. Roy. Soc. 224A, 336-347 (1954)
McDONALD, B.J., STUART, W.I.: Crystal structures of some plutonium borides, Acta Cryst. 13, 447-452 (1960)
MENTH, A., BUEHLER, E., GEBALLE, T.H.: Magnetic and semiconducting properties of SmB_6. Phys. Rev. Lett. 22, 295-297 (1969)
MERCURIO, J.P.: Synthèse, stabilité thermique et propriétés électriques et magnétiques des hexaborures de terres rares. Thèse Doct. Sci. Phys. Univ. BORDEAUX I (1974)
MERCURIO, J.P., ETOURNEAU, J., NASLAIN, R., HAGENMULLER, P., GOODENOUGH, J.B.: Etude comparée des propriétés électriques des hexaborures d'europium et d'ytterbium. Mat. Res. Bull., 8, 837-844 (1973)
MERCURIO, J.P., ETOURNEAU, J., NASLAIN, R., HAGENMULLER, P., GOODENOUGH, J.B.: Propriétés électriques et magnétiques des solutions solides $La_xEu_{1-x}B_6$. J. Solid State Chem. 9, 37-47 (1974)
MOISSAN, H., WILLIAMS, P.: Sur la préparation et les propriétés des borures de calcium, de strontium et de baryum. C.R. Acad. Sci. (Paris) 125, 629-634 (1897)
NASLAIN, R., ETOURNEAU, J.: L'hexaborure de potassium. C.R. Acad. Sci. (Paris) 263, 484-487 (1966)
NASLAIN, R., GUETTE, A., BARRET, M.: Sur le diborure et le tétraborure de magnésium. Considérations cristallochimiques sur les tétraborures. J. Solid State Chem. 8, 68-85 (1973)
NICKERSON, J.C., WHITE, R.M., LEE, K.N., BACHMANN, R., GEBALLE, T.H., HULL, G.W.: Physical properties of SmB_6. Phys. Rev. B3(6), 2030-2042 (1971)
NIEMYSKI, T., KIERZEK-PECOLD, E.: Crystallization of lanthanum hexaboride. J. Cryst. Growth 3-4, 162-165 (1968)
NIIHARA, K.: The preparation and non-stoichiometry of samarium hexaboride. Bull. Chem. Soc. Japan 44, 963-967 (1971)
PADERNO, Yu.B., GORYACHEV, Yu.M., GARF, E.S.: Polyborides, their structure and properties. Electron Techn. 3(1/2), 175-183 (1970)
PADERNO, Yu.B., NOVIKOV, V.I., GARF, E.S.: Poroshkovaya Met. Akad. Nauk Ukr. SSr., 11(83), 70 (1969)
PADERNO, Yu.B., POKRZYWNICKI, S.: Magnetic properties of some heavy rare earth tetraborides. Sov. Phys. Solid State 24, K11-12 (1967)
PADERNO, Yu.B., POKRZYWNICKI, S., STALINSKI, B.: Magnetic properties of some rare earth hexaborides. Sov. Phys. Solid State 24, K73-76 (1967)
PAULING, L., WEINBAUM, S.: The structure of calcium boride CaB_6. Z. Kristallogr. 187, 181-182 (1934)
POST, B., MOSKOWITZ, D., GLASER, F.W.: Borides of some rare-earth metals. J. Amer. Chem. Soc. 78, 1800-1802 (1956)
RABENAU, A., KAUER, E., KLOTZ, H.: Sur l'hexaborure de lanthane LaB_6 in: Propriétés thermodynamiques, physiques et structurales des dérivés semi-métalliques, Proc. Int. Symp. CNRS, ORSAY, Sept.28 - Oct.1, 495-498 (1967)
RUDERMAN, M.A., KITTEL, C.: Indirect exchange coupling of nuclear magnetic moments by conduction electrons. Phys. Rev. 96, 99-102 (1954)
SAMSONOV, G.V., PADERNO, Yu.B., FOMENKO, V.S.: Hexaborides of rare earth metals. Poroshkovaya Met. Akad. Nauk Ukr. SSR 6(18), 24-31 (1963)
SCHWETZ, K. LIPP, A.: Homogeneity region of the europium hexaboride phase: J. Less Common Metals 33, 295-303 (1973)
SHANNON, R.D., PREWITT, C.T.: Effective ionic radii in oxides and fluorides. Acta Cryst., 25, 925-928 (1969)
SPEAR, K.E.: Phase equilibria and related properties of the dysprosium boron-system. In: Boron, preparation, structure and properties. Ed. Science, (Moscow), 207-215, 1974
STURGEON, G.D.: Personal communication (1973)
STURGEON, G.D., MERCURIO, J.P., ETOURNEAU, J., HAGENMULLER, P.: Single crystals of SmB_6 for conductivity and other measurements. Mat. Res. Bull. 9, 117-120 (1974)
STACKELBERG, M. Von, NEUMANN, F.: Die Kristallstruktur der Boride der Zusammensetzung MeB_6. Z. Phys. Chem. B19, 314-320 (1932)
YAJIMA, S., NIIHARA, K.: The non stoichiometry of rare earth hexaborides. 9th Rare-earth Res. Conf. Blacksburg, Virginia, October 10-14 (1971)

YAMAZAKI, M.: Group-theoretical treatment of the energy bands in metal borides MeB_6. J. Phys. Soc. Japan, 12, 1-6 (1957)
YOSIDA, K.: Magnetic properties of Cu-Mn alloys. Phys. Rev., 106, 893-898 (1957)
ZALKIN, A., TEMPLETON, D.H.: The crystal structures of CeB_4, ThB_4 and UB_4 J. Chem. Phys. 18(3), 391 (1950)
ZALKIN, A., TEMPLETON, D.H.: Acta Cryst. 6, 269 (1953)

X. Crystal Chemistry of Boron and of Some Boron-Rich Phases; Preparation of Boron Modifications

R. NASLAIN

1. Introduction

For many years, most physico-chemical studies of boron and its compounds were made with elemental boron which was of low purity and usually amorphous or at least very poorly crystallized. Such a type of boron was generally obtained either by reduction of a boron compound using a highly electropositive metal (reduction of $B_2O_3$by magnesium or of KBF_4 by sodium for example), or by electrolysis of molten salts containing a borate or a fluoborate as a boron source. In all cases, the boron resulting from the reduction was separated from by-products by prolonged washing in aqueous solutions, so that its purity generally was not higher than 80 to 95%. WEINTRAUB and KAHLENBERG even made the hypothesis that in certain cases (for example MOISSAN boron obtained by reduction of B_2O_3 by magnesium or alkali metals) the reaction product was a boron sub-oxide, with a formula close to B_6O, rather than elemental boron. Although these materials have low purity, they are easily prepared, and their purity can be greatly improved by outgasing in high vacuum and at a temperature near to the melting point of boron. The products resulting from chemical or electrochemical reduction of boric oxide, boron trifluoride, borates or fluoborates, are now utilized as starting raw materials in obtaining highly pure boron when subjected to purification cycles involving boron halides as principal intermediates (MOISSAN, 1892 a,b; WEINTRAUB, 1909; KAHLENBERG, 1925; ANDRIEUX, 1929; COOPER, 1945, a, b; SOWA, 1949; COOPER, 1951 a, b; SPEVACK, 1954; ANDRIEUX and DEISS, 1955; MIKHEEVA et al., 1957; a, b, c; HAAG, 1957; COOPER et al., 1959; YANNACAKIES and NIES, 1960; FINN et al., 1960; SAMSONOV et al., 1960; COOPER 1961; SCHOTT et al., 1963).

The first samples of highly pure crystallized boron were probably prepared in 1909 by WEINTRAUB, by reduction of BCl_3 with hydrogen in an electric arc (WEINTRAUB, 1909).

For more than 60 years, the work of WEINTRAUB has been the basis of much research on the synthesis of boron modifications by decomposition (or by reduction with hydrogen) of boron halides either in an electric arc, or in contact with different hot substrates. These studies have led to the publication of a large number of X-ray diffraction patterns which were attributed, all too readily, to new and different boron modifications. HOARD and NEWKIRK (1960) made a critical study of published diffraction patterns, from which they concluded that only a few boron allotropes could be accepted as definitely established, namely those which have been investigated by single crystal technique. Thus, they suggested that the other crystallized products, assumed to be also true boron modifications, could be most often either a mixture of phases or boron-rich non-stoichiometric borides. These products could be the result (1) of an initial boron-substrate interaction which leads, under specific growth conditions, to phases with broad homogeneity range extending almost to pure boron, or (2) of a stabilization of new boron frameworks by traces of impurity atoms from the gas phase.

The most recent work on products resulting from hydrogen reduction of boron halides, with special attention given to the purity of starting materials, to the use of boron itself as the substrate and relying on precise structure determination from single crystals, indicates that there may be only four, or perhaps three, crystalline forms of boron.

The α-rhombohedral variety (α-rh.)(a = 5.075 Å; α= 58°06; space group R$\bar{3}$m), also called the low-temperature form or red boron, is obtained either by precipitation from melts of boron-saturated precious metals, or by low-temperature (800-1200°C) reduction of BX_3 boron halides (X = Br or I) (McCARTY et al., 1958; DECKER and KASPER, 1959; HORN, 1960; McCARTY and CARPENTER, 1960; NIEMYSKI and ZAWADZKI, 1962; AMBERGER and DIETZE, 1965; HAGENMULLER and NASLAIN, 1965; SITARIK and ELLIS, 1966; WALD, 1970; NASLAIN et al., 1970; NASLAIN, 1970).

The β-rhombohedral variety (β-rh.)(a = 10.145 Å; α = 65°17'; space group R$\bar{3}$m, also called the high temperature form, is prepared either by solidification of liquid boron, or by annealing α-rh. boron beyond the α-rh. → β-rh. transformation point, or by reduction of a boron halide with hydrogen on a substrate heated to a temperature above approximately 1400°C (POWELL et al., 1960; BEAN and MEDCALF, 1960; STARKS and MEDCALF, 1960; GREINER, 1960; HAGENMULLER and NASLAIN, 1965; HINZ and WIRTH, 1965; ARMINGTON et al., 1965; NIEMYSKI et al., 1965; KING et al., 1965; NASLAIN, 1967; ARMINGTON et al., 1970).

Between 1000 and 1400°C approximately, the reduction of boron halides by hydrogen leads to two types of crystalline tetragonal phases, while the identification of α-rh. and β-rh. phases as real allotropic boron forms has not been questioned, this is not the case for the tetragonal phases, which are still the subject of much controversy.

The first crystals of the α-tetr. phase (a = 8.75 Å; c = 5.06 Å; space group $P4_2/nnm$ or P$\bar{4}$2m), also called tetr. I by some authors, were obtained by LAUBENGAYER et al. (1943) by reduction of boron tribromide with hydrogen on a tantalum filament heated to 1300°C approximately. HOARD et al. (1958) published what they considered to be the first completely established structure for an allotropic form of pure boron, based on their structure analysis of single crystals. However, the latest studies on the reduction of boron halides by hydrogen at 1000-1300°C all seem to indicate that the α-tetr. phase does not occur when highly pure reagents are utilized, but that it can be obtained in a reproductible manner in presence of impurities such as carbon, nitrogen or even some metals. Thus, the α-tetr. phases would seem not to correspond to an allotropic form of pure boron, but in fact, could be boron-rich borides, comprising a boron framework of tetragonal symmetry, stabilized by impurity atoms (OLEMPSKA et al., 1966; MIERZEJEWSKA, 1966; AMBERGER and PLOOG, 1968; BOIRET, 1969; APPENHEIMER et al., 1970; AMBERGER et al., 1970; AMBERGER and PLOOG, 1971; PLOOG and AMBERGER, 1971; AMBERGER und SCHMIDT, 1971; PLOOG et al., 1972).

The β-tetr. variety (a = 10.12; c = 14.14 Å; space group $P4_122$), also called tetr. II or tetr. III, could be, on the contrary, a real allotropic form of pure boron, since it has been obtained by several authors in a reproductible manner and from highly pure reagents. The crystal structure of this boron form has not yet been established but it could be closely related to those of lithium, beryllium and aluminum borides, characterized by lattice parameters very close to those of β-tetr. boron and by the same space group (TALLEY et al., 1960 a, c; AMBERGER and PLOOG, 1968; BOIRET, 1969; AMBERGER et al., 1970; AMBERGER and PLOOG, 1971; PLOOG et al., 1972).

Boron can also occur in an amorphous or vitreous state with remarkable mechanical properties. An important application of these properties, described elsewhere, is the utilization of fibers (or filaments) of amorphous boron as reinforcing elements for certain types of composite materials mostly used in the aerospace industry (TALLEY et al., 1960b; WANER et al., 1967).

Crystalline boron forms are most interesting from a physical as well as from a chemical point of view. From the physical point of view, the structural differences of various forms of boron, particularly the nature of bonding (in α-rh. and β-rh. forms especially) are reflected in quite different physical properties. From the chemical point of view, the crystalline forms of boron are much purer, because of the methods required to prepare them, than amorphous products resulting from reduction of boric oxide, borates and fluoborates, by metals; hence, they appear to be most suitable materials for the synthesis of boron compounds (e. g. borides), for which further physical studies are planned. Furthermore, the various cristalline boron modifications can differ markedly in their chemical behaviour, because of their different crystal-chemical characteristics. For example, NASLAIN, ETOURNEAU and HAGENMULLER showed that under the conditions of their experiments, only the α-rh. variety combined with potassium to form KB_6. Thus, it can be useful to have reproductible preparation methods for each of the various crystalline forms of boron (NASLAIN and ETOURNEAU, 1966; NASLAIN et al., 1967).

After some general considerations regarding the crystal chemistry of boron and boron-rich borides, we shall describe briefly the crystal structures of various modifications; we refer the reader to other critical reviews for further details (NEWKIRK, 1961, 1964; HOARD and HUGHES, 1967; NASLAIN, 1972). Most of this chapter will then be concerned with the preparation of high-purity α-rh. and β-rh. boron, the two crystalline forms which have been characterized most completely and which also seem to be the most important for practical application.

2. Boron Polymorphism

a) General Considerations of the Crystal Chemistry of Boron-Rich Phases

Boron varieties (α-and β-rh., α-tetr. phases), as well as a great number of boron-rich borides, are all characterized by the existence of three-dimensional skeletons of boron atoms built from a B_{12} elementary unit, itself consisting of twelve boron atoms at the vertex positions of a quite regular icosahedron. The presence of icosahedral clusters, though rather unusual in inorganic structures, is not all that surprising in the case of boron-rich phases with a covalent character, in view of the electron-deficient nature of boron and the efficient manner for distributing electrons provided by the compact grouping of twelve atoms characteristic of an icosahedron.

The B_{12} clusters are bound by *directed bonds*, either to one another or to isolated atoms, giving rise to a great variety of skeletal arrangements that is quite remarkable for an element. The variety and complexity of the crystal structures of boron and of boron-rich phases result from the special properties of B_{12} icosahedral units and from the rules that govern the way they combine.

α) The Boron B_{12} Icosahedron

The regular icosahedron is a polyhedron with a very large number of symmetry elements, namely: a center of symmetry, six five-fold axes, ten three-fold axes, fifteen two-fold axes and fifteen mirror planes. The most unique feature of this polyhedron is *the presence of five-fold axes*, which cannot lead to either a two-or a three-dimensional lattice. A regular icosahedron can be repeated within three-dimensional lattices only by elementary translations or by utilizing additional symmetry elements (Fig. 1a).

The B_{12} icosahedron is generally quite regular. It can be approximated for the purpose of packing considerations, by a sphere with a diameter of 5.1 Å.

β) Association Modes of B_{12} Icosahedra

In a B_{12} icosahedron, each boron atoms shares B-B bonds with the five closest neighbors which are related by a five-fold axis and which belong to the same icosahedron. In addition, each boron atom shares a sixth bond that is directed outwardly from the icosahedron and which is responsible for the formation of the many kinds of three-dimensional skeletons that occur in boron-rich phases. As a general rule, the average length of intra-icosahedral B-B bonds is greater than the length of the sixth outwardly directed bond; this property reflects the electron deficient character of the B_{12} unit (Fig.1b).

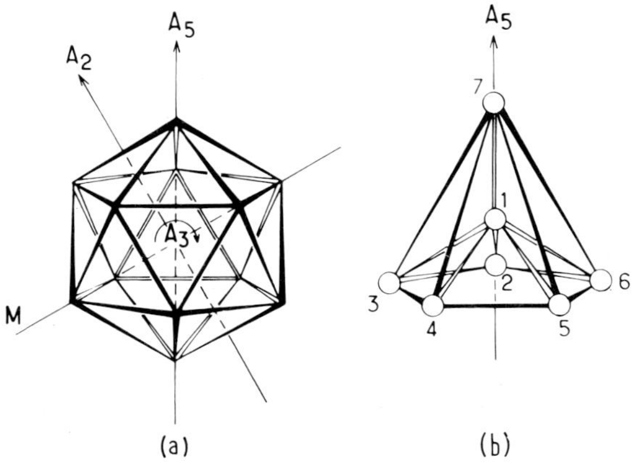

Fig. 1 (a). Some symmetry elements of the regular icosahedron. (b) The six-fold coordination boron polyhedron as it occurs in various allotropic forms of the element and in many boron-rich phases

According to HOARD and HUGHES, this sixth bond would be quite rigid and *preferentially* directed along the corresponding pseudo-five-fold axis. A result of this assumption would be that in boron-rich phases the polyhedron for the *ideal* coordination of boron would be a pyramid with a pentagonal base. It is to be noted that while the B_7 unit does not exist in an isolation in the various forms of boron or in boron-rich phases, it is however, the basic element of the crystal structure of MgB_4 magnesium tetraboride (HOARD and HUGHES, 1967; GUETTE et al., 1972; NASLAIN et al., 1973; GUETTE, 1974).

The existence of six five-fold axes in the B_{12} icosahedron and the preferential orientation of outwardly directed bonds along these axes, give rise, in fact, to the complexity of the crystal structures of boron-rich phases. Indeed, in the case of a three-dimensional lattice which would contain *only* B_{12} icosahedra, the assumption of HOARD and HUGHES would mean that each B_{12} unit would be itself icosahedrally surrounded by twelve B_{12} icosahedra and that this arrangement could be extended for a long distance. Such an arrangement is theoretically impossible. On the one hand, it would imply constructing a lattice with five-fold symmetry axes; on the other hand, the exclusive use of icosahedra would result in very poor space filling. The conflict between maintaining the ideal orientation for the sixth bond and the impossibility of extending for long distances the five-fold symmetry inherent in an icosahedron means that compromises must be made and that there must be distortions of the ideal coordination polyhedron for boron. The great variety of the structures of boron-rich phases reflects the very large number of ways in which these distortions can take place.

Analysis of actually known structures shows that, except for the particular case of the α-rh boron, which will be discussed later, two types of three-dimensional boron skeletons are found. In the first case, to which belong the α-tetr. phases and many borides such as NaB_{15}, $MgAlB_{14}$, AlB_{10}, the boron skeleton consists of B_{12} icosahedra bonded to each other either by direct B-B bonds, or through additional *isolated* atoms of boron or of elements that readily form covalent bonds (carbon and silicon for example). In the second case, the lattice formed by boron atoms can be described on the basis of *complex units* formed from B_{12} icosahedra, such as $B_{12}(B_6)_{12}$ or $B_{12}(B_{12})_{12}$ units in β-rh. boron or in YB_{66}-type phases respectively, that in fact correspond to attempts to extend the five-fold symmetry of the central icosahedron to longer distances. These complex units then combine through directed bonds to give complex three-dimensional structures.

In both cases, association through *directed bonds* according to the ideal orientation defined by the postulate of HOARD and HUGHES (1967) of B_{12} units or of more complex $B_{12}(B_6)_{12}$ and $B_{12}(B_{12})_{12}$ units, necessarily leaves voids in the structure whose size is related to the size of the basic unit of the boron framework itself. The stability of the crystal structure requires that these voids be filled with additional atoms, which are generally either (1) boron atoms isolated or belonging to more or less complex configurations according to the size of the voids, or (2) foreign atoms in the case of borides.

Boron Atom Frameworks Based on Simple B_{12} Icosahedral Unit. In many boron-rich phases the boron sublattice can be described from simple B_{12} icosahedral units. In this case, only some of the directed sixth bonds can be used to form inter-icosahedral linkages between adjacent units, because of the necessity to use additional symmetry elements that are not characteristic of the icosahedron. The remaining boron atoms of the icosahedral units have their coordination completed by sharing a sixth bond with *isolated atoms* (most often boron), localized in the interstices of the B_{12} framework. In all cases, these isolated atoms occur in such positions that the sixth bonds that they share with boron atoms of the icosahedra *fulfill as best as possible* the orientation condition of HOARD and HUGHES (1967). Such frameworks occur in phases derived from α-rh. and α-tetr. boron, in AlB_{10}, NaB_{15} or $MgAlB_{14}$; these two latter examples are particularly interesting for illustrating the principles that we wish to discuss in this section.

The Na_xB_{15} (with $0<x<1$) phases crystallize in the orthorhombic system with the space group Imam. The B_{12} icosahedra that form the boron sublattice have one of their mirror planes parallel to the bc plane (Fig.2). The structure can be described either as a stacking in the $\vec{c}$ direction of relatively compact planes of B_{12} units, or as an alignment of chains of icosahedra parallel to the $\vec{c}$ direction. In a chain, the B_{12} icosahedra have one of their pseudo-five-fold axes inclined in an alternating manner at an angle of ±13°23 to the $\vec{c}$ direction. The chains are bonded to each other by inter-icosahedral B-B bonds or by bonds involving isolated boron atoms (Fig. 2 and 3, NASLAIN and KASPER, 1970; GUETTE, 1974; NASLAIN et al., 1976).

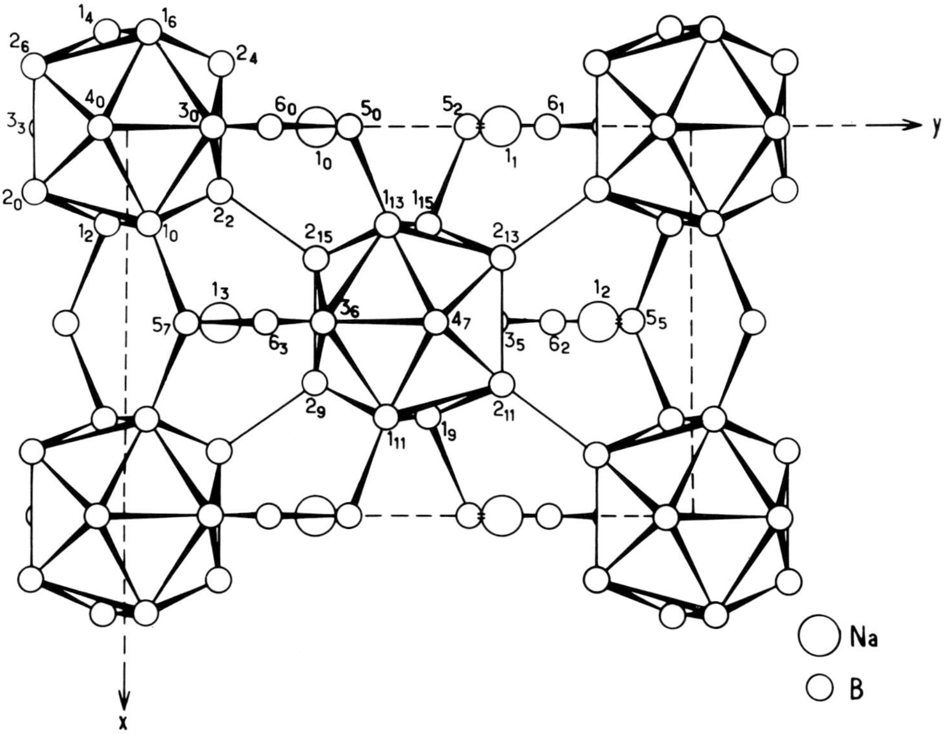

Fig. 2. Projection of the structure of NaB_{15} on the ab-plane (only atoms with $-1/4 < z < 1/4$ are represented). (From GUETTE, 1974, NASLAIN et al., 1976)

Figures 2 and 3 show that of the twelve external bonds shared by the atoms of an icosahedron, only six are direct inter-icosahedral bonds. The boron atoms of B_{12} units not involved in these bonds complete their coordination by forming directed bonds with isolated B(5) and B(6) boron atoms, which occupy a portion of the voids existing in the bc plane. The larger voids still available in the boron framework (sites-I with 16 coordination) are filled by sodium atoms.

A ternary boride with ideal formula $MgAlB_{14}$, has been studied at the same time by MATKOVICH and ECONOMY (1970). Its structure is isotypic with NaB_{15}, the metal atoms being distributed both in sites-I (4 e positions) which are occupied by sodium in Na_xB_{15} phases and also in the smaller sites-II (1/4, 1/4, 1/4) which are empty in the case of sodium boride. Since magnesium and aluminum atoms (r_{Mg} = 1.60 Å; r_{Al} = 1.43 Å) are far smaller than sodium atoms (r_{Na} = 1.90 Å), there is a contraction of the lattice in going from NaB_{15} to $MgAlB_{14}$. The largest contraction is for the b-parameter (8.415 Å for NaB_{15}, 8.115

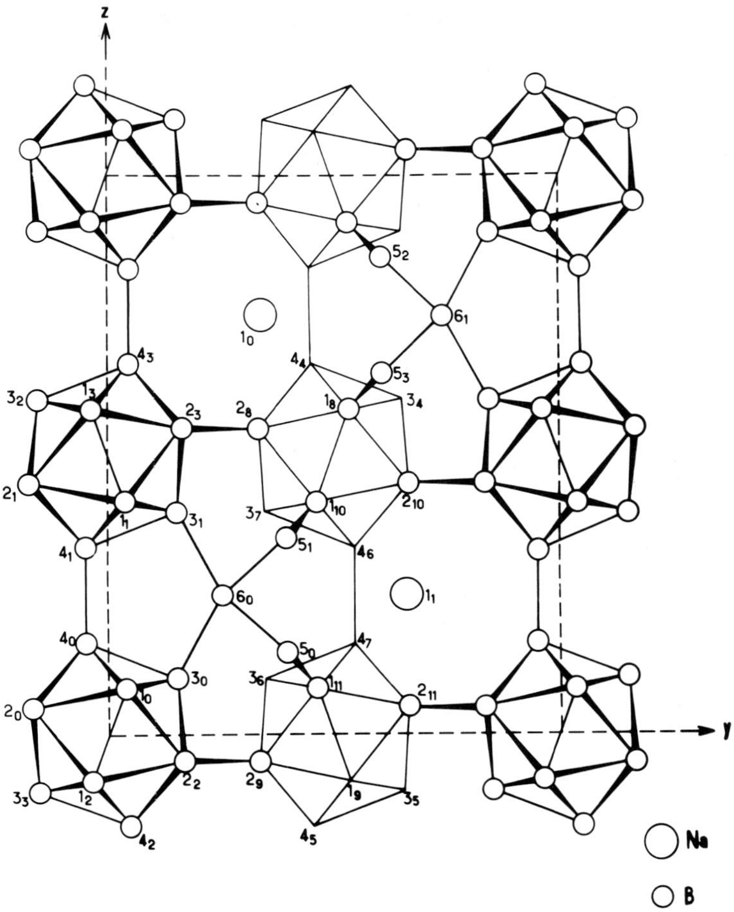

Fig. 3. Projection of the structure of NaB_{15} on the bc-plane (only atoms with $-1/4 < x < 1/4$ are represented). From GUETTE 1974, NASLAIN et al. 1976)

for $MgAlB_{14}$), so that the presence of isolated B(6) atoms is unnecessary in the case of $MgAlB_{14}$ to complete the external bonds of B_{12} units (Fig.4). Table 1 gives for both phases, the α-angle formed by each of the twelve external bonds with the corresponding pseudo-five-fold axis of the icosahedron; it shows that the deviations from the ideal orientation defined by HOARD and HUGHES are not equal for different cases, but that these deviations remain quite low (NASLAIN and KASPER, 1970; MATKOVICH and ECONOMY, 1970; GUETTE, 1974; NASLAIN et al., 1976).

In as much as the twelve external bonds of B_{12} units are entirely satisfied, for this kind of framework, by boron atoms, deviations from stoichiometry because of metal vacancies frequently occur. Thus, in $MgAlB_{14}$ crystals studied by MATKOVICH and ECONOMY the occupancy of sites-I and II by metal atoms was only 75% approximately. NASLAIN, GUETTE and HAGENMULLER showed that, while it is possible to obtain a stoichiometric NaB_{15} sodium boride when operating under sodium pressure, the pyrolysis of NaB_{15} at low temperature and in high vacuum leads to Na_xB_{15} sodium-deficient phases wher x can practically reach a value of zero, without the collapse of the boron skeleton. Although the metal atoms are not indispensable for the boron framework, they do however,

Table 1. α-angles between the pseudo five-fold axes and the B-B bonds outwardly directed from the B_{12} unit in $MgAlB_{14}$ and NaB_{15}. From GUETTE, 1974; NASLAIN et al., 1976

B-B bonds	Number of bonds	$\alpha_{MgAlB_{14}}$	$\alpha_{NaB_{15}}$
4_0 - 4_1	2	7°88	13°23
2_2 - 2_{15}	4	22°03	16°66
1_0 - 5_7	4	8°92	11°47
3_0 - 5_0	2	10°03	-
3_0 - 6_0	2	-	21°98

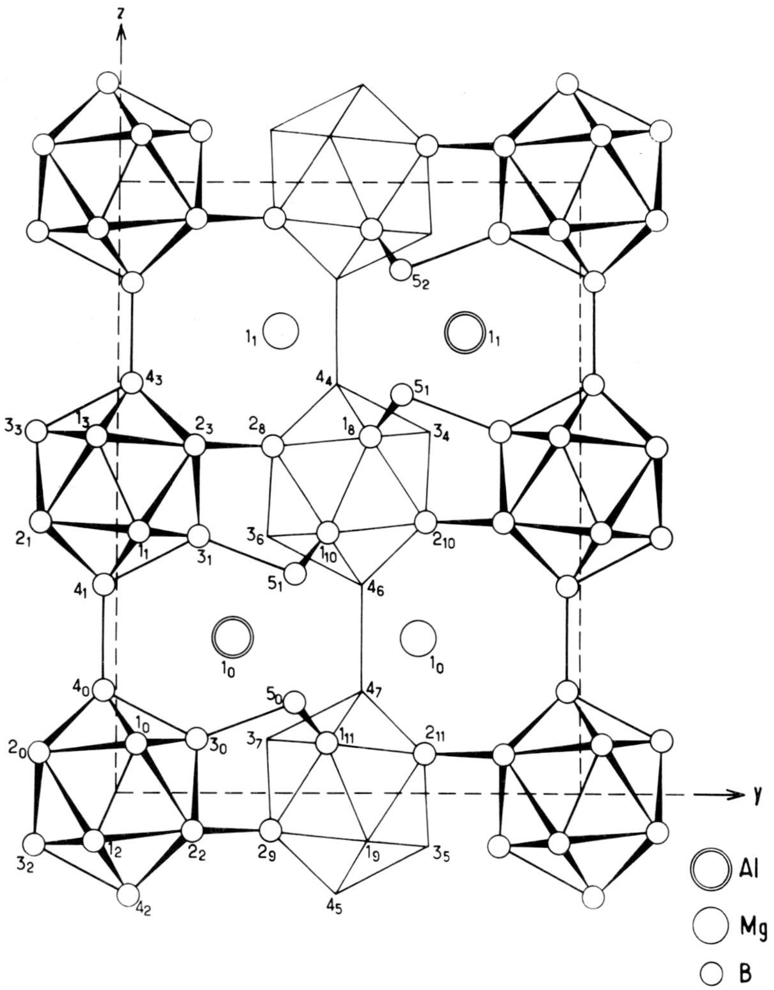

Fig. 4. Projection on the bc-plane of the structure of $MgAlB_{14}$. (Based on data of MATKOVICH and ECONOMY, 1970)

in this kind of phase, contribute to increasing its stability by filling totally or partially the voids and by contributing electrons to the boron framework ($Na_{\sim 0}B_{15}$ transforms into α-or β-rh. boron at 950°C whereas NaB_{15} is stable under sodium pressure at temperatures over 1200°C).

This example beautifully illustrates how the directed sixth bond is maintained as a rigid bond and, as a consequence, how isolated boron atoms might have to be introduced within the icosahedral framework. Furthermore it shows how, in particular circumstances, non-stoichiometric borides can lead to metastable boron monotropes which have sometimes been considered - wrongly - as new allotropic forms of elemental boron. The great structural variety of borides, together with the well-known difficulties of analysis in this field, is responsible for the proliferation in the literature of the many "new" crystalline boron modifications.

Boron Atom Frameworks Based on Complex Icosahedral Units. Even though it is not theoretically possible to have an icosahedral arrangement of B_{12} units continued for long distances, such a distribution can nonetheless occur *locally*. Thus, the crystal structure of β-rh. boron can be described on the basis of $B_{12}(B_6)_{12}$ polyhedron made up of a central B_{12} icosahedron surrounded icosahedrally by twelve pentagonal B_6 pyramids. Likewise, the structure of YB_{66}-type phases determined by RICHARDS and KASPER (1969), can be described on the basis of an even more complex polyhedron made up of a central B_{12} icosahedron surrounded locally by twelve B_{12} icosahedra. In these polyhedra the twelve bonds shared by boron atoms of the central icosahedron with atoms belonging to peripheral icosahedra have (within a few degrees) the ideal orientation defined by the postulate of HOARD and HUGHES. To compensate for the fact that the icosahedral arrangement of the B_{12} units poorly fills up space, and in order to make B-B bonds possible between atoms belonging to two peripheral icosahedra, it is necessary that the peripheral icosahedra be considerably larger than the central icosahedron. In the YB_{66}-type phases, the length of B-B bonds, and consequently the resulting effective radius of boron, progressively increase as the distance from the center of the $B_{12}(B_{12})_{12}$ unit increases, as shown in Figure 5. Thus, the average length of the B-B bond within the central icosahedron of the unit is only 1.726 Å whereas it is 1.808 Å within outer icosahedra. Furthermore, the dispersion of intra-icosahedral bond lengths, hence the distortion of the icosahedra, is much larger for peripheral ico-

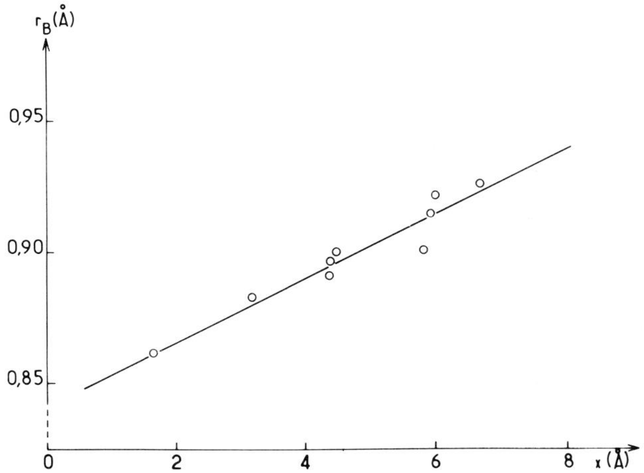

Fig. 5. Variation of the effective radius of boron, r_B, as a function of its distance, x, from the center of the $B_{12}(B_{12})_{12}$ unit in YB_{66}. (From RICHARDS and KASPER, 1969)

sahedra (from 1.752 to 1.863 Å) than for the central icosahedron (1.719 and 1.742 Å) (HUGHES et al., 1963; HOARD et al., 1970; RICHARDS and KASPER, 1969).

As already mentioned by RICHARDS and KASPER (1969), it is doubtful whether an even larger polyhedron would exist where the twelve outer icosahedra of the $B_{12}(B_{12})_{12}$ unit would themselves be icosahedrally surrounded by B_{12} units. Such an arrangement would involve, in the new outer shell of icosahedra, bonds far too long to correspond to usual B-B distances, and the cluster would lack stability. Thus, it appears that the 156-atom unit of YB_{66}-type phases very likely represents the limit beyond which it is no longer possible to maintain the pseudo five-fold axes of the central icosahedron. Figure 6, according

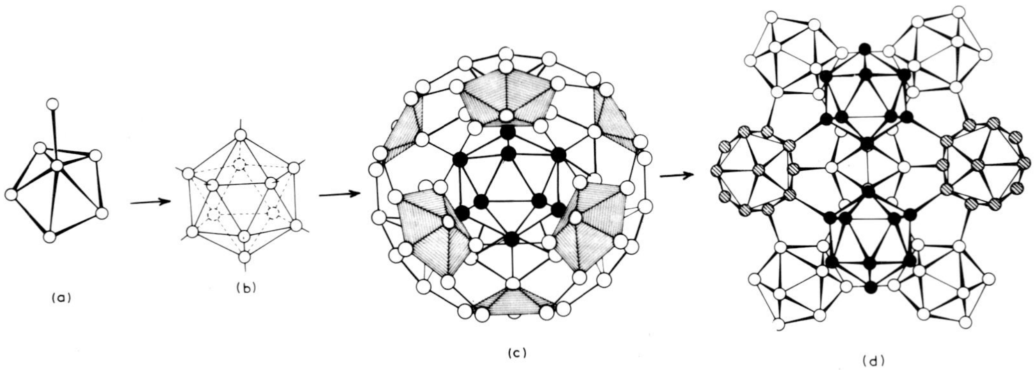

Fig. 6a-d. Evolution of the basic unit of the boron framework in some boron-rich phases: (a) MgB_4, (b) NaB_{15}, (c) β-rh. boron, (d) YB_{66}. (From NASLAIN et al. 1974)

to NASLAIN, GUETTE and KASPER, summarizes the successive stages in the progression from the pyramidal B_7 unit (strongly individualized in MgB_4) to the giant $B_{12}(B_{12})_{12}$ polyhedron of YB_{66} always maintaining *locally* one or more pseudo-fivefold axes (RICHARDS and KASPER, 1969; NASLAIN et al., 1972).

The giant $B_{12}(B_6)_{12}$ or $B_{12}(B_{12})_{12}$ units then associate by directed B-B bonds to yield three-dimensional skeletons of boron atoms of β-rh. boron (or of its solid solutions with transition elements) or of YB_{66} phases. Association, by means of rigid bonds, of these giant units, and maintaining as much as possible the symmetry elements of the icosahedron, have the consequence of producing large voids in the boron framework, the larger the basic unit the larger the voids. The stability of corresponding crystal structures requires that these voids be filled either by boron or by metal atoms, with the former contributing, as already observed, to completing the coordination of the boron atoms belonging to peripheral icosahedra of the giant units. Whereas in previous structures the interstitial voids were filled by a small number of isolated boron atoms (three per formula group in Na_xB_{15} phases, two in the case of $MgAlB_{14}$), in lattices with complex units, the void size requires considerably more boron atoms. Thus in β-rh. boron, the voids left by the association ob B_{84} polyhedra are occupied by B_{10}-B-B_{10} configurations (HUGHES et al., 1963; HOARD et al., 1970). Likewise, in YB_{66}-type phases additional boron atoms are grouped within two more complex configurations (since the basic unit is larger) with formulas YB_{36} and YB_{48} respectively. Figure 7 illustrates the association mode for $B_{12}(B_{12})_{12}$ polyhedra in YB_{66} (a) and represents one of these two interstitial configurations (b) (RICHARDS and KASPER, 1969).

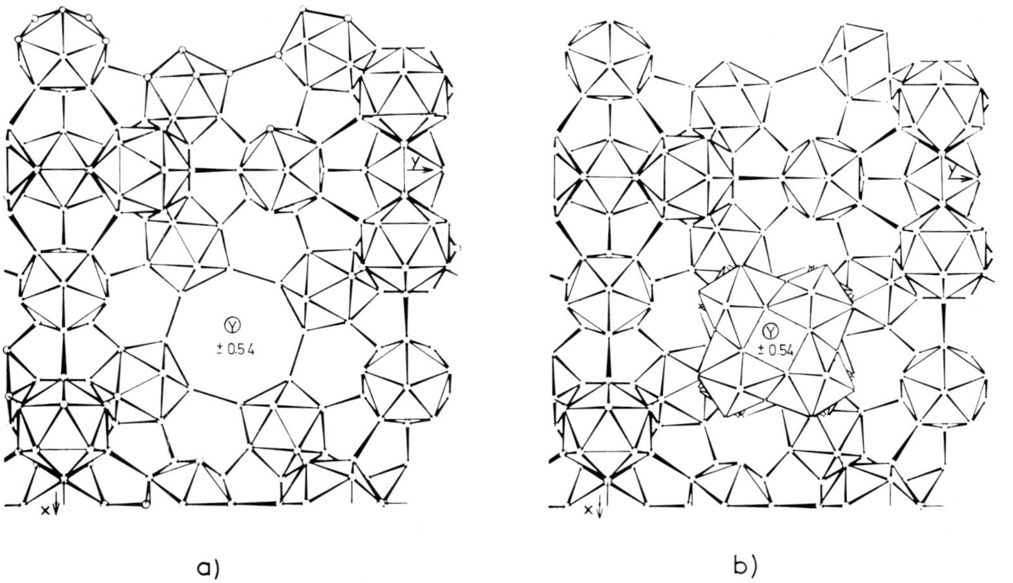

Fig. 7a and b. Association of $B_{12}(B_{12})_{12}$ polyhedra in YB_{66} showing: (a) the directed character of intericosahedral bonds, as well as the existence of large voids in the boron framework (b) the filling of voids by additional boron and yttrium atoms (partial projection on the ac-plane, according to RICHARDS and KASPER 1969)

Structural Classification of Boron Frameworks Containings Icosahedral Units. In noticing that the structures of boron forms and of most boron-rich phases could be described on the basis of nearly spherical boron clusters and much smaller isolated atoms, MATKOVICH, GIESE and ECONOMY proposed in 1965 that an analogy might exist between these phases and the Hägg interstitial compounds. On this basis they showed that all phases known up to that date could be deduced from three types of compact or pseudo-compact stackings of alomst spherical B_{12}(5.1 Å) or B_{84}(10.2 Å) units (MATKOVICH et al., 1965).

While their classification shows that the lattice parameters in these phases essentially depend on the association modes of the icosahedra (or of more complex $B_{12}(B_6)_{12}$ polyhedra), it does not however take into account the fact that these *covalent* units are themselves bonded to each other by *rigid bonds*, and it minimizes the fundamental role of isolated boron atoms. From a physical chemical point of view, the boron-rich, covalent phases are very far from HAGG interstitial compounds with metallic character, so that any parallelism between them can be only formal.

b) Crystal Structure of Rhombohedral Boron Modifications

The crystal structure of α-rh. boron can be described simply on the basis of a single icosahedral B_{12} unit. The structure of β-rh. boron also can be described in this manner but with the more complex $B_{12}(B_6)_{12}$ eighty four atom unit replacing the single unit of α-rh. boron and with an approximate doubling of the lattice parameter, as suggested by MATKOVICH and coworkers (1965). The bonds between units are of special kind in the case of α-rh. boron but are of the directed type that conforms to the postulate of HOARD and HUGHES (1967) for the high temperature form.

α) Crystal Structure of α-Rhombohedral Boron

The crystal structure of α-rh. boron determined by DECKER and KASPER (1959) is the most simple of structures based upon the B_{12} icosahedral unit. The rhombohedral cell (a = 5.057 ± 0.03 Å; α = 58.06° ± 0.05; space group R$\bar{3}$m) contains twelve boron atoms grouped within a B_{12} cluster. The B_{12} units, centered at the vertex of the rhombohedral cell, are oriented in such a way that one of their three-fold axes coincides with that of the space group, i.e. with the [111] direction of the rhombohedron. Three of the six pseudo-five-fold axes of each B_{12} unit coincide almost perfectly with the edges of the rhombohedron; consequently, adjacent B_{12} units are bonded to each other, along the edges, by direct B-B bonds with the ideal orientation defined by HOARD and HUGHES (DECKER and KASPER, 1959).

The structure may be considered as being formed by layers of B_{12} icosahedra parallel to the basal plane of the hexagonal cell and forming, in the classification of MATKOVICH and coworkers, a cubic close packing of icosahedra which is slightly deformed since the rhombohedral angle is 58.06°-instead of 60° for an ideal cubic close packing (Fig. 8a). The cohesion between layers is provided by inter-icosahedral bonds (1.71 Å) with ideal orientation directed along the edges of the

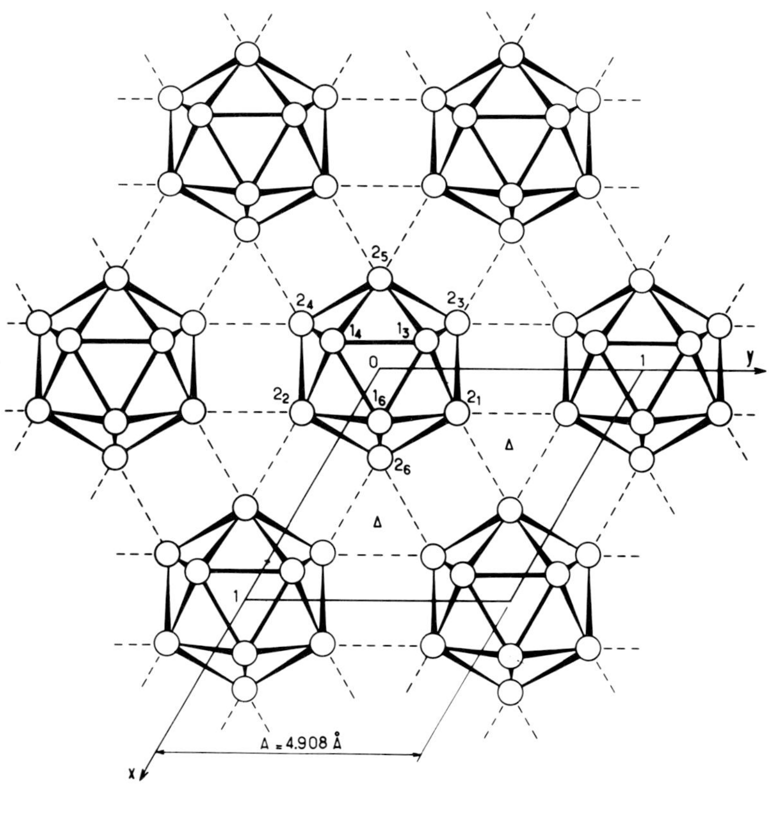

Fig. 8a. Crystal structure of α-rh. boron, (a) layer of B_{12} icosahedra centered in the (00.1) plane. The icosahedra are bonded to each other by three-center bonds situated in alternating parallel planes at z = ± 0.0245. The various B-B bonds have the following lengths: 1_3-1_8, 1.71 Å; 1_3-2_3, 1.79 Å; 1_3-1_4, 1.73 Å; 1_4-2_5, 1.79 Å; 2_1-2_3, 1.78 Å and 2_1-2_2, 2.03 Å

rhombohedron (each icosahedron having three bonds to a layer above and three bonds to a layer below) (Fig. 8b). In a given layer, boron atoms not involved in interlayer bonding and belonging to adjacent B_{12} units are close enough (2.03 Å) to be considered as weakly bonded, and in fact do form the novel type of three-center bond - such as frequently occurs in boron hydrides - where two electrons are shared between three boron atoms (Fig. 8a).

LONGUET-HIGGINS and ROBERTS (1955) showed that the stability of the B_{12} icosahedron requires 26 electrons for internal bonding. Six electrons per B_{12} unit are involved in the six intericosahedral directed covalent bonds, whereas four are involved in forming the six three-center bonds. The various bonds, internal and external, of the icosahedron then require a total number of 36 electrons per B_{12} unit: this is exactly the number of electrons yielded by the 12 atoms of the cluster.

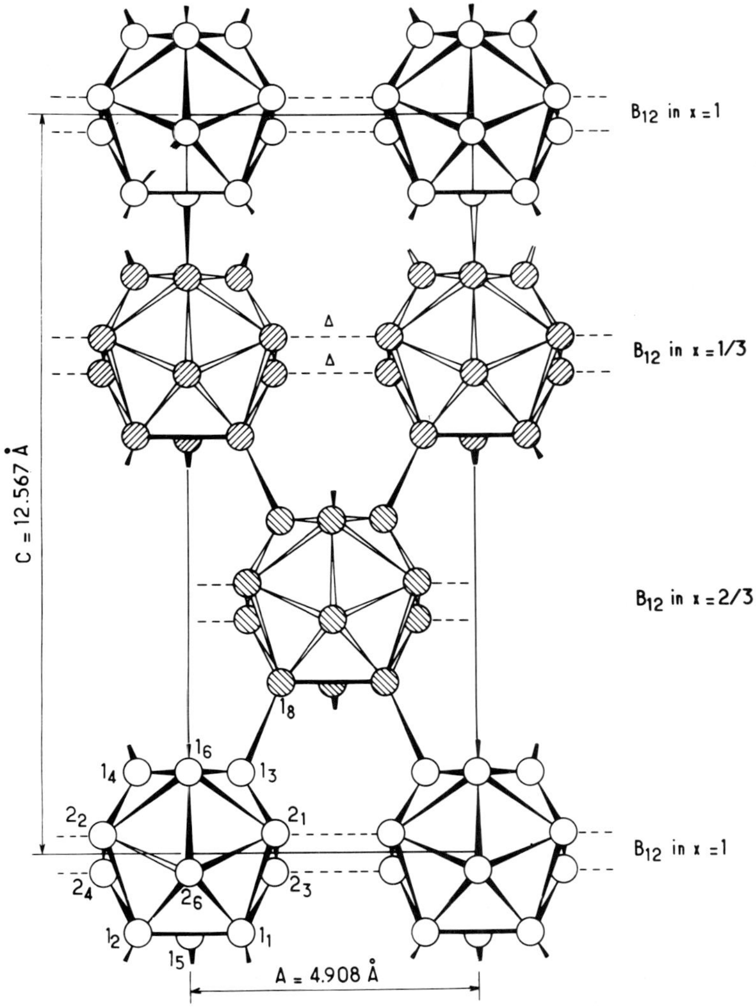

Fig.8b. View perpendicular to the bc-plane, showing the bonds along the c-axis between icosahedra at x = 1, x = 2/3 and x = 1/3 (icosahedra at x = 0 are not represented). (Based on data of DECKER and KASPER, 1959)

The presence of weak three-center bonds in α-rh. boron is responsible for its weak thermal stability: α-rh boron transforms into β-rh. boron around 1200°C. The presence of three-center bonds also confers to α-rh. boron a high reactivity, which makes this variety the most interesting material for boride preparations as demonstrated by NASLAIN ETOURNEAU and HAGENMULLER in the synthesis of alkali-metal borides, or by BECHER in the synthesis of BeB_{12} (NASLAIN et al., 1970; NASLAIN, 1967; BECHER, 1963).

While the proximity of B_{12} units in pseudo-compact planes of α-rh. boron provides, through three-center bonds, a pattern of inter-icosahedral bonding not involving isolated atoms, the resulting framework is, however, quite unstable since 50% of the boron atoms are not in their ideal coordination. In compounds formed by boron with carbon, silicon, phosphorus, arsenic, oxygen and sulfur, and that have the same space group and similar lattice and positional parameters, the foreign atoms, on being inserted in the "octahedral interstices" of the deformed cubic close packing of B_{12} units, take the place of the three-center bonds of α-rh. boron. They form chains, C-B-C or P-P for example, in such a way that *covalent bonds* can occur between the terminal atoms of the chain and the neighboring icosahedra; furthermore, these bonds are in the *ideal orientation* with respect to the HOARD and HUGHES (1967) postulate (Fig. 9) (CLARK and HOARD, 1943). Thus, in boron compounds derived from α-rh. boron, all the external bonds of a B_{12} unit have the ideal orientation. Consequently, these compounds, characterized by a framework of extremely rigid covalent bonds, are very stable and very hard; $B_{13}C_2$ (the hardest known material after diamond and borazon) for example, melts around 2500°C, whereas α-rh. boron transforms at 1200°C. This example again shows the important role of isolated atoms and of directed bonds that they form with icosahedral units, in this type of structure.

β) Crystal Structure of β-Rhombohedral Boron

The crystal structure of the β-rh. boron variety was determined by HOARD et al. (1970) and confirmed by GEIST et al. (1970). The primitive rhombohedral cell (a = 10.145 ± 0.015 Å; α = 65°17 ± 8'; space group $R\bar{3}m$) contains 105 atoms in the model proposed by HOARD et al. with a corresponding theoretical density of 2.29, compared to a value of 2.35 measured by floatation for a single crystal (Fig. 10) (HUGHES et al., 1963).

The β-rh. boron framework can be described in terms of a P polyhedron which corresponds to a $B_{12}(B_6)_{12}$ unit with 84 boron atoms, and which maintains to a good approximation the symmetry elements of the central B_{12} icosahedron (Fig. 11). In this polyhedron the B-B bonds within the B_6 pentagonal pyramids are, on an average, longer (1.85Å) than the internal bonds of the central icosahedron (1.767 Å). The longest bonds (1.88 and 1.92 Å) are situated in the hexagonal rings that surround the threefold symmetry axes of the P polyhedron. The shortest bonds are those from the central icosahedron to the twelve B_6 pyramids (1.62 to 1.72 Å). The observed variations in bond lengths within the $B_{12}(B_6)_{12}$ unit illustrate - as mentioned previously - the distortions resulting from the extension of the icosahedral symmetry of the central B_{12} unit (Fig. 5).

The B_{84} units are centered at the vertices of the rhombohedral cell and form - as in the case of α-rh. boron - a deformed cubic close packing (α = 65° 17' instead of 60°) (Fig. 10b). They are oriented in

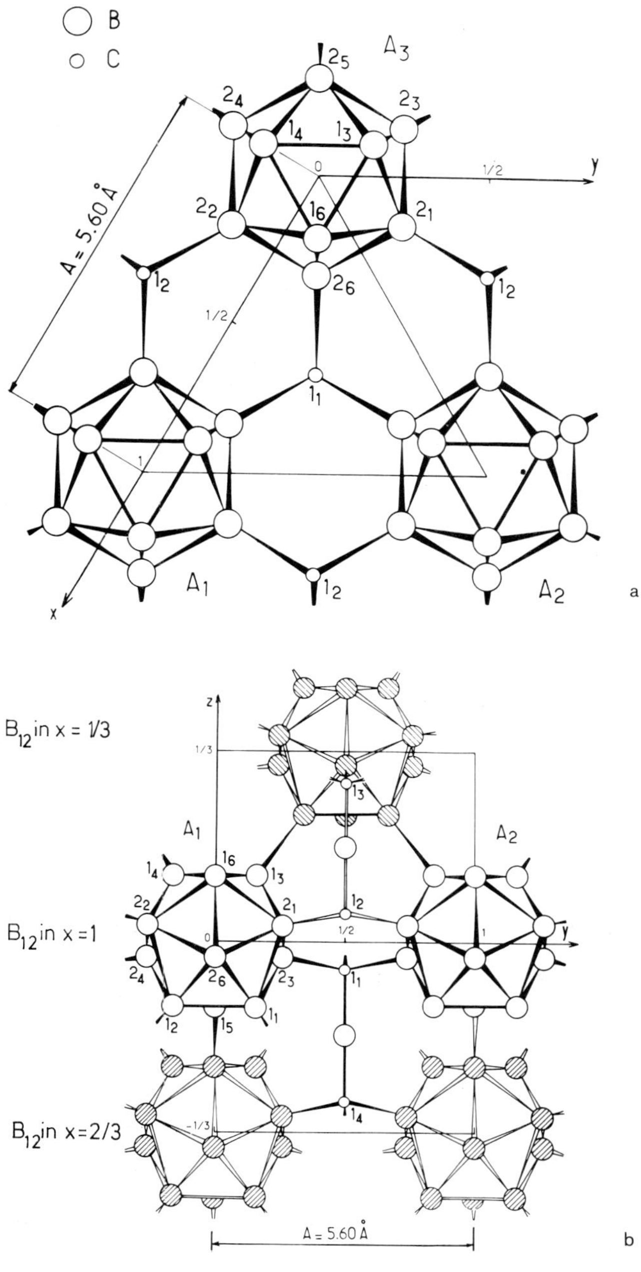

Fig. 9a and b. Crystal structure of $B_{13}C_2$ boron carbide: (a) a layer of B_{12} icosahedra centered in the (00.1) plane. The icosahedra are bonded to each other through terminal carbon atoms of C-B-C chains that are parallel to [001] (atoms 1_1 have a negative z value, atoms 1_2 a positive one). (b) View perpendicular to the (100) plane. B-B bonds between icosahedra are parallel to edges of the rhombohedral cell. C-B-C chains oriented parallel to [001], occupy octahedral voids of the deformed cubic-close packing of icosahedra. The 1_1-1_4 chain is situated at the back of the (100) plane and the 1_2-1_3 chain at the front (icosahedra at x = 0 are not represented) (Based on data of CLARK and HOARD, 1943)

such a way that three of their pseudo-five-fold axes are directed along the egdes of the rhombohedron. Consequently, in the middle of each edge, two B_6 pyramids (p_2 and p'_2, for example) belonging to two adjacent P polyhedra (P_1 and P_3, for example) form a B_{12} icosahedron (A_2, for example). Thus there exists in the network, chains of B_{12} icosahedra along the edges of the rhombohedron, in which the icosahedra are bonded to each other by direct B-B bonds with the ideal orientation (Fig. 10a, b).

The structure can also be described - as in the case of α-rh. boron - in terms of pseudo-compact layers of B_{84} units that are stacked perpendicularly to the $\vec{c}$ axis of the hexagonal cell. The large size of the almost spherical B_{84} units leads to octahedral voids in the deformed

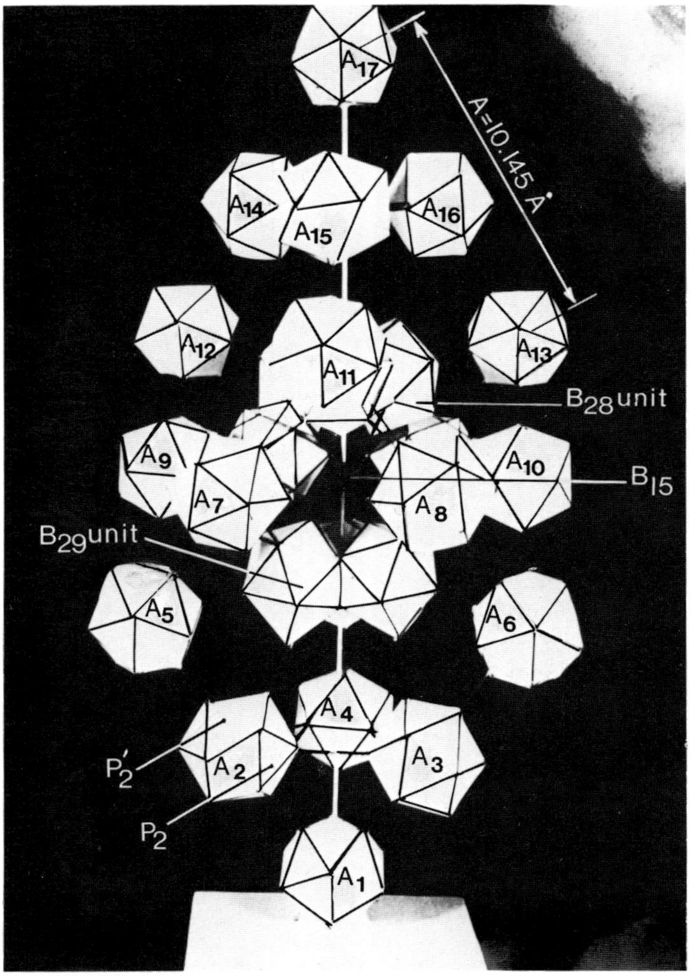

Fig. 10a. Crystal structure of β-rh. boron (a) the primitive rhombohedral cell with the structural formula $B_{84} + 2B_{10} + B$. The isolated B_{15} atom, situated at the center of the cell, is represented by a sphere (from a model constructed by M. BOURDEAU)

cubic close packing that are far larger than those in α-rh. boron and its derivatives. Whereas the octahedral voids are completely filled by only one C-B-C chain in the case of boron carbide, those of β-boron require a large number of additional atoms. These additional atoms form two new P' polyhedra, with 10 boron atoms, bonded together by the only isolated boron atom of the structure, B(15), which is situated at the center of the cell. The P' polyhedra result from three pentagonal pyramids condensing around a common (B_{14}) vertex, as shown in Figure 12a. Thus, the B_{10}-B-B_{10} configuration formally plays a role analogous to that of a C-B-C chain in boron carbide. In a P' polyhedron the bond length varies from 1.72 to 1.81 Å.

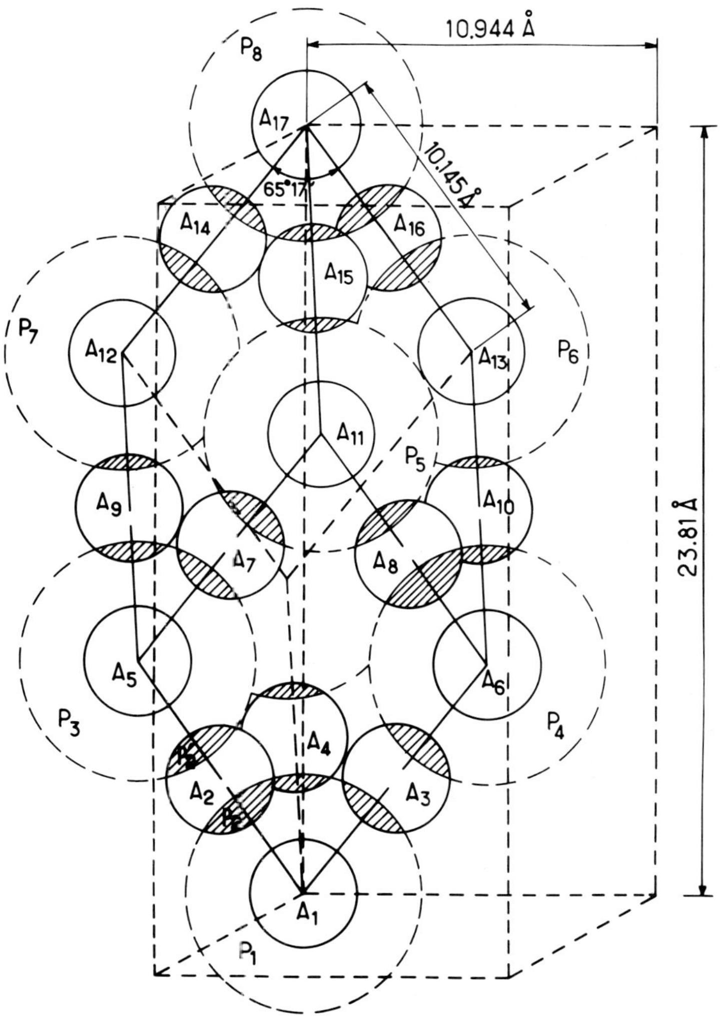

Fig.1ob. The arrangement of P polyhedra, with 84 boron atoms, in the primitive cell. The P_i polyhedra of $B_{12}(B_6)_{12}$ formulas are schematized by large diameter circles. The A_i icosahedra situated at the centers of edges are represented by small circles; they are formed from two pentagonal B_6 pyramids, belonging to two adjacent P_i polyhedra, and represented by hatched zones (p_2 and p'_2) (partial representation)

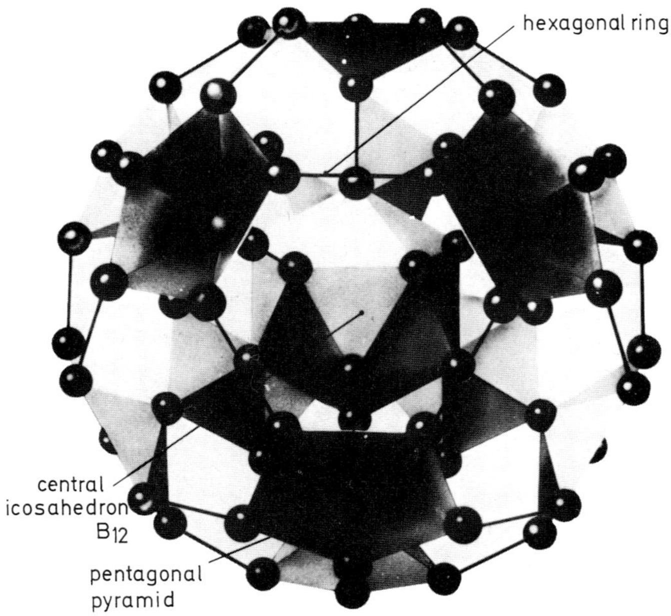

Fig. 11. Crystal structure of β-rh. boron. The P polyhedron with the formula $B_{12}(B_6)_{12}$. (From MATKOVICH et al., 1965)

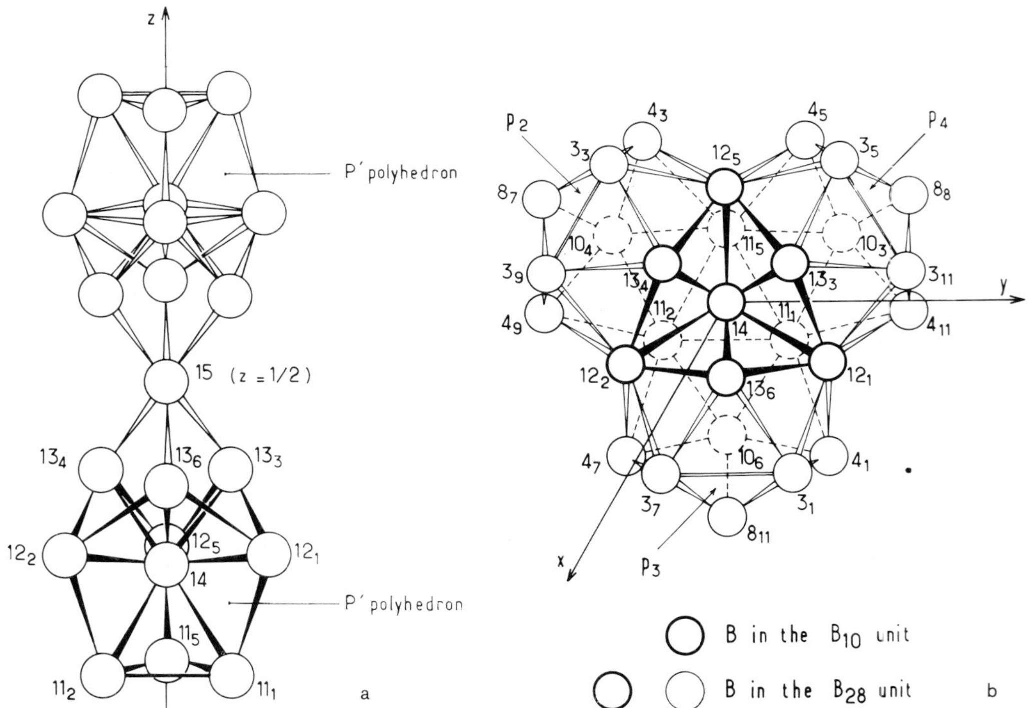

Fig. 12a and b. Crystal structure of β-rh. boron.
(a) the B_{10}-B-B_{10} chain occupying the octahedral interstices of the pseudocubic close packing of $B_{12}(B_6)_{12}$ units, (b) the B_{10} and B_{28} subunits represented by projection on the (00.1) plane of the hexagonal cell. The atoms have the values $0.3 < z < 0.5$. (Established from data of HOARD et al., 1970)

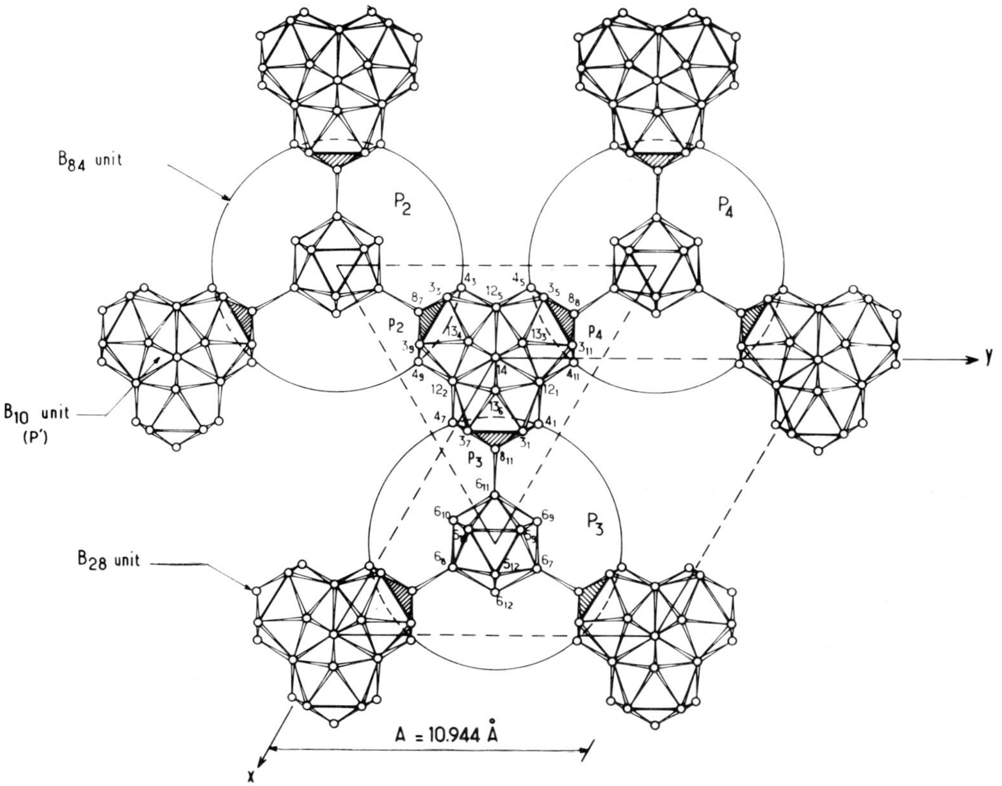

Fig. 13. Crystal structure of β-rh. boron: partial projection on (00.1) showing how the $B_{12}(B_6)_{12}$ units are bonded to each other through intermediary B_{10} sub-units in planes parallel to basal plane of the hexagonal cell. The P polyhedra corresponding to the $B_{12}(B_6)_{12}$ units are schematized by circles, including the central B_{12} icosahedron and three of the twelve B_6 pentagonal pyramids. Condensation of a B_{10} sub-unit with three pentagonal pyramids (e.g. p_2, p_3 and p_4) gives a B_{28} unit constituted of three condensed icosahedra. The boron atoms that are shown in the figure have the values $0.25 < z < 0.45$. (Based on data of HOARD et al., 1970)

In a pseudo-compact layer, three P polyhedra with 84 atoms (P_2, P_3 and P_4) are bonded to each other through a P' polyhedron with 10 boron atoms, which, in the general bonding scheme of basic units, plays a role which is formally analogous to that of the three-center bonds of α-rh. boron or of the C-B bonds of boron carbide (Fig. 13). Figure 13 shows that each P' polyhedron forms with three B_6 pentagonal pyramids (p_2, p_3 and p_4) belonging to adjacent P polyhedra (P_2, P_3 and P_4), three icosahedra condensed around the common (B_{14}) vertex, thus defining a new $B_{10} + 3B_6$ unit with 28 boron atoms (Fig. 12b). According to this description, the ideal structural formula associated with the primitive cell of β-rh. boron is: $B_{84} + 2B_{10} + B$.

In the model proposed by HOARD et al. (1970), the skeleton formed by boron atoms is determined by 15 crystallographically independent boron atoms. All these positions are considered to be fully occupied except for B(13) type atoms (in the 18(h) positions in the multiple hexagonal cell) which are represented in Figure 12a by the six atoms surrounding

Table 2. The nature, coordination number and % occupancy of the interstitial sites in β-rh. boron and the solid solutions formed by it with chromium, manganese and copper. The crystallographic positions are given for the triple hexagonal cell; Cu (2) corresponds in fact to two neighboring positions: (2') in 18 (h) and (2'') in 36 (i). From HOARD et al., 1970; ANDERSSON and LUNDSTRÖM, 1970; ANDERSSON and CALLMER, 1974; LUNDSTRÖM, 1974

Site Desi-gnation	Crystal-lographic Designation	Coordination Number	Positional Coordinates			Site Occupancy in %			
			x	y	z	CrB_{41}	MnB_{23}	CuB_{28}	β-rh. Boron
A (1)	6 (c)	12	0	0	0.135	Cr(1)71.9	Mn(1)25.6	Cu(1)6.1	-
A (2)	18 (h)	12	0.108	0.216	0.100	-	-	-	-
A (3)	36 (i)	12	0.279	0.248	0.046	-	-	-	-
D	18 (h)	15	0.203	0.406	0.174	Cr(2)18.0	Mn(2)43.1	Cu(2)43.1	-
E	3 (c)	15	0	0	0.230	-	Mn(3)66.2	Cu(3)50.5	-
F (1)	18 (h)	7 - 9	0.065	0.130	0.245	-	-	-	-
F (2)	18 (h)	3 -16	0.117	0.233	0.252	-	-	-	-
G	18 (h)	8 -14	0.055	0.109	0.117	-	-	B(16)20.9	B(16)33.3

the B(15) atom. The occupancy of 18(h) sites by B(13)-type atoms was indicated by HOARD et al. to be approximately 2/3 which would mean that the three 13_3, 13_4 and 13_6 positions of Figure 12a would, in fact, be occupied in a statistical manner by only two boron atoms. This hypothesis is corroborated by the presence of very short B-B distances (1.684 Å) between B_{13}-type atoms and the B(15) atom situated in the center of the cell (in the case of a full occupancy, the octahedral surrounding of the B(15) atom would involve quite longer B-B distances).

Without making a detailed study of the orientation of bonds directed outwardly from the icosahedra, it may be noted that about 80% of the boron atoms in the framework have an almost ideal coordination polyhedron. This characteristic is responsible for the high thermal stability of β-rh. boron; the structure is stable up to the melting point (about 2200°C under a pressure of one bar) and up to 1800°C under a pressure of 80kb (WENTORF 1965).

γ) Interstitial Solid Solutions Derived from β-Rhombohedral Boron

The structure of the β-rh. boron is more open than that of α-rh. boron, as evidenced by the density values of the two phases (2.35 and 2.46 respectively). By giving the boron atoms a uniform radius of 0.88 Å, ANDERSSON and LUNDSTRÖM (1970) calculated that the space-filling by boron atoms was only 36% for β-rh. boron. Consequently, the network formed by icosahedra contains many interstitial sites, some of which are large enough to accommodate either boron atoms or foreign atoms. Thus, β-rh. boron gives interstitial solid solutions with various elements such as lithium, beryllium, magnesium, aluminum, silicon, scandium and most transition elements (GIESE et al., 1965; ANDERSSON and LUNDSTRÖM, 1970; MATTES et al., 1970; CARLSSON and LUNDSTRÖM, 1970; WÄPPLING et al., 1972; ANDERSSON and CALLMER, 1974; LUNDSTRÖM, 1974; PITON et al., 1974; MATTES et al., 1976).

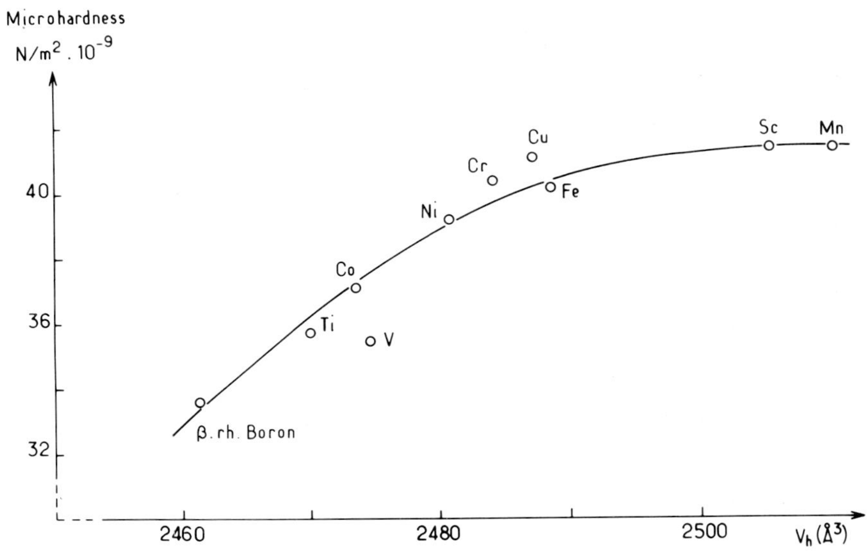

Fig. 14. Variation of the Vickers microhardness (measured under a load of 50g.) as a function of the hexagonal cell volume for solid solutions of 3d transition metals in β-rh. boron. (From CARLSSON and LUNDSTRÖM 1970)

The nature of interstitial sites and their degree of occupancy have been ascertained by LUNDSTRÖM and coworkers, from a series of studies with single crystals of CrB_{41}, MnB_{23} and CuB_{28} (Table 2). At the same time this work confirms the model of HOARD et al. (1970) for β-rh. boron.

In the structure model proposed by HOARD et al., for what is supposedly non-doped β-rh. boron, the G-type site (in the notation of LUNDSTRÖM) is occupied at approximately 33% by interstitial B(16)-type atoms. These atoms also occur in CuB_{28} (20.9% site occupancy) but not in CrB_{41} nor in MnB_{23}.

In solid solutions formed by β-rh. boron with 3d-elements, the metal atoms (more or less ionized depending on the specific element) occupy two or three types of interstitial sites, as shown in Table 2. In all cases occupancy of *interstitial sites* is never complete, in the same way that the occupancy of 18(h) positions by B(13) boron atoms in the B_{10} sub-unit is only partial (for which the following occupancies have been found: CrB_{41}:71.7%; MnB_{23}:65.0%; CuB_{28}:61.1% and β-rh. boron: about 2/3). It seems that this fractional occupancy is the result of both geometrical and electronic factors. Mössbauer studies carried out on the solid solution of iron in β-rh. boron have confirmed that iron, in both oxidation states Fe^{2+} and Fe^{3+} occupies several kinds of interstitial sites (ANDERSSON and LUNDSTRÖM, 1970; WÄPPLING et al., 1972; ANDERSSON and CALLMER, 1974; LUNDSTÖM, 1974).

Presence of interstitial metal atoms in the voids formed by icosahedra causes a large expansion of the lattice parameter of β-rh. boron. This is accompanied by an increase in cohesion, as indicated by the variation of the microhardness with the unit cell colume (Fig. 14; CARLSSON and LUNDSTRÖM 1970).

From a practical point of view, ease with which foreign atoms can be inserted in interstitial sites of β-rh. boron makes for great difficulty in achieving high purity and also accounts for the great range of variability of some of its physical properties, electrical properties, for example (HAGENLOCHER, 1960; GEIST, 1965).

c) Crystal Structures of Tetragonal Phases

Although it was naturally assumed by HOARD et al. (1958) that the α-tetr. (or tetr.-I) structure was that of an allotropic form of pure boron, it now appears, after considerable thorough further research, that small amounts of impurity atoms are required for stabilizing the tetragonal arrangement of B_{12} icosahedra. The structure of the second tetragonal variety (referred to variously as β-tetr., tetr.-II or tetr.-III) has not yet been determined in spite of extensive research on the part of many investigators.

α) On the Existence and the Crystal Structure of α-Tetragonal Boron

The HOARD, HUGHES and SANDS Structure. The crystal structure of a tetragonal phase considered initially to be pure boron, was described in detail by HOARD et al. (1958), on the basis of an X-ray study of single crystals prepared previously by LAUBENGAYER et al. (1943). The preparative method, which has now become standard, involves reduction of BBr_3 by hydrogen on a hot tantalum filament. The crystals were obtained either in the shape of needles (type-I) or as hexagonal plate (type-II) and, interestingly enough, their diffraction patterns showed significant differences both in spacing and in intensity for corresponding reflections.

The lattice parameters showed considerable variation from one crystal to another, with average values of: a = 8.75 Å c = 5.06 Å. The space group was taken to be $P4_2/nnm$, although the authors noted the existence of a few spots of very low intensity that were inconsistent with this space group, and which were attributed to parasitic double reflection phenomena. The density measured by floatation (d = 2.29) indicated 50 boron atoms per unit cell, assuming that this phase did not contain appreciable amounts of impurity.

The unit cell, represented in projection on the ab plane in Figure 15, mainly contains four quite regular B_{12} icosahedra, centered at 4(c) positions with symmetry 2/m (1/4, 1/4, 1/4; 3/4, 3/4, 1/4; 3/4, 1/4, 3/4 and 1/4, 3/4, 3/4) and having one of their pseudo-five-fold axes directed parallel to c-axis. The icosahedra are bonded to each other either by direct B-B bonds, or through isolated atoms occupying 2(b) positions (0,0,1/2; 1/2, 1/2, 0) considered as boron atoms by HOARD et al. (1958). The three-dimensional skeleton resulting from association of the icosahedra contains several kinds of interstitial voids, some of which are sufficiently large to accomodate either boron or metal atoms (2(a) sites at 0 0 0, 1/2 1/2 1/2, for instance).

The icosahedra are quite regular in size, as evidenced by the weak variations in intraicosahedral bond length (from 1.785 to 1.846 Å for a type-I crystal). Each icosahedron has 12 outwardly directed bonds: 10 towards boron atoms belonging to adjacent icosahedra and two towards isolated atoms in 2(b) positions. Of these twelve bonds: four are abnormally long (B_1-B_1:1.858 Å) for *intericosahedral* bonds; six are, as expected, considerably shorter ($4B_2$-B_2:1.658 Å and $2B_4$-B_4:1.709 Å) than the average length (1.806 Å) of intraicosahedral bonds; and two, that are directed towards atoms in 2(b) positions, are quite short (B_3-B_5:1.601 Å) for B-B bonds. The external bonds of the icosahedron do not equally fulfill the ideal orientation criterion defined by HOARD and HUGHES (1958). Thus, eight of the twelve bonds are oriented satisfactorily close to the corresponding pseudo-five-fold axes - including those directed towards isolated atoms - but four of them deviate by more than 20°.

HOARD et al. (1958) noticed on Fourier sections residual electron densities that they attributed to a partial occupancy of interstitial sites in the network formed by B_{12} icosahedra. Thus, an excess of electron density (relative to the value expected for a boron atom) occurred in projection on the bc plane in positions corresponding to projections of isolated boron atoms occupying 2(b) sites. It was attributed by HOARD et al. to a partial occupancy of 2(a) interstitial sites (0, 0, 0; 1/2, 1/2, 1/2) or 4(g) sites (0, 0, z; 0, 0, $\bar{z}$; 1/2, 1/2, 1/2 + z; 1/2, 1/2, 1/2 - z with z $\leqslant$ 0.183) by boron atoms. Likewise, weak electron density peaks at 0, 1/2 and 1/2, 0 were attributed to a partial occupancy by boron atoms of 4(d) sites (0, 1/2, 1/4; 1/2, 0, 1/4; 0, 1/2, 3/4 and 1/2, 0, 3/4) or 4(c) sites (0, 1/2, 0; 1/2, 0, 0; 0, 1/2, 1/2 and 1/2, 0, 1/2). The degree of occupancy of interstitial sites by boron atoms was determined from difference Fourier sections ρ_o-ρ_c. The resulting values was: type-I crystals, 1/4 at 2(a) or 1/8 in 4(g), 1/16 in 4(c); type-II crystals, 1/16 in 4(g), 1/8 in 4(c). Taking into account the interstitial atoms, the reliability factor R was 11.4% (type-I crystal) and 13.5% (type-II crystal).

Thus, the structural formula associated with the model of HOARD et al. is: $4B_{12} + 2B_2 + 0.37$ to 0.75 B, taking into account interstitial atoms.

Critical Study of the Model of J.L. HOARD et al. (1958). A surprising feature for the structural results reported by HOARD et al. for the α-tetr. phase, is that the lattice parameters of an element, and hence its density, should show such a large variation from one crystal to another. For exemple, the c-parameter of type-I crystal varies from 5.030 ± 0.001 to 5.090 ± 0.003 Å.

Furthermore, the tetrahedral surroundings of B_5 atoms situated in 2(b) positions is unusual for boron. The length (1.60 Å) of the four B_3-B_5 bonds is also particularly short for a B-B bond, although NASLAIN et al. (1976) found a length of the same order for B_3-B_6 bonds in NaB_{15}, again between an isolated B_6 boron atom and boron atoms belonging to icosahedra (Fig. 3).

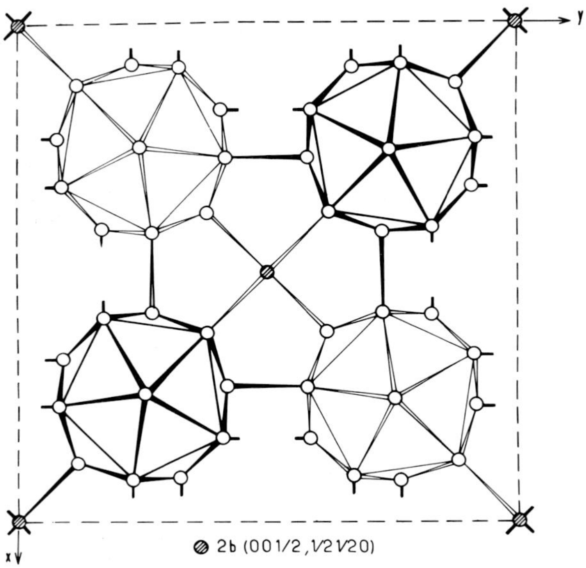

Fig. 15. Crystal structure of α-tetr. phases in projection on (001). (Based on data of HOARD et al., 1958)

LONGUET-HIGGINS and ROBERTS (1955), and also LIPSCOMB and BRITTON (1960), showed that 160 electrons per unit cell were necessary to satisfy: the internal bonding within the icosahedra (4 x 26 = 104 electrons); their external bonds (4 x 12 = 48 electrons); and the bonds shared by atoms in 2(b) positions (2 x 4 = 8 electrons). In the case of the formula providing the largest number of electrons, $4B_{12} + 2B_2 + 0.75$ B, there are still only 152 electrons, which means a deficiency of eight electrons per cell.

These observations, together with the existence of residual electron density peaks noted by HOARD et al. (1958), suggest that the atoms occupying 2(b) positions could actually be more electron-rich than boron atoms, for example, carbon, nitrogen or oxygen, which are common impurities of boron that are difficult to eliminate.

The Structural Investigations of PLOOG et al. OLEMPSKA et al. (1966) were among the first to express doubts about the existence of an α-tetr. pure boron variety, by showing that the crystals of the α-tetr. phase were easily obtained between 1300 and 1800°C only in very particular deposition conditions (slow deposition rate and addition of traces or air to the gas phase).

In 1968, AMBERGER an PLOOG showed that the reduction of BBr_3 by hydrogen on a hot substrate (between 950 and 1350°C) never resulted in the formation of α-tetr. boron, when it was carried out with highly pure reagents. On the other hand, an α-tetr. *phase* could be prepared reproducibly as large crystals by doping the gas phase with small amounts of methane; furthermore, the crystal quality and size improved with increasing CH_4 content (AMBERGER and PLOOG, 1968). These findings were confirmed later by PLOOG and AMBERGER (1971), and by AMBERGER and SCHMIDT (1971), they demonstrated that between 900 and 1600°C, the hydrogen reduction of highly pure boron halides produced only the α-rh., β-rh. and β-tetr. forms of boron. α-tetr. phases, $B_{48}B_2N_{1-2}$, $B_{48}B_2C$ and $B_{48}B_2C_2$, were obtained reproducibly by using BBr_3-H_2 mixtures doped with small quantities of nitrogen, CH_4 or $CHBr_3$. PLOOG et al. (1972) extended the results of the chemical analysis with a complete single crystal structure determination of $B_{48}B_2C_2$ and $B_{48}B_2N_2$, thereby identifying the interstitial atoms and proving definitely that the α-tetr. icosahedral framework was stabilized by carbon and nitrogen atoms occupying 2(b) positions.

$B_{48}B_2C_2$ and $B_{48}B_2N_2$ crystallize in the tetragonal system with well established lattice parameters, given in Table 3. Both structures, although assigned the space group $P4_2/nnm$, could in fact belong to space group $P\bar{4}2m$, as had been considered already by HOARD et al. (1958). The measured crystal densities are in good agreement with values calculated for the $B_{48}B_2X_2$ formulation (with X = C or N) and differ markedly from that reported by HOARD et al. (Table 3; PLOOG et al., 1972; WILL and KOSSOBUTZKI, 1976).

Table 3. Crystallographic data for the tetragonal phases $B_{48}B_2X_2$ (X = C, N) and for "α-tetr. boron". (From HOARD et al., 1958; PLOOG et al., 1972)

	Temperature for synthesis (°C)	Lattice parameters		Density		Chemical
		a (Å)	c (Å)	obs.	calc.	formula
$B_{48}B_2C_2$	1200	8.722±0.001	5.080±0.001	2.43	2.426	$B_{50}C_{1.9}$
$B_{48}B_2N_2$	1400	8.646±0.001	5.127±0.001	2.46	2.463	$B_{50}N_{1.8}$
"$B_{48}B_2$"	1300	8.74-8.77	5.03-5.09	2.31	2.34-2.29	

PLOOG et al. (1972), on the basis of precise measurements of diffracted intensities of $B_{48}B_2C_2$ and $B_{48}B_2N_2$ crystals, calculated the absolute electron density at the location of the different interstitial sites and determined, consequently, the kind of atom and its fractional occupancy in these sites (Table 4). The atoms in 2(b) positions, that play an important role in stabilyzing the B_{12} icosahedral framework, were found to be only carbon or nitrogen atoms and not boron; this could account for the anomalies described in the previous paragraph. Whereas in $B_{48}B_2C_2$ the isolated boron atoms statistically occupy three types of sites, 2(a), 8(c) and 8(h); in $B_{48}B_2N_2$, on the contrary, only one position 8(i) is occupied.

Table 4. Absolute electron density in the interstitial sites of $B_{48}B_2C_2$ and $B_{48}B_2N_2$. (From PLOOG et al., 1972)

Positions	Atoms	Coordinates x	y	z	Absolute electron densities $B_{48}B_2C_2$	$B_{48}B_2N_2$
2 (a)	B	0	0	0	0.64	0
8 (i)	B	0.4330	0	0	0.49	-
"	B	0.4299	0	0	-	1.15
8 (h)	B	0	1/2	0.8557	0.56	-
2 (b)	C or N	0	0	1/2	5.44	6.54

Conclusion. These recent studies clearly establish then, that the α-tetr. arrangement of B_{12} icosahedra cannot exist without the presence of stabilizing atoms such as carbon and nitrogen. The work of PLOOG et al. (1972) relative to the stabilizing effect of carbon and nitrogen suggests that the same conclusions could be made for BeB_{12}, (Be, Si)B_{12} and NiB_{25} phases that have the same space group and similar parameters (BECHER, 1960; DECKER and KASPER, 1960; BECHER, 1963; BECHER, 1965; BECHER and NEIDHARD, 1968).

β) β-Tetragonal Boron

TALLEY et al. (1960a) reported the existence of another tetragonal form of boron, obtained by reduction at 1270°C of highly pure boron tribromide on tungsten or rhenium filaments. The spectrographic analysis of the resulting polycristalline material indicated the presence of less than 0.005% aluminum and a total content less than 0.02% for other impurities which were: Ba, Sr, Ca, Cu, Fe, Mg and Si. The Debye-Scherrer pattern was indexed in the tetragonal system with lattice parameters (a = 10.12 Å; c = 14.14 Å) close to those of α-AlB_{12} (a = 10.16 Å; c = 14.28 Å; space group $P4_22_12$?). The density of the material, d = 2.364 ± 0,005, corresponds to 192 boron atoms in the unit cell (TALLEY et al., 1960a, 1960b; KOHN, 1960; HALLA and WEIL, 1939).

Just as for α-tetr. phases, there are other boron rich borides (e.g. BeB_6 and $LiB_{10.85}$, in addition to α-AlB_{12}) whose unit cells are very similar to that of the phase studied by TALLEY et al. In view of the outcome of the thorough studies of α-tetr. phases, it is an open question whether the TALLEY et al. boron is a real allotropic form of the pure element or whether it is another boron framework stabilized by small amounts of foreign atoms (HALLA and WEIL, 1939; KOHN, 1960; SANDS et al., 1961; SECRIST, 1967; SULLENGER et al., 1969).

In fact, AMBERGER and PLOOG (1968) showed that the reduction of BBr_3 by hydrogen in contact with a tantalum filament heated to 950 - 1350°C, leads mainly to the formation of a β-tetr. phase when highly pure reagents are utilized. The product is microcrystalline when prepared at low temperatures but is obtained in the form of large crystals at higher temperatures. A study of single crystals confirmed the space group and parameters proposed by TALLEY et al., and furthermore implied the presence of B_{12} icosahedra in the structure. Simultaneously, BOIRET, using similar preparative conditions established that the β-tetr. and vitreous boron are the main products of the hydrogen reduction of highly pure boron tribromide at 1200°C (AMBERGER and PLOOG, 1968; AMBERGER and PLOOG, 1971; BOIRET, 1969).

SULLENGER et al. (1969) noted the presence of crystals of a tetragonal phase in products resulting from passing pulverized boron through an argon plasma at very high temperatures, the main component being the β-rh. boron. X-ray analysis of these crystals confirmed the findings of TALLEY et al.: a = 10.061 ± 0.005 Å; c = 14.210 ± 0.005 Å; possible space groups: $P4_12_12$ and $P4_32_12$; measured density: 2.367 ± 0.002.

All of this work seems to indicate that a β-tetr. form of boron can indeed exist, since it has been prepared by several investigators under different conditions and apparently in a state of high purity. However, SULLENGER et al. (1969) noted that when several successive passes of boron in the plasma torch were made, eliminating the β-tetr. phase after each pass (as well as crystals belonging to a cubic phase, which is discussed later), the resulting product was only β-rh. boron. This experiment would give reason to suspect the presence of foreign atoms in the β-tetr. form; only a very precise determination of the crystal structure of this phase will make it possible to determine whether this is really an allotropic variety of elemental boron.

Single crystals of β-tetr. boron have a metallic aspect; on the contrary, thin sheets are red in transmitted light. According to TALLEY et al., this form is stable well above 1400°C; however, it transforms to β-rh. boron after melting and further solidification. On the other hand, the studies of SULLENGER et al.(1969) suggest that β-tetr. boron could be stable at very high temperatures, but could be obtained at room temperature only by drastic quenching. In all cases, this form seems to be obtained only under very special experimental conditions. Up to the present, no one has reported the existence of a reversible solid state transformation, which would definitely establish a thermodynamic stability range for β-tetr. boron.

d) Other Crystalline Forms of Boron

Among the many phases described as being new crystalline forms of boron, two seem particularly noteworthy: that obtained by WENTORF (1965) under high pressure, and that prepared by SULLENGER et al. (1969) at very high temperatures in an argon plasma.

Among the various crystalline forms of boron, α-rh. boron has the highest density ($2.46 g/cm^3$); however, even in this case, the percentage of space occupied by atoms is very low (37%), which results from the peculiar bonding characteristics of the icosahedral units which in part have strong inter-icosahedral covalent bonds and also weak three-center bonds with large interatomic spacing. It would seem reasonable that application of high pressure would produce, in analogy to the case of hexagonal boron nitride and graphite, a boron form that is considerably denser. Experiments with the view towards achieving this result were made by WENTORF (1965).

By submitting either amorphous or β-rh. boron to pressures higher than 80kb and to temperatures above 1800°C, WENTORF obtained a product with a density of 2.46 - 2.52 which transforms into β-rh. boron when annealed at 1500°C under a pressure of 30kb. Although the possibility of contamination by materials from the container (e.g. titanium, tantalum or magnesia) cannot be eliminated, a priori, the fact that the high pressure phase reversibly transforms to β-rh. boron might indicate that it does correspond to an elemental form of boron. Further experiments of this kind would be necessary and desirable.

At the same time as they obtained β-tetr. phase crystals, SULLENGER et al. (1969) also abserved the presence of cubic crystals in boron passed through the plasma torch. From a study of single crystals they assigned to this new phase the space group Pn3n with an a - parameter of 23.472 ± 0.006 Å. The measured density (2.367), close to that of β-tetr. and β-rh. boron crystals, called for 1708 boron atoms in the unit cell. A comparison with the YB_{66}-type phases studied by KASPER et al. (a = 23.440 ± 0.006 Å, space group Fm3c, 1584 boron atoms and 24 yttrium atoms per unit cell) obviously suggest itself (RICHARDS and KASPER, 1969; SULLENGER et al., 1969, NASLAIN et al., 1971).

The fact that both these phases are cubic with pratically the same parameter implies that they have the same boron skeleton in spite of a difference in space groups (RICHARDS and KASPER also noted for YB_{66} the presence of diffraction spots of very low intensity, inconsistent with a face-centered cell). It is quite likely that the structural differences concern the way the voids resulting from the association of $B_{12}(B_{12})_{12}$ giant clusters are filled by additional boron or metal atoms. According to SULLENGER et al. (1969), the powder pattern of this cubic form of boron ressembles that of the PuB_{100} cubic hecto-boride studied by EICK (1965) more than that of YB_{66}.

Here again, as with the forms discussed before, further work involving single crystal structure determination coupled with trace analysis would be needed to establish whether this cubic phase is a new boron allotrope or whether traces of foreign atoms are required for stabilizing the boron framework of $B_{12}(B_{12})_{12}$ clusters.

e) Conclusions Regarding Boron Polymorphism

This review emphasizes the complexity of the problems in boron polymorphism and the necessity for coupling single crystal structure determination with analysis of impurities present in trace amounts. Actually, it seems that only α-and β-rh. modifications, and possibility the β-tetr. form, are generally accepted as being real allotropes of pure elemental boron.

The problem of boron polymorphism is better considered in the more general framework of the crystal chemistry of boron-rich phases. It is readily apparent that for each boron modification there is a family of closely related boron-rich borides and it is probable that reciprocally, for each boron-rich phase there can occur a boron "monotrope" which is more or less stable.

Thus, corresponding to α-rh. boron, there is the family of binary compounds of boron with carbon, silicon, phosphorus, arsenic, oxygen and possibly sulfur. Whereas the thermal stability of α-rh. boron is low, the corresponding boron-rich phases are, on the contrary, very stable. This increase in stability of the boron framework is due to the formation of additional covalent bonds - as pointed out previosuly - with ideal orientation, between some of the boron atoms of the icosahedra and the foreign atoms occupying the voids within the icosahedral skeleton.

Likewise, corresponding to β-rh. boron, there are many interstitial solid solutions of the transition elements. Figure 14 shows that the microhardness, and consequently the cohesive energy, is considerably increased by insertion of foreign atoms into the β-rh. boron lattice.

Conversely, it is likely that, under special experimental conditions, there may occur B_{12} icosahedral networks totally - or nearly totally in certain cases - free of the foreign atoms which stabilized them and which correspond to the various structure-types of boron-rich borides. This situation occurs for Na_xB_{15} ($0<x\leq 1$); for α-AlB_{12} and the

related β-tetr. boron; for YB_{66} and the cubic "boron" of SULLENGER et al. (1969) and finally for the α-tetragonal $B_{48}B_2M_2$ (M = C, N).

At this point of our conclusion, we wish to assert that any kind of B_{12} icosahedral framework, be it with orthorhombic tetragonal, rhombohedral or cubic symmetry, is always stabilized by the introduction of foreign atoms even in trace amounts. The readiness with which foreign atoms may be inserted depends on the number and size of the sites created by the association of icosahedra with their directed covalent bonds. The stabilizing effect of foreign atoms can be understood from the fact that in all cases, they contribute either in completing the coordination of the boron atoms in the icosahedra, or in giving the lattice a stable electronic structure by means of electron transfer.

Foreign atoms play only a minor role in the β-rh. network of B_{12} icosahedra, which is probably the most stable framework formed by boron atoms in the absence of foreign atoms. On the contrary, in the case of α-rh. (e.g. $B_{13}C_2$), orthorhombic (e.g. NaB_{15} or $MgAlB_{14}$) and α-tetr. (e.g. $B_{48}B_2C_2$) networks, the stabilizing effect of foreign atoms is determinative. YB_{66} and α-AlB_{12} could be regarded as intermediate cases in as much as the corresponding arrangements of icosahedra can be obtained at high temperatures, *apparently* without any stabilizing atoms (for example, β-tetr. boron).

The stabilizing effect of impurity atoms upon B_{12} icosahedra arrangements has the following immediate consequence: it will be very difficult to obtain the element boron itself in a state of high purity, because of the strong affinity of boron for a vast number of elements.

3. Methods of Preparing Crystalline Boron

Many methods have been proposed to prepare boron, but only those based on thermal decomposition or hydrogen reduction of a volatile boron compound (halide or borane) are dependable for obtaining crystalline boron of high purity.

Methods based on the reduction of B_2O_3 by a very electropositive metal, magnesium for example (MOISSAN process), or on the reduction of an alkali-metal fluoborate (KBF_4) by an alkali-metal yield a very impure boron, that is partly oxidized, but that can be utilized as a starting material for preparation of boron halides, due to its great chemical reactivity (MOISSAN, 1892; WEINTRAUB, 1909; KAHLENBERG, 1925; SOWA, 1949; SPEVACK, 1954; MIKHEEVA et al., 1957 a, b, c; SHAMRAI et al., 1957; HAAG, 1957; YANNACAKIS and NIES, 1960; FINN et al., 1960; SCHOTT and HANSLEY, 1963).

Likewise, the cathodic reduction of melts containing a borate (ANDRIEUX process) or a fluoborate (COOPER process) gives an amorphous or poorly crystallized boron, that is too impure (B% 95 to 97) for most applications (ANDRIEUX, 1929; COOPER, 1945a, b; COOPER, 1951 a, b; ANDRIEUX and DEISS, 1955; COOPER, 1959; SAMSONOV et al., 1960; COOPER, 1961).

Purification of these products by outgasing in high vacuum, at a temperature close to the melting point of boron, eliminates only the most volatile impurities and leaves a sintered or partly melted residue of β-rh. boron, which is chemically unreactive and difficult to pulverize because of its hardness (SOWA, 1949; SPEVACK, 1954; HAAG, 1957; FINN et al., 1960; SCHOTT and HANSLEY, 1963). These methods of preparing boron have been reviewed in great detail by NEWKIRK (1961, 1964) and more recently by the author (NASLAIN, 1972).

Here we shall study particularly the preparation of crystalline forms of boron from the halides BX_3 (X = Cl, Br, I) or through solidification from a liquid phase.

a) Preparation of Crystalline Boron from Boron Halides

α) *Advantages of Using Halides as a Boron Source*

We have discussed how certain elements (C, N, Si, O, Al, Mg, etc ...) are able to stabilize the various boron icosahedral frameworks that characterize the different boron modifications. One of the advantages of using BX_3 halides for boron synthesis is precisely that one avoids the introduction of such foreign elements into the reaction environment. Furthermore, BX_3 halides are liquid at temperatures close to room temperature; this property enables one to achieve a very high degree of purity for them by well-known standard techniques, such as

Table 5. Some physico-chemical properties of boron halides. (From URRY, 1967)

BX_3	Boiling point at 1 bar (°C)	Melting point (°C)	Weight ratio B/BX_3
BF_3	- 99.9	- 127.1	0.1594
BCl_3	12.5	- 107	0.0923
BBr_3	91.3	- 46	0.0432
BI_3	210	49.9	0.0276

fractional distillation, gas phase chromatography or zone melting (Table 5).

The boron halide method is based on a gas phase reduction, according to the following reaction:

$$2BX_{3\,(g)} + 3H_{2\,(g)} \longrightarrow 2B_{(S)} + 6HX_{(g)} \quad (1)$$

The thermodynamic study of this reaction was made by STERN and LYNDS (1958), and more recently by AUGUSTIN and VUILLARD (1974). The calculation of the free energy of the reaction as a function of temperature shows that the reduction of BF_3 by hydrogen would require temperatures that are too high for a technically feasible method of preparing boron (Fig.16). On the other hand, the reduction of BCl_3 and BBr_3 is thermodynamically possible at temperatures that are more accessible. The reduction of BBr_3 starts at 650°C, although that of BCl_3 requires temperatures above 1000°C. For BI_3, it is generally preferable to utilize simple thermal decomposition in high vacuum rather than reduction by hydrogen.

At the temperatures of reduction all chemical species inside the reaction vessel are gaseous except boron which is solid. Another advantage of the halide method is then the extreme ease with which the boron product can be separated. The reaction is frequently carried out in contact with a substrate maintained at the reaction temperature; it can also be carried out within a fluidized bed of boron granules or even in an argon-hydrogen plasma.

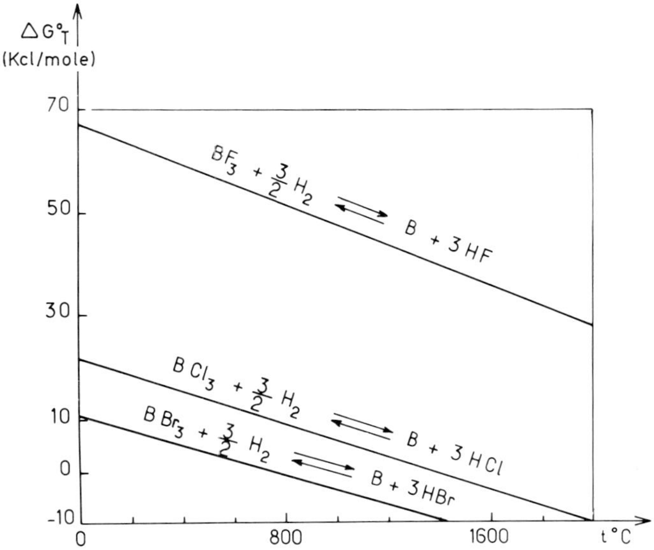

Fig. 16. Variation of the standard free energy of the reactions: $BX_3 + 3/2\ H_2 \rightleftharpoons B + 3HX$ as a function of temperature. (From STERN and LYNDS, 1958)

Another advantage of this method is the capability of varying the parameters that control the boron deposition, thereby optimizing the conditions for preparing a desired variety of boron. The most important parameters are the substrate temperature and the deposition rate. In general, α-rh. and amorphous boron are formed at lower temperatures ($t < 1000$ to $1200\,^\circ C$), whereas β-rh. boron is predominant at higher temperatures ($t \gtrsim 1300$ to $1400\,^\circ C$). For intermediate temperatures ($1000 < t < 1400\,^\circ C$ approximately), the reduction of boron halides by hydrogen leads mainly to the β-tetr. form when highly pure reagents are utilized, and to α-tetr. phases stabilized by impurity atoms otherwise, as we have already indicated. However, it is to be noted that the β-rh. form has been found as a minor component in deposits of amorphous boron or of α-rh. boron, for much lower reduction temperatures (NASLAIN et al., 1970).

Practical application of the reaction can be made in various ways according to the nature of the substrate and to the method of heating. The first experiments were carried out with an electric arc maintained between two cooled electrodes (WEINTRAUB, 1909; KAHLENBERG, 1925; HACKSPILL et al., 1931; CUEILLERON, 1944; HACKSPILL and CUEILLERON, 1944). Utilization of a quartz tube heated externally gives a product of reasonable purity if the reduction is carried out at low temperatures, but that limits the use of this technique to BBr_3 (KIESSLING, 1948; BECHER and SCHÄFER, 1960; NIEMYSKI and OLEMPSKA, 1962; HAGENMULLER and NASLAIN, 1965; NASLAIN et al., 1970). The reduction can also be carried out in a bed of porous boron granules kept in a state of fluidization by the BX_3/H_2 gaseous mixture, or with a plasma torch (MURDOCH and HAMBLYN, 1969; NASLAIN et al., 1972; NASLAIN et al., 1973). But the method more generally used and that yields the purest boron is the reduction of BBr_3 or BCl_3 halides on a filament heated by the Joule effect (DUPUY and HACKSPILL 1923; FREYMANN and STIEBER, 1934; LAUBENGAYER et al., 1943; STERN and LYNDS, 1958; POWELL et al., 1960; TALLEY et al., 1960 b; HINZ and WIRTH, 1965; HAGENMULLER and NASLAIN, 1965).

Impurities eventually present in boron obtained by hydrogen reduction of BX_3 halides have their origin in the starting reagents themselves and in the substrate used for boron deposition.

Except for hydrogen, which is an easily purified gas, the main impurity source is the boron halide itself. However, BCl_3, and mostly BBr_3 and BI_3, are compounds that can be made of very high purity by fractional distillation, by chromatography in gas phase, or by zone melting. Some impurities, such as carbon, for example, however, are very difficult to eliminate from boron halides.

The substrate on which the boron deposition is carried out is another important source of boron contamination. In the case of deposition in quartz tubes, it is minimized by working at the lowest possible temperature and in large diameter tubes. Contamination by reactor walls can also be reduced, by carrying out the reaction in contact with porous boron granules in a fluidized bed, or even totally eliminated by working with a plasma torch. In the case of deposition upon filaments, contamination by the substrate is minimized by using wires of very small section or even totally avoided by making the filaments or rods of boron.

The hydrogen reduction of boron halides is without doubt one of the best methods of preparing boron, when carried out with highly pure reagents and in an appropriate apparatus. Purity of better than 99% is being obtained currently and samples containing less than 500 ppm carbon and only a few ppm aluminum, silicon, calcium and iron were obtained by ARMINGTON et al. (1965) by using the method of deposition upon filaments.

However interesting, the use of boron halides is not without some disadvantages. They are very corrosive and expecially sensitive to moisture which makes it necessary to carry out operations in a strictly dry atmosphere. Furthermore, BX_3 halides are characterized, as shown in Table 5, by very low B/BX_3 weight ratios, especially for the most interesting cases of BBr_3 (0.0432) and BI_3 (0.0276). In the most favorable case of a quantitatively complete reduction, the preparation of 100g boron from BBr_3 requires 2300g of BBr_3.

Other volatile boron compounds that do not present this drawback can be used in preparing highly pure boron. Thus boron hydrides, and especially B_2H_6, are characterized by B/B_nH_m weight ratios much more favorable than the B/BX_3 ratios (0.781 for B_2H_6 for example). Furthermore, B_2H_6 is readly converted into boron by simple thermal decomposition on a hot filament. With this method, HINZ and WIRTH obtained boron samples that after zone melting apparently contained only 30ppm carbon and no impurities detectable by spectrographic analysis. However, the toxicity and the risks of explosion in manipulating large quantities of B_2H_6, make this method unattractive in spite of its real advantages.

β) Preparation and Purification of Boron Halides

Choice of a Boron Halide. BCl_3 is seldom utilized in preparing highly pure boron although it has one of the highest boron contents among the trihalides (Table 5). BCl_3 is reduced by hydrogen only above $1000^{\circ}C$ and its reduction often leads to the formation of chloroboranes that can cause explosions. Moreover, purifying BCl_3 by fractional distillation means using either refrigerated apparatus or cold rooms (B.P. = $12.5^{\circ}C$ under normal pressure). And one of the main impurities in BCl_3, carbon, in the form of $COCl_2$ (B.P. = $8.3^{\circ}C$) is practically impossible to eliminate by distillation due to the proximity of boiling points.

For these reasons, it is generally preferable to use BBr_3 or BI_3. Boron tribromide is easier to work with than the tri-iodide, since it is a liquid over a large temperature range on either side of room temperature (Table 5). On the other hand, BBr_3 is properly purified only by fractional distillation, whereas BI_3 can be purified also by sublimation or zone melting. It is generally accepted that the purity of boron obtained from BI_3 is quite higher than that from BBr_3, especially in regard to carbon content. In both cases, the reduction of boron halides by hydrogen can be carried out at temperatures far below 1000°C.

Preparation of BBr_3 and BI_3 by Direct Synthesis. BBr_3 can be made similarly to BCl_3 by reacting the corresponding halogen with B_2O_3 in the presence of carbon, or by reacting BF_3 with $AlBr_3$ (GAMBLE et al., 1940; NASLAIN, 1972). The boron tri-iodide was prepared by W.C. SCHUMB et al. (1949), by reacting iodine with $LiBH_4$. The most appropriate method, for obtaining the great amounts of BBr_3or BI_3 that are necessary for the preparation of boron in a state of very high purity, is, however, direct synthesis from the elements.

Preparation of BBr_3 or BI_3 by direct synthesis is based upon the reaction of the corresponding halogen vapor with boron heated to a moderate temperature within a vertical reactor, generally made of silica. It can be done either with a fixed bed of boron pellets, or by the fluidized layer technique. The halogen is placed in contact with boron either by distillation in vacuum, or by means of a carrier gas (helium or argon) at atmospheric pressure.

The yield of the synthesis is quantitatively complete for BBr_3, whereas even under the most favorable conditions, it is only 60% for BI_3 (SCHUMB et al., 1949; McCARTY et al., 1958; McCARTY and CARPENTER, 1960; AMBERGER and DIETZE, 1965; HAGENMULLER and NASLAIN, 1965; ARMINGTON et al., 1965; NASLAIN, 1972). The reaction rate at a given temperature depends on the nature of the boron and on the technique used. Amorphous boron prepared by the MOISSAN process, for example, is a prime product for the synthesis of boron halides because of its great chemical reactivity. Thus, ARMINGTON et al. (1965) report that for the same temperatures, using the fixed-bed technique, the yield of BI_3 increases from only 10% for crystalline boron to 50 or 60% for amorphous boron. Use of crystalline boron is, however, possible if working with a fluidized bed.

Although the reaction of bromine with boron takes place at 420°C, the synthesis of BBr_3 is generally carried out between 700 and 850°C. For BI_3, the yield of the reaction of iodine with boron depends on the temperature and goes through a maximum around 850-900°C (Table 6) (McCARTY and CARPENTER, 1959; ARMINGTON et al., 1965). The action of iodine on boron leads, even under optimum conditions, to a triiodide mixed with a relatively large amount of free halogen.

Table 6. Yield of the synthesis of BI_3 and silicon content of the resulting product as a function of temperature. (From ARMINGTON et al., 1965)

Temperature (°C)	650	750	800	850	900	1000	1100
Yield (%)	8	43	55	62	57	47	35
Silicon content (ppm)	100	300	100	300	600	1000	1000

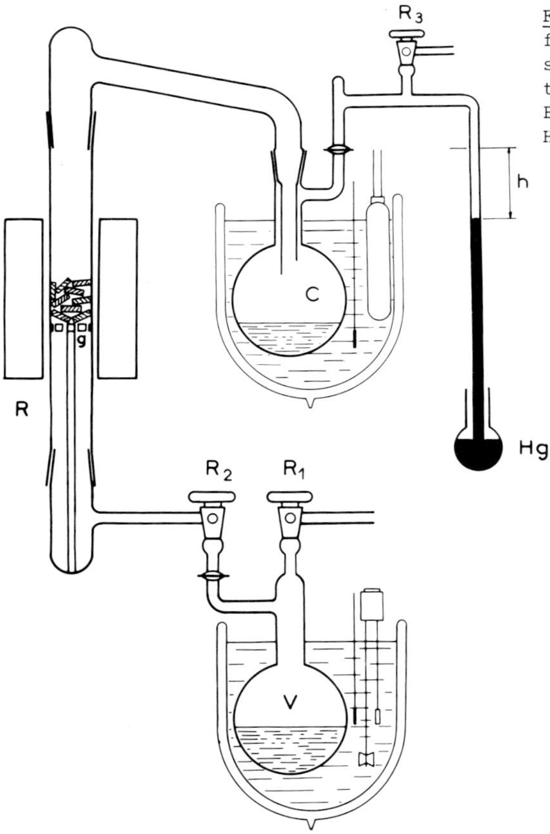

Fig. 17. Apparatus for BBr_3 synthesis from low purity boron and bromine: R, silica tube; g, silica grid supporting the boron; V, bromine vaporizer; C, BBr_3 condenser kept at -40°C. (From HAGENMULLER and NASLAIN, 1965)

Since the contamination is lower, the lower the temperature of the boron, it is best to utilize amorphous boron which reacts at lower temperatures than the crystalline forms of boron.

Figure 17 gives an illustration of the apparatus used by HAGENMULLER and NASLAIN (1965) for the synthesis of BBr_3 from amorphous boron. The halogen is distilled under reduced pressure (100 torr) through a fixed layer of boron pellets kept at 700-750°C on a silica grid (g). In order to obtain a total conversion of the bromine into BBr_3, a large boron excess is used: two to three times the stoichiometric quantity. BBr_3 is collected in a condenser (C) kept at -40°C. The rate of distillation of bromine is adjusted so that the product condensing in (C) stays colorless, i.e. devoid of free bromine. The crude product generally contains some solid impurities in suspension, which are easily eliminated by distillation under reduced argon pressure (NASLAIN, 1967; NASLAIN, 1970; NASLAIN, 1972).

Purification of Boron Halides. Purification of the boron halide is a fundamental step in the process of preparing pure boron by thermal decomposition or hydrogen reduction of BBr_3 or BI_3. Indeed, the purity of the resulting boron depends greatly on the efficiency of the methods for purifying the boron halide.

Table 7. Boiling points of boron halides and of the impurities that they may contain according to Weast, 1968. The boiling points are given for a pressure of 760 torr unless indicated by a number for the pressure given in parenthesis; (a) sublimation, (b) decomposition, (c) sublimation at high temperature

	BX_3	X_2	AlX_3	SiX_4	CX_4	COX_2	MgX_2	CaX_2	FeX_3	CrX_3	NiX_2	TiX_4	SnX_4	HgX_2	ZnX_2	SOX_2	CuX
BCl_3	12.5	-34.6	178[a]	57.6	76.8	8.3	1412	>1600	310[a]	1300[a]	973[a]	136.4	114.1	302	732	78.8 (746)	1490
BBr_3	91.3	58.8	263.3 (747)	154	189.5	64.5	(c)	810	(a)	(c)	(c)	230	202	322	650	138 (773)	1345
BI_3	209	184.4	360	287.5	170[b]	-	700[b]	-	(b)	350[a]	(c)	377.1	364.5	354	624	-	1290

The main impurity contents in boron halides prepared by direct synthesis are of two types: varying amounts of free halogen and traces of halides or oxi-halides originating from impurities in the starting boron or from chemical attack of the reaction vessel (Table 7).

Since the synthesis of BBr_3 is quantitatively complete the crude product generally contains only traces of free bromine that can be easily eliminated by simply stirring at room temperature, in the presence of mercury (HAGENMULLER and NASLAIN, 1965; NASLAIN, 1970). The crude triiodide, on the contrary, contains considerably more free iodine, due to the poor yield of the synthesis. SCHUMB et al. (1949) recommend reduction of free iodine by tin or mercury after dissolving the crude product in carbon disulfide. McCARTY and CARPENTER (1960) use fractional distillation, while ARMINGTON et al. (1964) use reduction by tin followed by distillation. The normal boiling points of iodine (184.35 °C) and of BI_3 (208-210 °C) being quite different, it appears to us that the most satisfactory method, which particularly allows the recycling of iodine that has not reacted, is fractional distillation with a very high performance column.

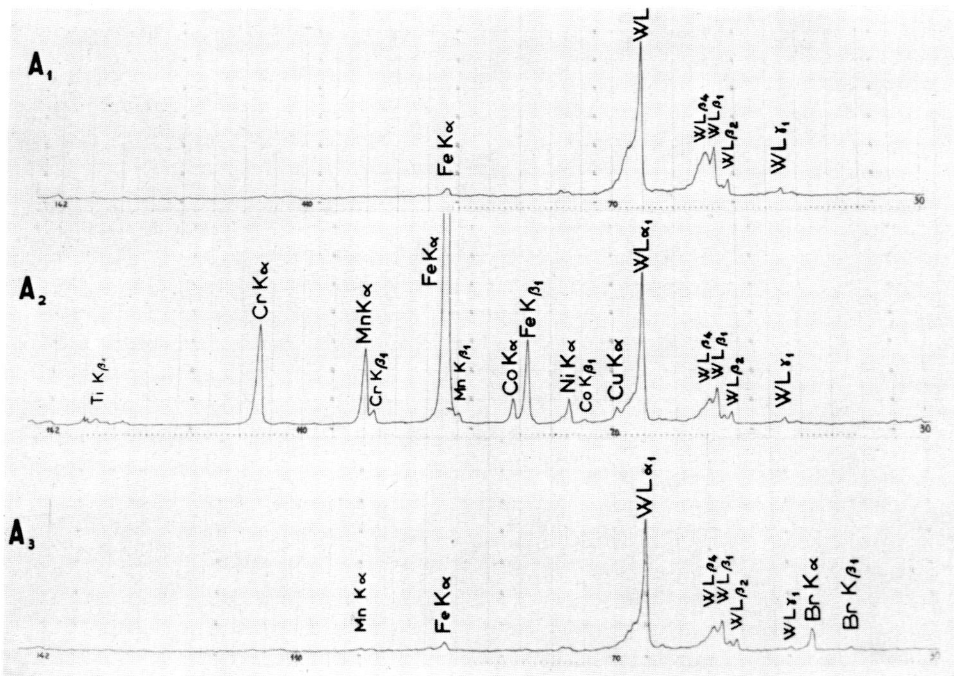

Fig. 18. X-ray fluorescence spectra of boron prepared by hydrogen reduction of BBr_3 upon filaments (A_1), in fused-silica tube at low temperature (A_3) and of low purity boron used for the synthesis of BBr_3 (A_2). The analysis was made with an X-ray tube containing a tungsten target. (From NASLAIN et al., 1970)

Boron utilized in the synthesis of BBr_3 or BI_3 generally contains many impurities which may form with bromine or iodine compounds of high volatility (Fig. 18). These compounds are condensed with the boron halides and constitute the main impurities of the crude products. Table 7 shows that most of these impurities can be separated from BBr_3 or BI_3 by fractional distillation except for carbon and silicon. The elimination of carbon and silicon, which can induce the formation of certain icosahedral frameworks and change profoundly the electrical properties of boron, is the main problem in the purification of boron halides.

Purification of boron halides is generally carried out by fractional distillation in an inert gas atmosphere within silica columns with 20 to 30 theoretical plates, with a reflux ratio of approximately 20 to 1. The central portion of the distillate then contains only a few ppm of the major impurities, including silicon.

It is interesting to study the case of carbon in as much as carbon is by far the most plentiful impurity of boron. It is well known that carbon is mainly present as COX_2 compounds in crude boron halides. $COBr_2$ is stable at the boiling point of BBr_3 (91.3 C under 1 bar). Its boiling point is sufficiently different from that of BBr_3 to allow its concentration significantly in the first fractions of the distillation. The situation is still more favourable in the case of BI_3, since no compound of iodine with carbon is stable at the normal boiling point of BI_3 (Table 7). Fractional distillation of BI_3 at atmospheric pressure leads to a rapid elimination of carbon which is concentrated in the lightes fractions of the distillate, as shown by ARMINGTON et al. 1970 (Table 8).

BI_3 can also be purified by sublimation or zone melting. The efficiency of these methods, with respect to elimination of silicon and carbon, does not seem to be much higher than that of fractional distillation (ARMINGTON et al., 1965; ARMINGTON et al., 1970; SCHUMB et al., 1949; AMBERGER and DIETZE, 1964).

γ) Low Temperature Hydrogen Reduction of BBr_3

The last stage in the process of obtaining pure boron from boron halides is reduction by hydrogen (or thermal decomposition in vacuum in the case of BI_3). As KIESSLING showed (1948), one of the simplest methods of realizing this reaction is to use a quartz tube externally heated by an electric furnace. When the BX_3-H_2 gaseous mixture passes into the hot zone, boron is deposited onto the internal wall of the tube that serves as a substrate. Since boron reduces silica at high temperatures this method of preparation can lead to pure boron only if it is carried out at low temperatures, which practically limits its appli-

Table 8. Carbon content of BI_3 submitted to two fractional distillations. In each case the product has been fractionated into two halves: the distillate and the residue. (From ARMINGTON et al., 1970)

	First distillation: carbon content in ppm.	Second distillation: carbon content in ppm.
Distillate	930	140
Residue	450	65

cation to the one halide, BBr_3 (NASLAIN, 1972; NASLAIN, 1970; KIESSLING, 1948). The reduction of BBr_3 by hydrogen in quartz tubes was studied in detail by NASLAIN, ETOURNEAU and HAGENMULLER; it can produce very pure boron of the α-rh. type under specific conditions (HAGENMULLER and NASLAIN, 1965; NASLAIN, 1967; NASLAIN et al., 1970; NASLAIN, 1970; NASLAIN, 1972).

In order to minimize the risk of contamination by the internal wall of the reactor, and in order to increase the rate of deposition, NASLAIN et al. (1973a, b) have recently proposed to carry out the reduction of BBr_3 within a fluidized bed of porous granules of boron.

Hydrogen Reduction of BBr_3 in Fused-Silica. The reaction starts at $600^\circ C$, but at this temperature the yield is low and boron is poorly crystallized. Above $850^\circ C$, the rate of boron deposition becomes significant, but boron begins to reduce silica. The temperature range from 700 to $850^\circ C$ has been chosen as a compromise between obtaining a good yield and avoiding extensive contamination by silicon.

Since the rate of reduction of BBr_3 remains low in this temperature range, only a small fraction of the boron tribromide in the vapor phase is reduced in the hot zone of the tube. To reach the highest possible yield, relative to the starting amount of BBr_3, the boron tribromide that has not been reduced is condensed at the exit of the silica tube and recycled.

At the beginning of a run, boron deposits only onto the inner wall of the quartz tube, as porous granules of very high specific area. After a few hours of reaction, the substrate on which boron deposits is no longer the silica wall, but rather the first deposited boron granules themselves acting as seeds: their very high specific area gives the reaction a significant rate despite the use of low temperatures. The hot zone becomes progressively filled with boron and in order to continue the reduction the hot zone must be shifted along the tube axis as a function of time by displacing the electrical furnace. The time spent by boron in the hot zone τ - or the rate of displacement of the furnace - is an important parameter. For τ less than a critical value τ_c, which depends on the experimental apparatus, the boron deposit consists only of discrete α-rh. granules that have very few contacts with themselves and with the silica wall. For $\tau > \tau_c$, the deposit is a compact powder consisting of a mixture of α-rh. and β-rh. phases, in which the proportion of β-phase is the greater the higher the temperature and the longer the deposition time (NASLAIN et al., 1970; NASLAIN, 1972).

The apparatus used in our laboratory is represented schematically in Figure 19. Commercial hydrogen (99%) is dried with CaH_2 (A); oxygen and nitrogen are removed by passing through a platinum-base catalyst (B), and by liquid nitrogen traps (P, P_1 and P_2) associated with furnaces containing magnesium (F_1, F_2). Purified hydrogen bubbles in the BBr_3 vaporizer (V), and then is admitted into the fused silica reduction tube (Q) where boron deposits. The unreacted boron tribromide is condensed at $-40^\circ C$ (C), and then recycled. The hydrogen bromide is absorbed, after passing through a drying column, by a soda solution. The BBr_3 vapor pressure is low at room temperature (70 torr at $25^\circ C$); in order to obtain a BBr_3-rich gaseous mixture then, it is necessary to heat the vaporizer (V) to about $60^\circ C$ (280 torr BBr_3). Consequently, the tubing between V and C, without stopcocks or ground glass joints, has to be heated to about $80^\circ C$ in order to avoid condensation of BBr_3.

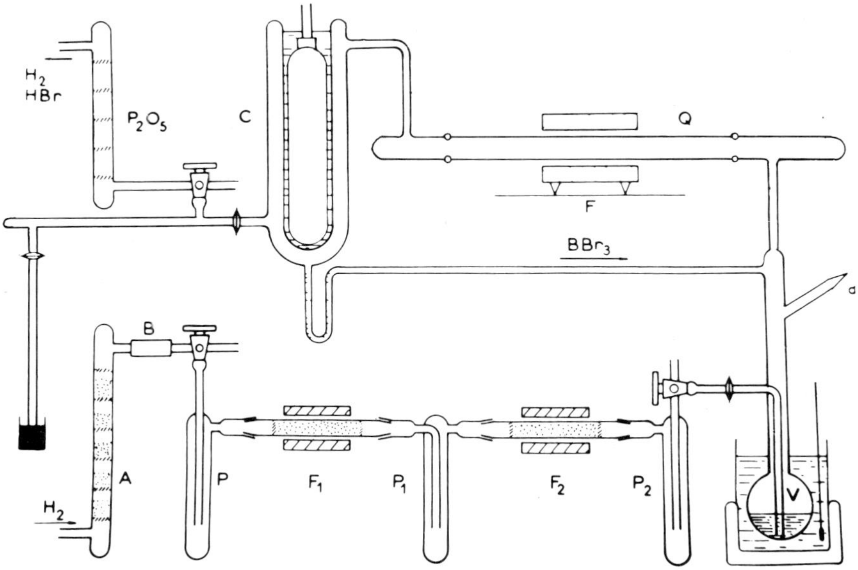

Fig. 19. Apparatus for the preparation of α-rh. boron in fused-silica tubes. (From NASLAIN et al., 1970)

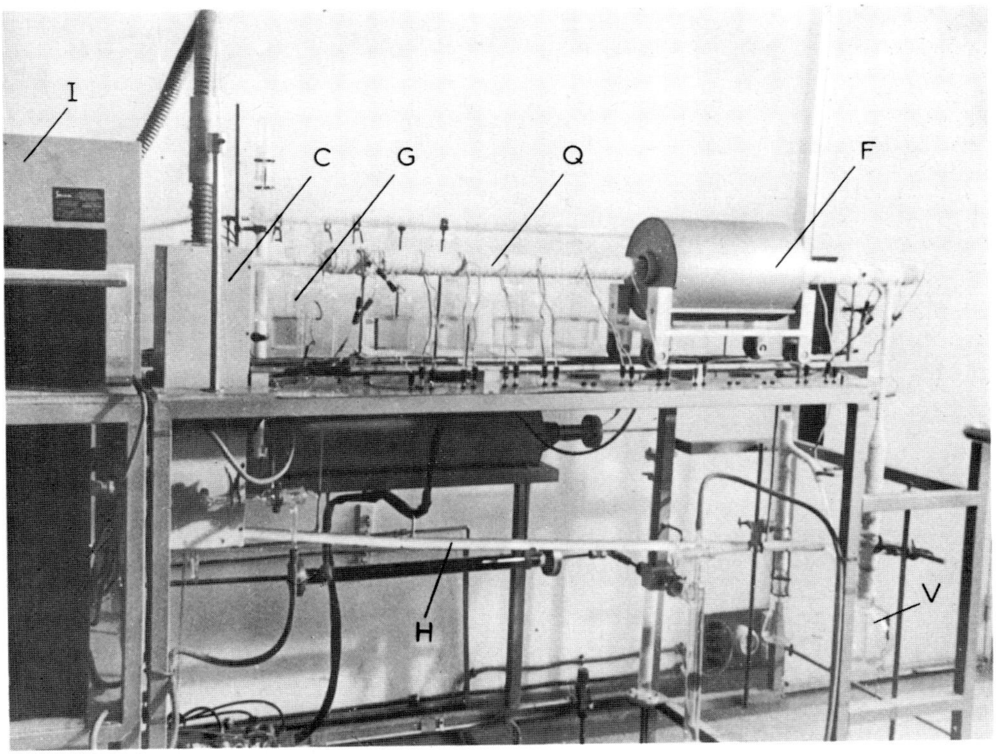

Fig. 20. Unit for the synthesis of finely divided α-rh. boron by hydrogen reduction of BBr_3; V: BBr_3 vaporizer; F: reducing furnace; Q: fused-silica tube; C: BBr_3 condenser; I: cryostat for maintaining condenser C at $-40^{\circ}C$; H: recycling of unreacted BBr_3; G: HBr absorber. (From NASLAIN et al., 1972)

The unit which actually operates in this laboratory is shown in Figure 20; it gives 50-100g of boron per run with a yield of 90% and a deposition rate of about 250mg/h.

Boron prepared at low temperatures according to this process is in the form either of a brown powder or as coarse porous granules of a diameter of 3 to 6mm. The latter are easily pulverized without risk of contamination: in this respect, this method is superior to the technique of deposition on filaments, whereby hard and massive products are the result.

The product is mainly α-rh. boron if $\tau < \tau_c$; it should be noted, however, that the lines of the Debye-Scherrer pattern are broad, indicating that the material is fine-grained(Fig. 21). If, on the contrary, $\tau > \tau_c$, a mixture of α-rh. and β-rh. forms results with α-rh. predominating. Boron prepared at low temperatures is very reactive chemically, because of its small particle size, its large specific surface area (about 4-5m^2/g) and the presence of weak three-center bonds in the α-rh. icosahedral network. This type of boron is slowly oxidized in air, necessitating its storage in glass flasks sealed in an argon atmosphere (NASLAIN, 1967; NASLAIN et al., 1970; NASLAIN, 1970; NASLAIN, 1972).

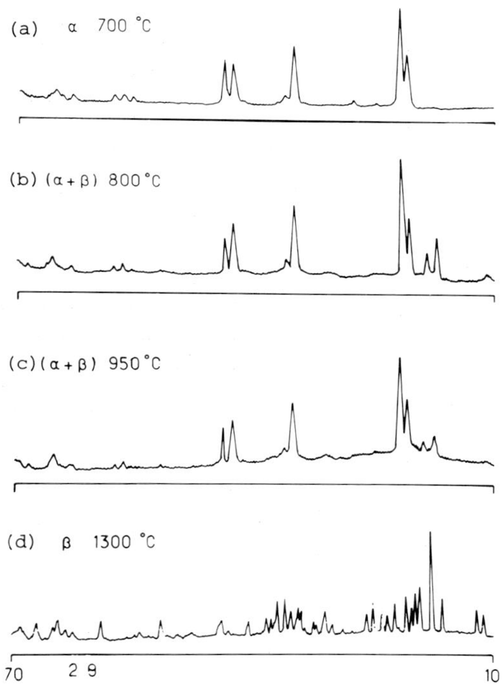

Fig. 21a-d. X-ray powder diagrams of synthesized boron: (a) at 700°C in a fused-silica tube; (b) at 800°C in a fused-silica tube; (c) at 950°C in a fluidized bed of porous boron granules; (d) at 1300°C on a metal filament

It has often been stated in the literature, that the use of silica tubes for the reduction of boron halides should be avoided because of the risk of contamination by the silicon resulting from the reduction of silica. Analysis of boron prepared in our laboratory (which has been used as the starting material for all the studies of borides that have been made in this laboratory) gives an elemental boron content of 99%. The X-ray fluorescence spectrum of this product is given in Figure 18; it shows that the only impurity present in significant amount is bromine (about 0.3%). Emission spectrography indicates that the three other significant impurities are: silicon (250ppm), copper (50ppm) and aluminium (25ppm). The analysis for light elements (H, C, N, O) has not been made.

These results show that the reduction of BBr_3 by hydrogen can be carried out effectively in fused-silica tubes by working at low temperatures with tubes of large cross-section. Under these conditions, the boron granules themselves act as a substrate and they make very few contacts with the silica inner wall.

Hydrogen Reduction of BBr_3 in Fluidized Beds. The main drawback of the previous method is that the deposition rates (250mg/h) are too low because of the requirement that the temperature of reduction must be kept low. For example, preparation of 100g of boron necessitates over two weeks' deposition at 750-800°C. While such a low deposition rate is acceptable for laboratory work, it is not practical for industrial use.

In order to increase the deposition rate without lowering considerably the purity of boron, NASLAIN et al. (1973a, b) proposed recently to carry out the reduction of BBr_3 within a fluidized bed of porous boron granules. Under these conditions, boron is deposited upon the boron granules, kept in suspension by the gas current, which, as a result make very little contact with the reactor wall. The reaction can be then carried out at higher temperatures with a higher deposition rate.

In the first experiments with this technique a deposition rate of about 1g/h at 950°C was obtained with a fluidized bed containing 20g of porous granules with an average diameter of 0.5mm. The minimum fluidization velocity was 1.6cm/sec and the conversion rate of BBr_3 into boron for one pass through the layer was 60%. As for the previous method, the boron product was a mixture of both rhombohedral forms with α-rh. boron predominant, the density of the product (2.465) being very close to that of α-rh. boron (2.46). The elemental boron content was over 99%, the main impurity still being bromine (0.4%).

Improvements in Boron Purity. Boron prepared at low temperatures from BBr_3 contains significant amounts of bromine, an impurity that is not harmful in many applications, but which it is still better to eliminate. Moreover, due to the porosity of the granules and their great specific surface area, this kind of boron probably contains hydrogen and other volatile impurities (for example HBr) in spite of the vacuum outgassing carried out at the reduction temperature before opening the reactor.

After being pressed into pellets, boron undergoes another outgassing in high vacuum (10^{-6} torr). The improvement in purity becomes appreciable only when the thermal treatment is carried out above the α-rh. → β-rh. allotropic transformation temperature (1200°C); this suggests that bromine is chemically bonded to boron in the α-rh. modification. If the temperature is limited to 1600-1700°C, boron does not sinter much and can be easily pulverized; its elemental boron content is then between 99.5 and 99.9% (NASLAIN et al., 1970).

Unfortunately, this improvement in purity is accompanied by a decrease of reactivity, since β-rh. boron is chemically more inert than α-rh. boron; however, this type of material is still quite suitable for the synthesis of borides of elements having quite a strong affinity for boron.

δ) Hydrogen Reduction of Boron Halides on Metal Filaments

The Apparatus. In the hot filament method, the reduction of the boron halide by hydrogen is carried out in contact with a refractory filament heated by the Joule effect. In fact, it is only at the very beginning of the experiment that boron is deposited directly onto metal; as soon as the boron is deposited, it itself becomes the substrate. To obtain large quantities of boron, the substrate temperature must be kept constant during all the experiment. The electrical conductivity of boron is very low at room temperature but rises very rapidly with temperature reaching $10\Omega^{-1}cm^{-1}$ at $900^{\circ}C$ for example. Because of this semi-conducting behavior, a boron deposit on a metal filament can be kept at a constant temperature by progressively increasing, as a function of its growth, the electrical power that is applied to it.

In order to limit the contamination of boron by the metal substrate, utilizable filaments must have small diameters: 80 to 200μ. On account of the very small surface area of the substrate in this method, appreciable quantities of boron are obtained only if the deposition rate is high, which involves a substrate temperature of at least 1200 to $1400^{\circ}C$. At such a temperature the three halides, BCl_3, BBr_3 and BI_3 are equally utilizable, and the boron deposits are homogeneous and compact, which is essential for their further use as resistance heaters.

Since 1923 the hot filament method has been employed by many investigators for obtaining small amounts of boron for the physical and chemical characterization of the element (FREYMANN and STEIBER, 1934; HOARD et al., 1958). Further development of this method was related to the work after 1950 on the semi-conducting properties of boron. Various types of apparatus allowing to prepare several hundred grams of crystalline boron per operation, have been described in detail in the proceedings of the First International Symposium on Boron. Here we shall give, as an illustration, the technique of POWELL et al. (1960) also used by HAGENMULLER and NASLAIN (1965). (See also NASLAIN, 1967; NASLAIN et al., 1970; NASLAIN, 1972).

In the case of BBr_3, the reduction can be carried out in an apparatus of the type schematically depicted in Figure 22. The experimental device is similar to that represented in Figure 19, but the fused-silica tube is replaced by the hot filament reactor. The reactor can be isolated by two valves R_1 and R_2, so that several deposits may be obtained from the same BBr_3 charge. The reactor is shown in Figure 23; it is made of boro-silicate glass and cooled with a water spray. The refractory filament is held in the axis of the glass flask, between two molybdenum electrodes and subjected to an adjustable tension to prevent sagging as the filament grows. Teflon O-rings (or mercury wells) permit the electrodes to slide, while isolating the inside of the reactor from the atmosphere. The filament is fed with a low-voltage current through B_1 and B_2.

The first few minutes of deposition are the most critical period in the growth of a boron rod, as vibrations induced by the heating current in the filament, which is still fragile, can cause its rupture if the tension is not rapidly adjusted. The temperature of the filament, measured with an optical pyrometer, is kept constant by regulating the applied tension. As in previous methods, the boron deposition rate is a function of the temperature of the substrate, of the gas flow in the reactor and of the temperature of the vaporizer that determines the partial pressure of BBr_3 in the vapor phase.

POWELL et al. (1960) report an average deposition rate of 8 to 10g/h and a capacity of about 100g per operation, for a filament 20cm in length heated to $1300\text{-}1400^{\circ}C$, with a hydrogen flow of 2 to 4l/min and a vaporizer temperature of 50 to $70^{\circ}C$. As in the case of the fused-silica tube apparatus, the condensation and recycling of non-reduced boron halide make it possible to achieve a final yield of 90% (relative to BBr_3).

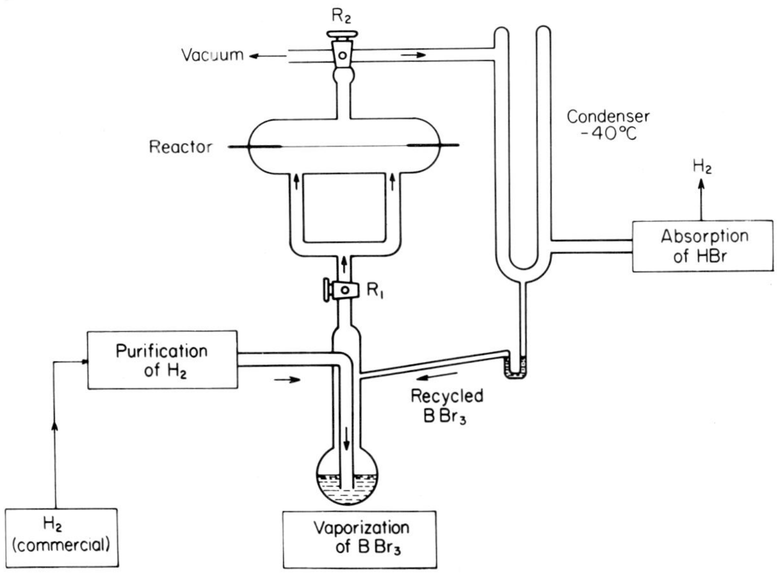

Fig. 22. Schematic representation of the principle of an apparatus for preparing boron on a hot metal filament. (From BOIRET, 1969 and NASLAIN, 1972)

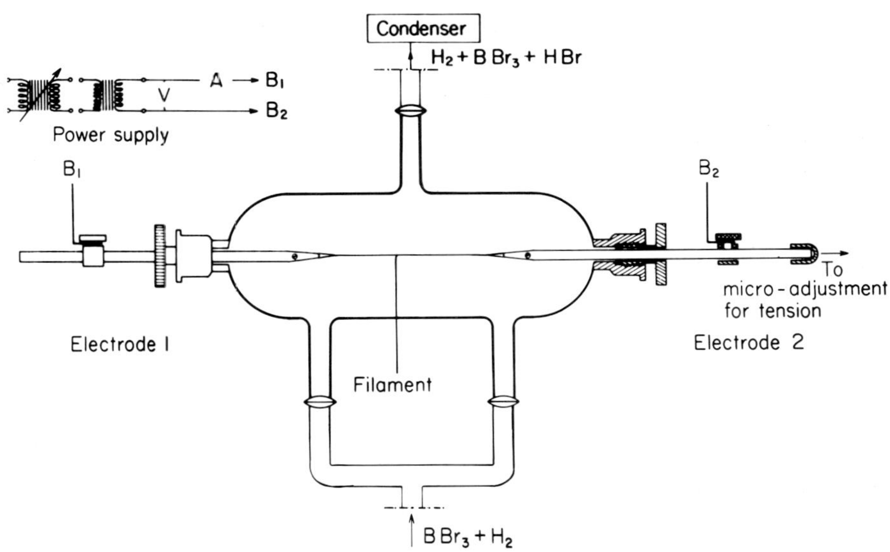

Fig. 23. Hot filament reactor used in preparing boron by high temperature hydrogen reduction of BBr_3. (From BOIRET, 1969 and NASLAIN, 1972)

Nature of the Metal Filament: Diffusion Phenomena at the Boron-Metal Substrate Interface. The choice of the substrate metal is important. The metal must not form low-melting eutectics with boron; this eliminates noble metals such as platinum, and metals of the iron family. Also, it must not diffuse into boron and must have sufficient mechanical strength at the temperatures of boron deposition. Most of the common metals react with boron at 1000°C; the ones that are most compatible with boron are the highly refractory transition metals of columns IV, V and VI in the periodic table.

During the first few minutes of deposition, boron atoms diffuse into the metal substrate and form a borided layer around the metal filament, the thickness of which increases with time. Depending on the specific metal, the diffusion leads to the formation of one or several borides. For zirconium, niobium and tantalum, only one phase is observed, the MB_2 boride. On the contrary, for molybdenum and tungsten, the diffusion results in the formation of both M_2B_5 and $MB_{\sim 4}$ phases at 1000 - 1400°C. The boriding of the substrate is quite rapid at the usual temperatures of boron halide reduction. For example, a tantalum filament, 200μ in diameter, is entirely transformed into TaB_2 after a deposition of 2 h at 1200°C (BOIRET, 1969).

Diffusion of boron into the metal is essentially a one-way phenomenon. It gives rise to a considerable increase in volume of the metal core, which induces a state of stress that causes radial cracks in the boron rod. Accompanying the flux of diffusing boron atoms is a significant flux in the opposite direction of vacancies which can result in the formation of macroscopic pores around the substrate - deposit interface. The presence of such pores decreases the cohesion between the boron deposit and the metal core and makes it easier to extract the latter. Tantalum, molybdenum and tungsten make the best substrates for preparing boron by the hot filament method. The effects of diffusion are minimized for filaments of very small diameter.

Purity of Boron Prepared on a Metal Filament. The main impurities are of two kinds; those that have their source in the starting materials and those that result from contamination by the metal filament; the contamination by the substrate is the main drawback of the hot filament method. However, for deposits of about 100g, made on thin filaments (80μ), it is not necessary to separate the metal core which represents only a small portion of the total mass of the product. Thus, POWELL et al. (1960) reported that the tantalum content in their samples was only 200ppm, without any separation of the substrate.

Very often, the amount of deposited boron is much smaller and the metal core can no longer be neglected. In this case, the substrate must be extracted by either physical or chemical methods. Sometimes, the metal core breaks off easily into tiny cylinders when one tries to reduce the boron deposits into fragments before the grinding operation; in this case, the substrate may be separated under a microscope. Generally, however, separation of the metal core is more difficult. HAGENMULLER and NASLAIN (1965) have proposed a method which is based on the differences between the densities of boron (2.35) and the borides present in the metal core (d = 12 for TaB_2 and d = 10 for $WB_{\sim 4}$) allowing an easy separation of the substrate in bromoform (d = 2.9). Chemical separations have also been suggested; they are generally not very selective, for the metal is not in a free state, but within compounds that have chemical properties similar to those of boron (STERN and LYNDS, 1958; FORMSTECHER and RYSKEVIC, 1945).

Crystallinity of Boron Deposits on a Metal Filament. The crystalline nature of boron deposited upon a filament depends on experimental conditions. Parameters that have the largest effect on the crystallinity of deposited boron are: the substrates temperature, the boron deposition rate, the nature of the substrates and the presence of certain impurities in the reaction environment. Depending on the choice of these parameters, the deposited boron may be amorphous or crystalline.

Deposition of amorphous boron upon metal filaments has been the subject of many studies concerning fabrication of boron filaments (or fibers) for use as reinforcing elements in composite materials. For such applications, the diameter of the deposits is limited, however, to 100-150μ, the substrate being made of a very thin tungsten filament (13μ) (WAVNER, 1967). But only a few studies have been made concerning the preparation of amorphous boron deposits having a large diameter (TALLEY et al., 1960 b; BOIRET, 1969).

Two parameters essentially control the formation of amorphous boron: the substrate temperature and the boron deposition-rate. Boron is amorphous if the deposition rate is high and if the substrate temperature is sufficiently low to prevent formation and growth of crystal nuclei. The first condition is fulfilled if the vapor phase is rich in boron halide and the hydrogen flow is rapid (0.5 to 1 l/min). The deposition temperature must be moderate and 1200°C appears to be the highest limit above which boron is crystalline. Even if these conditions are fulfilled, crystal nuclei will form at the surface of the boron rod after some time and will grow at the expense of amorphous boron. According to TALLEY et al. (1960b) and to BOIRET (1964) it is difficult to obtain amorphous boron rods with a diameter greater than 2-3mm from a thin metal filament. Deposits of amorphous boron have a characteristic appearance; the surface ressembles ears of corn and consist of nodules which result from a radial growth into cones, similar to deposits of pyro-carbons (Fig. 24).

The hydrogen reduction of boron halides, on a filament, results in a great variety of crystalline forms of boron. At low temperatures ($T < 1100^\circ$C approximately), the only cristalline phase present in considerable amounts with amorphous boron, is the α-rh. modification (Fig. 22) (AMBERGER and DIETZE, 1965; AMBERGER and PLOOG, 1971; BOIRET, 1969). AMBERGER and PLOOG observed that in boron deposits obtained at 900-950°C from BBr_3, the amount of α-rh. boron increases - relative to that of amorphous boron - as a function of the deposition time (composition of vapour phase: 9 moles % BBr_3; hydrogen flow: 200cm^3 per minute) (AMBERGER and PLOOG, 1971). This result suggests that the α-rh. boron might have grown within amorphous boron itself by a recrystallization process. The α-rh. boron occurs as red crystals which can be extracted from the vitreous of microcrystalline mass by treating with nitric acid (McCARTY and CARPENTER, 1960). By decreasing the content of BBr_3 in the vapor phase and by increasing the hydrogen flow (0.66 mole = BBr_3; 400cm^3/min), no amorphous boron formation is observed and the deposit then consist mainly of α-rh. boron (95%) and a small amount of β-rh. boron (5%) (AMBERGER and PLOOG, 1971).

At high temperatures ($T > 1400^\circ$C approximately), the boron deposits generally consist of only β-rh. boron (HAGENMULLER and NASLAIN, 1965; POWELL et al., 1960; BEAN and MEDCALF, 1960; AMBERGER and SCHMIDT 1971). OLEMPSKA et al. however, reported the presence of α-tetr. crystals in boron deposits prepared from BCl_3 between 1300 and 1800°C, but no information was given concerning the carbon and nitrogen content of these products (OLEMPSKA et al., 1966).

For intermediate temperatures ($1100 < T < 1400^\circ$C), the phenomena are more complex, the purity of reagents being a determining factor, and various crystalline forms of boron may result. In the absence of detectable amounts of impurities in the starting materials, hydrogen reduction of boron halides at 1100 to 1400°C produces the β-tetr. and β-rh. modifications, that eventually coexist at low temperatures with amorphous boron (Fig. 24). The β-tetr. form can be obtained pure quite readily, as microcrystals, by working at 1150-1250°C with a tantalum

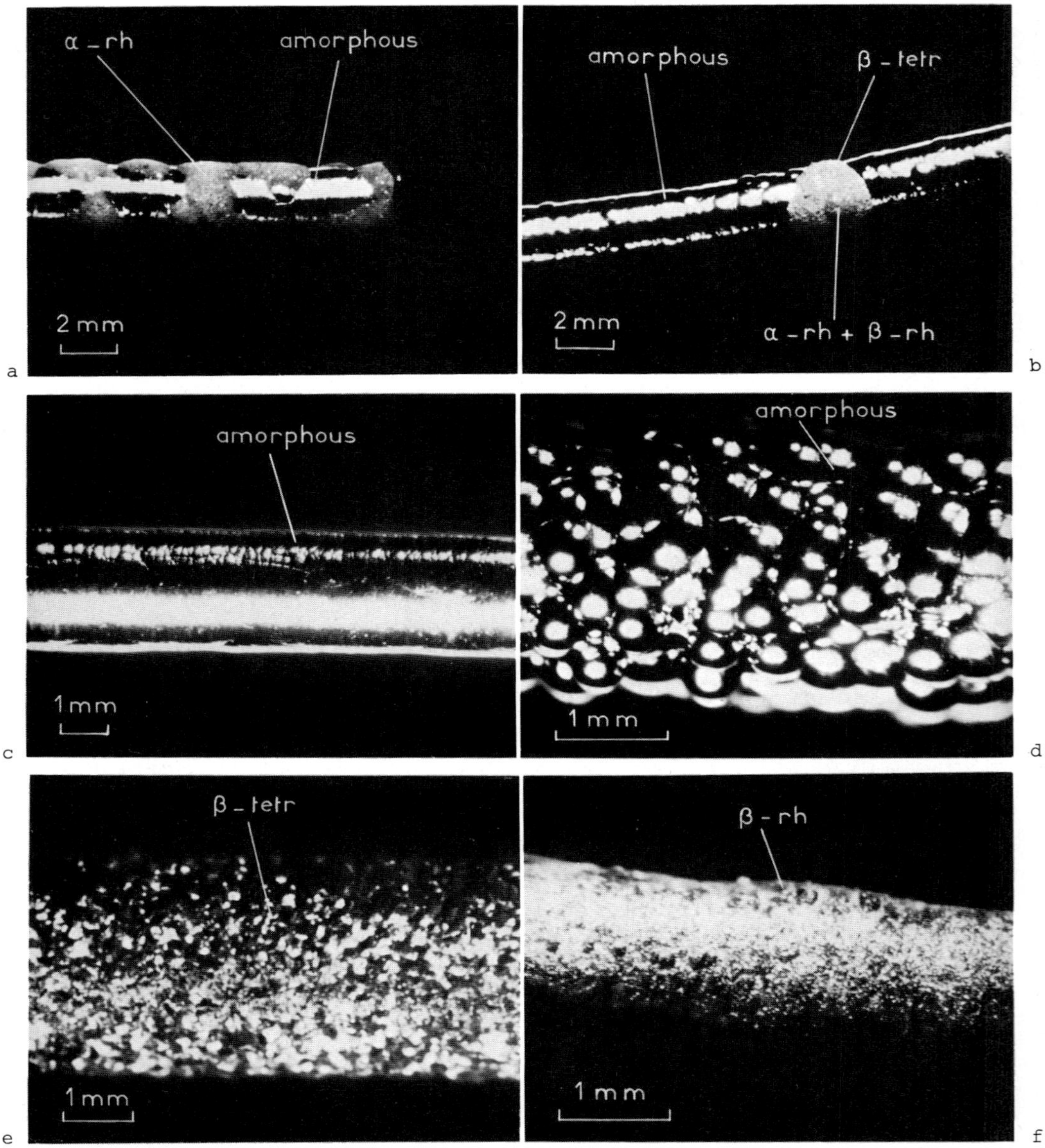

Fig. 24. Characteristic morphologies of boron as a function of deposition temperature. (From BOIRET, 1969 and NASLAIN, 1972). (a) 1000°C; (b-d) 1100°C; (e) 1200°C; (f) 1300°C

or rhenium filament (composition of vapor phase: 9 mol % BBr_3; hydrogen flow: 200 to 500 cm^3/min) (AMBERGER and PLOOG, 1971; BOIRET, 1969; TALLEY et al., 1960 a).

When the BX_3-H_2 gas mixture is doped with small amounts of nitrogen, CH_4 or $CHBr_3$, the reduction of the boron halide yields $B_{48}B_2N_{\sim 2}$ and $B_{48}B_2C_{\sim 2}$ α-tetr. phases, that correspond, as previously discussed to tetragonal lattices of boron icosahedra, stabilized by carbon or nitrogen atoms (AMBERGER and PLOOG, 1968; PLOOG and AMBERGER, 1971; AMBERGER and SCHMIDT, 1971; PLOOG et al., 1972).

Table 9. Nature of the phases formed by hydrogen reduction of BCl_3 or of BBr_3 on a BN substrate, as a function of substrate temperature and of vapor phase composition. (From AMBERGER and SCHMIDT, 1971)

	Temperature of deposition (^oC)					
	1200	1200	1400	1400	1200-1400	1500
H_2O (mol.%)	90-94	75-40	94	90-40	40	40
N_2 "	1-5	20-55	1	5-55	55	55
BCl_3 "	5	5	5	5	-	5
BBr_3 "	-	-	-	-	5	-
Phases formed	α-Tetr. β-Tetr. β-rh.	α-Tetr.	α-Tetr. 10% β-rh. 90%	α-Tetr.	α-Tetr.	α-Tetr. BN

Table 10. Influence of carbon on the nature of the phases formed at 1150^oC by hydrogen reduction of BBr_3 on a tantalum substrate. (From PLOOG and AMBERGER, 1971)

	CH_4 content of the vapor phase (mol.%)					
	0	0.4	1.0	1.8	2.5	3.6
BBr_3 (mol.%)	9.1	9.1	9.1	9.1	9.1	9.4
H_2 "	90.9	90.5	89.9	89.1	88.4	87.0
Phases formed	β-Tetr. boron	β-rh. boron	$B_{48}B_2C$ (α-tetr.)	$B_{48}B_2C_2$ (α-tetr.)	$B_{13}C_2$ (rh.)	B_4C (rh.)

According to AMBERGER and PLOOG (1968), the $B_{48}B_2N_{\sim 2}$ phase is obtained quantitatively when a boron nitride substrate heated to 1200-1600^oC is used with boron chloride or bromide mixed with nitrogen. As indicated in Table 9, 1% nitrogen in the BCl_3-N_2 gas phase is enough to induce at 1400^oC the formation of detectable amounts of $B_{48}B_2N_{\sim 2}$. This result corroborates the observation of OLEMPSKA et al. (1966), that an accidental air leak into the deposition apparatus favoured the formation of crystals of the α-tetr. phase (AMBERGER and SCHMIDT, 1971).

The effect, on the nature of the crystalline deposits, of doping with small amounts of CH_4 or $CHBr_3$ has also been determined by PLOOG and AMBERGER (1971). Table 10 shows that 0.4 mol% of methane added to the gas mixture is enough to inhibit the formation of β-tetr. boron at 1150^oC and only β-rh. boron occurs in the deposits. More doping (1 to 1.8 mol.% CH_4) makes it possible to prepare highly pure α-tetr. phases with the formula $B_{48}B_2C_{1-2}$. And finally, for an even higher methane content (2.5 to 3.6 mol%) rhombohedral boron carbide results.

The influence of carbon on the formation of the α-tetr. lattices of boron icosahedra was confirmed in this laboratory by M. BOIRET (1969). Transition metals may have an influence similar to that of nitrogen or carbon (DECKER and KASPER, 1960; AMBERGER et al., 1976).

The results discussed in this section and the observations made during structural studies now provide a better understanding of the active controversies that have existed relative to the identity of α-tetr. phases. The "α-tetr. boron" used in the first investigations was made from boron halides that probably contained large amounts of carbon. Indeed, we have seen how prevalent this impurity is, and how difficult it is to eliminate it. Furthermore only very recently have there been developed precise methods of analyzing boron for carbon; thus it remains unknown how much carbon was present in the products considered by the first investigators to be α-tetr. boron. Finally, since nitrogen has an influence analogous to that of carbon, the effect of this impurity (which can arise from insufficiently purified hydrogen, from uncomplete flushing of the deposition apparatus, or from an accidental air leak) has an additional influence to that of carbon or even that of the transition metal substrate (LAUBENGAYER et al., 1943; OLEMPSKA et al., 1966; BECHER and SCHÄFER, 1960; UNO, 1958). Thus, it is probable that pure α-tetr. *boron* does not exist, but that boron icosahedral arrangements, with space group $P4_2/nnm$ or $P\bar{4}2m$, are commonly observed in the presence of traces of stabilizing impurities acting as electron donors (AMBERGER and PLOOG, 1968; BOIRET, 1969; AMBERGER et al., 1970; AMBERGER and PLOOG, 1971; PLOOG and AMBERGER, 1971; PLOOG et al., 1972).

ε) *Hydrogen Reduction of Boron Halides on a Boron Substrate*

The reduction of boron halides with metal filaments presents two drawbacks: its application is delicate when rods of large diameter are to be prepared from very thin filaments; furthermore, contamination of boron by the substrate is still possible. These drawbacks can be overcome by using boron itself as a substrate in the form of rods which are more stable mechanically at high temperatures than metal filaments. This technique requires, first of all, the preparation of the substrate, then the reduction of a boron halide in contact with the boron rod kept at a high temperature either by Joule heating or using high frequency currents.

Preparation of Boron Substrates. BEAN and MEDCALF (1960, 1962) showed that boron deposits on straight tantalum filaments could, under certain conditions, be separated from the metal core. They occur as hollow cylinders (Øext.: 4mm, Øint.: 1.5mm, length: 150mm) and can be transformed into dense rods by zone melting. The possibility of extracting the metal substrate from the boron deposit depends on the adhesion of the deposit to the filament, which results itself from diffusion phenomena at the boron-substrate interface. If the deposition is carried out at low temperatures (1100-1175°C), the filament is only slightly borided and the layer of TaB_2 is regular: under these conditions, the filament does not strongly adhere to boron after quenching and can be extracted with its superficial TaB_2 film, without breaking the boron deposit itself. The latter then serves as a satisfactory substrate.

Boron rods, generally with larger cross-sections, can also be prepared by zone melting from sintered boron cylinders (SIRTL, 1964).

Heating of Boron Substrate. The high electrical conductivity of boron at high temperatures allows for heating boron substrates either by the Joule effect, or by high-frequency induction. These methods, however, require that the substrates be preheated with a separate furnace of with a metallic susceptor, since boron is an insulator at room temperature ($\rho \sim 10^6\ \Omega$ cm at 25°C).

Figure 25 illustrates two types of vertical chambers for boron deposition upon a boron substrate. In the apparatus of BEAN and MEDCALF (1960), the boron rod is mounted along the axis of a silica tube, between two graphite electrodes cooled by circulating water. The top electrode is equiped with a device for absorbing the linear dilatation of the boron rod.

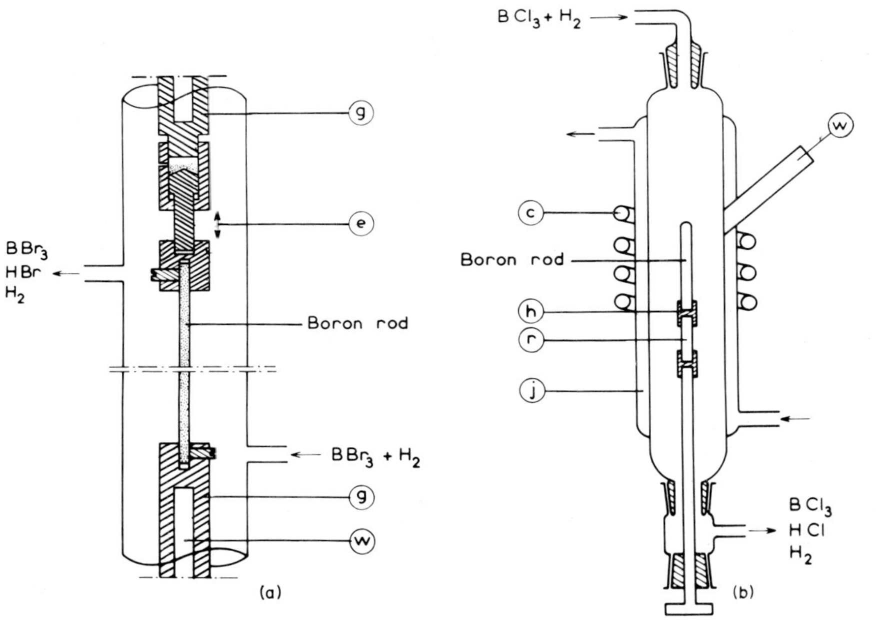

Fig. 25. Apparatus for the reduction of BBr_3 in contact with a boron substrate (a) graphite piston-electrode; g: graphite electrode; e: zone of expansion; w: cooling water circulation (according to BEAN and MEDCALF 1960); (b) substrate heated by a high frequency current; c: induction coil; h: molybdenum holder; r: thermal insulator; j: water cooling (From OLEMPSKA et al., 1966)

Heating the boron rod by a coil which is fed by a high frequency current and which is outside the reaction chamber, is a particularly simple way to keep the substrate temperature constant (Fig. 25b; OLEMPSKA et al., 1966).

These two variants of a reactor can be inserted in a general scheme of the type given in Figure 22. BEAN and MEDCALF (1960) obtained with this technique, boron deposits up to 2cm in diameter and 15cm in length.

Nature of Boron Deposited on Boron Substrate. When utilizing a boron substrate which eliminates all risks of contamination, the hydrogen reduction of boron halides can be carried out at high temperatures and consequently at high deposition rates. Boron obtained by this method can be spectrographically free of impurities if the boron halide and the hydrogen have been carefully purified. The main impurities are still carbon and silicon. Hence, this method is especially suitable for the preparation of massive β-rh. boron samples for studies of physical properties.

ζ) Thermal Decomposition of BI_3 by the Hot Filament Method

Thermal decomposition of boron halides leads to the formation of elemental boron and free halogen according to the following reaction:

$$2BX_3 \longrightarrow 2B + 3X_2\uparrow \text{ (with X = Br; I)} \qquad (2)$$

Although there has been some work with BBr_3, this method is applied most often to BI_3 whose low thermal stability we have already noted (Table 6; NARAY-SZABO and TOBIAS, 1949; McCARTY and CARPENTER, 1960; ARMINGTON et al., 1964; AMBERGER and DIETZE, 1974).

Thermal dissociation of boron halides is generally done in vacuum, in hot filament reactors similar to those used when they are reduced by hydrogen; since the method does not necessitate a carrier gas, the apparatus is quite simple (Fig. 26). The substrate can be either a metal filament, or a boron rod. Since the entire apparatus is under high vacuum, the boron halide is easily transferred by distillation or sublimation from the vaporizer V to the filament F. Due to the small deposition surface, in comparison with the reactor volume, only part of the boron halide is converted into boron in contact with the filament, so that when leaving the reactor, a mixture of non-decomposed BI_3 and of free iodine is condensed in C (AMBERGER and DIETZE, 1964).

The main advantage of this method in comparison with the hydrogen reduction of boron halides, is in achieving boron of very high purity. Since the dissociation is carried out in high vacuum, the amounts of hydrogen, oxygen, nitrogen and halogen in boron are low. Furthermore, it has been observed that it is easier to obtain a boron iodide with low carbon content than a bromide, this result being related to the fact that at the normal boiling point of BI_3 (208-210 $^\circ$C), there are no stable carbon and iodine compounds (Table 8). According to ARMINGTON et al. (1965, 1970) the carbon content of boron from BI_3 is about three times lower than that of boron from BBr_3 the lowest reported values being in the range of 50 and 100ppm.

The crystallinity of boron deposits obtained by thermal dissociation of BI_3 depends on the temperature and nature of the substrate, on the boron deposition rate and on the presence of impurities in the gas phase. The method is especially recommended for the preparation of α-rh. crystals (McCARTY and CARPENTER, 1960; AMBERGER and DIETZE, 1964; AMBERGER and DIETZE, 1965). According to AMBERGER and DIETZE, α-rh. boron crystals are formed at 950-1100 $^\circ$C by crystallization of vitreous boron (initially produced by dissociation of BI_3) from crystal nuclei which are later deposided at the surface (Fig. 27). For a deposit of a given diameter the amount of α-rh. boron crystals is greater for a tantalum filament with a relatively large cross-section (for example, 0.5 to 2mm) than for a very thin filament. This result can be explained in terms of the temperature gradient that is created in the deposit, between the surface and the metal core; indeed, the temperature of the core has to be progressively increased so that during the growth of the boron deposit, the external surface stays at constant temperature. Since this gradient is lower in the case of a large-diameter metal filament, it will favour the transformation of vitreous boron into α-rh. boron rahter than into a high temperature variety (AMBERGER and DIETZE, 1965).

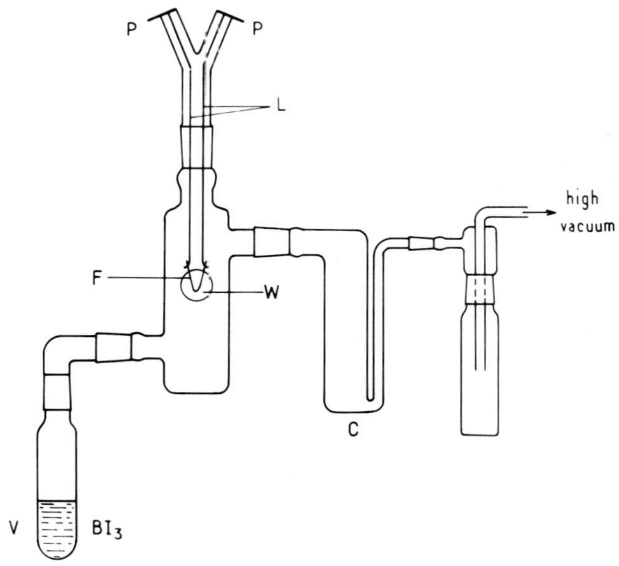

Fig. 26. Boron preparation by thermal dissociation of BI_3; V: vaporizer; F: tantalum filament; L: tantalum conductors; P: connectors to electric generator; C: condenser; W: window for pyrometer measurements. (From AMBERGER and DIETZE, 1964)

η) *Conclusions on Utilizing Boron Halides for Boron Preparation*

Boron halides, BCl_3 and particularly BBr_3 and BI_3, appear to be excellent intermediates in preparing highly pure boron, due to the fact that they can be made very pure by fractional distillation. Even if the initial boron used for the synthesis of the halide itself contains large amounts of impurities, these are generally all eliminated during the distillation, except for silicon and carbon, which remain as main impurities. When boron substrates are used, the reduction (or the thermal decomposition) does not introduce other impurities, and the boron product is generally free of any impurities detectable by the usual methods of emission spectroscopy, except for a few ppm silicon. The choice between BBr_3 and BI_3 is dictaded by considerations of purity and the ease of practical experimentation. BI_3 yields a less carbon-rich boron, whereas BBr_3, which is liquid at room temperature, is particularly easy to work with.

According to HINZ and WIRTH (1965), the thermal decomposition of B_2H_6 yields boron of higher purity than that obtained from halides (C : 30 to 50; Ta $<$ 0.3; Cu $<$ 0.01; As $<$ 0.0001ppm). Even if this result was confirmed by other investigators, it would not, however, compensate for the danger in manipulating great quantities of diborane.

Hydrogen reduction (or thermal decomposition) of boron halides gives boron samples which occur in a great variety of forms and conditions: finely divided powders, granules, massive vitreous or crystalline deposits, single crystals. By judicious choice of experimental parameters, it is possible to adapt the method of preparation to the desired application.

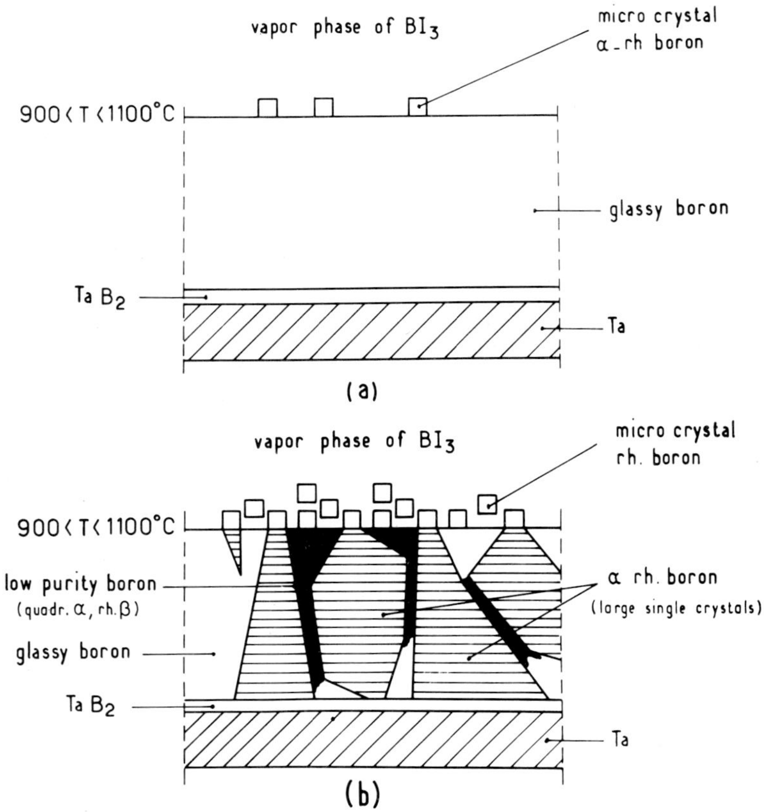

Fig. 27a and b. Mechanism of formation of single crystals of α-rh. boron: (a) initial formation of a thick vitreous boron deposit by dissociation of BI_3. The filament is covered with a TaB_2 film that acts as a diffusion barrier. (b) from nuclei created at the surface of the deposit, the vitreous boron changes into single crystals of highly pure α-rh. boron. The impurities rejected by these crystals induce the formation of crystalline phases of lower purity (α-tetr. or β-rh.). (From AMBERGER and DIETZE, 1965)

b) Preparation of Crystallized Boron by Solidification of a Liquid Phase

Both rhombohedral forms of boron can be prepared as single crystals by solidification of liquid phases of appropriate composition. Some metals have the property of dissolving, in the liquid phase, a relatively great quantity of boron, likely to reprecipitate upon cooling. Among these metals, platinum is especially interesting, for the boron-platinum diagram shows an eutectic with a very low melting point, which is the basis of a method of preparing small-sized crystals of α-rh. boron. Large single crystals of β-rh. boron have been obtained also by solidification of boron in a floating-zone or by the CZOCHRALSKI method. Since the crystal growth of boron is the subject of a particular chapter in this book, only a few data will be given here regarding the methods of preparing crystalline boron from a liquid phase.

α) Preparation of α-Rh. Boron by Solidification of Metal Alloys with Low Melting Points

Certain metals, such as copper, gold, silver and particularly platinum, form with boron eutectics with low melting points. This property has been utilized by various investigators in attempts to prove that α-rh. boron was a thermodynamically stable allotrope at low temperatures, and for obtaining crystals in order to define its physical properties.

The Platinum-Boron Phase Diagram. The first attempts at establishing the equilibrium phase diagram of the platinum-boron system were made by F.H. HORN (1960), who reported the existence of an eutectic with 40 at.% boron and with a very low melting point (830°C). No stable platinum boride was found at higher boron concentration (Fig. 28a).

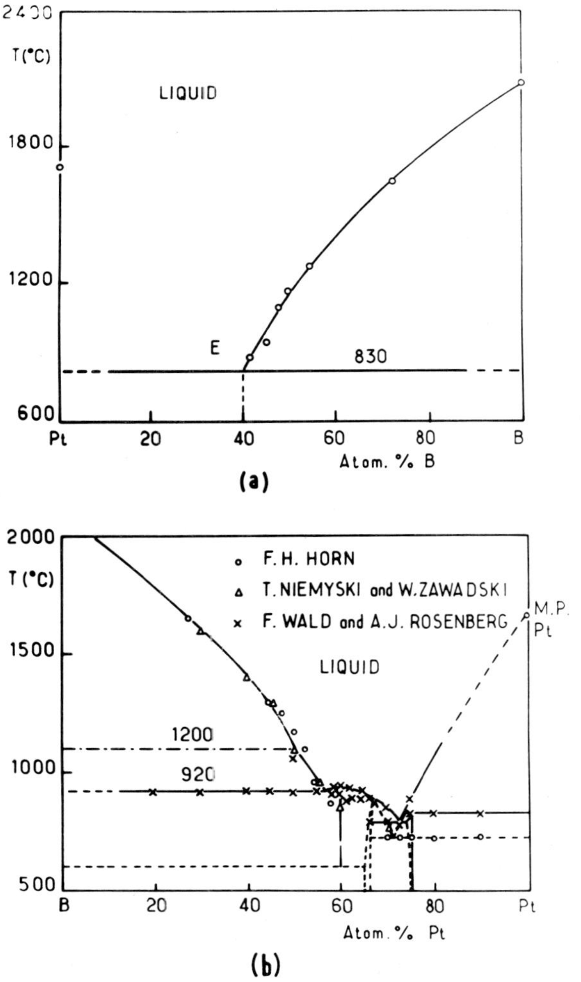

Fig. 28a and b. Equilibrium phase diagram of the boron-platinum system. (a) From HORN, 1960, (b) as established by WALD, 1970, on data of HORN, of NIEMYSKI and ZAWADSKI and of WALD and ROSENBERG

The phase diagram was completed later by NIEMYSKI and ZAWADSKI (1962) as well as by WALD and ROSENBERG (1965). Figure 28b, from WALD (1970), sums up all the results of the various investigators. While the phase diagram proposed by WALD differs noticeably from that of HORN (1960), especially by the presence of intermediate compounds, both phase diagrams show a very pronounced minimum (about at 800 °C) in the liquidus.

On the Stability of α-Rh. Boron at Low Temperatures. The possibility of obtaining α-rh. boron by solidification of alloys is dependent on knowledge of its thermodynamical stability range, which is still a controversial matter.

According to McCARTY et al. (1958), α-rh. boron microcrystals produced by thermal decomposition of BI_3 are thermally stable at 1100 °C, but start to transform into β-rh. boron at 1200 °C. According to NASLAIN et al. (1970), α-rh. boron prepared by hydrogen reduction of BBr_3 at 700-750 °C, undergoes this transformation at 1180-1200 °C. On the contrary, the inverse transformation which might eventually take place, by a prolonged annealing at temperatures below 1200 °C has never been observed.

It should be noted also that the thermal decomposition of BI_3 or the hydrogen reduction of BBr_3 lead to α-rh. boron only for deposition temperatures below 1100-1200 °C. At higher temperatures, as we have already noted, the β-tetr., β-rh. or even α-tetr. forms are obtained depending on boron halide purity.

If α-rh. boron was a true allotrope, thermodynamically stable below 1200 °C, the liquidus of the platinum-boron phase diagram should show an inflexion point at this temperature. Moreover, the solidification of Pt-B alloys with 40-50 at.% boron should produce α-rh. boron crystals. Figure 28b shows that even if it is not possible, in the actual state of our knowledge of the platinum-boron phase diagram, to affirm that the liquidus does have an inflexion point at 1200 °C, the data of NIEMYSKI and ZAWADSKI (1962) would however suggest it to be so.

Nonetheless, many authors have observed the formation, at temperatures below 1200 °C, of α-rh. boron crystals, not only in the platinum-boron system, but also in many other binary systems (B-Cu; B-Ag; B-Au), and even in ternary systems (B-Cu-Au) (WALD 1970). Likewise, whiskers of α-rh. boron were prepared by SITARIK and ELLIS (1966) by reducing, at 1000 °C, BCl_3 by hydrogen, in contact with a silica or boron substrate on which small platinum particles were deposited. The whiskers, 2mm long and 25μ in cross-section, had a rounded end - where the platinum had accumulated - which is a typical growth pattern when working with a vapor-liquid-solid (V.L.S.) system.

All these experimental results suggest that the α-rh. boron could be thermodynamically stable at low temperatures, in spite of the fact that β-rh. boron has been observed as a minor component in deposits obtained by chemical vapo-deposition at temperatures as low as 800 °C.

Preparation of α-Rh. Boron Crystals by Solidification of B-Pt Alloys. If it is admitted that α-rh. boron is thermodynamically stable below 1200 °C, Figure 28b shows that controlled cooling of B-Pt alloys with 40-50 at.% boron, must lead to primary precipitation of α-rh. boron crystals from 1200 to 920 °C.

Following this idea, α-rh. boron crystals have been prepared in a reproductible manner by HORN et al. (HORN, 1959; HORN, 1960; HORN et al., 1965). The method developed by HORN consists of pulling vertically a boron nitride crucible containing a liquid B-Pt alloy with 40-50 at.% boron, from a heating coil fed by a high frequency current. The rate of translation of the crucible was about 0.04mm/mn. The solidification starts at the bottom of the crucible and progressively extends into the whole mass.

α-rh. boron crystals are mainly concentrated in the top part of the platinum matrix. They can be extracted either by crushing the matrix and then separating the boron phase in bromoform (d(α-rh. B) = 2.45; d (Pt) = 21.4; d (bromoform) = 2.9), or by electrolytic dissolution of the platinum in a concentrated solution of HCl.

Crystals obtained in this manner are small in size (maximum 0.2 to 0.5mm) and in the form of rhombohedra. They are of an orange-red colour and show pronounced anisotropy and dichroism (HORN et al., 1965).

The presence of carbon inhibits the formation of α-rh. boron, as was shown by HORN et al. (1965). This observation agrees with the conclusions of AMBERGER and DIETZE (1965) and PLOOG and AMBERGER (1971) that obtaining α-rh. boron crystals between 900 and 1100°C from BI_3 or BBr_3 is possible only by utilizing highly pure products that are especially free from carbon.

β) Preparation of β-rh. Boron by Solidification of Liquid Boron

Many investigations have been made of the preparation of β-rh. boron by solidification of liquid boron with zone-melting methods or by involving pulling techniques of the CZOCHRALSKI type, for example. These studies, motivated by the interest in the semi-conducting properties of boron, aimed either at improving the purity of boron, or at obtaining bulk or even monocrystalline materials from crude starting materials.

The high melting point of boron (~2200°C) and more over its extreme chemical reactivity with most refractory materials make it delicate to use conventional crystal-growth methods. Few materials are chemically compatible with liquid boron; the one used most frequently, namely hexagonal boron nitride, probably introduces contamination of boron by nitrogen and the impurities that it contains even when it has been out-gassed at high temperatures (HORN, 1960; STARKS and MEDCALF, 1960; NIEMYSKI et al., 1965).

These observations indicate why the preferred melting techniques are either the use of a "self-crucible" or floating-zone methods, and also make clear why floating methods are favored over crystal-pulling techniques for growing crystals of β-rh. boron (HORN, 1960; STARKS and MEDCALF, 1960; GREINER, 1960; GAULE et al., 1960; HINZ and WIRTH, 1965; ARMINGTON et al., 1965; NIEMYSKI et al., 1965; KING et al., 1965; ARMINGTON et al., 1970; DIETZ and HERRMANN, 1965, 1970).

Floating-Zone Melting of Boron. In the floating-zone technique, melting takes place locally along a vertical bar made of the material to be treated. The equilibrium of the liquid zone results from two types of opposing forces: the weight of the liquid drop and the forces of surface tension. This equilibrium is more easily maintained if the material is light and the surface tension of the liquid is great, which is precisely the situation for boron. So, excluding contamination problems, floating-zone melting is particularly recommended in the case of boron.

In order to increase the stability of the liquid zone, its height, which is generally about the diameter of the rod to be treated, has to be reduced by using special devices to concentrate the heat energy. Sometimes, electron gun techniques are used for heating (ARMINGTON et al., 1965; KING et al., 1965). Most often, however, the heating is done with a high frequency current. The latter method arises some problems, due to the fact that boron is a good electrical conductor only at high temperatures. Direct coupling between the induction coil and a boron bar can be achieved only above about 1000 °C, and so it is necessary to preheat boron locally with a graphite or tantalum susceptor, which is frequently one of the holders of the bar.

Zone melting of boron is carried out either in vacuum or in a purified rare-gas atmosphere, which is sometimes doped with a small amount of hydrogen. However, melting in a gas atmosphere seems to cause the bars to be fragile.

Floating-zone melting requires the use of boron bars, that are already sufficiently compact to be heated by direct coupling with an induction coil. These bars can be produced by melting in horizontal boats or by sintering with a binding substance (B_2O_3) which is later eliminated at high temperatures, involving, however, a certain risk of contamination (GREINER, 1960; NIEMYSKI et al., 1965). Methods of boron preparation by the hot filament technique yield rods that can be mounted directly in a zone melting apparatus. If the deposition has been carried out on metal filaments, these must be extracted mechanically or dissolved chemically as already mentioned. The best bars are, of course, those that have been prepared on boron filaments (STARKS and MEDCALF, 1960).

Distribution Coefficients of Impurities between Liquid and Solid Boron. Zone melting not only allows densification of the treated material, but can also result in a significant improvement in purity, as has been proven in the case of silicon, germanium and many metals. The distribution coefficients, K, of the various impurities between liquid and solid phases have been determined in the case of boron and are listed in Table 11. The elements for which $K < 1$ are concentrated at the bottom end of the bar, whereas carbon, aluminium and titanium, which have K values above 1, are concentrated at the top end of the bar. Silicon, copper and titanium, which have K coefficients close to 1, are difficult to separate by zone melting. According to HINZ and WIRTH, zone melting could allow, however, separation of carbon after many passes (HINZ and WIRTH, 1965). But not all investigators are in agreement regarding the effect of zone melting for this impurity (ARMINGTON et al., 1965).

Zone melting of boron, especially when carried out in vacuum, favours a considerable decrease in the content of volatile impurities. This is the case particularly for magnesium, oxygen and hydrogen.

Crystalline Nature of Boron Obtained by Floating-Zone Melting. Solidification of liquid boron yields exclusively the β-rh. modification. Boron bars that have undergone several zone-melting passes generally consist of a core that is mostly monocrystalline, surrounded by a polycrystalline shell, 1 to 2mm thick. In the monocrystalline part, the c-axis of the crystal lattice tends to be oriented parallel to the axis of the bar.

Conclusions on the Use of Zone Melting. The purification that can be expected from zone melting in the case of boron is relatively limited in regard to the major impurities, silicon and carbon. On the contrary, it is highly beneficial in removing volatile impurities, especially hydrogen. Table 12 gives the contents of various inpurities in boron samples made from BBr_3 and BI_3 and then zone-melted, compared to those

Table 11. Values of the partition coefficient K for different impurities in liquid and solid boron

	Si	Cu	Fe	V	Ca	Ni	Cr	Mg	Al	Ti	C
Starks and Medcalf, 1960	<1	<1	<1	<1	<1	-	-	>1	-	-	-
Horn, 1960	<1	<1	≪1	-	-	<<1	<<1	-	>>1	>1	>>1
Hinz and Wirth, 1965	-	-	-	-	-	-	-	-	-	-	4

Table 12. Impurity content (in ppm) of boron samples prepared from BBr_3 or BI_3 and submitted to zone-melting, and impurity content of the boron that has been used for the synthesis of the halides. (From ARMINGTON et al. 1965)

Elements	Emission spectroscopy			Mass spectroscopy	
	Initial boron	Bromide process boron	Iodine-process boron	Bromide-process boron	Iodide-process boron
Al	99	ND	ND	600(?)	1
Si	16,000	1-5	3	100(?)	11
Ca	ND	ND	ND	10	100(?)
Cu	358	ND	2	1.5	-
Ni	96	ND	ND	1	-
Mg	3,000	ND	ND	-	-
Fe	2,000	ND	ND	130	3.6
Mn	2,000	ND	ND	-	-
Sn	9	ND	ND	-	-
Cr	14	ND	ND	200	-
Ti	12	ND	ND	-	-
Co	11	ND	ND	100	-
Mo	-	-	-	4	-
W	-	-	-	30	13
C	-	-	-	500	>500

of the raw boron used in the preparation of halides. This table shows that even taking into account the uncertainties inherent in methods of trace analysis, the BI_3 boron iodide is probably a better intermediate than BBr_3 for obtaining very pure boron.

Floating-zone melting, in as much as it does not require a crucible, appears to be an easily applied method for the preparation of massive highly cristalline boron samples.

c) Conclusions Regarding Methods of Preparing Boron

The problems of boron crystal chemistry and of the preparation of various boron forms are now practically all solved, except for the crystal structure of β-tetr. modification which remains unknown, and for obtaining boron free of carbon.

There are two aspects to the problem of preparing boron, depending on whether it is to be used for the synthesis of its compounds (borides, for example) or for the study of its physical properties. In the first case, it seems to us that α-rh. boron prepared by hydrogen reduction of BBr_3 in fused-silica tubes or in fluidized beds is actually the best material available to the chemist. This product has an elemental boron content greater than 99% and the main impurity, bromine, is not harmful in many applications inasmuch as it is eliminated by volatilization during syntheses carried out at high temperatures. The fact that it is finely divided and essentially only the α-rh. modification gives it a high reactivity that can be most advantageous when it is desired to react boron with elements for which it does not have a great affinity. β-rh. boron obtained by the allotropic transformation of α-rh. boron is more pure since it is free of bromine (> 99.7%) and is specially recommended for solid-state preparations and for studying the solid solutions that it forms with many elements.

As to boron specimens intended for the study of the physics of the solid state, decisive progress has been made in the last ten years in the characterization and preparation of highly pure boron in various crystalline forms. Single crystals, the size of which still remains limited, belonging to α-rh., β-tetr. allotropes or to α-tetr. phases, can be obtained in a reproductible manner, by thermal decomposition of BI_3 or by hydrogen reduction of BBr_3, either pure or doped with carbon or nitrogen. Controlled solidification of alloys of boron with metals such as platinum allows for preparing small single crystals of α-rh. boron. The development of this method provides hopes that larger crystals might be obtained quite easily. And finally, monocrystalline bars of high purity can be prepared by chemical vapor deposition in the vapor phase upon boron substrates followed by floating-zone melting.

Acknowledgments. The author would like to thank Dr. J.S. KASPER of the General Electric Research and Development Center for many discussions concerning the crystal chemistry aspects of boron-rich phases and for his kind help in the writing of the final form of this review.

References

AMBERGER, E., DIETZE, W.: Zur Bildung von α-rhomboedrischem, rotem Bor. Z. anorg. allg. Chem., 332, 131-139 (1964)

AMBERGER, E., DIETZE, W.: Synthesis of red, α-rhombohedral boron. In: Boron (Vol. II), Preparation, Properties and Applications. (G.K. GAULÉ, ed.) pp. 1-7. New-York: Plenum Press 1965

AMBERGER, E., DIETZE, W., DRUMINSKI, M., PLOOG, K.: Formation of modifications of boron and boron carbides. In: Boron (Vol. III), Preparation and Properties (T. NIEMYSKI, ed.) pp. 133-141. Warsaw: Pol. Sci. Publ. 1970

AMBERGER, E., PLOOG, K.: Zur Bildung der tetragonalen Modifikation des Bors. J. Less Common Metals 15, 240-241 (1968)

AMBERGER, E., PLOOG, K.: Bildung der Gitter des reinen Bors. J. Less Common Metals, 23, 21-31 (1971)

AMBERGER, E., POLBORN, K., RAUH, P.A.: The boron-rich boron nitride, boron phosphide, boron arsenide and transition borides. Proc. 5th. Int. Symp. Boron and Borides (Bordeaux 1975); to be published in J. Less Common Metals (1976)

AMBERGER, E., SCHMIDT, H.: Stickstoffinduziertes Gitter beim Bor: I-tetragonales Bornitrid $(B_{12})_4B_2N_{1-2}$. Z. Naturforschg. 26b (7), 641-646 (1971)

ANDERSSON, S., CALLMER, B.: The solubilities of copper and manganese in β-rhombohedral boron as determined in $CuB_{\sim 28}$ and $MnB_{\sim 23}$ by single crystal diffractometry. J. Solid State Chem. 10 (3), 219-231 (1974)

ANDERSSON, S., LUNDSTRÖM, T.: The solubility of chromium in β-rhombohedral boron as determined in $CrB_{\sim 41}$ by single crystal diffractometry. J. Solid State Chem., 2, 603-611 (1970)

ANDRIEUX, L.: Recherches sur l'électrolyse des oxydes métalliques dissous dans l'anhydride borique ou dans les borates fondus. Nouvelles méthodes de préparation du bore amorphe, des borures et de quelques métaux. Ann. Chimie (Paris), 10 (12), 423-507 (1929)

ANDRIEUX, J.L., DEISS, W.J.: Sur l'obtention du bore par électrolyse ignée. Bull. Soc. Chim., 838-841 (1955)

APPENHEIMER, S., NIEMYSKI, T., JABLONSKI, J.: Investigation of hydrogen absorption in boron. In: Boron (vol. III), Preparation and Properties (T. NIEMYSKI ed.) pp. 29-33 Warsaw: Pol. Sci. Publ. 1970

ARMINGTON, A.F., BUFORD, J.T., STARKS, R.J.: Ultrapure boron halide intermediates. In: Boron (Vol. II), Preparation, Properties and Applications (G.K. GAULÉ, ed.) pp. 21-33. New-York: Plenum Press 1965

ARMINGTON, A.F., DILLON, G.F., MITCHELL, R.F.: The preparation of high-purity boron via the iodide. Trans. Metallurg. Soc. A.I.M.E., 230, 350-354 (1964)

ARMINGTON, A.F., STARKS, R.J., BUFORD, J.T.: The determination of carbon in iodide boron. In: Boron (Vol. III), Preparation and Properties (T. NIEMYSKI, ed.) pp. 21-28. Warsaw: Pol. Sci. Publ. 1970

AUGUSTIN, F., VUILLARD, G.: personal communication (1974)

BEAN, K.E., MEDCALF, W.E.: Utilization of boron filaments in vapour-phase deposition of boron. In: Boron (Vol. I), Synthesis, Structure and Properties (J.A. KOHN, W.F. NYE, G.K. GAULÉ, eds.) pp. 48-57. New York: Plenum Press 1960

BEAN, K.E., MEDCALF, W.E.: Boron deposition method. U.S. Pat. 3, 053, 636 (Sept. 11, 1962)

BECHER, H.J.: Über das Berylliumboride BeB_{12} mit β-Bor-Struktur. Z. anorg. allg. Chem. 306, 266-272 (1960)

BECHER, H.J.: Über die Bildung der Tetragonalen Bormodifikation durch Substitution der Berylliumatome im Gitter des BeB_{12}. Z. anorg. allg. Chem. 321, 217-223 (1963)

BECHER, H.J.: Tetragonal boron and borides with similar structures. In: Boron (Vol. II), Preparation, Properties and Applications (G.K. GAULÉ, ed.) pp. 89-95. New York: Plenum Press 1965

BECHER, H.J., NEIDHARD, H.: Siliciumhaltiges β-AlB_{12} und $AlBeB_{24}$ vom Typ des BeB_{12}. Acta Cryst. B24, 280-281 (1968)

BECHER, H.J., SCHÄFER, A.: Über die Modifikationen des Bors. Versuche zur Einlagerung von Beryllium in das Borgitter. Z. anorg. allg. Chem. 306, 260-265 (1960)

BOIRET, M.: Reduction du tribromure de bore par l'hydrogène sur substrat métallique.

Structure des dépots de bore et phénomènes de diffusion dans le substrat. Thèse 723 Univ. Bordeaux (1969)
CARLSSON, J.O., LUNDSTRÖM, T.: The solution hardening of β-rhombohedral boron. J. Less Common Metals. 22, 317-320 (1970)
CLARK, H.K., HOARD, J.L.: The crystal structure of boron carbide. J. Amer. Chem. Soc. 65, 2115-2119 (1943)
COOPER, H.S.: Method of manufacturing boron trichloride. U.S. Pat. 2,369, 212 (Feb. 13, 1945a)
COOPER, H.S.: Method of purifying boron trichloride. U.S. Pat. 2,369, 214 (Feb. 13, 1945b)
COOPER, H.S.: Electrolytic method of making boron. U.S. Pat. 2,572, 248 (Oct. 23, 1951a)
COOPER, H.S., Electrolytic production of elemental boron. U.S. Pat. 2,572, 249 (Oct. 23, 1951b)
COOPER, H.S., SHAEFER, J.C.: Production of boron by fused salt bath electrolysis. U.S. Pat. 2, 918, 417 (Dec, 22, 1959)
COOPER, H.S.: In: Rare Metal Hand Book (C.A. HAMPE, ed). Chap. 5. Princeton (N.J.): v. Nostrand-Rheinhold 1961
CUEILLERON, J.: Contribution à l'étude du bore. Ann. Chim. (Paris) 19. 459-486 (1944)
CUEILLERON, J., CRUIZIAT, B.: Préparation du bore de pureté élevée par reduction du chlorure de bore par hydrogène dans un plasma d'induction d'argon. Bull. Soc. Chim. France 4, 1207-1208 (1973)
DECKER, B.F., KASPER, J.S.: The crystal structure of a simple rhombohedral form of boron. Acta Cryst. 12 (7), 503-506 (1959)
DECKER, B.F., KASPER, J.S.: The crystal structure of a nickel boride related to tetragonal boron. Acta Cryst. 13, 1030-1031 (1960)
DIETZ, W., HERRMANN, H.: Conductivity, Hall effect, Optical absorption and band gap of very pure boron. In: Boron (Vol. II), Preparation, Properties and Applications (G.K. GAULÉ, ed.) pp 107-118. New York: Plenum Press 1965
DIETZ, W.H., HERRMANN, H.A.: Investigations on the conduction mechanism in boron single crystals. In: Boron (Vol. III), Preparations and Properties (T. NIEMYSKI, ed.) pp 195-205. Warsaw: Pol. Sci. Publ. 1970
DUPUY, E., HACKSPILL, L.: Sur la loi de dilatation thermique du bore. C.R. Acad. Sci. (Paris) 197, 229-230 (1923)
EICK, H.A.: Plutonium borides. Inorg. Chem. 4 (8), 1237-1239 (1965)
FINN, J.M. Jr., NELSON, E.N., MERCURI, R.A.: Improvements in and relating to the production of boron. Brit. Pat. 857, 060 (Dec. 29, 1960)
FORMSTECHER, M., RYSKEVIC, E.: Preparation du bore cristallisé pur et mesure de sa dureté. C.R. Acad. Sci. (Paris) 221, 747-749 (1945)
FREYMANN, R., STIEBER, A.: Effet de la température et des radiations visibles et infrarouges sur la résistance électrique du bore. C.R. Acad. Sci. (Paris) 199, 1109-1110 (1934)
GAMBLE, E.C., GUILMONT, P., STIEF, F.: The reaction of boron fluoride with aluminium chloride or bromide. J. Amer. Chem. Soc. 62, 1257-1258 (1940)
GAULÉ G.K., BRESLIN, J.T., PASTORE, J.R., SHUTTLEWORTH, R.A.: Optical and electrical properties of boron and potential application. In: Boron, Synthesis, Structure and Properties (J.A. KOHN, W.F. NYE, G.K. GAULÉ, eds.) pp 159-174. New York: Plenum Press 1960
GEIST, D.: Electron paramagnetic resonance, electrical conductivity, and impurity diffusion in doped boron. In: Boron (Vol. II). Preparation, Properties and Applications (G.K. GAULÉ, ed.) pp. 203-214. New York: Plenum Press 1965
GEIST, D., KLOSS, R., FOLLNER, H.: Verfeinerung des β-rhomboedrischen Bors. Acta Cryst. B26 (11), 1800-1802 (1970)
GIESE, R.F. Jr., ECONOMY, J., MATKOVICH, V.I.: Interstitial derivatives of β-boron. Z. Kristallgr. 122 (S), 144-147 (1965)
GREINER, E.S.: The floating-zone melting of boron and the properties of boron and its alloys. In: Boron, Synthesis, Structure and Properties (J.A. KOHN, W.F. NYE, G.K. GAULÉ, eds.) pp. 105-109. New York: Plenum Press 1960
GUETTE, A: Etude cristallochimique des borures de magnésium riches en bore. Les phases $MM'B_{14}$ (M = Mg, Na; M' = Mg, Al, B). Thèse 462 Univ. Bordeaux-I (1974)

et Centre Documentation du CNRS Paris A.O. 10799 (1974)

GUETTE, A., NASLAIN, R., GALY, J.: Structure cristalline du tétraborure de magnésium. C.R. Acad. Sci. (Paris) 275 (c), 41-44 (1972)

HAAG, H.: Method for the production of a substantially pure boron. U.S. Pat 2, 794, 708 (June 4, 1957)

HACKSPILL L., CUEILLERON, J.: Bore colloidal. C.R. Acad. Sci (Paris) 218, 458-461 (1944)

HACKSPILL, L., STIEBER, A., HOCART, R.: Sur le bore cristallisé. C.R. Acad. Sci. (Paris) 193, 776-778 (1931)

HAGENLOCHER, A.K.: Semiconductor properties of boron. In: Boron, Synthesis, Structure and Properties (J.A. KOHN, W.F. NYE, G.K. GAULÉ, eds.) pp. 128-134. New York: Plenum Press 1960

HAGENMULLER, P., NASLAIN, R.: Sur la préparation de bore pur sous les formes rhomboédriques α et β. Rev. Hautes temp. et Réfract. 2, 225-236 (1965)

HALLA, F., WEIL, R.: X-ray investigation of "crystallized boron". Z. Kristallogr. 101, 435-450 (1939)

HINZ, I., WIRTH, H.: Purity of boron produced by the decomposition of diborane and subsequent zone-melting. In: Boron (Vol. II), Preparation, Properties and Applications (G.K. GAULÉ, ed.) pp. 9-20. New York: Plenum Press 1965

HOARD, J.L., HUGHES, R.E.: Elemental boron and compounds of high boron content: structure, properties, and polymorphism. In: The Chemistry of Boron and its Compounds (E.L. MUETTERTIES, ed.) Chap. 2, pp 25-154, New York, London, Sydney: J. WILEY 1967.

HOARD, J.L., HUGHES, R.E., SANDS, D.E.: The structure of tetragonal boron. J. Amer. Chem. Soc. 80, 4507-4515, (1958)

HOARD, J.L., NEWKIRK, A.E.: An analysis of polymorphism in boron based upon X-ray diffraction results. J. Amer. Chem. Soc. 82, 70-76 (1960)

HOARD, J.L., SULLENGER, D.B., KENNARD, C.H.L., HUGHES, R.E.: The structure analysis of β-rhombohedral boron. J. Solid State Chem. 1, 268-277 (1970)

HORN, F.H.: The crystallization of a simple rhombohedral boron from platinum J. Electrochem. Soc. 106, 905-906 (1959)

HORN, F.H.: Simple rhombohedral boron, preparation and properties. In: Boron, Synthesis, Structure and Properties (J.A. KOHN, W.F. NYE, G.K. GAULÉ, eds.) pp. 110-115. New York: Plenum Press 1960

HORN, F.H., TAFT, E.A., OLIVER, D.W.:On the birefringence of simple rhombohedral boron. In: Boron (Vol. II). Preparation, Properties and Applications (G.K. GAULÉ, ed.) pp 231-234. New York: Plenum Press 1965

HUGHES, R.E., KENNARD, C.H.L., SULLENGER, D.B., WEAKLIEM, H.A., SANDS, D.E., HOARD, J.L.: The structure of β-rhombohedral boron. J. Amer. Chem. Soc. 85, 361-362 (1963)

KAHLENBERG, H.H.: Boron and boron suboxide. Trans. Amer. Electrochem. Soc. 47, 23-61 (1925)

KIESSLING, R.: A method for preparing boron of high purity. Acta Chem. Scand. 2, 707-712 (1948)

KING, I.R., WAVNER, F.E. Jr., TAYLOR, G.R., TALLEY, C.P.: Preparation and characterization of single-crystal boron. In: Boron (Vol. II), Preparation, Properties and Applications (G.K. GAULÉ, ed.) pp 45-62. New York: Plenum Press 1965

KOHN, J.A.: Crystallography of the aluminum borides. In: Boron, Syntheses, Structure and Properties (J.A. KOHN, W.F. NYE, G.K. GAULÉ, eds.) pp 75-82. New York: Plenum Press 1960

LAUBENGAYER, A.W., HURD, D.T., NEWKIRK, A.E., HOARD, J.L.: Boron I. Preparation and properties of pure crystalline boron. J. Amer. Chem. Soc. 65, 1924-1931 (1943)

LIPSCOMB, W.N., BRITTON, D.: Valence structure of the higher borides. J. Chem. Phys. 33 (1), 275-280 (1960)

LONGUET-HIGGINS, H.C., ROBERTS, M. de V.: The electronic structure of an icosahedron of boron atoms. Proc. Roy. Soc. A 230, 110-119 (1955)

LUNDSTRÖM, T.: Crystal structure of the solid solutions of transition metals in β-rhombohedral boron. In: Boron, Synthesis, Structure and Properties (F.N. TAVADZE, ed.) pp 44-51. Moscow: Nauka 1974

McCARTY, L.V., CARPENTER, D.R.: The preparation of a new crystalline modification of boron, and notes on the synthesis of boron triiodide. J. Electrochem. Soc.

107, 38-42 (1960)
McCARTY, L.V., KASPER, J.S., HORN, F.H., DECKER, B.F., NEWKIRK, A.E.: A new crystalline modification of boron. J. Amer. Chem. Soc. 80, 2592 (1958)
MATKOVICH, V.I., ECONOMY, J.: Structure of $MgAlB_{14}$ and a brief critique of structural relationships in higher borides. Acta Cryst. B26, 616-621 (1970)
MATKOVICH, V.I., GIESE, R.F. Jr., ECONOMY, J.: Packing of B_{12} groups in boron and boride structures. Z. Kristallogr. 122, 116-130 (1965)
MATTES, R., MAROSI, L., NEIDHARD, H.: Darstellung und Kristallstruktur von $Cu_2Al_{2.7}B_{104}$. J. Less Common Metals 20, 223-228 (1970)
MATTES, R., TEBBE, K.F., NEIDHARD, H., RETHFELD, H.: The crystal structure of BeB_3. Proc. 5th.Int. Symp. Boron and borides (Bordeaux 1975), J. Less Common Metals 47, 29-32 (1976)
MIERZEJEWSKA, S., NIEMYSKI, T.: Vapour growth of boron crystals. J. Less Common Metals 10, 33-37 (1966)
MIKHEEVA, V.I., MARKINA, V. Yu, KRYUKOVA, O.N.: Preparation of amorphous boron of a high degree of purity. II. Physicochemical analysis for the reaction of magnesium with boric anhydride. Russ. J. Inorg. Chem. 2, 1232-1241 (1957b)
MIKHEEVA, V.I., SHAMRAI, F.I., KRYLOVA, E. Ya.: Preparation of amorphous boron of a high degree of purity. I. Reduction of boric anhydride by light metals. Russ. J. Inorg. Chem. 2 (6), 1223-1231 (1957a)
MIKHEEVA, V.I., SHAMRAI, F.I., KRYLOVA, E. Ya.: Preparation of amorphous boron of a high degree of purity. IV. Evaluation of the quality of amourphous boron Russ. J. Inorg. Chem. 2, 1248-1253 (1957c)
MOISSAN, H.: Action des métaux alcalins sur l'acide borique. Etude critique des procédés de préparation du bore amorphe. C.R. Acad. Sci. (Paris) 114, 319-324 (1892a)
MOISSAN, H.: Préparation du bore amorphe. C.R. Acad. Sci.(Paris) 114, 392-397 (1892b)
MURDOCH, H.D., HAMBLYN, S.M.L.: Reaction chimique dans un plasma d'induction. Brev. Fr. 1 582, 154 (Aug. 18, 1969)
NARAY-SZABO, S.V., TOBIAS, C.W.: X-ray powder patterns of boron coated Mo and W filaments. J. Amer. Chem. Soc 72, (2), 1882-1883 (1949)
NASLAIN, R.: Préparation de bore pur sous les formes rhomboèdriques α et β. Les borures alcalins. Thèsis Univ. Bordeaux n° 188 (1967) et Centre Doc. du CNRS, Pairs, n° AO 1332 (1967)
NASLAIN, R.: Alpha-rhombohedral boron. In: Inorganic Syntheses, vol XII (R.W. PARRY, ed.) Chap. 4, pp. 145-152. New York: Mc. Graw Hill 1970
NASLAIN, R.: Boron Syntheses. In: Preparative Methods in Solid State Chemistry (P. HAGENMULLER, ed.) pp. 439-485. New York and London: Academic Press 1972
NASLAIN, R., BARRET, M.: Sur la solubilité du magnésium dans le bore rh-β(personal communication) 1972
NASLAIN, R., ETOURNEAU, J.: L'hexaborure de potassium. C.R. Acad. Sci. (Paris) 263 (c), 484-487 (1966)
NASLAIN, R., ETOURNEAU, J. HAGENMULLER, P.: Le système bore-potassium. Bull. Soc. Chim Fr. 2529-2538 (1967)
NASLAIN, R., ETOURNEAU, J., HAGENMULLER, P.: Preparation of pure α and β-rhombohedral boron: The alkali borides. In: Boron (Vol. III), Preparation and Properties (T. NIEMYSKI, ed.) pp. 35-47. WARSAW: Pol. Sci. Publ. 1970
NASLAIN, R., ETOURNEAU, J., KASPER, J.S.: Structure cristalline de la phase δ du système bore-thorium. Un nouveau borure à structure type YB_{66}. J. Solid State Chem. 3, 101-111, (1971)
NASLAIN, R., GUETTE, A., BARRET, M.: Sur le diborure et le tétraborure de magnésium. Considérations cristallochimiques sur les tétraborures. J. Solid State Chem. 8, 68-85 (1973)
NASLAIN, R., GUETTE, A., HAGENMULLER, P.: Crystal Chemistry of some boron-rich phases. Proc. 5th. Int. Symp. Boron and Borides (Bordeaux 1975), J. Less Common Metals 47, 1-16 (1976)
NASLAIN, R., GUETTE, A., KASPER, J.S.: Some aspects of the crystal chemistry of the higher borides. In: Boron, Synthesis, Structure and Properties (F.N. TAVADZE, ed.) pp. 160-171. Moscow: Nauka 1974.
NASLAIN, R., KASPER, J.S.: The crystal structure of the φphase in the boron-sodium system. J. Solid State Chem. 1, 150-151 (1970)

NASLAIN, R., SOUM-POUYALET, B., ETOURNEAU, J., HAGENMULLER, P.: Un nouveau procédé de préparation du bore pur: la réduction de BBr_3 par l'hydrogène en lit fluidisé. Mater. Res. Bull. 8, 825-836 (1973)
NASLAIN, R., SOUM-POUYALET, B., ETOURNEAU, J., HAGENMULLER, P.: Procédé pour l'obtention de bore pur et dispositif pour sa mise en oeuvre. Brevet Fr. 7, 212, 260 (apr. 1972)
NEWKIRK, A.E.: Preparation and chemistry of elementary boron. In: Borax to Boranes. Advances in Chemistry Series 32 (R.F. GOULD, ed.) pp. 27-41 - Washington: Amer. Chem. Soc. 1961
NEWKIRK, A.E.: Elemental Boron. In: Boron, Metallo-boron Compounds and Boranes (R. ADAMS, ed.) Chap. 4, pp. 233-299. New York, London, Sidney: Wiley Interscience 1964
NIEMYSKI, T., OLEMPSKA, Z.: The preparation of pure boron for semi-conductor investigations. J. Less Common Metals 4, 235-243 (1962)
NIEMYSKI, T., PRACKA, R., SZCZERBINSKI, R., FRUKACZ, Z.: Studies on crystallization of pure boron: Boron (Vol. II), Preparation, Properties and Applications (G.K. GAULÉ, ed) pp; 35-43. New York: Plenum Press 1965
NIEMYSKI, T. and ZAWADZKI, W.: Some properties of pure polycristalline boron Phys. Letters, 2 (1), 30-1, (1962)
OLEMPSKA, Z., BADZIAN, A., PIETRZAK, K., NIEMYSKI, T.: Crystalline modifications of boron deposited on boron substrates. J. Less Common Metals 11, 351-359 (1966)
PESHEV, P., ETOURNEAU, J., NASLAIN, R.: Sur le système bore-scandium. Mater. Res. Bull. 5, 319-328 (1970)
PITON, J.P., VUILLARD, G., LUNDSTRÖM, T.: Sur les cristaux de borure de cuivre. C.R. Acad. Sci. (Paris) 278 (c), 1495-1496 (1974)
PLOOG, K., AMBERGER, E.: Kohlenstoff-indurzierte Gitter beim Bor: I-tetragonales $(B_{12})_4B_2C_2$ und $(B_{12})_4B_2N_2$. J. Less Common Metals 23, 33-42 (1971)
PLOOG, K., SCHMIDT, H., AMBERGER, E., WILL, G., KOSSOBUTSKI, K.H.: $B_{48}B_2C_2$ und $B_{48}B_2N_2$, Zwei Nicht metallboride mit der Struktur des Sog. I-tetragonalen Bors. J. Less Common Metals 29. 161-169 (1972)
Powell, C.F., ISH C.J., BLOCHER, J.M. Jr.: The preparation of high-purity boron by hot-wire technique. In: Boron, Synthesis, Structure and Properties (J.A. KOHN, W.F. NYE, G.K. GAULÉ, ed.) pp. 7-14. New York: Plenum Press 1960
RICHARDS, S.M., KASPER, J.S.: The crystal structure of YB_{66}.Acta Cryst. 25, 237-251 (1969)
SAMSONOV, G.V., OBOLONCHIK, V.A. KOULICHINA, G.N.: Electrolytic production of boron. Zh. Prikl. Khim. 33. 1365-1368 (1960)
SANDS, D.E., CLINE, C.F., ZALKIN, A., HOENIG, C.L.: The beryllium-boron system. Acta Cryst. 14 (3), 309-310 (1961)
SCHOTT, S., HANSLEY, V.L.: Boron production. U.S. Pat. 3, 086, 847 (Apr. 23, 1963)
SCHUMB, W.C., GAMBLE, E.L., BANUS, M.D.: Lower iodides of boron. J. Amer. Chem. Soc. 71, 3225-3229 (1949)
SECRIST, D.R.: Compound formation in the systems: Lithium-carbon and lithium-boron, J. Amer. Ceram. Soc. 50 (10), 520-523 (1967)
SHAMRAI, F.I., MIKHEEVA, V.I., KRYLOVA, E. Ya.: Preparation of amorphous boron of a high degree of purity. III. Refining of amorphous boron. Russ. J. Inorg. Chem. 2, 1242-1247 (1957)
SIRTL, E.: Process of producing compact boron, particularly in monocrystalline form. U.S. Pat. 3,160, 476 (Dec. 8, 1964)
SITARIK, J.P., ELLIS, W.C.: Preparation and morphology of boron filamentary crystals grown by the vapor-liquid-solid mechanism. J. Appl. Phys. 37 (6), 2399-2401 (1966)
SOWA, F.J.: Process for producing elemental boron. U.S. Pat. 2, 465, 989 (Apr. 5, 1949)
SPEVACK, J.S.: Process for preparing boron. U.S. Pat. 2, 685, 501 (Aug. 3, 1954)
STARKS, R.J., MEDCALF, W.E.: Growth of boron crystals by the czochralski and floating-zone methods, In: Boron, Synthesis, Structure and Properties (J.A. KOHN, W.F. NYE, G.K. GAULÉ, eds.) pp. 59-69. New York: Plenum Press 1960
STERN, D.R., LYNDS, L.: Pure crystalline boron. J. Electrochem. Soc. 105, 676-682 (1958)

SULLENGER, D.B., PHIPPS, K.D., SEABAUGH, P.W., HUDGENS, C.A., SANDS, D.E., CANTRELL, J.S.: Boron modifications produced in an inductioncoupled argon plasma. Science 163, 935-937 (1969)

TALLEY, C.P., LA PLACA, S., POST, B.: A new polymorph of boron. Acta Cryst., 13 (3), 271-272 (1960a)

TALLEY, C.P., LINE, L.E.Jr., OVERMAN, Q. D.: Preparation and properties of massive amorphous elemental boron. In: Boron, Synthesis, Structure and Properties (J.A. KOHN, W.F. NYE, G.K. GAULÉ, eds.) pp. 94-104. New York: Plenum Press 1960b

TALLEY, C.P., POST, B., LA PLACA, S.: A new modification of elemental boron. In: Boron, Synthesis, Structure and Properties (J.A. KOHN, W.F. NYE, G.K. GAULÉ, eds.) pp. 83-85. New York: Plenum Press 1960c

UNO, R.J.: On the electrical properties of polycristalline boron. J. Phys. Soc. Japan 13 (7), 667-675 (1958)

URRY, G.: Boron halides. In: The Chemistry of Boron and its Compounds (E.L. MUETTERTIES, ed.) Chap. 6, pp. 325-375. New York, London, Sidney: Wiley and Sons 1967

WALD, F.: On the stability of the red α-rhombohedral boron modification. In: Boron (Vol. III), Preparation and Properties (T. NIEMYSKI, ed.) pp. 103-108. Warsaw: Pol. Sci. Publ. 1970

WALD, F., ROSENBERG, A.J.: Constitutional investigations in the boron-platinum system. Trans. AIME 233, 796 (1965)

WÄPPLING, R., HÄGGSTRÖM, L., DEVANARAYANAN, S.: Mossbauer investigation of iron in boron. Physica Scripta 5, 97-98 (1972)

WAVNER, F. E. Jr.: Boron filaments. In: Modern Composite Materials(L.J. BROUTMAN, R.H. KROCK, eds.) Chap. 10, pp. 244-269. Reading (Mass.): Addison-Wesley Publish. Co. 1967

WEAST, R.C.: Handbook of Chemistry and Physic,49th Edition. Cleveland (Ohio): Chem. Rubber Publ. Co. 1968

WEINTRAUB, E.: Preparation and properties of pure boron. Trans. Amer. Electrochem. Soc. 16, 165-184 (1909)

WENTORF, R.H. Jr.: Boron-another form. In: Boron (Vol. II), Preparation, Properties and Applications (G.K. GAULÉ, ed.) pp. 97-100. New York: Plenum Press 1965

WILL, G., KOSSOBUTZKI, K.H.: X-ray diffraction analysis of $B_{50}C_2$ and $B_{50}N_2$ crystallizing in the "tetragonal boron" lattice. Proc. 5th. Int. Symp. on Boron and Borides (Bordeaux 1975), J. Less Common Metals 47, 33-38 (1976)

YANNACAKIS, J., NIES, N.P.: Preparation of crystalline boron. In: Boron, Synthesis, Structure and Properties (J.A. KOHN, W.F. NYE, G.K. GAULÉ, eds.) pp. 38-41. New York: Plenum Press 1960

C. Preparations and Properties

I. Chemical Properties of Boron

J. CUEILLERON and F. THEVENOT

1. Introduction

It is impossible to describe in detail the chemical properties of elementary boron in a few pages. The object of this article is to give general information. A more complete study of the reactions of boron with the other elements of the periodic table is presented in Chapters C VI-XIII of this volume.

Borax was used as early as the beginning of the Middle Ages, particularly since the XVIIth century, as a flux for fusion of welding of metals. The nature of borax and boric acid was particularly studied. In 1808, boron was isolated simultaneously by DAVY in England and by GAY-LUSSAC and THENARD in France, each by an independent method.

Soon after its discovery, the chemical properties of boron were rigorously studied, yet even now, in spite of a large number of studies, they are relatively poorly known. Boron in fact presents very complex properties, and, except for its high reducing power, shows only few analogies with the neighboring elements of the periodic system (aluminum, carbon and silicon).

While barely reactive at room temperature, boron reacts strongly at high temperatures with almost all the elements of the periodic system. Moreover, its reducing properties allow it to react with numerous compounds and, as in the case of oxygenated or halogenated compounds, sometimes in a violent manner.

The experimental and theoretical study of the chemistry of boron is made difficult by some of its physical and chemical properties. The following points must be particularly considered.

Due to its high reactivity, the preparation of pure boron is extremely difficult. Purification of boron samples is further complicated by the fact that impurities are mainly inserted in the crystal lattice, combined with boron or combined together.

Due to its extreme hardness, pulverization of boron introduces impurities from the crushers and causes a superficial oxidation of the grains.

In general, boron reacts only at high temperatures, attacking the refractory compounds containing oxides, and combining with graphite and refractory metals. In order to avoid such reactions, crucibles made of hexagonal boron nitride may be used or metallic crucibles (e.g. copper or silver) cooled by water circulation. Yet even under very high vacuum or under controlled atmosphere boron reacts with traces of oxygen or nitrogen or with carbonaceous residues from the oils of vacuum pumps. Sometimes in order to limit these drawbacks, the preparation of solid boron compounds is performed by sintering. In such processing it is very difficult to control the homogeneity of the products and particularly their stoichiometry, because of differences between vapor pressures of boron (which is volatile) and the other elements. It has been shown in many cases that impurities have a very important influence on the properties of boron and on its derivatives (reactivity, stabilization of crystalline species, deformation of the crystal lattice, etc.).

The original studies of the chemical properties of boron were done with samples containing nearly fifty % impurities. Gradually, boron of increasing purity was made available. Today, workers mainly use "Moissan's boron" (amorphous boron prepared by the reduction of boric anhydride with, at best, 92% purity) and "Cooper's boron" (β-rhombohedral boron, obtained by electrolysis of molten salts which can reach a purity of 99%). Some workers use boron obtained by thermal decomposition of boron dihydride. This product, obtained at relatively low temperatures, is very fine and reactive, and absorbs significant quantities of ambient gases in contact with air. Other workers use boron prepared by reduction of boron halides or by dissociation of boron iodide. These samples are of high purity which is in most cases unverified. They are generally well crystallized and their reactivity is small.

In the literature, often only the commercial origin of boron used is indicated; more rarely, the boron content of the samples is given; exceptionally, the main impurities are determined. This lack of accurate data on the purity of the products can be explained by the analytical difficulties, (analysis of elemental boron, of bonded boron and of the impurities, THEVENOT and VIALA, 1975; THEVENOT and CUEILLERON, 1977). Furthermore, to provide an accurate description of the chemical properties of a boron sample, the physical state of samples must be considered. Yet in the literature, the granulation or the crystalline state of the boron used is seldom indicated. Boron reactivity depends upon the grain size and changes greatly from one allotropic variety to the other (amorphous boron being generally more reactive than crystallized boron, α-rhombohedral boron being more reactive than β-rhombohedral boron) (NASLAIN et al., 1967).

These remarks provide some understanding as to why the thermodynamic data relating to boron reactions are not very accurate, why reaction temperatures are poorly defined and reaction kinetics poorly established. There is considerable controversy about the existence as well as the exact formula of a great number of boron derivatives.

Because its ionization energy is high, boron tends to form compounds with a well-marked covalent character. Yet using the classical theories of valency, the electron deficiency does not permit understanding of the existence of many of its compounds. Thus, the complex theory of bonding with three or more bond centers had to be introduced. The electron deficiency which exists in numerous compounds results in the tendency of boron atoms to be grouped together in chains, in hexagonal cycles or in triangles or squares; these geometrical figures are also found in more complex assemblies, such as octahedrons, cubo-octahedrons, and icosahedrons. These assemblies behave as entities to form crystals of boron or its derivatives. Furthermore, the icosahedrons (or fragments of icosahedrons) are found in numerous hydrogenated derivatives of boron.

The low atomic number of boron introduces some difficulties in the study of its derivatives by X-rays. The presence of two important isotopes in natural boron necessitates, in certain cases, synthesis of compounds using boron enriched with a mass 10 isotope, in order to study their structure.

The preceding remarks may help in understanding the difficulties in establishing the varied and usually very complex structures of boron and its derivatives.

2. Reactions of Elementary Boron

a) With Hydrogen

Even though the chemistry of boron hydrides and carboranes has been extensively studied, the direct combination of boron with hydrogen is possible only at very high temperatures and at very low efficiency even after soaking the gases after reaction. In fact boron hydrides are not stable at temperatures at which the reaction takes place (CUEILLERON and PICHAT, 1964; DEVYATYKH and YUSHIN, 1969; BIRYUKOV et al., 1970).

In suspension in water, even in a colloidal state, boron does not react with native hydrogen.

b) With IVth B-Group Elements

In this group, only carbon and silicon react with boron. Compounds of boron with germanium, tin or lead do not seem to exist.

A considerable amount of research has been carried out on boron and carbon compounds. These compounds have some practical applications; because they are refractory, they have particularly good mechanical properties, are very hard and are good chemical reducers.

The combination of boron with carbon takes place at above 1000°C, under vacuum, in controlled atmosphere or in molten metals. Carbides can be obtained by sintering or by working with a plasma torch (injection of boron powder and methane, propane or butane). The synthesis of carbides has also been studied under high temperatures and pressures. After reaction, it is very difficult to eliminate an excess of either boron or carbon.

In addition to the unstable groups detected by mass spectrometry, the existence of more than twelve compounds of boron and carbon has been reported. The formulae expressed in terms of combination of one carbon atom with a number of boron atoms vary from one to at least twenty-four. Many of these species, imprecisely characterized, seem to be only mixtures or solid solutions. In fact, (BOSCH, 1962; AMBERGER et al., 1971; KIEFFER et al., 1971; REKOV and SAMSONOV, 1971; PERKOVA et al., 1973; PLOOG, 1973, 1974a and b; BECHER and THEVENOT, 1974), on the basis of numerous physico-chemical determinations (phase-diagrams, X-ray studies, infra-red studies, electrical and mechanical properties) and of theoretical considerations (e.g. electronic equilibrium, geometrical possibilities), only two well-determined compounds seem to exist. The first one, $B_{13}C_2$, melts without decomposition at around 2450°C, the other, $B_{12}C_3$, often referred to as B_4C, melts in an incongruent way at around 2350°C. This substance ($\Delta H^0_{(298)<B_4C>}$ = -17 Kcal/mole) might simply be a solid solution of carbon in $B_{13}C_2$.

These two substances can be derived from the structure of α-rhombohedral boron. With some lattice deformation a chain of three carbons atoms, or, more probably, a C-B-C chain may be inserted along the three-fold axis of the rhombohedron of boron. Simultaneously, a substitution may occur in the B_{12} icosahedron where a boron atom is replaced by a carbon atom.

The compounds of boron and silicon have been studied less than those of carbon and boron. However, there are considerable similarities in preparation and properties (SAMSONOV and SPLETSOV, 1964; LAMIKHOV, 1968; ETTMAYER et al., 1969).

As with carbides, in addition to the species detected by mass spectrometry, the existence of a number of boron silicides has been reported. In fact, only the following compounds appear to exist: SiB_3 which decomposes above 1400°C, SiB_6 resulting from this decomposition, and SiB_{14}.

SiB_{3-4} is rhombohedral and isomorphous with boron carbide. However, because the silicon atom is larger than the carbon atom, this structure corresponds to the insertion of only two silicon atoms along the three-fold axis in α-rhombohedral boron. Substitution of boron atoms by silicon atoms in icosahedrons also occurs.

c) With Vth B-Group Elements

Boron combines with nitrogen, phosphorus and arsenic. Compounds with antimony or with bismuth do not seem to exist.

With nitrogen, only one compound is obtained directly at high temperature, i.e., boron nitride. This substance has an exothermal heat of formation ($\Delta H = -61$ Kcal/mole). It decomposes above 2000°C. Different parameters of the direct synthesis (temperature, pressure, use of different types of boron, etc. and kinetics of the reaction) have been studied (BOSCH, 1962; BAHN, 1969; AMBERGER and SCHMIDT, 1971; BOROVINSKAYA and MERZHAKOV, 1971).

Boron nitride is a white powder which shows significant similarities with graphite. Its usual form consists of hexagonal $(BN)_3$ cycles instead of the C_6 cycles found in graphite. However, boron nitride does not conduct electricity. It is used as a particularly stable and inert refractory, especially in contact with boron.

A high temperatures boron reacts with different types of nitrogen compounds (ammonia, cyanhydric acid, cyanogen etc.) to give boron nitride. It also reacts with different thiocyanides (potassium or ammonium) producing boron nitride and sulphide. At high temperatures and pressures, boron reacts with different nitrides (lithium, magnesium) and gives boron nitride in the cubic form, the structure and properties of which are similar to those of diamond (DE VRIES and FLEISCHER, 1972).

Boron reacts with phosphorus vapor at about 1000°C. The effects of temperature and pressure upon this reaction have been studied. For boron phosphides, many different formulae have been reported. It appears that during the reaction of boron with phosphorus, the boron derivative BP is formed first. It can be amorphous or cubic and has an exothermal heat of formation ($\Delta H_{cryst} = -22$ to -100 kcal/mole). Upon further heating it decomposes to give $B_{12}P_2$, which is very stable and inert at high temperatures. The structure of this substance is derived from the α-rhombohedral boron form by insertion of two phosphorus atoms along the three-fold axis of the rhombohedron of boron (PERET, 1964; GRAYON et al., 1965; GOOD and MANSON, 1966; THEVENOT, 1968; THEVENOT and CUEILLERON, 1968; GARDNER, 1969; GROSS et al., 1969; AMBERGER and RAUH, 1972; GAL'CHENKO et al., 1972; BECHER and THEVENOT, 1974).

Boron phosphide is also obtained (PERET, 1964; THEVENOT and CUEILLERON, 1968) by reaction of boron upon phosphine or upon certain metal phosphides (zinc, aluminum).

There is little information about arsenides. As with phosphorus the compound BAs is obtained at about 800°C. It decomposes at high temperatures to give $B_{12}As_2$. The influence of temperature and pressure on the synthesis reaction has been studied (ELISEEV et al., 1964; GRAYON et al., 1965; OSUGI et al., 1966).

The structure of boron arsenides are similar to those of phosphides; their thermal stability is lower than those of phosphides.

At high temperatures, boron does not appear to react with arsine or with metal arsenides.

d) With VIth B-Group Elements

Boron combines with oxygen, sulfur and selenium but does not appear to react with tellurium.

Exposed to air at room temperature, boron surface is oxidized. It must be preserved under vacuum or in inert atmosphere (THEVENOT and VIALA, 1975).

The combustion of boron (which gives off a green flame) results normally in boric anhydride (B_2O_3). This vitreous and colorless mass consists of an irregular assembly of planar and triangular figures (BO_3), each boron atom being bound to three oxygen atoms and each oxygen atom to two boron atoms. In a massive form boron burns poorly in oxygen, even under pressure (for example, in a calorimetric bomb). This is caused by the fact that boron is enveloped during the combustion with a liquid layer of B_2O_3 which vaporizes only above 800°C. Consequently, there are many different combustion kinetics dependent upon temperatures. On the contrary, in the powdery state, boron can burn (above a few hundred degrees) and the combustion can become explosive. The possible use of powdered boron as a combustible in aerospace applications (the heat of formation of B_2O_3 is around 300 Kcal/mole) has given rise to numerous studies (ignition temperature under various pressures, and in various gaseous mixtures, techniques of ignition, induction time of reaction, combustion kinetics as a function of granulation, the nature of impurities, crystallinity, and treatments to which it has been submitted (GOOD and MANSON, 1966; RHEIN, 1966; MACEK et al., 1970; McLAIN et al., 1970; BRYANT and SIPPEL, 1971b; KING, 1972; MOHAN and WILLIAMS, 1972; SCHADOW, 1972; SAVELEV et al., 1973b; ZOLOTAR and OZEROV, 1973).

Besides boric anhydride (B_2O_3), many suboxides have been described, whose formulas vary from B_8O to BO. The existence of many of these is uncertain. Using boron and oxygen, it is perhaps possible to obtain the suboxide $(BO)_2$. However, it is well established that suboxide of composition B_6O can be prepared at temperatures above 1000°C, either under pressure, or in the presence of platinum (ZUBOVA and BURDINA, 1971; HOLCOMBE and HORNE, 1972; PETRAK et al., 1972; JEANBLAIN and CUEILLERON, 1973). This boron derivative is harder than boron carbide and is unreactive. It corresponds to the insertion of a chain of two oxygen atoms in the α-rhombohedral boron along its three-fold axis (PETRAK et al., 1972; BECHER and THEVENOT, 1974).

Boron, which is a powerful reducing agent, can react with oxygenated compounds. The reactions generally occur more readily with amorphous boron than with crystalline boron and even more readily if boron is not pure (one can "induce" in certain cases such reactions by adding, for example, a little magnesium to boron).

Cold water seems to react with boron very slowly. In sunlight, boron reacts with boiling water. At red-hot temperatures, water vapor is reduced, accompanied by formation of boric acid and emission of hydrogen. The action upon hydrogen peroxide when concentrated can become very violent.

Boric anhydride at elevated temperature gives, along with other species detectable by mass spectrometry, sub-oxide $(BO)_2$ (in which boron seems to be trivalent) (BLACKBURN et al., 1969). This amber-colored substance is vitreous and amorphous. Under very high pressures, the reaction of boron on boric anhydride leads to suboxide B_6O.

Carbonic anhydride and carbon monoxide are reduced at red-hot temperatures by boron. Molten silica seems to give a complex reaction with boron which, depending upon the experimental conditions, may result in the formation of silicon, SiB_3 or SiB_4 mixed with silicon.

At red-hot temperatures, boron reduces nitrogen oxides N_2O and NO with the formation of boric anhydride and boron nitride. Phosphorous and arsenic oxides react at lower temperatures with the emission of phosphorus and arsenic; sulfur dioxide gives off sulphur.

Generally, metal oxides are reduced by boron at high temperatures resulting in borides or (in the case of an oxide excess) borates. In the presence of carbon, one may obtain borocarbides.

With a certain number of poorly volatile oxides (e.g. titanium, zirconium, hafnium, aluminum, magnesium) the reduction by boron leads to the formation of the suboxide B_2O_2 (BARTON and NICHOLLS, 1966; PESHEV, 1967; FAVRE and THEVENOT, 1971; BESOMBES-VAILHE and TRICHE, 1972; ERMOLENKO et al., 1973).

Certain oxides (lead, tin, antimony, etc.) give a reaction with boron that can be very violent at room temperature in conjunction with grinding or a shock.

Alkaline hydroxides which, in boiling solution, have a slight action on boron, react at high temperatures to form borates. Sodium peroxide reacts at high temperatures in the same way.

Sulphuric acid, like sulpho-chromic acid, reacts slowly with boron. Sulphuric acid gives ill-defined products (DONNET and KOBEL, 1967a and b; DONNET and MESSIET, 1967). Pure nitric acid (or in the presence of sulfuric acid or hydrogen peroxide), as well as nitrohydrochloric acid, reacts with boron. Depending upon the nature of the sample boron, its particle size, the concentration of acid and the temperature, these reactions may become violent.

In solution, phosphoric acid attacks boron only slightly. At high temperatures it is reduced with the emission of phosphorus. Chloric acid is reduced to chlorous acid. Iodic acid is reduced by boron giving off iodine.

At red-hot temperatures, alkaline and alkaline-earth sulphates are reduced by boron to sulphides. (The reactions of cerium sulfate and ammonium persulphate with boron are used for boron analysis.)

While alkaline nitrates are not reduced by boron up to about 400°C, lead nitrate gives a very brisk reaction at room temperature, as also potassium nitrite.

Depending upon the electrode potential (which is difficult to define) boron reduces (in solution) different metal nitrates or chlorides with metal precipitation (e.g. Pt^{4+}, Ag^{+}, Cu^{2+}, Hg^{+}, Pb^{2+}). The reaction is believed to lead, in case of gold, to formation of a suboxide or, in other cases, to a reduction of the metal valence (Fe^{3+}) (DEL DUCA, 1972).

In solution, the alkaline carbonates react with boron slowly. Unlike alkaline-earth carbonates. they are reduced by boron at red-hot temperatures. A brisk reduction takes place for a great number of oxygenated salts (chlorates, arsenites, arsenates, chromates, permanganates, etc.). Some of these reduction can take place at room temperature (permanganate by grinding) or even in solutions (iodates, periodates, etc.).

The reactions of boron with sulphur are less well known than the reactions of boron with oxygen. At high temperature, boron with sulphur produces B_2S_3 ($\Delta H \simeq -57$ Kcal/mole). It is a vitreous compound similar to B_2O_3 and extremely hygroscopic. In addition to the numerous species observed by mass spectrometry, B_2S_5 and BS_2 seem to exist. At high temperatures, boron and B_2S_3 react to produce $(BS)_2$ similar to $(BO)_2$ (HAGENMULLER and CHOPIN, 1962; KIRK and TIMMS, 1967; MEDVEDEVA et al., 1968; ROSSO and DUBUSC, 1969).

Generally, boron reacts with sulphur compounds and gives B_2S_3. Hydrogen sulphide (sometimes with addition of hydrogen) reacts at red-hot temperature and gives B_2S_3 and $(BS_2H)_3$. Certain sulphides (carbon, tin, arsenic, antimony, lead, alkaline metals, alkaline-earth metals) react with boron to give the element (preliminary bounded with sulfur) or to give borides (CUEILLERON et al., 1971).

Our knowledge of the reactions of boron with selenium are rather fragmentary (CUEILLERON and HILLEL, 1967; BORYAKOVA et al., 1969; HILLEL and CUEILLERON, 1972; KORYAZHKIN, 1972).

Aside from unstable groups observed at high temperatures, selenium combines with boron at intermediate temperatures, producing B_2Se_3 which, by dissociation in vapor state, leads to a gaseous mixture of BSe_2 and BSe. The compounds B_4Se and B_2Se also possibly exist.

Reactions of boron with selenium compounds have been given little attention. However, hydrogen selenide (sometimes with addition of hydrogen) reacts at elevated temperatures with boron to give B_2Se_3.

e) With VIIth B-Group Elements

The elements of the group VII B react with boron to give BX_3-type compounds which are particularly sensitive to water action.

While fluorine, even when diluted with air or nitrogen reacts with boron at low temperatures (like ClF_3) (GAL'CHENKO et al., 1972; GARDNER, 1969; GROSS et al., 1969; BRYANT and SIPPEL, 1971a), chlorine, bromine and iodine have to be used between 400°C and 900°C. The reaction temperature varies with the crystalline state, purity, grain size of boron and the purity of the halogenide which, due to the attack of refractories, is not very high when used at high temperatures.

The heats of formation decrease from fluorine to iodine. Boron fluoride is very stable while the iodine can be decomposed at lower temperatures.

The boron trihalogenides are colorless. Fluoride and chloride are gaseous at room temperatures. Bromide is liquid and iodide is solid.

The molecules have the shape of an equilateral triangle in the center of which is a boron atom.

These substances can produce very numerous additional compounds (boron being acceptor). This is particularly the case for fluoride which has catalytic properties similar to those of aluminum chloride.

Since the reaction of fluorine with boron is particularly exothermal (ΔH = -270 Kcal/mole), it has been considered for use in aerospace applications and has given rise to numerous studies, such as dissociation of boron fluoride at elevated temperatures, kinetics and mechanisms of the reaction of boron with molecular or atomic fluorine, with mixed halogenides or with mixtures containing fluorinated derivatives (NF_3, NOF etc., NOF + FH) and influence of temperature and pressure on various reactions (ROSNER and ALLENDORF, 1971).

Bromine water attacks boron slowly at low temperatures, but iodine water seems to be non-reactive.

Because of its affinity for halogens, boron reacts with numerous halogenides.

While the hydracides (such as the non-oxidant) in solution do not react with boron, they are reduced at red-hot temperature and give the boron trihalogenide with the release of hydrogen (SAVELEV et al., 1973a). With the exception of alkaline and the alkaline-earth fluorides, the fluorides are reduced at more or less high temperature with the formation of boron fluoride. A certain number of fluorides give very violent reactions (silver, lead, etc.).

Phosphorus trifluoride and nitrogen trifluoride are reduced at red-hot temperature with the formation of boron trifluoride and the emission of phosphorus or nitrogen.

Except for alkaline and alkaline-earth chlorides the chlorides are reduced at high temperatures with formation of boron chloride. The same is the case for carbon tetrachloride and a certain number of organic chlorides.

Various reactions with bromides and iodides have been reported. Above 1000°C reaction of boron with boron fluoride, chloride and bromide leads to formation of subhalogenides of BX type (KIRK, 1969; KRENEV and EVDOKIMOV, 1972; SAVELEV et al., 1973b).

f) Metals

Boron has less affinity for metals than for non-metals. Though the heats of formation of borides are little known (because their combustion in a classical calorimeter is often impossible) it has been reported that, for the borides of MB type, the enthalpies of formation vary from -20 to -40 Kcal/mole.

Borides are obtained, in most cases, by direct combination of boron with the metal at high temperatures; this must be done under vacuum or under controlled atmosphere. In order to avoid the use of refractories, sintering is often used. Such samples undergo a variety of treatments, such as long stress relieving, hardenings, homogenization, grinding, etc. Yet many borides are obtained by an indirect method, such as heating of boron or of boron carbide with oxide or with an oxygenated compound or the metal sulfide, reduction (particularly by carbon) of a mixture of boron anhydride and metal oxide, electrolysis of a mixture of molten salts, Van Arkel process, precipitation in solution.

Boron combines with almost all metals. While some ten metals give only one boride, the others form a number of compounds. Up to twelve borides have been reported for certain metals.

For all the metals, the existence of more than 200 borides has been reported, though about half of them either do not exist or represent mixtures and solid solutions. Some metal borides exist only when stabilized by impurities while others exist only at high temperatures. These species are represented by about twenty formulae ranging from M_4B (or even M_6B) to about MB_{100}. The uncertainty as to the existence of a great number of borides and their composition is not difficult to understand if one considers the difficulties in synthesis, analysis and purification as well as those in obtaining a single crystal or in determining phase diagrams at high temperatures.

In fact, among the hundred borides which really seem to exist, about three quarters correspond to the formulas MB, M_3B_4, MB_2, MB_4, MB_6 and MB_{12}.

3. Conclusion

In the period of over 150 years of boron chemistry, considerable information on the chemical nature of boron has been obtained. A rather large number of workers have dealt with very difficult experimental conditions to establish a basis for understanding the complicated chemistry of this element.

Needless to say, much remains to be done to refine the accuracy and correctness of data, and numerous systematic and elaborate studies are required in this interesting field of chemistry.

References

AMBERGER, E., DRUMINSKI, M., PLOOG, K.: Boron carbon system in the kinetic range of formation: Pyrolytic formation of carbon - rich boron - carbon phases. J. Less Common Metals 23, 43-52 (1971)

AMBERGER, E., RAUH, P.: Phosphorous incorporated lattice with boron - monoclinic boron - phosphorus phase $B_{14}P$. J. Cryst. Growth 16, 259-64 (1972)

AMBERGER, E., SCHMIDT, H.: Nitrogen - induced lattice of boron - I tetragonal boron nitride $(B_{12})_4$ B_2N_{1-2}. Z. Naturforschg. 26, 641-46 (1971)

BAHN, G.: Investigation of the concept of nitrogen reaction in Ramjet engines, West States Sec. Combust. Inst. WSS. CI 69-12 (1969)

BARTON, L., NICHOLLS, D.: The hydrogenation of boron monoxide to diborane and the reactions of boron and boron carbide with titanium and zirconium dioxide. J. Inorg. Nucl. Chem. 28, 1367-72 (1966)

BECHER, H., THEVENOT, F.: IR spectra of boron carbide and some other compounds of this structural type. 2nd IME Boron Meeting Leeds (1974) and Z. anorg. allg. Chem 410, 274-286 (1974)

BESOMBES-VAILHE, J., TRICHE, H.: Spectrographic analysis of boron monoxide bands and calculation of rotation temperature from one of these bands during titanium diboride formation in a plasma torch. Analusis 1, 135-9 (1972)

BIRYUKOV, V., VINOGRADOV, B., PEREZHOGIN, G., KHOLOPOVA, M.: Gas-phase composition during electric discharge alloying elektron. Tekh. Sb. Upr. Kach. Stand. 5, 96-9 (1970)

BLACKBURN, P., BUCHLER, A., STAUFFER, J.: Thermodynamics of vaporization in the aluminum oxide-boron oxide system, J. Phys. Chem 70, 2469-74 (1969)

BORYAKOVA, V., GRINBERG, Ya.K., ZHUKOV, E., KORYAZKIN, V., MEDVEDEVA, Z.: Phase diagram of the boron-selenium system. Izv. Akad. Nauk SSSR. Neorg. Mater 5, 477-80 (1969)

BOROVINSKAYA, I., MERZHAKOV, A.: Self-propagating high temperature synthesis of refractory inorganic compounds metalloterm. Protsessy, Khim. Met. Mater. Konz. 58-65 (1971)

BOSCH, F.: Synthesis of boron carbide and nitride-silicates. Ind. 27, 587-90 (1962)

BRYANT, J., SIPPEL, N.: Ignition of metal powders in gaseous chlorine trifluoride. J. Spacecr. Rockets 8, 192-3 (1971a)

BRYANT, J., SIPPEL, N.: Amorphous Boron Dust Flames - Combust. Sci. Technol. 3, 145-51 (1971b)

CUEILLERON, J., HILLEL, R.: Preparation of boron selenide B_2Se_3. Bull. Soc. Chim. Fr. 2973-6 (1967)

CUEILLERON, J., LAHET, G., THEVENOT, F., PARIS, R.: New preparation of metal borides, J. Less Common Metals 24, 317-22 (1971)

CUEILLERON, J., PICHAT, P.: Action of hydrogen on boron. Bull. Soc. Chim. Fr. 10, 2547-8 (1964)

DEL DUCA, B.: Electrochemical behavior of the boron anode in aqueous solutions. N.A.S.A. Tech. Note T.N.D. 6589 (1972)

DE VRIES, R., FLEISCHER, J., Phase equilibrium pertinent to the growth of cubic boron nitride. J. Cryst. Growth 13-14, 88-92 (1972)

DEVYATYKH, G., YUSHIN, A.: Reaction of boron with hydrogen. Zh. Neorg. Khim. 14, 2015-8 (1969)

DONNET, J., KOBEL, L.: Superficial oxidation of amorphous boron by sulfuric acid. Rev. Chim. Miner. 4, 249-67 (1967a)

DONNET, J., KOBEL, L.: Superficial oxidation of amorphous boron with sulfuric acid. Rev. Chim. Miner. 4, 399-412 (1967b)

DONNET, J., MESSIET, J., Oxidation of amorphous boron, Bull. Soc. Chim. Fr. 2963-7 (1967)

ELISEEV, A., BABITSYNA, A., MEDVEDEVA, Z.: X-ray study of the boron - arsenic system. Zh. Neorgan. Khim., 9, 1158-62 (1964)

ERMOLENKO, N., TABULINA, L., EFROS, M.: Composition of a compound formed during the heat treatment of the aluminum oxide-boron system. Vestsi. Akad. Navuk. Belarus SSR, Ser. Khim. Navuk. 2, 5-10 (1973)

ETTMAYER, P., HORN, H., SCHWETZ, K.: Electron-probe microanalysis in the silicon - boron system. Mikrochim. Acta. Suppl. 4, 87-95 (1969)

FAVRE, R., THEVENOT, F.: Synthesis and study of titanium, zirconium and hafnium diborides. Bull. Soc. Chim. Fr. 3911, 16 (1971)

GAL'CHENKO, G.L., TIMOFEEV, B.I., GRINBERG, Y.K.: Heat of formation of boron phosphide. Izv. Akad. Nauk SSSR, Neorg. Mater. 8, 634-638 (1972)

GARDNER, P.J.: The standard heat of formation of boron phosphide. Inorg. Chem. 8, 1187-1188 (1969)

GOOD, W., MANSON, M.: Thermochemistry of boron and some of its compounds, J. Phys. Chem. 70, 97-104 (1966)

GRAYON, P., BUFORD, J., ARMINGTON, A.: Pressure synthesis of boron phosphide and boron arsenide. Electrochem. Technol. 3, 338-9 (1965)

GROSS, P., HAYMAN, C., STUART, M.C.: Heat of formation of boron phosphide. Trans. Faraday. Soc. 65, 2628-2632 (1969)

HAGENMULLER, P., CHOPIN, F.: Boron-sulfur system. C.R. Acad. Sci. (Paris) 255, 2259-2260 (1962)

HILLEL, R., CUEILLERON, J.: Boron selenide B_2Se_3, Bull. Soc. Chim. Fr 98-103 (1972)

HOLCOMBE, C., HORNE, O.: Preparation of boron suboxide B_7O. J. Amer. Ceram. Soc., 55, 106 (1972)

JEAN-BLAIN, H., CUEILLERON, J.: Synthesis of boron suboxide B_7O. C.R. Acad. Sci. (Paris) Ser. C 277, 977-78 (1973)

KIEFFER, R., GUGEL, E., LEIMER, G., ETTMAYER, P.: Boron-carbon system. Ber. Deut. Keram. Ges. 48, 358-359 (1971)

KING, M.: Boron ignition and combustion in air-augmented rocket afterburners. Combustion Sci. Techn. 5, 155-164 (1972)

KIRK, R.: Synthetic chemistry of several high temperature boron-containing species. Univ. Microfilms, Ann Arbor, Michigan, Order No. 70-17, 594 (1969)

KIRK, R., TIMMS, P.: Synthesis of thioborine, HBS. Chem. Commun. 1, 18-19 (1967)

KORYAZHKIN, V.: Electronic absorption spectrum of the boron selenide (BSe_2) molecule. Vestn. Mosk. Univ. Khim. 13, 719-720 (1972)

KRENEV, V., EVDOKIMOV, V.: Kinetics of the reaction of boron with boron trichloride. Zh. Neorg. Khim. 17, 327-330 (1972)
LAMIKHOV, L.: Preparation, properties and uses of refractory alloys of the silicon-boron system. Izv. Akad. Nauk. SSSR, Neorg. Mater. 4, 1076-1080 (1968)
MACEK, A., SAMPLE, J., Mc KENZIE: Combustion of boron particles at elevated pressures. 13th Symp. Combust. 859-868 (1970)
McLAIN, W., WILLIAMS, R., WALTER, L.: Ignition and combustion of single boron particles in air at ambient pressure. West States Sect. Combust. Inst. No. 70-13 (1970)
MEDVEDEVA, Z., BORYAKOVA, V., GRINBERG, YA.Kh., ZHUKOV, E.: Phase diagram of the boron disulfide - sulfur system. Zh. Neorg. Khim. 13, 1440-1443 (1968)
MOHAN, G., WILLIAMS, F.: Ignition and combustion of boron in oxygen - inert atmosphere. Am. Inst. Aeronaut. Astronaut. J. 108, 776-783 (1972)
NASLAIN, R., ETOURNEAU, J., HAGENMULLER, P.: Boron-potassium system. Bull. Soc. Chim. Fr. 2529-2538 (1967)
OSUGI, J., SHIMIZU, K., TANAKA, Y.: Reaction of boron and arsenic at high temperature and high pressure. Proc. Japan. Acad. 42, 48-53 (1966)
PERET, J.: Preparation and properties of the boron phosphiedes. J. Amer. Ceram. Soc. 47, 44-46 (1964)
PERKOVA, G., DEMIN, A., CHALYKH, E., CUNDOROVA, N.: Solubility of boron in a graphite lattice. Tsvet. Metal. 4, 42-45 (1973)
PESHEV. P.: Thermodynamic study of some reactions used for the preparation of borides of groups II-VI. Rev. Int. Hautes Temp. et Refract. 4, 289-296 (1967)
PETRAK, D., RUH, R., GOOSEY, B.: Preparation and characterization of boron suboxide. Nat. Nur. Stand. U.S. Spec. Publ. No. 364, 605-611 (1972)
PLOOG, K.: Preparation of a rhombohedral boron carbide $B_{13}C_2$. J. Less. Common Metals 31, 177-180 (1973)
PLOOG, K.: Tetragonal and orthorhombic boron carbide. J. Less Sommon Metals 75, 115-130 (1974a)
PLOOG, K.: Rhombohedral boron carbide. J. Less Common. Metals 35, 131-145 (1974b)
REKOV, A., SAMSONOV, V.: Boron solid solutions in boron carbide. Khim. Tverd. Topl 2, 140-142 (1971)
RHEIN, R.: Ignition of ultrafine powder boron in air, nitrogen carbon dioxide and in mixtures of nitrogen and carbon dioxide Pyrodynamics 4, 55-6 (1966)
ROSNER, D., ALLENDORF, H., Kinetics of the attack of refractory solids by atomic and molecular fluorine. J. Phys. Chem. 75, 308-317 (1971)
ROSSO, J., DUBUSC, M.: Boron sulfur and compounds boron sulfur metals. Chimie Industrie 102, 409-415 (1969)
SAMSONOV, G., SPLETSOV, V.: Manufacture of boron silicon alloys. Poroshkovaya Met. Akad. Nauk. Ukr. S.S.R. 4, 58-67 (1964)
SAVELEV, B., KRENEV, V., EVDOKIMOV, V.: Interaction of hydrogen bromide with elemental boron. Th. Neorg. Khim. 18, 317-319 (1973a)
SAVELEV, B., KRENEV, V., EVDOKIMOV, V.: Reaction of boron tribromide with elemental boron. Zh. Neorg. Khim. 18, 1413 (1973b)
SCHADOW, K.: Boron combustion characteristics inducted rockets. Combust. Sci. Technol. 5, 107-117 (1972)
THEVENOT, F.: Heat of formation of crystalline boron phosphide. Bull. Soc. Chim. 98-101 (1968)
THEVENOT, F., CUEILLERON, J.: Preparation of crystallized boron phosphide. Bull. Soc. Chim. 3204-3206 (1968)
THEVENOT, F., VIALA, J.: Comparison of methods of determination of impurities in boron. Analysis 3, 76-84 (1975)
ZOLOTAR, E., OZEROV, E.: Calculation of the inflammability limit of a boron particle. Fiz. Goreniya. Vzryva 9, 515-521 (1973)
ZUBOVA, E., BURDINA, K.: Synthesis of B_6O at high pressures. Dok. Akad. Nauk. S.S.S.R. 197, 1055-1056 (1971)

II. Methods of Preparation of Amorphous Boron

A. F. ZHIGACH and D. C. STASINEVICH

1. Introduction

Preparation of pure elemental boron presents certain difficulties. The conditions necessary for its preparation must be carefully maintained because boron reacts strongly with many substances at high temperatures required for reduction of its compounds. Any by-products formed during the reduction are difficult to separate.

Properties of elemental boron depend to a large extent upon the method of its preparation. The first of these conditions is temperature. At comparatively low temperatures, about 600-800°C, boron is usually obtained in the form of amorphous powder. At temperatures above 1000°C crystalline boron is produced, in many cases, as a mixture of various modifications. Crystalline modifications of elemental boron are often produced by special thermal treatment of amorphous boron obtained at lower temperatures. It is difficult to draw a clear distinction between methods of preparing amorphous and crystalline boron because they are based on the same reaction. In early works the structure of boron was not usually investigated. The reactions resulting in a mixture of amorphous and crystalline boron without distinctly marked structure will be discussed in this Chapter. Naturally, the main attention will be focused on the processes of current practical application, and processes capable of development. Information about investigations on boron preparation, which generally did not lead to high purity can be found in MELLOR (1924), LAUBENGAYER et al. (1943), GMELIN'S Handbook (1954), SAMSONOV et al. (1960b), NEWKIRK (1961), OLEMPSKA (1967) and BOWER (1970).

The known methods of boron preparation may be divided into four groups. Three of them are based on reduction of boron compounds by metals, hydrogen and other agents, or by electrolysis. The fourth method consists of thermal decomposition of boron compounds.

2. Reduction of Boron Compounds by Metals

Boron oxide, borax, metal borates, the halogenides (chloride; less frequently bromide, fluoride and iodide) and metal tetrafluorborates are types of boron compounds which can be used for the practical production of elemental boron.

As was shown by thermodynamic calculation (EFREMKIN, 1957) isobaric potentials of boron oxide reduction by magnesium, calcium, and aluminum are negative in a temperature range of 298 to 2000°K, and therefore, these metals can readily reduce B_2O_3 to boron. Silicon reduces weakly at temperatures up to 1900°K, but does not reduce at higher temperatures. Sodium cannot reduce boron oxide to elemental boron in stoichiometric proportions of reagents in the above range of temperatures.

From the metals listed above only magnesium has found practical application in boron production. Powdered magnesium is readily available and is inexpensive in comparison with such reducers as alkaline metals and calcium.

The thermomagesium method of boron production developed by MOISSAN (1892) is based on a reaction which is usually written as:

$$B_2O_3 + 3Mg = 2B + 3MgO \qquad (1)$$

Successful reaction depends on the use of an excess of boron oxide which binds the formed magnesium oxide in the form of borates. Thus, the following reactions take place:

$$2B_2O_3 + 3Mg = 2B + 3MgO \cdot B_2O_3 \qquad (2)$$

$$4B_2O_3 + 3Mg = 2B + 3(MgO \cdot B_2O_3) \qquad (3)$$

Optimum correlation corresponds to about 1.8 moles of boron oxide per three atoms of magnesium (MARKOVSKII and BELOV, 1964, 1967; MARKOVSKII, 1970).

The reduction process of boron oxide by magnesium metal is strongly exothermal. The thermal effect of reaction (1) is equal to 136 Kcal., and is estimated that temperatures as high as 2540°C can be developed in the reactive mixture, if there is no heat loss (EFREMKIN and MIKULINSKII, 1957).

Energetic interaction, characterized by a fast temperature increase and blaze appearance, takes place when the finely-divided reagents are heated to the reaction temperature. Reduction occurs within several seconds. Thermographical studies have shown (MIKHEEVA and MARKINA, 1957) that exothermal reaction begins at 700-880°C, i.e., above the melting points of boron oxide and magnesium metal. At the initial stage, the reaction takes place between two liquid phases. At higher temperatures the evaporation of magnesium begins promoting the loosening of a film of reaction products at the interface of boron oxide with magnesium. When this occurs, the speed of reaction is sharply increased. Probably the boron oxide reduction (EFREMKIN and MIKULINSKII, 1957) occurs by stages. The first stage is apparently the formation of boron suboxide:

$$B_2O_3 + Mg = 2BO + MgO \qquad (4)$$

$$2BO + 2Mg = 2B + 2MgO \qquad (5)$$

RAY (1914) gives other formulas for boron suboxides present in boron obtained by the thermomagnesium method. Investigations show that a reaction with magnesium oxide forms other subborates, such as $MgB_{30}O_{10}$. Their presence in thermomagesium boron was proved experimentally (MARKOVSKII and BELOV, 1967; MARKOVSKII, 1970). The interaction of magnesium metal with boron results in the simultaneous formation of borides MgB_2, MgB_4 and MgB_{12}. These borides are thermally unstable and dissociate into boron and magnesium at the high temperatures developed during the reduction (MARKOVSKII et al., 1955; MARKOVSKII and KAPUTOVSKAYA, 1962). This last fact is extremely important because it explains why preparations with a high percentage of boron cannot be obtained by using other metals as reducers. Calcium and aluminum, for example, form stable borides.

If the boron reduction is carried out in air, the formation of BO and boron nitride is possible due to interaction of boron with oxygen and nitrogen. In this case, excessive boron anhydride and magnesium borates, produced during the reaction, form a film on the surface of boron particles. This film prevents further oxidation and nitridation of boron. KROLL (1918) and MIKULINSKII et al. (1957a) proposed adding sulfur to the reaction charge. It was believed that a film of melted magnesium sulfide would then prevent boron interaction with air.

Investigations showed that the effect of the cooling rate on the oxidation of boron products results in decreased oxygen content in boron. Sulfur usage was rejected by MIKULINSKII et al. (1957a). Its use does not considerably improve the quality of the product, and it requires and additional comsumption of magnesium. Besides, the product contains an impurity of sulfur up to 0.3% which forms hydrogen sulfide during the acid treatment.

The substitution of boric acid for boron anhydride has also been attempted (MIKULINSKII et al., 1957a). In this process water is formed during the dehydration of boric acid. This water reacts with magnesium at higher temperatures, consuming additional magnesium and evolving hydrogen. Because presence of hydrogen may lead to formation of an explosive mixture, the use of boric acid was abandoned when a convenient method of preparing boron anhydride was developed (MARON and VAL'SHCHIKOV, 1962; MIL'MAN and BOBKOV, 1962).

For preparation of pure boron, use of raw materials in as pure a form as possible is recommended, since all the impurities tend to gather in the end product. Aluminum, which forms stable borides, is especially undesirable. Fine crushing of reagents is not necessary. Magnesium with particle size of 0.3-0.8 mm is recommended (MIKULINSKII, et al., 1957a).

Use of borax as a raw material for boron production (MARCHESINI, 1968) has been found acceptable. Small amounts of boron were produced in crusibles of chamotte or stainless steel (MOISSAN, 1892, MIKULINSKII et al., 1957a), using powdered or briquetted charge (MIKULINSKII et al., 1957a). For reductions carried out on a production scale, vacuum retort or tunnel ovens are used (MARON and VAL'SHCHIKOV, 1962; MIL'MAN and BOBKOV, 1962).

There are some indications that the quality of the product is higher when the charge is spread in a thin layer as reduction then takes place at a lower temperature (MARCHESINI, 1968). Under laboratory conditions this reaction can be carried out in an argon atmosphere without external heating. Potassium chlorate is added to the charge as an oxidizer (PERMINOV and NERONOV, 1969).

The cake thus produced is rapidly cooled in air, crushed and treated with acid for dissolution and decomposition of impurities, such as magnesium oxide, magnesium borates, and magnesium borides. Early workers used prolonged boiling with hydrochloric acid for the decomposition of magnesium borides (MOISSAN, 1892). (It should be noted that during the acid decomposion of magnesium borides, formation of small amounts of highly toxic boron hydrides may occur.) Acid treatment was followed by alkaline treatment and finally by boiling with hydrofluoric acid and drying in vacuum. After such treatment, the product contained 94-95% of boron, 2-4% magnesium and 1.2-1.5% undissolved residue.

Later, the treatment of the product was simplified (MIKULINSKII et al., 1957a). After 2-4 h of treatment with 12-15% hydrochloric or 20-25% sulfuric acid (POLYAK et al., 1957), washing and drying, the product is subjected to oxidative roasting in air at temperatures of 540-560°C for ten minutes. During this treatment, magnesium borides are oxidized, but boron remains unchanged (MIKULINSKII et al., 1957b; SEREBRENNIKOVA, 1962). Treatment of the boron cake with superheated steam at 500°C produces the same effect.

After roasting, the product is treated with an acid for the second time, washed with water, dried at 80-90°C, and packed into tightly sealed packages (EFREMKIN and SHEVELEVA, 1957). It will gradually oxidize if stored in air. The product obtained by this method is a brown powder with 94.5-95.5% boron (MIKULINSKII et al., 1957a). The yield of boric acid is 40% and that of magnesium is 57%. Some of the boric acid may be regenerated from the wash liquid (MIKULINSKII et al., 1957a).

An X-ray study of thermomagesium boron shows that it is amorphous. The only lines seen on the film are those given by the impurities of magesium boride MgB_{12} (MARKOVSKII et al., 1955).

Magnesium boride impurities may be separated from boron by fusing the mixture with a large amount of boron oxide (MOISSAN, 1892; MAZZA et al., 1958). After acid treatment the product contained 98.3% of base substance, 0.4% magnesium and 1,2% of a residue undissolved in nitric acid (apparently boron nitride). It is possible to obtain preparations containing 99.2-99.6% B, when the reaction is carried out in an inert atmosphere. Boron is also fused with a KBF_4 + KHF_2 mixture at 950 to 1000°C for purification from magnesium (YANNACAKIS and NIES, 1960). Impurities of metals contained boron, are turned into volatile chlorides during treatment by boron chloride at 1200-1350°C (CUEILLERON and PICHAT, 1965a)

Magnesium removal and an increase of boron content is accomplished by annealing the product in a vacuum (5.10^{-2}-5.10^{-10}mm Hg). However, for complete removal of magnesium, temperatures of 1400-1800°C are insufficient (SHAMRAI, MIKHEEVA, 1957). Only at 1900-2000°C is it possible to achieve a product containing 99.5-99.8% boron. That is comparable to the degree of purification obtained in the best samples using other methods (ZINTL et al., 1940; MARKOVSKII et al., 1955; HAAG, 1957; SWIFT and WHITE, 1957). Floating with oleic acid in a water medium may be used to remove some impurities such as boron carbide, iron, etc., from boron (SCHAEFFER and SCHMIDT 1964; 1965; 1967).

If it were possible to produce pure boron by reduction of the boron oxide with metallic aluminum, this would be the most economical process. However, the primary product of the interaction of these substances is aluminum boride (WOHLER and St. CLAIR DEVILLE, 1857). There is evidence that boron can be produced by metallic aluminum reduction of melted borax when the metallic aluminum is in such a ratio as to produce a low-melting $B_2O_3 \cdot Al_2O_3$ slag. However, such product is of low quality (89% B, 7% Al, 0.8% Fe and 2.3% of an undissolved residue).

Boron oxide reduction by alkaline metals, lithium, sodium and potassium (ANDRIEUX, 1929; MARKOVSKII, KONDRASHEV, 1957), as well as their hydrides, results in products with low boron content. Calcium boride (CaB_6) is usually produced by reduction of boron oxide by calcium, calcium hydride and calcium carbide (HAGENMULLER and DE PAPE, 1959).

Thermodynamic calculations have shown that boron can be produced by reduction of its halogenides (BCl_3, BF_3) by different metals (EFREMKIN, 1957). These processes, however, have no practical application since halogenides are very expensive, and the equipment needed is very complicated.

Laboratory experiments were carried out to investigate reduction of boron halogenides by sodium (KAMENAR, 1962; PICHAT and FOREST, 1967; HOBBS et al., 1964; SCHOTT and HANSLEY, 1964), potassium and potassium-sodium alloy (PICHAT and FOREST, 1967), magnesium (HOBBS et al., 1964), zinc (FEDOROV et al., 1960; HIGASHI, 1961; ITO et al., 1962, 1963a,b,c; HOBBS et al., 1964; CUEILLERON and PICHAT, 1965; NISEL'SON et al., 1965; PICHAT and FOREST, 1967), aluminum and tin as well as by lithium hydride and calcium hydride (HAGENMULLER and DE PAPE, 1959).

The reaction of boron chloride with a liquid alloy of K-Na is interesting, because it permits the production of elemental boron at very low temperatures (about 50°C) (PICHAT and FOREST, 1967). The reaction with molten sodium and its vapors is carried out at 440-460°C.

Boron chloride reduction by zinc vapors at 850-950°C attracted attention because zinc is one of the few metals which does not form borides (FEDOROV et al., 1960; NISEL'SON et al., 1965).

3. Electrolytic Method of Boron Production

Boron is produced by electrolysis of melts, containing metal borates or metal fluoroborates (KAHLENBERG, 1925; ANDRIEUX, 1927, 1929, 1948; ANDRIEUX and DEISS, 1965). The majority of workers believe that boron formation is the result of a secondary chemical reaction between alkaline metals, initially evolved on a cathode, and boron compounds contained in the melts. There are indications, however, that boron is produced in the primary electrode process (MEERSON and SMIRNOV, 1955). Boron is formed as loose powder on a cathode. This layer is impregnated with melt and can be isolated after cooling and treatment with hot water.

A high quality product was obtained by use of an electrolyte containing potassium fluoroborate mixed with potassium cloride and potassium fluoride (COOPER, 1945, 1951a,b; COOPER and SCHAEFER, 1959; SAMSONOV et al., 1959, 1960a). The approximate composition of the electrolyte is as follows:

KBF_4 - 17.8%, KCl - 70%, KF - 12%, H_2O - 0.2%

Electrolysis was carried out at 750-800°C (ANDRIEUX 1927, 1948; DEISS and ANDRIEUX, 1959; PICHAT, 1967). The walls of the elctrolysis vessel were made of fine-grained graphite which was used as an anode. The metal cathode was made of Monel metal. Current density was 80 A/dm^2, Δ adn the voltage of electrolysis was 5-6 volts. Yield of current was 74% and electric energy consumption was 55 kw/h/kg. The product containing 94-95% boron was obtained after washing and drying. This process permitted the production of several kilograms of boron per month (COOPER and SCHAEFER, 1959).

Other compositions of electrolytes may be used for electrolytic production of boron, for example: $B_2O_3 + KBF_4 + KCl$; $B_2O_3 + KF + KCl$; $B_2O_3 + K_2O + KF$; $MgO{\cdot}B_2O_3 + MgF_2$ (KAHLENBERG, 1925; ANDRIEUX, 1927, 1948; ANDRIEUX and DEISS, 1955; NIES et al., 1958; MILLER, 1959, 1960) $B_2O_3 + Na_3AlF_6$ (MONIER et al., 1965).

X-ray studies of electrolytic boron do not give simple results. Some samples appeared to be amorphous (MILLER, 1959, 1960). However, the films of other samples, showed lines, not always distinct, that indicated a crystalline structure (COOPER, 1945, 1951a,b; COOPER and SCHAEFER, 1959).

Another electrolytic method for preparation of pure boron is based on the dissolution of a low purity boron or boron carbide at the anode and the deposition of pure boron on the cathode. Electrolysis is carried out at 800°C. The electrolyte consists of KBF_4 (20%), NaCl (40%) and KCl (40%). After acid treatment, washed and dryed, the product contained 99.8% boron (STERN and LYNDS, 1958; STERN, 1960a,b).

The electrolytic boron isolated on a cathode can be subjected to refining. In primary electrolysis where the graphite walls of the vessel are used as a cathode, the anode is also made of graphite. After deposition of boron sediment on the walls of the vessel, a rod of low-carbon steel is introduced. As this rod becomes a cathode, and the walls of the vessels become an anode, the electrolysis can be repeated (YANO et al., 1961, 1962).

4. The Preparation of Amorphous Boron by Reduction of Boron Compounds by Hydrogen

Thermometal and electro-chemical methods of boron preparation have many disadvantages. Among others, they produce a low yield and a low-quality product. The complicated procedures are difficult to control. There are difficulties in material selection for large installations, use of non-standard reactors, complicated purification of the end-product, etc. For these reasons, boron production by hydrogen reduction of boron compounds is preferred (SAMSONOV et al., 1960a).

Among the investigated boron compounds only boron halogenides are suitable for boron production. Hydrogen reduction of copper borate (DEB and KHUNDAR, 1958), nickel borate (HUG et al., 1958) and cobalt borate (JAHAN and KHUNDAR, 1961) results only in formation of metals and boron oxides.

Comparative thermodynamic calculations of the reduction of boron halogenides (BF_3, BCl_3, BBr_3) by hydrogen have shown (CUEILLERON, 1944; EFREMKIN, 1957; STERN and LYNDS, 1958) that boron fluoride may be reduced only at very high temperatures (2000°C), thus excluding the possibility of its practical application. Hydrogen reduction of boron chloride to elementary boron takes place at temperatures above 1000°C, according to the following reaction:

$$2BCl_3 + 3H_2 = 2B + 6HCl \tag{6}$$

An increase of ratio of BCl_3 to H_2 in the reactice mixture and a decrease of pressure increases the yield of boron (CUEILLERON, 1944; EFREMKIN et al., 1957; STERN and LYNDS, 1958; NISEL'SON et al., 1965; KRENEV et al., 1972). At lower temperatures (700-900°C) interaction of boron chloride with hydrogen leads more readily to formation of dichlorborane (ZHIGACH and STASINEVICH, 1968) according to equation:

$$BCl_3 + H_2 = BHCl_2 + HCl \tag{7}$$

Boron bromide is reduced by hydrogen more easily than boron chloride. The reaction occurs at 750-850°C.

Bromide and iodide of boron are inaccessible and expensive products and can be used only under laboratory conditions. For these reasons only boron trichloride has practical significance for boron production from boron haligenides.

Hydrogen reduction of boron halogenides is endothermic, in constrast to reduction by metals and the reaction must be carried out by use of an external heat supply.

There are three means for heating reactive gases: the plasma of electric arc, discharge in tubular electric ovens, or incandescent heated filaments or rods.

Because the quality of the product is strongly dependent upon the purity of the initial materials (FETTERLY, 1960), the available grades of boron halogenides and hydrogen are subjected to thorough purification.

Boron halogenides are easily boiled and readily separated during rectification from metal and nonmetal impurities including boron suboxides. Consequently, a product of any required degree of purification can be obtained.

a) Reduction in a Plasma Reactor

With the plasma method of heating, boron formation takes place in plasma volume or in heated parts of the reactor. By feeding the mixture of boron trichloride and an excess of hydrogen into a powerful high-voltage arc of alternating current, boron of 99.8-100% purity was obtained in the form of layers between the cooling copper electrodes. Highly dispersed powder, containing up to 99% boron after washing by water was also found on the walls of the reactor (WEINTRAUB, 1907, 1913). In these experiments, however, the yield of boron was low, and the coefficient of energy consumption and productivity of the equipment was also small.

A better yield from boron trichloride conversion (up to 50%) results when the process is carried out in a glow discharge at pressures of up to 150 mm Hg (at about 4000 volts) and a frequency of 1000 Hz on the cooling copper electrodes. The gas mixture, containing a large excess of hydrogen, is fed into a discharge through the channel inside the electrode (HOBBS et al., 1964). Boron is deposited on the walls of the discharge tube as a black, dispersed powder (containing traces of Si, Fe, and Al).

It was recommended by CUEILLERON and CRUIZAT (1973, 1974) that reduction of boron trichloride by hydrogen be carried out in argon plasma, created in a cooling quartz tube. They use alternating current of a 12 KW generator with a frequency of 5-10 Mhz. Gas was supplied in the following quantities: 15 l/min of argon, 1.0-1,5 l/min of hydrogen, and 0.11-0.22 l/min of boron trichloride.

The axis of the plasma reactor registered 16000°K and on periphery, 8000°K. A thin coating of elemental boron, forming at a rate of 4 g/h was observed on the walls of the tube. This coating was found to consist of β-rhombohedral boron structure.

The boron yield, on the basis of consumed boron trichloride, was 70-75%. Emission spectroscopy analysis of the product has shown the presence of silicon, carbon, aluminum, oxygen and other impurities. The total of the impurities amounted to less than 0.4%. Si (from the tube walls) was 0.09% and C (formed from phosgene impurities in boron trichloride) was also 0.09%.

The use of a plasma reactor for chemical reactions at high temperatures has been proposed by ITO et al. (1964). Hydrogen reduction of boron trichloride in a plasma burner, capable of producing 250 g/h with 25-30 KW., has been described also (HAMBLYN et al., 1973). A product has been obtained with a particle size of 0.5 μ and 99% of boron content. Pure elemental boron in the form of rods was obtained on electrodes when a mixture of boron tribromide and hydrogen was heated in plasma (CUEILLERON, 1944). In this case, black highly dispersed "colloidal" boron was deposited on the walls of the reactor.

Boron of 98.6-99.5% purity was also obtained by passing the mixture of boron tribromide and hydrogen through a high frequency arc, formed between tungsten electrodes (HACKSPILL et al., 1931). X-ray studies have shown that boron produced in the arc, has a structure ranging from amorphous to crystalline (CUEILLERON, 1944). The latter was reported to be of rhombohedral form.

b) Reduction in Tubular Electric Ovens

To obtain elemental boron, a mixture of boron halogenides with hydrogen is passed through a quartz tube inside an electric oven.

Initially, this process produced a low yield of elemental boron contaminated by silicon and other impurities from the reactor walls. Later, boron of 98.9% purity was obtained by a more thorough purification of the initial products and hydrogen reduction of boron trichloride in a tube of high-quality quartz (KIESSLING, 1948).

It has been reported (EPELBAUM and STAROSTINA, 1958) that a product containing 98.9% boron and the following impurities: Al - 0.4%; Si - 0.1%; Mg - 0.2%; O_2 - 0.1% and H - 0.1% was obtained during the hydrogen reduction of boron tribromide in a quartz tube at 800°C. The same results were obtained on reduction of boron trichloride in a quartz tube at 1200°C (ITO et al., 1964). A 63% yield of high purity boron can be produced when the molecular ratio of hydrogen to boron trichloride equals 60.

A still higher degree of purity in boron was reported by NIEMYSKI and OLEMPSKA (1963a, b). They passed the pure boron trichloride mixture with purified hydrogen through a quartz tube heated externally up to 1000-1250°C. Amorphous boron with 99.9999% B purity and a polycrystalline product with 99.99% B content was deposited on the walls of the tube. The amorphous boron contained the following impurities: Si - 10^{-4}% and Mg - 10^{-5}%. The impurities in the crystalline product were: Si - 10^{-2}% and Mg - 10^{-5}%.

c) Reduction of Boron Halogenides in the Reactor with Incandescent Filament or Rods

This method was proposed for the first time in 1922 (KOREF, 1922) and described in detail by BELYAEV et al. (1969) and POWELL et al. (1960).

In this process, the reduction of boron halogenides occurs on the surface of a heated filament or rod. Simultaneous intermediate hydrogenation reactions of boron halogenides are possible since chloroborane and diborane are found in the condensed boron trichloride which did not participate in the reaction.

Amorphous boron produced by this method is the result of the reduction of boron trichloride at 600-1000°C (LAUBENGAYER et al., 1943; TALLEY et al., 1960). Reduction of boron trichloride and boron tribromide on a heated filament at higher temperatures (1075-1475°C), leads to various crystalline modifications of boron. In comparison, by reduction in a heated tube, crystalline boron can be prepared at lower temperatures (750-800°C). This may be related to lower concentrations of boron compounds in the reaction zone and, consequently, lower velocities of the atoms forming elemental boron. Consequently, the atoms have time to take their places in a crystalline lattice. Such slow crystal growth may not occur in reactions where the boron deposition is rapid.

Kinetic process data on reduction of boron tribromide by hydrogen on tungsten filament were reported by LAUBENGAYER et al. (1943).

The conditions of preparation of amorphous boron include a rate of hydrogen flow of 500 cc/min at 1 atm and 25°C. Boron tribromide content in the reactive mixture is 8-10 mol %. Samples of tungsten wire of 25 mm in diameter and 5 cm in length are heated to 1130°C. Under these conditions the rate of deposition is 4.5 mg $B/cm^2/min$. The diameter of the wire is increased by 1 mm in 25 min. Before deposition the wire is conditioned in hydrogen at 1430°C for 5-10 min. The sample temperature during deposition must not exceed 1230°C. At higher temperatures boron is deposited in a polycrystalline form. The rate of deposition is relatively low at temperatures below 830°C.

A horizontal position for the sample is preferred to the vertical one, since is decreases temperature differences along the length of the sample, and simplifies the process control. The rate of deposition is increased as the consumption of gas mixture increases, and as the ratio of boron tribromide to hydrogen approaches the stoichiometric proportions. Boron is obtained in the form of small cylinders, approximately 2 mm in diameter and 5-10 cm in length, as well as in the form of spheres and cones. The samples contain 98.8% (wt) boron. The major part of the impurities comes from the tungsten core and consists of 0.5% on the average.

The type of boron obtained in this manner is referred to as amorphous, because X-ray studies show the presence of two diffuse haloes, typical for glassy materials. Heating at 1260°C and above can convert "amorphous" boron into another modification.

5. Production of Amorphous Boron by Thermal Decomposition of Boron Compounds

In practical applications, preparation of boron by the thermal decomposition method utilizes only boron halogenides and hydrides. Borides, such as nitrides, sulphides, phosphides and metal boron hydrides form products of unknown composition when heated to 1200-1500°C (WEINTRAUB, 1909; TALLEY et al., 1960).

a) Decomposition of Boron Halogenides

Fine crystalline boron (ARMINGTON et al., 1964; WIEBERG et al., 1966) or red α-rhombohedral boron (McCARTY and CARPENTER, 1960; AMBERGER and DIETZE, 1964) can be formed by thermal decomposition of boron triiodide. In this case amorphous boron may be formed only as an intermediate phase, which quickly turns crystalline at temperatures of about 1000°C. Other crystalline boron modifications can also be formed during thermal decomposition in the arc discharge (MEYER and ZAPPNER, 1921) or on the incandescent tungsten wires (ARKEL 1934; UNO, 1958).

A mixture of polycrystalline and amorphous boron is deposited when tungsten filament is at about 1200°C (McCARTY and CARPENTER, 1960).

b) Decomposition of Boron Hydrides

Amorphous boron of a varying degree of purity can be formed by decomposition of diborane, pentaborane and decaborane in a temperature range from 300 to 900°C (TODD and MILLER, 1946; APKER et al., 1948; SCHLESINGER and SCHAEFFER, 1950; JOHNSTON et al., 1951; LAGRENAUDIE, 1953; DEVYATYKH, 1966). Decomposition is carried out by passing a gaseous hydride, usually diluted with hydrogen, through the heated quartz tube. Similar results were also obtained by heating solid boron hydrides at 800-900°C in a small boat (made of boron nitride).

Pyrolysis on an incandescent tungsten and tantalum (WIRTH and DIETZE, 1964; HINZ and WIRTH, 1965) filament results in formation of boron rods or films (SCHRANGLE and DIETZE, 1972) with crystalline structure.

References

AMBERGER, E., DIETZE, W.: Formation of α-rhombohedral red boron. Z. Anorg. Allg. Chem. 332, 131 (1964)

ANDRIEUX, L.: New process for the preparation of boron. C.R. Acad. Sci. (Paris) 185, 119-20 (1927)

ANDRIEUX, L.: The electrolysis of metallic oxides dissolved in fused boric anhydride or borates. New methods for preparation of amorphous boron, borides and some metals. Ann. Chem. Phys. 12, 432-507 (1929)

ANDRIEUX, L.: Making of metallic powders by electrolysis of fused salts. Rev. Metallurg. 45, 49-59 (1948)

ANDRIEUX, L., DEISS, W.J.: Obtaining boron by electrolysis of melts. Bull. Soc. Chim. 838, (1955)

ARKEL, A.E.: Production of high melting metals by thermal decomposition of their compounds. Metallwirtschaft 13, 405-8 (1934)

APKER, L., TAFT, E., DICKEY, J.: Photoelectric emission and contact potentials of semiconductors. Phys. Rev. 74, 1462-74 (1948)

ARMINGTON, A.F., DILLON, G.F., MITCHELL, R.F.: Preparation of high purity boron via the iodide. Trans. Metallurg. Soc. AIME 230, N 2, 350 (1964)

BELYAEV, A.I., ZHEMCHUZHINA, E.A., FIRSANOVA, L.A.: Metallurgiya chistych metallov i elementarnych poluprovodnikov (Metallurgy of pure metals and elemental semiconductors), Chapt.15, Moskow: Metallurgiya 1969

BOWER, J.G.: Progress in Boron Chemistry. Vol. II, pp. 231-271. Oxford: Pergamon Press 1970

COOPER, H.S.: BCl_3, US Patent 2369212 (1945)

COOPER, H.S.: Electrolytic production of boron. US Pat. 2572248 (1951a)

COOPER, H.S.: Electrolytic production of boron. US Pat. 2572249 (1951b)

COOPER, H.S., SCHAEFER, C.: Production of boron by fused salt-bath electrolysis. US Pat. 2,918,417 (1959)

CUEILLERON, J.: Purification of elemental boron. Ann. Chim. 19, 459 (1944)

CUEILLERON, J., PICHAT, P.: Manufacture of pure boron. C.R. Acad. Sci. (Paris) 260, 190 (1965)

CUEILLERON, J., CRUIZAT, B.: Reduction of copper borate by hydrogen. Bull. Soc. Chim. France, 1207 (1973)

CUEILLERON, J., CRUIZAT, B.: Preparation of high purity boron by reduction of boron chloride with hydrogen in an argon induction plazma. In: Boron, Preparation, Structure and Properties, p. 15-17 (F.N. TAVADZE, Ed.) Moscow: Nauka 1974

DEB, S.K., KHUNDKAR, M.H.: Boridation of metallic powder by B chloride. J. Indian Chem. Soc. 23, 555 (1958)

DEISS, W.J., ANDRIEUX, L.: The boridation of metallic powders by boron chloride. Bull. Soc. Chim. France, 178-82 (1959)

DEVYATYKH, G.G.: In: Poluchenie i analiz veshchestv osoboi chistoty (Preparation and Analysis of Super Pure Substances), p. 15, Moskow: Nauka 1966

EFREMKIN, V.V.: Trudy Ural Nauch-Issledovatel. Chimi. Inst. (Trans Ural's Sci. Res. Inst. Chem.), Vol. V, p. 152, Moskow: Goschimizdat 1957

EFREMKIN, V.V., MIKULINSKII, A.S.: Mechanism of the reduction of boric anhydride with magnesium. Trans. Ural's Sci. Res. Inst. Chem. Vol. V, p. 211, Moskow: Goschimizdat 1957

EFREMKIN, V.V., SHEVELEVA, S.S.: Preparation of amorphous boron. Trans. Ural's Sci. Res. Inst. Chem. Vol. V, p. 232, Moskow: Goschimizdat 1957

EPELBAUM, V.A., STAROSTINA, M.T.: In: Bor. Trudy konferencii po chimii Bora i ego soedinenii (Trans. Conf. Boron and its Compounds Chemistry) p. 97, Moskow: Goschimizdat 1958

FEDOROV, T.F., SHAMRAI, F.I., NISELSON, L.A., PETRUSEVICH, I.V.: Preparation of elemental boron. Zhurnal neorganicheskoi Chimii (Inorg. Chem.) 5, 226 (1960)

FETTERLY, G.H.: In: Boron (Ed. G.K. Gaulê), Vol. I, p. 15, New York: Plenum Press 1960

GMELIN'S Handbuch der anorganischen Chemie. 8, Vol. XIII Bor. Weinheim: Chemie Verlag (1926) Supplementary Vol. (1954)

HAAG, H.: High purity boron. US Pat. 2794708 (1957)

HACKSPILL, L., STIEBER, A., HOCART, R.: Crystallized boron. C.R. Acad. Sci. (Paris) 193, 776-8 (1931)

HAGENMULLER, P., PAPE, R. de: Reduction of B-trifluoride by Ca-hydride. C.R. Acad. Sci. (Paris) 249, 114 (1959)

HAMBLYN, S.M.L., REUBEN, B.G., THOMPSON, B.: Hydrogen reduction of boron trichloride to boron in an R.F. plasma. Spec. Ceram. N 5, 147 (1973)

HIGASHI, KEI: Preparation of B by reducing BCl_3 with Zn. (ii) Preparation end purification of BCl_3. (iii) Factors affecting the deposition of B. Kyushu Kozan Gakkai-Shi 29, 209, 368 (1961)

HINZ, I., WIRTH, H.: Purity of B produced by the decomposition of diboran and subsequent zone melting. In: Boron (Ed. G.K. Gaulê), Vol. II, p. 9, New York: Plenum Press 1965

HOBBS, D.Z., CAMPBELL, I.T., BLOCHE, F.E.: Methods used in preparing boron. Rep. Invest. Bur. Mines US Dept. Int. N 6456 (1964)

HUG, A.K., DEB, S.K., KHUNDKAR, M.H.: Separation of heavy metals ions from mixtures with H_3BO_3. Indian J. Appl. Chem. 21, 81 (1958)

ITO, HISASHI, YANAGASE TSUTOMU, HIGASHI, KEI: Purity and crystallinity of boron by reduction of BCl_3 with Zn. Nippon Kogyo Kaishi 78, 667 (1962)

ITO, HISASHI, HIGASHI, KEI, MARUTA, KENJI: On preparation of boron by reducing boron trichloride with zinc. IV. Reduction reaction under low pressure. Kyushu Kozan Gakkai-Shi 31, 390 (1963a)

ITO, HISASHI, AJUMA, KEI, MARUTA, KENJI: Nippon Kogyo Kaishi 79, 886 (1963b)

ITO, HISASHI, YANAGASE, TSUTOMU, HIGASHI, KEI: Preparation of boron by reducing BCl_3 with zinc. Trans. Japan Inst. Metals, 4, 205 (1963c)

ITO, HISASHI, HIGASHI, KEI, MARUTA, KENJI: The preparation of boron by reducing boron trichloride with hydrogen. Kyushu Kozan Gakkai-Shi 32, 128 (1964)

JAHAN, R., KHUNDKAR, M.H.: Reduction of cobalt borate by hydrogen. J. Indian Chem. Soc. 38, 7 (1961)

JOHNSTON, H.L., HERSH, H.N., KERR, E.C.: Low temperature heat capacities of inorganic solids, V.J. Am. Chem. Soc. 73, 1112-7 (1951)

KAHLENBERG, H.H.: Boron and boron suboxide. Trans. Am. Electrochem. Soc. 47, 23 (1925)

KEMENAR, B.: Preparation of elemental boron. Z. Anorg. Allg. Chem. 342, 108 (1962)

KIESSLING, R.: Preparing boron of high purity. Acta Chem. Scand. 2, 707-12 (1948)

KOREF, F.: Further growth of metal crystals by separation from the gaseous phase. Z. Electrochem. 28, 511-7 (1922)

KRENEV, V.A., SAVEL'EV, B.A., EVDOKIMOV, V.I.: In: Termodinamika i kinetika processov vosstanovleniya metallov (Thermodynamics and Kinetics of Metal Reduction Processess), p. 122, Moskow: Nauka 1972

KROLL, W.: Preparation of amorphous boron. Z. Anorg. Chem. 101, 1-33 (1918)

LAGRENAUDIE, J.: Semiconducting properties of boron. J. Chem. Phys. 50, 629 (1953)

LAUBENGAYER, A.W., HURD, D.T., NEWKIRK, A.E., HOARD, H.: Preparation and properties of pure crystalline boron. J. Am. Chem. Soc. 65, 1924 (1943)

McCARTY, L.V., CARPENTER, D.R.: Preparation of crystal modifications of boron - Synthesis of B-tri-iodide. J. Electrochem. Soc. 107, 38 (1960)

MARCHESINI, L.: Reduction of boric anhydride with magnesium. Techn. Ital. 33, 139 (1968)

MARKOVSKII, L.Ya.: Chemistry of magnesiothermal preparation of boron. Electron. Technol. 3, 95 (1970)

MARKOVSKII, L.Ya., BELOV, M.P.: Chemical reactions occuring in the reduction of boric anhydride and magnesium. Zhur. Prik. Chim. (J. Appl. Chem.) 37, 1658 (1964)

MARKOVSKII, L.Ya., BELOV, M.P.: Source of oxygen in amorphous boron, prepared by magnesium thermal method. Zhur. Prik. Chim. (J.Appl.Chem.) 40, 727 (1967)

MARKOVSKII, L.Ya., KAPUTOVSKAYA, G.V.: Properties of Mg borides and their role in the preparation of boron by the Mg thermal method. Zhur. prikladnoi chimii (J. Appl. Chem.) 35, 723 (1962)

MARKOVSKII, L.Ya., KONDRASHEV, Ju.D.: Composition and properties of the borides of Groups I and II of the periodic system. Zhur. Neorg. Chim. (J. Inorg. Chem.) 2, 34 (1957)

MARKOVSKII, L.Ya., KONDRASHEV, Ju.D., KAPUTOVSKAYA, G.V.: Zhurnal obshchei chimii. (J. Gen. Chem.) 25, 433 (1955)

MARON, F.S., VAL'SHCHIKOV, V.I.: Tezisy dokladov konferencii po Boru, ego Soedineniyam i splavam (Proc. Conf. Boron, its Alloys and Compounds) p. 12, Kiev 1962

MAZZA, H., SAWYER, D.L., BAIER, R.W.: Amorphous high-purity Boron. US Pat. 2866688 (1958)

MEERSON, G.A., SMIRNOV, M.P.: Mechanism of electrolytic preparation of borides of difficultly fusible metals. Chemistry of rare elements (Chimiya redkich elementov), Acad. Nauk S.S.S.R., 2, 130 (1955)

MELLOR, J.W.: Comprehensive Treatise on Inorganic and Theoretical Chemistry, Vol. V, p. 1-147, London: Longmans Green 1924

MEYER, F., ZAPPNER, R.: The preparation of boron by the dissociation of boron bromide. Ber. 54 B, 550-5 (1921)

MIKHEEVA, V.I., MARKINA, V.Yu.: Physical chemical analysis for the reaction of magnesium with boric anhydride. Zhur. Neorg. Chim. (J. Inorg. Chem.) 2, 1232 (1957)

MIKULINSKII, A.S., EFREMKIN, V.V., SHEVELEVA, S.S., SACHKO, A.P.: Preparation of amorphous B by magnesium thermal method. Trudy Ural. Nauch. Issledovatel Chimi. Inst. (Trans. Ural's Sci. Res. Inst. Chem.) Vol. V, 5, 166 (1957a)

MIKULINSKII, A.S., YAKUNCHIKOV, V.N., VAL'SHCHIKOV, V.I., EFREMKIN, V.V.: Refining of amorphous B by oxidative roasting in a fluidized bed. Trudy Ural. Nauch. Issledovatal. Chim. Inst. 5, 206 (1957b)

MILLER, G.T.: Electrolytic production of boron. J. Electrochem. Soc. 106, 815-9 (1959)

MILLER, G.T.: Production of boron. J. Electrochem. Soc. 107, 817 (1960)

MIL'MAN, D.N., BOBKOV, V.H.: Tezisy dokladov konferencii po boru, ego soedineniyam i splavam. (Proc. Conf. Boron, its Compounds and Alloys) p. 15, Kiev: 1962

MOISSAN, H.: Preparation du bor amorphe. C.R. Acad. Sci. (Paris) 114, 392 (1892)

MONIER, R., TISSOT, P., PEARSON, R.: Preparation of boron by electrolysis of boron sesquioxide solution in cryolite. Helv. Chem. Acta 49 (1), 67-72 (1965)

NEWKIRK, A.E.: Preparation and chemistry of elementary boron. Adv. Chem. Series 32, 27-41 (1961)

NIEMYSKI, T., OLEMPSKA, Z.: High grade boron. Polish Pat. 46818 (1963a)

NIEMYSKI, T., OLEMPSKA, Z.: High grade boron. Polish Pat. 46819 (1963b)

NIES, N.P., FAJANS, E.W., THOMAS, L.L., HILBERT, L.E., MORGAN, N.: Electrolytic production of boron. US Pat. 2832730 (1958)

NISEL'SON, L.A., PETRUSEVICH, I.V., SHAMRAI, F.I., FEDOROV, G.V.: In: Vysokotemperaturnye neorganicheskie soedineniya (High Temperature Inorganic Compounds) p. 362, Kiev: Naukova Dumka 1965

OLEMPSKA, Z.: Current problems of the chemistry and technology of boron. Wiadom. Chem. 21, 169 (1967)

PERMINOV, V.P., NEBONOV, V.A.: Magnesiothermic furnace-less production of boron from boric oxide. Poroshkovaya metallurgiya (Powder Metallurgy) 15 (1969)

PICHAT, P.: β-Rhombohedral from of boron. C.R. Acad. Sci. (Paris) C 264, 1121 (1967)

PICHAT, P., FOREST, D.: Preparation of powdered boron. Bull. Soc. Chim. 10, 3825 (1967)

POLYAK, A.M., SHEVELEVA, S.S., USPENSKAYA, Z.P.: Substitution of sulfuric acid by hydrochloric acid in the production of elemental B. Trudy Uralskogo Nauchno-Issledovatelskogo Chimicheskogo Instituta (Trans. Ural's Sci. Res. Inst. Chem.), Vol. 5, p. 222, Moskow: Goschimizdat (1957)

POWELL, C.F., ISH, C.J., BLOCHER, J.M.: Preparation of high purity boron. In: Boron (Ed. G.K. Gaulé), Vol. I, p. 7, New York: Plenum Press 1960
RAY, R.C.: Magnesium boride and amorphous boron. J. Chem. Soc. (London) 105, 2162-8 (1914)
SAMSONOV, G.V., OBOLONCHIK, V.A., KULICHKINA, G.N.: MP diagram of the KBF_4 - KCl system. Chimicheskaya Nauka i promyshlennost (Chemical Science and Industry) 4, 804 (1959)
SAMSONOV, G.V., OBOLONCHIK, V.A., KULICHKINA, G.N.: Electrolytic production of boron. Zhurnal prikladnoi chimii (J. Appl. Chem.) 33, 1365 (1960a)
SAMSONOV, G.V., VALYASHKO, M.G., MARKOVSKII, L.Ya., ZHIGACH, A.F.: Bor, ego splavy i soedineniya (Boron, its Alloys and Compounds) Kiev: Naukova Dumka 1960b
SCHAEFFER, J.C., SCHMIDT, E.C.: Purifying Boron. Brit. Pat. 1064093 (1967); Fr. Pat. 1388928 (1965); US Pat. 3137647 (1964)
SCHLESINGER, H.I., SCHAEFFER, G.W.: Boron coatings. US Pat. 2528514 (1950)
SCHOTT, S., HANSLEY, T.: Boron. US Pat. 3086847 (1964)
SCHRANGLE, W., DIETZE, W.: Boron Coating of Metal Fibers by Vacuum Pyrolytic deposition from Boron Compounds. Ger. Pat. 1521199 (1972)
SEREBRENNIKOVA, M.A.: Tezisy dokladov konferencii po boru, ego splavam i soedineniyam (Proc. Conf. Rep. Boron, its Alloys and Compounds), p. 14, Kiev: Naukova Dumka 1962
SHAMRAI, F.I., MIKHEEVA, V.I.: Refining of Amorphous Boron. Zhurn. Neorg. Khim. (J. Inorg. Chem.) 2, 1242 (1957)
STERN, D.R., In: Boron (Ed. G.K. Gaulé), Vol. I, p. 27, New York: Plenum Press 1960a
STERN, D.R.: Preparation of boron from boron carbide. J. Electrochen Soc. 107, 441 (1960b)
STERN, D.R., LYNDS, L.J.: Pure crystalline boron. J. Electrochen Soc. 105, 676 (1958)
SWIFT, R.M., WHITE, D.: Low temperature heat capacities of magnesium diboride (MgB_2) and magnesium tetraboride (MgB_4). J. Amer. Chem. Soc. 79, 3641-4 (1957)
TALLEY, C.P., LINE, L.E., OVERMAN, O.D.: Preparation and Properties of massive amorphous elemental boron. In: Boron (Ed. G.K. Gaulé), Vol. I, p. 94, New York: Plenum Press 1960
TODD, B.J., MILLER, R.R.: Heat of formation of B_2O_3. J. Am. Chem. Soc. 68, 530 (1946)
UNO, R.: Electric properties of boron. J. Phys. Soc. Japan 13, 667 (1958)
WEINTRAUB, F.: Boron. Trans. Am. Electrochem. Soc. 14, 167 (1907)
WEINTRAUB, F.: Preparation and properties of pure boron. Trans. Am. Electrochem. Soc. 16, 165 (1909)
WEINTRAUB, F.: The properties and preparation of the element boron. Ind. Eng. Chem. 3, 2 (1911); 5, 106 (1913)
WIEBERG, E., AMBERGER, E., DIETZE, W.: Process for the preparation of boron with an α-rhombohedral crystal structure. Ger. Pat. 1231221 (1966)
WIRTH, H., DIETZE, W.: High purity boron. Ger. Pat. 1174750 (1964)
WOHLER, F., SAINT CLAIR DEVILLE, H.: Über das Bor. Ann. Chim. Phys. 100, 635 (1857)
YANNACAKIS, J., NIES, N.P.: Preparation of crystalline boron. In: Boron (Ed. G.K. Gaulé), Vol. I, p. 38, New York: Plenum Press 1960
YANO; TOMASABURO; KOMATSU; AKITOSHI: Electrolytic preparation of boron. Japan. Pat. 3856 (1961); Electrolytic preparation of high purity metallic boron; Japan. Pat. 7005 (1961), 6109 (1961), 6110 (1961); Electrolysis of boron, Japan. Pat. 12503 (1962)
ZHIGACH, A.F., STASINEVICH, D.S.: Chimiya gidridov (Hydride Chemistry), Leningrad: Chimiya 1968
ZINTL, E., MORAWITZ, W., GASTINGER, E.: Boron monoxide. Z. Anorg. Allg. Chem. 245, 8 (1940)

III. Methods of Preparation of α-Rhombohedral Boron

J. A. UGAI and N. E. SOLOVIEV

1. Introduction

The preparation of a crystalline form of boron, later referred to as α-rhombohedral[1] was reported first in 1958 by McCARTY et al. (1958), who had found red boron crystals in the deposit formed under the pyrolysis of boron triiodide. However, as DECKER and KASPER (1959) stated, α-boron had earlier been prepared by the decomposition of boron hydride and had also been found in some commercially prepared boron samples.

The methods of α-boron preparation from boron-containing compounds and alloys at present known are mainly reduced to the decomposition of boron out of the vapor phase and to the crystallization from melt-solutions forming low-melting eutectic. The preparation of α-boron from the vapor phase can be obtained by decomposition of diborance B_2H_6 as well as by the decomposition or reduction of boron halogenides, BX_3.

To grow α-boron crystals from the solutions, the alloys of boron with platinum have been used. Other solvents can also be used to this end. Furthermore, α-boron can be prepared according to the vapor-liquid-solid technique (VLS process) in which the process of deposition from the vapor phase is combined with the crystallization from the melt solution.

The production of α-boron during the crystallization of amorphous boron should also be regarded as an independent method.

In spite of the numerous ways of preparation, the problem of growing sufficiently large α-boron crystals necessary for the investigation of the physico-chemical nature of this material has not yet been solved. Therefore, the authors of this present review have attempted to generalize the available data and analyze the causes preventing the growth of large α-boron crystals.

2. α-Boron Preparation from the Vapor Phase

As has already been shown (McCARTY et al., 1958) among the boron halogenides, BI_3 is the least stable and at 1000 °C can be decomposed into iodine and boron. During the halogenide hydrogen reduction which follows the reaction $2BX_3 + 3H_2 \rightarrow 2B + 6HX$, the low-temperature boron preparation can be accomplished with boron tribromide

[1] Since we know two rhombohedral modifications of boron, two different terms are used to distinguish between them. They are: low-temperature and high-temperature modifications; simple rhombohedral and complex rhombohedral; α-rhombohedral and β-rhombohedral modifications. Further, we shall denote these modifications as α-boron and β-boron respectively.

beginning with 600°C and above. The reduction of BCl_3 and BF_3 is observed at higher temperatures but hydrogen reduction of BI_3 is carried out with difficulty because of unstability of hydrogen iodide.

Furthermore, α-boron can be prepared by decomposition of diborane at temperatures below 1000°C.

The equipment used for α-boron preparations from the vapor phase is similar to that used for the preparation of polycrystalline silicon in semiconductor manufacturing. Boron decomposition takes place in evacuated quartz chambers on substrates heated by electrical current passing through them. As a material for substrates, one can successfully use tantalum, boron nitride as well as boron filaments prepared by CHOKHRALSKYI's method combined with the zone melting or by depositing a thin layer of crystalline boron on the tantalum filament followed by the removal of tantalum and a treatment of the boron residue (BEAN and METCALFE, 1960). Substrate temperatures are usually read with an optical pyrometer. It is necessary to bear in mind that in this case, the values obtained indicate the temperature of brightness compared to which the real values must be somewhat higher. In the opinion of TALLEY et al. (1960), this difference in temperature range in the preparation of α-boron is approximately 50°C.

The vapor pressure of highly volatile boron compounds is defined by the temperature of the container with the original substance.

a) Decomposition of Boron Triiodide

To produce α-boron, original BI_3 was carefully purified either by distillation in a glass column with the efficiency of 30 theoretical plates (McCARTY and CARPENTER, 1960) or by the method of zone melting (AMBERGER and DIETZE, 1964; WIBERG et al., 1967) placing the samples in an evacuated glass tube, passed through the system of alternate heating and cooling cells. The hot zone was created by passing water of about 90°C through copper tubes, and a cold zone created by passing water at 10°C. Having passed through all the zones, the tube with BI_3 was automatically placed in the initial position and was again subjected to the same temperature treatment. The tube velocity was 5 cm/h. The efficiency of the purification can be estimated through the melting temperature of BI_3. The melting point of the original, almost colorless, BI_3 was 48.4°C. After zone melting 50 times, the tube was 3/4 filled with BI_3 with the melting point at 49.0°C, and no impurities could be found by spectral analysis.

Purified BI_3 was pyrolytically decomposed on the heated substrate at partial vapor pressure 0.1-1.0 mm of Hg (McCARTY and CARPENTER,1960). During the boron deposition on tantalum wire (0.075 mm diameter) the surface temperature of which was kept at 1000°C, the 1.8 mm diameter deposit of boron was obtained in 4.5 h. On breaking this deposit, the accumulation of red crystals became clearly visible to the naked eye. Among them, there was a wedge-shaped crystal widening to the outer surface of the deposit, which, in the authors' opinion, testifies to the influence of tantalum upon the character of the deposit. Tantalum is likely to act as an absorber of impurities and assists in this manner in the local growth of α-boron crystals. The preparation of large diameter deposits on these filaments was difficult because of filament failures under the weight of boron deposits. Therefore, the subsequent experiments have been carried out on tantalum rods of 3.2 mm diameter. The results of these experiments have been presented in Table 1.

Table 1. Thermal decomposition of BI_3 on the tantalum rod 91.5 cm long

Decomposition temperature $^{\circ}C$	BI_3 Temperature $^{\circ}C$	Decomposition time h	Deposit weight g	Yield %
800	0 - ~ 25	73.5	10.3	Not determined
900	20	23	11	43
1000	~ 25	64.3	56.6	78

At 800°C a black-scaled deposit was formed which arbitrarily separated from the rod, in the course of the experiment. X-ray powder patterns of the product obtained indicate the presence of α-boron and a substantial quantity of tantalum boride in it. X-ray emission analysis of this sample shows the presence of 78% of tantalum.

X-ray powder photographs of the deposits obtained at 900°C also disclosed the presence of α-boron and tantalum boride. However, tantalum boride is in a far smaller quantity than in the deposit prepared at 800°C. The exterior of the deposit has a dull texture with a brownish shade. On the exterior of the deposit, hemispherical growths of 1.5 mm in diameter can be observed. They contain a dark glassy layer 0.1 mm thick which is a clear red in transmitted light. Because the glassy layer is observed only on the deposit exterior, its appearance, in contrast to wedge-shaped crystals, is in no way connected with the tantalum presence. The analysis of the deposit for the content of impurities shows the following results:
carbon - 0.007 ± 0.002%, tantalum - 0.03%, nickel < 0.01%. Apart from this, the presence of iodine has also been revealed.

X-ray photographs of the deposit obtained at 1000°C differ to some extent in the intensity of the lines from those of the sample prepared at 900°C. However, the α-boron is also the major phase. After the samples have been treated in a warm concentrated nitric acid, three different phases were observed under the microscope; red powder, red microcrystals ~0.25 mm long, red glassy material. According to the X-ray emission analysis, the red crystals contain about 0.04% iodine and not more than 0.001% other impurities with an atomic number ⩾ 13. The analysis of the deposit for oxygen content has not been carried out. The authors, however, assume that the oxygen content would not exceed 0.001%.

The largest tantalum content in the samples prepared at 800°C contrasted to a higher temperature preparation contradicts the diffusion laws. Therefore, the abnormal penetration of tantalum into boron deposit at 800°C is explained by the authors (McCARTY et al., 1960) by the presence of iodine in the vapor phase, where it is likely to form compounds with tantalum at this temperature. At higher temperatures the intermediate compound does not form because of its instability.

The decomposition of purified boron triiodide carried out by AMBERGER and DIETZE (1964); AMBERGER et al. (1970) differs slightly in experimental technique from the work described above. But the more accurate investigation of the character of the deposits made it possible for the authors to draw some additional conclusions. Boron deposition was being carried out on the tantalum wire 0.2-2 mm in diameter (optimal diameter being 0.5 mm) at 850°-1100°C for 4-25 h. The vessel with BI_3 was kept at room temperature. The boron deposits increased the wire diameter to 8 mm. The boron yield was approximately 65%. To reveal the structure of the boron deposits obtained, they were treated in the warm concentrated nitric acid. Apart from this, the structure could be observed, even without etching, in the polarized reflected light.

In the temperature range between 850°C and 950°C the formation of microcrystal α-boron and tetragonal boron was observed.

First, in the range 950°C and 1100°C, a thin layer of TaB_2 was formed. Under stable experimental conditions, a glassy boron mass was deposited on this layer. When experimental conditions are slightly changed, crystals of α-boron or β-boron were formed in the surface of the glassy boron. The crystallization of glassy boron occurs in the presence of microcrystals of α-boron and sufficiently large α-boron crystals are formed (Figs. 1, 2). The largest size of crystals was 2 mm. When β-boron appears, the formation of α-boron is fully suppressed. It has been observed that β-boron does not contain any appreciable impurities. The content of oxygen, nitrogen and carbon were not accounted for. The growth of the α-boron crystals directly on the tantalum substrate was not observed in either of the experiments.

At temperatures above 1100°C and under the prolonged boron deposition, tantalum wire becomes brittle and breaks. The boron deposits contain tantalum impurities found in the wire.

On the basis of these observations, AMBERGER and DIETZE came to the following conclusions: (1) the glassy boron and microcrystalline α-boron arises in the vapor phase; (2) microcrystals of α-boron serve as centers of crystal growth in glassy boron and assist in the growth of large α-boron crystals; (3) tantalum boride does not affect the process of α-boron crystallization.

In their work AMBERGER and DIETZE (1965) observed that α-boron crystals always have a high level of purity. They concluded that α-boron is the only boron modification in which crystallographic vacancies are too small to allow accommodation of foreign atoms. Therefore, the presence of foreign atoms at boron crystallization decreases the probability of α-boron formation.

On the basis of this observation the purified BI_3 has an average purity. The purities of the various products may be listed as follows:

Glassy boron	-	average purity
α-boron	-	high purity
Tetragonal boron	-	low purity
β-boron	-	low purity

Thus, according to AMBERGER and DIETZE the process of α-boron formation by BI_3 pyrolysis can be described in the following manner.

At 950°C-1100°C and under stable pyrolysis conditions, a layer of tantalum boride is formed on the tantalum. Glassy boron, having the same degree of purity as the initial BI_3, is deposited on the tantalum boride. Casual changes in pyrolysis conditions are likely to cause the emergence of α-boron microcrystals on the surface of glassy boron with a purity higher than that of glassy boron. Microcrystals of α-boron serve as crystallization centers for α-boron of high purity, formed from glassy boron, of less purity.

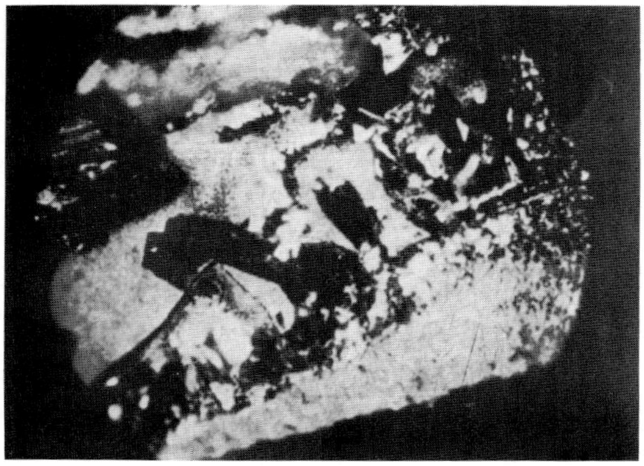

Fig. 1. Cross-section of the boron sample deposited on a tantalum wire -10^x (AMBERGER and DIETZE, 1964)

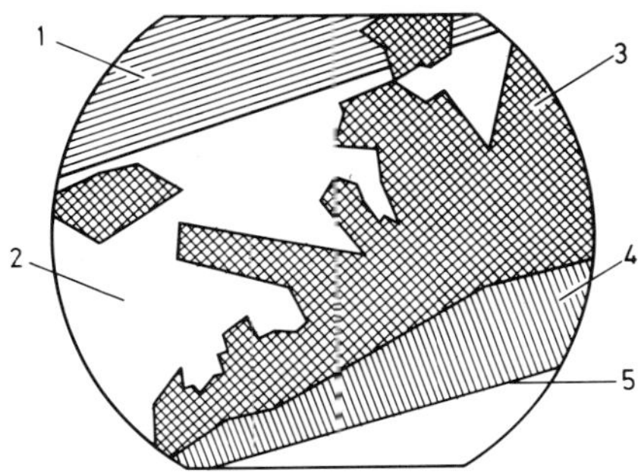

Fig. 2. The outline to Fig. 1. *1*: tantalum wire *2*: amorphous boron *3*: macrocrystalline α-boron *4*: microcrystalline α-boron *5*: the deposit surface

In this case the impurities deposit on the boundaries of α-boron crystals and cause crystallization of tetragonal and β-forms with low purity. Under further pyrolysis, glassy boron is no longer formed. Instead, the formation of microcrystals of α-boron and β-boron takes place on the surface of the deposit.

b) Boron Halogenide Reduction

As already mentioned above, boron preparation by the BBr_3 reduction in hydrogen can be accomplished at temperatures as low as 600°C. NASLAIN et al. (1970) implemented this reaction in a quartz tube, heated in a resistance furnace. In the temperature range of 700-850°C, they obtained α-boron. The temperature range of 700-850°C has been chosen as a compromise permitting deposition of boron at a noticeable rate while avoiding appreciable contamination by silicon.

The device in which the reduction process took place enables a high degree of purification of original substances. Because of low temperatures the boron yield is small, and unreacted BBr_3, subjected to fractional distillation, is returned to a vaporizer. The temperature of the vaporizer is kept at 45-50°C, which corresponds to the partial vapor pressure of BBr_3 at 160 mm of Hg.

In the course of the experiment it was found that the factor defining the deposit type was the residence time of the deposit in the reduction zone. There was a critical time τ_c below which ($\tau < \tau_c$) α-boron deposit was formed as grains not containing β-boron. No interaction between boron and quartz was observed. When $\tau > \tau_c$, the deposit looks like a solid powder containing α- and β-boron modifications. With the increase of temperature and the prolongation of time, the amount of β-boron formed is increased. It was also noted that a slight interaction of boron and quartz occurred in that case.

The critical time depends upon the geometrical dimensions of the reaction chamber and upon temperature. At constant chamber dimensions, τ_c is lower when higher temperatures are used (Table 2). The effect of the deposition conditions upon the boron structure is shown in Table 3.

Table 2. The values of τ_c found experimentally for the reaction chamber of 30 mm diameter and the hot zone of 60 mm

Reduction temperature (^{0}C)	P BBr_3 (mm of Hg)	Deposition rate (mg/h)	τ_c (h)
850	150	250	10
750	160	150	13
700	160	120	15
675	160	80	25

Table 3. Experimental parameters for BBr_3 low-temperature reduction in hydrogen

Deposit structure	Reduction temperature ^{0}C	BBr_3 pressure mm of Hg	Hot zone length cm	τ_c h	τ h	Deposition rate mg/h
	750	160	6	13	12	150
α + β	800-850	170	12	10	40	250

Boron content in samples was 99-99.5%.

X-ray fluorescent analysis has shown that in the course of hydrogen reduction the deposits are contaminated by bromine which can be removed by heating in high vacuum (10^{-6} mm of Hg) at 1200^{0}C. On these grounds the suggestion was made that bromine was chemically bonded to boron. The samples obtained also contained W, Mn, Fe and, furthermore, traces of Si, O_2 and H_2 were found.

Comparing the results of low-temperature reduction of BBr_3 with the investigation of α→β-transformation (the temperature of which is found to be 1200^{0}C), it was suggested (NASLAIN et al., 1970) that the β-boron appearance at the temperatures <1200^{0}C (after the critical time) is due to the kinetic factors and that α→β-transformation at temperatures <1200^{0}C are hardly probable. The process of deposit formation has two steps: (1) the spontaneous nucleation and the growth of α-boron microcrystals from the vapor phase with the further development into granules; (2) β-boron deposition from the vapor phase in the intervals between the granules "which could result from a gradual modification *in situ*" of the kinetics of β-boron formation, which would require a lower activation energy in the presence of porous α-boron granules.

TALLEY et al. (1960a) as well as BEAN and MEDCALF (1960), using the process of hydrogen reduction of BBr_3 (purified by the distillation method) did not make it their special task to prepare α-boron. However, all of them observed its formation: the former - under investigation of the preparation conditions of glassy boron, the latter - under investigation of boron deposition on boron rods. The conditions under which α-boron formation is observed in these works are given in Table 4.

Table 4.

Conditions of α-boron formation by hydrogen reduction of BBr_3	TALLEY et al. (1960a)	BEAN and MEDCALF (1960)
Substrate material	W + 1% ThO_2	β-boron
Substrate size	d = 0.025 mm l = 5 cm	d = 1.6 mm
Substrate temperature	1200 °C	1000 °C
Hydrogen stream velocity	500 ml/min	970 ml/min
Vaporizer temperature (rate of evaporation) of BBr_3	25 °C -	- 60 ml/min
BBr_3 concentration in hydrogen	8-10 M %	20 M %
Deposition rate of boron	4.5 mg/cm^2 min (0.01 g/h)	2-3 g/h

α-Boron, prepared by the method described by BEAN and MEDCALF (1960) contained, according to the spectrographic analysis, Si-5.10^{-5} percent, Mg-10^{-6} percent. Cu-0.5-10^{-6} percent. As mentioned by TALLEY et al. (1960a), α-boron formation did not always take place and when it did, it appeared only as a transition layer between the glassy boron and polycrystalline (probably β-rhombohedral) boron. In the same work it has been pointed out that the glassy boron heated above 1260 °C turns into a crystalline state, including α-boron. Taking into account the fact that the mentioned temperature is a brightness temperature, and the real temperature must be about 50 °C higher, the transformation of glassy boron to α-rhombohedral form takes place at temperatures above 1310 °C. As will be shown below, the authors of the present review have observed the transformation of powdered amorphous boron into α-boron at approximately these temperatures.

AMBERGER et al. (1970) have reported the possibility of α-boron preparation by hydrogen reduction of BBr_3 and BCl_3 of technical purity by the use of a graphite susceptor of high frequency current. Preliminarily, the susceptor was covered with pyrolytic film. This process was carried out by the use of CH_4-BX_3 mixture at about 1200°C. BX_3 reduction occurred in the temperature range 1000-1100°C. Molar ratio of BX_3 to hydrogen was 1:10. The flow rate was not higher than 10 l/h. The boron yield proved to be higher with BBr_3 than with BCl_3.

In the deposits, small α-boron crystals and glassy boron were found. α-Boron crystals up to 1 mm in size have been disclosed by etching glassy boron in a warm concentrated nitric acid. These crystals have been found only in contact with amorphous glassy boron.

3. α-Boron Preparation by Diborane Reduction (ROBB and LANDAUER, 1959)

Diborane was subjected to preliminary purification by passing its vapor through a molecular sieve. However, methane and carbon oxide are not completely removed by this method, and the boron prepared from the diborane purified in such a way, contained 0.1 percent carbon. The best purification of diborane was achieved by the use of distillation. In this case, a boron sample of up to 99.99 percent purity was obtained from diborane.

For preparation of elementary boron the purified diborane was subjected to thermal decomposition on a heated substrate. Diborane decomposition took place at temperatures below 1000°C and general pressure in the decomposition chamber was no higher than 10 mm of Hg (the vapor pressure close by the substrate was seldom higher that 0.1 mm of Hg). The crystalline deposits containing α-boron was obtained at deposition rate of 0.5 g/h cm^2.

4. α-Boron Crystal Growth from Melt Solution

It is well known that the process of melt-solution crystallization makes it possible to obtain crystals at temperatures much lower than the melting point of the substance under investigation. This process is especially suitable for growing crystals of modifications whose upper temperature limit of stability is limited. In the case of boron, the existence of irreversible polymorphous transformation of α-boron to β-boron (DECKER and KASPER, 1959) has been ascertained. When using the method of crystallization of melt solutions the emergence of α-boron crystals may be expected at temperatures below α→β-transition temperature.

HORN (1959, 1960) was the first to prepare α-boron by crystallizing melt solutions. He studied the boron-platinum system where there is a low-melting eutectic, and reported an absence of compounds between pure boron and eutectic composition up to 40 percent boron content. The temperature of eutectic horizontal was established to be 825 ± 5°C.

Further investigations of the phase diagram were developed later (NIEMYSKI et al., 1962; WALD, 1970; WALD and ROSENBERG, 1965) and the phase diagram of B-Pt system is given here (WALD, 1970) and is derived from the combined data (Fig. 3). Without considering this phase diagram in detail, we shall only note that its form depends greatly upon the presence of outside impurities. For instance, 1.1 at. percent of silicon prevents the formation of Pt_3B and leads to the emergence of eutectic between Pt_2B and Pt (WALD, 1970).

For the boron crystallization to occur, the composition of boron-platinum alloys was selected so that the starting temperature of alloy crystallization should not exceed the temperature of α-boron

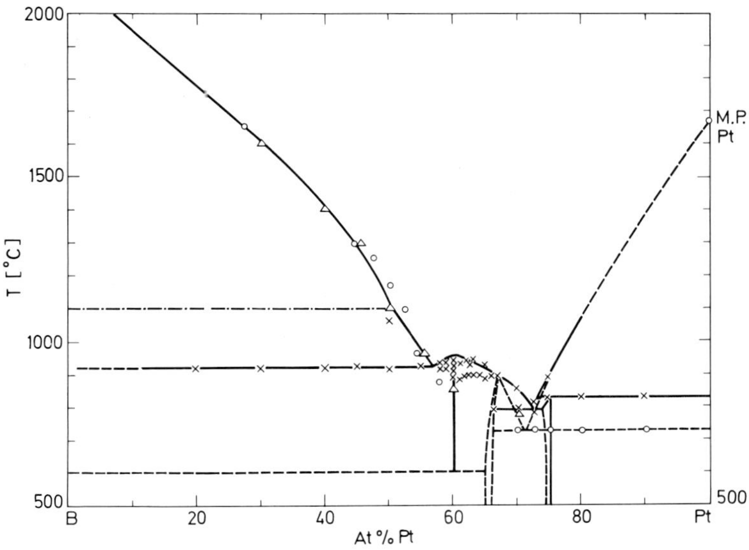

Fig. 3. Diagram of the state of platinum-boron system derived from the works of: o: HORN (1959), Δ : NIEMYSKI and ZAWADZKI (1972), x: WALD and ROSENBERG (1965), −·−: WALD (1970)

formation from the vapor phase. This condition satisfies the boron composition with 50 percent (at.) which melts at 1200°C. The alloy was prepared from the commercial platinum and boron, purified by zone melting. After removing the slag initially formed by melting from the exterior of the bar, the bar was put into a crucible made of boron nitride with the conical bottom. The melting was carried out in a vacuum in which a high frequency generator was used as a heating device. When the eutectic temperature was reached, the alloy was abruptly cooled. The formation of α-boron crystals was observed in the upper part of the bar. The crystals could have been isolated either by crushing the bar with the subsequent separation in bromoform (HORN, 1959) or by separating platinum matrix by electrolysis in concentrated hydrochloric acid followed by washings and decantation (HORN et al., 1965). The largest of the crystals observed reached 0.5 mm in size.

Attempts have been made to prepare α-boron crystals from other solutions. Encouraging results were obtained by WALD (1970), who observed α-boron emergence from the solution of boron in platinum, germanium, palladium, copper, silver, as well as in alloys of Au-Ge and Cu-Ni.

Utilizing a method of crystal preparation by a vapor-liquid-solid phase technique and using platinum as a liquid-forming agent, SITARIK and ELLIS (1966) developed α-boron whiskers 2 mm long and 25 μm in cross-section. Transportation of boron to the substrate was accomplished by the reduction of BCl_3 in hydrogen at about 1000°C. The melted quartz and amorphous boron covered with finely dispersed platinum were used as substrates. The substrates on the graphite heater were introduced into a quartz tube and a flow of purified hydrogen was passed through it at a flow rate of 1.5 l/min. The graphite was heated by high frequency current. When the necessary temperature was reached, purified boron halogenide was added to hydrogen. BCl_3 flow rate was 1.4 ml/min.

The experiments resulted in translucent α-boron whiskers of red color, having spherical thickenings at the ends, characteristic of VLS process. The direction of the largest part of the whiskers growth was determined by X-ray methods as <$\bar{4}$401> (hexagonal presentation). Apart from the straight whiskers, bent ones were observed with growth direction - <$\bar{4}$401> at the base and - <$\bar{4}$041> at the peak. These directions make an angle of approximately 70° and both are in the plane (0114). These directions also have an angle close to 70° with c axis of α-boron. The cause of such a growth is not clear.

5. α-Boron Preparation by Amorphous Boron Crystallization

The authors of the present review investigated the crystallization process of powdered amorphous boron and after high frequency heating of amorphous boron of 99.6 percent purity, prepared by pressing, found a thin layer of red color between the areas of original boron and the gray crystalline mass (Fig. 4). X-ray photograph investigations indicate that this material belongs to α-boron. In view of the fact that in a number of works on boron preparation from the vapor phase, mentioned above, α-boron emergence was observed on the boundary with the amorphous glassy boron, and in view of the fact that the powdered amorphous boron used by us was also prepared from the vapor phase, it was of interest to find out the crystallization conditions of powdered boron in detail in order to compare the results obtained with those found in the literature.

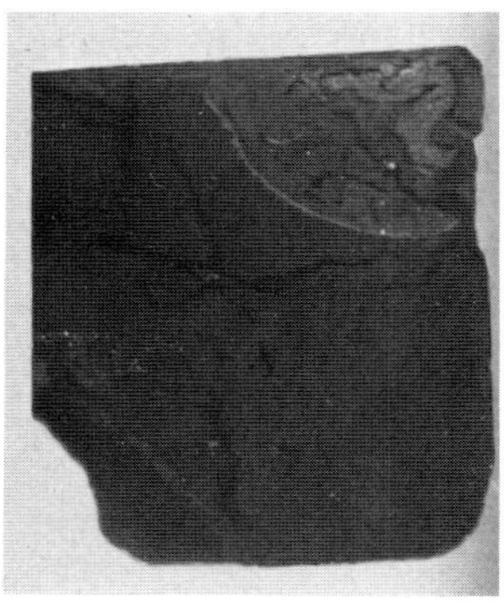

Fig. 4. Photo pattern of amorphous boron, subjected to high frequency heating. Between the amorphous boron (*lower part*) and β-boron (*right hand top corner*) there is α-boron vein (*thin light line*)

To find out the conditions of crystallization, the pressed samples of powdered boron were heated in argon atmosphere with the simultaneous recording of thermo-effects by the method of differential thermoanalysis. Thermographs were taken in the quartz tubes at the temperature range of 800-1400°C and a heating rate of 3-40° per min. PtRh-Pt thermocouples were used.

When amorphous boron are heated at atomspheric pressures, exothermic effect was found in the range of temperatures between 1270°C and 1330°C (Fig. 5a). No effects were observed on the cooling curves. X-ray pattern investigation of samples after cooling pointed to α-boron content in them as the main component as well as β-boron and suboxide $B_{13}O_2$. The interpretation of X-ray patterns was made by comparing them with powdered patterns of different boron modifications (HOARD and NEWKIRK, 1960; TALLEY et al., 1960b) and borides of some elements (LA PLACA and POST, 1961; PASTERNAK, 1959). The color of α-boron was yellow to

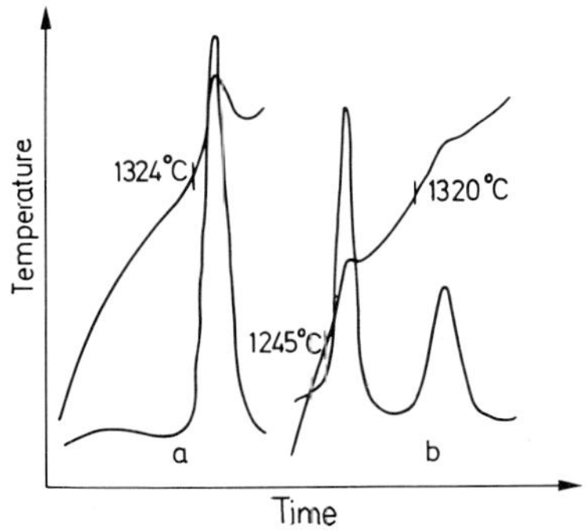

Fig. 5. Thermograms of boron samples:
a: amorphous boron *b*: amorphous boron with higher B_2O_3 content

red. Judging by the electron microscopy results, the size of the crystallites was 1-2 μ.

Thermographs were also taken of the samples of amorphous powdered boron with the addition of boron oxide (B_2O_3), which brought out one more exothermic effect preceding the effect of α-boron formation (Fig. 5b). As the results of X-ray analysis show, the samples, cooled abruptly after the first effect, contain amorphous boron and suboxide $B_{13}O_2$. Hence, it is possible to assert that the first exothermic effect was caused by the formation of suboxide $B_{13}O_2$.

The presence (Fig. 5a) of a diffused maximum on the differential curve and a rather sharp bend on the simple curve is also likely to indicate the formation of $B_{13}O_2$. However, this exothermic effect is not observed when the B_2O_3 content in the initial amorphous boron is small.

The heating of analogous samples of amorphous powdered boron at low pressure of argon (10-3 mm of Hg) did not result in observation of both $B_{13}O_2$ and α-boron, which is probably caused by a decrease in B_2O_3 content in amorphous boron in the process of heating due to partial evaporation of B_2O_3 (SPEISER et al., 1950) and formation of the highly volatile compound B_2O_2 (SCHEER, 1958). At heating the amorphous boron in a vacuum the β-boron formation has been observed, as mentioned earlier (GILLESPIE, 1966).

Thus, the following:
1. Thermoheating amorphous powdered boron in argon atmosphere at atmospheric pressures results in α-boron formation starting at 1270°C.
2. Some amount of suboxide $B_{13}O_2$ which can be detected by X-ray patterns is always present in the prepared α-boron samples.
3. $B_{13}O_2$ formation occurs at lower temperatures than α-boron formation, i.e., $B_{13}O_2$ emergence precedes α-boron emergence.
4. At low argon pressures neither suboxide $B_{13}O_2$ nor α-boron is formed.

On grounds of the above enumerated facts and isostructure of α-boron and suboxide (LA PLACA and POST, 1961), the conclusion was reached that the presence of suboxide $B_{13}O_2$ in amorphous boron not only does not interfere with the α-boron formation but is likely to be beneficial, as $B_{13}O_2$ plays a part in crystal forming nuclei.

It may thus be expected that other boron compounds, isostructural with it, will also assist in α-boron formation.

The fact that small additions of some impurities, particularly carbon, do not prevent formation of α-boron has been mentioned earlier (AMBERGER et al., 1970). In the same work it was shown that at temperatures below 1000°C the simultaneous reduction of BI_3 and CI_4 in hydrogen resulted in formation of crystalline boron carbide $B_{96}C_{12}$, which was grown on an α-boron layer (Fig. 6). According to the phase diagram of boron-carbon, given in this work, $B_{96}C_{12}$ compound (whose structure has not been fully studied) has the largest boron content among carbonic boron compounds, which form in the same temperature range as α-boron. As may be observed in Figure 6, among the grown crystals of $B_{96}C_{12}$ there are α-boron crystals. Hence, we may conclude that not only crystals of $B_{96}C_{12}$ can grow on α-boron, but, also *vice versa*. If this conclusion is true, the compound $B_{96}C_{12}$ as well as $B_{13}O_2$ can act as crystal-forming centers under conditions of α-boron formation.

On the basis of the above investigations, a method of α-boron preparation from the amorphous boron has been worked out, by which pressed amorphous boron powder of 99 percent purity can be crystallized (AVERBAKH et al., 1973). Fig. 7 demonstrates a photo of the sample prepared by this technique.

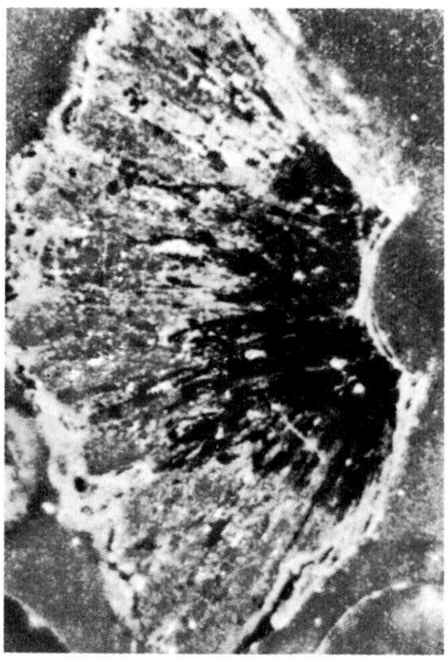

Fig. 6. Photograph of the cross-section of carbide boron sample $B_{96}C_{12}$ (*left*), formed on the α-boron (*right*) (AMBERGER et al., 1970)

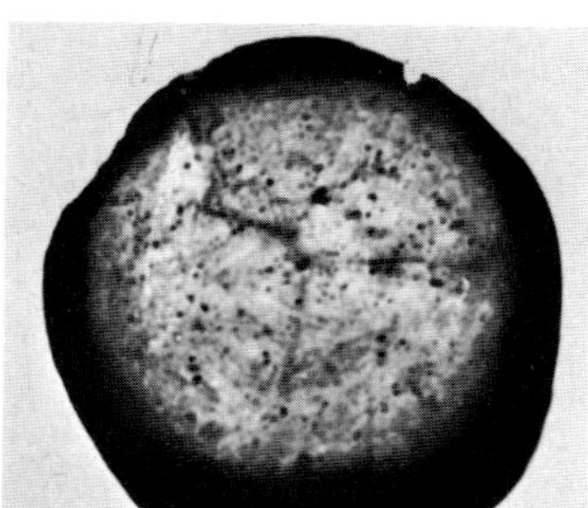

Fig. 7. Cross section of α-boron sample. *Presurface dark layer*: β-boron.

When pieces of glassy boron are placed in amorphous boron powder and heated, formation of α-boron scales up to 2 mm in size can be observed. Because the crystallization of glassy boron took place under the same conditions as crystallization of powdered boron and since these amorphous materials have identical X-ray patterns, a conclusion about the similarity of these materials can be drawn. The difference between them is probably in that bonding between atoms of glassy boron extends farther, i.e., powdered boron is like finely dispersed glassy boron. Therefore, processes observed at crystallization of powdered boron also take place at glassy boron crystallization.

The process of α-boron formation from amorphous boron may be analogous to the process of crystallization of glassy substances where, besides definite temperature conditions, the presence of crystallization centers is necessary. Depending upon the nature of these centers, α-boron formation may be expected in different temperature ranges. This observation is confirmed by reports in which experimental conditions differed greatly. For instance, in our experiments, amorphous boron crystallization in the presence of suboxide occurs at 1270 °C and above. The formation of α-boron crystals in glassy boron, when its exterior is coated with microcrystals of the same modification, is observed in the range 950-1100 °C (AMBERGER and DIETZE, 1964).

Thus, in the process of glassy boron crystallization some impurities can play a positive part in forming crystal nuclei. However, as soon as the process of crystallization has been initiated, the presence of impurities has a negative effect on further growth of α-boron crystals. This has been described by AMBERGER and DIETZE (1965) and has been discussed earlier. It should be added that among impurities there can also be atoms of starting materials from gas-transport reactions (McCARTY and CARPENTER, 1960; NASLAIN et al., 1970) due to incomplete process of these reactions. Such atoms may not only cause the distortion of the α-boron lattice but also reduce the crystal growth of this modification.

The phenomenon of the impurity segregation during α-boron crystal growth and the higher purity of the α-boron formed in comparison with the starting glassy boron, resembles the phenomenon of segregation at crystallization of melts on a liquid-crystal boundary. In this case, however, the glassy boron, formed from the vapor phase, plays the part of a "liquid". As the directed crystallization of melts has successfully been used for the preparation of monocrystals of different substances, the preparation of homogeneous glassy boron samples and their controlled crystallization seems worthy of some attention for the purpose of growing comparatively large crystals of the α-rhombohedral boron modification.

References

AMBERGER, E., DIETZE, W.: Zur Bildung von α-rhomboedrischem, rotem Bor. Z. anorg. allg. Chem., 332, 131 (1964)

AMBERGER, E., DIETZE, W.: Synthesis of red, α-rhombohedral boron. In: Boron - Preparation, Properties and Applications. 2, p. 1, New York: Plenum Press 1965

AMBERGER, E., DIETZE, W., DRUMINSKY, M., PLOOG, K.: Formation of Modifications of Boron and Boron Carbides. In: Boron - Preparation and Properties, 3, p. 133, Warszawa: PWN, (1970)

AMBERGER, E., DIETZE, W., JAUMANN, J. RUNOW, P.: Preparation of α-rhombohedral boron and electron microscope examination. Phys. Solid State A., (Germany), 2, 2, K 59, (1970)

AVERBAKH, E.M., SOLOVIJEV, N.E., UGAI, Ya.A.: The method of α-rhombohedral boron formation. Author's Certificate No. 331661, cl. C01b, 35/00, claimed 1970, published January 16, 1973

BEAN, K.E., MEDCALF, W.E.: Utilization of boron filaments in vapor-phase deposition of boron. In: Boron - Synthesis, Structure and Properties. 1, p. 48. New York: Plenum Press 1960

DECKER, B.F., KASPER, J.S.: The crystal structure of a simple rhombohedral form of boron. Acta. Cryst. 12, 7, 503, (1959)

GILLESPIE, J.S.: Crystallization of massive amorphous boron. J. Amer. Chem. Soc. 88, 11, 2423, (1966)

HOARD, J.L., NEWKIRK, A.E.: An analysis of polymorphism in boron based upon X-ray diffraction results. J. Amer. Chem. Soc., 82, 1, 70, (1960)

HORN, F.H.: On the crystallization of simple rhombohedral boron from platinum. J. Electrochem. Soc., 106, 10, 905, (1959)

HORN, F.H.: Simple rhombohedral boron - preparation and properties. In: Boron - Synthesis, Structure and Properties, 1, p. 110. New York: Plenum Press 1960

HORN, F.H., TAFT, E.A., OLIVER, D.W.: On the birefringence of simple rhombohedral boron. In: Boron - Preparation, Properties and Applications, 2, p. 231. New York: Plenum Press 1965

LA PLACA, S., POST, B.: The boron carbide structure type. Planseeber Pulvermet. 9, 109, (1961)

McCARTY, L.V., CARPENTER, D.R.: The preparation of a new crystalline modification of boron and notes on the synthesis of boron triiodide. J. Electrochem. Soc., 107, 1, 38, (1960)

McCARTY, L.V., KASPER, J.S., HORN, F.H., DECKER, B.F., NEWKIRK, A.E.: A new crystalline modification of boron. J. Amer. Chem. Soc. 80, 2592, (1958)

NASLAIN, R., ETOURNEAU, J., HAGENMULLER, P.: Preparation of pure α- and β-rhombohedral boron, the alkali borides. In: Boron - Preparation and Properties. 3, p 35. Warszawa: PWN 1970

NIEMYSKI, T., ZAWADZKI, W.: Some properties of pure polycrystalline boron. Physics Letters. 2, 1, 30, (1962)

PASTERNAK, R.A.: Crystallographic evidence for the existence of B_7O. Acta. Cryst. 12, 612, (1959)

ROBB, W.Z., LANDAUER, L.C.: Production of boron from diborane. Abst. Boston Meet. ACS, p. 53 Division of Inorganic Chemistry April 1959

SCHEER, M.D.: The molecular weight and vapor pressure of gaseous boron suboxide. J. Phys. Chem. 62, 4, 490, (1958)

SITARIK, J.P., ELLIS, W.C.: Preparation and morphology of boron filamentary crystals grown by the vapor-liquid-solid mechanism. J. Appl. Phys. 37, 6, 2399, (1966).

SPEISER, R., NAIDITCH, S., JOHNSTON, H.L.: The vapor pressure of inorganic substances II, B_2O_3. J. Amer. Chem. Soc., 72, 6, 2578, (1950)

TALLEY, C.P., LINE, L.E. OVERMAN, Q.D.: Preparation and properties of massive amorphous elemental boron. In: Boron - Synthesis, Structure and Properties, 1, p. 94. New York: Plenum Press 1960

TALLEY, C.P., POST, B., LA PLACA, S.: A new polymorph of boron. In: Boron - Preparation, Properties and Applications, 1, p. 83. New York: Plenum Press 1960

WALD, E.: On the stability of the red α-rhombohedral boron modification. In: Boron - Preparation and Properties, 3, p. 103. Warszawa: PWN 1970

WALD, F., ROSENBERG, A.: Constitutional investigations in the boron-platinum system Trans. Met. Soc. AIME, 233, April, 796, (1965)

WIBERG, E., AMBERGER, E. DIETZE, W.: Verfahren zum Herstellen von Bor mit α-rhomboedrischer Kristallstruktur. Pat. GFR., No. 1231221, cl. 12i, 35/00 (C01b), claimed 1963, publ. 1967

IV. Characterization of Localized States in β-Rhombohedral Boron

M. PRUDENZIATI

1. Introduction

Boron is an elemental semiconductor with very fascinating properties, namely low density, high melting point (2300 °C), a hardness close to that of diamond and remarkable mechanical and nuclear properties.

These features could also play an important role in electronic devices such as neutron detectors (SHAW et al., 1957; GAULÉ et al., 1965), high-power and high-temperature devices (DIETZ and HELMBERGER, 1965; PRUDENZIATI and MAJNI, 1973) and switching devices (DIETZ and HELMBERGER, 1965; PRUDENZIATI et al., 1973).

Among the different polymorphs, β-rhombohedral boron is the only one grown from the melt and actually prepared in suitable size for electrical investigation. This is why our knowledge of semiconducting properties of cristalline boron is limited to this polymorph and in this article, the term "boron" always refers to β-rhombohedral boron.

A prediction of the basic semiconducting parameters - such as the energy gap, band structure and effective masses of the charge carriers - on a theoretical basis has been hampered so far by the complex structure of this material. In fact, the unit cell contains 105 atoms, among which 91 have the coordination number six, 12 the coordination number eight and two atoms the coordination number nine.

Failing theoretical computations we turn to qualitative approaches. With respect to the density of electronic states SOSNOWSKI (1970) suggests that the atoms with coordination number six should give electronic bands with a width of several electron volts and a high density of states in the order of $10^{22}/cm^3$ eV (such as in more simple structures), while atoms with coordination numbers eight and nine could contribute subbands or parts of bands with a state density 10^{20}-$10^{21}/cm^3$ eV. These low-density bands can be considered localized states and may form the top of the valence band or parts of excited band states.

These states will be referred to as "intrinsic localized states" and juxtaposed to "extrinsic localized states" which, of course, exist in the boron gap, due to impurities and defects.

The density of these extrinsic states depends, of course, on the quality of the material. Unfortunately, the growth of relatively defect-free and pure single crystals of boron is a difficult problem and the researchers have, up to now, had to study samples of poor quality.

In fact, the best single crystals actually available to our knowledge (Wacher Chemie, West Germany and Alfa Inorganics, U.S.A.) contain a high amount of carbon (≥50 ppm) and near equal concentrations of nitrogen and oxygen[1], so a density of extrinsic localized states to the order of $10^{19}/cm^3$ could be expected. However, it is difficult to say if we can speak in terms of concentration for these impurities, since they can be nonuniformly distributed into the bulk, as suggested by the detection of inhomogeneities by microanalysis (WERHEIT et al., 1972) and of boron-carbide clusters by electron microscope (KLEINHENZ and RUNOW, 1968), together with dislocations and other defects.

Then we could suspect effects of heterogeneities and potential fluctuations in the transport phenomena of boron, while, as is well known, these effects are less relevant in optical phenomena.

On the basis of these considerations, we see that there are different possibilities in order to justify the "anomalous" behavior of some properties of boron, that is, its complex structure with the possible existence of intrinsic localized states, and extrinsic localized states and heterogeneities, which can (in principle at least) be removed by use of a careful method of preparation.

An answer to the question of whether intrinsic or extrinsic characteristics of the material are dominant in boron, is of primary interest both from the physical and the practical point of view, since it can determine the methods of the workers who prepare the material and who hope to use boron in electronic devices.

In reviewing the experimental data on boron we will focus our attention on such experiments as can yield information on the energy location and density of the localized states and thus help to solve the problem already stated, since intrinsic high-density localized states would be detected consistently in different boron samples, with different techniques, at the same energy depths.

2. Optical Absorption

The most detailed investigation of optical absorption in boron was done by JAUMANN and WERHEIT (1969). The data on optical absorption coefficient, for a wide range of temperature ($-160^\circ C$ - $+600^\circ C$) and wavelength (0.7 μm - 8 μm), can be interpreted with a model which involves two indirect allowed transitions with phonon emission and one direct allowed transition. The threshold energies for the three transitions are $\Delta E_1 = 1.56$ eV, $\Delta E_2 = 1.33$ eV and $\Delta E_3 = 0.37$ eV respectively at $T = 0^\circ K$.

In the first transition, electrons are excited by the valence to the conduction band and the $\Delta E_1 = 1.56$ eV is the energy gap, in agreement with other investigations (BORCHERT et al., 1965; GAULÉ et al., 1965).

In the second transition, electrons are transferred from extended to localized states or vice versa. Electrical conductivity and thermopower data reported for the same sample by WERHEIT and LEIS (1970) indicate that the first case is most plausible. Thus, there exists a level at a distance of $\Delta E_1 - \Delta E_2 = 0.23$ eV above the valence band. The level has a great density of states, although considerably lower

[1]Measurements performed by F.G. SATKIEWICZ, CGA Corp. Mss. supported by the John Hopkins University, Maryland. Unpublished results.

than the VB states, and it is occupied by electrons at low temperatures, thus behaving as a hole-trapping level.

The third transition is explained as due to electrons which are excited from electron traps at a minimum distance $\Delta E_3 = 0.37$ eV to the conduction band. The occupation density of the electron traps increases with the factor exp $(-E_0/KT)$ with $\Delta E_0 = 0.96$ eV for thermal excitation of charge carriers from the VB.

The three threshold energies are temperature-dependent according to

$$\Delta E(T) = \Delta E(0) - \beta T^2 \qquad \text{where } \beta = 3.27 \times 10^{-7} \text{ eV/deg}^2. \tag{1}$$

At low temperatures the absorption spectrum strongly depends on previous illumination (GAULÉ et al., 1965; DZHAMAGIDZE et al., 1968; WERHEIT, 1970; ANISIMOV et al., 1973). This additional "photoabsorption" is caused by photons ionizing the localized states, having been occupied by photo-electrons at low temperatures. A direct recombination via localized electron and hole traps is very improbable due to their spatial separation, while the recombination via extended states is delayed by the improbable thermal excitation of carriers from the traps due to their relatively high activation energies.

WERHEIT (1970) has treated the measured optical spectrum obtained in this condition with the theory of EAGLES (1960), which involves the ratio m_e^+/m_h^+ of electron and hole effective masses. The fit is quite good when $m_e^+/m_h^+ = 0.09$ and electron depth is 0.44 eV. This energy, somewhat greater than the minimum threshold energy obtained at higher temperatures, suggests that electron traps are distributed on various levels with a spread of depths. The concentration of electron traps is estimated to be about $5\times10^{18}-10^{20}/cm^3$.

It is difficult to extract the optical excitation spectrum for holes from the data; however, there is support for the 0.23 eV level in photoconductivity measurements (Sect. 4a).

In conclusion, if we take the activation energies of localized states from pure optical measurements and anticipate the results of transport properties which suggest their relative position with respect to the allowed bands, the picture in Fig. 1 results.

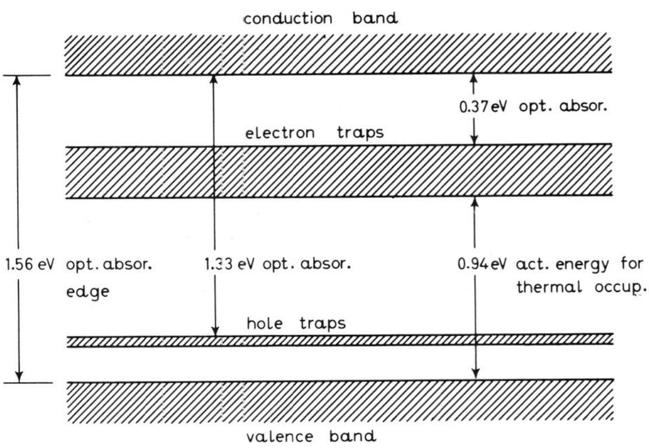

Fig. 1. Energy levels in the gap of boron, according to the results of optical and electrical measurements of JAUMANN and WERHEIT (1969); WERHEIT (1970); WERHEIT and LEIS (1970)

3. Dark Transport Measurements

Transport mechanisms in boron have been and are the subject of intense debate but our understanding of the simplest phenomena and the knowledge of the fundamental parameters are still quite incomplete. For example, we do not know the electron and hole mobilities and effective masses, the role of polaron effects, and of various scattering effects, and so on. This comes partly from the fact that the most informative experiments, namely Hall effect, resistivity and thermopower, have even now given only a very little information on boron. This is due to the concurrence of three facts, namely, the high resistivity of non-intentionally doped samples, the lack of a pure extrinsic region in the temperature dependence of resistivity at low temperatures, and the difficulty of treating Hall and thermopower data with classical formulae.

At high temperatures these properties are perhaps influenced by both electrons and holes, but it is not clear if the true intrinsic conductivity has been evidenced by the measurements even up to 1300 °K. Figure 2, for example, shows the Hall constant and thermopower measured by DZHAMAGIDZE et al. (1968) and GOLIKOVA et al. (1970) and here we note that the two sets of data are consistently different, so either one or both do not represent the plain intrinsic properties of the material.

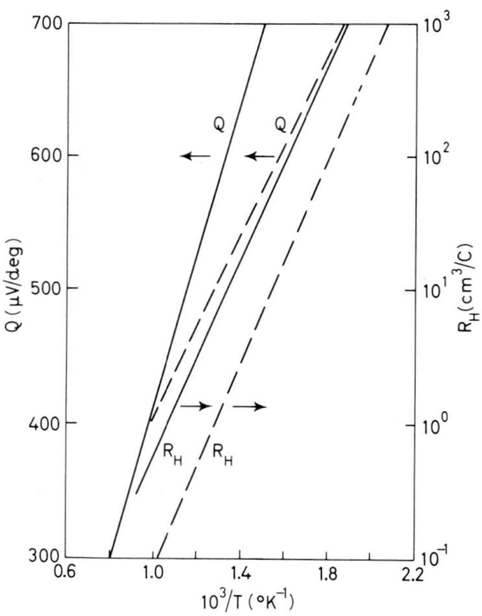

Fig. 2. Temperature dependence of the Hall coefficient (R_H) and of thermopower (Q) at high temperatures. *Continuous lines*: data of DZHAMAGIDZE et al. (1968), *broken lines*: data of GOLIKOVA et al. (1970)

WERHEIT and LEIS (1970) propose the assumption that only holes contribute to the conductivity up to 1000 °K. Under this assumption, the comparison of σ vs T with Q vs T according to BOSMANN and CREVECOEUR (1965) gives a hole mobility which increases exponentially with the temperature with an activation energy 0.19 eV, in rough agreement with the value 0.23 eV obtained in optical measurements (see Sect. 2), but no figure for this mobility is given.

At low temperatures the Hall signal is overwhelmed by noise and, assuming a one-carrier band-like transport, an appraisal of the maximum value of mobility is given around 10^{-1}-1 cm^2/V sec (BORCHERT et al., 1965; NEFT and SEILER, 1965; GOLIKOVA et al.,1971). The thermopower is always positive down to temperatures where the measurements are made difficult by the high resistivity of the material and appear to be affected by the sample preparation conditions (WERHEIT et al., 1970).

This uncertainty in the knowledge of the electron and hole mobilities has grave consequences in the interpretation of all the transport properties of boron and hence in the determination of depth and density of localized states by means of electrical measurements.

a) Hopping Conduction Near E_F

The temperature dependence of resistivity in non-intentionally doped boron at low temperatures can be described (WERHEIT and LEIS, 1970; MAJNI and PRUDENZIATI, 1971; BEREZIN et al., 1973a; BEREZIN et al., 1973b) by the relationship

$$\sigma = \sigma_0 \exp - \left(\frac{T_0}{T}\right)^{1/4} \quad (2)$$

derived by MOTT (1971) for thermally assisted hopping of carriers near the Fermi level E_F.

From the "slope" T_0 in Eq. (2), information can be obtained on the localized state density at the FERMI level, g_0; according to HILL (1971):

$$T_0 = \frac{16\alpha_0^3}{g_0 K} \quad (3)$$

where α_0 is the damping constant of the wave-function of the localized states. α_0^{-1} should be in the range 2-5 Å (MOTT and DAVIS, 1971). Moreover, after BRODSKY and GAMBINO (1972) the pre exponential factor σ_0 in Eq. (2) can be written

$$\sigma_0 = \left(\frac{g_0\alpha_0^{-1}}{KT}\right)^{1/2} \left(\frac{3e^2\nu_{ph}}{2}\right)\left(2\pi\right)^{-1/2} \quad (4)$$

We plot $\ln\sigma T^{1/2}$ vs $1/T^{1/4}$ in Figure 3, with the data reported in the literature and calculate the g_0 value with Eq. (3), assuming α_0^{-1} = 2-5 Å. In Table 1 the maximum and minimum of the calculated g_0 are reported. The quite reasonable values, ranging from 8.10^{18} to $10^{20}/cm^3$ eV, seem more appropriate for extrinsic than for intrinsic localized states. However, when we take into account also σ_0 and try to solve simultaneously Eqs. (3) and (4), in order to get both α_0 and g_0, we obtain very unreasonably high values for both quantities in all cases. In other words, the experimental σ_0 factor is too high in comparison to that expected from Eq. (4), with g_0 deduced by Eq. (3).

b) AC Conductivity

Sporadic data have been reported on ac conductivity measurements in boron, showing the behavior:

$$\sigma_{ac}(\omega) \propto \omega^s \quad (5)$$

where s is less than 1. This behavior could be taken as a support to the hypothesis of thermally assisted hopping of carriers among states near E_F, since in this case, we should have, according to AUSTIN and MOTT (1969):

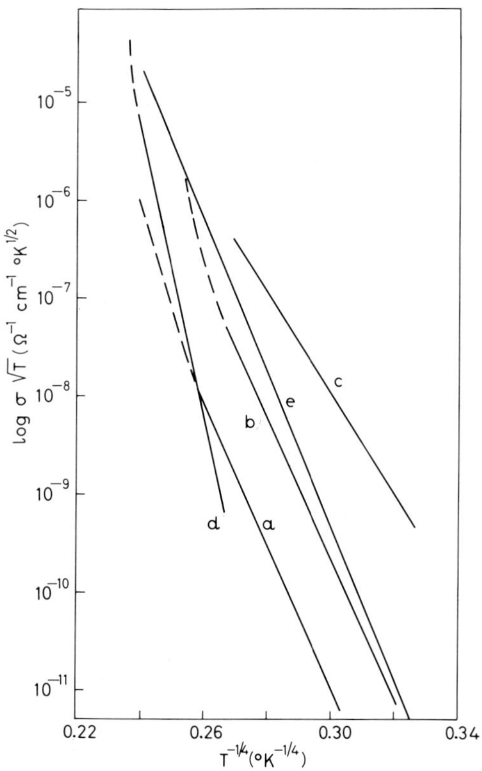

Fig. 3. $\ln \sigma T^{1/2}$ vs $1/T^{1/4}$ for boron at low temperatures. The data have been taken from a: WERHEIT and LEIS (1970) b: GEIST (1965) c: DZHAMAGIDZE et al. (1968) d: MAJNI and PRUDENZIATI (1971) e: BEREZIN et al. (1973b), as indicated on each line

$$\sigma(\omega) = \frac{\pi}{3} e^2 KT \alpha_0^{-5} g_0^2 \omega \{\ln(\nu_{ph}/\omega)\}^4 \qquad (6)$$

Unfortunately, a frequency-dependence of the type expressed in Eq. (5) is not a unique identification of the hopping process, as pointed out by FRITZSCHE (1972) but it results quite generally from any relaxation process which involves a wide and flat distribution of relaxation times.

Nevertheless, we can try to check if the sporadic data on boron are in rough agreement with the hopping theory {Eq. (6)} and to get an appraisal of the order of magnitude of g_0. This evaluation is possible with the data of AVDEENKO et al. (1969) at 77°K; assuming α_0^{-1} = 3-8 Å we obtain $g_0 = 10^{18} - 10^{20}/\text{cm}^3$ eV which are reasonable values for localized states even if these figures do not allow to settle on intrinsic or extrinsic nature of these states. The data are not sufficient to check the linear temperature - dependence predicted in Eq. (6) and the small s = 0.4 at 300°K suggests a multi-phonon effect (GOLIN, 1963).

NEFT and SEILER (1965) obtain $0.54 \leq s \leq 0.86$ around 300°K but also in this case the linear temperature-dependence in Eq. (6) cannot be accurately checked because of the narrow temperature range investigated. Moreover, the data are reported in terms of conductance instead of conductivity, so no information of g_0 can be derived.

Measurements of DIETZ and HERMANN (1970) point out that a frequency dependence of the type in Eq. (5) can be due to crystallographic

Table 1. Density and location of localized centers in boron

	$N_t\,(cm^{-3})$	$g_0\,(cm^{-3}eV^{-1})$	E(eV)	Reference
Optical absorp.	<<VB states		0.23 above VB	JAUMANN and WERHEIT (1969).
	$5\ 10^{18}$-10^{20}		0.37 ÷ 0.44 below CB	WERHEIT (1970).
Hopping conduct.		10^{18}-10^{19}		DZHAMAGIDZE et al. (1968).
		$8\ 10^{18}$-10^{20}	-	MAJNI and PRUDENZIATI (1971).
ac conductivity		10^{19}-10^{20}		AVDEENKO et al. (1969).
SCLC	$<3\ 10^{19}$		0.23 above VB[a]	PRUDENZIATI et al. (1972)
	$<10^{20}$		0.05 above VB[a]	PRUDENZIATI et al. (1972)
			0.36 above VB	
High field cond.	$\sim 10^{17}$-10^{18}[b]		0.05 above VB[b]	BEREZIN et al. (1973).
		$\sim 10^{22}$-10^{24}[b]	"deep"[b]	BEREZIN et. al. (1973).
Photocurr. yield	-		0.42	WERHEIT (1970).
	-		0.23	WERHEIT (1970).
Photoc. time-depend.	$<10^{18}$		0.27 ÷ 0.45	NADOLNY (1972).
Photocond. and EPR	$2.4\ 10^{19}$ donor		0.23 above VB	GEIST (1973).
	$2.4\ 10^{21}$ accept.		0.23 above VB	GEIST (1973).
	$2.4\ 10^{20}$ traps		0.36 below CB	GEIST (1973).
Trans. photocond.	-		0.25 below CB	PRUDENZIATI et al. (1972).
			0.23 above VB[a]	PRUDENZIATI et al. (1972).
			0.36 above VB[a]	PRUDENZIATI et al. (1972).
Magnetic suscept.	10^{22}		0.51 below CB	GEWINNER et al. (1973). KUBLER et al. (1973).
	10^{21}-10^{22}		0.23 above VB	GEWINNER et al. (1973). KUBLER et al. (1973).

[a] Different results in different samples.
[b] Different results with different models.

defects since $\sigma \alpha \omega^s$ can be shifted to lower frequency and higher temperatures by plastic deformation in boron samples.

c) Non-ohmic Conduction at High Fields

Information on the depth and distribution of localized centres in boron have been obtained by PRUDENZIATI et al. (1972) under space charge limited (SCL) conditions. The samples have contacts made by gold evaporated layers on etched surfaces or by platinum-boron eutectic alloy (PRUDENZIATI and MAJNI, 1973).

At low temperatures and high electric fields (10^3 up to 10^5V/cm at $T \leq 200^\circ$K) the injected carriers' concentration becomes greater than the normal equilibrium density and the current exceeds the ohmic behavior, observed at low fields.

The I-V characteristics assume different shapes in different classes of sample (Figs. 4a, b) as expected in SCL conditions for different distributions of traps near the Fermi level. In the simplest case the non-ohmic part of the characteristics consists in a region where the current is a quadratic function of the voltage, so according to LAMPERT (1956) part of injected carriers are captured by N_t empty traps located on a single level at a depth E; this latter can be derived from the temperature dependence of SCL currents according to the relation:

$$J = \frac{9}{8} \varepsilon\varepsilon_o \frac{V^2}{W^3} \frac{N_t}{N_v} \mu_o \exp\left(-\frac{E}{KT}\right) \tag{7}$$

with the usual meaning of symbols.

From the experimental data and Eq. (7) it follows: E = 0.23 eV, $N_t/\mu_o = 3\ 10^{17} cm^{-5}$Vsec, which implies a maximum value for N_t of $10^{19}/cm^3$, assuming a maximum value for the microscopic mobility $\mu_o = 10^2 cm^2/V$ sec.

The comparatively less simple characteristics of Fig 4b are fitted with the SCLC theory which deals with an exponential distribution of traps with a cut-off energy E_1 and a discrete shallow level E_o. It results: E_1 = 0.36 eV, E_o around 0.05 eV-0.02eV and a concentration of traps less than $10^{11}/cm^3$ for the distribution and less than $10^{20}/cm^3$ for the shallow level, taking a maximum microscopic mobility equal to $10^2 cm^2/V$ sec.

TPM measurements (see Sect. 4e) on the same samples strongly support the validity of the interpretation outlined above of the I-V characteristics and indicate that holes are the injected carriers.

The J-E characteristics (Fig. 5) analyzed by BEREZIN et al. (1973b) differ from the preceding ones essentially in two aspects: a non-ohmic behavior is present even at fields as low as 10 V/cm and at high temperatures, and the current is a relatively weak function of the field up to very high field strength where the conductivity shows a tendency to saturation. These differences could probably be ascribed to the different contacts (silver contacts in this case) and surface treatment (etching with concentrated nitric acid) given to the samples.

BEREZIN et al. discuss their results with three different models: Poole-Frenkel, hopping, and small polaron model.

In the Poole-Frenkel model, charge carriers move in a band; their density changes as an effect of the field which lowers the activation energy of donor (or acceptor) levels, while the microscopic mobility

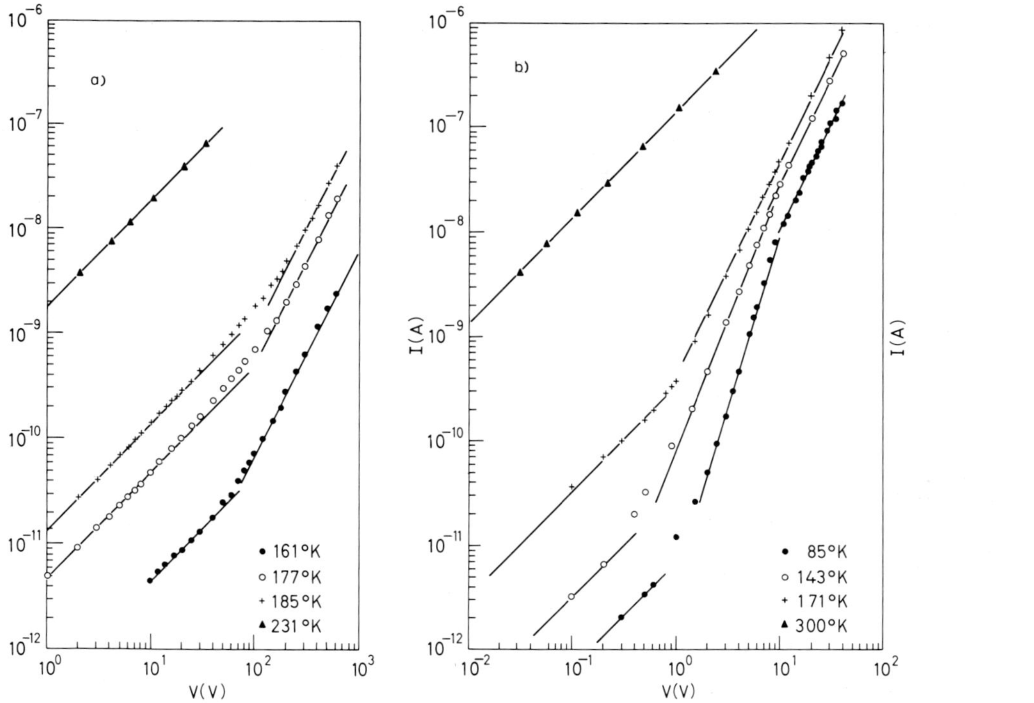

Fig. 4a and b. (a) Current-voltage characteristics at different temperatures in boron single-crystals (P 221 samples); $S = 7\ 10^{-2}\text{cm}^2$, $W = 2.5\ 10^{-2}\text{cm}$. (After PRUDENZIATI et al., 1972). (b) Current-voltage characteristics at different temperatures in boron single-crystals (P 225 samples), $S = 7\ 10^{-2}\text{cm}^2$, $W = 1.5\ 10^{-2}\text{cm}$. (After PRUDENZIATI et al., 1972)

is assumed independent of the electric field. A satisfactory agreement between the results and theory is obtained if the screening of Coulomb centers by free carriers is included, but drift mobility values ($10^{-5} \div 10^{-9}\text{cm}^2/\text{Vsec}$) are estimated which seem too low for band-like transport. The depth and density of traps are inferred from the tendency of the conductivity to saturate at high fields and considering this saturation as a consequence of the depletion of traps. Using merely the arguments of BOLTZMANN statistics, an activation energy of 0.05 eV and a density of the order of 10^{17}-$10^{18}/\text{cm}^3$ results.

In the hopping model, the jumps of the carriers among localized states are facilitated at high fields by the lowering of the site's energy in the direction of the field. According to this model, MOTT (1971) proposes the relation;

$$\sigma(E)\ \alpha\ \exp\left(-\frac{E_0}{E}\right)^{1/3} \quad \text{where } E_0 = \frac{8\alpha_0{}^4}{g_0 e} \tag{8}$$

while according to SHKLOVSKII (1973) this latter has to be changed to the following:

$$\sigma(E)\ \alpha\ \exp\left(-\frac{E_0}{E}\right)^{1/4} \quad \text{where } E_0 = \frac{16\alpha_0{}^4}{g_0 e} \tag{9}$$

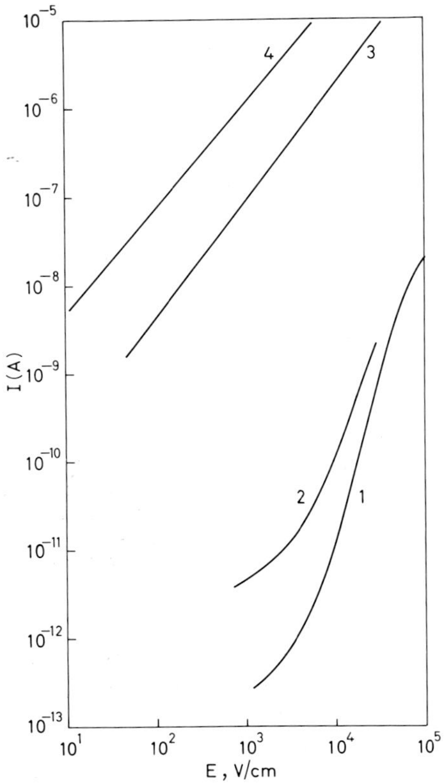

Fig. 5. J-V characteristics of zone-molted boron for different temperatures: 1: 77°K, 2: 113°K, 3: 205°K, 4: 293°K, $S = 10^{-2} cm^2$, $W = 10^{-2}$cm. After BEREZIN et al. (1973b)

Both these relations fit the data of BEREZIN et al. well, while in the two cases the derived hopping site densities are: $g_o = 10^{23}$-8 $10^{24}/cm^3$ eV from Eq. (8), and $g_o = 10^{22}$-7 $10^{23}/cm^3$ eV from Eq. (9). Then, in contrast to the POOLE-FRENKEL model, the hopping model gives exceedingly high densities of localized states, which, moreover must be located deep in the gap if Eqs. 8 or 9 are to be satisfied, as pointed out by SHKLOVSKII (1973).

Finally the small-polaron model does not fit the data satisfactorily. The authors propose a mixed type of conduction in which both hopping and small-polaron are involved.

4. Photoconductivity and Related Phenomena

a) Wave-length-Dependence of Photoconductivity

Boron exhibits a peak in photoconductivity at about 1.55 eV (BRUNGS and JACOBSMEYER, 1964; GAULÉ et al., 1965; WERHEIT, 1970), according to the onset of inter-band electronic transitions and a fall-off on the high-energy side of the edge, probably due to the increased role of surface recombination for carriers generated by strongly absorbed light. WERHEIT (1970) observed on well-polished samples another peak, not so far explained, at 3.5 eV.

On the low-energy side of the edge, excitation bands in photoconductivity, due to electronic transitions from localized states to extended states, can be easily measured in boron only after the localized states have been populated at low temperatures by illumination with $h\nu \geqslant 1.55$ eV (in strict analogy with the enhancing of optical absorption described in Sect. 3). The photoconductivity spectrum, measured under these conditions (WERHEIT, 1970) shows broad structures at a wave-length corresponding to transitions requiring 0.42 eV and 0.23 eV, in agreement with the absorption measurements in the same samples (Sect. 2).

It is not possible to infer the density of the localized states responsible for these transitions, since photocurrent intensity depends on too many quantities (BUBE, 1960) either unknown or imperfectly known in boron, such as mobility and life-time of carriers and quantum efficiency for photogeneration.

b) Time-Dependence of Photoconductivity

The parameters of localized states in photoconductors can easily be studied in suitable cases, measuring the time-dependence of the photocurrent intensity. For example, in the simplest cases where one carrier only contributes to the photocurrent, in the presence of a discrete trapping level, the decay of the photocurrent is merely a time-exponential function with a time-constant depending on the trap depth (BUBE, 1960).

On the contrary, concurrent effects of many trapping levels produce quite complicated rise and decay curves whose analysis often requires arduous work in curve fitting, in an attempt to derive the energy depths and density of the involved levels. This is, unfortunately, the case in boron, where moreover, the measurements are tiresome because of the exceedingly long times (hours at low temperatures) required for reaching stationary values of the photoconductivity (NEFT and SEILER, 1965; NADOLNY et al., 1966; ADIROVICH and GOL'DSHTEIN, 1969; GAULÉ et al., 1965).

In the last two works cited it is reported that a function sum of two time-exponential terms fits the experimental data well, while NADOLNY et al. (1966) showed that a time-logarithmic function fits the data better. In any case, no information on the trap parameters has been obtained from these studies until a recent work by NADOLNY (1972), where the photoconductivity is examined with a series of assumptions, namely: photocurrent is due to holes only, retrapping is negligible, there exists a distribution of electron traps with various emptying energies, the concentration of trapped electrons is equal to that of excess holes.

With this model the fitting parameters are: electron traps energies distributed over a range from 0.27 eV to 0.45 eV, and a concentration not higher than $10^{18}/cm^3$ assuming that at strong illumination and saturated photocurrent (over sufficiently long periods) the electron traps are completely filled and taking $\mu_p = 10^{-4} cm^2/V$ sec at $273^{o}K$.

c) Photoconductivity and Electron Spin Resonance

In an attempt to simplify and make the interpretation of photoconductivity experiments in boron more reliable, GEIST and MEYER (1970) and ADIROVICH et al. (1971) performed simultaneous measurements of photoconductivity and electron paramagnetic resonance (EPR). The density of the unpaired spins detected by EPR is believed to be correlated to the density of carriers on occupied localized states.

This contention is supported by experiments of GEWINNER et al. (1973) which moreover indicate a trapping level with a depth 0.23 eV.

A quantitative description of the photoconductivity and photo-EPR has been carried out by GEIST (1973) with the aid of the energy levels scheme of WERHEIT (see Sect. 2.) and suitable assumptions made on the magnetic behavior of the donors, acceptors and traps. The time-, illumination- and temperature-dependence of photoconductivity and photo-EPR may be described in this model by the combined action of traps in a concentration 2.4 $10^{20}/cm^3$ with an acceptor character at 0.36 eV below the conduction band and a level 0.23 eV above the valence band, made up simultaneously of donors (2.4 $10^{21}/cm^3$) and acceptors (2.4 $10^{19}/cm^3$).

The agreement between the data and the computation appears quite striking, considering the sophistication of the model.

d) Thermally Stimulated Conductivity (TSC)

TSC experiments are a well-known, simple method for the detection of traps in high-resistivity materials. However, it cannot be taken as a precise method of determining trap parameters since many assumptions are involved in data reduction (NICHOLAS and WOODS, 1964). On the contrary, TSC measurements can provide a useful check of consistency of some models, in conjunction with other independent experiments.

TSC data have been reported for boron by many authors but the information gathered is limited due to a lack of critical analysis of these data.

ZAREBA and LUBOMIRSKA (1969) obtained two trap levels with depths 0.05 eV and 0.08 eV in the same sample, with the "initial rise method" i.e. from the slope of the plot ln i vs 1/T where i is the TSC current in the lower temperature side of a peak. This method is independent of strong and weak retrapping cases and does not lose its validity when samples contain a distribution of various trapping levels (BUBE, 1965). However, in this latter case the method gives the more shallow level of the distribution which can be evidenced by the repeated use of the method on the data of the different cycles of decayed TSC curves (BUBE, 1965). This can be made with the data reported by ZAREBA (1971) and we obtain different activation energies ranging from 0.08 eV to 0.17 eV.

DZHAMAGIDZE et al. (1969) and ANISIMOV (1971) calculate the trap depths with the "Fermi level method" i.e. from the relation:

$$E = KT_m \ln\left(\frac{N_c e\mu}{\sigma_m}\right) \qquad (10)$$

where T_m is the temperature at which the maximum conductivity σ_m is observed. Some assumptions for the charge carrier mobility are made. Activation energies for traps ranging from 0.1 to 0.3 eV have been obtained. However, it is worthwhile pointing out that the "Fermi level" method is misleading when applied to samples containing distribution of trapping levels, such as those investigated in the two works cited. In fact, the decayed TSC curve reported by ANISIMOV et al. (1971) analyzed with the "initial rise method" gives activation energies ranging from 0.09 eV to 0.18 eV.

TSC measurements have been used as a check of consistency of a model in comparison with TPM (Sect. 4.e) and SCLC (Sect. 3.c) by PRUDENZIATI et al. (1972). The decayed TSC curves at lower temperatures have shown that the 0.25 eV depth for electron traps obtained in TPM measurements is the cut-off energy of a distribution containing more shallow levels; the "variation of heating rate method" (NICHOLAS and WOODS, 1964) has been applied for a peak at higher temperatures, the same activation energy (0.36 eV) being obtained for hole traps with TPM and SCLC.

The trap density cannot be obtained by TSC measurements in boron essentially since the calculation requires the knowledge of the photoconductivity gain (BUBE, 1960), which is a quantity not yet known in this material.

e) Transient Photoconductivity Measurements (TPM)

The experimental apparatus and basic concepts of TPM are the same as in the well-known "transit time" or "time of flight" technique (SPEAR, 1969; MARTINI et al., 1972). This latter technique is the clearest method to study electron and hole drift velocity and trapping effects separately in the same sample of high resistivity material.

A short flash of strongly absorbed light is projected on a contact of the biased sample, so pairs of charge carriers are created to a depth much shorter than the sample thickness. The current pulse is mainly due to the type of carriers whose sign is equal to the relative sign of the potential on the illumined contact.

In the time of flight method the width of the current pulse allows calculation of the drift velocity and the temperature-dependence of the pulse width is simply related to the activation energies of traps. A net kink is the current pulse, when the drifting carriers are collected on the back contact - and then a pulse width - cannot be observed in boron samples (Fig. 6) owing to recombination phenomena. Hence drift mobility cannot be measured in boron.

However, information on the location of trapping centers can be obtained from the temperature-dependence of the current pulse amplitude. In fact, the amplitude of the current signal is (HARTKE, 1962):

$$i = N_o e \mu_{eff} V/W^2 \qquad (11)$$

where N_o is the total number of created charges and μ_{eff} their trap-controlled drift mobility. This latter is correlated to trapping parameters by the relation:

$$\mu_{eff} = \mu_o \gamma \exp(-E/KT) \qquad (12)$$

when the free carriers are much less than the empty traps μ_o is the microscopic mobility, E the depth of a discrete trapping level ($\gamma = N_t/N_c$ in this case) or the cut-off energy of a distribution of trapping levels ($\gamma = E/KT$ for a uniform or linear distribution). γ is temperature-independent or it has a very weak temperature-dependence, negligible in respect to the exponential term in Eq. (12). So the plot ln i vs 1/T gives the energy depth of a discrete level or the cut-off energy of a distribution.

These considerations imply that the free carriers are in thermodynamic equilibrium with the trapped ones. In this case the current signal should be of constant amplitude until the collection of the drifting carriers. In most cases however, some other phenomenon is present, by which carriers are subtracted from the band. If this phenomenon is characterized by a long time-constant compared with the time required to reach the thermodynamic equilibrium, the slow decay of the current signal does not affect the measurements if we take the amplitude at its maximum. This maximum is called i(O) in Figure 6 a and b.

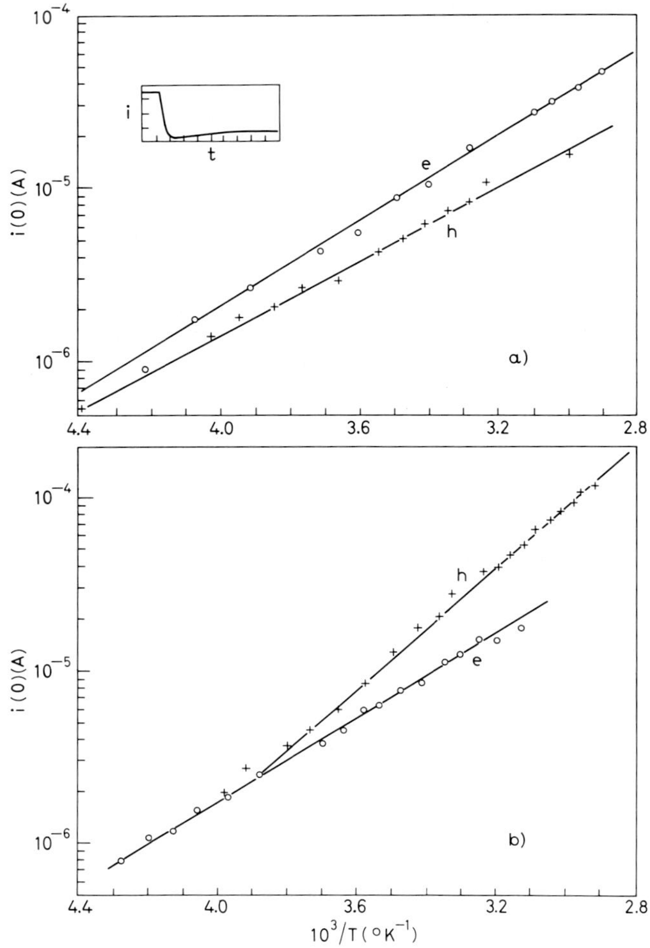

Fig. 6a and b. Temperature-dependence of the maximum amplitude of the current pulse i (O) for excitation with a light flash on the cathode (e) and on the anode (h) in TPM after PRUDENZIATI et al. (1972). (a) P211 sample, E_e = 0.25 eV, E_h = 0.23 eV, (b) P225 sample, E_e = 0.25 eV, E_h = 0.36 eV. In the insert a typical current wave shape is shown, (2 μA/div, 50 μsec/div)

It is convenient to note explicitly that TPM experiments have some relevant advantages with respect to the near steady-state photoconductivity experiments described in the previous sections, besides the possibility already mentioned of discriminating between electron and holes. Since the temperature-dependence of the current signal is analyzed over a relatively narrow temperature range, the temperature-dependence of the microscopic mobility (e.g. $\mu_0 \alpha T^{-3/2}$) and of the surface and bulk recombination has little effect on the investigated position of the trapping centers. Moreover, the knowledge of absolute values for the microscopic mobilities and N_0 (provided it is constant during the measurements) is not necessary for the determination of the activation energies of the trapping centers. On the contrary, these quantities should be well-known in order to determine the concentration of the localized states.

TPM measurements have been performed by PRUDENZIATI et al. (1972) on boron samples of the best quality actually available. The results (Fig. 6a and b) show electron traps at about 0.25 eV from CB in all the investigated samples. The trapping centres of the holes are at 0.36 eV from VB in some samples and at 0.23 eV in other samples.

These results, supported and enriched by the information provided by SCLC and TSC measurements, allow us to sketch the position of localized trapping centers in the boron gap for the two different classes of samples as in Figure 7a and b. The results obtained from TPM measurements in different samples show a heavy effect of extrinsic localized states in boron. We note in passing that the above results are the first ones showing transport for electrons in boron without ambiguity.

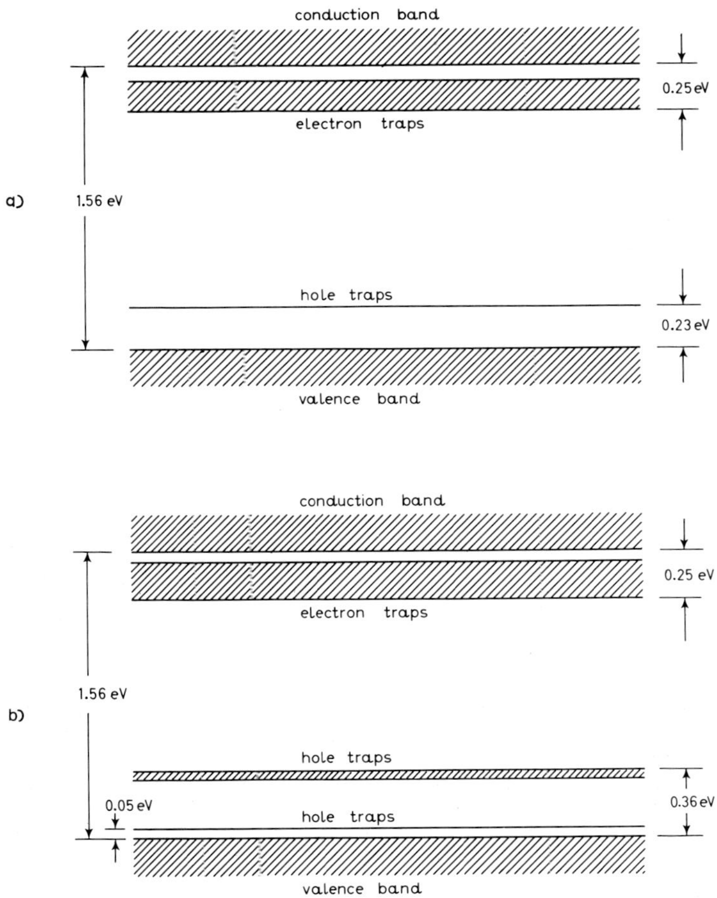

Fig. 7a and b. Scheme of the energy levels in boron single-crystals according to TPM, SCLC and TSC measurements. (After PRUDENZIATI et al., 1972) (a) P 221 samples, (b) P 225 samples

5. Magnetic Susceptibility

Magnetic susceptibility of solids depends in general in a complicated way on the electronic structure, and a comparison of experimental data with theory is usually not straightforward. However, these measurements may provide a possible test of consistency of a model.

Here it seems interesting to cite the most recent results of magnetic susceptibility measurements in boron at high temperatures (KUBLER et al., 1973) and to compare them with the results of experiments reported in previous sections.

In fact, an unusual paramagnetic component is observed above $500^{\circ}K$, with an experimental number of centers dependent on the temperature of the form $N = N_0 \exp(-\Delta E/KT)$. The obtained experimental results are: $\Delta E = 0.41$ eV and $N_0 = 10^{22}/cm^3$. This component of the magnetic susceptibility is independent of the carbon content up to 1% and of the sample annealing.

These experimental evidences have been interpreted assuming that thermal excitation creates paramagnetic centers in two narrow bands separated by an energy gap $2\Delta E$ (Fig. 8). At low temperatures N_1 is empty and P_1 is doubly occupied, so they do not contribute to the paramagnetism. The lower centers P_1 are supposed to be at 0.23 eV from the VB as suggested by photo-EPR measurements (GEWINNER et al., 1973) in the same material.

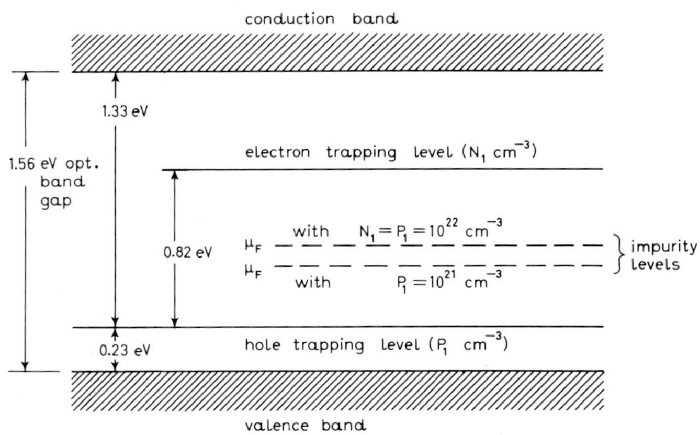

Fig. 8. Presumed band model of boron according to photo-EPR and magnetic susceptibility measurements after GEWINNER et al. (1973) and KUBLER et al. (1973). Here μ_f represents the Fermi level

It is noted by the authors that the level at 0.23 eV is common to other models, namely, to those obtained from optical absorption (Sect. 2.) and SCLC (Sect. 3.c) and TPM (Sect. 4.e) measurements.

The two high-density bands have been associated with "intrinsic localized states typical of pure crystalline boron". This statement seems to be in disagreement with some transport properties of boron. In fact, from the picture in Figure 8 and the postulated state of charge (N_1 empty and P_1 doubly occupied at $T = 0$) it follows that we should have too small TPM signals and too low SCLC. For example, TPM current pulses for electrons should be 10^{-2} times lower than the observed ones even with a microscopic mobility at 300°K equal to $10^4 cm^2/V$ sec, which is a very improbable high value for electron mobility in boron. Similarly the interpretation of SCLC on the basis of the situation depicted in Figure 8 requires a too high hole mobility, i.e. of the order of 10^4-$10^5 cm^2/V$ sec.

Since we cannot assume that these localized states exist without acting as trapping centers (they show this behavior in the experiments mentioned above) we have to argue that their density has been strongly overestimated.

6. Remarks and Conclusion

Different experiments which provide information on localized states in boron have been reviewed in this article. It has been shown that some experiments - optical absorption, wave length-dependence of photoconductivity, SCLC and TPM - provide a reliable picture for the activation energies of the localized states, since the interpretation of the data needs neither a detailed knowledge of the transport properties of boron nor a large number of simplifying assumptions. On the other hand, only TPM gives unambiguous information on the absolute position of the localized states within the gap. However, from the experiments cited above we can infer only an upper limit for the density of localized states, until some quantities (e.g. matrix elements for optical transitions and charge carriers mobility) become known.

On the other hand, the other techniques suffer from more severe disadvantages.

Hopping conduction near E_F and ac conductivity methods rest on the assumption of pure hopping transport and give no information on the energy depth of the localized states. The interpretation of high-field conduction is ambiguous, since different models fit the data with different physical pictures. Finally the data provided by other experiments - time-dependence of photoconductivity, photoconductivity and EPR, magnetic susceptibility - cannot be handled in a straightforward way, and require a lot of prior assumptions. From this picture it therefore results that, in the present state of the knowledge of boron, different methods should be used on the same samples, in order to check the consistency of an assumed model. This has not always been done. Nevertheless, let us now try to summarize and discuss the results actually available in order to elucidate the role played by extrinsic and intrinsic localized states in the semiconducting properties of boron.

These results are summarized in Table 1, where we consistently have the presence of high densities for states located at the same energy depth in the case of intrinsic localized states. The results in Table 1 are not so clear-cut on this point as to give a certain reply to our problem. In fact only some experiments - photoconductivity and EPR, magnetic susceptibility and high-field conduction interpreted with the hopping model - give such a high concentration of localized states ($\geq 10^{21}/cm^3$) that one is tempted to assign them to intrinsic properties of the boron lattice. But if this is really the case, these states should be observed also in optical and in transport experiments, where on the contrary, much lower densities are assigned to the same levels.

On the other hand, it can be seen that all the experiments indicate high densities and relatively high ionization energies for the localized states. In some cases also, equal activation energies have been obtained, e.g. the level at 0.23 eV above VB (detected in optical, photoconductive, TPM, and SCLC measurements) and the 0.05 eV level above VB (detected in SCLC and high-field conduction). However, these localized states, which act as hole trapping centers, are not detected in all the samples even with the same techniques. On the other hand, the activation energies of localized states acting as electron traps are more scattered.

In any case, according to Table 1, we can argue that what we observe in many experiments is a heavy influence of extrinsic localized states, due to impurities and defects, because we have different results in different samples, even using the same techniques.

Our feeling, which of course requires experimental confirmation, is that if intrinsic localized states really exist in boron, they are confined to narrow tails of the valence and conduction bands and can explain the low mobilities of the charge carriers and other properties, such as the magnetoconductivity, as suggested by SOSNOVSKI and ZAREBA (1971). These conclusions can be confirmed for example by photoemission and detailed field-effect measurements, mainly if (and we would express our hope in this sense) careful methods of preparation result in single crystals of higher purity to be studied and used for future investigations.

Acknowledgements. The constant interest and support of Prof. A. Alberigi Quaranta are sincerely acknowledged. Encouragement, discussion and help of many colleagues at this laboratory - Prof. G. Ottaviani, Dr. C. Jacoboni, C. Canali, G. Majni - are appreciated. The paper was partially supported by the Consiglio Nazionale delle Ricerche, Italy.

List of Symbols

CB	Conduction Band
e	Electronic charge, C
E	Energetical position of a trap level, referred to the CB for electron traps and to the VB for hole traps, eV.
E_F	Fermi level
E_g	Energy gap, eV.
g_o	Density of localized states per unit energy, eVcm^3.
i(O)	Amplitude of current pulse
J	Steady state current density, A/cm^2
k	Boltzmann constant, eV/^{0}K
m_e^+	Electron effective mass, g.
m_h^+	Hole effective mass, g.
N_o	Total number of electrons and holes created by a light flash
$N_{c(v)}$	Effective density of states in the conduction (valence) band/cm^3.
N_t	Concentration of traps in a discrete energy level/cm^3.
Q	Thermoelectric power, mV/^{0}K

R_H	Hall constant, cm^3/C
S	Electrode area, cm^2
SCLC	Space charge limited current
T	Absolute temperature, ^oK
T_o	Characteristic temperature in hopping conductivity, ^oK
T_m	Temperature at the maximum of the current peak, ^oK
TSC	Thermally stimulated currents
V	Applied voltage, V
VB	Valence Band
W	Sample thickness, cm
α_o	Radial decay of the localized state wavefunction/cm.
ε_o	Dielectric constant of vacuum, F/cm.
ε	Relative dielectric constant.
$\mathcal{E}$	Electrical field, V/cm.
λ	Optical wavelength, cm.
μ_o	Microscopic mobility, cm^2/V sec.
σ	Conductivity, mho/cm.
σ_o	Pre-exponential factor in temperature-dependent conductivity, mho/cm.
ν	Frequency of light/sec.
ν_{ph}	Phonon frequency, (10^{13})/sec.
ω	Frequency of a voltage signal/sec.

References

ADIROVICH, É.I., GOL'DSHTEIN, L.M.: Photoconductivity of boron single crystals and films. Sov. Phys. Semiconduct. 3, 196 (1969)

ADIROVICH, É.I., BENDERSKII, V.A., BRIKENSHTEIN, V.Kh., KOROSTELEV, Yu.A.: Investigation of the EPR signals observed during the illumination of boron single crystals. Sov. Phys. Semicond. 5, 13 (1971)

ANISIMOV, B.B, DZHAMAGIDZE, Sh.Z., SHVANGIRADZE, R.R.: Investigation of the effects caused by preliminary illumination of boron crystals. Sov. Phys. Semiconduct. 5, 991 (1971)

ANISIMOV, B.B., DZHAMAGIDZE, Sh.Z., SHVANGIRADZE, R.R.: Anomalous behavior of the electrical resistance and optical absorption of β-boron. Sov. Phys. Semiconduct. 6, 1873 (1973)

AUSTIN, I.G., MOTT, N.F.: Polarons in crystalline and non-crystalline materials. Adv. Phys. 18, 41 (1969)

AVDEENKO, B.K., GOLIKOVA, O.A., ZHUBANOV, M.Zh., KISKACHI, A.Yu., MATVEEV, V.N.: Conductivity of boron as a function of frequency and temperature. Sov. Phys. Solid State. 10, 2007 (1969)

BEREZIN, A.A., GOLIKOVA, O.A., ZAITSEV, V.K.: Nature of hopping conduction in β-rhombohedral boron. Sov. Phys. Solid State. 15, 1237 (1973a)

BEREZIN, A.A., GOLIKOVA, O.A., KAZANIN, M.M., TKALENKO, E.N., ZAITSEV, V.K.: Studies of a conductivity mechanism of β-rhombohedral boron in a strong electric field. Phys. Stat. Sol (a). 20, 447 (1973b)

BORCHERT, W., DIETZ, W., HERMANN, H.: Untersuchungen über die Halbleitereigenschaften von Bor. Z. Ang. Phys. 19, 485 (1965)

BOSMANN, A.J., CREVECOEUR, C.: Mechanism of the electrical conduction in Li-doped NiO. Phys. Rev. 144, 763 (1965)

BRODSKY, M.H., GAMBINO, R.J.: Electrical conduction in evaporated amorphous silicon films. J. Non-Cryst. Solids. 8-10, 739 (1972)

BRUNGS, R.A., JACOBSMEYER, V.P.: Activation energies of monocrystalline β-rhombohedral boron. J. Phys. Chem. Solids. 25, 701 (1964)

BUBE, R.H.: Photoconductivity in Solids. New York: Wiley and Sons 1960

BUBE, R.H.: Photoelectronic Materials and Devices. Larach Ed, p. 100. Princeton: Van Nostrand (1965)

DIETZ, W., HELMBERGER, H.: Boron Semiconductor Devices. In: Boron Vol. II, p. 301, (G.K. GAULÉ, ed.). New York: Plenum Press 1965

DIETZ, W., HERMANN, H.: Investigations on the Conduction Mechanism in Boron Single Crystals. In: Boron Vol. III, p. 195, (T. NIEMYSKI, ed.). Warsaw: Pol. Sci. Publ. 1970

DZHAMAGIDZE, Sh. Z., MAL'TSEV, Yu.A., SHVANGIRADZE, R.R.: Investigation of the electrical properties of boron. Sov. Phys. Semiconduct. 2, 320 (1968)

DZHAMAGIDZE, Z., MAL'TSEV, A., SHVANGIRADZE, R.R.: Trapping levels and temperature dependence of carrier mobility in boron. Sov. Phys. Semiconduct. 3, 80 (1969)

DZHAMAGIDZE, Sh. Z., SHVANGIRADZE, R.R., GVILAVA, M.F.: Investigation of the optical absorption of boron. Sov. Phys. Semiconduct. 2, 316 (1968)

EAGLES, D.M.: Optical absorption and recombination radiation in semiconductors due to transition between hydrogen-like acceptor impurity levels and the conduction band. J. Phys. Chem. Solids. 16, 76 (1960)

FRITZSCHE, H.: Electronic phenomena in amorphous semiconductors. Ann. Rev. Mat. Science. 2, 697 (1972)

GAULÉ, G.K., BRESLIN, J.T., PATTY, R.R.: Trap dominated electrical and optical effects in crystalline boron. In: Boron Vol. II, p. 169 (G.K. GAULÉ, ed.). New York: Plenum Press 1965

GAULÉ, G.K., ROSS, R.L., BLOOM, J.L.: B^{10}/B^{11} Thermistor pairs and their application. In: Boron Vol. II, p. 317 (G.K. GAULÉ, ed.). New York: Plenum Press 1965

GEIST, D.: Electron paramagnetic resonance, electrical conductivity and impurity diffusion in doped boron. In: Boron Vol. II, p. 203 (G.K. GAULÉ, ed.). New York: Plenum Press 1965

GEIST, D.: The mechanism and quantitative description of the photoconductivity and photo-EPR in β-rhombohedral boron single crystals at 77°K. Z. Natur. 28 a, 953 (1973)

GEIST, D., MEYER, J.: Carrier Mobility in High Purity Boron Single Crystals. Proc. 10th Int. Conf. Phys. Semic., Cambridge 1970

GEWINNER, G., KUBLER, L., KOULMANN, J.J., JAÉGLÉ, A.: EPR investigation of paramagnetic centres in β-rhombohedral boron. Phys. Stat. Sol. (b)., 59, 395 (1973)

GOLIKOVA, O.A., KISKACHI, A.Yu., KHOMIDOV, T.: Electrical properties of β-rhombohedral boron at high temperatures. Sov. Phys. Semiconduct. 4, 683 (1970)

GOLIKOVA, O.A., NIKITIN, E.N., TKALENKO, É.N.: Electrical properties of doped boron. Sov. Phys. Semiconductors., 4, 1193 (1971)

GOLIN, S.: Polarization conductivity in p-type germanium. Phys. Rev. 132, 178 (1963)

HARTKE, J.L.: Drift mobilities of electrons and holes and space-charge-limited currents in amorphous selenium films. Phys. Rev. 125, 1177 (1962)

HILL, R.M.: Hopping conduction in amorphous solids. Phyl. Magazine XXIV, 1307 (1971)

JAUMANN, J., WERHEIT, H.: Optische und photoelektrische Eigenschaften von β-rhombohedrischem Bor. Phys. Stat. Sol. 33, 587 (1969)
KLEINHENZ, K., RUNOW, P.: Herstellung dünner Borfolien und Elektronenmikroskopische Beobachtung ihrer Baufehler. Phys. Stat. Sol. 29, 627 (1968)
KUBLER, L., GEWINNER, G., KOULMANN, J.J., JAÉGLÉ, A.: Magnetic susceptibility of pure boron. Phys. Stat. Sol. (b). 60, 117 (1973)
LAMPERT, A.: Simplified theory of space-charge-limited currents in an insulator with traps. Phys. Rev. 103, 1648 (1956)
MAJNI, G., PRUDENZIATI, M.: Thermoelectric power in crystalline boron. Phys. Stat. Sol. (a). 5, K129 (1971)
MARTINI, M., MAYER, J.W., ZANIO, K.R.: Applied Solid States Science Vol. 3, (R. Wolfe, ed.). New York: Academic Press 1972
MOTT, N.F.: Conduction in non-crystalline systems VII Non-Ohmic behaviour and switching. Phyl. Mag. 24, 911 (1971)
MOTT, N.F., DAVIS, E.A.: Electronic Processes in Non-Crystalline Materials. Oxford: Clarendon Press 1971
NADOLNY, A.J.: Electron Trapping in β-Rhombohedral Boron. IV Int. Conf. on Boron, Tbilissi, Oct. 1972
NADOLNY, A.J., OSTROWSKI, J.W., PRZEGALIŃSKA-MIESKOWSKA, M.Z.: Kinetics of photoconductivity in boron. Phys. Stat. Sol. 16, K133 (1966)
NEFT, W., SEILER, K.: Semiconductor properties of boron. In: Boron Vol. II, p. 143, (G.K. GAULÉ, ed.). New York: Plenum Press 1965
NICHOLAS, K.H., WOODS, J.: The evaluation of electron trapping parameters from conductivity glow curves in cadmium sulphide. Brit. J. Appl. Phys. 15, 783 (1964)
PRUDENZIATI, M., LANZI, A., MAJNI, G., MALAVASI, G.: Switching effect in β-rhombohedral boron. Phys. Stat. Sol. (a). 18, 651 (1973)
PRUDENZIATI, M., MAJNI, G.: Boron thermistors for high temperature measurements. IEEE Trans. Ind. Elec. Control Instrum. IECI 20, 30 (1973)
PRUDENZIATI, M., MAJNI, G., ALBERIGI-QUARANTA, A.: Determination of trapping centers in β-rhombohedral boron. J. Phys. Chem. Solids, 33, 245 (1972)
SHAW, W.C., HUDSON, D.E., DANIELSON, G.C.: Electrical properties of boron single crystals. Phys. Rev. 107, 419 (1957)
SHKLOVSKII, B.I.: Hopping conduction in semiconductors subjected to a strong electric field. Sov. Phys. Semiconduct. 6, 1964 (1973)
SOSNOWSKI, L.: The present state of research on boron. In: Boron Vol. III, p. 15, (T. NIEMYSKI, ed.). Warsaw: Pol. Sci. Publ. 1970
SOSNOWSKI, L., ZAREBA, A.: Influence of magnetic field on conductivity due to electron tunnelling between localized states. Phys. Stat. Sol. (b). 47, 63 (1971)
SPEAR, W.E.: Drift mobility techniques for the study of electrical transport properties in insulating solids. J. Non-Cryst. Solids. 1, 197 (1969)
WERHEIT, H.: Optical and photoelectrical properties of β-rhombohedral boron. Phys. Stat. Sol. 39, 109 (1970)
WERHEIT, H., LEIS, H.G.: On the conductivity mechanism of β-rhombohedral boron. Phys. Stat. Sol. 41, 247 (1970)
WERHEIT, H., HAUSEN, A., BINNENBRUCK, H.: Optical anisotropy of β-rhombohedral boron (from 0.4 to 16 µm). Phys. Stat. Sol. (b). 51, 115 (1972)
WERHEIT, H., RUNOW, P., LEIS, H.G.: On boron-suboxide surface layers and surface states of β-rhombohedral boron. Phys. Stat. Sol. (a). 2, K125 (1970)
ZAREBA, A.: On thermally stimulated conductivity of boron. Phys. Stat. Sol. (a). 5, K121 (1971)
ZAREBA, A., LUBOMIRSKA, A.: Thermally stimulated current in boron. Phys. Stat. Sol. 34, K41 (1969)

V. Alkali Metal Borides

R. NASLAIN, J. ETOURNEAU, and P. HAGENMULLER

1. Introduction

The first studies relevant to alkali metal borides were carried out in the nineteenth century as an incidental part of the effort to isolate elemental boron by reduction of several boron compounds (boron oxide, boron fluoride and fluoborates by alkali metals) (GAY-LUSSAC and THÉNARD, 1808; DAVY, 1811; WÖHLER and SAINTE-CLAIRE DEVILLE, 1858; RAWSON, 1888).

MOISSAN (1892) characterized the nature of the products obtained by the first experimenters. He observed that even after repeated washing with hydrochloric acid, the boron samples still contained small amounts of sodium or potassium. He hypothesized that the alkali metal borides were probably mixed with amorphous boron and he attributed the emitted gases - observed during the acid washing - to partial hydrolysis of the borides. The observations of MOISSAN were confirmed later by KROLL (1918) and by KAHLENBERG (1925).

ANDRIEUX and BARBETTI (1932) tried to isolate alkali metal borides by submitting borates and corresponding fluorides to fused salt electrolysis; however, they succeeded only in preparing ternary compounds with a $M_xM'_{1-x}B_6$ formula where M was an alkali metal and M' an alkaline earth or a rare earth metal. From this result they concluded that alkali metals could form hexaborides, isomorphous with those of the alkaline earth or rare earth metals.

Subsequently BERTAUT and BLUM determined, by fused salt electrolysis, that only sodium among the alkali metals could form ternary borides of composition $Na_xBa_{1-x}B_6$ or $Na_xTh_{1-x}B_6$, with the limits of x corresponding to 0.43 and 0.77 respectively. These ternary borides were indeed isostructural with CaB_6. It was noted by these investigators that the partial substitution by alkali metal, while not constituting a firm basis for the existence of binary alkali metal borides, did prove that a stable CaB_6-type structure could be achieved for less than two conduction electrons per metal atom (BERTAUT and BLUM, 1952; BLUM and BERTAUT, 1954).

More recently, NASLAIN, ETOURNEAU and HAGENMULLER have made a critical study of the earliest methods used in the attempts to synthesize alkali-metal borides. They established that failures of these attempts could be attributed to two factors: (1) the lack of reactivity of alkali metals with boron at low temperatures and (2) the thermal dissociation at higher temperature of any borides that might otherwise be formed. They proposed a method that utilized the enhanced activity resulting from the use of a particularly reactive boron with an excess of alkali metal under pressure and at sufficiently high temperature. With this method, they succeeded in preparing, for the first time, three binary alkali metal borides: NaB_6, KB_6 and NaB_{15} and they noted the existence of lithium borides (HAGENMULLER and NASLAIN, 1963; NASLAIN and ETOURNEAU, 1966; NASLAIN, 1967; NASLAIN et al., 1967; HAGENMULLER et al., 1967; NASLAIN et al., 1970).

Parallel to this work, SECRIST and CHILDS, by an analogous method, also prepared a lithium boride to which they attributed the formula $LiB_{10.85}$. But so far as we know, there have been no published results concerning the boron-rubidium and boron-cesium systems (SECRIST and CHILDS, 1962; SECRIST, 1967).

2. Preparation of Alkali Metal Borides

The best method for preparing alkali metal borides is the direct synthesis from the elements. They are other methods that can be resorted to - for example, utilization of lithium hydride or reduction of fluoborates - but they lead to incomplete reactions or to less pure products.

a) Preparation of Alkali Metal Borides by Direct Synthesis

α) *Principle*

Since metals generally react with boron only at temperatures above approximately 700°C and since alkali metal borides have only a weak thermal stability, the preparation of alkali metal borides by direct synthesis can be carried out at high temperatures, with an excess of metal and in completely sealed containers:

$$M + x\,B \longrightarrow MB_x \ (M = Li,\ Na,\ K) \quad (1)$$

The formation of alkali metal borides by direct synthesis requires temperatures between 750 and 1200°C, depending on the variety of boron utilized and the specific metal. The high temperature requirements involve two important consequences:

1. Since the synthesis temperatures are higher than the dissociation temperatures of the phases that are likely to be formed (at a pressure of 1 bar), it is necessary to make the alkali metal react with boron under pressure. This pressure is obtained by utilizing metal in excess of the stoichiometric proportions defined by reaction (1), inside sealed vessels. At reaction temperatures, the alkali metal is present in a liquid state in equilibrium with a vapor phase, the pressure of which is fixed by the temperature of the coldest point in the reactor. It is this pressure, generated by the excess metal (the more volatile the metal, the higher the pressure), that favors the synthesis reaction relative to the reverse dissociation reaction.

2. The reactivity of alkali metals with boron decreases when their atomic number increases: lithium reacts completely with boron at 700°C, whereas with potassium the reaction is not complete until 1200°C. Figure 1 gives the equilibrium vapor pressures of liquid alkali metals in this temperature range, and shows that the pressure within the reactors can attain 20 bar at 1200°C, in the most unfavorable cases. The pressure values demand that reactors of a special type with thick walls be used.

By using a large metal excess, which can also act as a flux, well crystallized borides can be obtained. After reaction, the excess metal is eliminated either by distillation at low temperature (sodium: 350°C, potassium: 300°C) and in high vacuum (10^{-5} to 10^{-6} torr), or possibly by washing with liquid ammonia.

β) *Application of the Method*

Alkali metal boride synthesis can be achieved with a reactor of the type schematically represented in Figure 2a. The boron-alkali metal

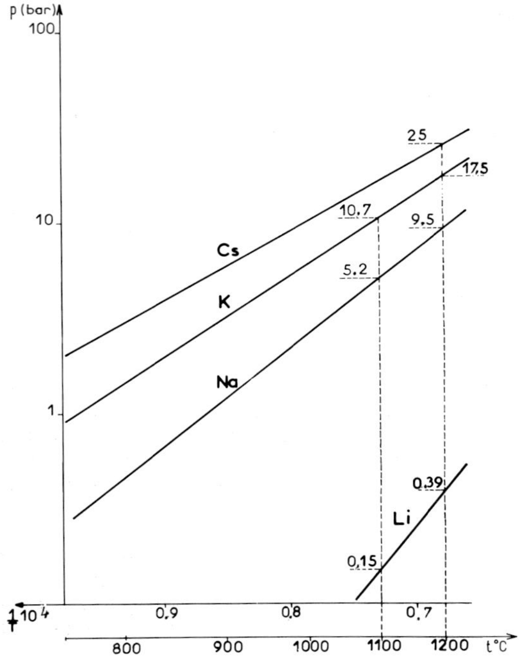

Fig. 1. Alkali metal saturated vapor pressure versus temperature. (From EWING et al., 1965a, b; STONE et al., 1965; BOGROS, 1930a, b; HICKS, 1963)

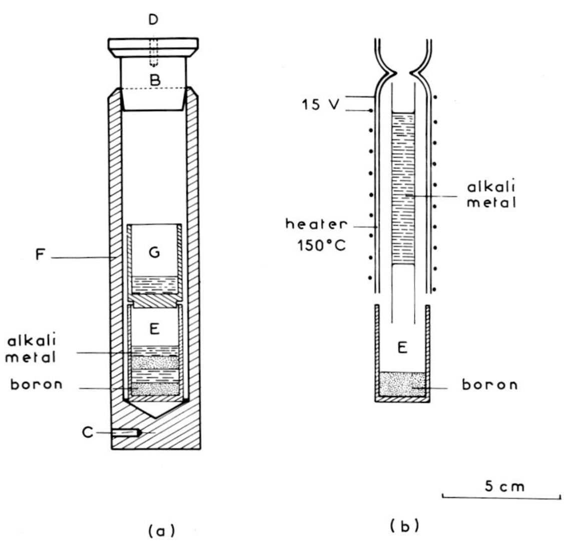

Fig. 2 a and b. Apparatus for the preparation of alkali-metal borides by direct synthesis: (a) The reactor. E: molybdenum crucible G: getter; F: thick wall container B: tight stopper C and D: holes for thermocouples. (b) introduction of liquid alkali metal into crucible E

mixture is placed in a molybdenum crucible (E) within a container (F) with thick walls made of iron or molybdenum, according to the reaction temperature.

In the case of sodium and potassium, the alkali metal from a prior distillation in glass flasks, is transferred in a liquid state to the crucible, as shown in Figure 2b. All manipulations of products containing an alkali metal are carried out, in a dry box, in purified argon atmosphere.

After the cap (B) is pressed tightly, it is welded to the body of the reactor; a getter (G) made of a small quantity of alkali metal is placed above the crucible (E); it is to protect the reaction mixture against any possible air entering during the welding (Fig. 2a).

The reactor is heated to synthesis temperature in a resistance furnace or by a high frequency current. The temperature can be followed by two thermocouples (C) and (D) and by an optical pyrometer. A metal liner (T) protects the outer tube and the furnace against possible alkali metal leakage under pressure at high temperature (Fig. 3).

γ) Role of Some Important Parameters

The most important parameters controlling the synthesis are the temperature, the vapor pressure of the alkali metal and the crystallographic boron variety. On the contrary, the reaction is hardly affected by the atomic ratio M/B, provided it is much larger than the M/B ratios characteristic of the phases that are to be prepared; values of about 1/2, 1 or 2, for example, are quite satisfactory.

Effect of Temperature and of the Crystallographic Nature of Boron. Since the reactivity of alkali metals towards boron decreases when the atomic number increases, the effect of temperature and of the crystallographic nature of boron is weak in the case of lithium, but becomes strong in the case of potassium.

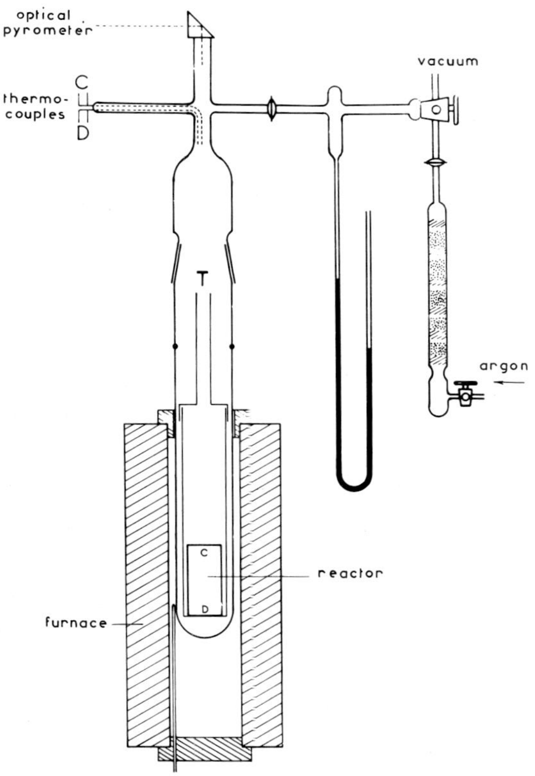

Fig. 3. Equipment for the preparation of alkali metal borides by direct synthesis: apparatus for heating to high temperatures, in a controlled atmosphere, of the thick-walled reactor shown in Fig. 2a

The reaction of lithium with boron - noticeable at 680°C according to NASLAIN (1967) and at 600°C according to SECRIST and CHILDS (1962) - is very rapid at 750°C and different boron samples are equally attacked regardless of crystallographic form or particle size.

The reaction of sodium with boron, detectable at 850°C, becomes complete at 900°C after a treatment of 24 h. At 1000°C the reaction is practically instantaneous and always complete. The reaction rate is independent of the variety of boron (α and β-rh.), if. the particle size is the same.

Liquid potassium, in equilibrium with its vapor, does not react with boron up to 1100°C (p_K^o = 10.7 bar). Above this temperature, a selective reaction is observed, leading to the formation of a hexaboride. Whereas lithium and sodium react equally with both α and β-rh. boron prepared at 800°C by reducing BBr_3 with hydrogen in a quartz tube (according to a technique described by NASLAIN, 1967; NASLAIN et al., 1970; NASLAIN, 1972), potassium combines only with the α-rh. variety. The reaction of lithium and sodium with boron is pratically instantaneous; that of potassium is on the contrary far slower: to be complete, it requires about 18 h at 1100°C (p_K^o = 10.7 bar) and 5 h ar 1200°C (p_K^o = 17.5 bar). Above 1200°C the reaction of potassium on α-rh. boron leads to products that contain simultaneously potassium hexaboride and β-rh. boron; this latter phase can result either from the early stages of decomposition of the hexaboride, or from a α-rh. ⟶ β-rh. transformation.

Thus the preparation of potassium boride by direct synthesis is possible only in a narrow range of temperature (1100-1200°C). To obtain a complete reaction, it is necessary to use only a highly reactive boron (α-rh.), i.e. prepared at the lowest possible temperature (for example by reduction of BBr_3 with hydrogen at 700°C) (NASLAIN, 1972).

The selective reaction of potassium with the α-rh. boron could be explained by the metastable character of this variety at high temperature and by the existence in the structure of weak three center B-B bonds, that correspond to particularly long B-B distances (2.03 Å, whereas a normal covalent B-B bond is 1.76 Å).

Effect of Alkali Metal Vapor Pressure. As we shall observe further on, the thermal stability of alkali metal borides is relatively low, which is to be expected from the especially high vapor pressures of the corresponding metals at high temperatures. Consequently, the alkali metal vapor pressure in the containers used for the synthesis also is a very important parameter.

In most cases alkali metal boride synthesis is done in isothermal reactors. This procedure permits the use of maximum alkali metal pressures and hence of optimum conditions of preparation.

The boron-sodium system include two phases with different thermal stabilities. By direct synthesis, either of these two borides can be obtained, by adjusting the alkali metal pressure in the vapor phase. Thus, the preparation of NaB_6 can be carried out in isothermal reactors where an equilibrium $Na_{liq.} \rightleftharpoons Na_{vap.}$ exists, up to 1100 C (p_{Na}^o = 4.5 bar). Above this temperature, the sodium pressure is no longer sufficient to prevent the thermal dissociation of NaB_6 and syntheses under such conditions lead to the second phase with a lower sodium content, NaB_{15}. However NaB_{15} can be prepared at temperatures below 1100°C, as long as the sodium pressure in the vapor phase is kept low (by providing a cold wall into the reactor or by substituting a sodium-potassium alloy for sodium, for example (NASLAIN, 1967).

b) Preparation of Alkali Metal Borides from Fluoborates

While the preparation of lithium and sodium borides by direct synthesis does not involve any particular difficulties because of the relative affinity of these metals for all forms of boron, the synthesis of KB_6 is more delicate since it requires a very reactive boron specimen, whose preparation, in a condition of high purity, remains a difficult operation.

Several investigators have taken out patents for a method of preparation of boron by low temperature reduction of alkali metal fluoborates with an alkali metal. NASLAIN and BARRET (1975) extended this technique to the preparation of alkali metal borides, by applying it in a sealed reactor.

Alkali metals reduce alkali metal fluoborates at low temperatures according to an exothermic reaction, which leads to the corresponding fluorides and to a finely divided boron powder that is amorphous or poorly crystallized. The reaction most often referred to is the reduction of potassium fluoborate by sodium:

$$KBF_4 + 3Na \longrightarrow B + KF + 3NaF$$

This reaction is carried out between 250 and 600°C in air and under a protective layer of sodium chloride (according to SOWA, 1949), or under reduced pressure (10^{-1} torr) according to HAAG (1957).

The reduction of alkali metal fluoborates, performed in two stages and with a large metal excess within sealed reactors similar to those used in direct syntheses, can lead to alkali metal borides through the following series of reactions:

$$6KBF_4 + 18\,Na \xrightarrow{300\,-\,600^\circ C} 6B + 6KF + 18\,NaF$$

$$6B + Na \xrightarrow{900^\circ C} NaB_6$$

$$6KBF_4 + 19\,Na \longrightarrow NaB_6 + 6KF + 18\,NaF$$

or, in case of a reduction by potassium:

$$6KBF_4 + 18K \xrightarrow{300\,-\,600^\circ C} 6B + 24KF$$

$$6B + K \xrightarrow{1100^\circ C} KB_6$$

$$6KBF_4 + 19K \longrightarrow KB_6 + 24KF$$

This method presents two advantages: it involves only commonly available starting materials (which eliminates the relatively delicate preparation of α-rh. boron in the case of potassium) and it permits the use of a boron specimen prepared in situ at a low temperature, i.e. one of high reactivity.

On the other hand, the method has a drawback: it yields an alkali metal boride mixed with fluorides in the final product. Furthermore, the boride synthesis in the presence of fluorides is seldom complete, so that the borides always coexist with a small quantity of residual fluoborate. Thus, alkali metal borides prepared by this method have to be isolated from the other components by prolonged washing in aqueous solution, which makes their purity much lower than that of phases obtained by direct synthesis. In spite of this drawback, the method can be utilized for preparation of KB_6 if a very high purity is not required (NASLAIN and BARRET, 1975).

c) Other Preparation Methods

In the case of lithium, the low volatility of the metal (p^o_{Li} = 1 bar at 1330°C) and the high thermal stability of its hydride, make it possible to use different variants of the general direct synthesis method previously described.

Thus, lithium boride can be obtained from elementary boron and lithium hydride; the reaction is carried out near the melting temperature of the hydride (688°C). Similarly boric oxide is reduced by LiH; the reaction can lead to the formation of a boride if the proportion of hydride is sufficient. In all cases, lithium boride must be separated from the hydride excess (and from the borates in the reduction of B_2O_3) by washing with dilute hydrochloric acid; this treatment lowers its purity appreciably. Because of its hygroscopicity, lithium hydride must be worked with in a glove box (just as lithium in the metallic state) so that there is no decisive advantage in using it rather than the metal (NASLAIN and BARRET, 1975).

Lithium boride seems to have been prepared by reaction of B_4C with lithium or its hydride at temperatures between 700 and 1000°C. The lithium boride is separated from the metal excess by aqueous treatment and contains quite a large amount of carbon (Elektroschmelzwerk Kempten 1964).

3. The Lithium-Boron System

As previously observed, lithium is the alkali metal that reacts most readily with boron. Thus, massive grains of β-rh. boron (considered to be the most chemically inert form of boron) are pulverized by a simple reaction at 600-1000°C, with lithium. In spite of this great affinity of lithium for boron, the lithium-boron system is still the least well-known of the alkali-boron systems (SECRIST and CHILDS, 1962).

All workers are in agreement about the existence, within the lithium-boron system of one or several phases, but there is a wide variation in the B/Li ratios from chemical analysis (from 3 to 11) depending on the author and the method of preparation.

The confusion in the lithium-boron system is partly the result of the difficulty in obtaining reliable chemical analysis of lithium borides. It seems that lithium borides are partially degradated while being washed with liquid ammonia or aqueous solutions, an operation carried out before the analysis in order to eliminate free residual lithium from the synthesis products. Moreover, it is difficult to interpret the elementary analysis because of heavy impurities that are frequently present (for example oxygen) and because of the low atomic weight of lithium. Since a large variety of formulas is suggested for lithium borides, it appears that it is not possible to give precise formulas, from elementary analyses alone, and that single crystal structure determination is required for a proper formulation.

According to BARRET and NASLAIN (1972) the lithium-boron system may include three phases. The reaction of lithium vapor with β-rh. boron at high temperature leads to a solid solution of lithium in β-rh. noron, B(Li) in analogy to the behavior of transition metals, as reported by LUNDSTRÖM (1974), except that for lithium the solution is more metal rich. At 1200°C for example, the composition limit of the solid solution established by chemical analysis, would correspond to a B/Li atomic ratio near to 7.5 (B%: 91.3; Li%: 7.7). The reaction of an excess of lithium with boron (α- or β-rh.) leads to the formation of two new phases γ and δ, characterized by X-ray diffraction analysis. The γ-phase, of a brown-yellowish color, is a relatively lithium-rich phase, whereas the δ-phase, which is usually well crystallized and of a brick-red color, is a more boron-rich compound. The formulas of these two phases have not yet been established exactly.

SECRIST and CHILDS (1962) also report the existence of a lithium boride to which chemical analysis attributes a B/Li atomic ratio near 10. It seems, because of similarities in their Debye-Scherrer pattern, that this boride and the γ-phase were the same phase; but the chemical compositions appear to be quite different. SECRIST (1967) showed that the Debye-Scherrer pattern of his phase could be indexed in the tetragonal system, with the same parameters (a = 10.16 and c = 14.28 Å) as those of "BeB_6" and α-AlB_{12} which have been studied as single crystals, and consequently with parameters near to those of β-tetragonal boron (a = 10.12 and c = 14.14 Å). It seems probable that all these phases have the same boron skeleton, built from B_{12} icosahedra or B_n clusters of a similar type, the interstitial holes of which could be filled, more or less completely according to preparation conditions, by metal or even boron atoms. This hypothesis, whose credibility is reinforced by the fact that lithium, beryllium and aluminium atoms are small (r_{Li} = 1.55; r_{Be} = 1.12; r_{Al} = 1.43 Å), would agree with the conclusions that NASLAIN et al. (1976) made after studying the $MM'B_{14}$ phases (where M = Na, Mg and M' = B, Al, Mg) for which an analogous description is possible.

Lithium borides, and particularly the γ and δ phases, are obtained as powders of low densities. They react with oxygen above 400°C and vigorously (especially the γ-phase) with concentrated oxidizing acids. Halogens (chlorine, bromine and iodine) react with lithium borides only above 450°C yielding the corresponding lithium halides and a quantity of boron halides which increases as the temperature rises.

4. The Sodium-Boron System

According to NASLAIN and HAGENMULLER, the boron-sodium system contains two borides with the formulas NaB_6 and NaB_{15} (HAGENMULLER and NASLAIN, 1963; NASLAIN, 1967; HAGENMULLER et al., 1967).

a) NaB_6-Sodium Hexaboride

The best method for preparing NaB_6 is direct synthesis at 1000°C in a sealed isothermal reactor from the elements in the B/Na atomic ratio of 1, followed by distillation in vacuum of the excess sodium. Washing with dilute hydrochloric acid eliminates soluble impurities without affecting the hexaboride.

α) Chemical Identification

The results of chemical analysis, summarized in Table 1, show that within the limits of precision, the boride obtained under these conditions does agree with the formula NaB_6. No variation has ever been found in the chemical composition of this phase.

The sodium hexaboride is obtained as a bluish-black powder, finely divided, with a density of 2.12 ± 0.01.

Table 1. Chemical analysis of NaB_6. (FROM NASLAIN, 1967)

	B%	Na%	(B+Na)%	(B/Na)at.
Before HCl washing	70.3	25.5	95.8	5.86
After HCl washing	73.8	25.7	99.5	6.11
NaB_6 (theor.)	73.84	26.16	100	6

β) X-ray Diffraction Identification

The crystalline state of NaB_6 is evidenced by the sharpness of the many lines in the Debye-Scherrer pattern (Fig. 4). In spite of their repeated attempts, HAGENMULLER and NASLAIN (1963) did not succeed in obtaining NaB_6 single crystals of a sufficient size to establish the crystal structure.

Inasmuch as there do exist $Na_xTh_{1-x}B_6$ and $Na_xBa_{1-x}B_6$ ternaries, isostructural with CaB_6 and containing a high percentage of sodium (x = 0.77 in the case of thorium), it could be expected that NaB_6 should also crystallize in the cubic system with the CaB_6-structure and with a parameter near to 4.16 Å, assuming that the $Na_xTh_{1-x}B_6$ solid solutions follow VEGARD'SLAW up to x = 1 (Fig. 5; BERTAUT and BLUM, 1952; BLUM and BERTAUT, 1954).

Comparing the Debye-Scherrer patterns of NaB_6 and CaB_6 shows that this does not happen: the sodium hexaboride obtained by reaction of sodium with boron is not isostructural with calcium hexaboride. This is the case whatever the preparation temperature and the B/Na atomic ratio in the starting mixture, even if the material undergoes annealing subsequently.

Recently, a physical study has been published by RUPP and HODGES (1974) for sodium boride synthesized as single crystals by flux growth in excess metal, by S. LA PLACA et al., According to these authors, the

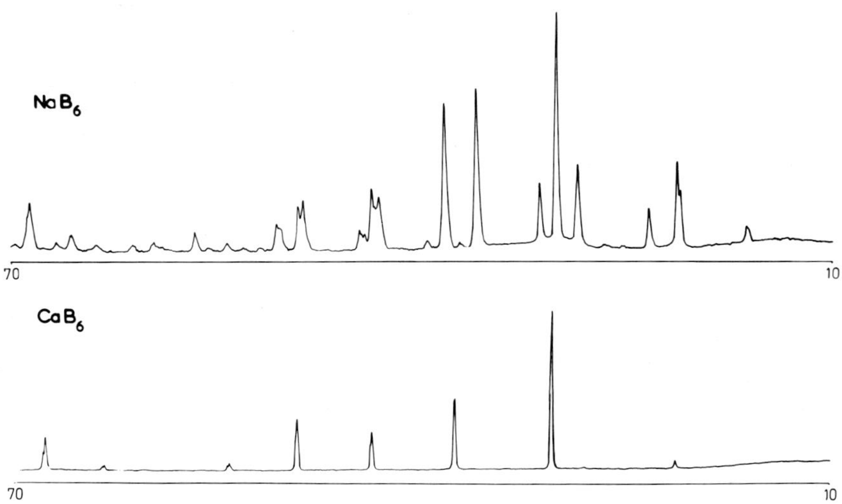

Fig. 4. Debye-Scherrer patterns of NaB_6 and CaB_6. (From NASLAIN, 1967)

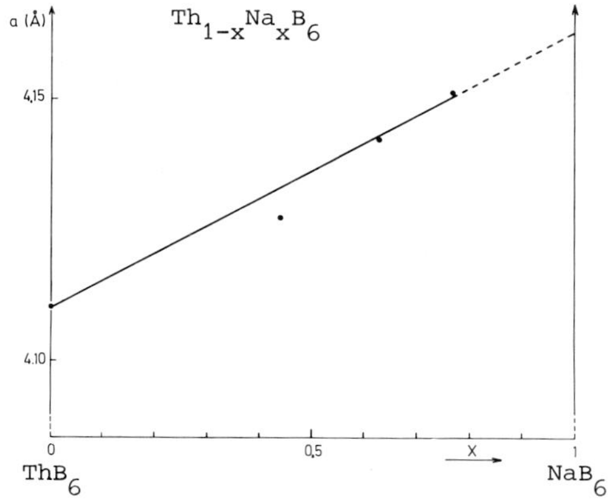

Fig. 5. Variation with x of the lattice parameter of $Th_{1-x}Na_xB_6$ cubic solid solutions. (From BERTAUT and BLUM, 1952; BLUM and BERTAUT, 1954)

sodium hexaboride crystallizes as the CaB_6-type; however no details are given about chemical analysis, purity and the lattice parameter of this phase.

On the light of these contradictions it would seem worthwhile to examine fully the evidence for the possible existence of a sodium hexaboride of the CaB_6-type. First, the work of ANDRIEUX and BARBETTI (1932), and later that of BERTAUT and BLUM (1952), showed that sodium can occupy, at least partially, the cages with a 24-coordination of the B_6 octahedron framework. Generally it is admitted that in this structure the metal atom is ionized and yields all or part of its valence electrons to the covalent boron skeleton. Because of the volume of the metal site, the inserted cations must be relatively large (Chap. B.IX of this vol.).

In the family of $M^{II}B_6$ hexaborides (M = Ba, Sr, Eu, Yb, Ca), the smallest cation likely to stabilize the cubic boron cage is calcium, the ionic radius of which is 1.12 Å according to SHANNON and PREWITT (1969) for VIII-coordination. In the rare-earth family, the existence of hexaborides for the elements of smaller size (Ho, Er, Tm, Lu) is questionable and it seems that Y^{3+} is the smallest trivalent cation ($r_{Y^{3+}}$ [VIII] = 1.06 Å) able to give a cubic hexaboride.

Assuming that the previous observations are applicable to $M^{I}B_6$ hexaborides, it appears that the Na^+ cation, whose radius for coordination VIII (r_{Na^+} = 1.16 Å) is larger than previous limit values, does agree with geometrical conditions for insertion into the [24]-coordination sites of the CaB_6-type structure. From Figure 6 (which is obviously only suggestive), cubic NaB_6 would have a corresponding cell parameter of 4.15 to 4.16 Å, in sufficiently good agreement with the value derived from the parameters of $Th_{1-x}Na_xB_6$ solid solutions (4.162 Å).

The fact that sodium has only one conduction electron to be transferred to the sublattice cannot be taken as an argument against the existence of a cubic hexaboride, since its potassium analog KB_6 is isostructural with CaB_6. Thus, the existence of a sodium cubic binary hexaboride seems to be possible a priori; however, it remains to be experimentally established.

γ) Thermal Stability of NaB_6

NaB_6 is stable up to 650°C in a high vacuum (10^{-6}torr). Between 650 and 750°C, its pyrolysis leads to a second boride: NaB_{15}. Unter an argon pressure of 1 bar, the pyrolysis becomes noticeable above 850°C:

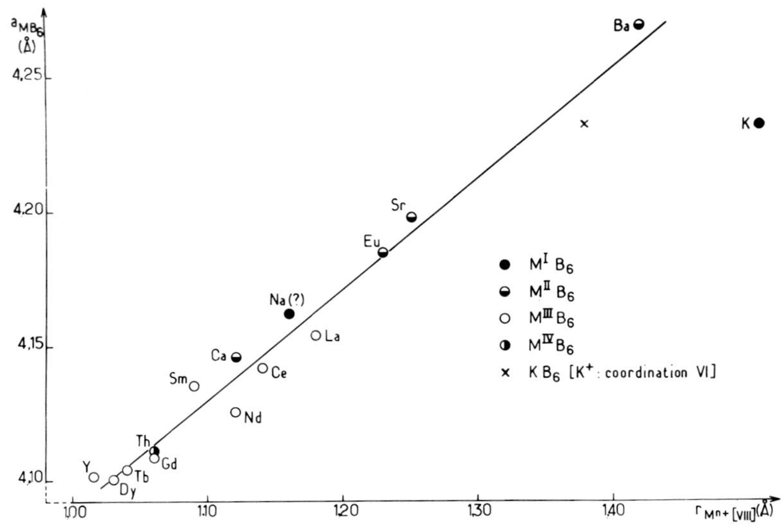

Fig. 6. Variation of the lattice parameter of cubic hexaborides with the size of inserted metal ion (the ionic radii, chosen for coordination VIII, are those of SHANNON and PREWITT, 1969); the value proposed for K^+ is certainly too large; the values of lattice parameters are taken from NASLAIN, 1967; ETOURNEAU, 1970; MERCURIO, 1974; that of a cubic hypothetical NaB_6 was derived from the parameters of the $Na_xTh_{1-x}B_6$ solid solutions. (From BERTAUT and BLUM, 1952; BLUM and BERTAUT, 1954)

$$5NaB_6 \longrightarrow 2\ Na_xB_{15} + (5 - 2x)\ Na\nearrow$$

Therefore, its preparation at 1000°C is made possible only by the presence, within synthesis reactors, of a sodium vapor pressure related to the $Na_{liq.} \rightleftharpoons Na_{vap.}$ equilibrium. Above 1100°C approximately, the sodium vapor pressure (p^o_{Na} = 4.5 bar) is no longer sufficient to prevent the decomposition reaction and NaB_6 gives Na_xB_{15}. The reaction, very slow at 1120°C, becomes rapid at 1200°C.

By substitution of sodium-potassium alloys with different composition (sodium borides are more stable than potassium one) for sodium, or by providing a cold wall into the reactor, the pressure of sodium in the vapor phase can be decreased, and thus the decomposition pressures of NaB_6 at temperatures below 1100°C can be determined. Figure 7 gives, schematically, the stability ranges of NaB_6 and of Na_xB_{15} phases resulting from its decomposition. The XX' straight line represents the variation with 1/T of the sodium vapor pressure in equilibrium with the metal, the YY' curve gives that of the decomposition pressure from NaB_6 to Na_xB_{15}.

In range I, the stable phase is the hexaboride and the syntheses with pure sodium lead to NaB_6. In range II, where the stable phase is Na_xB_{15}, the reverse phenomenon occurs and the hexaboride, initially formed during the temperature rise of the synthesis furnace, dissociates and gives Na_xB_{15}. In range III, the sodium pressure in the vapor phase is not sufficient to ensure the stability of NaB_6 (NASLAIN, 1967).

b) Na_xB_{15} Phases ($0 < x \leq 1$)

The Na_xB_{15} phases (with $0 < x \leq 1$) are obtained by thermal decomposition of NaB_6. The best method for preparing stoichiometric NaB_{15} is

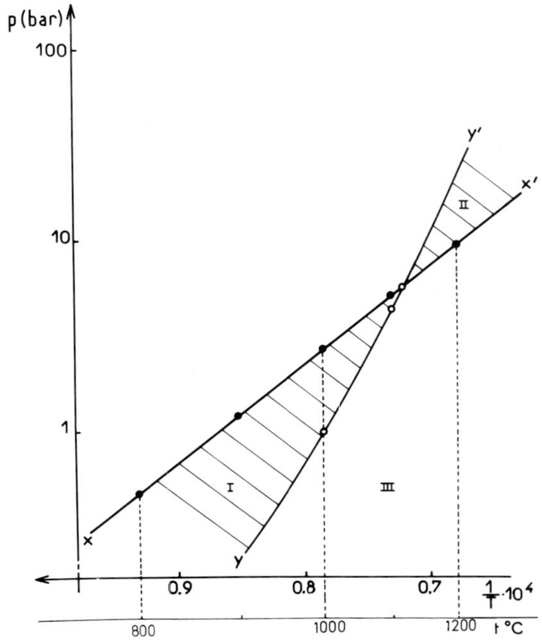

Fig. 7. Stability ranges of sodium borides versus temperature and the pressure of sodium in the vapor phase (I: stability range of NaB_6 II and III: stability ranges of Na_xB_{15} phases)

by direct synthesis, at 1150-1200°C, in a sealed isothermal reactor, in the presence of a large excess of sodium acting as a flux. Na_xB_{15} is obtained under these conditions as grey-black crystals with a metallic aspect and a density of 2.44 ± 0.01 (Fig. 8a) (NASLAIN, 1967; HAGENMULLER et al., 1967).

α) Chemical Identification

Table 2 gives the results of chemical analyses carried out on some samples obtained by thermal decomposition of NaB_6 within pure sodium or within sodium-potassium alloys.

Table 2. Chemical analysis of boron-rich sodium boride. (From NASLAIN et al., 1970)

Experimental conditions	B%	Na%	(B+Na)%	(B/Na)at.	d_{exp} (20°C)
Pure sodium (1200°C; B/Na = 2)	87.50	11.12	98.62	16.7	2.444
Pure sodium (1200°C; B/Na = 1.75)	87.52	11.58	99.10	16.0	-
0.25 Na-0.75 K (1050°C)	87.82	10.86	98.68	17.2	2.436
NaB_{15} (theor.)	87.58	12.42	-	-	2.428

Because of the low sodium content, the chemical analysis is not precise enough to allow the assignment of an exact formula, and the identification had to be completed by X-ray diffraction methods.

β) X-ray Diffraction Identification and Crystal Structure of NaB_{15}

The crystal structure of the boron-rich sodium boride was determined from a single crystal study by NASLAIN and KASPER (1970). The lattice

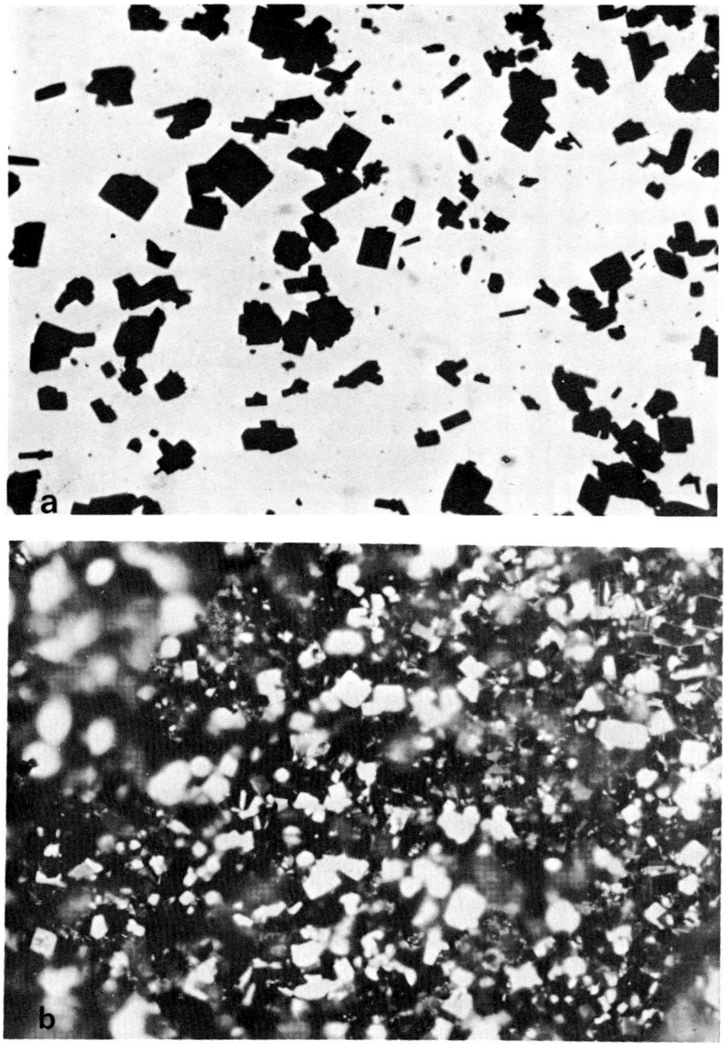

Fig. 8a and b. Micro-crystals of alkali metal borides (a) NaB_{15} (b) KB_6. (From NASLAIN, 1967)

is orthorhombic with space group Imam and parameters a = 5.847, b = 8.415 and c = 10.298 Å. The elementary cell contains four formula groups, and the best agreement between calculated and observed densities was obtained for the formulation NaB_{15} ($d_{theor.}$ = 2.428).

The structure of NaB_{15} was established from that of a ternary phase of ideal formula $MgAlB_{14}$, studied by MATKOVICH and ECONOMY (1970).

Table 3 gives the fractional coordinates and the isotropic thermal parameters of boron and sodium atoms, assuming 100% occupancy of the different site types by the corresponding atoms.

Table 3. Reduced coordinates and isotrope thermal parameters in NaB_{15}. (From NASLAIN and KASPER, 1970)

Atoms	Sites	x	y	z	B(Å^2)
Na	4 (e)	0	0.3386	3/4	1.44
B (1)	16 (j)	0.2475	0.0364	0.0820	0.89
B (2)	16 (j)	0.1609	0.8343	0.0484	0.81
B (3)	8 (h)	0	0.1529	0.1011	0.98
B (4)	8 (h)	0	-0.0472	0.1641	0.84
B (5)	8 (h)	0	0.3941	0.1450	0.93
B (6)	4 (e)	0	0.2541	1/4	1.57

Nature of the Boron Sublattice within Na_xB_{15} Phases. The structure of Na B_{15} is represented in Figures 9 and 10 in projections on xOy and yOz planes. Formally, it can be viewed as a stacking of rather compact planes of quasi-spherical icosahedral clusters of boron atoms with a packing diameter of about 5.1 Å (Fig. 9).

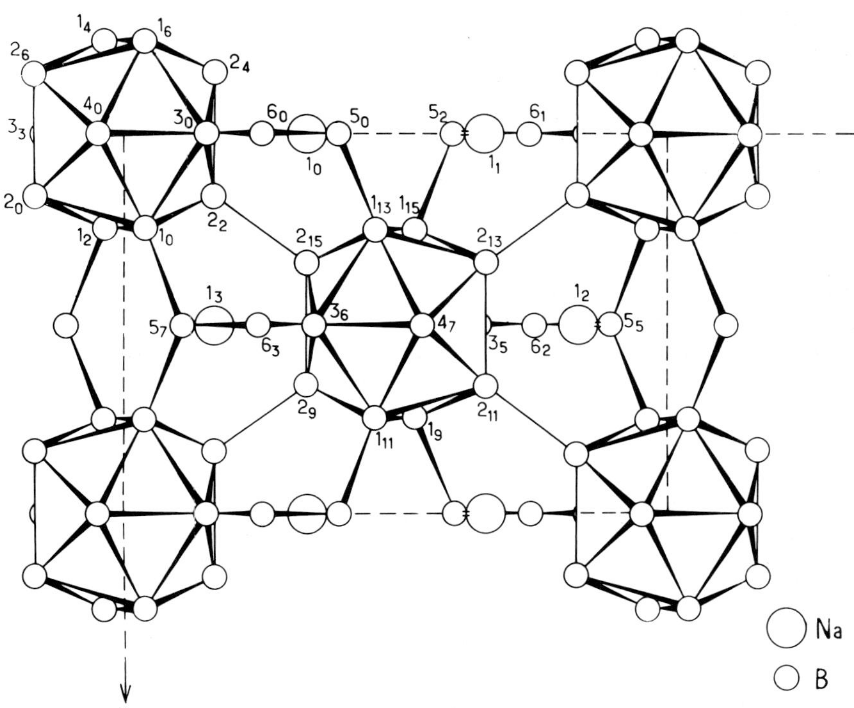

Fig. 9. The projection of the NaB_{15} crystal structure on the xOy plane. (From GUETTE, 1974; NASLAIN et al., 1976; only atoms with $-1/4 \leq z \leq 1/4$ are represented)

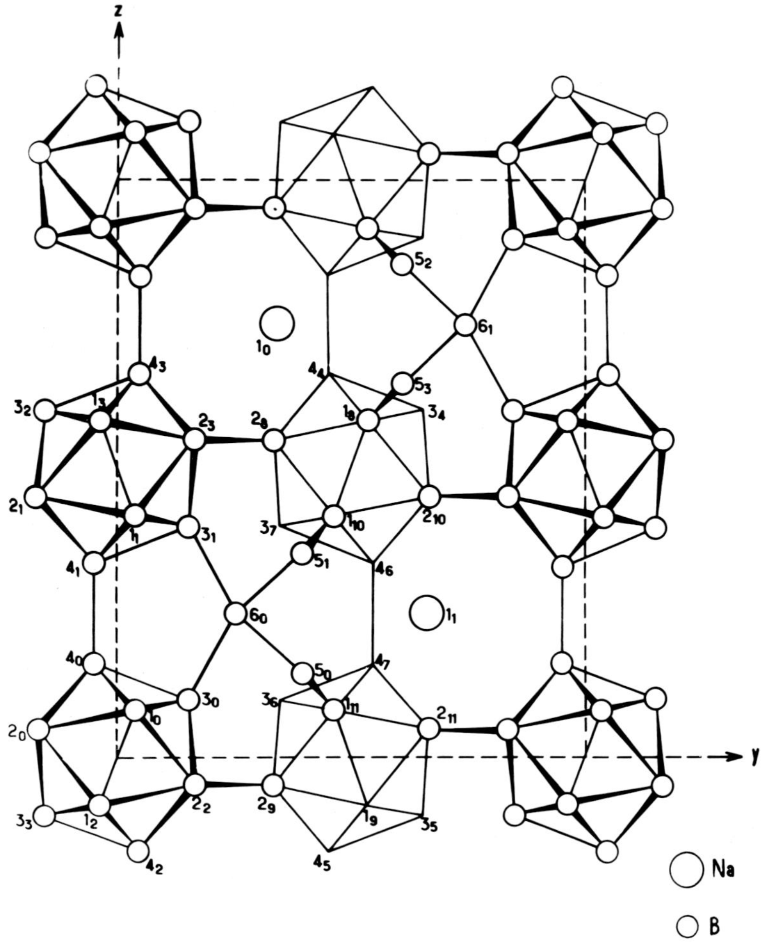

Fig. 10. The projection of the NaB_{15} crystal structure on the yOz plane. (From GUETTE, 1974; NASLAIN et al., 1976; only atoms with $-1/4 \leq x \leq 1/4$ are represented)

It can also be described as an alignment of icosahedron chains parallel to the $\vec{c}$ axis and centered at (x = 0, y = 0) and (x = 1/2, y = 1/2) (Fig. 10). In a chain, for example that centered at (x = 0, y = 0), the icosahedra are oriented in such a way that one of their mirror planes is parallel to the yOz plane, one of their pseudo-five-fold axes being inclined alternately at ± 13°23 to the $\vec{c}$ axis. Inside a chain, the icosahedral units are bonded to each other by B-B bonds of 1.769 Å parallel to the $\vec{c}$ axis (4_0-4_1 for example; Fig. 10). Cohesion between B_{12} icosahedron chains is secured either by direct B-B bonds between icosahedra (2_2-2_{15} for example) or by bonds involving isolated B(5) or B(6)-type boron atoms.

Within a cluster, the 12 boron atoms are bonded by 30 bonds ranging from 1.787 to 1.847 Å. The average value (1.806 Å) is close to that observed in the outer icosahedra of the giant $B_{12}(B_{12})_{12}$ clusters of YB_{66}-type borides, but is quite higher than those that are characteristic of α-rh. boron (1.77 Å) or those of the inner icosahedron of the B_{156} units of YB_{66} (1.726 Å) (DECKER and KASPER, 1959; RICHARDS and KASPER, 1969).

Each B_{12} unit has twelve bonds outwardly directed, which consequently give rise to a three-dimensional skeleton formed by the boron atoms. In NaB_{15}, there are four types of such bonds, depending on whether they are directed towards a boron atom belonging to an adjacent icosahedron (example: 2_2-2_{15} and 4_0-4_1) or whether they involve B(5) or B(6) isolated boron atoms. This average length (1.745 Å) is far less than the average length of intra-icosahedral bonds (1.806 Å), which is a property of all boron-rich phases with a boron sublattice made of boron atom clusters.

HOARD and HUGHES (1967) have postulated that in phases with icosahedral units, the twelve external bonds are preferentially oriented along the corresponding pseudo-five-fold axes of the icosahedron. In NaB_{15} this rule is followed to within about 15° for ten of the twelve external bonds, and only two, of the 3_0-6_0 type, deviate markedly from this rule (by 21°98).

Isolated Boron Atoms in NaB_{15}. Two types of isolated atoms, B(5) and B(6), are present in the structure of NaB_{15}, whereas only B(5) atoms occur for $MgAlB_{14}$ (which has the same boron icosahedral skeleton). These isolated boron atoms complete the pattern of external bonding of the B_{12} units of the boron sublattice with their participation in directed B-B bonds.

In Table 4 one notices that most of the B-B bonds, involving isolated boron atoms, are short in comparison with other B-B bonds and consequently correspond to strong interactions. Thus the isolated boron atoms are really a part of the covalent boron skeleton.

Table 4. B-B bonds involving B(5) and B(6) isolated atoms in NaB_{15} and $MgAlB_{14}$. (From MATKOVICH and ECONOMY, 1970; NASLAIN et al., 1976)

B-B bonds	NaB_{15}	$MgAlB_{14}$
5_0 - 1_{11}	1.715	1.754
5_0 - 1_{13}	1.715	1.754
5_0 - 3_0	(2.079)	1.797
5_0 - 6_0	1.599	-
6_0 - 3_0	1.754	-

The B(5) type atoms have a trigonal surrounding: in $MgAlB_{14}$ the three bonds are directed towards boron atoms belonging to three adjacent icosahedra, the 5_0-3_0 bond being longer than the 5_0-1_{11} and 5_0-1_{13} ones. On the contrary, in NaB_{15}, the larger b-parameter (8.415 Å in NaB_{15} vs. 8.115 Å in $MgAlB_{14}$) due to the presence of a larger atom (sodium), results in very weak 5_0-3_0-type bonds (2.079 Å in NaB_{15}) but allows the B(5) atoms to form a very strong bond (1.599 Å) with B(6)-type atoms.

The B(6)-type atoms occupy a very special kind of site, since they have four B-B bonds situated in the same bc plane. These bonds are of two types, those involving boron atoms of the icosahedra being longer (1.754 Å) than those between B(6) and B(5) atoms (1.599 Å). One could expect the B(6) atoms, which seem necessary to complete the pattern of external bonding to icosahedral units, to occupy their sites at 100%. Electron density sections and results of refinement of the NaB_{15} structure indicate the opposite: according to NASLAIN et al. (1976) the occupancy of sites 4(e) by B(6) atoms is only about 85%.

Metal Atoms in the NaB_{15} Lattice. The covalent skeleton formed by boron atoms has two types of interstitial sites, of different sizes, which can accomodate metal atoms. The role of sodium atoms is more difficult to understand than that of isolated boron atoms. They can be assumed to be at least partially ionized and to act as electron donors, since the boron sublattice of boron-rich phases is generally electron-deficient. Furthermore, the metal atoms contribute to stabilizing the structure since they fill the voids that necessarily result from any kind of packing of B_{12} icosahedral clusters involving bonds oriented along the five-fold axes.

The site I, corresponding to a 4(e) position is 100% filled by sodium atoms in the case of phases prepared in the presence of liquid sodium in equilibrium with its vapor, the latter giving rise to a high pressure within the reactor where the synthesis is being carried out. Site I corresponds to a cage of 16 boron atoms, with a mirror plane passing though 4_0, 4_1, 5_0 and 5_1 boron atoms for the 1_1 sodium atom for example (Fig. 11). This high coordination site can accomodate relatively large atoms: sodium atoms in the case of NaB_{15} and magnesium atoms (or aluminum atoms) in the case of $MgAlB_{14}$ (Table 5).

Table 5. M-B bonds involving metal atoms (of M_1 type) occupying sites I in NaB B_{14} and $MgAlB_{14}$ phases. (From MATKOVICH and ECONOMY, 1970; GUETTE, 1974; NASLAIN et al., 1976)

M - B bonds	Number	M = Mg or Al in $MgAlB_{14}$ (Å)	M = Na in NaB B_{14} (Å)
$M-B4_0$	2	2.811	2.607
$M-B2_0$	4	2.660	2.704
$M-B5_0$	2	2.361	2.496
$M-B2_{11}$	4	2.784	2.871
$M-B1_{11}$	4	2.736	2.818

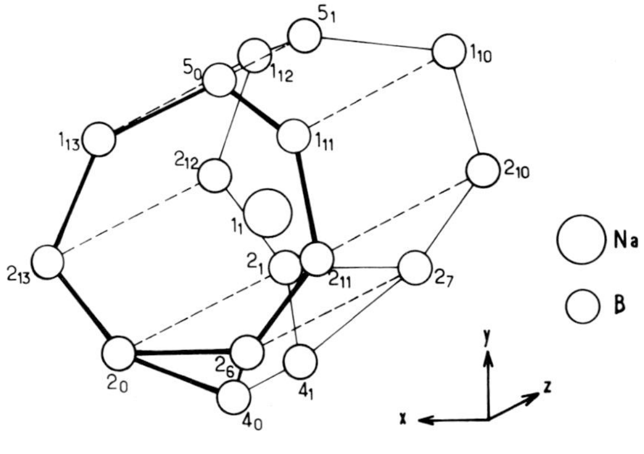

Fig. 11. Sodium environment in NaB_{15} (site I). (From GUETTE, 1974; NASLAIN et al., 1976)

The site II, corresponding to a 4(d) position (1/4, 1/4, 1/4), smaller than the previous one, is surrounded by 12 boron atoms only. It is not used in NaB_{15}, whereas it is 75% occupied by aluminum atoms in $MgAlB_{14}$.

It emerges from this structural study that the ideal formula for boron-rich sodium boride is NaB_{15}, which it is appropriate to write as NaB_3B_{12} to emphasize the occurrence of two distinct kinds of boron atoms: icosahedral B_{12} units and isolated B(5) and B(6) boron atoms. Alternatively, the formula may be written as $NaBB_{14}$ in analogy with $MaAlB_{14}$, although the sites occupied by the B(6) atoms and the Al atoms are different.

γ) Deviations from Stoichiometry in NaB_{15}

In NaB_{15}, as in $MgAlB_{14}$, the 12 external bonds of the icosahedra are completely satisfied by boron atoms, isolated or belonging to adjacent icosahedra. From this, one could conclude that in the bonding pattern of the icosahedra the metal atoms have only a secondary role, which would consequently allow large deviations from stoichiometry because of metal vacancies. Thus, for example, the occupancy of sites I and II by magnesium and aluminum atoms was found to be only about 75% for the $MgAlB_{14}$ crystal studied by MATKOVICH and ECONOMY (1970).

The high volatility of sodium suggested the hypothesis that far larger deviations from stoichiometry should be possible in NaB_{15} (NASLAIN, 1967; NASLAIN et al. 1976).

Preparation of Non-Stoichiometric Na_xB_{15} Phases ($0 < x < 1$). Na_xB_{15} phases are obtained by thermal decomposition of stoichiometric NaB_{15}. NaB_{15} starts losing sodium at 750°C when heated in high vacuum (10^{-6} torr). The loss of sodium takes place without any apparent modification of crystal morphology; in this condition, there is observed a slight variation of lattice parameters and quite a large change in the relative intensities of some of the Debye-Scherrer lines (Fig. 12). Such phenomena are typical of interstitial compounds where the inserted element occupies the voids of a rigid three-dimensional skeleton; they make it possible to assume that the loss of sodium leaves the boron sublattice intact.

The pyrolysis of NaB_{15} is slow at 750-850°C, but owing to the great volatility of sodium, it does lead to a large loss of alkali metal,

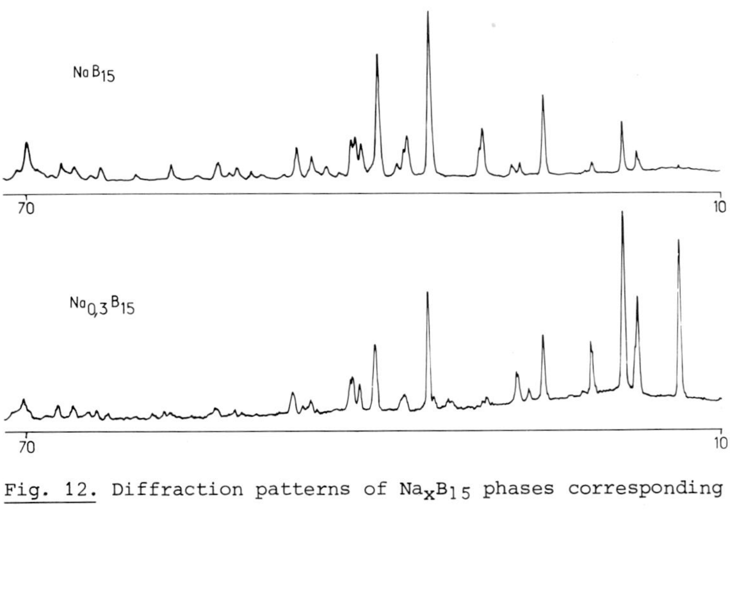

Fig. 12. Diffraction patterns of Na_xB_{15} phases corresponding to two values of x

the composition of the limit phase obtained at 850°C, for example, being near $Na_{0.3}B_{15}$.

Above 950°C the last traces of sodium are removed and the residue consists of a mixture of α and β-rh. boron and of a new boron monotrope with orthorhombic symmetry isostructural with NaB_{15}, but apparently without any sodium.

At 1200°C the only identified phase is β-rh. boron; the new boron monotrope is consequently a metastable form, which proves that while sodium atoms do not seem absolutely necessary in the crystal network; they do however contribute in stabilizing it since NaB_{15} is stable at 1200°C under a pressure of sodium vapor (NASLAIN, 1967).

Crystal Structure of Na_xB_{15}: Nature of Deviations from Stoichiometry. Two single crystals corresponding to two stages of the pyrolysis have been isolated and studied by X-ray diffraction by NASLAIN et al. (1976). Their structural formulas and lattice parameters are given in Table 6, as well as those of the stoichiometric boride.

Table 6. Lattice parameters of some single crystals of formula Na_xB_{15}. (From GUETTE, 1974; NASLAIN et al., 1976)

Formulas	a (Å)	b (Å)	c (Å)
$Na_{0.72}B_{0.84}B_{14}$	5.847	8.401	10.285
$Na_{0.85}B_{0.84}B_{14}$	5.852	8.408	10.300
$Na\ B_{0.8}\ B_{14}$	5.847	8.415	10.298

The refinement of the crystal structure of $Na_{0.72}B_{0.84}B_{14}$ and $Na_{0.85}B_{0.84}B_{14}$ phases showed that the sodium given off, only partially for the two phases, does not significantly alter the covalent boron skeleton and that observed changes of the intensities of Debye-Scherrer lines are related only to the decreased occupancy of sites I by sodium atoms. Consequently, the pyrolysis of stoichiometric NaB_{15} proceeds as follows:

$$NaB_{15} \xrightarrow{750\ -\ 850°C} Na_x \square_{1-x}B_{15} + (1-x)\ Na\nearrow$$

It is verified in Table 6 that the loss of sodium does not alter (within the limits of precision) the a - parameter but one observes, as could be expected, a regular decrease of the b-parameter and to a lesser extent a decrease of the c-parameter.

The behavior of NaB_{15} somewhat resembles that of clathrates formed by silicon and germanium with alkali metals (KASPER et al., 1965; CROS, 1970; CROS et al., 1970).

c) Properties of Sodium Borides

NaB_6 and NaB_{15} can be worked with in air and they are not attacked significantly by dilute hydrochloric solutions at room temperature. The action of oxygen on NaB_6 is detectable only above 500°C. The oxidation is slow at this temperature, but becomes rapid at 800°C; however a protective layer of glass is formed at this temperature and slow the reaction down. Sodium borides are very sensitive to the action of oxidants in aqueous solutions, just as boron itself.

Concerning the electrical properties, NaB_6 is an insulator ($\rho = 10^6\,\Omega cm$ at 25°C for a sintered sample). NaB_{15} sintered at 650°C under a pressure of 50kb. has a semiconductor behavior; in the intrinsic conduction range, which is reached near 300 K, the width of the forbidden gap, calculated by the following formula:

$$\mathrm{Log}\,\frac{1}{R} = -\frac{Eg}{2kT} + \text{constant}$$

is about 0.3 eV, a value which is quite lower than that reported for β-rh. boron, for example (1.4 to 1.6 eV) (Fig. 13).

5. The Potassium-Boron System

According to NASLAIN, ETOURNEAU and HAGENMULLER, the boron-potassium system has only one intermediate phase, the KB_6 hexaboride (NASLAIN and ETOURNEAU, 1966; NASLAIN, 1967; NASLAIN et al., 1967; HAGENMULLER et al., 1967; NASLAIN et al., 1968).

KB_6 is the most difficult alkali metal boride to prepare, due to the high vapor pressure of potassium at high temperatures, the necessity of utilizing only pure and highly reactive boron, and the low thermal stability of this phase, all of which considerably limit the temperature range in which the synthesis is possible (1100 to 1200°C).

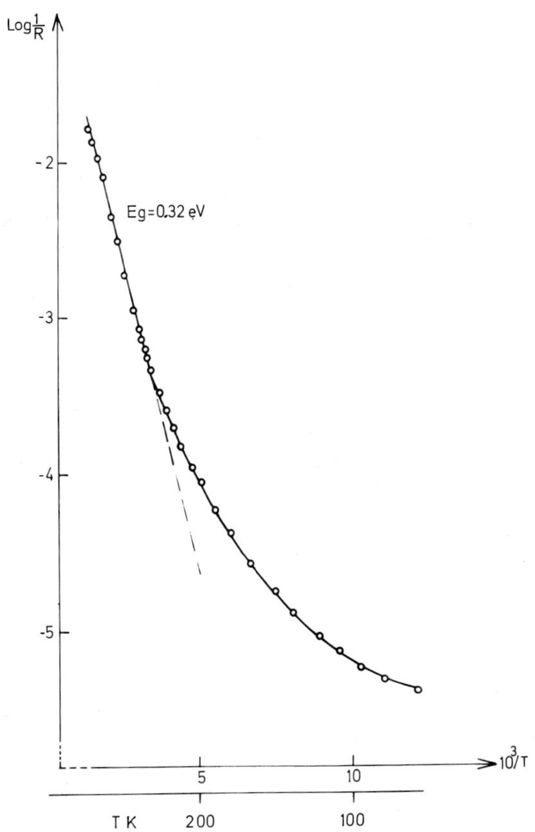

Fig. 13. Electrical behavior of a NaB_{15} sample-sintered at 650°C under a pressure of 50kb. (From GUETTE, 1974)

The most satisfactory preparative method for KB_6 is direct synthesis at 1130-1150°C, in the presence of an excess of potassium (K/B at. = 1 to 2) acting as a flux, followed by slow cooling and distillation of excess potassium. Prepared in these conditions, KB_6 is obtained as cubic microcrystals, with a density of 2.29 ± 0.01 (Fig. 8b).

a) Chemical Identification of KB_6

Table 7 gives results of chemical analyses carried out after washing with dilute hydrochloric acid, of several products resulting from the action of potassium on boron. It illustrates the importance of temperature and the crystallographic nature of boron for the realization of the reaction; it shows that even under optimum conditions, the potassium hexaboride does not seem strictly stoichiometric or contains sligh amounts of free boron.

Table 7. Chemical analysis of products resulting from the action of potassium on boron. (From NASLAIN, 1967; NASLAIN et al., 1967)

Experimental conditions				Chemical analyses			Phases obtained
Temperature (°C)	Duration (h)	P_K^o (bar)	Boron varieties used	K%	B%	(K+B)%	
1125	16	12.0	α-rh.	33.1	64.5	97.6	KB_6
1150	16	13.5	α-rh.	34.6	63.4	98.0	KB_6
1150	16	13.5	α-rh.+β-rh.	15.3	81.0	96.3	$KB_6+B_{\beta\text{-rh.}}$
1200	5	17.5	α-rh.	31.4	67.7	97.1	KB_6+traces $B_{\beta\text{-rh.}}$
1260	3	23.0	α-rh.	13.4	84.0	97.4	$KB_6+B_{\beta\text{-rh.}}$
				37.58	62.42	100	KB_6 (theor.)

b) X-Ray Diffraction Identification and Crystal Structure

Potassium hexaboride can easely be characterized by its Debye-Scherrer pattern which contains only a small number of lines and is of high quality if the sample is prepared within a potassium flux.

The KB_6 and CaB_6 patterns represented in Figure 14 show that KB_6, unlike NaB_6, is evidently isostructural with calcium hexaboride. The lines of the KB_6 pattern can indeed be indexed in the cubic system, with a parameter of 4.232 Å and space group Pm3m, O_h^1. The density (2.29 ± 0.01), measured in diethyl-orthophtallate, allows for one KB_6 formula unit per elementary cell; it is in good agreement with the X-ray density (d = 2.28) (RABARDEL et al., 1971).

The structure of KB_6 was established as that of CaB_6, determined in 1932 by VON STACKELBERG and NEUMANN (1932) and further refined by PAULING and WEINBAUM (1934). According to the former authors, if the calcium atom is situated at the origin, the boron atoms can be situated only in (1/2, 1/2, ± x) (1/2, ± x, 1/2) and (± x, 1/2, 1/2) with x = 0.20 ± 0.02.

The only unknown parameter of the structure x was determined from the powder pattern, by a graphical intensity comparison method. The value obtained for KB_6 (x = 0.208) is very near to that proposed by VON STACKELBERG and NEUMANN (1932) for CaB_6. In describing the CaB_6-struc-

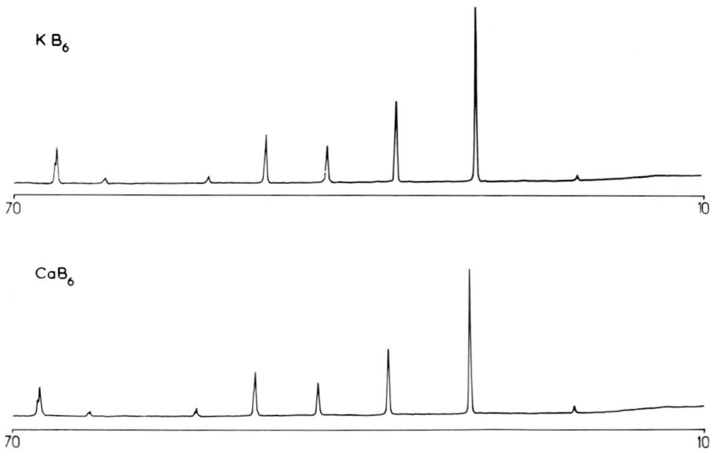

Fig. 14. X-ray diffraction patterns of potassium and calcium hexaborides

ture, PAULING and WEINBAUM (1934) placed the origin of the coordinate axes in the center of the B_6 octahedron, so that the boron atoms are situated at (O, O, ± u), (O, ± u, O) and (± u, O, O) with u = 1/2-x. The value of the u-parameter in the case of KB_6 is then 0.292 (Fig.15a).

The crystal structure of KB_6 is represented in Figure 15b; the structure consists of chains of B_6 octahedra forming a rigid skeleton. The holes of the B_6 skeleton are occupied by potassium. Boron has a five-fold coordination within its sublattice, whereas potassium occupies centers of large cages formed by 24 boron atoms.

Each boron atom shares with its neighbors five B-B bonds: four of them are orientated towards boron atoms belonging to the same octahedron, while the fifth one bonds to a boron atom of an adjacent octahedron. While the first four B-B bonds have the same length, due to symmetry, the fifth is crystallographically independent. The value of u, obtained graphically for KB_6 (u = 0.292) is however very near to the theoretical value, u = 0.2929, which would give the same length for the five B-B bonds. The interatomic distances in KB_6 are:

B-B = 1.747 Å, B-K = 3.119 Å and K-K = 4.232 Å

BLUM and BERTAUT (1954) thought that the impossibility of preparing $K_xM_{1-x}B_6$ phases by fused salt electrolysis was due to the size of the potassium atom, which is large relatively to M atoms that were known to form cubic hexaborides. On the contrary, Figure 6 shows that the lattice parameter of KB_6 is intermediate between that of SrB_6 and BaB_6, which agrees with the fact that the atoms inserted within the boron sublattice cages are ionized and that the size of the K^+ ion is intermediate between that of Sr^{2+} and Ba^{2+} (the radius of the K^+ ion given by SHANNON and PREWITT (1969) for VIII-coordination seems over estimated; according to the same authors, the value of this radius for the VI-coordination is 1.38 Å and that indicated by AHRENS (1952) for the same coordination is 1.33 Å).

Several authors have recently established, after refining the structure of some MB_6 borides, that the two types of crystallographically independent B-B bonds do not have usually the same length, contrary to the conclusions of VON STACKELBERG and NEUMANN (1932) or of PAULING

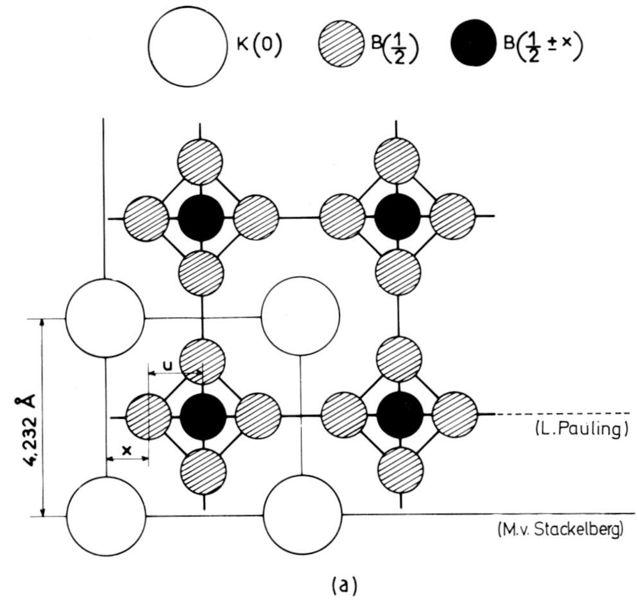

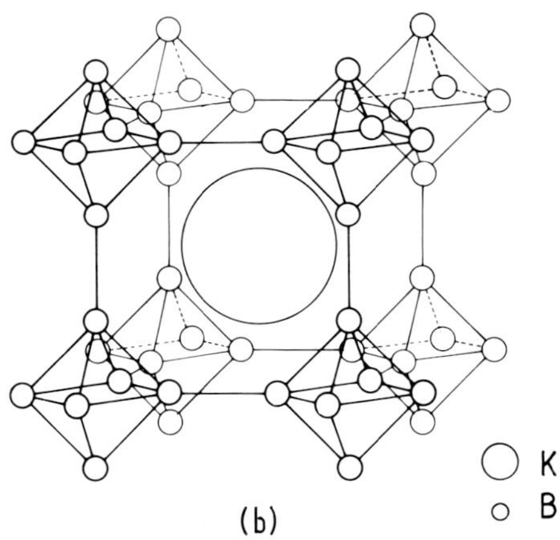

Fig. 15a and b. Crystal structure of KB_6 (a) in projection on the (100) plane (b) the [24]-coordination boron cage

and WEINBAUM (1934). Table 8 gives the length of these intra- and inter-octahedral B-B bonds, as well as the unique length of the B-B bond $[a/(1 + \sqrt{2})]$, assuming equivalence of B-B distances (in this case $x = a/2(1 + \sqrt{2}) = 0.207$). Using these values, the variation of the two different B-B bond lengths is represented in Figure 16, as a function of the radius of the inserted ion (the value of 1.35 Å for potassium has been deduced from Fig. 6) (LA PLACA, 1969; ETOURNEAU et al., 1971; BARRET and NASLAIN, 1973).

Several interesting observations can be made from Figure 16. In all cases, the intra-octahedral bond is longer than the inter-octahedral one; this property, that all B_n clusters seem to have in common, corresponds to the elctron defficiency of condensed boron atom units. On the other hand, the variation of the inter-octahedral bond length is more understandable than that of the intra-octahedral bond: when

Table 8. Lengths of various B-B bonds in some cubic hexaborides (for comparison, the $\frac{a}{1+\sqrt{2}}$ length is given, corresponding to the supposition of equidistance of boron atoms)

	a(Å)	B-B (Å) (hyp.equid.)	B-B (Å) intraoct.	B-B (Å) interoct.	References
YB_6	4.113	1.704	1.82	1.54	LA PLACA (1969)
$Th_{0.78}B_6$	4.112	1.703	1.767	1.612	ETOURNEAU et al. (1971)
CaB_6	4.146	1.717	1.752	1.667	LA PLACA (1969)
LaB_6	4.154	1.720	1.764	1.659	LA PLACA (1969)
SrB_6	4.200	1.739	1.757	1.715	BARRET and NASLAIN (1973)
KB_6	4.232	1.753	-	-	NASLAIN and ETOURNEAU (1966)
BaB_6	4.269	1.768	1.780	1.752	LA PLACA (1969)

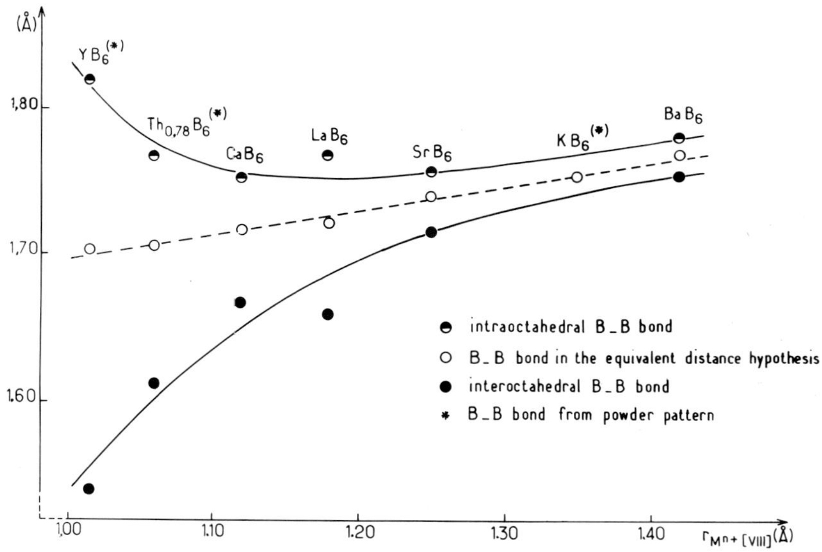

Fig. 16. Variation of the length of intra- and inter-octahedral B-B bonds in hexaborides as a function of the radius of inserted cation. (From data of LA PLACA, 1969; ETOURNEAU et al., 1971; BARRET and NASLAIN, 1973). Ionic radii for coordination VIII are those given by SHANNON and PREWITT (1969), except for K^+ for which the value of 1.35 Å was derived from Fig. 6

the size of the cation increases, the modification of the boron skeleton affects principally the inter-octahedral B-B bonds, that progressively become longer and consequently weaker. However, except for the marginal case of YB_6, it is observed that the size of the B_6 octahedron remains quite unchanged from ThB_6 to BaB_6 in spite of increasing size of the cation. Finally the difference between both bond types, which was large for small cations, decreases progressively as the cation increases, and becomes very small in the limiting cases of KB_6 and BaB_6. This could explain why the graphic method (with its limited precision), used in determining the value of x in the case of KB_6, leads to a value near to that corresponding to equal distances.

c) Thermal Stability and Deviation from Stoichiometry of KB_6

KB_6 is stable under an argon pressure of 1 bar up to about 800°C. Above this temperature, its dissociation is characterized by loss of metallic potassium and formation of β-rh. boron.

$$KB_6 \longrightarrow 6B + K\nearrow$$

The pyrolysis of KB_6 in high vacuum (10^{-6} torr) is of some interest inasmuch as, similarly to the case of NaB_{15}, the high volatility of potassium could lead, at low temperatures, to non-stoichiometric $K_x\square_{1-x}B_6$ phases, the limit of which might be a new cubic boron monotrope made of B_6 octahedra stabilized by traces of potassium.

In high vacuum, KB_6 is stable up to 750°C. Above this temperature a great loss of metal is observed: the quantity of residual potassium decreases, quite rapidly at the beginning of the pyrolysis, then more slowly down to a quite constant value which is about 19 weight % at 900°C (Fig. 17).

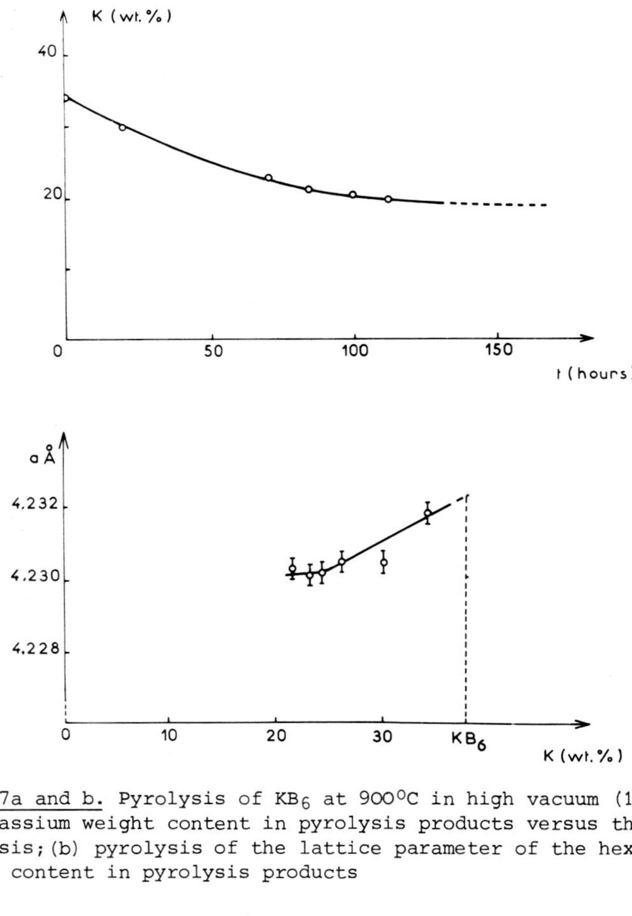

a)

b)

Fig. 17a and b. Pyrolysis of KB_6 at 900°C in high vacuum (10^{-6}torr); (a) variation of potassium weight content in pyrolysis products versus the duration of the pyrolysis; (b) pyrolysis of the lattice parameter of the hexaboride versus potassium weight content in pyrolysis products

These observations could lead to the conclusion that a non-stoichiometric phase with metal vacancies has been formed, the limiting composition of which at 900°C would be $K_{0.40}B_6$. Large deviations from stoichiometry with metal vacancies have been already reported for other cubic hexaborides such as ThB_6, LaB_6 or SmB_6. In fact, the absence of significant variation in the relative intensities of the Debye-Scherrer pattern, and the discrepency in densities ($d_{exp.} = 2.30$ instead of 1.76 calculated for the $K_{0.40}B_6$ formula) show that this is not what happens (JOHNSON and DAANE, 1961; ETOURNEAU et al., 1971; NIHARA, 1971).

It can be accepted that the very first stages of the pyrolysis of KB_6 correspond effectively to the formation of a small number of potassium vacancies as suggested by the decrease of the lattice parameter with the amount of residual potassium (Fig. 17b). Beyond a limiting value, the B_6-sublattice, which is no longer electronically stable (in the case of KB_6, potassium yields a maximum of one electron to the boron framework whereas in LaB_6 or ThB_6 the electron contribution of the metal atom is 3 or 4 per MB_6 formula), collapses to give amorphous boron because of the low temperature of the pyrolysis. The density of the pyrolysis product (d = 2.30) is compatible with that for a mixture of amorphous boron (2.35 to 2.45) and potassium deficient hexaboride (NASLAIN, 1967; NASLAIN et al., 1967).

KB_6 is the least stable of the cubic MB_6 hexaborides. According to ETOURNEAU et al. (1972, 1976) the thermal stability of MB_6 borides can be interpreted with two parameters: (1) the vapor pressure of the M metal which tends to favor greater dissociation, the higher its value for a given temperature, and (2) the stability of B-B bonds which increases as these bonds grow shorter and consequently as the lattice parameter decreases (MERCURIO, 1974; Chap. V IX of this vol.).

From this point of view, the fact that potassium is the most volatile of the metals forming hexaborides and that the lattice parameter of KB_6 is one of the largest explains the weak thermal stability of KB_6.

d) Physical Properties of KB_6

The elements that form cubic hexaborides generally belong to columns IIA (Ca, Sr, Ba), IIIA (the lightest rare - earths) and to the actinide family (Th, Pu). KB_6 is the first cubic binary hexaboride involving a monovalent element. Because of this peculiar circumstance, KB_6 has distinctive physical properties that contrast with those of the other members of the CaB_6-type structure family.

Although no electric measurements have been made on single crystals, conduction measurements carried out on sintered samples show that KB_6 has a weak conductivity at room temperature ($10^{-4}\Omega^{-1}cm^{-1}$ at 300 K) which increases strongly with temperature (Fig. 18). The log $\sigma = f(1/T)$ curve is typical of non-stoichiometric or doped semiconductors. The activation energies, $W_I = 0.29$ and $W_{II} = 0.008$ eV, corresponding to the straight parts of the curve, were calculated from the following relations:

$$\sigma = \sigma_o \exp\left(-\frac{W_I}{2kT}\right) \quad \text{and} \quad \sigma = \sigma_o \exp\left(-\frac{W_{II}}{kT}\right)$$

KB_6 has a weak paramagnetism at temperatures between 100 and 600 K, practically independent of temperature (Fig. 19).

The physical properties of KB_6 can be interpreted most successfully with band structure models (developed in other chapters of this book).

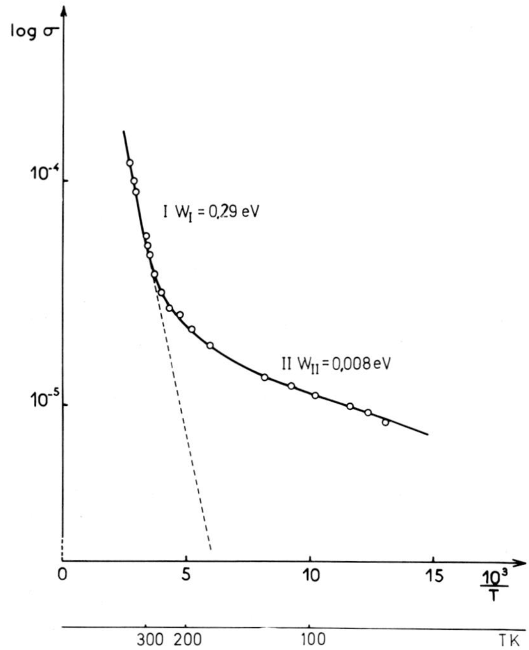

Fig. 18. Electrical conductivity of a sintered KB_6 sample versus reciprocal absolute temperature

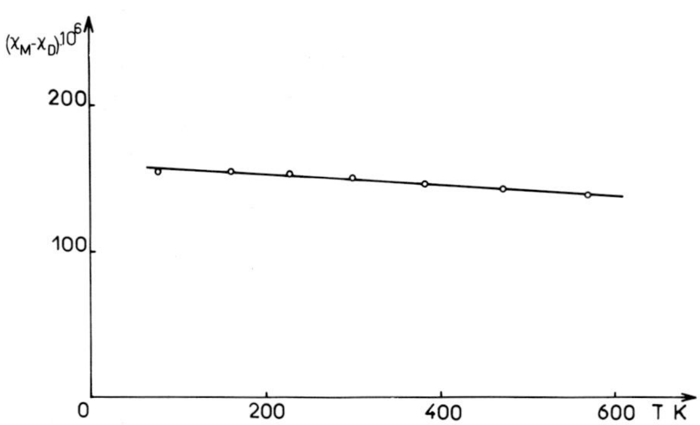

Fig. 19. Magnetic susceptibility of potassium hexaboride versus temperature

After the theoretical work of LONGUET-HIGGINS and ROBERTS (1954) and that of YAMAZAKI (1957), it has been generally assumed that the band structure of CaB_6 contains bonding bands typical of the boron sublattice and capable of accomodating 20 electrons per CaB_6 formula, separated from antibonding bands by a relatively narrow gap (from 1.5 to 4.4 eV). The boron atoms of the B_6 octahedron yield only 18 electrons: from this, it has been concluded that a transfer of two electrons from the metal to the boron sublattice is necessary to stabilize the crystalline framework. The semi-conductor properties of $M^{II}B_6$ phases and the metallic ones of $M^{III}B_6$ or $M^{IV}B_6$ phases directly result from this model (Chap. B.IX of this vol.).

The validity of previous models may be questionable, because of the existence and stability of $Na_xB_{1-x}B_6$ solid solutions and moreover of KB_6, since they prove that the CaB_6-type structure is still stable when the electron contribution of the inserted atom is less than two. As has already been proposed for EuB_6, a detailed description of physical properties of hexaborides probably involves not only the bonding and anti-bonding boron bands considered by LONGUET-HIGGINS and ROBERTS (1954) or by YAMAZAKI (1957), but also bands originating in atomic orbitals of inserted metal (MERCURIO, 1974; MERCURIO et al., 1974).

The existence of KB_6 is accounted for very adequatly by a model of band structure presented very recently by PERKINS and SWEENEY (1976), a result of which is that the distribution of charges in KB_6 corresponds to the formula $K^{+0.74}B_6^{-0.74}$. Thus, a much smaller amount of electron transfer, namely 0.74 instead of 2, suffices to stabilize the structure. It would follow from this result that potassium hexaboride can be stable, even with the presence of an appreciable number of potassium vacancies as long as the charge transfer requirements are fullfilled. According to these authors. KB_6 should have electrical properties close to those of CaB_6, i.e., a weak conductivity and a semi-conductiong behavior.

The similarity of KB_6 and alkaline-earth hexaborides, in regard to their electrical properties (as made clear by the most recent measurements), leads to the following hypothesis:

The valence electrons of potassium and of alkaline-earth elements, far from being fully transferred onto the boron sublattice, could belong to a narrow band that is not completely filled in the case of KB_6. At low temperatures, the conductivity would be of a metallic type, the low values obtained for σ being explained by the narrowness of the conduction band. At high temperatures, the nature of the conductivity curve could be explained by progressive excitation, with increasing temperatures, of 4s potassium electrons from their original level onto the anti-bonding band of the boron lattice, which has evidently an acceptor character. This hypothesis could also explain the temperature-independent paramagnetism of KB_6, i.e. it could be a PAULI paramagnetism of a magnitude comparable to that of transition metals which also have narrow bands. This interpretation would be consistent with the existence of a degenerate conduction electron gas and of a PAULI paramagnetism, established by conduction electron spin resonance, for a cubic sodium hexaboride by RUPP and HODGES (1974).

e) Chemical Properties of KB_6

The chemical properties of KB_6 are similar to those of sodium borides; KB_6 is not hydrolyzed, particularly in hydrochloric aqueous solutions. The oxidation of KB_6 by oxygen starts at 500°C and is more energetic than for NaB_6, but glass formation prevents it from being complete.

The KB_6 hexaboride, just as boron when finely divided, reacts very strongly with oxidizing acids, even at room temperature.

References

AHRENS, L.H.: The use of ionization potentials. Part 1. Ionic radii of the elements. Geochim. Cosmochim. Acta 2, 155-169 (1952)

ANDRIEUX, L., BARBETTI, D.: Sur les borures alcalins. C.R. Acad. Sci. (Paris) 194, 1573-1574 (1932)

BARRET, M., NASLAIN, R.: personal communication (1972)

BARRET, M., NASLAIN, R.: personal communication (1973)

BERTAUT, F., BLUM, P.: Étude des hexaborures et de la substitution alcaline. C.R. Acad. Sci. (Paris) 234, 2621-2623 (1952)

BLUM, P., BERTAUT, F.: Contribution à l'étude des borures à teneur élevée en bore. Acta Cryst. 7, 81-86 (1954)

BOGROS, A.: Pression de vapeur saturante du lithium. C.R. Acad. Sci. (Paris) 191, 322-324 (1930a)

BOGROS, A.: Pression de vapeur saturante du lithium. C.R. Acad. Sci. (Paris) 191, 560-561 (1930b)

CROS, C.: Sur quelques nouveaux siliciures et germaniures alcalins à structure clathrate: étude cristallochimique et physique. Thesis 291 Univ. Bordeaux (1970). Centre Doc. CNRS Paris n°AO 4489 (1970)

CROS, C., POUCHARD, M., HAGENMULLER, P.: Sur une nouvelle famille de clathrates minéraux isotypes des hydrates de gaz et de liquide. Interpretation et résultats obtenus. J. Solid State Chem. 2, 570-581 (1970)

DAVY, H.: Versuche über die Zersetzung und Zusammensetzung der Boraxsäure J. Chem. Physik 2, 48-57 (1811)

DECKER, B.F., KASPER, J.S.: The crystal structure of a simple rhombohedral form of boron. Acta Cryst. 12, 503-506 (1959)

Elektroschmelzwerk Kempten: Procédé de préparation de borures alcalins. French patent 1.461.878 (1965)

ETOURNEAU, J.: Les hexaborures de type CaB_6. Le système bore-thorium. Thesis 285 Univ. Bordeaux (1970). Centre Doc. CNRS Paris AO 4312 (1970)

ETOURNEAU, J., NASLAIN, R., LA PLACA, S.: L'hexaborure de thorium non-stoechiometrique $Th_{1-x}B_6$. J. Less Common Metals 24, 183-193 (1971)

ETOURNEAU, J., MERCURIO, J.P., NASLAIN, R.: Etude de la stabilité thermique des hexaborures de terres rares. C.R. Acad. Sci. (Paris) 275, 273-276 (1972)

EWING, C.T., STONE, J.P., SPANN, J.R., STEINKULLER, E.W., WILLIAMS, D.D., MILLER, R.R.: High temperature Properties of Potassium. NRL Report 6233 (1965a) (available from CFSTI, Springfield, Virginia, USA)

EWING, C.T., STONE, J.P., SPANN, J.R., STEINKULLER, E.W., WILLIAMS, D.D., MILLER, R.R.: High temperature properties of cesium. NRL Report 6246 (1965b) (available from CFSTI, Springfield. Virginia, USA)

GAY-LUSSAC, THENARD: Sur la décomposition et la recomposition de l'acide borique. Ann. Chim. 68, 169-174 (1808)

GUETTE, A.: Etude cristallochimique des borures de magnésium riches en bore. Les phases $MM'B_{14}$ (M = Mg, Na; M' = Mg, Al, B). Thesis 462 Univ. Bordeaux (1974). Centre Doc. CNRS Paris AO 10799 (1974)

HAAG, H.H.: Method for the production of a substantially pure boron. U.S. Pat. 2, 794, 708 (1957)

HAGENMULLER, P., NASLAIN, R.: Un borure alcalin: l'hexaborure NaB_6. C.R. Acd. Sci. (Paris) 257, 1294-1296 (1963)

HAGENMULLER, P., NASLAIN, R., POUCHARD, M., CROS, C.: New binary compounds of the alkali metals with boron, silicon and germanium. In: The Alkali Metals. SP 22, pp. 207-221. London: The Chemical Society 1967

HICKS, W.T.: Evaluation of vapour pressure data for mercury, lithium, sodium and potassium. J. Chem. Phys. 38, 1873-1880 (1963)

HOARD, J.L., HUGHES, R.E.: Elemental boron and compounds of high boron content: structure, properties and polymorphism. In: The Chemistry of Boron and its Compounds (E.L. MUETTERTIES ed.) Chap. 2, pp. 25-154. New-York, London, Sydney: Wiley and Sons (1967)

JOHNSON, R.W., DAANE, A.H.: The lanthanum-boron system. J. Phys. Chem. 65, 909-915 (1961)

KAHLENBERG, H.H.: Boron and boron suboxide. Trans. Amer. Electrochem. Soc. 47, 23-63 (1925)

KASPER, J.S., HAGENMULLER, P., POUCHARD, M., CROS, C.: Clathrate Structure of silicon Na_8Si_{46} and Na_xSi_{136} ($x < 11$). Science 150 (3704), 1713-1714 (1965)

KROLL, W.: Über die Darstellung des amorphen Bors. Z. anorg. allgem. Chem. 102, 1-33, (1918)

LA PLACA, S.: personal communication (1969)

LA PLACA, S., CASTELLANO, R.N., SCHMIDT, P.H. and COOPER, A.S.: Cystal Growth and Identification of Several New Alkali Metal Borides (to be published)

LONGUET-HIGGINS, H.C., DE V. ROBERTS, M.: The elctronic structure of the borides MB_6. Proc. Roy. Soc. 224 (A), 336-347, (1954)

LUNDSTRÖM, T.: Crystal structure of the solid solutions of transition metals in β-rhombohedral boron. In: Boron: Synthesis, Structure and Properties (F.N. TAVADZE, ed.) pp.44-51. MOSCOW: Nauka 1974

MATKOVICH, V.I., ECONOMY, J.: Structure of $MgAlB_{14}$ and a brief critique of structural relationships in higher borides. Acta Cryst. B26, 616-621, (1970)

MERCURIO, J.P.: Synthese, stabilité thermique et properiétés électriques et magnétiques des hexaborures de terres rares. Thesis 434 Univ. Bordeaux (1974). Centre Doc. CNRS Paris AO 10000 (1974)

MERCURIO, J.P., ETOURNEAU, J., NASLAIN, R., HAGENMULLER, P., GOODENOUGH, J.B.: Propriétés électriques et magnétiques des solutions solides $La_xEu_{1-x}B_6$. J. Solid State Chem. 9, 37-47 (1974)

MOISSAN, H.: Action des métaux alcalins sur l'acide borique. Etude critique des procédés de préparation du bore amorphe. C.R. Acad. Sci. (Paris) 114, 319-324 (1892)

NASLAIN, R.: Préparation de bore pur sous les formes rhombohedriques α et β. Les borures alcalins. Thesis 188 Univ. Bordeaux (1967). Centre Doc. CNRS Paris AO 1332 (1967)

NASLAIN, R.: Boron syntheses. In: Preparative Methods in Solid State Chemistry (P. HAGENMULLER, ed.) pp. 439-485. New York and London: Academic Press 1972

NASLAIN, R., BARRET, M.: personal communication (1975)

NASLAIN, R., ETOURNEAU, J.: L'hexaborure de potassium. C.R. Acad. Sci. (Paris) 263, 484-487 (1966)

NASLAIN, R., ETOURNEAU, J., HAGENMULLER, P.: Le système bore-potassium. Bull. Soc. Chim. France, 2529-2538 (1967)

NASLAIN, R., ETOURNEAU, J., HAGENMULLER, P.: Preparation of pure α and β-rhombohedral boron. The alkali borides. In: Boron Vol. III ,Preparation and Properties (T. NIEMYSKI, ed.) pp. 35-47. Warsaw: Pol. Sci. Publ. 1970

NASLAIN, R., ETOURNEAU, J., BARRET, M.: personal communication (1970)

NASLAIN, R., KASPER, J.S.: The crystal structure of the Øphase in the boron-sodium system. J. Solid State Chem. 1, 150-151 (1970)

NASLAIN, R., GUETTE, A., HAGENMULLER, P.: Crystal chemistry of some boron-rich phases. In: Proc. 5th. Int. Symp. on Boron and Borides (Bordeaux 1975). J. Less Common Metals 47, 1-16 (1976)

NIIHARA, K.: The preparation and non-stoichiometry of samarium hexaboride. Bull. Chem. Soc. Japan 44, 963-967 (1971)

PAULING, L., WEINBAUM, S.: The structure of calcium boride. Z. Kristallogr. 187, 181-182 (1934)

PERKINS, P.G., SWEENEY, A.V.J.: The electronic structure of metal borides. In: Proc. 5th. Int. Symp. on Boron and Borides (Bordeaux 1975), J. Less Common Metals 47, 165-173 (1976)

RABARDEL, L., POUCHARD, M., HAGENMULLER, P.: Détermination précise des masses spécifiques par poussée hydrostatique. Mat. Res. Bull. 6, 1325-1336 (1971)

RAWSON, S.G.: On the preparation of boron. Chem. News 58, 283 (1888)

RICHARDS, S.M., KASPER, J.S.: The crystal structure of YB_{66}. Acta Cryst. 25, 237-251 (1969)

RUPP, L.W.Jr., HODGES, D.J.: Conduction-electron spin resonance in alkali hexaborides. J. Phys. Chem. Solids 35, 617-619 (1974)

SECRIST, D.R., CHILDS, W.J.: Lithium-boron carbide reaction studies. U.S. Atomic Energy Comm. Report TID 17149 (1962) (available from CFSTI, Springfield, Virginia, USA) 1962

SECRIST, D.R.: Compounds formation in the systems lithium-carbon and lithium-boron. J. Amer. Ceram. Soc. 50 (10), 520-523 (1967)

SHANNON, R.D., PREWITT, C.T.: Effective ionic radii in oxides and fluorides. Acta Cryst. B25, 925-946 (1969)

SOWA, F.J.: Process for producing elemental boron. U.S. Pat. 2,465,989 (1949)

STACKELBERG, M. VON, NEUMANN, F.: Die Kristallstruktur der Borid der Zusammensetzung MeB_6. Z. Phys. Chem B19, 314-320 (1932)

STONE, J.P., EWING, C.T., SPANN, J.R., STEINKULLER, E.W., WILLIAMS, D.D., MILLER, R.R.: High temperature properties of sodium. NRL Report 6241 (1965) (available from CFSTI, Springfield, Virginia, USA)

WÖHLER, F., SAINTE-CLAIRE DEVILLE, H.: Du bore. Ann. Chim. Phys. 52 (3), 63-92 (1858)

YAMAZAKI, M.: Group theoretical treatment of the energy bands in metal borides MeB_6. J. Phys. Japan 12, 1-6 (1957)

VI. Ib and IIb Subgroup Borides

V. N. GURIN and M. M. KORSUKOVA

Ib and IIb subgroup elements are essentially the only representatives among the metals of the periodic system whose ability to form borides remains questioned. The products of interaction between Ib and IIb subgroup metals with boron are ordinarily considered to be solid solutions. Two factors make this problem particularly difficult. First, information on boron systems is scarce. Second, there is a question as to what constitutes a chemical compound in general, and whether these systems will fit the definition when it is established. In light of these factors, it is worthwhile to apply the well-known criteria for interactions between elements to the boron Ib and IIb subgroup systems.

1. A Comparative Evaluation of the Existence of Interaction Between Boron and Ib, IIb Elements

Many criteria may be employed to establish interaction which result in the formation of chemical compounds between elements in binary systems. We will use here those which we believe to be most suitable for the boron systems considered.

Computers were used by SAVITSKY and GRIBULYA (1970) to predict the formation of AB, A_2B and A_3B chemical compounds (A and B being various elements). They show boron as capable of forming compounds AB with all IB and IIb elements, compounds A_2B with all but mercury and A_3B with all but copper, zinc and cadmium. Unfortunately, tables predicting more complex compounds (e.g. $A_{22}B$, etc.) have not yet been compiled. It is hoped that the results will clarify possible patterns of interactions in these systems.

Within the framework of the configuration model developed by SAMSONOV et al. (1971), interacting atoms should tend to form stable electronic configurations. Indeed, boron may form such configurations, sp^3 in particular with Ib and IIb elements. The electronic interaction between their atoms may be written as follows:

Ib Subgroup

(a) univalent state:

xB + y(I)b $\to B_x(I)b_y$

$s^2p + s \to s^2p^2 \to sp^3$

(b) divalent state:

$s^2p + (s + d) \to s^2p^3 \to sp^4 \to sp^3 + p$

(c) trivalent state:

$s^2p + (s + d^2) \to s^2p^4 \to sp^5 \to sp^3 + p^2$

IIb Subgroup

x B + y (II)b $\to B_x(II)b_y$

$s^2p + s^2 \to s^2p^3 \to sp^4 \to sp^3 + p$

It should be noted that both the divalent and trivalent states of the Ib elements are energetically favored. For example, the single-electron energies of the d- and s- orbitals by HERMAN and SKILLMAN (1963) are as follows; $U_s^{Cu} = 0.5091$; $U_d^{Cu} = 0.7431$; $U_s^{Ag} = 0.4724$; $U_d^{Ag} = 0.9300$; $U_s^{Au} = 0.4775$; $U_d^{Au} = 0.9200$ R, where IR = 13.595 eV. According to NEKRASOV (1970) copper and gold become diand trivalent even more easily than silver. For example, the energies required to remove 2 and 3 electrons derived from single-electrons are: $U^{Cu2+} = 1.2522$; $U^{Ag2+} = 1.4024$; $U^{Au2+} = 1.3975$; $U^{Cu3+} = 1.9953$; $U^{Ag3+} = 2.3324$; $U^{Au3+} = 2.3175$R.

On the basis of the concept of stability in electron configurations, one may conclude that the formation of stable configurations in B-Ib and IIb systems is possible. Therefore, with boron acting as an electron acceptor, the formation of relatively stable compounds or solid solutions should be possible. Furthermore, as the principal quantum number increases, the possibility of formation of such compounds should decrease. This is related to the decreasing stability of sp^3 configurations for the ρ-elements i.e. the elements of the Ib and IIb subgroups, as supported by the well-known experimental fact of the decreasing reactivity of boron in direction from copper to gold, and from zinc to mercury.

GURIN and OBUKHOV (1966) proposed to use the difference in the sum energies of their valence bands (SEVB) as a measure of interaction of elements in binary systems. The SEVB are quantities proportional to sums of successive atomic ionization potentials, and represent sums of products of single-electron energies by the number of the valence orbital electrons. The SEVB differences are found under the conditions of maximum saturation of positive atomic valencies. A positive difference leads to the formation of chemical compounds, while a negative one involves the absence of interaction. The formation of unstable compounds and solid solutions is indicated when the difference is close to zero. In boron systems with Ib and IIb elements, the SEVB differences vary from +0.4821 (B-Zn) to +0.9204R (B-Ag). For di- and trivalent Ag the differences are +0.4680 and 0.0052R. Thus, the formation of borides or solid solutions in all of these systems is possible.

By comparing these three different approaches to the problem of interaction between boron and Ib and IIb elements, we can draw the conclusion that compounds and solid solutions can form in all these systems. However, a difficulty arises here in classifying the interaction products as either chemical compounds or solid solutions. The structural factors which usually represent the distinctive feature of chemical compounds, and solid interaction products in particular, are of minor importance here. The lattice type and period cannot serve as a measure of the "qualitativeness" of a new compound, especially if its formula is XB_{22} or XB_{66}, etc. and its structure is similar to that of boron. However, one should know the arrangement of impurity atoms in the boron structure which will define an increment in the lattice period. The period changes by this amount indicate a qualitatively new product of interaction between the elements, that is, a new chemical compound or a new solid solution is formed. This new qualitativeness is revealed to a greater extent in the physical and chemical properties than in structural features. Unfortunately, these ideas have not yet been fully explored. This is partially why the problem of classification of the interaction products in the systems in question remains unsolved.

2. Factors Accounting for the Specific Features of Boron Interaction with Ib and IIb Elements

The specific features of interaction stem from the electronic structure of interacting atoms. They are shown in both the mechanism of formation and in the structure and properties of interaction products. Compared with other elements, the electronic structures of boron and of its solid phase, in particular, as well as of the Ib and IIb elements have some distinctive features.

NEWKIRK (1964) pointed out that the main elements of the boron framework in the solid phase are icosahedra representing electron-deficient arrangements. In the existing crystal modifications the existence of α-, β-rhombohedral (α-rh, β-rh) and tetragonal (α-Q, β-Q) structures have been clearly established. HOARD and HUGHES (1967) have shown the electronic density between the atoms located at different icosahedral sites and outside the icosahedra to be different. This may be attributed to different coordination of these atoms and the distance between them. The difference in the electron characteristics of different boron atoms may affect the mechanism and the kinetics of their interaction with the atoms of other elements.

At the same time, the removal of electrons from the Ib and IIb elements, whether the elements are in the form of isolated atoms or in the condensed phase, does not require large energy expenditures. Examples of this include their single-electron energies (see above) or the effective work function in condensed phase given by SCHULZE (1967). Furthermore, the ionic size of the atoms makes it possible for them to occupy icosahedron openings in the boron structure as pointed out by HOARD and HUGHES (1967). Hence, when the ionic size and donor properties of the Ib and IIb elements are combined with the acceptor properties of boron, conditions are favorable for the formation of new phases on the basis of the boron structure, rather than of new structural species. This is what occurs in actual fact. However, there is one possibility for an exception to this rule. When the Ib subgroup elements are in the di- or trivalent state, the latter may change their donor to acceptor properties. This modification is due to the participation of d-orbitals in the bonds. In this case, we could assume the formation of new structural modifications. This assumption is confirmed to a certain extent by some experimental data on the Ag-B and Au-B systems.

Thus, on the basis of general considerations, one may conclude that the interaction of boron with the Ib elements may result in the formation of both individual structures and of interaction products based on the boron structure. The latter probably may also occur with IIb elements.

3. Borides of the Ib Subgroup

The most comprehensive review papers on boron systems with copper, silver, and gold was written by SAMSONOV et al. (1960). This work contains the relevant data available. Data on these systems can also be found in works by MELLOR (1946), SAMSONOV and MARKOVSKY (1956) and MARKOVSKY and KONDRASHEV (1957). The original works reviewed in these papers can be arranged in two parts, those confirming and those denying the existence of interaction in the systems considered.

a) Cu-B System

MARSDEN (1880) should probably be credited with the first report on the products of interaction in the Cu-B system. He made an attempt to initiate direct reaction of boron with metallic copper by heating them for three to four hours at a temperature considerably exceeding the melting point of copper. The end product contained 79.75 wt% Cu, 9.41 wt% B and 10.84 wt% Si on the average.

From this composition, the author derived the formula Cu_3B_2. However, since the amount of silicon is so pronounced in this alloy (larger than that of boron), and since subsequent studies contain no confirmation of the existence of this compound, these data may be questionable. Some other early works also indicated the formation of copper alloys or compounds with boron. COWLES et al. (1885) produced a Cu-B alloy by passing current through a mixture of H_3BO_3 with copper. HUNT (1886) reported a preparation of boron bronze by reduction of B_2O_3 with liquid Cu. WARREN (1891) also obtained an alloy of copper with boron by electrolyzing B_2O_3 melt with liquid copper as a cathode. NISCHK (1923) produced boron alloys containing 98-99 wt% Cu by reducing the mixture of Cu and B_2O_3 with carbon. HACKSPILL (1933) reported using the aluminothermic reaction, under plant operating conditions, to produce a so-called copper boride with 12-15 wt% boron and at least as much aluminum.

Other early papers deny the possibility of interaction between copper and boron as in the data presented by TUCKER and MOODY (1901, 1902). BINET du JASSONEIX (1909) succeeded in isolating inclusions in the copper matrix by its electrolytic dissolution in KCN. However, he deduced the absence of interaction between copper and boron because the product isolated by him did not reveal any traces of H_3BO_3 after dissolution in HNO_3, using a Moissan apparatus. GIEBELHAUSEN (1915) found that amorphous boron was not trapped in the copper melt and the melting point of copper did not change when sections contain no crystals of a second phase. LOSKIEWICZ (1929) also reported no traces of interaction of Cu with boron.

The negative results in these works may be explained by the possible presence of other elements which form boron compounds and thus prevent the boron from reacting with copper or yielding ternary compounds. Moreover, we do not yet know how to isolate the products of boron interaction with copper from the copper matrix.

Thus, the results of early studies showed that, while direct interaction of boron with copper is inhibited, they can form alloys. At the same time, no individual phase of a Cu-B compound was isolated from the alloys obtained. Finally, no data are listed concerning the properties, structure and other characteristics of these alloys.

Since the mid-fifties a series of works has been published containing studies of the Cu-B system. The results of these studies are recorded in Table 1. On the basis of these data, Figure 1 represents a tentative phase diagram for the Cu-B system. Interaction within this system is confirmed in nearly all these works. However, there is disagreement in the interpretation of some results and in the classification of the interaction products. The question is whether the interaction products should be considered a chemical compound (copper boride) or a solid solution of copper in boron. With these considerations in mind, the following is an analysis of the main results of these works.

The classic work of LIHL and FEISCHL (1954) on the copper-boron system is a thorough study of several ways to obtain copper-boron alloys. The authors claim to be the first to isolate a compound from the alloys. This dark, powdered compound was obtained by anodic dissolution of the copper matrix in an acidic solution of $CuSO_4$ at a current density

Table 1. Composition, structure and properties of the phases from Cu-B system

Properties \ Phase		CuB_{22} LIHL and FEISCHL (1954)	$CuB_{\sim 34}$ REXER and PETZOV (1970)	$CuB_{\sim 40}$ CARLSSON and LUNDSTRÖM (1970)	$Cu_2Al_{2.7}B_{104}$ MATTES et al. (1970)
Composition (weight %)		Cu:20.7 B :76.8 Fe: 2.18	Cu:14.6	-	Cu: 9.7 B :84.8 Al: 5.5
Space Group		-	$R\bar{3}m$	$R\bar{3}m$	$R\bar{3}m$
Cell	a,A°	-	11.08	10.9657	10.99
param.	c,A°	-	24.11	23.8847	23.98
a. Volum	V, A^{o3}	-	2563.28	2487.20	2508.3
Density, g	g/cm^3	-	2.57	-	2.64
Micro-hardness,	kg/mm^2	1115 (P=120g) 1260 (P=54g)	-	4180 (P=50g)	2500 (P=50g)

0.3 A/dm^3. In sections the compound can be seen as fairly large blue-grey crystals arranged in an irregular polyhedra. Chemical analysis attributes to it a formula CuB_{22}. The authors admit that the compound may undergo a partial change in composition during the process of anodic dissolution of the matrix. This paper was the first to confirm the formation of a eutectic between the solid solution of boron in copper and CuB_{22}. The solubility of boron in copper in the solid state is very low, constituting 0.06 wt% at room temperature, and 0.09 wt% at $T_{eut.} = 1060 \pm 2^{\circ}C$. As the boron content increases, the hardness of copper increases from 50 to 103 Kg/mm^2 at its eutectic concentration. This paper also presented for the first time the part of the copper-boron phase diagram with the eutectic (2 wt% boron).

The authors obtained X-ray powder patterns of the copper boride CuB_{22} which exhibited a large number of lines. Unfortunately, they did not specify their interplanar spacings and reflection intensities. Moreover, the lines are unindexible in both tetragonal and hexagonal structures, and do not fit the set of lines of any known borides. The authors believe that this indicates the formation of an individual low symmetry structure. However, when these data were published, no standard X-ray powder diffraction patterns were known for various crystallographic modifications of boron. Therefore, the X-ray powder patterns of CuB_{22} might be similar to one of the boron modifications. SMIRYAGIN and KVURT (1965), using purer starting material, have confirmed the existence of CuB_{22}. They obtained a final product with a microhardness of 2500 Kg/mm^2 (P = 50g), which is a factor of two higher than the value measured in the preceding work. They also established the existence of a eutectic and a lower solubility of boron in copper in the solid state. However, their results are somewhat incomplete. They did not isolate the boride from the copper matrix and did not carry out its chemical and X-ray diffraction analysis.

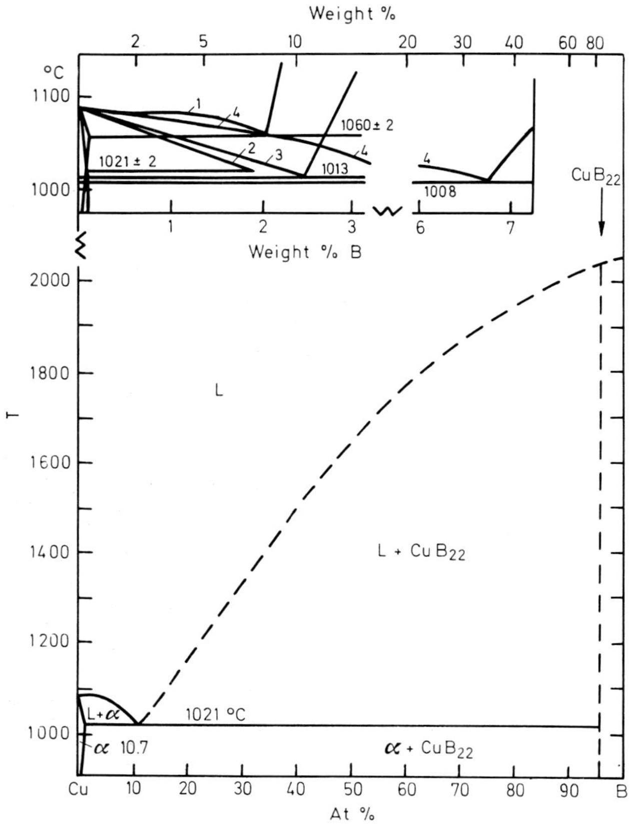

Fig. 1. Phase diagram of the copper-boron system and summary of data for its copper-rich portion (at top)

Later works indicate the existence of a boron-rich phase close in composition to CuB_{22} representing, in the opinion of the authors, a solid solution of copper in boron. WALD and STARMONT (1965) obtained alloys of copper with zone-melted boron by reaction of the metal and boron in the liquid state. These authors confirmed the existence of the eutectic and showed that part of it forms α-rh boron. Also, confirming the low solubility of boron in copper in the solid state, the lattice period of copper revealed almost no change ($a = 3.614 \pm 0.002$ Å). The authors disclaim the existence of the compound CuB_{22}, while attributing the effects observed in the thermograms at 1022°C to a polymorphic transformation of boron and the associated change in the solubility of copper in boron. The X-ray powder patterns of the alloys obtained by the authors indicate a transformation of the original β-rh boron into α-rh and β-QB. The latter reveals an increase of the lattice period a (by 1.22 Å) at a constant c. In our opinion, the conclusions of the authors (WALD and STARMONT, 1965) are too categorical. First, the formation of a boron-rich phase with the β-QB structure and an increased parameter may indicate inclusion of copper. As shown by HOARD and HUGHES (1967) this modification can exist only in the presence of interstitial atoms in its structure. Second, the formation of copper boride, given actual conditions of the work in question, seems highly unlikely since the processes leading to the inclusion of copper in boron should proceed more easily in the presence of solid boron and liquid copper rather than when they are mixed in the liquid state. Finally we should mention a discrepancy in this work. The authors state a eutectic at >2 wt% boron while the phase diagram presented in the work indicates the eutectic as containing ~32 at % boron (which is ~7 wt% boron).

ROSSO (1969) studied the interaction of copper with α-rh B and established the existence of an allotropic high-temperature transition of α-rh into β-rh boron induced by copper. The author rejects the existence of CuB_{22}. After boiling in 30%-HNO_3, the powder isolated from the copper matrix contained 0.1-0.3 wt% copper, while its X-ray powder diffraction lines revealed only boron lines: β-rh and β-QB phases in samples produced at 1300-1400°C, and β-rh phase at T>1400°C. One should point out that dissolution of the copper matrix in boiling HNO_3 could have resulted in a considerable decrease of the copper content in the reaction product. This was found by LIHL and FEISCHL as early as 1954. Therefore, ROSSO was unable to isolate CuB_{22}.

REXER and PETZOV (1970) produced copper boron alloys by arc-melting under purified argon. They established the existence of solid solutions of copper in boron and a eutectic between the copper and boron phase, which was isolated from the matrix. Unfortunately, they did not specify their method of isolation. The results obtained seem to imply that this phase represents a solid solution of copper in β-rh boron, with an ordered inclusion of 3 copper atoms in a unit cell made up of 104-105 boron atoms. The cell parameter increases in relation to the copper content. It is of interest that the copper content in the phase, as determined with a microprobe, is higher than that established chemically. The authors attribute this to inaccurate quantitative calibration in the microprobe technique. However, this difference may also have been due to partial dissolution of the phase during its isolation from the matrix. Nevertheless, chemical analysis shows that the formula CuB_{34} can be ascribed, which is sufficiently close in composition to CuB_{22}.

Information on the boron-rich phase can also be found in other works. RAO and ANDERSON (1971) consider the copper-boron system as a conventional eutectic. From the data obtained by electron microprobe analysis, they found the alloys to be composed of nearly 100% boron with traces of copper and carbon. In the authors' opinion, this does not permit its positive identification. CARLSSON and LUNDSTRÖM (1970) established the existence of a CuB_x phase representing a solid solution of copper in β-rh-boron. This causes an increase both in the parameters a and c of the hexagonal unit cell and in its microhardness.

The solid solution in question is assumed to be saturated and to have the composition $CuB_{\sim 40}$. LUNDSTRÖM (1974) also reported obtaining a solid solution of copper in β-rh boron of composition $CuB_{\sim 28}$[1]. MATTES et al. (1970) prepared a boron-copper melt containing aluminum, cooled it slowly and were able to isolate blue-grey crystals by dissolving the matrix in 50% HNO_3. From the data of chemical and electron microprobe analysis, they determined the composition of the crystals to be $Cu_2Al_{2.7}B_{104}$. In the authors' opinion, the existence of the CuB_{22} is quite probable, however, it should be regarded as a solid solution of copper in β-rh boron rather than as a stoichiometric boride.
PERRY et al. (1973) have apparently obtained a solid solution of copper in β-rh boron in the form of inclusions in the copper matrix. These were isolated by dissolving the copper matrix in concentrated HNO_3. The

[1]Some new data on the composition and structure of such solid solutions are recently reported by PITON et al. (1974), ANDERSSON and CALLNER (1974) and LUNDSTRÖM and TERGENIUS (1976).

authors suggest a reduction of copper content in the solid solution during the isolation process. They also consider the process to be responsible for a slight increase in the unit cell parameters of solid solution of copper in β-rh boron obtained as compared with that of β-rh B.

Finally, we have obtained blue-grey hypidiomorphic crystals by cooling a melt containing 55g Cu, 45g Zn and 2g B (see Fig. 2a). The crystals possessed a high microhardness close to that of CuB_{22} (H_m = 2710 ± 130 kg/mm^2) (P=30g). However, we did not succeed in isolating them by dissolving the matrix in 30% HNO_3. Thus, their composition was not established.

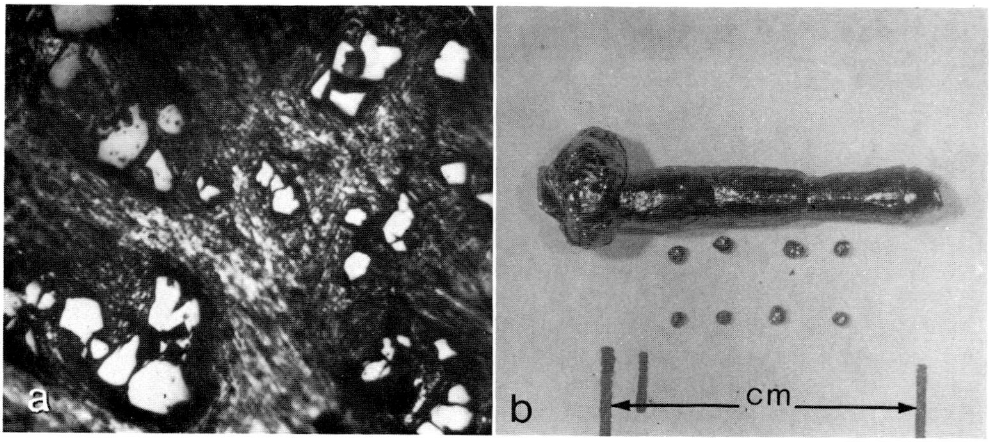

Fig. 2a and b. Compounds in Cu-B and Zn-B systems.
(a) Microstructure of Cu-Zn-B alloy containing crystals (white) of boron-rich phase (x 450). (b) Photograph of zone-melted β-rhombohedral boron rod and ZnB_{22} particles obtained from it (x4)

In summary, an analysis of the above works permits us to draw the following conclusions:

1. The products of interaction of copper with boron are:
a) A eutectic with the most probable content of:
B = 2 wt% and T_{eut} = 1021 °C (see Fig. 1)
b) A chemical compound with a structure of interstitial solutions with a broad range of homogeneity; its upper limit corresponds approximately to composition CuB_{22}, and the lower limit to $CuB_{\sim 35}$. This compound is close in structure to β-rh boron and contains statistically ordered interstitial copper atoms (see Table 1).
c) Solid solutions of boron in copper containing 0.01-0.06 wt% boron at room temperature, and 0.05-0.09 wt% boron at T_{eut}.

2. Except for CuB_{22}, the copper-boron system can form phases with a lower copper content CuB_x (when x ⩾ 35-40).

3. Formation of CuB_{22} and CuB_x phases may take place by copper atoms entering both the interstitial and substitutional mechanisms.

It should be stressed that optimum conditions apparently exist for the formation of the CuB_{22} and CuB_x phases. These conditions consist of high temperatures (but not exceeding the boron melting point) and a definite size of crystalline boron grains. When grains of this size are present, the process of their saturation by copper proceeds most intensively and completely with the formation of the CuB_{22} or CuB_x type of phases.

The practical application of the products of interaction in the Cu-B system can be envisaged in the development of composites. Indeed, PERRY et al. (1973) report the production of boron-fiber reinforced copper alloys by directional solidification of the copper-boron eutectic (with 2 and 2.54 wt% boron). The authors showed that the fiber-like inclusions of boron formed in the alloy possess high VTS (roughly about 185 kg/mm^2). The inclusions cause the properties of the composite to be changed and strengthened. For example, conductivity is reduced to 83% and the strength is increased by 75% (in comparison with pure copper). The data of this work support the suggestion of CARLSSON and LUNDSTRÖM (1970) that it is possible to alloy boron fibers and whiskers with copper to increase their mechanical strength. Furthermore, BREDZS and SCHWARTZBART (1961) established that alloys of 78% Cu, 20% Zn, 1.5% B, 0.5% Li can be used for welding metal in air without fluxes. It may also be assumed that boron alloyed copper, particularily those close to the eutectic composition, can possess high corrosion resistance. Practical applications may also be found for CuB_{22} when its properties are better known.

b) Ag-B System

Most of the authors who have studied this system agree that silver and boron never interact. Others point out the possibility of obtaining alloys, while only one claims obtaining a compound.

Interaction between silver and boron was first mentioned by WARREN (1890) who succeeded in adding up to 6 wt% boron to metallic silver. Ingots of this alloy exhibited greater chemical inertness to reagents than pure silver. BREDZS and SCHWARTZBART (1961) obtained alloys which did not wet steel by alloying pure silver with commercial ferroboron containing 11% boron. However, neither WARREN nor BREDZS and SCHWARTZBART presented any information on the actual form of boron in the alloy. According to the authors of both studies, the presence of a fairly large amount of impurities in the alloys is favorable for the formation of alloys.

Only OBROWSKI (1961) has obtained the compound AgB_2 at a high temperature. X-ray measurements showed it to have a C32 type hexagonal lattice with a period a = 3.14 Å and c/a = 1.12. However, the author does not present any information on the actual experimental conditions, nor does he explain the technique used to isolate the boride crystals. Furthermore, he does not disclose the composition of the starting material and the data obtained from the chemical analysis of the compound. It may be added that MOISSAN (1892) already noted the ease with which silver can be melted with boron.

Some of the workers disclaim flatly any melting or interaction of silver with boron. Among those works are the studies of BINET du JASSONEIX (1909) and LOSKIEWICZ (1929). Other workers reports are conflicting and do not seem to provide conclusive proofs of the absence of interaction between silver and boron. GIEBELHAUSEN (1915) noted the absence of solubility of amorphous boron in liquid silver at 1500-1600 $^{\circ}$C. Yet the reddish color of the samples may indicate the presence of a second phase. However, examination of the sample sections did not reveal its presence. ROSSO (1969) reported the absence of any compounds or eutectic, but chemical analysis of the inclusions isolated from the silver matrix showed them to consist of boron containing 0.1-0.2 wt% silver. At $T<1400^{\circ}$C this boron is β-QB, while at $T>1400^{\circ}$C it is β-rh B. Since the method of isolating inclusions from the silver matrix by boiling in 30% HNO_3 could result in a considerable reduction of the silver content, his results could be erroneous. WALD and STARMONT (1965) noted almost complete immiscibility in the liquid, and very low solubility in the solid state. In their opinion, the phase diagram of the Ag-B system is monotectic. They believe that the AgB_2 compound does not exist in the system, since annealing an alloy containing 66-2/3 wt% boron at $900^\circ C$ for one month did not yield any results. However, as already pointed out for the Cu-B system, the mixing liquid silver and boron could inhibit formation of compounds or solid solutions between them, thus leading the authors to erroneous conclusions.

In conclusion, the data of most works indicate that Ag-B is monotectic with immiscibility of the components in the liquid, and very low solubility in the solid state. Existence of boron compounds or solid solutions cannot be clearly determined without further study.

Since knowledge of their properties is inadequate the practical applications of products of interaction in this system receive little attention in the literature. BREDZS and SCHWARTZBART (1961) proposed the use of a silver-based alloy with lithium and boron as an air resistant solder.

c) Au-B System

All the authors who have dealt with this system agree that gold and boron do interact.

LOSKIEWICZ (1929) was the first to discover an alloy of gold and boron and to suggest that they exhibited two-component phases. He also presented data on microhardness which differed over the ingot, but produced no other studies.

MULLER and MERL (1959) confirmed that boron was soluble in gold at high temperatures. In the solid state the solubility was found to be ~0.5 wt% boron. Except for the crystals of the α-solid solutions, they observed in the sections crystals of unknown type which they believed to be either boron or an Au-B compound. The wire they drew from the alloys revealed an enhancement of VIS by more than a factor of two, microhardness was increased by about a factor of 7, and electrical resistivity was increased by more than a factor of 6 as compared with pure gold.

As we have seen, OBROWSKI was the only author who obtained a compound in the Ag-B system. He repeated this success in the Au-B system. Using X-ray data, he prepared the formula of AuB_2 with C32- type hexagonal lattice, and the parameter a = 3.00Å, a/c = 1,08. However, just as for AgB_2, he does not specify the natural conditions of synthesis, the method of isolation, and the analytical procedure used with the boride obtained. In later works, no one succeeded in obtaining AuB_2.

WALD and STARMONT (1965) established formation of a eutectic at $T = 1056^\circ C$ and Au concentration above 95 at%. At boron content ~50 at% in the liquid state, there exists a region of miscibility between boron and gold, while the solubility of boron in gold in the solid state is very low. The X-ray diffraction data on the inclusions isolated from the gold matrix by dissolving it in mercury are used to draw a conclusion on the low solubility of gold in boron in the solid state, and on the absence of the compound AuB_2.

On the basis of the above works one may conclude that the system Au-B belongs to the type of simple eutectics with a low solubility of components in one another in the solid state. However, it should be pointed out that the formation of compounds in this system is not impossible, so that this problem awaits further investigation.

The practical applicability of Au-B alloys was reported by MÜLLER and MERL (1959). They also proposed to use them in semi-conductor technology for the connection of p-n junction semi-conductor devices. They also suggested their use in other areas, for example, in the electrochemical industry and the jeweller's art.

In conclusion, with the exception of the above original and review papers, one may find information on the B-Ib systems in the handbooks by HANSEN and ANDERKO (1958), VOHL (1959), ELLIOTT (1965) as well as in the theses by TSAGAREISHVILI (1971) and KUZ'MA (1973).

4. Borides of the IIb Subgroup

Since little is known about the B-IIb systems, there are no reviews on them in the literature.

a) Zn-B System

MARSDEN (1880) and GIEBELHAUSEN (1915) claimed that boron did not interact with liquid and boiling zinc even though the final products were not analyzed and the purity of the starting material was not specified. The authors of these works did not succeed in detecting and isolating a compound of boron with zinc. MARKOVSKY and KONDRASHEV (1957) heated a mixture of elemental boron with pure zinc powder for several hours in sealed quartz ampoules at 800-1100°C. After cooling, the X-ray powder patterns of the product revealed a change in the intensity of the zinc lines, which could indicate boron entering the zinc lattice. However the authors did not attempt to separate the reaction products. FEDOROV et al. (1960) have proposed a method of obtaining pure boron by reducing BCl_3 with zinc. They did not observe any interaction between boron and zinc but they did mention that purified boron contained traces of zinc, aluminum and silicon. The possible existence of a Zn-B compound was first pointed out by NEWKIRK and HURD (1955) who studied the formation of boron hydrides in the reaction of H_2 with various borides. They concluded that when Zn reacts with B_2O_3 at 250°C, it forms a boride which interacts with H_2. This interaction of the boride with H_2 forms boranes by analogy with the boride interaction of other elements with H_2. However, zinc boride was not isolated in this case.

The first person to synthesize and isolate a compound of zinc with boron was GURIN (1966). The precise composition of the compound, the conditions required for the synthesis, its properties, and some aspects of its practical applications have been studied by KORSUKOVA et al. (1971, 1972). These authors used relatively pure starting materials: zinc in granules, 99.99%, amorphous boron (not less than 99%), and zone-melted β-rh B (rods with metallic impurity content not more than 10^{-5} wt%).

The conditions required for the formation of the compound of zinc with boron were studied by heating mixtures of boron with excess zinc (Zn: B $\geqslant$ 6). The heating process took place in sealed quartz ampoules or quartz tubes, under purified dry argon with mechanical agitation at 500-1000°C, for 1 min to 96 h. The mixture was subsequently quenched in water, or cooled slowly. The final product represented a blackish-brown powder outside the zinc ingot. To remove excess zinc, a section was prepared from the ingot and treated with HCl (1:1). The powder was filtered, washed and dried and then subjected to the X-ray and chemical analyses. The technique of chemical analysis and the chemical properties of the compound, as well as the intensities and interplanar spacings derived from the powder X-ray reflections, are described by

KORSUKOVA et al. (1971). A metallographic study of the zinc sections revealed an absence of any eutectic and a noticeable solubility of boron in solid zinc (the microhardness of Zn did not change). One could observe only a few inclusions. They were of bluish color, 5-12 μ in size and a microhardness of ~2880 ± 350 kg/mm^2. These findings suggested to KORSUKOVA that the impurities were of the boride type. Since their amount was very small, no chemical analysis of their composition was carried out.

On the basis of the chemical analysis performed (Zn-24.58÷18.55 wt%, B-75.41÷81.18 wt%, Zn+B-99.99÷99.73 wt%) we ascribed to the resulting compound the formula $ZnB_{\sim 18-26}$ or ZnB_{22} on the average. This compound was produced by heating amorphous boron in a temperature range of T_{boil}. Zn - 700°C for several hours. We also obtained this compound from rods of zone-melted β-rh B under the same conditions (see Fig. 2). The rods became dispersed into granular bright black particles ranging from 2 to 2.000 μ in size (Fig. 2b). After treating the particles preliminarily with HCl(1:1), our chemical analysis revealed the overall content of zinc to be 20-23 wt% which corresponds to the formula ZnB_{20-24}, for an average of ZnB_{22}.

As obtained from both amorphous boron and β-rh B rods the X-ray powder patterns of this compound are identical and consist of a set of lines different from those of boron. This set of lines is consistent while the zinc content in the product decreases down to ~13-14 wt% which corresponds to the formula $ZnB_{\sim 35-40}$. This indicates a broad range of homogeneity of ZnB_{22} whose structure, judging from the X-ray powder pattern, is of low symmetry. When X-ray diffraction measurements were carried out on single crystal granules of ZnB_{22}, the granules proved to be the same structural type as β-rh B. As a result of zinc atoms entering the boron lattices, parameters a and c of the hexagonal unit cell increase (see Table 2). Unfortunately, the actual arrangement of the zinc sites and the resulting distortions of the boron lattice were not studied.

The data on the structure and some of the physical properties of ZnB_{22} are listed in Table 2. The data shows the properties of ZnB_{22} to be close to those of pure boron. Moreover, ZnB_{22} decomposed in vacuum at 1000°C into zinc and a product containing 0.2-7 wt% zinc. The X-ray powder patterns of these products of thermolysis are identical with those of β-rh B.

We attempted to identify the zinc boride obtained. To do this we used a DS-4039 spectrophotometer to record the IR absorption spectra of powders of ZnB_{22} (particle size 1-3 μ), amorphous boron, and $ZnB_{\sim 40}$ (the lowest limit of homogeneity) in a KBr matrix. As seen from the spectra (Fig. 3), the strong absorption band of amorphous boron and $ZnB_{\sim 40}$, which is of less intensity than boron, practically disappears for ZnB_{22}. The same difference is observed in the spectra of β-rh B and the ZnB_{22} obtained from it. We believe this band to be associated with the lattice vibrations of β-rh B, since its disappearance in ZnB_{22} is connected with a change in the boron structure caused by zinc atoms entering the lattice.

The similarity of the ZnB_{22} and β-rh B structures is confirmed indirectly by the curves depicting zinc interaction with various crystal modifications of boron (see Fig. 4). The differences in the structure of these modifications result in different zinc content. α-rh B can contain only 6-7 wt% zinc (curve 3) while amorphous and β-rh B can contain up to 25 wt% (curves 1 and 2). The same amount of interstitial zinc may serve as an indirect proof of the similarity between the

Table 2. Structure and properties of ZnB_{22}, β-rh B and amorphous B

Properties \ Phase		ZnB_{22} crystalline	ZnB_{22} powder	β-rh B TSAGAREISHVILY (1972)	B amorph. SAMSONOV (1960)
Space Group		R $\bar{3}$ m	-	R $\bar{3}$ m	-
Cell	a, Å	11.00±0.04	-	10.94	-
param.	c, Å	24.04±0.08	-	23.81	-
a. volume	V, $Å^3$	2519.05	-	2467.8	-
Density,	g/cm^3	-	2.86±0.04	2.34	2.30
Microhardness	kg/mm^2	3510±210 (P=30g)	-	3500 (P=200g)	-
Resistivity, om·cm,	$T_{298}K$	$2.5 \cdot 10^3$	$2.5 \cdot 10^6$	$1 \cdot 10^6$	$1 \cdot 10^6$
Type of conduction		P	P	P	P
Magnetic susceptibility χ_g $10^6 cm^3 \cdot g^{-1}$, $T_{298}K$		-	-0.64±0.02	-	-0.62

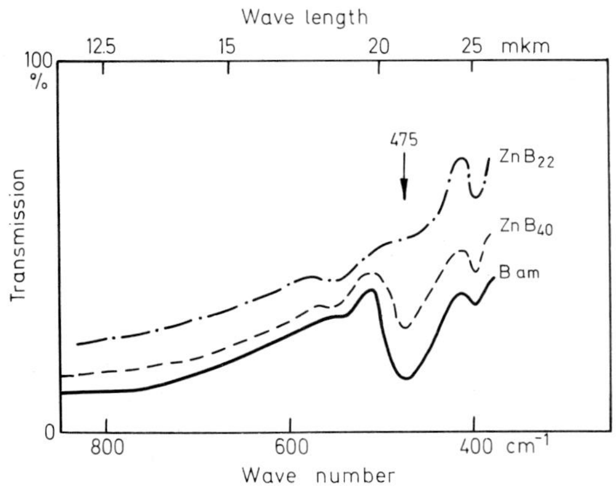

Fig. 3. Infrared absorption spectra of amorphous boron, ZnB_{22} and $ZnB_{\sim 40}$

structures of amorphous and β-rh B. This was suggested by BADZIAN (1970) who considered amorphous boron as a frozen stage in the transition from α-rh to β-rh B.

The highly dispersed state of the starting amorphous boron powder should facilitate heterophase reactions. The consequences of this highly dispersed state are greater interaction between amorphous boron and zinc, and a greater zinc content in the final product than in β-rh B.

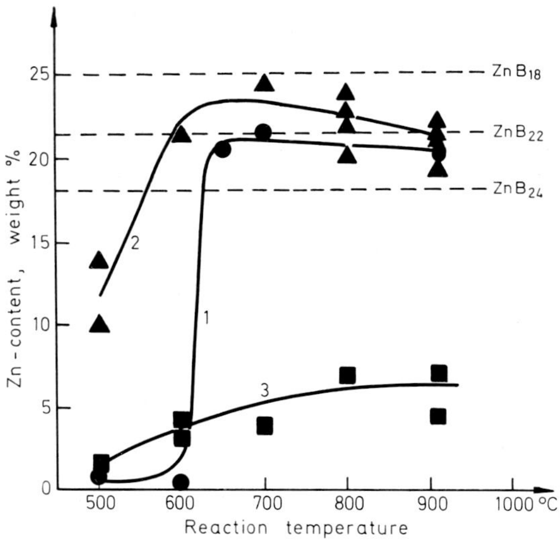

Fig. 4. Zn-content in various boron modifications vs reaction temperature. *Curves*: 1 - β-rh B, 2 - amorphous B, 3 - α-rh B

The data obtained permit the conclusion that zinc atoms enter the openings in the boron lattice. We recognize, however, the possibility of ordered occupation of interstitial sites or even substitution of boron atoms. The ZnB_{22} phase thus formed is similar to interstitial solutions such as CuB_{22} with upper and lower homogeneity range limits of ZnB_{22} and $ZnB_{\sim 40}$. This phase is a chemical compound because of its qualitative difference (judging from its X-ray powder patterns and the IR spectra) from the structure of boron and because the solid solutions of zinc in boron contains less zinc than the lower limit of the zinc boride homogeneity range. Also, the compounds CuB_{22} and ZnB_{22} can be considered as interstitial structures which are derivatives of β-rh B, as was suggested for SiB_{14} and AlB_{12} by GIESE et al. (1965), and by ECONOMY and MATKOVICH (1970). The details of this analogy may be refined when we ascertain the arrangement of the ZnB_{22} structure[2].

The possible practical application of ZnB_{22} has not been studied adequately. KORSUKOVA et al. (1972) investigated its possible use as resistor material in connection with its high electrical resistivity. However, its fairly large negative temperature coefficient of resistivity, which is not reduced by alloying ZnB_{22} with metals (Al, Cu, Mg, Ag, Mn) prevents it from being employed in resistor technology. Furthermore, investigation of its properties may reveal its use in semiconductor technology, heterophase catalysis, etc.

In conclusion, the existence of a chemical compound ZnB_{22}, and of solid solutions of zinc in boron was established in the Zn-B system, and the solubility of boron in zinc in the solid state is practically zero.

b) Cd-B System

Only two papers deal with the interaction in this system.

---

[2]Some new data on the interaction of zinc with various modifications of boron are reported by KORSUKOWA et al. (1977).

MARKOVSKY and KONDRASHEV (1957) studied the Cd-B system in a manner similar to that used by them in their work on the Zn-B system. They did not detect any new compound or a solubility in the solid state in the Cd-B system.

KORSUKOVA et al. (1972) established that zone-melted β-rh B does not react with molten cadmium, while amorphous boron forms a blackish-brown powder containing about 5 wt% cadmium at T = 800°C. At 400°C cadmium enters boron at less than 1 wt%. The analysis and chemical properties of this product are similar to those of ZnB_{22} (see above). Its X-ray powder patterns showed it to be practically amorphous, while the IR absorption spectrum is the same as the starting amorphous boron. This may indicate a stability in the boron structure while cadmium atoms enter the lattice.

In short, no compounds seem to exist in the Cd-B system. In the solid state, boron is practically insoluble in cadmium, while amorphous boron dissolves (as soon as boiling point of Cd is reached) about 5 wt% cadmium. It should be stressed that the data obtained do not provide answers to all questions concerning the interaction in this system. These questions undoubtedly deserve further study.

No areas of practical application of the CdB_x powder produced have been found. Its use as resistor material like ZnB_{22} turned out to be unpromising.

c) Hg-B System

The first and apparently only study of interaction in this system was carried out by DAVY (1809). He proposed the absence of reaction between boron and mercury when heated. At the present time no other information exists on this system.

5. Conclusion

The experimental data presented above support the theoretical predictions concerning the interaction in the B-Ib and IIb system. In other words all Ib and IIb subgroup elements may form chemical compounds, solid solutions, or eutectics with boron. The only exception appears to be mercury but the Hg-B system remains practically unexplored. The experimental data show that the interaction intensity of the Ib and IIb elements with boron decreases as we proceed from copper to gold, and from zinc to mercury. An exception in the Ib subgroup is that silver interacts less actively with boron than do copper and gold. Higher energy is required to remove two or three electrons from a silver atom than from copper and gold. At the same time, the divalent state which afterwards is more easily reached by copper and gold than by silver favors the formation of compounds with boron. LIPSCOMB and BRITTON (1960) proposed that to stabilize the boron framework in the formation of borides requires a transfer of two electrons from the atom of metal to the boron atoms. As for the IIb subgroup elements, experiments have confirmed that the intensity of their interaction with boron decreases as we progress from zinc to mercury. This finding is at variance with the amount of energy required to remove two electrons from their atoms ($U^{Zn^{2+}}$ = 1.2370, $U^{Cd^{2+}}$ = 1.1340, and $U^{Hg^{2+}}$ = 1.1323R). This may be attributed to the reverse course of variation of their atomic and ionic radii. They increase from zinc to mercury while for cadmium and mercury, they are much larger than for zinc. This may strongly inhibit their diffusion in boron. Consequently, cadmium and especially mercury are expected to exhibit weaker interactions with boron.

References

ANDERSSON, S., CALLNER, B.: The solubilities of copper and manganese in β-rhombohedral boron as determined in $CuB_{\sim 28}$ and $MnB_{\sim 23}$ by single-crystal diffractometry. J. Solid State Chem. 10, 219-231 (1974)

BADZIAN, A.R.: Radial distribution function of non-crystalline boron. Electron Technol. 3, 143-146 (1970)

BINET DU JASSONEIX, A.: Recherches sur les combinaisons du bore avec metaux. Ann. Chim. Phys. 17, 145-216 (1909)

BREDZS, N., SCHWARTZBART, H.: Recent progress in development of self-fluxing, air proof brazing alloys. Welding J. 40, 123S-129S (1961)

CARLSSON, J.-O., LUNDSTRÖM, T.: The solution hardening of β-rhombohedral boron. J. Less-Common Metals 22, 317-320 (1970)

COWLES, E.H., COWLES, A.H., MABERY, C.F.: On the electrical furnace and the reduction of the oxides of boron, silicon, aluminum and other metals by carbon. Am. J. Sci. 30, 308-312 (1885)

DAVY, H.: An account of some new analytical researches on the nature of certain bodies, particularly the alkalies, phosphorus, sulphur, carbonaceous matter and the acids hitherto undecompounded; with some general observations on chemical theory. Phil. Trans. 99, 39-104 (1809)

ECONOMY, J., MATKOVICH, V.I.: Crystal chemistry of higher borides. Electron Technol. 3, 159-165 (1970)

ELLIOTT, R.P.: Constitution of Binary Alloys. New York: McGraw-Hill 1965

FEDOROV, T.F., SHAMRAY, F.I., NISELSON, L.A., PETRUSEVICH, I.V.: O poluchenii elementarnogo bora. Zh. Neorgan. Khim. 5, 226-228 (1960)

GIEBELHAUSEN, H.: Über das Verhalten des Vanadins zu Silicium, Nickel, Kupfer und Silber und das Verhalten des Bors zu Nickel. Z.anorg.Chem. 91, 251-262 (1915)

GIESE Jr., R.F., ECONOMY, J., MATKOVICH, V.I.: Interstitial derivatives of β-boron. Z. Kristallogr. 122, 144-147 (1965)

GURIN, V.N.: Issledovanie protsessa sintesa silitsidov i drugikh tugoplavkikh soedinenii perekhodnykh metallov IV-VI grupp periodicheskoi sistemy elementov v rasplave zinka. Thesis, Leningrad 1966

GURIN, V.N., OBUKHOV, A.P.: Interaction of elements of the periodic table. Poroshkovaya Met. N 12, 72-83 (1966)

HACKSPILL, L.: Recherches recentes sur les borates et sur le bore. Helv. Chim Acta 16, 1099-1118 (1933)

HANSEN, M., ANDERKO, K.: Constitution of Binary Alloys. New York: McGraw-Hill 1958

HERMAN, F., SKILLMAN, S.: Atomic Structure Calculations. New Jersey: Prentice-Hall 1963

HOARD, I.L., HUGHES, R.E.: Elemental boron and compounds of high boron content: structure, properties and polymorphism. In: The Chemistry of Boron and Its Compounds. (E.L. Muetterties, ed.) pp. 25-154. New York: Wiley and Sons 1967

HOARD, I.L., NEWKIRK, A.E.: An analysis of polymorphism in boron based upon X-ray diffraction results. J. Am. Chem. Soc. 82, 70-76 (1960)

HUNT, S.: An electrical furnace for reducing refractory ores. Chem. News, 53, 64-65 (1886)

KORSUKOVA, M.M., GURIN, V.N., KOSLOVA, I.R., TERENT'EVA, S.P., NIKANOROVA, L.E., OBUKHOV, A.P.: Poluchenie i nekotorye svoistva soedineniya bora s zinkom. Izv. Akad. Nauk SSSR, Neorg. Materialy 7, 1518-1522 (1971)

KORSUKOVA, M.M., GURIN, V.N., SOROKIN, V.N., YUSOV, Yu.P., TERENT'EVA, S.P.: Svoistva i nekotorye perspektivy ispol'sovania vysshikh boridov podgruppy medi i zinka. In: Bor. Poluchenie, Struktura i Svoistva (F.N. Tavadze, ed.), pp. 235-239. Moscow: Nauka 1974

KORSUKOVA, M.M., GURIN, V.N., KUZ'MA, Ju.B., KISKATSHY, A.Ju., SOLOVJEV, N.E.: Structurnie ismenenija v modifikatsijakh bora pod deistviem zinca. Kristallographija 22, N4 (1977), to be published

KUZ'MA, Yu.B.: Issledovaniya po kristallokhimii boridov. Thesis, L'vov 1973

LIHL, F., FEISCHL, O.: Herstellung und Konstitution von Kupfer-Bor-Legierungen. Metall 8, 11-19 (1954)

LIPSCOMB, W.W., BRITTON, D.: Valence structure of higher borides. J. Chem. Phys. 33, 275-280 (1960)

LOSKIEWICZ, W.: Cementowanie miedzi, srebra i slota berilem, krzemen i borem. Przeglad Gorniczo-Hutniczy 21, 583-609 (1929)

LUNDSTRÖM, T., TERGENIUS, L.E.: On the solid solution of copper in β-rh boron. J. Less-Common Metals 47, 23-28 (1976)

LUNDSTRÖM, T.: Strukturnye issledovaniya rastvorimosti v tverdom sostoyanii perekhodnykh metallov v β-rhombohedricheskom bore. In: Bor. Poluchenie, Struktura Svoistva (F.N. Tavadze, ed), pp. 44-51. Moscow: Nauka 1974

MARKOVSKY, L., KONDRASHEV, D.: O sostave i svoistvakh boridov elementov pervoi i vtoroi grupp periodicheskoi sistemy. Zh. Neorgan. Khim. 2, 34-41 (1957)

MARSDEN, R.S.: On a new boro-copper compound of the formula B_2Cu_3. J. Chem. Soc. (L) 37, 672-674 (1880)

MATTES, R., MAROSI, L., NEIDHARD, H.: Darstellung und Kristallstruktur von $Cu_2Al_{2,7}B_{104}$. J. Less-Common Metals 20, 223-228 (1970)

MELLOR, J.W.: A Comprehensive Treatise on Inorganic and Theoretical Chemistry, Vol V, pp. 23-24. London-New York-Toronto: Longmans, Green and Co. 1946

MOISSAN, H.: Etude des proprietes du bore amorphe. C.R. Acad. Sci. (Paris) 114, pt. 1, 617-622 (1892)

MULLER, K., MERL, W.: Untersuchungen an einigen in der Halbleitertechnik verwendeten Edelmetal-Legierungen. Elektrotechn. Z. 80, 515-518 (1959)

NEKRASOV, B.V.: Osnovy obshei khimii, Vol. 3, pp. 39-44. Moscow: Isdatel'stvo Khimiya 1970

NEWKIRK, A.E.: Elemental boron. In: Boron, Metallo-Boron Compounds and Boranes. (R.M. Adams.), pp. 233-299. New York: J. Wiley and Sons 1964

NEWKIRK, A.E., HURD, D.T.: The direct synthesis of boron hydrides. J. Am. Chem. Soc. 77, 241-242 (1955)

NISCHK, K.: Studien über Umsetzungen zwischen Metallen oder Metalloxyden und Kohlenstoff. Z. Elektrochem. 29, 373-390 (1923)

OBROWSKI, W.: Die Struktur der Diboride von Gold und Silber. Naturwissenschaften 48, 428-428 (1961)

PERRY, A.I., NICOLL, A.R., PHILLIPS, K., SAHM, P.R.: The copper-boron eutectic-unidirectionally solidified. J. Mater. Sci. 8, 1340-1348 (1973)

PITON, J.-P., VILLARD, G., LUNDSTRÖM, T.: Sur les cristaux de borure de cuivre. Compt. Rend. Acad. Sci. Paris C278, 1495-1496 (1974)

RAO, M.V., ANDERSON, R.N.: Reassessment of the copper-boron phase diagram. J. Less-Common Metals 25, 427-430 (1971)

REXER, J., PETZOW, G.: Über den Aufbau und einige Eigenschaften von Kupfer-Bor-Legierungen. Metall 24, 1083-1086 (1970)

ROSSO, J.C.: Action du cuivre et de l'argent sur le bore à haute temperature. Chim. Ind., Gen. Chim. 101, 1156-1161 (1969)

SAMSONOV, G.V., MARKOVSKY, L.Ya.: Khimiya boridov. Uspekhi Khim. 25, 190-241 (1956)

SAMSONOV, G.V., MARKOVSKY, L.Ya., ZHIGACH, A.F., VALYASHKO, M.G.: Bor, ego soedineniya i splavy. Kiev: Izd. Akad. Nauk Ukr. SSR 1960

SAMSONOV, G.V., PRYADKO, I.F., PRYADKO, L.F.: Konfiguratsionnaya model' veschestva, p. 26, Kiev: Naukova Dumka 1971

SAVITSKY, E.M., GRIBULYA, V.B.: Opyt prognozirovaniya sostava i svoistv soedinenii s pomosch'yu EVM. Dokl. Akad. Nauk SSSR 190, 1147-1150 (1970)

SCHULZE, G.E.R.: Metallphysik, p. 312-312. Berlin: Akad. 1967

SMIRYAGIN, A.P., KVURT, O.S.: Diagramma sostoyaniya med'-bor. Trudy gos. nauchno-issled. i proektnogo instituta splavov i obrabotki tsvetnykh metallov 24, 7-11 (1965)

TSAGAREISHVILY, G.V.: Issledovanie uslovii obrasovaniya, struktury i svoistv β-rhombohedricheskogo bora. Thesis, Kiev 1971

TUCKER, S.A., MOODY, R.: The production of some new metallic borides. Proc. Chem. Soc. (L) 17, 129-130 (1901)

TUCKER, S.A., MOODY, R.: The production of hitherto unknown metallic borides. J. Chem. Soc. (L) 81, 14-17 (1902)

VOHL, A.E.: Stroenie i svoystva dvoynykh metallicheskikh sistem Vol 1, Moscow: Fizmatgiz 1959

WALD, F.: On the stability of the red α-rhombohedral boron modification. Electron Technology 3, 103-108 (1970)

WALD, F., STARMONT, R.W.: Investigations on the constitution of certain binary boron-metal systems. J. Less-Common Metals. 9, 423-433 (1965)

WARREN, H.N.: Boronised silver. Chem. News 61, 231-232 (1890)

WARREN, H.N.: Electrolytic method of preparing metallic alloys. Chem. News 64, 302-302 (1891)

VII. Borides of the IVb Group

G. N. Makarenko

1. Compounds of Boron with Carbon

a) Phase-Diagram of the B-C System

The first attempts to prepare compounds of boron with carbon were described by MUHLHAUSER (1894), MOISSAN (1899) and PODSZUS (1933). The products obtained were attributed to compositions of BC, B_2C_2, B_3C, B_4C and B_6C. The results were based on bulk chemical analysis only.

RIDGWAY (1934) was the first to construct the approximate phase diagram on the basis of X-ray data and establish the existence of boron carbide with B_4C composition and determine some of its physico-chemical characteristics.

The phase diagram of the B-C system was studied by ORMONT (1950) and MEERSON and SAMSONOV (1953). MEERSON and SAMSONOV (1953) confirmed the existence of the B_4C compound ($B_{12}C_3$) by using the data of thermal, X-ray, and metallographic analysis to construct a phase diagram of this system (Fig. 1) and B_4C was assigned a variable composition.

The crystal structure of boron carbide was investigated by ZHDANOV and SEVAST'YANOV (1941, 1943), CLARK and HOARD (1943), YAMAZAKI (1957), MATKOVICH (1961a, b) and LIPP (1966). The boron carbide lattice was established as belonging to the $D3d^5$ - $R\bar{3}m$ space group with rhombohedral constants of a = 5.19 Å and $\alpha = 66^\circ18'$ or corresponding hexagonal dimensions of a = 5.60 Å, c = 12. 12°Å. The B_4C unit cell contains 15 atoms distributed among the following positions: $6h_1 + 6h_2 + 2c + 1b$ (Fig. 2).

ZHDANOV et al. (1953) suggested the possibility of substituting boron atoms for carbon atoms in the 1b position and in 1954 established the

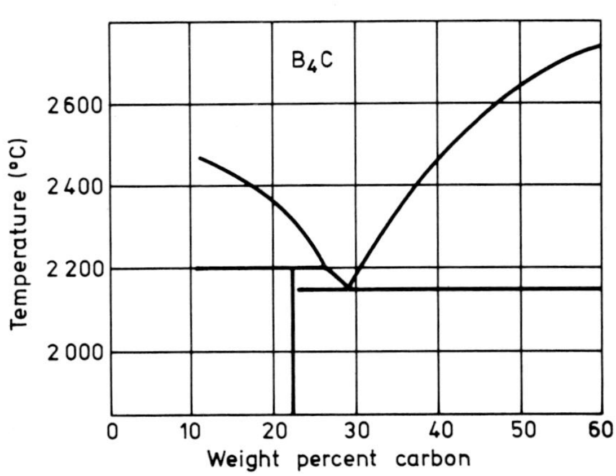

Fig. 1. The plot of phase-equilibrium diagram of B-C system (MEERSON and SAMSONOV, 1953)

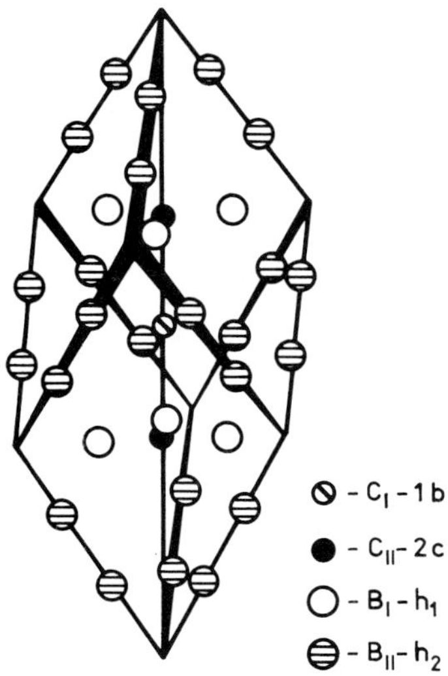

Fig. 2. B_4C structure (ZHDANOV and SEVAST'YANOV, 1941)

existence of $B_{13}C_2$ carbide. This carbide is formed by diffusion-method substitution of boron atoms in $B_{12}C_3$ for carbon atoms. Formation of $B_{13}C_2$ was also confirmed by DAWIHL (1952) and KRANZ (1963). Rhombohedral carbides $B_{13}C_2$ and $B_{12}C_3$ were prepared by PLOOG and AMBERGER (1971) using the method of precipitation from the gaseous phase. AMBERGER et al. (1971) showed that on precipitation from the gaseous phase at temperatures between 900 and 1200°C the $B_{13}C_2$ phase was predominant and thermodynamically more stable. Furthermore, an investigation of boron carbide dislocation structure by ASHBEE and DuBOSE (1972) assumes the stoichiometric composition of the carbide phase in the boron-carbon system to be $B_{13}C_2$. Only by using high-temperature annealing can hyperstoichiometric composition $B_{12}C_3$ be obtained.

In a paper by PLOOG (1973), the widely used boron carbide B_4C ($B_{12}C_3$) is assumed to be a solid solution of carbon in the thermodynamically stable compound $B_{13}C_2$. On the other hand, GLASER et al. (1953) point out the presence of a wide homogeneity range of B_4C (from 4 to 28 wt% of carbon). They showed that in this composition range the lattice period and specific gravity continuously increase while electrical resistance decreases. On this basis the authors described boron carbide as a solid solution of carbon in boron lattice. ALLEN (1953) pointed out the possibility of substitutional solid solutions in boron carbide.

Using X-ray analysis of numerous samples of various compositions of boron carbide, KUDRIAVTSEV and SAFRONOV (1960) established that the homogeneity range was due to exchange of boron and carbon atoms in the $B_{12}C_3$ lattice. According to these data, two boron atoms can be introduced into the boron carbide lattice forming, in the limiting case, a carbide with the formula $B_{14}C_3$ (19.2 wt% of carbon), and with further lattice distortion, a composition of $B_{15}C_2$ (12.8 wt% of carbon). In this, as in other papers, it is assumed that the formation of a solid solution takes place by replacing carbon atoms with boron atoms in a C-C-C linear chain. With carbon excess, the possible introduction of two additional carbon atoms in $B_{12}C_3$ leads to an arbitrary formula of $B_{12}C_5$ (31.15 wt% of carbon). Such extensive solubility of carbon into boron carbide can occur because of the presence of impurities.

By investigation of purer samples of the B-C system SAMSONOV et al. (1956) showed that carbon can dissolve in boron carbide up to 24 wt%. They also established the presence of a carbon-rich carbide at 72.5 wt% carbon. According to ZHURAVLEV, this carbide, which was called γ-phase, contains 68-69 wt% of carbon and is formed by a peritectic reaction $B_4C + C \rightleftharpoons B_nC_m$.

Systematic investigation of the B-C system alloys in the composition range of 4.5 to 90.0 wt% of carbon was carried out using pure samples by ZHURAVLEV et al. (1961). The investigation of compositions versus pyncometric density, microhardness, electrical resistivity, thermal emf., and X-ray analysis established the existence of compounds of the $B_{12}C$, $B_{13}C_2$, and $B_{12}C_3$ compositions. Figure 3 represents the phase diagram constructed on the basis of these results.

Another version of the diagram is given by DOLLOFF (1960) and ELLIOT (1970). Based on differential thermal analysis, X-ray and metallographic analysis, and microhardness measurements in the B-C system the authors established the existence of a unique compound corresponding to B_4C composition (Fig. 4, 5). The results of investigation by KIFFER et al. (1971) agree with the above findings.

Investigation by LOWELL (1967) carried out in the carbon-rich region of the phase diagram shows limited dissolution of boron in carbon. The maximum solubility comprises 2.35% at 2300°C.

Thus, at present, there is no commonly accepted definition of either phase compositions or of ranges of homogeneity in the boron-carbon system.

b) Preparation of Boron-Carbon Alloys

Boron carbide is prepared using the following methods:

1. Reduction of boron anhydride with carbon:

$$2B_2O_3 + 7C = B_4C + 6CO$$

2. Reduction of boron anhydride with magnesium in the presence of carbon:

$$2B_2O_3 + 6Mg + C = B_4C + 6MgO$$

3. Reduction of boron trichloride with hydrogen in the presence of carbon:

$$4BCl_3 + 6H_2 + C = B_4C + 12HCl$$

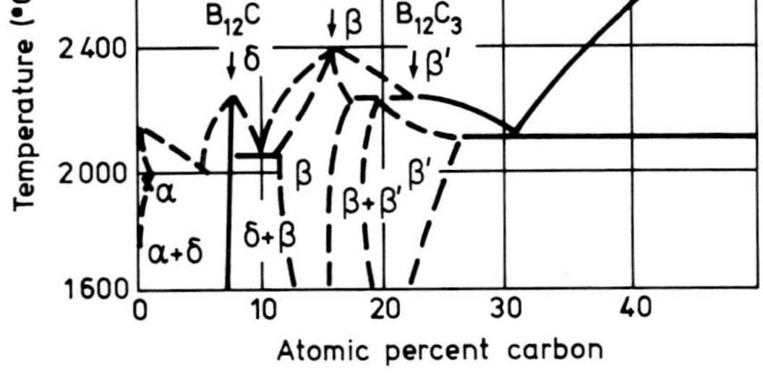

Fig. 3. Phase-diagram of the B-C system (ZHURAVLEV et al., 1961)

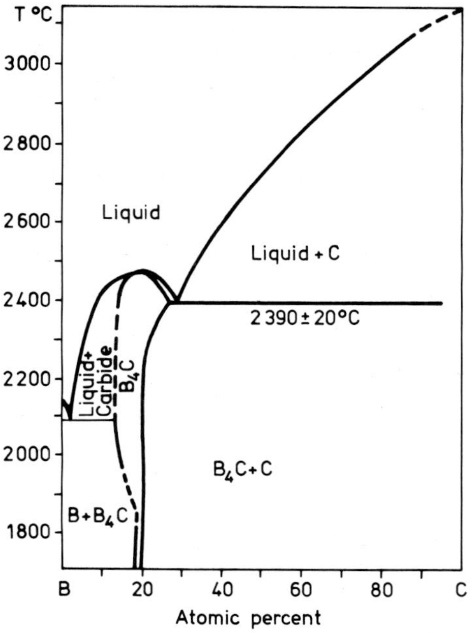

Fig. 4. Phase-diagram of the B-C system (DOLLOFF, 1960)

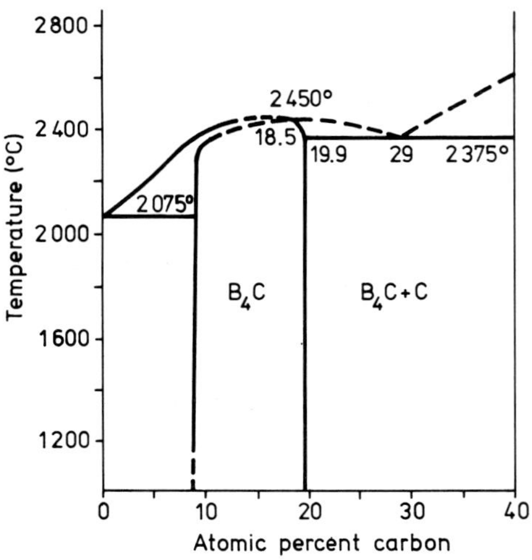

Fig. 5. Phase-diagram of the B-C system (ELLIOT, 1970)

4. Precipitation from gaseous phase:

$$BCl_3 + CCl_4 + 8H_2 = B_4C + 16HCl$$

5. Synthesis from elements:

$$4B + C = B_4C$$

Reduction of boron anhydride with carbon is used in the commercial production of boron carbide. The process takes place in two stages (SAMSONOV et al., 1960).

$$B_2O_3 + 3CO = 2B + 3CO_2$$

$$4B + C = B_4C$$

The temperature at the beginning of boron anhydride reduction with carbon monoxide (determined from pressure dissociation of B_2O_3) is 1400°C. However, as pointed out by MARKOVSKII et al. (1952), the temperature may be increased to 2000 - 2500°C to enhance the process rate.

Production techniques for boron carbide using carbon reduction differ according to the type of furnace in which the process is carried out. Boric acid and coal or petroleum coke (previously mixed and dehydrated at 800°C) are used as starting materials.

Production of boron carbide in arc furnaces is carried out at 2500°C (MARKOVSKII et al., 1952) and is based on B_4C crystallization from the melt. However, at such high temperatures, boron carbide decomposes peritectically; boron becomes volatile; and the resulting product is enriched with free carbon, the content of which reaches 10-15 wt%.

In boron carbide production (in core-type and coreless-type furnaces) (REKOV and SAMSONOV, 1961) thermal reduction of boron anhydride by carbon takes place without melting or with only partial melting. In such furnaces the charge is distributed at three temperature zones. The product of the central high temperature zone of the coreless-type furnace represents boron carbide, which after grinding and refining contains no more than 0.6-0.8% free carbon (SAMSONOV et al., 1960). Temperature increase to above 2200°C in the central zone of the core-type furnace leads to an increase in free carbon content in boron carbide because of its peritectic decomposition. Products of the second and third zones are used as semiproducts for subsequent charges of the furnace.

An advantage of boron carbide production in graphite-tube furnaces is uniform heating over the whole volume at any given temperature. These furnaces reach temperatures of 1700-1800°C, and the process can be carried out in a protective atmosphere. Fine-crystal boron carbide with particle size of 0.5-5 μ produced by this method possesses a composition close to the stoichiometric one; the disadvantage of this method is low productivity.

The *magnesium-thermal method* of boron carbide production is based on the reaction:

$$2B_2O_3 + 6Mg + C = B_4C + 6MgO$$

In the papers by MEERSON and SAMSONOV (1958), SAMSONOV (1958) it was shown that boron carbide of stoichiometric composition can be formed by heating definite charges at temperatures of 1000-1200°C for one hour. The product obtained is freed from impurities of magnesia by treatment with hydrochloric or sulphuric acid and subsequent washing with hot water. Boron carbide powder has a bulk density of 0.1-0.2 g/cm^3 and is composed of fine particles in the order of 0.5-1.5 μ. This method is used in laboratory experiments.

Production of B_4C by *reduction of boron trichloride with hydrogen in the presence of carbon* is carried out by heating a charge of boron anhydride and carbon in a flow of chlorine at 600°C. The resulting BCl_3 and CO are mixed with hydrogen and passed over the incandescent tungsten filament on which deposition of the boron carbide layer takes place (SAMSONOV et al., 1960).

The method of precipitation from the gaseous phase consists of the interaction of boron halogenides with carbon-containing compounds in the presence of hydrogen. Formation of boron carbide takes place on hot substrate in the form of a dense pyrolytic precipitate of lamellar columnar structure (NESHPOR et al., 1970; PLOOG and AMBERGER, 1971) or mono-crystals (MIERZIEYEWSKA and NIEMYSKI, 1965). Under conditions of low temperature plasma, this process results in super-fine powder of boron carbide with a particle size of 0.01-0.1 μ (CRUIZIAT, 1970).

Amorphous boron and lamp black are used as starting materials for the production of boron carbide by *synthesis from elements*. To obtain boron carbide powder this process is carried out in vacuum at 1800°C in electric resistance furnaces. Directly compacted samples may be obtained by hot pressing in a protective atmosphere of argon at 2000°C (MAKARENKO and SAMSONOV, 1958). In this case, boron carbide is formed by mutual diffusion of boron and carbon where boron diffusion in carbon is predominant (KISLII and SAMSONOV, 1960). The method of synthesis from elements results in comparatively pure B_4C of strictly stoichiometric composition, but economically, it is inefficient and finds no practical application.

c) Sintering of Boron-Carbon Alloys

Production of dense material from boron carbide using sintering of previously compacted blanks is a rather difficult problem. Boron carbide's low plasticity and high slip resistance on grain boundaries (up to temperatures close to melting point) combined with low-surface tension in the solid state impedes the sintering process (SAMSONOV et al., 1960). The use of various additives including metals, glass, or alumina, to intensify this process does not enable the formation of articles with more than 78% theoretical density (JANES and NIXDORF, 1966; GLASSON and JONES, 1969).

To obtain material of greater density, sintering is carried out by simultaneous application of high temperatures and pressure. Porosity in the order of 0-2% can be obtained by hot pressing at 2000-2100°C for 5-7 min, under pressure of 150-200 kg/cm^2. To avoid contamination with oxygen and nitrogen, hot pressing is performed in a protective atmosphere of argon (ZHURAVLEV et al., 1961). A detailed description of boron carbide compacting is found in a report by SAMSONOV and KOVAL'CHENKO (1962)

d) Properties of Boron-Carbon Alloys

Boron carbide is a dark grey to black crystalline powder with crystals in the form of rhombohedral needles or hexagonal plates.

The heat of formation of boron carbide determined calorimetrically by combustion in oxygen (SMITH et al., 1955) in fluorine (DOMALSKY and ARMSTRONG, 1967) and chlorine (GAL'CHENKO et al., 1970) is 13.8 ± 2.7 kcal/mole, 17.1 ± 2.7 kcal/mole and 13.7 ± 3.5 kcal/g - atom of C respectively (in the latter case boron carbide had a composition of $B_{4.233 \pm 0.043}C$).

Thermodynamic characteristics of boron carbide were evaluated in a paper by STULL (1965) based on experimental data of low-temperature heat capacity (KELLY, 1941) and temperature-dependence enthalpy of B_4C (KING, 1949). According to these data, standard enthalpy of boron carbide is 6.47 ± 0.07 kcal/mole deg. The temperature dependence of enthalpy is described by equation:

$$H^{o}_{T} - H^{o}_{298.16} = 22.22\ T + 2.70 \times 10^{3} \cdot T^{2} + 10.72$$

$$\times\ 10^{5} T^{-1} - 10690\ \text{cal/mole}$$

Boron carbide, possessing chemical bonding of a predominantly covalent type, is semiconductor, as shown earlier by RIDGWAY (1934) and GLASER et al. (1953). Theoretical evaluation of energy bank parameters in boron carbide carried out by YAMAZAKI (1957) showed that the gap between valence and conduction bands in the center of the BRILLOUIN zone is 2.5 eV, and at the zone boundary in the direction of the (111) wave vector, about 1 eV. The width of the B_4C forbidden band (experimentally determined) is 0.7 eV as given by LAGRENAUDIE (1953).

The effect of temperature on resistivity of pure samples of boron carbide is represented in Figure 6 (SAMSONOV and SINEL'NIKOVA, 1961). As noted, transition to the conductivity proper starts at 1100-1150°C and the width of the forbidden band evaluated from these data is 1.64 eV. This agrees with values which were calculated by YAMASAKI (1957) using a quantum mechanical method.

Investigation of the relation of B_4C thermal emf and porosity shows that thermal emf of green samples decreases approximately linearly with decrease in porosity, and after annealing increases in density (ZHURAVLEV et al., 1961). This effect is caused by a highly developed imperfection of boron carbide crystal structure, which is especially noticeable in porous samples. After annealing, porosity decreases as a consequence of a greater uniformity of composition and a more even distribution of impurities.

The dependence of thermal emf and specific electrical resistance on the composition of alloys of the B-C system is represented in Figure 7 (SAMSONOV et al., 1960; ZHURAVLEV et al., 1961). High values of thermal emf are observed for $B_{13}C_2$ and $B_{12}C_3$ phases, maximum values of specific electrical resistance correspond to the same phases, and also to the $B_{12}C$ carbide. Compound $B_{12}C$ is formed by replacing a boron atom in boron structure where each atom possesses five bonds by a carbon atom which, in turn, is able to form only four bonds. Consequently, $B_{12}C$ carbide has a p-type conductivity and relatively high thermal emf. $B_{13}C_2$ composition, which is characterized with a C-B-C linear chain, shows electron conductivity, while $B_{12}C_3$ composition with C-C-C chain shows hole-type conduction.

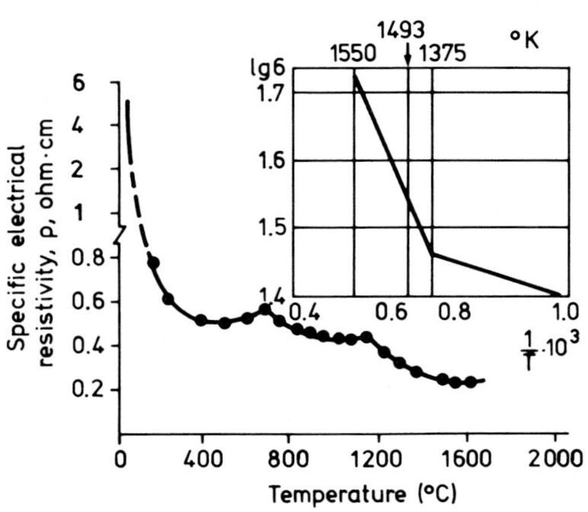

Fig. 6. Temperature versus specific electrical resistivity of B_4C (SAMSONOV and SINEL'NIKOVA, 1961)

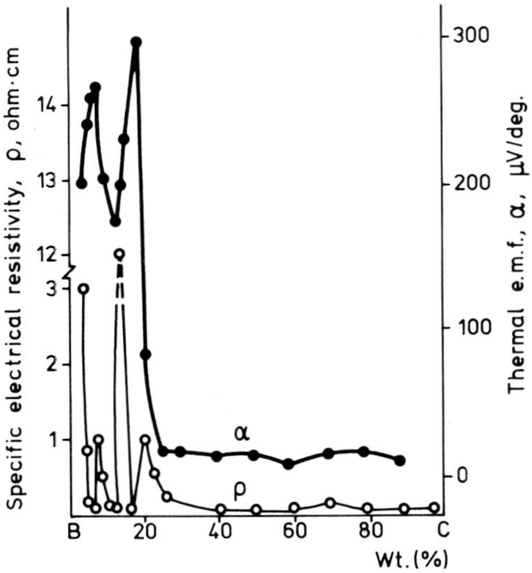

Fig. 7. Electrical resistivity and thermal emf of B-C alloys versus composition (SAMSONOV et al., 1960; ZHURAVLEV et al., 1961)

SAMSONOV et al. (1960) noticed that specific resistivity of commercial-grade boron carbide is about one order of magnitude lower than that of pure B_4C.

NESHPOR et al. (1970) reported a Hall coefficient for samples of pyrolitic boron carbide with theoretical density and lamellar structure as + 2.19 ± 0.01 x 10^{-2} cm^3/coulomb. The value of concentration of current carriers is 3.4 x 10^{20} cm^{-3}, and the value of Hall's mobility is 0.13 cm^2/V sec.

KONDAKOV and FEDOROVA (1955) pointed out that a hardness, inferior only to that of diamond, is a distinctive feature of boron carbide. FITZGERALD (1963) showed that boron carbide, in contrast to transition metal carbides, preserves its hardness up to temperatures of 1800°C. This observation illustrates the tendency of B_4C to preserve the stability of its crystal lattice at high temperatures. The high degree of hardness of boron carbide enables friction sliding of B_4C - B_4C pairs, and a low 0.43 friction coefficient which decreases slowly to the value 0.23 at 1800°C (SAMSONOV et al., 1970).

The dependence of flexural strength upon temperature is also an indication of high strength of bonds in the B_4C crystal lattice at high temperatures (LIDMAN and HAMJIAN, 1952; see Table 1).

Because of its great hardness and strength, boron carbide is inferior in abrasive resistance only to diamond. According to studies by MARKOVSKII et al. (1952) the abrasive resistance of diamond is 0.613, that of boron carbide 0.4-0.422, that of silicon carbide 0.314, and that of electrocorundum 0.145-0.160, in arbitrary units. A detailed study of abrasive properties of boron carbide is found in paper by ZHDANOV et al. (1954).

The wear resistance of boron carbide was investigated by KIFFER and KOLBE (1949). According to their data, wear of boron carbide, cast tungsten carbide and WC-based hard metal containing 12% cobalt, determined under similar conditions, result in 4.5, 28, and 68 mm^3/ unit of time respectively.

Table. 1. Properties of boron carbide

Characteristics	B_4C
Boron content, wt%	78.30
Carbon content, wt%	21.70
Molar weight, g	55.28
Density g/cm^3	2.52 ± 0.01
Crystal lattice	rhombohedral
Lattice periods, Å	
a	5.60-5.61[a]
c	12.10-12.12[a]
c/a	2.165
Melting point, °C	2350[b] 2450-2470[c]
Boiling point, °C	3500
Standard heat of formation, kcal/g.-atom C	-13.7 ± 3.5[e]
Heat capacity at 400-1730°K, cal/mol. deg	$22.22T + 2.7010^{-3}T^2 + 10.720.10^{-5}T^{-1} - 10690$[f]
Standard entropy, cal/g deg	6.47 ± 0.07[g]
Heat conductivity, cal/deg. cm. sec	
at °C 100	0.29 [g]
300	0.22 [g]
500	0.18 [g]
700	0.15 [g]
Microhardness, kg/mm^2	4950[b]
Coefficient of thermal expansion, deg^{-1}	2.6×10^{-6}[g]
25-800°	4.5×10^{-6}[g]
Electrical resistivity, ohm. cm, at 20°C	0.9[i]
Thermal emf, microvolts/deg at room temperature	80[i]
Compressive strength, kg/mm^2 at 20°C	196[h]
Flexural strength, kg/mm^2	
at 20°C	21 – 28[h]
	31 – 34 [j]
870°C	24.6 [j]
1095°C	20.9 [j]
1316°C	19.5 [j]
Tensile strength, kg/mm^2 at 20°C	16.3
Abrasion resistance in arbitrary units	0.4-0.422
Radiation coefficient at 1800°C λ = 0.655	0.91[g]

[a] ZHDANOV and SEVAST'YANOV (1941); CLARK and HOARD (1943);
[b] MEERSON and SAMSONOV (1953); [c] DOLLOFF (1960); [d] ELLIOT (1970);
[e] GAL'CHENKO et al. (1970); [f] STULL (1965); [g] SAMSONOV et al. (1960);
[h] RIDGWAY (1934); [i] ZHURAVLEV et al. (1961); [j] ZHDANOV et al. (1954)

High bond strength in B_4C crystalline lattice results in a low thermal expansion coefficient of 4.5×10^{-6}/degree (SAMSONOV et al., 1960).

The basic properties of boron-carbon alloys are presented in Table 1.

Boron carbide is chemically one of the most stable compounds. On heating in vacuum, B_4C dissociation is observed at the relatively high temperatures of 2350-2600°K and is described by the equation: $1/4\ B_4C_{solid} = B_{gas} + 1/4\ C_{solid}$. The heat of dissociation is $\Delta H^o_{298} = 138.0 \pm 1.0$ kcal/mole. Equilibrium vapor pressure of boron can be described by the equation: $\lg P_{B\ atom} = 7.506 - \frac{29630}{T}$ (ROBSON and GILLES, 1964; HILDENBRAND and HALL, 1964; BOLGAR, 1965).

Boron carbide is not dissolved in mineral acids and their mixtures, or in alkali solutions of various concentrations. It is not corroded under action of acid-oxidizer mixtures and hot concentrated solutions of hydrofluoric and nitric acid and their mixtures (RIDGWAY, 1934; SHAFRAN, 1938). The most aggressive media for boron carbide are mixtures of concentrated nitric, sulphuric, and hydrofluoric acids (SHAFRAN and PAVLOVA, 1938). The most complete study of the chemical properties of boron carbide was carried out by NAZARCHUK (1959; 1961).

One technique of B_4C refinement from free carbon is based on the difference in oxidation between carbon and boron carbide. This technique consists of heating a weighed amount of boron carbide in a mixture of 10 ml of H_2SO_4, 10 ml of HNO_3, 5 ml of $HClO_4$, and 20 mg of $K_2Cr_2O_7$.

Oxidation of boron carbide starts at 600°C and takes place in stages. The formation and behavior of a B_2O_3 film limits the oxidation rate at various temperatures. The oxidation of B_4C will take place at lower temperatures in the presence of free carbon. This principle is used to determine free carbon in B_4C as described by MEERSON and SAMSONOV (1950).

Investigations (LITZ and MERCURY, 1963) of the interaction of boron carbide with dry and wet air showed that at 250°C, B_4C is oxidized more rapidly with aqueous vapor than with air; at higher temperatures oxidation rate with dry air exceeds oxidation rate with vapor - air mixtures.

At 1200°C, boron carbide does not interact with sulphur, phosphorus or nitrogen. It interacts with chlorine at 1000°C forming BCl_3 and graphite. Bromine and iodine do not interact with B_4C.

Heating boron carbide with transition metal oxides results in reduction of the oxides by carbon and the corresponding formation of borides. This principle can be used as a method of preparing refractory borides (SAMSONOV et al., 1960).

The wetability of B_4C with liquid metals has been studied by SAMSONOV et al. (1960, 1969) and JANES and NIXDORF (1966). The results of these investigations are presented in Table 2.

e) Applications of Boron-Carbon Alloys

Applications of boron carbide are based on the utilization of its high degree of hardness, abrasive resistance, wear resistance, semiconductive properties and ability to absorb thermal neutrons. B_4C is used both, as powder and as sintered and cast parts.

As an abrasive powder, boron carbide is widely used for polishing glass, hard metals, and natural and man-made minerals (TIHONOV, 1940; SAMSONOV et al., 1960). The grinding power of boron carbide cutting

Table 2. Contact angles of boron carbide with liquid metals

Metal	Temperature, ^{o}C	θ, degrees
Cu	1130	133 ± 1.5
Zn	540	121.5
Al	670	118.0
Pb	395	113.5
Cr	1820	0
Mn	1300	0
Fe	1550	16.0 ± 2.0
Co	1500	64.0 ± 0.5
Ni	1450	85.0 ± 1.5
Ni	1470	41.0 ± 1.0
Cr-Ni (1 : 1)	1400	0

tools used to grind hard alloys was found to decrease proportionally with the grain size and to rise with decreasing grinding rates. Grinding powders of boron carbide are utilized successfully in grind-in applications (SAMSONOV et al., 1960).

Boron carbide powders are also utilized to prepare polished specimens of very hard materials (SAMSONOV and KIPARISOV, 1958). It should be mentioned that the abrasive resistance of boron carbide powder is sharply increased in mixtures containing 1-5% diamond powder of smaller grain size.

In powder form, boron carbide is further used to prepare transition metal borides by the carbide-boron method of SAMSONOV et al. (1960), and it is an active component of diffusion-boronizing parts of steels, refractory metals and alloys (EPIK et al., 1968).

As an additive in some types of tool steels and copper alloys, powdered boron carbide increases their strength and heat conductivity (MARKOVSKII et al., 1952). The introduction of boron carbide in steels and cast iron (in the form of intermediate alloys containing 7.5-8% nickel) to increase wear, corrosion resistance and viscosity was reported by SAMSONOV et al. (1960).

Being highly wear-resistant, boron carbide can be used for sand blasting nozzles and sliding friction pairs in the cast form and very often in sintered form. SAMSONOV et al. (1960) indicated that although the cost of sand blasting nozzles produced from sintered boron carbide may be up to seventeen times that of those produced from cast iron, their lifetime in practical application is about 300 times longer.

Boron carbide is used to manufacture various gauges, in particular those controlling aluminum parts in aeronautics; matrixes for drawing abrasive rods; draw plates; studs for drilling spherical pits in oil cups of ruby bearings and indentors for measurements of hardness at up to 1850°C (SAMSONOV et al., 1960, DZHEMELINSKII et al., 1970). Indentors of boron carbide, alloyed with titanium, can measure the hardness of hard and refractory materials in a temperature range of 300 to 2000°C (DZHEMELINSKII et al., 1973).

Utilization of boron carbide as a material for adjusting instruments instead of diamond or diamond metallic pencils is also promising (SAMSONOV et al., 1960).

High chemical stability of B_4C, combined with its high degree of hardness and wear resistance makes it possible to fabricate boron carbide mortars used for grinding hard and brittle materials, and various chemical vessels for use with aklalies and acids (SAMSONOV, 1957).

Boron carbide is used in the production of igniters, wireless resistors and semi-conductive thermocouples able to operate over long periods in melts of non-ferrous metals, steels, cast irons and salts (SAMSONOV and KISLII, 1959; KISLII and SAMSONOV, 1959).

Because of the large cross section for the capture of thermal neutrons, boron carbide finds applications in nuclear technology as protective material and for control rods.

Boron carbide also finds wide applications in the structure of superalloys (SAMSONOV et al., 1960).

2. Compounds of Boron with Silicon

a) Phase-Diagram of the B-Si System

MOISSAN and STOCK (1900) were first to report compounds of boron with silicon. They assignet to them the chemical compositions of SiB_3 and SiB_6. Subsequent attempts by BREWER et al. (1951) to obtain silicon borides were not successful but they indicated the existence of a compound containing about 90% boron. SAMSONOV and LAT'ISHEVA (1955) reported an SiB_3 composition with tetragonal symmetry. Silicon boride with SiB_3 composition was also reported by PETRUSEVICH et al. (1967).

ZHURAVLEV (1956) reported the existence of an SiB_6 isomorphous with CaB_6. However X-ray analysis of crystal structure carried on SiB_6 single crystals by CLINE (1958), ADAMSKY (1958) indicates that this boride is crystallized in an orthorhombic cell with the following parameters:

$a = 14.392 \pm 0.001$ Å,

$b = 18.267 \pm 0.0015$ Å,

$c = 9.885 \pm 0.001$ Å; space group - Pnn2 or Pnnm

$Z = 40$

A new chemical compound of boron and silicon of composition SiB_4, isomorphous with B_4C was reported almost simultaneously by RIZZO and BIDWELL (1960), CLINE and SANDS (1960), BROSSET and MAGNUSSON (1960), MATKOVICH (1960) and COLTON (1961a). This boride possesses a rhombohedral lattice with constants: a = 5.52 Å and α = 69.1°; the corresponding hexagonal cell dimensions are: a = 6.35 Å, c = 12.69 Å. The possible space groups are R32; R3m and R$\bar{3}$m; Z = 3 (COLTON, 1961a). According to MAGNUSSON and BROSSET (1962) the structure of silicon tetraboride can be described as a B_4C structure in which two silicon atoms occupy positions of two carbon atoms of the C-C-C linear chain, the remaining silicon atoms statistically replacing boron atoms in icosahedron. Like boron carbide, SiB_4 possesses a wide homogeneity range ($SiB_{2.82-4.0}$). Apparently, the SiB_3 phase found by MOISSAN and STOCK (1900), SAMSONOV and LAT'ISHEVA (1955), and PETRUSEVICH et al. (1967) was in fact silicon boride SiB_4.

NOWOTNY et al. (1957a, b) reported a boron rich phase, SiBn (n ~ 10) referred to later (NOWOTNY et al., 1958) as SiB_{12}. This phase is isotypical with AlB_{12} and is isomorphous with one of the modifikations of elemental boron. Boride of the SiB_{14} composition (GIESE et al., 1965) is isostructural with β-boron and crystallized in a rhombic lattice. Parameters of its hexagonal cell are: a = 11.13 Å, c = 23.83 Å. Thus, the exact composition of the boron-rich phase (SiB_{10-14}) is still uncertain.

A phase of unknown composition with hexagonal lattice, isostructural with AlB_{12} (C_{32}-type) and unit cell dimensions of a = 3.17 Å, c = 3.54 Å was reported by LIPP and RODER (1966). This contradicts the results of SAMSONOV and SLEPTSOV (1964) and LAMIHOV et al. (1972), who found only one boride phase in the Si-SiB_4 range.

Solubility of boron in silicon was studied by HORN (1955), NOWOTNY et al. (1957a), EPEL'BAUM et al. (1959), SAMSONOV and SLEPTSOV (1963), and ELLIOT (1970).

The most reliable results are found in a paper by SAMSONOV and SLEPTSOV (1963), who show that the solubility of boron in silicon at room temperature corresponds to 0.81 at %, and 3.6 at % at 1375°C. According to HESSE (1968) in the temperature interval of 900 to 1200°C, a solid solubility region exists in the range of 0.2 to 0.85 at % of boron.

The solubility of silicon in boron (according to X-ray studies at 1750°C) is 8.7 at %. Based on microscopic investigations of the cast alloy structures ELLIOT (1970) reported that solubility of silicon in boron does not exceed 1 at %.

A phase diagram of the B-Si system was first proposed by SAMSONOV and SLEPTSOV (1962). A more exact phase diagram of this system is presented in Figure 8.

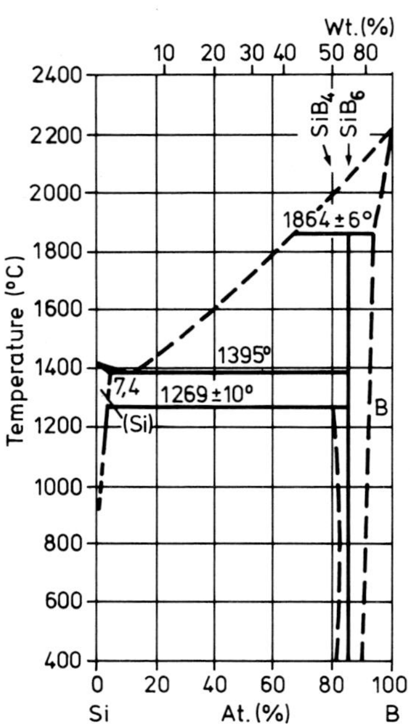

Fig. 8. Phase-diagram of the B-Si system (ELLIOT, 1970)

b) Preparation of Boron-Silicon Alloys

The following methods may be used to prepare silicon borides:

1. Synthesis from elements (sintering, melting, hot pressing).
2. Reduction of silicon oxide with boron or boron anhydride with silicon.
3. Thermal reduction of boron anhydride and silicon oxide mixtures with magnesium or calcium.
4. Precipitation from gaseous phase.
5. Electrolysis of salt melts containing silicon and boron.

Synthesis from elements and thermal reduction of silicon oxide by boron are the most widely utilized methods in practical use in the manufacture of silicon borides. For sintering boron and silicon mixtures, optimum temperatures for the preparation of the SiB_4 phase are 1200 to 1370°C. Above 1370°C a peritectic decomposition of SiB_4 takes place forming SiB_6 and silicon (BROSSET and MAGNUSSON, 1960; MATKOVICH, 1960; COLTON, 1961a; SAMSONOV and SLEPTSOV, 1964). According to SAMSONOV and SLEPTSOV (1964) formation of the SiB_4 phase starts at 800°C and is practically complete at 1200°C in about 120 min. According to some data, formation of SiB_6 takes place from stoichiometric mixtures of boron and silicon powders at 1300 - 1500°C.

Silicon borides obtained by melting boron and silicon must be given an additional treatment with acid to extract excess boron or silicon (BROSSET and MAGNUSSON, 1960; MEERSON et al., 1965).

Practically poreless SiB_3 samples are produced by sintering boron and silicon mixtures using hot pressing at 1600 - 1800°C (SAMSONOV et al., 1960). SiB_6 can be prepared with residual porosity of the order of 2% by hot pressing of boron and silicon mixtures at 2150°C for a short time under pressure of 100 kg/cm^2 (SAMSONOV et al., 1960).

Thermal reduction of silica by boron is based on the reaction proposed by SAMSONOV and SLEPTSOV (1969):

$$SiO_2 + 8\ B = SiB_6 + 2BO\ (B_2O_2)$$

However, they did not succeed in obtaining a SiB_6 phase using this method. At 1400°C a compound was formed which corresponds to the composition of $SiB_{2.93-3.3}$ and was found to be isostructural with silicon boride SiB_4. Analogous results were reported by RIZZO and BIDWELL (1960) on thermal reduction of boron anhydride by silicon in air. LAMIHOV et al. (1972) studied thermal reduction of boron anhydride by silicon in the presence of $6Si + Al_2(SO_4)_3$. The mixtures were preheated to 600°C and reaction in the mixtures was induced by electric arc. The resulting product was found to represent a mixture of $SiB_{2.87-3.11}$ and silicon.

Thermal reduction of a mixture of boron anhydride and quartz by magnesium and calcium also resulted in a compound of SiB_3 composition only (SAMSONOV and LAT'ISHEVA, 1955).

Silicon boride of SiB_3 composition was also formed by precipitation from $SiCl_4$, BBr_3 and H_2 vapor-gas mixtures on tantalum wire heated to 1150°C (PETRUSEVICH et al., 1967). Lattice parameters of this phase were close to those reported for $SiB_{2.89}$ (MAGNUSSON and BROSSET, 1962).

ALLEONARD (1961) obtained alloys of the B-Si system by using electrolysis of KBF_4, K_2SiF_6 and KF salt melts in argon atmosphere. The resulting compound was identified as SiB_3.

c) Properties of Boron-Silicon Alloys

From a practical point of view, one of the most valuable properties of alloys of the boron-silicon system is their oxidation and high thermal shock resistance (MOISSAN and STOCK, 1900; RIZZO et al., 1960; COLTON, 1961b; MEERSON et al., 1965; LAMIHOV, 1968).

According to RIZZO et al. (1960) samples containing 35 wt% of boron showed no significant weight increase after exposure to air for 24 h at 980°C. Samples of SiB_4 and SiB_6 exposed to air for 60 h at 1375°C showed a gain of 62 mg/cm^2 and 55 mg/cm^2 for SiB_4 and SiB_6 respectively (COLTON, 1961a). MEERSON et al. (1965) reported that oxidation kinetics of SiB_4 boride has a parabolic diffusion character. A gradual decrease in oxidation rate takes place due to the formation of a dense, closely attached, glass-like film. COLTON (1961c) reported that oxidation resistance of silicon borides is higher than that of ZrB_2, TiB_2, BN, and B_4C compounds. Exceptionally high thermal shock resistance of articles produced from these borides was also noted. No cracks were observed in the samples after 50 cycles of heating and cooling from 1100° to 20°C, or by quenching in water.

Alloys of the B-Si system possess a high degree of strength, both at room and elevated temperatures. The strength of sintered articles depends upon sintering conditions (COLTON, 1961c). Samples of SiB_6 sintered in air for 24 h at 1375°C have tensile strength 8.95 kg/mm^2 at 25° and 7.44 kg/mm^2 at 1100°C. Flexural strength is 8.27 kg/mm^2 and 25° and 7.25 kg/mm^2 at 1100°C. The flexural strength of SiB_6 specimens with 2% residual porosity, produced by hot pressing in vacuum at 1600°C (DUTTA and GAZZA, 1973) is 19.7 kg/mm^2 at room temperature, and the modulus of elasticity for the sample with 6% residual porosity is 28.8×10^3 kg/mm^2. According to the data by COLTON (1961c), SiB_6 has approximately the same abrasive properties as silicon carbide.

The electrical properties of B-Si system alloys are of some practical interest. The introduction of 0.0005% boron into silicon decreased its electrical resistance 100-fold and addition of 1% boron 1000-fold (SAMSONOV et al., 1960). In this case, the temperature coefficient of resistance (which is negative with very small boron additions and positive with 1% boron content), decreases. Additions of 0.01-1% boron to silicon induces hole-type conductivity. Alloys of boron with silicon containing from 5 to 60 wt% of boron have electrical resistance at room temperature from 1 to 150 ohm.cm (RIZZO et al., 1960). At 1300°C electrical resistance decreases from 22.9×10^3 to 4.8×10^3 ohm.cm with an increase in boron content from 0.1 to 10% (SAMSONOV and SLEPTSOV, 1964). Figure 9 represents the results of thermal emf for boron-silicon alloys. Some properties of boron-silicon compounds are given in Table 3.

Silicon borides possess a rather high chemical stability (MOISSAN and STOCK, 1900). Oxidation in air as pointed out above, is difficult to induce. Borides interact with fluorine on heating and with chlorine only at red head (~ 700°C). Bromine does not interact with these compounds. Nitrogen does not act on them up to 1000°C. They are not decomposed with halogenohydrogen acids, but dissolve in boiling concentrated sulphuric acid. Boiling HNO_3 reacts mainly with SiB_4. Melts, such as alkali carbonates or their mixtures with nitrates, readily decompose both of these compounds at red heat.

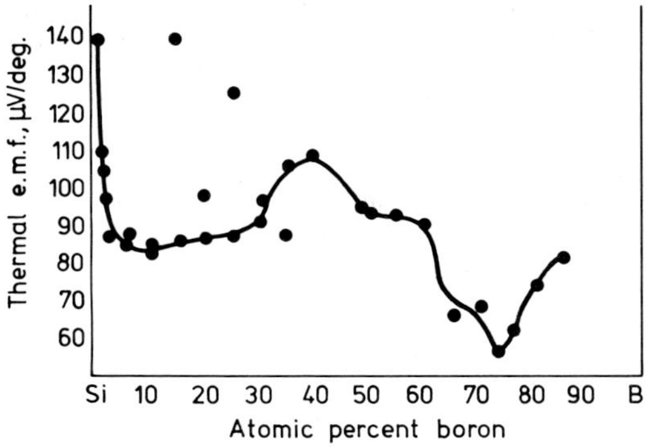

Fig. 9. Thermal emf of the B-Si system versus composition (SAMSONOV and SLEPTSOV, 1962)

Table 3. Properties of boron-silicon alloys

Characteristics	SiB_4	SiB_6
Boron content, wt%	60.6	69.9
Silicon content, wt%	39.4	30.1
Molar weight, g	71.34	82.96
Crystal lattice	rhombohedral	rhombic
Lattice periods, Å		
a	6.35[a]	14.392[b]
b	-	18.267[b]
c	12.69[a]	9.8852[b]
Density, g/cm^3 X-ray	2.41[a]	2.39[b]
pycnometric	2.44[a]	2.43[b]
Melting point, $^{\circ}C$	1269[c]	1864[c]
	decomposes	decomposes
	1830[d]	1950[d]
Microhardness, kg/mm^2	2000 - 2500[e]	3200 - 3500[e]
	1870 - 2290[f]	2520 - 2870[f]
Electrical resistivity, ohm-cm, at $20^{\circ}C$	1.75[g]	0.2[a]

[a]COLTON (1961a)
[b]ADAMSKY (1958)
[c]ELLIOT (1970)
[d]SAMSONOV and SLEPTSOV (1962)
[e]SAMSONOV and SLEPTSOV (1964)
[f]RIZZO and BIDWELL (1960)
[g]MEERSON et al. (1965)

d) Application of Boron-Silicon Alloys

Because of their oxidation and thermal shock resistance, boron-silicon alloys find wide applications in compositions of heat-resistant and fire-resistant "cermet"-type alloys (RIZZO et al., 1960; LAMIHOV, 1968). According to RIZZO et al. (1960) fire-resistant materials consist of unreacted silicon and silicon boride (SiB_4) scattered in a boron-silicate matrix. This material, stable in air up to 1550°C, possesses high thermal stability and low density.

Articles made of SiB_4 and SiB_6 are used for work at high temperature in oxidizing media. Heating rods including SiB_4 in their composition are stable in oxidizing media up to 1600°C (LAMIHOV, 1968).

Semiconductive properties of solid solutions of boron in silicon are applied in the production of solar batteries used to convert solar into electrical energy (SHCHEKIN, 1958; CHECHIK, 1965).

Silicon borides can be used as control rods and protective devices in nuclear reactors (COLTON, 1961b, c).

3. Compounds of Boron with Germanium, Tin and Lead

SAMSONOV et al. (1960) investigated the preparation of boron compounds with germanium using a number of methods:

1. Synthesis from elements attempted by heating boron and germanium mixtures at 800-1200° in argon atmosphere and by hot pressing at 2000°C.

2. Vacuum-thermal reduction of germanium dioxide with a mixture of boron carbide and carbon at 1000-1400°C.

3. Reduction of germanium dioxide with boron in vacuum at 1250°C. Chemical and X-ray analysis showed that in mixtures with boron, germanium is formed as a result of reduction reactions. Direct interaction of boron and germanium powders did not result in the formation of new phases or mutual solid solutions. These results were confirmed by WALD (1965).

Investigation of boron diffusion in germanium by KRASIUK and GRIBOV (1961) showed that the diffusion coefficient equals 1.8×10^{-9} cm²/sec, the value of activation energy being 105 kcal/mole.

Based on results of X-ray, metallographic, thermal analysis and electron microscopic investigations of sintered and melted alloys, BIDWELL (1970) proposed a schematic phase-diagram of B-Ge system which is shown in Figure 10. This system may be characterized as a classical system of monotectic eutectic without intermediate phase formation. Insignificant solubility of germanium in solid boron and boron in liquid germanium is observed. Upon solution of boron in germanium the lattice of germanium is reduced by 0.01 - 0.02 vol %.

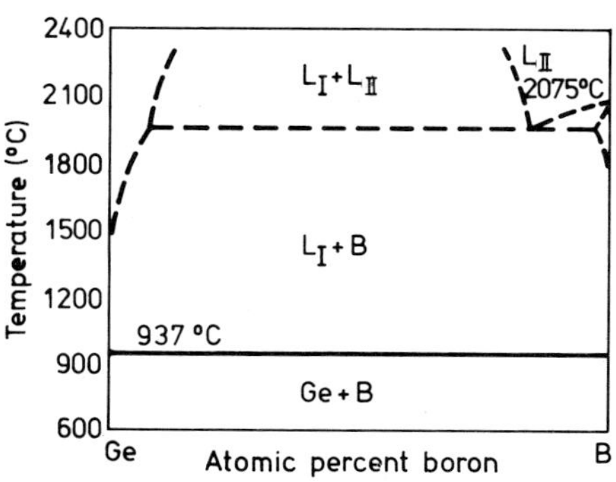

Fig. 10. Phase-diagram of the B-Ge system (BIDWELL, 1970)

These results show that germanium can be used as a diffusion barrier for boron fibers in composites.

Attempts to obtain boron compounds with tin and lead also showed the absence of solubility or formation of boride phases in these systems (SAMSONOV et al., 1960; WALD, 1965).

References

ADAMSKY, R.: Unit cell and space group of orthorhombic SiB_6. Acta Cryst. 11, pt 10, 744-755 (1958)

ALLEN, R.: Solid solution of the boron in boron carbide. J. Am. Chem. Soc. 75, 3582-3583 (1953)

ALLEONARD, S.: Preparation de composes bore-silicium par electrolyse ignee, Bull. Soc. Chim. France Sec 5, 34-36 (1961)

AMBERGER, E., DRUMINSKI, M., PLOOG, K.: Das B-C System im kinetischen Bildungsbereich pyrolitischer Bildung kohlenstoffreicher B-C Phase. J. Less Common Metals 23, 43-52 (1971)

ASHBEE, K. H. G., DuBOSE, C.K.H.: Dislocation nodes in boron carbide with special reference to non-stoichiometry. Acta. Metallurica 20, 241-247 (1972)

BIDWELL, L.R.: The boron-germanium system. J. Less Common Metals 20, 19-27 (1970)

BOLGAR, A.S.: Thesis IPM, Acad. Sci., USSR 1965

BREWER, L., SAWYER, D., TEMPLETON, D., DAUBEN, C.: A study of the refractory borides. J. Am. Ceram. Soc. 34, 173-174 (1951)

BROSSET, C., MAGNUSSON, B.: The silicon-boron system. Nature 187, 54-55 (1960)

CHECHIK, P.O.: New radio sources. USSR M-L., Gosenergoizdat (1965)

CLARK, H., HOARD, J.: Crystal structure of B_4C. J. Am. Chem. Soc. 65, 2115-2119 (1943)

CLINE, C.F.: Preliminary investigations of the silicon boride SiB_6. Nature 181, 446-477 (1958)

CLINE, C.F., SANDS, D.E.: A new silicon boride SiB_4. Nature 185, 456 (1960)

COLTON, E.: Two boron silicides. Mat. Design Eng. 53, 9-11 (1961a)

COLTON, E.: On the boron-silicon reaction. J. Inorg. Nuclear Chem. 17, 108-111 (1961b)

COLTON, E.: Hexaboron silicide: Promising material in nuclear control devices. Ceramic Industry 76, 70-71 (1961c)

CRUIZIAT, B.: Realisation de quelques reactions chimiques a l'aide d'un Chalumeaner a Plasa Inductit. Thesis, Univ. Lyon, 1970

DAWIHL, W.: Über die Gleitschwindung bei Metall-Metalloxyd-Gemischen und bei Borkarbid. z. Metallkunde 43, 138-146 (1952)

DOLLOFF, R.T.: Boron-carbon binary system - Investigation of the phase equilibrium in some non-metallic carbides at high temperatures. Air Force Report, WADD, Techn. Rep. 60-143, 1, 1-22 (1960)

DOMALSKY, E.S., ARMSTRONG, G.T.: The heat of formation of boron carbide. Phys. Chem. A 72, 133 (1967)

DUTTA, S.K., GAZZA, G.E.: Properties of hot-pressed SiB_6. Am. Ceram. Soc. Bull. 52, 552-554 (1973)

DZHEMELINSKI, V.V., KOVAL'CHENKO, M.C., BORISENKO, V.A., MAKARENKO, G.N.: Indentors for measurements of hardness of materials at high temperatures. In: Refractory Carbides (ed. G.V. Samsonov) pp. 233-236. Kiev: Naukova Dumka 1970

DZHEMELINSKI, V.V., KOVAL'CHENKO, M.C., MAKARENKO, G.N.: Indentor materials for measurements of hardness at high temperatures. Poroshkovaia Metallurgia 2, 97 to 100 (1973)

ELLIOT, R.P.: Structure of binary alloys. Metallurgia (1970)

EPEL'BAUM, V.A., GUREVICH, M.A., STAROSTINA, M.I.: Solubility of boron in silicon. ZhNH. 4, No. 8, 1881-1884 (1959)

EPIK, L.P., KOTLIARENKO, L.A., DERKACH, V.D.: Protective Coating on Metals. Kiev: Naukova Dumka, 1968

FITZGERALD, L.M.: The hardness at high temperatures of some refractory carbides and borides. J. Less Common Metals 5, No. 4, 356-364 (1963)

GAL'CHENKO, G.L., TIMOFEEV, B.I., MAKARENKO, G.N., SAMSONOV, G.V.: Heat of reaction of boron carbide with chlorine ZhFH 44, No. 10, 2469-2472, USSR (1970)
GIESE, R., ECONOMY, J., MATKOVICH, V.: Interstitial derivatives of boron. Z. Kristallogr. 122, No. 1-2, 144-147 (1965)
GLASER, F., MOSKOWITZ, D., POST, B.: Investigations of boron carbide. J. Appl. Phys. 24, No. 6, 731-733 (1953)
GLASSON, D.R., JONES, J.A.: Formation and reactivity of borides, carbides and silicides. II. Production and sintering of boron carbide. J. Appl. Chem. 19, No. 5, 137-140 (1969)
GMELINS, L.: Bor. In: Handbuch der anorganischen Chemie. System No. 13 8th ed. Erg. band. Weinheim: Verlag Chemie 1954
HESSE, J.A.: Löslichkeit und Ausscheidungskinetik von Bor in polikristallinem Silizium. Z. Metallkundt 59, 499-503 (1968)
HILDENBRAND, D.L., HALL, W.F.: Pressure decomposition of B_4C and heat sublimation of the boron. J. Phys. Chem. 68, 989-994 (1964)
HORN, F.: Densitometric and electrical investigations of boron in silicon. Phys. Rev. 97, 1521-1525 (1955)
JANES, S., NIXDORF, J.: Herstellung von Sinterkorpern aus Borkarbid und Bornitrid. Ber. Deut. Keram. Ges. 43, 136-145 (1966)
KELLY, K.K.: The specific heat at low temperature of crystalline boric oxide, boron carbide and silicon carbide. J. Am. Chem. Soc. 63, 1137 (1941)
KIFFER, R., KOLBE, F.: Tungsten-carbide-free-band metals. Powd. Metall. Bull. 4, 4-12 (1949)
KIFFER, R., GUGEL, E., LEIMER, G., ETTMAYER, P.: Untersuchungen zum System Bor-Kohlenstoff. Ber. Deut. Keram. Ges. 49, 385-389 (1971)
KING, E.G.: High-temperature heat content of boron carbide. Ind. Eng. Chem. 41, 1298 (1949)
KISLII, P.S., SAMSONOV, G.V.: Thermocouples of high temperature semiconductors. Acad. Sci. USSR, OTH, ser. Metallurgia Toplivo 6, 133-137 (1959)
KISLII, P.S., SAMSONOV, G.V.: Diffusion of boron in carbon. Physics of solid state. II, 8, 1729-1732 (1960)
KONDAKOV, G.V., FEDOROVA, N.J.: Determination of microhardness of artificial abrasive materials. In: Collection of Scientific-Technical Information, VNIIASH. Abrasivi. Bull. 12, 10-14 (1955)
KRANZ, R.: Darstellung und Eigenschaften von Borcarbid. Keram. 8, 459-464 (1963)
KRASIUK, B.A., CRIBOV, A.P.: Semiconductors, germanium and silicon. Metallurgizdat (1961)
KUDRIAVTSEV, V.I., SAFRONOV, G.V.: Bulletin from Seminar on: Heatresistant materials. Abrt. 5, Kiev: Acad. Sci. USSR (1960)
LAGRENAUDIE, J.: Composes semi-conducteurs du bore avec le carbone et l'auminium. J. Chem. Phys. 50, 352-355 (1953)
LAMIHOV, L.K.: Methods of preparation, properties and applications of refractory alloys of the silicon-boron system in unorg. materials. Ed. Acad. Sci., USSR 4, 1076-1080 (1968)
LAMIHOV, L.K., NERONOV, V.A., SAMSONOV, G.V.: Silicothermic method of preparation of boron silicide, SiB_4: metallothermic method of preparation of compounds and alloys, pp. 11-13 Novosibirsk: Nauka (Siberian Dept. Acad. Sci.) 1972
LIDMAN, W., HAMJIAN, H.: Reactions during sintering of a zirconium carbide-niobium cermet. J. Am. Cer. Soc. 35, 236-240 (1952)
LIPP, A.: Elementary cell of boron carbide. Ber. Deut. Keram. Ges. 43, 60-62 (1966)
LIPP, A., RODER, M.: Verbindungen im System B-C-Si bzw. B-Si. Z. Anorg. Allg. Chem. 344, 5-6, 225-229 (1966)
LITZ, L.M., MERCURY, R.A.: Oxidation of boron carbide by air, water and air-water mixtures at elevated temperatures. J. Electrochem. Soc. 110, 921-926 (1963)
LOWELL, C.: Solid solution of boron in graphite. J. Am. Cer. Soc. 50, 142-144 (1967)
MAGNUSSON, B., BROSSET, C.: The crystal structure of $B_{2.89}Si$. Acta Chem. Scand. 16, 449-455 (1962)
MAKARENKO, G.N., SAMSONOV, G.V.: Preparation of alloys of boron and carbon. Inf. Let. No. 20. IMSS Acad. Sci. UkUSSR, Kiev (1958)
MARKOVSKII, L.Ya., ORSHANSKII, D.L., PRIANISHNIKOV, V.: Electrothermic Chemistry, M.-L. USSR: Goshimizdat (1952)

MATKOVICH, V.I.: A new form of the boron silicide. Acta. Cryst. 13, 679-680 (1960)
MATKOVICH, V.I.: Interstitial compounds of boron. J. Am. Chem. Soc. 83, 1804-1806 (1961a)
MATKOVICH, V.I.: Unit cell, space group and composition of a lower boron phosphide. Acta Cryst. 14, pt 1, 93-194 (1961b)
MEERSON, G.A., KIPARISOV, S.S., GUREVICH, M.A., DEN-FEN-VYAN: Investigations of conditions of preparation and some properties of hard alloys of pseudobinary systems B_4C-B_4Si. Poroshkovaia Metallurgia 3, 62-68 (1965)
MEERSON, G.A., SAMSONOV, G.V.: Determination of free carbon content in boron carbide. Zav. Lab. 16, 1423-1428 (1950)
MEERSON, G.A., SAMSONOV, G.V.: Investigations of the phase diagram of the boron-carbide system and the natural phase in that system. Rpts Sector for Phys-Chem. Anal. 22, 92-104 (1953)
MEERSON, G.A., SAMSONOV, G.V.: The questions of conditions of preparation of boron carbide. Boron, Proc. Conf. on boron and its compounds, 52-59, M. GHI, USSR (1958)
MIERZIEYEWSKA, S., NIEMYSKY, T.: Preparation of crystalline boron carbide by vapor-phase reaction. J. Less Common Metals 8, 368-374 (1965)
MOISSAN, H.: Preparation et proprietes de borure de carbone, C.R. 118, 556-560 (1899)
MOISSAN, H., STOCK, A.: Preparation et proprietes de deux borures de silicium SiB_3 et SiB_6. C.R. 131, 139-143 (1900)
MULHAUSER, D.: Über Borcarbid. Z. Anorg. Chem. 5, 92-93 (1894)
NAZARCHUK, T.N.: Chemical stability of boron carbide. ZhNH 4, 2665-2670 (1959)
NAZARCHUK, T.N.: Boron carbide, chemical properties and methods of analysis; proc. Heat-resistant coatings No. 6, pp. 30-35, Kiev, Ed. Acad. Sci., UkSSR (1961)
NAZARCHUK, T.N.: MEHANOSHINA, L.N.: On the question of oxidation resistance of boron carbide, Poroshkovaia Metallurgia 2, 46-50 (1964)
NESPHOR, V.S., NIKITIN, V.D., ROBOTNOV, V.V.: Some physical properties of boron carbide: Refractory carbides, Kiev: Naukova Dumka, 1970
NOWOTNY, H., DIMAKOPOULOU, KUDIELKA, H.: Untersuchungen in dem Dreistoffsystem: Molybdan-Silizium-Bor, Wolfram-silizium-Bor und in dem System: VSi_2 - $TaSi_2$. Monatsheft für Chemie 88, 180-192 (1957a)
NOWOTNY, H., KIFFER, R., BENESOVSKY, F.: Silicoboride der Übergangsmetalle Vanadium, Niobium, Tantalum, Molybdenum und Wolfram. Planseeber, Pulvermet. Bd. 5, 86-93 (1957b)
NOWOTNY, H., PIEGGER, E., KIFFER, R., BENESOVSKY, F.: Das Dreistoffsystem Chrom-Silizium-Bor. Wolfrom Silizium Bor und in dem System: VSi_2 - $TaSi_2$. Monatsheft für Chemie 89, 611-617 (1958)
ORMONT, B.F.: Structure of Inorganic Matter. USSR Gostehizdat (1950)
PETRUSEVICH, I.V., NILL'SON, L.N., BELIAEV, A.I., GUREVICH, M.A.: On preparation of silicon boride by deposition from gaseous phase. Ed. Acad. Sci. USSR, Inorganic Materials 3, 1389-1392 (1967)
PLOOG, K.: Darstellung eines rhomboedrischen Borcarbids der Zusammensetzung $B_{13}C_2$. J. Less Common Metals 31, 177-180 (1973)
PLOOG, K., AMBERGER, E.: Kohlenstoffinduzierte Gitter beim Bor: I-Tetragonales $(B_{12})_4$ B_2C und $(B_{12})_4$ B_2C_2. J. Less Common Metals 23, 33-42 (1971)
PODSZUS, E.: Die Darstellung von Bor, sowie einiger seiner Verbindungen mit Stickstoff und Kohle. Z. Anorg. Chem. 221, 41-48 (1933)
REKOV, A.I., SAMSONOV, G.V.: A carbothermic method of preparation of boron carbide. Poroshkovaia Metallurgia 5, 80-91 (1961)
RIDGWAY, B.: Boron carbide - a new crystalline abrasive and wearresisting product. Trans. Am. Electr. Soc. 66, 117-133 (1934)
RIZZO, M.F., BIDWELL, L.R.: Formation and structure of SiB_4. J. Am. Ceram. Soc. 43, 550-552 (1960)
RIZZO, M.F., WEBER, B.C., SCHWARTZ, M.A.: Refractory composition based on silicon-boron-oxygen reactions. J. Am. Ceram. Soc. 43, 497-504 (1960)
ROBSON, H.E., GILLES, P.W.: High-temperature properties of the boron carbide and heat sublimation of boron. J. Phys. Chem 68, 983-989 (1964)
SAMSONOV, G.V.: Articles made of boron carbide. Vestnik Mashinostroenia 3, 24-27 (1957)
SAMSONOV, G.V.: Magnesio-thermic preparation of boron carbide. Ukr. Chem. J. 24, 799-805 (1958)

SAMSONOV, G.V., KIPARISOV, S.S.: Techniques of metallographic examination of boron carbide - Technology of colored metals. Metallugizdat, 367-371 (1958)
SAMSONOV, G.V., KISLII, P.S.: Technology of manufacture of tubes and rods from molybdenum disilicide. Ognewpori 6, 276-278 (1959)
SAMSONOV, G.V., KOVAL'CHENKO, M.S.: Hot-pressing. Kiev: Gostehizdat 1962
SAMSONOV, G.V., LAT'ISHEVA, V.P.: On the question of chemical compounds of boron and silicon. Kiev: Ed. Acad. Sci. 105, 499 (1955)
SAMSONOV, G.V., MARKOVSKII, L.Ya., ZHIGACH, A.F., VALIASHKO, M.T.: Boron, its compounds and alloys. Kiev: Ed. Acad. Sci., Uk. SSR 1960
SAMSONOV, G.V., PANASIUK, A.D., KOZINA, G.K.: Contact interaction of silicon carbide and boron. Acad. Sci. Uk.SSR 5, 468-472 (1969)
SAMSONOV, G.V., SINEL'NIKOVA, V.S.: On the question of the spread of the homogeneity range in boron carbide. Ukr. Physical J. 6, 687-690 (1961)
SAMSONOV, G.V., SLEPTSOV, V.M.: Forefront version of the phase diagram in the boron-silicon system. Acad. Sci. Uk. SSR 8, 1066-1067 (1962)
SAMSONOV, G.V., SLEPTSOV, V.M.: Investigation of solubility of boron in silicon. ZhNH 8, 2009-2011 (1963)
SAMSONOV, G.V., SLEPTSOV, V.M.: Preparation of boron-silicon alloys. Poroshkovaia Metallurgia 6, 58-67 (1964)
SAMSONOV, G.V., TKACHENKO, Yu.G., DOBROVOL'SKII, A.G., Temperature-dependence of the coefficient of friction of carbides of boron and silicon in vacuum. Poroshkovaia Metallurgia 7, 79-81 (1970)
SAMSONOV, G.V., ZHURAVLEV, N.N., AMNUEL, I.G.: On the question of physicochemical properties of boron-carbon alloys. Fizika Metallov i Metallovedenie 3, Abstr. 2, 309-313 (1956)
SAMSONOV, G.V., ZHURAVLEV, N.N., SHULISHOVA, O.I., PADERNO, Yu.B., SEREBRIAKOVA, T.I.: Interaction of gallium, indium, tallium, germanium, tin and lead with boron. ZhSH 1, 458-463 (1960)
SHAFRAN, I.G.: Oxidation of oxidation-resistant materials in superheated systems. ZhPH 11, 561-563 (1938)
SHAFRAN, I.G., PAVLOVA, M.V.: Analysis of boron and boron carbide. Zav. Lab. 7, 1245-1348 (1938)
SHCHEKIN, V.: Sun batteries. Radio. 8, 29-31 (1958)
SMITH, D., DWORKIN, A., van ARTSDALEN, E.R.: The heat of formation of boron carbide. J. Am. Chem. Soc. 77, 2654-2661 (1955)
STULL, D.R.: JANAF Thermochemical Tables, PB-168370, Clearing House, U.S. Dept. Comm. Natl Bur. of Std. 1965
TIHONOV, A.: Application of boron carbide as an abrasive. Novosti Tehniki 9, 9-11 (1940)
WALD, F.: Investigations on the constitution of certain binary boron metal-systems. J. Less Common Metals 9, 423-435 (1965)
YAMAZAKI, I.M.: Electronic band structure of boron carbide. J. Chem. Phys. 27, 746-751 (1957)
ZHDANOV, G.S., MEERSON, G.A., ZHURAVLEV, N.N., SAMSONOV, G.V.: On the question of solubility of boron and carbon in boron carbide. $B_{12}C_3$, (B_4C). ZhFH 28, Abstr. 6, 1076-1082 (1954)
ZHDANOV, G.S., SEVAST'YANOV, N.G.: Crystal structure of B_4C. Acad. Sci. USSR 32, 432-435 (1941)
ZHDANOV, G.S., SEVAST'YANOV, N.G.: X-ray investigations of boron carbide structure. ZhFH 52, 326-335 (1943)
ZHDANOV, G.S., ZHURAVLEV, N.N., ZEVIN, D.S.: X-ray determinations of formation of solid solutions in boron carbide. Acad. Sci. USSR 92, 767-768 (1953)
ZHURAVLEV, N.N.: X-ray determinations of the structure of SiB_6. Kristallographia 1, 666-668 (1956)
ZHURAVLEV, N.N., MAKARENKO, G.N., SAMSONOV, G.V., SINEL'NIKOVA, V.S., TSEBULIA, G.G.: On the question of properties and composition of phases of boron-carbon alloys. Ed. Acad. Sci. USSR, OTN, Metallurgia i Toplivo 1, 133-141 (1961)

VIII. Borides of Group VIb Elements

G. L. GAL'CHENKO, E. G. LAVUT, E. A. LAVUT, and L. M. VIDAVSKY

1. Boron-Oxygen

Reference is made in the literature to the following boron oxides: B_7O, $B_{13}O_2$, B_6O, B_5O, B_3O, B_4O_2, B_2O, B_4O_3, $(BO)_x$, B_4O_5, B_2O_3 and BO_2. Among these compounds only B_2O_3 has been investigated in detail, while no data are available in the recent literature on the existence of B_5O, B_3O, B_4O_2, B_4O_3, and B_4O_5.

a) Boron Oxide, B_2O_3

In laboratory practice, B_2O_3 is generally used in the vitreous state in the form of a tough, solid mass which begins to soften at a temperature above 200°C. The only practical method of obtaining pure boric oxide is provided by the dehydration of boric acid at high temperature. Low-temperature techniques of boric acid dehydration in a fluidized bed have been developed (BERLIN et al., 1969; MOLLARD, 1970). These techniques make it possible to obtain amorphous B_2O_3 in the form of fine powder.

Studies on the relationship of temperature to diverse physico-chemical properties of boric oxide in the vitreous state prompted numerous workers to conclude that B_2O_3 is a polymer (at least up to a temperature of about 800°C). However, the polymer structure of B_2O_3 is a matter of controversy. For example, BARTENEV and EREMEEVA (1960), TARASOV (1964), and TARASOV et al. (1971) consider B_2O_3 to be a linear polymer having a very small number of bonds between discrete molecules. On the other hand, SPERRY and MACKENZIE (1968) hold a point of view which appears to agree more closely with the behavior of boric oxide in a broad temperature range. They postulate the existence of a strongly cross-linked network at low temperature and the gradual disappearance of the cross-links as the temperature increases. In this way, the polymer acquires linear structure. Further temperature elevation results in complete degradation of polymer chains into discrete B_2O_3 molecules linked with one another by B-O bonds. (The trigonal bi-pyramidal structure once ascribed to the B_2O_3 molecules has been abandoned.) In another model (OTTAR and RUIGH, 1962), discrete B_4O_6 molecules are assumed to be bonded by a temperature-dependent number of commutating B-O bonds, but this model appears to offer no advantages and at the same time disagrees with the NMR and X-ray analysis data. Moreover, the model in question implies the existence of an equilibrium state for each temperature. This assumption is at variance with both theoretical concepts and available experimental results in this field. For example, at temperatures corresponding to the highly elastic state of B_2O_3, BARTENEV et al. (1972) and BARTENEV and ABROSIMOVA (1973) observed enhancement of structure formation in vitreous boric oxide subjected to heat treatment.

Studies involving the use of X-ray structural analysis (PORAI-KOSHITZ and WARREN, 1971; WASEDA and OHTANI, 1972), NMR analysis (KLINE et al., 1968), and Raman and IR spectroscopy (KRISHNAN, 1971; BROADHEAD and NEWMAN, 1971a, b; SHUKER and GAMMON, 1971) warrant the conclusion that vitreous B_2O_3 is essentially composed of interconnected and more or less randomly arranged triangular groups $(BO_3)^{3-}$, but these authors do not agree on details of the structure. Some authors suggest that, apart from planar triangular groups, $(BO_4)^{5-}$ groups in tetrahedral configuration are present, while other authors agree that all the boron atoms have an identical tetrahedral configuration with distorted tetrahedra, so that the boron atom is strongly displaced towards a face. Another group of workers believe that some of the boron atoms are represented by -B=O groups and that the number of such groups increases as the temperature rises.

Under normal conditions, the cooling of molten B_2O_3 results only in the formation of vitreous boric oxide. Crystalline trigonal, B_2O_3 (I) can be prepared by subjecting cubic HBO_2 (I) to prolonged heating at 220°C. The heating time may be decreased in the presence of a seed crystal (McCULLOCH, 1937; REYNAUD, 1967; STANTON, 1971). Another cooling method is to crystallize the vitreous form under 10-20 kbar pressure. (DACHILLE and ROY, 1959; MACKENZIE and CLAUSSEN, 1961; UHLMANN et al., 1967). This method produces single crystals which are large enough for X-ray studies. The single crystals are characterized by a trigonal lattice cell of the following parameters: a = 4.3358; c = 8.3397 Å; space group, $P3_1$, and density, 2.44 g/cm^3 (STRONG and KAPLOW 1968; GURR et al., 1970; STRONG et al., 1971).

Under pressure higher than 20 kbar, orthorhombic B_2O_3 (II) is obtained having the following parameters: a = 6.613, b = 7.803 and c = 4.129 Å; space group, $Ccm2_1$, and density, 3.11 g/cm^3 (PREWITT and SHANNON, 1968). Optimum conditions for preparing this phase are the heating of thoroughly dehydrated B_2O_3 to 1200°C at 65 kbar pressure, followed by cooling to 400°C at a rate of 100°C/h and quenching to room temperature. When the starting mixture contained water, the product contained B_2O_3 (II) crystals. Some colorless, water soluble crystals were assumed to be cubic HBO_2 (I).

The B_2O_3 (II) phase displays kinetic stability under standard pressure up to the melting point. This behavior is corroborated by X-ray studies in a temperature range of up to 525°C. Melting point measurement by means of a warm-stange microscope showed this phase to melt at 525°C, while on the DTA diagram the peak temperature is 550°C. DACHILLE and ROY (1959) and MACKENZIE and CLAUSSEN (1961) plotted tentative high-pressure phase diagrams for B_2O_3, on which the B_2O_3 (I)-B_2O_3 (II) - melt ternary point is approximately 1100°C at 70 kbar.

The gaseous B_2O_3 molecule is known to exist in the monomeric form and among the suggested configurations, namely an angular and a trigonal bipyramidal structure. The first configuration appears to be more likely, although the structure of this molecule is far from fully explained (ROSS and EDWARDS, 1967; MAL'TSEV, 1971).

Boric oxide, B_2O_3, is an acid anhydride and combines vigorously with water to yield boric acid. Even in the molten state at a high temperature, boric anhydride is capable of binding substantial amounts of water. For example, at a temperature of 1000°C and a water vapor pressure of 1 atm, the equilibrium concentration of water in boric anhydride was found to be 2 mole % (FRANZ, 1969).

Numerous borates are obtained by fusing B_2O_3 with basic oxides. Because of the low volatility, boric oxide when heated with salts of volatile acids also yields borates. The acidic nature of B_2O_3 likewise manifests itself in the reaction with alcohols which leads to the formation of esters:

$B_2O_3 + 6R\text{-}OH \rightleftharpoons 2B(OR)_3 + 3H_2O$

The equilibrium is usually shifted by combining the water chemically, but water elimination can also be effected by azeotropic distillation using a suitable organic solvent (NENCETTI and TARTARELLI, 1968).

Fluorine is the sole halogen with which B_2O_3 reacts in forming BF_3. At 300 - 400°C, B_2O_3 reacts with BF_3 and yields trimeric boron oxyfluoride. The reaction between B_2O_3 and carbon at temperatures above 2000°C results in the formation of boron carbide, but at temperatures between 1350° and 1550°C, there occurs the reaction $B_2O_3 + 3C = 3CO + 2B$. At higher temperatures the process of reduction proceeds according to the following equation: $B_2O_3 + C = B_2O_2 + CO$ (RENTZEPIS et al., 1960). The first reaction underlies the well-known process of B_2O_3 halogenation in the presence of carbon. This halogenation occurs at a temperature as low as that corresponding to red glow, because of high volatility of boron halides.

Several active metals are capable of reducing B_2O_3 to free boron. However, this reaction is complicated by the side process of boride formation, since the extent of the secondary reaction is dependent on experimental conditions and the thermal stability of a particular boride. Nonmetallic reducing agents, such as CaH_2 (ERB, 1959), may be used to reduce B_2O_3. However, in this case practically all the boron will be combined in the form of calcium hexaboride.

Boric anhydride, apart from predominant acidic properties, also exhibits to some degree, amphotheric properties, as evidenced by the ability of B_2O_3 to combine with pyrosulfuric acid and form $H[B(S_2O_7)^4]$, which is a strong acid in 45% oleum (DACRE, 1967). FLOWERS et al. (1956) and GREENWOOD and THOMPSON (1959) have reported the formation of a similar compound, $H[B(HSO_4)^4]$, as a reaction product of B_2O_3 or BCl_3 with 100% sulfuric acid. In these compounds, boron exhibits its general tendency to form complexes with four-fold coordination. On the other hand, in such compounds as bisulfate, $B(HSO_4)_3$ (GREENWOOD and THOMPSON, 1959) and perchlorate, $B(ClO_4)_3$ (MOSHER et al., 1963), B_2O_3 functions as a basic oxide in the "classical" sense, inasmuch as the systems in question are anhydrous.

Thermodynamic properties of boric oxide have been studies in detail, and reliable thermodynamic values are reported in relevant handbooks.

Boric oxide can be applied in the production of boron and borides. B_2O_3 additions improve the wear and heat resistance of some polymers, particularly rubbers. Boric oxide is used as a catalyst in organic reactions. B_2O_3 is employed extensively in the manufacture of ceramics, glass, sitalls, high-temperature paints and adhesives, and enamels and fluxes.

b) Boron Oxides B_7O, $B_{13}O_2$ and B_6O

After reducing B_2O_3 with magnesium, PASTERNAK (1959) made X-ray studies of the resulting product. He indexed the powder analysed as $B_{6.6}O$ in terms of an orthorhombic cell (a = 8.20; b = 5.35; c = 5.13 Å). Assuming that Z=4 and using the measured density of 2.64 g/cm^3, the molecular weight of this oxide was calculated as 89.7. However, because the molecular weights of B_7O and $B_{6.5}O$ ($B_{13}O_2$) are 91.7 and 86.3 respectively, the author assigned the compound a formula of B_7O assuming the presence of some B_2O_3 as impurity.

Later LA PLACA and POST (1961) demonstrated that the powder pattern obtained by PASTERNAK could be indexed more satisfactorily on the basis of the rhombohedral cell of the B_4C type having the parameters a = 5.14 Å and α = 62°56'. These parameters correspond to the hexagonal cell parameters of a = 5.37; c = 12.31 Å. The same idea was advanced by MATKOVICH (1961) and LIPP (1966). These authors assumed the phase composition to be $B_{13}O_2$, in which case the calculated density would be 2.80 g/cm^3. More recently, POST (1964) suggested the composition $B_{12}O_2$ and obtained nearly the same values for the calculated density (2.65 g/cm^3) and the measured density (2.64 g/cm^3).

RIZZO et al. (1962) synthesized an oxide having a composition closely corresponding to the formula B_6O (chemical analysis involved B and O determinations). The procedure consisted in reacting stoichiometric amounts of B_2O_3 and boron at 1500°C for one hour in a helium atmosphere (2 atm). The reaction product in the powdered form was red-brown in color and the density of some specimens was as high as 2.62 g/cm^3. For hot-pressed specimens the density was 2.588 g/cm^3.

The hardness of B_6O is inferior only to that of diamond or borazon; the VICKERS microhardness under a load of 100 g is 3.820 kg/mm^2. The oxide is noted for its high chemical stability. It resists the attack of hydrofluoric, hydrochloric or dilute nitric acid, and dissolves with difficulty only in hot concentrated HNO_3. When heated to a temperature of 1850 - 1950°C, the oxide apparently loses a part of oxygen and becomes black in color.

Several papers should be mentioned which are related to this oxide. AHMAD and CAPSIMALIS (1969) subjected boron calcination at a temperature of 1850 - 1950°C in an inert atmosphere containing a mixture of oxygen. They obtained filamentary single crystals, identified as B_7O on the basis of X-ray powder patterns. HOLCOMBE and HORNE (1972a, b) reported preparation of boron suboxide, B_7O, by the reaction $ZnO + 7B = B_7O + Zn$. The procedure involved calcining a pressed mixture of stoichiometric amounts of finely ground reactants at a temperature of 1500°C in an argon atmosphere. However, they presented no strong proof for the suggested formula B_7O.

ZUBOVA and BURDINA (1971) and PETRAK et al. (1972) synthesized a boron suboxide and ascribed to it the formula B_6O (or $B_{12}O_2$ according to the classification suggested by MATKOVICH et al. (1965). In both cases, the synthesis consisted in subjecting a B_2O_3 + B mixture to diverse pressures and temperatures. The authors presented mass-spectrometric evidence to show that the suboxide, on being heated, decomposes and forms boron and B_2O_2. ZUBOVA and BURDINA (1971) did not substantiate the suggested formula B_6O, while PETRAK et al. (1972) cited analysis data which might be ascribed to compounds in the composition range from B_4O to B_8O.

It is quite likely that the suboxide structure consists of closely packed, slightly distorted B_{12} icosahedra with each icosahedron bonded to two discrete oxygen atoms located in interstices among them (HOARD and HUGHES, 1967). If we assume that some kind of disorder in the oxygen is possible the sublattice inconsistencies in the specimen compositions reported by different workers could be clarified to a degree. At any rate, it is reasonable to suggest that the composition range under consideration could be assigned to a single phase.

c) Boron Oxide B_2O

Boron oxide, B_2O, in the form of a red-brown solid material, was prepared and studied by HALL and COMPTON (1965) (cf. also HALL, 1965). The synthesis involved either the reduction of B_2O_3 with boron or lithium, or the oxidation of boron with $KClO_3$ at a temperature range of 1200° to 1800°C under 50 - 75 kbar pressure. The resulting B_2O is thermodynamically stable only under high pressures, so that when the reaction is completed (a heating period of 5 sec and a 3 min dwelling time), the product can be cooled with simultaneous specimen quenching, followed by releasing the pressure.

The crystal lattice of B_2O is of the rhombohedral (graphite) type (alternating layers ....ABCABC....). The lattice parameters are as follows:

$a = 7.98$; $c = 9.09$ Å; $z = 18$;

$d_{calcd.} = 2.24\ g/cm^3$; $d_{meas.} = 2.24 \pm 0.04\ g/cm^3$.

The interatomic distances in the layer are 1.54 Å, while the interlayer distances are 3.03 Å. HALL and COMPTON (1965) suggested that B_2O could be prepared in the diamond-like form, since B_2O and diamond are isoelectronic. So far, attempts in this direction have failed, but X-ray difraction patterns of B_2O specimens obtained under a pressure of 120 kbar showed feeble lines which could be assigned to the cubic diamond-like B_2O.

MAL'TSEV and SHEVEL'KOV (1971) examined regularities in vibrational frequencies and bond lengths of M_2O and MO suboxide molecules for boron subgroup elements. They then evaluated these parameters for the hypothetical B_2O molecule and obtained the following values:

$\nu_1 = 1450$, $\nu_2 = 420$, and $\nu_3 = 1870\ cm^{-1}$; $R_{B-O} = 1.29$ Å.

No IR spectrum of gaseous B_2O could be obtained, presumably due to B_2O instability under low pressure.

d) Boron Oxide $(BO)_x$

ZINTL et al. (1940) were the first to synthesize $(BO)_x$ by calcining a pressed mixture of boron and ZrO_2 at 1800°C in a vacuum. The composition of the resulting light-brown sublimated material approximated that of boron monoxide. Later, RAY and SINHA (1941) reported the preparation of $(BO)_x$ by leaching the reacted mixture Mg + B_2O_3 with aqueous ammonia. Next, they evaporated the solution and subjected the dry residue to thermal decomposition in a vacuum. When decomposition was caused by "mild" heating, the product was a colorless compound, readily soluble in water. Though the formula of B_4O_5 was ascribed to the product, it is reasonable to assume that the product in question was an incompletely dehydrated white modification of boron monoxide. The same was described by WARTIK and APPLE (1955). These authors modified the RAY and SINHA method by decomposing sub-boric acid, $B_2(OH)_4$, in vacuum, which was prepared by the action of water vapor on diboron tetrachloride at room temperature. This material began to lose water in vacuum at about 90°C, but to remove the last traces of water it was necessary to heat the product for four hours at 220°C. The resulting boron monoxide was white in color and appeared substantially unchanged at temperatures up to about 500°C. Only at 650°C, in vacuum, was the well-known light brown modification obtained. When the white, solid boron monoxide was not completely dehydrated, its conversion to the brown form proceeded at a lower temperature and was accompanied by heat evolution. Thus, samples having the approximate composition $BO \cdot 1/10\ H_2O$, when briefly heated to about 400°C in vacuum, underwent spontaneous heating to incandescence. Also, the conversion to the brown form was accompanied by the evolution of a small amount of hydrogen and the formation of minute quantities of a white solid.

Both forms of the monoxide yield aqueous solutions with reducing properties. However, the brown form is much less soluble than the white form. WARTIK and APPLE (1955) suggested that boron monoxide, in both modifications, is polymeric and assigned the following formula:

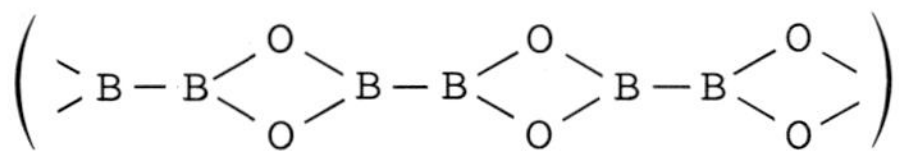

(white form)

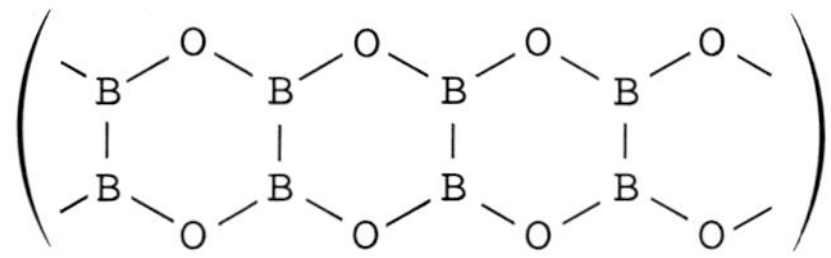

(brown form)

The observed coloration of the second form may be explained by the contribution of quinoid structures such as:

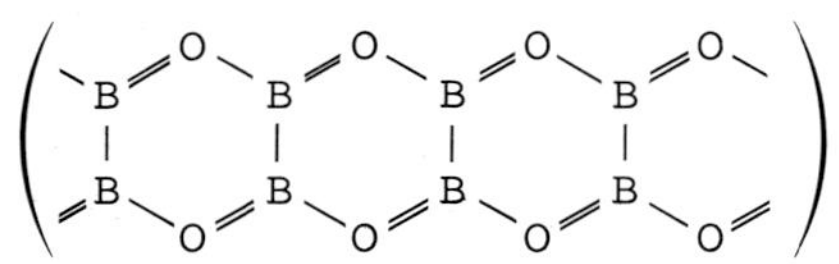

Later, McCLOSKEY et al. (1961a, b) assumed the existence of three-dimensional molecules, apart from linear molecules, and ascribed to the white modification less strained six-membered cycles (boroxol-type cycles), instead of four-membered cycles:

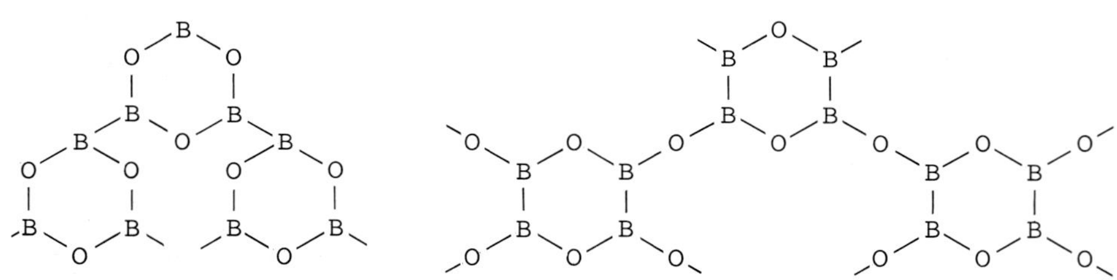

(white form) (brown form)

It is interesting to note that NIKITIN et al. (1963), in the course of studying the IR spectra of boron monoxide powders, found the spectrum of the brown modification to be quite similar to the IR spectrum of B_2O_3. From this observation they concluded that the brown form of boron monoxide is actually a mixture of B_2O_3 and boron. This inference, however, can hardly be reconciled with the data of other workers obtained by studying the properties of brown boron monoxide.

McCLOSKEY et al. (1961a, b) found that the white modification absorbed readily atmospheric moisture. As a result it underwent reconversion to subboric acid within 1.5 h at room temperature and a humidity of 50%. These workers carried out an interesting experiment by reacting boron monoxide with BCl_3 vapors at a temperature of about 200°C and obtaining diboron tetrachloride. These findings provide convincing proof that B-B bonds in boron monoxide remained intact.

Methods of synthesizing boron monoxide similar to the Zintl method have found widespread use. For example, KANDA et al. (1956); NICHOLLS (1965) prepared boron monoxide by reducing B_2O_3 with boron in vacuum at temperatures of 1050° to 1400°C. The reaction product collected in the cooler part of a reactor is a vitreous, amber-like sublimate. From X-ray data, Kanda and co-workers found the sublimate to be amorphous and the density of the material to be 1.765 ± 0.001 g/cm³. Attempts to induce crystallization by heating to 400°C, failed, while at temperatures above 450°C the product underwent disproportionation and yielded B_2O_3 and boron. Boron monoxide reacted vigorously with water and alcohols, evolving hydrogen with traces of boranes. The resulting solution exhibited reducing properties, as did the solution reported by McCLOSKEY et al. (1961a, b). These workers ascribed the dissimilarity in the properties of boron monoxide samples obtained by various investigators to the fact that the samples sublimed in vacuum contain up to 30% of boron in a form different from boron monoxide. This species is responsible for the formation of boranes in contact with water.

In subsequent syntheses of boron monoxide, CLARK et al. (1963) also used diverse borates, while NICHOLLS (1965) employed C, B, B_4C, BN, Ti, TiC and TiB_2 for the reduction of B_2O_3. BARTON and NICHOLLS (1964, 1966) also employed B_4C for reducing TiO_2 and ZrO_2 in the synthesis of boron monoxide.

An interesting application of boron monoxide, either as such or in the form of reaction mixtures that yield boron monoxide, is associated with the synthesis of boranes from BO in contact with hydrogen at high temperatures. The high-temperature syntheses of boron monoxide is for the most part related, in one way or another, to the research in the field of boranes.

According to INGHRAM et al. (1956) and SCHEER (1958), gaseous boron monoxide exists up to temperatures of 1500°C. It exists mainly in the form of dimeric molecules, and is assigned a linear confirguation of O = B = B = O. The concentration of monomeric molecules becomes significant at higher temperatures only (UY and DROWART, 1970; SHRIVASTAVA et al., 1971).

e) Boron Oxide BO_2

Studies of the so-called "Boric Acid Fluctuation Bands" in the 3700 - 6800 Å wave length region made it possible to suggest the existence of BO_2. Subsequent investigations (KASKAN and MILLIKAN, 1960a, b; KASKAN et al., 1961; MAL'TSEV et al., 1961; MATVEEV et al., 1961; RUSIN and TATEVSKY, 1961) provided convincing proof that "Boric Acid Fluctuation Bands", (BAFB) do owe their origin to BO_2. The following arguments were advanced to substantiate this conclusion:

1. The intensity of BAFB increases in the presence of oxygen and drops to zero in the presence of hydrogen or water; these changes in intensity indicate an enhanced oxygen content in the molecule under examination as compared to B_2O_3;

2. In samples having different B^{10}; B^{11} ratios, a single isotopic shift of BAFB edges was observed, so that we can only be certain of presence of one boron atom in the spectrum-producing molecule;

3. Studies of the relationship between BAFB intensity of gaseous mixtures of boron compounds either in the burner flame or during explosion and equilibrium gaseous mixtures obtained by combustion or explosion, indicate BO_2 molecules as the exclusive source of BAFB.

KASKAN and MILLIKAN (1960a, b) using equilibrium data for

$$HBO_2 + OH \rightleftharpoons H_2O + BO_2$$

calculated the enthalpy of BO_2 (g) formation to be -84 kcal/mole. Later, RUSIN and TATEVSKY (1963, 1965) showed this value to be excessively high, since they ignored a pronounced temperature dependence of absorption by BO_2 vapors in the 1400-2200 K temperature range. They reported the value of $\Delta H^o_{f,298}$ (BO_2, g) = -68 ± 7 kcal/mole. RUSIN and TATEVSKY (1963, 1965) found that at a temperature of 3000°K, the principal products of B_2H_2 combustion in H_2, O_2, CO, Ar and N_2 mixtures are HBO_2 and BO_2 in an approximate ratio of 2 : 1.

Although KASKAN and MILLIKAN (1960a, b) also employed IR spectroscopy, the majority of the above papers report the results of studies based on absorption in the visible part of the spectrum (predominantly, in the 5470 Å band). In later investigations, BO_2 was likewise studied by using IR absorption spectra in a solid argon matrix (SOMMER et al., 1963) and fluorescence spectra obtained with the mixtures of BCl_3 (RUSSEL et al., 1973) or $B(C_2H_5)_3$ (SHARMA and JOSHI, 1972) with atomic oxygen, the reaction product being BO_2.

Among other investigations, mention should be made of the mass spectrometric study of the equilibrium:

$$HBO_2\ (g) + OH(g) \rightleftharpoons H_2O(g) + BO_2(g)$$

The reaction was conducted by passing H_2O vapors into a Knudsen cell containing molten B_2O_3 at a temperature of from 1160 to 1400°K. In contrast to the results of INGHRAM et al. (1956), UY et al. (1971) found by ionizing BO_2 molecules at relatively low ionizing voltage, found the BO_2 ions formed in the mass spectrum. They calculated the $\Delta H^o_{f,0}$ (BO_2,g) to be - 77.2 ± 2.5 kcal/mole. The atomization energy D_0^o = 327.0 ± 5.0 kcal/mole was estimated from this value. Hence, the bonding energy of oxygen in BO_2 lies between respective values for B-O (133 kcal/mole) and B=O (191.8 ± 1.2 kcal/mole).

2. Boron-Sulfur

The boron-sulfur system includes binary compounds B_2S_3, BS, B_2S_2, BS_2, B_2S_5, $B_{12}S$, $(B_2S_3)_m$ and $(BS_2)_n$. In this group, only diboron trisulfide has been studied in detail in both the solid and gaseous states. In general, boron sulfides are not well understood. Furthermore, the very existence of B_2S_5 and B_4S, in particular, are disputed.

Investigations in the field of boron sulfides are hampered by their high sensitivity towards moisture. Only $B_{12}S$ is stable when flushed in water and other reagents.

The initial important contribution to the study of boron sulfides occured at the turn of the century, but further progress in the field has only been made in the last 15 years. The results of earlier investigations pertaining to boron sulfides are conveniently presented in the section "Boron" of Gmelin's handbook (GMELIN, 1926). MUETTERTIES (1967) reviewed the vast array of studies on the chemistry of boron and sulfur compounds, including their organic derivatives. In addition, CRAGG (1968) compiled abstracts of paper on the chemistry of B-S system compounds for the years 1950 - 1967. Selected boron sulfides and ternary boron-sulfurmetal compounds are discussed in the review published by ROSSO and DUBUSK (1969).

All currently known boron sulfides are described below.

a) Diboron Trisulfide

B_2S_3, is the most important compound of the boron-sulfur system. It is the principal product in high-temperature reactions of either boron with sulfur, hydrogen sulfide, carbon desulfide, or metal (arsenic, antimony, tin, lead, zinc or iron) sulfides, or of metal borides and alloys (ferroboron and manganoboron) or boron halides (BCl_3, B_2Cl_4, BBr_3) with hydrogen sulfide, as well as boron oxide with aluminum sulfide. Diboron trisulfide prepared in this way can be used for the synthesis of other boron sulfides for research purposes.

These pathways of B_2S_3 synthesis were discovered in the early period of research concerned with boron sulfides (GMELIN, 1926). Recent studies showed that the more promising synthetic routes can be improved. These are: (1) the synthesis of B_2S_3 from boron and hydrogen sulfide, and (2) thermal decomposition of metathioboric acid.

The synthesis of B_2S_3 according to the first method involves the formation of metathioboric acid (HBS_2) as an intermediate product and can be represented by the following sequence of reactions:

$$2B\ (s) + 4H_2S\ (g) = 2HBS_2(g) + 3H_2\ (g) \quad (1)$$

$$HBS_2\ (g) = HBS_2\ (s) \quad (2)$$

$$2HBS_2\ (s) = B_2S_3\ (glass) + H_2S\ (g) \quad (3)$$

Reaction (1) is begun in a quartz tube at 500°C. Towards the end of the process the temperature is raised to 600 - 800°C. Hydrogen sulfide is carefully dried, and then mixed with H_2 to prevent hydrogen sulfide from decomposing by the reaction $xH_2S\ (g) = x\ H_2(g) + S_x(g)$. This hydrogen sulfide mixture is then fed to the quartz tube in an inert carrier (helium, argon), and the metathioboric acid obtained condenses in the cooler part of the tube [Eq.(2)].

Reaction (3) is usually conducted in a separate quartz or graphite container in vacuum. The process is begun at a temperature of 100°C and then raised to 300°C. Boron sesquisulfide, B_2S_3, is obtained in the glassy state, and the analysis of a typical sample is 95% B_2S_3, 0.7% sulfur, 2.7% SiO_2, and 1.4% oxygen (GREENE and GILLES, 1964). The sulfur can be removed in part by reaction with H_2, but so far no effective method for purifying B_2S_3 has been elaborated. In this regard, THOMAS and TRIDOT (1964) emphasized the advantages offered by the low-temperature method (2) based on the employment of metathioboric acid. They synthesized the acid by reacting BBr_3 with H_2S at 55°C in a carbon disulfide medium according to the following equation:

$$3BBr_3 + 6H_2S \longrightarrow (HBS_2)_3 + 9HBr \quad (4)$$

The acid was crystallized at -40°C in a stream of H_2S as white needle crystals (trimer). At 95°C, the $(HBS_2)_3$ underwent rapid and complete decomposition to form H_2S and B_2S_3. The resulting diboron trisulfide was a highly pure white amorphous powder. The preparation gives no absorption bands in IR spectra in the 2 - 15 micron region, displays appreciable volatility at 300°C in a stream of H_2, and undergoes sublimation, without decomposition, in vacuum at 570°C (THOMAS and TRIDOT, 1964). One disadvantage of the above method of preparing B_2S_3 is that the reaction (4) proceeds slowly and takes about 12 days. EDWARDS et al. (1966) studied the vaporization and thermal decomposition of metathioboric acid with a mass spectrometer, and found HBS_2 decomposition to be a complicated process which, in contrast to the decomposition of HBO_2, involves a number of intermediate metastable compounds.

Direct synthesis of B_2S_3 from boron and sulfur calls for high temperatures. The reaction begins at 600°C (MOISSAN, 1892), but terminates because a B_2S_3 coating on the boron surface forms (KAMSU KOM 1964). Thus, a temperature as high as 1200°C is required to bring the reaction to completion (MOISSAN, 1892). The samples of B_2S_3 prepared by this method contain an excess of sulfur and other impurities arising from the contact of reactants with the material from which the reactor is made (quartz or graphite). Stoichiometric B_2S_3 containing no sulfur impurity can be obtained by reacting boron with ZnS at 700°C (SOMMER et al., 1960), or with FeS at 650°C (CHEN and GILLES, 1970). SNYDER (1958) patented a method for the preparation of B_2S_3 involving the reaction of H_2S or sulfur vapors at 400 - 700°C in a liquid medium with boron or calcium and iron borides:

$$FeB_2 + 4H_2S \xrightarrow{700^\circ C} B_2S_3 + FeS + 4H_2$$

The crystallization of B_2S_3 proceeds with difficulty. HAGENMULLER and CHOPIN (1962) prepared crystalline B_2S_3 by heating a mixture of boron and sulfur taken in an atomic ratio of 2:3, in a sealed quartz tube at 590°C. They were the first to obtain an X-ray powder pattern of B_2S_3. CHEN et al. (1970) prepared B_2S_3 white crystals (0.1 × 0.1 × 0.2 mm) by heating FeS and boron at 700 - 730°C in a graphite tube contained inside a quartz tube. The crystalline B_2S_3 samples were 99.1% pure (the impurity being carbon) and were characterized by a density of 1.71 - 1.72 g/cm^3 and a melting point of 563 ± 3°C. CHEN et al. (1970) obtained the crystal patterns for B_2S_3 and X-ray powder patterns identical with those reported by HAGENMULLER and CHOPIN (1962). From these data they determined the monoclinic lattice parameters of B_2S_3 to be: a = 4.048 ± 0.004 Å; b = 37.09 ± 0.04 Å; c = 10.73 ± 0.01 Å; Δ = 90° ± 1.5°; space group, C2, Cm or C 2/m, and the number of molecules in a unit cell $\underline{Z}$ = 14.

Using mass spectrometric data, CHEN and GILLES (1972) found the relationship between the vapor pressure and temperature for B_2S_3 (cryst.) in the 722 - 840°K range to be $\log P_{atm.} = -(12.04 \pm 0.13) \cdot 10^3/T + (10.26 \pm 0.17)$ and for B_2S_3 (1) in the 840 - 939°K range, $\log P_{atm.} = -(9.51 \pm 0.37) \cdot 10^3/T + (7.28 \pm 0.42)$. They calculated the enthalpy of melting ΔH_m = 11.6 ± 1.8 kcal/mole and the entropy of fusion ΔS_m = 13.7 ± 2.1 cal/mole. deg. for the melting temperature of 567 ± 10°C. The standard enthalpy of formation, ΔH_f°, 298.15 kcal/mole, for B_2S_3(glass) was found to be - 58.4, and for B_2S_3 (cryst.), - 60.4 (SABATIER, 1891a, b) and - 60.3 (MOROSOVA and RYBAKOVA, 1968).

According to GREENBERG et al. (1969) diboron trisulfide undergoes the disproportionation in the 700 - 950°C range as follows:

B_2S_3 (g) = BS_2(g) + BS(g), For this reaction:

ΔH° = 34.9 ± 1.5 kcal/mole, ΔS = 35.5 ± 1.4 cal/mole/deg, log K

= - (7630 ± 300/T) + (7.76 ± 0.30).

SOMMER et al. (1960) studied mass spectra and IR emission spectra of the vapors of boron sulfide prepared by reacting ZnS with B at 1000°C. From the data obtained they estimated that for B_2S_3 (g), $\Delta H_{f,0}^\circ$ = -11 kcal/mole, and that the bond energies for B-S and S-S bonds equal 125 kcal/mole and 100 kcal/mole, respectively.

AKISHIN and SPIRIDONOV (1959) assessed the geometric configuration of the B_2S_3 molecule by the electronographic method. From the feasible models they selected the angular model of symmetry C_{2v}, and obtained the following parameters:

$r(B=S) = 1.65 \pm 0.03$ Å; $r(B-S) = 1.81 \pm 0.02$ Å;

$B-S=B$ angle is $96^\circ \pm 5O$, and $S=B-S$ angle is 180°.

GREENE and MARGRAVE (1959) studied IR emission and absorption spectra of B_2S_3 at 600 - 1000°C and assigned a twisted zigzag model of C_2 symmetry for the B_2S_3 molecule.

Chemically, diboron trisulfide is a thioanhydride and hydrolyzes readily according to the equation:

$$B_2S_3\ (s) + 6H_2O\ (l) = 2H_3BO_3\,(s) + 3H_2S\ (g).$$

For the most part, chemical activity of B_2S_3 has been studied in relation to various classes or organic compounds, such as alcohols, aldehydes, carboxylic acids, mercaptans, etc. (CRAGG, 1968). Diboron trisulfide appears to offer promise as a valuable synthetic reagent.

b) Boron Sulfides BS (g) and B_2S_2 (g)

Boron sulfides BS and B_2S_2 are known to exist in the gaseous state only. Boron monosulfide, BS, has been identified in the examination of vapors of B_2S_3 at 700°C from emission spectra (ZEEMAN, 1950, 1951). Molecular constants of the BS molecule have been determined in the ground state and in several excited states (ZEEMAN, 1951; MAL'TSEV, 1965; KORYAZHKIN, 1967). For BS, the interatomic distance equals 1.8177 Å (ZEEMAN, 1951). The ratio of dissociation energies, D_0° (BO) / D_0° (BS), is 1.26 (MAL'TSEV, 1965). According to UY and DROWART (1970), D_0° (BS) = 137.9 ± 2.2 kcal/mole, so that $\Delta H^\circ_{f,298.15}$ BS(g) = 60 kcal/mole.

The sulfide B_2S_2 (g) is detected in the vapors resulting from the reaction of ZnS with boron at 700 - 900°C (SOMMER et al., 1960), as well as from the systems B_2S_3 + Zn or FeS + B + Zn at 650°C (CHEN and GILLES, 1970). From mass spectrometric data, the heat of formation of B_2S_2 (g) was estimated to be $\Delta H^\circ_{f,0} = 18 \pm 10$ kcal/mole (SOMMER et al., 1960).

c) Boron Sulfide BS_2

Crystalline BS_2 was prepared by MEDVEDEVA et al. (1968) by heating a mixture of amorphous boron and sulfur at 600 - 700°C for 24 h. The resulting white boron disulfide, BS_2, had a density of 1.96 g/cm^3. BS_2 reacts vigorously with water to produce hydrogen sulfide. Samples of BS_2 melt congruently at a temperature of 417 ± 2°C, and the heat of fusion, as assessed from thermogravimetric measurements, is ΔH_m = 8.9 kcal/mole, and the entropy of fusion was found to be ΔS_m = 12.9 cal/deg/mole.

KORYAZHKIN and MAL'TSEV (1966) analyzed the electronic absorption spectra of BS_2 in the 4100 - 5550 Å region at 1000°C. They concluded that the configuration of the BS_2 molecule should be slightly bent configuration rather than linear. The heat of formation $\Delta H^\circ_{f,0}$, for BS_2 (g) was estimated from mass spectrometric data to be 13 ± 10 kcal/mole (SOMMER et al., 1960).

d) Boron Sulfide B_2S_5

Only contradictory data are available on the existence of diboron pentasulfide. MOISSAN (1892c) reported the preparation of B_2S_5 by reacting sulfur with boron iodide in a CS_2 solution at 60°C.

HAGENMULLER and CHOPIN (1962) heated amorphous boron and excess sulfur in a sealed quartz vessel at 590°C. From the reaction mixture obtained they separated a yellowish powder of B_2S_5 composition. They found this compound to be distinct from B_2S_3 in both X-ray diffraction pattern and water reaction. When separated from H_3BO_3 and H_2S, B_2S_5 yields free sulfur. MEDVEDEVA et al. (1968) carried out the thermal analysis of a sample prepared by the same technique. They concluded that the material in question is not an individual compound of the formula B_2S_5, but a (BS_2 + α-S) eutectic. Also, X-ray diffraction patterns of this material indicated existence of the BS_2 and S phases.

e) Boron Sulfide B_4S

SABATIER (1891b) reported the synthesis of boron sulfide B_4S by heating amorphous boron to incandescence in a stream of hydrogen sulfide. Boron sulfide B_4S is a black product of considerable hardness. When reacted with water, it yields elemental boron, as well as boric acid and hydrogen sulfide. In other works on boron sulfides, the preparation of B_4S under similar reaction conditions is not mentioned. GMELIN (1926) cites early papers which report experiments similar to that of SABATIER, but they resulted in the formation of a mixture of boron and B_2S_3.

f) Boron Sulfide $B_{12}S$

This compound, the highest boron-content sulfide yet produced, was synthesized by MATKOVICH (1961). He fired a compressed, 1:1 ratio mixture of powdered amorphous boron and sulfur at 1600 - 1700°C. The reaction mixture was placed in a graphite crucible which was heated in an induction furnace. In order to avoid sulfur loss, it was essential that the sample be heated very quickly to the required temperature. Also, the sample must be maintained at the highest temperature for a short time (1 to 2 min), since the reaction product undergoes rapid decomposition at temperatures exceeding 1700°C. The reaction product is then kept at 1650°C until evolution of excess sulfur ceases, and the sample appears uniformly black. (ROSSO, 1968), suggested preparing $B_{12}S$ in an apparatus of a different design, in which a reaction mixture pellet is dropped into a crucible when the temperature of the crucible in a furnace has reached 1650°C. ROSSO used this technique to synthesize $B_{12}C_2S$, a compound related to $B_{12}S$.

The crystalline structure of the $B_{12}S$ sulfide is similar to that of $B_{12}C_3$ and $B_{13}C_2$ carbides and is characterized by a rhombohedral unit cell with parameters a = 5.19 Å and α = 67°56'. The sulfur atom is assumed to be placed in the center of the unit cell. The density of the $B_{12}S$ sample containing about 2% carbon as impurity was found to be 2.40 g/cm³, while the density calculated from the X-ray data equalled 2.33 g/cm³. MATKOVICH regarded $B_{12}S$ as an interstitial compound and did not exclude the possibility of a homogeneity region existing in the $B_{12}S$ phase.

Since $B_{12}S$ sulfide, like boron carbide, displays high chemical and thermal stability, it appears to hold much promise in semiconductor technology.

g) High-Molecular Weight Boron Sulfides $(B_2S_3)m$ and $(BS_2)n$

GREENE and GILLES (1962) found the mass spectrum of vapors over B_2S_3 glass containing sulfur to be complex. As evidence, they pointed out the presence of about 30 ions of boron sulfides ranging in composition from $B_2S_3^+$ to $B_{10}S_{17}^+$. The complexity of the mass spectrum was assumed to result from a large number of parent compounds, namely boron sulfide molecules of a high molecular weight and dissimilar structures (chains, networks, icosahedra, etc., like those observed in boranes). This finding attracted considerable interest and gave impetus to subsequent extensive studies of the boron-sulfur system. High temperature, mass spectrometric, spectroscopic and tensimetric techniques have been used. Further investigations, GREENE and GILLES (1964) concluded that the complex mass spectra of vapors over sulfur-rich B_2S_3 samples (glass at 300°C and a viscous liquid at 600°C) owe their origin to the ionization of polymeric B_nS_{2n} molecules (n = 8; 7; 10; 9 and 6 in the order of decreasing relative amounts of polymers in the vapors) in equilibrium with B_2S_3. The total pressure of polymeric boron sulfides was found to be equal to about 10% of the pressure of B_2S_3 vapors. GREENE and GILLES estimated the heat of vaporization of B_2S_3 as 36 ± 5 kcal/mole, and that of polymeric sulfides to be 55 ± 10 kcal/mole.

At temperature below 300°C, the principal vapor component over the sulfur-rich B_2S_3 glass samples is the B_8S_{16} polymer (EDWARDS et al., 1971). This polymer is assumed to have a ring structure consisting of a B-atom cycle located between two cycles built up by sulfur atoms.

It has not yet been determined how $(BS_2)_m$ is formed in sulfur-rich B_2S_3 samples, but two hypotheses have been advanced (GREENE and GILLES, 1964; EDWARDS et al., 1971), to explain this formation.

1. The sulfur dissolved in the sample reacted with B_2S_3.

2. The dissolved hydrogen sulfide in the sample reacted with B_2S_3. For example, B_8S_{16} may appear as a result of the decomposition of the intermediate mestastable $H_2B_2S_5$ species.

The presence of polymeric boron sulfides is responsible for high viscosity inherent in diboron trisulfide melts and enhances the tendency of B_2S_3 samples to form glasses.

GREENBERG et al. (1970a) investigated crystalline BS_2 samples and determined the relationship of temperature to equilibrium constants of the dissociation reaction:

$$(BS_2)_n\ (g) \rightleftharpoons nBS_2\ (g)$$

for the $(BS_2)_n$ dimer and tetramer at 550 - 950°C, and also calculated the heats and entropies of this reaction: for the dimer $\log (Kp)_2 = (-7.69 \pm 0.82) \cdot 10^{-3}/T + (9.19 \pm 0.81)$ (p-mm of mercury), $\Delta H_T = 35.2 \pm 3.8$ kcal/mole, and $\Delta S_T = 42.0 \pm 3.7$ cal/deg mole; for the tetramer, $\log (Kp)_4 = -(12.19 \pm 0.70) \cdot 10^{-3}/T + (15.18 \pm 0.66)$ (p-mm of mercury), $\Delta H_T = 55.8 \pm 3.2$ kcal/mole, and $\Delta S_T = 69.5 \pm 3.0$ cal/deg mole. In the temperature range of 590 to 765°C, the partial vapor pressures dependence on temperature and heat and entropy of volatilization were found to be as follows:

1. for the monomer, $\log P(BS_2) = (-4120 \pm 40)/T + (5.54 \pm 0.04)$ (p-mm of mercury), $\Delta H_T = 18.8 \pm 0.2$ kcal/mole, and $\Delta S_T = 25.3 \pm 0.2$ cal/deg mole;

2. for the dimer, $\log P(BS_2)_2 = (-900 \pm 50)/T + (2.29 \pm 0.06)$ (p-mm of mercury), $\Delta H_T = 4.1 \pm 0.2$ kcal/mole, $\Delta S_T = 10.5 \pm 0.3$ cal/deg mole; and

3. for the tetramer, log $P(BS_2)_4 = (-4240 \pm 1.30)/T + (6.94 \pm 0.14)$ (p-mm of mercury), $\Delta H_T = 19.4 \pm 0.6$ kcal/mole, and $\Delta S_T = 31.8 \pm 0.6$ cal/deg mole.

When B_2S_3 is synthesized by high-temperature reactions in the systems Zn + B and FeS + B, no polymeric sulfides, $(BS_2)_n$ (g), will be formed, and the gas phase contains only the sulfides B_2S_3 and B_2S_2 (SOMMER et al., 1960; CHEN and GILLES, 1970).

When crystalline B_2S_3 samples free of sulfur impurities are vaporized at 400 - 650°C, they yield in the gaseous phase only the $(B_2S_3)_{\underline{m}}$ species, where $\underline{m}$ = 1,2...5 (CHEN and GILLES, 1970). For the B_4S_6 ($\overline{g}$) dimer, in the temperature range of 722 - 840°K, the dependence of vapor pressure on temperature (over crystalline B_2S_3) was found to be (CHEN and GILLES, 1972) log P (atm., B_4S_6) $= -(13.52 \pm 0.27) \cdot 10^3/T + (10.62 \pm 0.34)$. At temperatures of 840 - 939°K (over molten B_2S_3) log P (atm., B_4S_6) $= -(9.21 \pm 0.41) \cdot 10^3/T + (5.44 \pm 0.45)$. The heat and entropy of dimerization for B_2S_3 were calculated to be $\Delta H^o = -47.8 \pm 1.6$ kcal/mole (dimer) and $\Delta S = -44.6 \pm 2.0$ cal/deg mole (dimer), respectively. CHEN and GILLES suggest that B_2S_3 dimerization leads to the formation of a four-membered cycle involving two B-S bonds of two B_2S_3 molecules. The molten diboron trisulfide is assumed to contain predominantly dimeric molecules $(B_2S_3)_2$.

3. Boron-Selenium, Boron-Tellurium, and Boron-Polonium

For the most part, the published data pertain to boron selenides. The formation of boron tellurides was not detected in either the condensed phase (MOISSAN, 1892a, 1895) or the gaseous phase (KORYAZHKIN, 1967; UY and DROWART, 1970). The boron-polonium system has not been investigated.

In the boron-selenium system, the only solid compound is B_2Se_3, while in the gaseous phase the BSe_2 and BSe molecules are formed. The existence of B_4Se (SABATIER, 1891a) has not been corroborated. According to BORYAKOVA et al. (1969) and HILLEL and CUEILLERON (1972), the phase diagram in the region of B_2Se_3 - Se presents an eutectic devoid of a discernible solid solution. BORYAKOVA et al. (1969) regards B_2Se_3 - Se as a degenerate eutectic, but HILLEL and CUEILLERON (1972) present evidence that the eutectic corresponds to about 90 at. % selenium and the eutectic temperature equals 50 ± 5°C.

SABATIER (1891a, b) and later MOISSAN (1892, 1895) synthesized boron selenide by reacting selenium or hydrogen selenide with boron; HOFFMAN (1911) substituted iron or manganese borides for boron. In these early investigations the composition and purity of the reaction product were not determined.

HUTCHINSON and EICK (1962) and CUEILLERON and HILLEL (1967) prepared boron selenide by reacting hydrogen selenide with boron. The synthesis of boron selenides from elements was described by CUEILLERON and HILLEL (1967), BORYAKOVA et al. (1969), MELUCCI and WAHLBECK (1970), GREENBERG et al. (1970), DEMBOVSKY et al. (1971), and HILLEL and CUEILLERON (1972). BORYAKOVA et al. (1969) carried out the synthesis in a rotary furnace at 650 - 700°C. The reaction product was a glass. HILLEL and CUEILLERON (1972) suggested heating boron and an excess of selenium in quartz ampules with an internal protective layer of carbon. Prolonged heating of the reaction product in vacuum results in selenium removal and in the formation of the crystalline stoichiometric boron selenide. Boron selenohalides (CUEILLERON and HILLEL, 1968) and, probably, copper selenoboride (KAMSU KOM, 1964), when subjected to thermal dissociation, yield B_2Se_3 containing no impurity of selenium which is hardly removable.

Pure crystalline B_2Se_3 samples are light grey in color (HILLEL and CUEILLERON, 1972), but impurity of elemental selenium imparts a red hue. Poorly crystalline powders are either orange or ochre in color (CUEILLERON and HILLEL, 1967), while the glassy B_2Se_3 samples are red (DEMBOVSKY et al., 1971, 1972). GREENBERG et al. (1970b) prepared cherry-red prismatic crystals. In yellow B_2Se_3 samples described by HUTCHISON and EICK (1962) and earlier workers, yellow coloration is presumably caused by presence of silicon diselenide. BORYAKOVA et al. (1969) determined the melting point of B_2Se_3 as $T_m = 753°K$, while HILLEL and CUEILLERON (1972) found the melting point to be $T_m = 752 \pm 3°K$. They also reported the value of the heat of fusion, $\Delta H_m = 2.5 \pm 0.3$ kcal/mole, so that the entropy of fusion can be estimated as $\Delta S_m = \Delta H_m/T_m = 3.3 \pm 0.4$ cal/mole deg. In the case of B_2Se_3, glass softening begins at 325°C (BORYAKOVA et al., 1969; DEMBOVSKY et al., 1971, 1972). From the X-ray evidence reported by HILLEL and CUEILLERON (1972), it follows that boron selenide is characterized by the monoclinic unit cell having the following parameters: a = 4.06 ± 0.01, b = 38.6 ± 0.1, and c = 10.43 ± 0.02 Å; β = 90°; space group, P_2P_m or $P_{2/m}$; Z = 8, and $d_{calcd.}$ = 2.10 g/cm³. The analogy between boron sulfide and boron selenide underlying the indexing of boron selenide X-ray patterns appears to lack complete justification in view of a substantial difference in the values of Z and the dissimilarity of space groups. According to HILLEL and CUEILLERON, the absorption maxima in the IR spectra of B_2Se_3 (glass) lie at 775, 838, and 861 cm^{-1}, and those of B_2Se_3 (cryst), at 768, 789, 804, 831, 868, and 883 cm^{-1}.

HUTCHINSON and EICK (1962), CUEILLERON and HILLEL (1967), and BORYAKOVA et al. (1969) presented IR spectroscopic data and X-ray diffraction patterns for B_2Se_3.DEMBOVSKY et al. (1972) also published the results of X-ray studies. It should be noted that the IR spectra and X-ray diffraction patterns of the samples reported by HUTCHINSON and EICK differ essentially from those reported by other workers.

Boron selenide, B_2Se_3, and selenium yield a mixture of the composition B_2Se_3 + 0.85 Se. This mixture is not separated readily; HILLEL and CUEILLERON (1972) assumed free selenium to occupy a part of interstices in the selenide structure. B_2Se_3 undergoes hydrolysis easily and forms boric acid and hydrogen selenide. At a temperature above 600°C, the reaction of B_2Se_3 with quartz yields silicon diselenide (CUEILLERON and HILLEL, 1967), while copper selenide, Cu_2Se, reacts with B_2Se_3 to form $CuBSe_2$ (KAMSU KOM et al., 1963; KAMSU KOM, 1964). Heating B_2Se_3 and thallium selenide at a temperature of up to 800°C results in the formation of glasses (DEMBOVSKY et al., 1971). The compounds of boron selenide with thallium selenide have been identified by DEMBOVSKY et al. (1972).

Thermal dissociation of boron selenide at 556 ± 10°C reported by HUTCHINSON and EICK (1962) has not been corroborated by subsequent investigations (BORYAKOVA et al., 1969; HILLEL and CUEILLERON, 1972; MELUCCI and WAHLBECK, 1970; GREENBERG et al., 1970). According to GREENBERG et al., (1970a, b) the process of disproportionation in the 675 - 1000°C range proceeds as follows: $B_2Se_3\,(g) \rightleftharpoons BSe_2\,(g) + BSe(g)$. In this reaction, enthalpy and entropy changes (948 - 1273°K) are equal to $\Delta H = 49.4 \pm 1.17$ kcal/mole and $\Delta S = 46.5 \pm 1.5$ cal/mole deg. respectively. HILLEL and CUEILLERON (1972) suggested B Se_2 (g) as the reaction product of liquid boron selenide and liquid selenium. KORYAZHKIN (1972) corroborated this contention, by observing no absorption spectra of BSe_2 (g) when they heated B_2Se_3, in the absence of selenium, to 1200°C. At temperatures above 2000°C UY and DROWART (1970) obtained BSe(g) by the gas-phase reaction of boron with yttrium selenide and determined the energy of dissociation (at 0°K) D_0° (BSe) = 109.4 ± 3.5 kcal/mole. For the hypothetical molecule BTe(g), the energy of dissociation was estimated to be D_0° (BTe) = 83.7 ± 4.8 kcal/mole.

No data are available on the use of boron selenides.

References

AHMAD, I., CAPSIMALIS, G.P.: Formation of filamentary single crystals of B, CaB_6, B_4C and B_7O. In: Reactivity of Solids, Proc. 6th Symp. pp. 447-486, 1968 (Pub. 1969)

AKISHIN, P.A., SPIRIDONOV, V.P.: Electronographic investigation of the structure of boron sulfide molecule. Dokl. AN USSR 129, 1317-1320 (1959)

BARTENEV, G.M., ABROSIMOVA, G.D.: Effect of high-temperature treatment on deformation properties of vitreous boric anhydride. Visokomol. Soed. B 15, 199-203 (1973)

BARTENEV, G.M., EREMEEVA, A.S.: Is boric anhydride a polymer? Visokomol. Soed 2, 1845-1849 (1960)

BARTENEV, G.M., TZYGANOV, A.D., ABROSIMOVA, G.D.: Effect of heat treatment on the electronic structure of the vitreous B_2O_3 system. Dokl. AN USSR 205, 375-378 (1972)

BARTON, L., NICHOLLS, D.: High-temperature routes to boron monoxide and diborane. Proc. Chem. Soc. 242 (1964)

BARTON, L., NICHOLLS, D.: The hydrogenation of boron monoxide to diborane and the reactions of boron and boron carbide. J. Inorg. Nucl. Chem. 28, 1367-1372 (1966)

BERLIN, L.E., LYKOV, M.V., FEDYUSHKIN, B.F., KOZLOVA, S.A., ODERBERG, A.S.: Preparation of powdered B_2O_3 in a three-zone fluidized bed unit. In: Collection of Papers Presented at the USSR Conference on Novel Means and Advanced Techniques in Drying Processes. Issue 2, pp. 59-63, Moscow (1969)

BORYAKOVA, V.A., GREENBERG, Ya.H., ZHUKOV, E.G., LORYAZHKIN, V.A., MEDVEDEVA, Z.S.: Phase diagram of B_2Se_3-Se system. Izv. AN USSR, Ser. Neorg. Materiali 5, 477-480 (1969)

BROADHEAD, P., NEWMAN, G.A.: The structures of orthoboric acid and its thermal decomposition products. Inorg. Macromol. Revs. 1, 191-201 (1971a)

BROADHEAD, P., NEWMAN, G.A.: The vibrational spectra of orthoboric acid and its thermal decomposition products. J. Mol. Struct. 10, 157-172 (1971b)

CHEN, H., GILLES, P.W.: High molecular weight boron sulfides V. Vaporization behavior of the boron-sulfur system. J. Am. Chem. Soc.92, 2309-2312 (1970)

CHEN, H., GILLES, P.W.: High molecular weight boron sulfides VIII. Vapor pressures of B_2S_3 (g) and B_4S_6 (g) over stoichiometric B_2S_3. J. Phys. Chem. 76, 2035-2038 (1972)

CHEN, H., CONRAD, B.R., GILLES, P.W.: High molecular weight boron sulfides VI. The unit cell of monoclinic diboron trisulfide. Inorg. Chem.9, 1776-1777 (1970)

CLARK, Ch.C., KANDA, F.A., KING, A.J.: (Olin Mathieson Chemical Corp.): Diborane. U.S. 3,021, 197 Feb. 13, 1962, Appl. Nov. 20, 1956, quot. from Chem. Abstr. 56, 138010 b (1963)

CRAGG, R.H.: The chemistry of boron-sulfur compounds for the period 1950-1967. In: Quarterly Reports on Sulfur Chemistry, pp. 1-80. Santa Monica, California: Intra-Science Research Foundation, 1968

CUEILLERON, J., HILLEL, R.: Preparation du seleniure de bore: B_2Se_3. Bull. Soc. Chim. France N8, 2973-2976 (1967)

CUEILLERON, J., HILLEL, R.: Etude des selenichalogenures de bor BSeX. Preparation, degradation thermique, proprietes spectrales. Bull. Soc. Chim. France N9, 3635-3640 (1968)

DACHILL, F., ROY, R.: A new high-pressure form of B_2O_3 and inferences on cation coordination from infrared spectroscopy. J. Am. Soc. 42, 78-80 (1959)

DACRE, B.: Solutions in oleum. Part 3. Cryoscopic behavior of acids, bases and salts in disulfuric acid. Trans. Faraday Soc. 63, 604-607 (1967)

DEMBOVSKY, S.A., KIRILENKO, V.V., BUSLAEV, Yu.A.: Vitrification in the Ti-B-S and Ti-B-Se systems. Izv. AN USSR, Ser. Neorg. Materiali 7, 510-511 (1971)
DEMBOVSKY, S.A., KIRILENKO, V.V., BUSLAEV, Yu.A.: The B_2Se_3 Ti_2Se system. Izv. AN USSR, Ser. Neorg Materiali 8, 237-241 (1972)
EDWARDS, J.G., LEITNAKER, J.M., WIEDEMEIER, H., GILLES, P.W.: High molecular weight boron sulfides VII. Lower temperature studies and mestastable decompositions. J. Phy. Chem. 75, 2410-2412 (1971)
EDWARDS, J.G., WIEDEMEIER, H., GILLES, P.W.: High molecular weight boron sulfides III. A mass spectrometric study of the vaporization and decomposition of metathioboric acid. J. Am. Chem. Soc. 88, 2935-2939 (1966)
ERB, A.: Contribution a la connaissance de la reduction de l'anhydride borique. Ann. Chimie. 4, 713-714 (1959)
FLOWERS, R., GILLESPIE, R.J., OUBRIDGE, J.V.: Solutions in sulfuric acid. Part XIX. The formation of tetra-(hydrogen sulphato) boric acid and its anion in solutions of boric acid and boric oxide. J. Chem. Soc. 1925-1938 (1956)
FRANZ, H.: Untersuchungen über die Aciditats-Basizitatsverhältnisse in oxydischen Schmelzen. 1. Infrarotspektroskopische Untersuchungen über die Löslichkeit von Wasserdampf in B_2O_3-Schmelzen. Glastechn. Ber. 38, 54-59 (1969)
GMELIN'S Handbuch der anorganischen Chemie: Bor. pp. 129-135, Leipzig-Berlin, 1926
GREENBERG, Ya.H., ZHUKOV, E.G., KORYAZHKIN, V.A.: Thermal dissociation of B_2S_3. Dokl. AN USSR 190, 847-849 (1969)
GREENBERG, Ya.H., ZHUKOV, E.G., KORYAZHKIN, V.A.: Thermodynamics of BS_2 evaporation. Dokl. An USSR 589-592 (1970a)
GREENBERG, Ya.H., ZHUKOV, E.G., KORYAZHKIN, V.A.: Thermal dissociation of B_2Se_3. Zh. Fiz. Khimmi 44, 2050-2051 (1970b)
GREENE, F.T., GILLES, P.W.: New Classes of high molecular weight boron sulfides. J. Am. Chem. Soc. 84, 3598-3599 (1962)
GREENE, F.T., GILLES, P.W.: High molecular weight boron sulfides II. Identification, relative intensites, appearance potentials, and origins of the ions. J. Am. Chem. Soc. 86, 3964-3969 (1964)
GREENE, F.T., MARGRAVE, J.L.: Infrared and visible spectra of gaseous B_2S_3 at high temperatures. J. Am. Chem. Soc. 81, 5555-5556 (1959)
GREENWOOD, N.N., THOMPSON, A.: The reactions of boron halides with anhydrous sulphuric acid: Boron tri(hydrogen sulphate)boric acid. J. Chem. Soc. 3643-3645 (1959)
GURR, G.E., MONTGOMERY, P.W., KNUTSON, C.D., GORRES, B.T.: The crystal structure of trigonal diboron trioxide. Acta Crystallogr B26, 906-915 (1970)
HAGENMULLER, P., CHOPIN, F.: Sur le system bore-soufre. Compt. Rend. Acad. Sci. 255, 2259-2260 (1962)
HALL, H.T.: Periodic compounds: Syntheses at high pressures and temperatures. Science 148, 1331-1333 (1965)
HALL, H.T., COMPTON, L.A.: Group IV analogs and high pressure, high temperature synthesis of B_2O. Inorg. Chem. 4, 1213-1216 (1965)
HILLEL, R., CUEILLERON, J.: Etude du seleniure de bore: B_2Se_3. Bull. Soc. Chim. France N1, 98-103 (1972)
HOARD, J.L., HUGHES, R.E.: Elemental boron and compounds of high boron content: structure, properties, and polymorphism. In: The Chemistry of Boron and its Compounds (ed.,E.L. Muetterties) pp. 25-154. New York-London-Sydney: John Wiley and Sons, 1967
HOFFMAN, J.: Die Gewinnung von Selenbor aus Ferro und Manganbor. Chemiker-Zeitung 35, 713 (1911)
HOLCOMBE, C.E., HORNE, O.J.: Preparation of boron suboxide, B_7O. J. Am. Ceram. Soc. 55, 106 (1972a)
HOLCOMBE, C.E., HORNE, O.J.: Us Atomic Energy Commission: Boron suboxide. U.S. 3,660,031 (Cl. 23/204R; C 01 b), 02 May 1972 (b) Appl. 122,506, 09 Mar. 1971, quot. from C.A. 77, 37152s (1972)
HUTCHINSON, W.E., EICK, H.A.: Preparation and properties of boron sesquiselenide. Inorg. Chem. 1, 434-436 (1962)
INGHRAM, M.G., PORTER, R.F., CHUPKA, W.A.: Mass spectrometric study of gaseous species in the B-B_2O_3 system. J. Chem. Phys. 25, 498-501 (1956)
KAMSU KOM, J.: Contribution a l'etude de quelques combinations ternaires sulfurees ou seleniees du bore ou du phosphere avec les metaux du groupe Ib ou avec l'iode. Ann. Chimie 9, 179-205 (1964)

KAMSU KOM, J., FLAHAUT, J., DOMANCE, L.: Sur deux series de composes ternaires, derives du bore: $CuBS_2$ et $CuBSe_2$, et derives du phosphore: $CuPS_2$. Compt. Rend. Acad. Sci. 257, 3919-3922 (1963)
KANDA, F.A., KING, A.J., RUSSELL, V.A., KATZ, W.: Preparation of boron monoxide at high temperatures. J. Am. Chem. Soc. 78, 1509-1510 (1956)
KASKAN, W.E., MILLIKAN, R.C.: Source of green bands from boron-containing flames. J. Chem. Phys. 32, 1273-1274 (1960a)
KASKAN, W.E., MILLIKAN, R.C.: Spectroscopic studies of flame gases containing boron. Symp. Combust., 8th, Pasadena, California pp. 262-275 (1960b) (Pub. 1962)
KASKAN, W.E., MACKENZIE, J.D., MILLIKAN, R.C.: Origin of the green bands in the boron-oxygen system. J. Chem. Phys. 34, 570-574 (1961)
KLINE, D., BRAY, P.J., KRIZ, H.M.: Structure of crystalline boron oxide. J. Chem. Phys. 48, 5277-5278 (1968)
KORYAZHKIN, V.A.: On spectra and structure of BO_2 BS_2 and BSe_2 molecules and CO_2^+ and CS_2^+ ions. Cand. Sci (Chem) Thesis, Moscow 1967
KORYAZHKIN, V.A., MAL'TSEV, A.A.: Electron absorption spectrum of BS_2. Vestnik Mosc. Univ., Ser. Khim. 21, 6-9 (1966)
KORYAZHKIN, V.A.: Electron absorption spectrum of BS_2. Vestnik Mosc. Univ., Ser. Khim. 21, 6-9 (1966)
KRISHNAN, R.S.: Raman spectrum of boric oxide glass. Indian J. Pure Appl. Phys. 9, 916-918 (1971)
LA PLACA, S., POST, B.: The boron carbide structure type. Planseeber. Plvermet. 9, 109-112 (1961), quot. from Chem. Abstr. 55, 24185h
LIPP, A.: Borcarbid-Elementarzelle. Ber. Deut. Keram. Ges. 43, 60-62 (1966)
MACKENZIE, J.D., CLAUSSEN, W.F.: Crystallization and phase relations of boron trioxide at high pressures. J. Am. Ceram. Soc. 44, 79-81 (1961)
MAL'TSEV, A.A.: Boron and aluminum sulfides. Dissertation Thesis, Chapter IX, pp. 228-240 Moscow 1965
MAL'TSEV, A.A.: High-temperature molecular spectroscopy. IR spectra of vapors of involatile compounds. In: Vebrational Spectra in Inorganic Chemistry. (ed., Yu. Ya. Kharitonov) pp. 65-88. Moscow: Nauka 1971
MAL'TSEV, A.A., SHEVEL'KOV, V.F.: Infrared absorption spectra of aluminum, gallium and indium suboxide vapors. Certain regularities in the frequencies of oscillations of Group III element suboxides and evaluation of molecular constants of B_2O. In: Vitrational Spectra in Inorganic chemistry. (ed., Yu.Ya., Kharitonov), pp. 89-92. Moscow: Nauka 1971
MAL'TSEV, A.A., MATWEEV, V.K., TATEVSKY, V.M.: On the nature of boric acid fluctuation bands. Dokl. AN USSR 137, 123-125 (1961)
MATKOVICH, V.I.: Interstitial compounds of boron. J. Am. Chem. Soc. 83, 1804-1806 (1961)
MATKOVICH, V.I., GIESE, R.F., ECONOMY, J.: Packing of B_{12} groups in boron and boride structures. Z. Kristallogr. 122, 116-130 (1965)
MATWEEV, V.K., MAL'TSEV, A.A., TATEVSKY, V.M.: Dependence of the intensity of boric acid fluctuation bands on oxygen pressure. Vestnik Mosc. Univ. Ser. Khim 51-53 (1961)
McCLOSKEY, A.L., BOONE, J.L., BROTHERTON, R.J.: The preparation of diboron tetrachloride from boron monoxide. J. Am. Chem. Soc. 83, 1766-1767 (1961a)
McCLOSKEY, A.L., BROTHERTON, R.J., BOONE, J.L.: The preparation of boron monoxide and its conversion to diboron tetrachloride. J. Am. Chem. Soc. 83, 4750-4754 (1961b)
McCULLOCH, L.: A crystalline boric oxide. J. Am. Chem. Soc. 59, 2650-2652 (1937)
MEDVEDEVA, Z.S., BORYAKOVA, V.A., GREENBERG, Ya.H., ZHUKOV, E.G.: Phase diagram of BS_2-S system. Zh. Fiz. Khimii 13, 1440-1443 (1968)
MELUCCI, R.C., WAHLBECK, P.G.: Mass spectrometric observations of gaseous boron selenides. Inorg. Chem. 9, 1065-1068 (1970)
MOISSAN, H.: Etude des proprietes du bore amorphe. Compt. Rend. Acad. Sci. 114, 617-622 (1892a)
MOISSAN, H.: Etude du trisulfure de bore. Compt. Rend. Acad. Sci. 115, 203-208 (1892b)
MOISSAN, H.: Sur le pentasulfure de bore. Compt. Rend. Acad. Sci. 115, 271-273 (1892c)

MOISSAN, H.: Etude du bore amorphe. Ann. Chim. Phys. 6, 296-320 (1895)
MOLLARD, P.: Boron oxide (B_2O_3) from orthoboric acid. Ger. Offen 1,951,830 (CL01b), May 6, 1970, quot. from Chem. Abstr. 73, 556 Oz. Oct., 17 (1968)
MOROSOVA, M.P., RYBAKOVA, G.A.: Enthalpy of formation of magnesium hexaboride. Vestnik Leningrad Univ. Ser. Fiz. i Khim 22, 161-163 (1968)
MOSHER, R.A., IVES, E.K., MORELLO, E.F.: Synthesis of boron chloroperchlorates and boron triperchlorate. J. Am. Chem. Soc. 85, 3037-3038 (1963)
MUETTERTIES, E.L.: Sulfur and selenium compounds of boron. In: The Chemistry of Boron and its Compounds (ed., E.L. Muetterties) pp. 647-667. New York-London-Sydney: John Wiley and Sons, 1967
NENCETTI, F., TARTARELLI, R.: Esteri borici da anidride borica per distillazione azeotropica dell'acqua con un mezzo organico. Chimica e industria 50, 1295-1298 (1968); quot. from Russ. Ref. J. Chim. (1959), 12H107
NICHOLLS, D.: The reduction of boric oxide to boron monoxide. J. Chem. Soc. 6644-6646 (1965)
NIKITIN, V.S., MAL'TSEV, A.A., PCHOLKINA, M.A., VINOGRADOVA, Z.F.: Infrared spectrum of diboron tetrahydroxide, $B_2(OH)_4$, and boron monoxide, $(BO)_x$. Vestnik Mosc. Univ., Ser. Khim 3, 14-27 (1963)
OTTAR, B., RUIGH, V.L.: The structure of boric oxide and borate glasses. A preliminary communication. Phys. Chem. Glasses 3, 95-98 (1962)
PASTERNAK, R.A., Crystallographic evidence for the existence of B_7O. Acta Crystallogr. 12, 612-613 (1959)
PETRAK, D.R., RUTH, R., GOOSEY, B.F.: Preparation and characterization of boron suboxide. Natl. Bur. Stand. (U.S.) Spec. Publ. No. 364, 605-611 (1972); quot.from Chem. Abstr. 77, 157316v (1972)
PORAI-KOSHITZ, E.A., WARREN, B.E.: New possibilities and objects of direct structural methods of investigation of glass-like substances. In: Collection of Papers "Glassy State," pp. 128-136. Leningrad: Nauka 1971
POST, B.: Refractory binary borides. In: Boron, Metalloboron Compounds and Boranes (ed., R.M. Adams) pp. 301-371. New York: Interscience, 1964
PREWITT, C.R., SHANNON, R.D.: Crystal structure of a high-pressure form of B_2O_3. Acta Crystallogr. B24, 869-874 (1968)
RAY, R.C., SINHA, P.C.: Two lower oxides of boron. J.Chem. Soc. 742-744 (1941)
RENTZEPIS, P., WHITE, D., WALSH, P.N.: The reaction between B_2O_3 (1) and C(s): Heat of formation of B_2O_2 (g). J. Phys. Chem. 64, 1784-1787 (1960)
REYNAUD, C.: Sur la cristallisation de l'anhydride borique. Compt. Rend. Acad. Sci. C264, 1115-1117 (1967)
RIZZO, H.F., SIMMONS, W.C., BIELSTEIN, H.O.: The existence and formation of the solid B_6O. J. Electrochem. Soc. 109, 1079-1082 (1962)
ROSS, V.F., EDWARDS, J.O.: The structural chemistry of the borates. In: The Chemistry of Boron and its Compounds (ed., E.L. Muetterties) pp. 155-207. New York-London-Sydney: John Wiley and Sons, 1967
ROSSO, J.C.: Preparation et identification du compose $B_{12}C_2S$. Compt. Rend. Acad. Sci. 266, 114-116 (1968)
ROSSO, J.C., DUBUSK, M.: Les sulfures de bore et des composes ternaires bore-soufre-metal. Chimie et Industrie-Genie chimique 102, 409-415 (1969)
RUSIN, A.D., TATEVSKY, V.M.: Nature of green luminescence and the composition of boron combustion products at 3100°K. Dokl. AN USSR 139, 630-633 (1961)
RUSIN, A.D., TATEVSKY, V.M.: On the heat of formation of BO_2 radical. Zh. Fiz. Khimii 37, 716-717 (1963)
RUSIN, A.D., TATEVSKY, V.M.: Composition of the products of boron combustion at 3000-3200°K and the heat of BO_2 formation. Teplofizika Visokikh Temperatur 3, 547-554 (1965)
RUSSEL, D.K., KROLL, M., DOWDS, D.A., BEAUDET, R.A.: The laser-excited fluorescence spectra of BO_2. Chem. Phys. Lett. 20, 153-155 (1973)
SABATIER, P.: Sur le sulfure de bore. Compt. Rend. Acad. Sci. 112, 862-864 (1891a)
SABATIER, P.: Sur le sulfure de bore. Bull. Soc. Chim. France 6, 215-218 (1891b)
SABATIER, P.: Sur le seneniure de bore. Bull. Soc. Chim. France 6, 218-219 (1891c)
SABATIER, P.: Sur le seleniure de bore. Compt. Rend. Acad. Sci. 117, 1000-1001 (1891d)
SCHEER, M.D.: The molecular weight and vapor pressure of gaseous boron suboxide. J. Phys. Chem. 62, 490-493 (1958)

SHARMA, A., JOSHI, J.C.: Fluorescent enhancement of the BO_2 band in the chemiluninescent reaction of triethyl boron and atomic oxygen. J. Quant, Spectrosc. Radiat. Transfer 12, 1074-1076 (1972)
SHUKER, R., GAMMON, R.W.: Low-frequency vibrational light scattering in visceous liquids. J. Chem. Phys. 55, 4784-4788 (1971)
SNYDER, P.E.: Preparation of boron trisulphide. U.S. Pat. 2,866,687 (1958)
SOMMER, A., WALSH, P.N., WHITE, D.: Mass spectrometric and infrared emission investigation of the vapor species in the B-S system at elevated temperatures. J. Chem. Phys. 33, 296-297 (1960)
SOMMER, A., WHITE, D., LIVENSYKY, M.J., MANN, D.E.: Infrared absorption spectra of B_2O_3, B_2O_2 and BO_2 in solid argon matrixes. J. Chem. Phys. 38, 87-98 (1963)
SPERRY, L.L., MACKENZIE, J.D.: Pressure dependence of viscosity of B_2O_3. Phys. Chem. Glasses 9, 91-95 (1968)
SRIVASTAVA, R.D., UY, O.M., FARBER, M.: Effusion-mass spectrometric study of the thermodynamic properties of BO and BO_2. Trans. Faraday Soc. 67, 2941-2944 (1971)
STANTON, L.S.: Crystalline hexagonal boric oxide. U.S. Pat. 3,582,272 (Cl. 23-149; C 01 b), 01 June (1971) Appl 11 July (1966)
STRONG, S.L., KAPLOW, R.: The structure of crystalline B_2O_3. Acta Crystallogr. B24, 1032-1036 (1968)
STRONG, S.L., WELLS, A.F., KAPLOW, R.: On the crystal structure of B_2O_3. Acta Crystallogr. B27, 1662-1663 (1971)
TARASOV, W.W.: Polymeric structure of boric anhydride and sodium borate glasses. In: Collection of Papers "Glassy State," Vol. III, No. 4, pp. 112-119, Minsk, 1964
TARASOV, W.W., BOZHKO, Yu.A., RATOBYL'SKAYA, V.A.: Temperature dependence of ultrasound veloscity and absorption in boric anhydrie and sodium borate glasses. In: Collection of Papers "Glassy State," pp. 274-278. Leningrad: Nauka 1971
THOMAS, D., TRIDOT, G.: Etude chimique et cristallographique de l'acide thioborique. Compt. Rend. Acad. Sci. 259, 3559-3562 (1964)
TIMMS, P.L., PHILLIPS, C.S.: The formation of some volatile hydrides from lower oxides. Inorg. Chem. 3, 606-607 (1964)
UHLMANN, D.R., HAYS, J.F., TURNBULL, D.: The effect of high pressure on B_2O_3; crystallisation, densification, and the crystallisation anomaly. Phys. Chem. Glasses 8, 1-10 (1967)
UY, O.M., DROWART, J.: Mass spectrometric determination of the dissociation energies of the boron monochalcogenides. High Temp. Sci. 2, 293-298 (1970)
UY, O.M., SRIVASTAVA, R.D., FARBER, M.: Mass spectrometric determination of the heats of formation of gaseous BO_2 and BOF_2. High Temp. Sci. 3, 462-468 (1971)
WARTIK, T., APPLE, E.F.: A new modification of boron monoxide. J. Am. Chem. Soc. 77, 6400-6401 (1955)
WASEDA, I., OHTANI, M.: Structure of vitreous SiO_2 and B_2O_3 by X-ray fourier analysis. Sci. Repts. Res. Inst. Tohoku Univ. A23, 201-210 (1972), quot. from Russ. Ref. J. Chim. 46 648 (1973)
ZEEMAN, P.B.: Spectrum of boron monosulfide (BS) molecule. Phys. Rev. 80, 902 (1950)
ZEEMAN, P.B.: Spectrum and structure of the BS molecule. Can. J. Phys. 29, 336-337 (1951)
ZINTL, E., MORAWIETZ, W., GASTINGER, E.: Boronoxyd. Z. Anorg. Chem 245, 8-12 (1940)
ZUBOVA, E.V., BURDINA, K.P.: Synthesis of B_6O under high pressures. Dokl. AN USSR 197, 1055-1056 (1971)

IX. Transition Metal Borides

T. LUNDSTRÖM

1. Crystal Structure

a) Structures of the CrB_4 and MnB_4 Type

The crystal structure of CrB_4 (ANDERSSON and LUNDSTRÖM, 1968) and MnB_4 (ANDERSSON and CARLSSON, 1970) are closely related, although the former is orthorhombic with a = 4.744 Å, b = 5.477 Å and c = 2.866 Å and the latter monoclinic with a = 5.503 Å, b = 5.367 Å, c = 2.949 Å and β= 122.71°. The relationship is, however, easily revealed if the MnB_4 structure is described using a non-conventional I-centered cell with a = 4.630 Å, b = 5.367 Å, c = 2.949 Å and β = 90.31 Å. From the projection of the CrB_4 structure along its short axis (see Fig. 1) it is seen that the boron atoms form squares in the ab plane. These boron squares are connected in the c direction as well, thus generating a three-dimensional network, intersected by channels in the c direction. The chromium atoms are situated in these channels, forming chains with a fairly long Cr-Cr distance of 2.87 Å. It is a characteristic feature of the structure that it contains voids (at 0,1/2,1/2 and 1/2,0,0), which are large enough to accommodate atoms with radii less than 1.0 Å without any appreciable distortion of the lattice. The MnB_4 structure may be regarded as a monoclinic distortion of the CrB_4 structure involving minor changes in the interatomic distances but no changes in coordination. No further representatives of these structure types are known at present. Attempts to prepare iso-structural phases in the V-B system were unsuccessful, while in the Mo-B and W-B systems a phase of a different structure occurs at the corresponding composition.

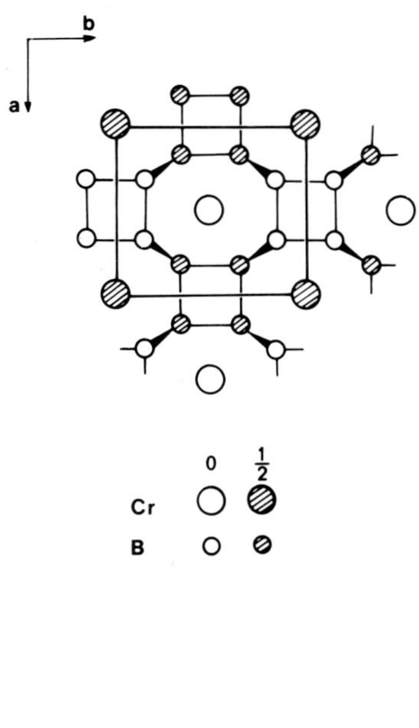

Fig. 1. The crystal structure of CrB_4

b) Structures Based on Two-dimensional Boron Networks

Transition metal borides in the composition range from about Me_2B_3 to MeB_4 are structurally characterized by planar or puckered boron layers interleaved with metal layers. The structure types and their representatives are collected in Table 1.

Table 1. Structure types and representatives with two-dimensional boron networks

Structure type	Stacking sequence	Representatives
AlB_2	AHAH...	TiB_2, ZrB_2, HfB_2, VB_2, NbB_2, TaB_2, CrB_2, MoB_2, WB_2, MnB_2 [a]
ReB_2	AK´BK´AK´BK´...	ReB_2, TcB_2
Ru_2B_{3-x}	AH´AK´BH´BK´AH´...	Ru_2B_3, Os_2B_3
W_2B_{5-x}	AHAK´BHBK´AH...	$WB_{2.0}$
Mo_2B_{5-x}	AHAKBHBKCHCKAH[b]...	$MoB_{2.3}$, W_2B_{5-x}
$Mo_{1-x}B_3$	A´HB´HA´H...	$MoB_{3.75}$, $WB_{\sim 4}$
RuB_2		RuB_2, OsB_2
Ir_2B_{3-x}		$IrB_{1.35}$

[a]For rare earth diborides see HOARD and HUGHES 1967
[b]Defective K layers

We can distinguish a group of structures with hexagonal symmetry based on close-packed layers of metal atoms (A, B, C) and four types of boron layers (K, K´, H, H´) as shown in Figure 2. The layers are planar with the exception of the K and K´ types of boron layers, which are slightly puckered. It should be noted that although layers of the K type are commonly used in descriptions of boride structures, their actual occurrence needs experimental confirmation. Layers of type H´ involve no boron-boron bonds, but isolated boron atoms. The stacking sequence of these layers in the c direction has been incorporated in Table 1 for the various structure types. The boron atoms have a trigonal prismatic metal environment (between AA layers) or a tetrahedral metal environment (between AB layers).

The simple AlB_2 structure type is the most common one and has been discussed at some length by various authors (POST, 1964; HOARD and HUGHES, 1967; RAMAN, 1967). The ReB_2-type structure (LA PLACA and POST, 1962) has only one further representative, TcB_2, (TRZEBIATOWSKI and RUDZINSKI, 1964) and contains puckered boron layers (K´) without boron-boron bonding in the c direction. The Mo_2B_{5-x}-type structure (KIESSLING, 1947; KUZ´MA et al., 1967) as well as the W_2B_{5-x}- and Ru_2B_3-type structures (KIESSLING, 1947; ARONSSON et al., 1962; LUNDSTRÖM, 1968) may be regarded as built up of structure elements from both the AlB_2 and ReB_2 structures. It is a characteristic feature of these structures that the boron layers are defective (H´ replaces H, vacancies in the K layers often result in a K´ layer), although this does not always correspond to a large range of homogeneity. A few of the AlB_2 type diborides, however, display a very extensive range of homogeneity, for instance NbB_2 and TaB_2.

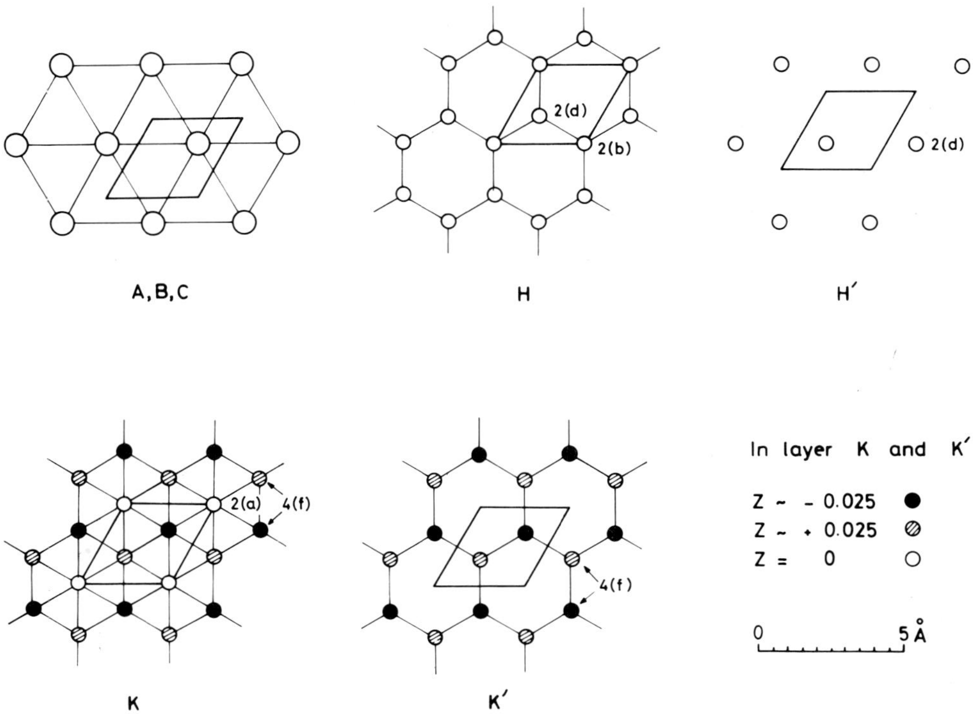

Fig. 2. Layers of metal (A, B, C) and layers of boron (H, H´, K, K´). These layers are used to describe hexagonal structures related to AlB_2

The earlier structure proposals for the highest boride of tungsten (ROMANS and KRUG, 1966; NOWOTNY et al., 1967) showed discrepancies as regards the number and positions of the boron atoms. It was noted that MoB_4 is isotypic with the tungsten compound (CHRÉTIEN and HELGORSKI, 1961; ARONSSON et al., 1968; GALASSO and PINTO, 1968). The structure of MoB_4 was recently determined and denoted $Mo_{1-x}B_3$ (LUNDSTRÖM and ROSENBERG, 1973). The structure may be described using defective metal layers of type A´B´ and boron layers of type H (see Fig. 3) with the stacking sequence A´HB´HA´H... The structure may be derived from the simple AlB_2 type by removing one-third of the metal atoms and distributing the vacancies in an ordered manner through the structure, which would correspond to the ideal formula MeB_3. However, the occupancy for one of the metal positions is only about 0.60, which leads to the composition $MoB_{3.75}$.

Two independent structure determinations have established the RuB_2 structure type (ROOF and KEMPTER, 1962; ARONSSON, 1963). The orthorhombic structure contains puckered metal and boron layers (see Fig. 4), which are pseudo-hexagonal in the projection along the c axis. The structure does not contain a three-dimensional boron network as claimed by ARONSSON (LUNDSTRÖM, 1969), but there is agreement on the occurrence of a void at 2(b) with z = 0.75, which would otherwise be sufficiently large to accommodate a boron atom.

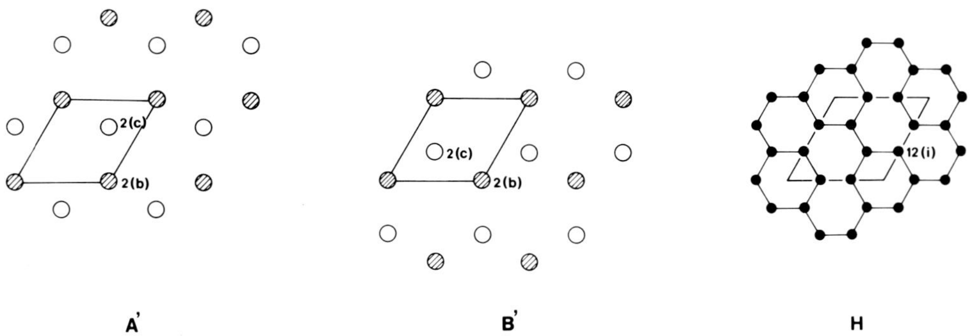

Fig. 3. Defective metal layers ($\underline{A}'$, $\underline{B}'$) and boron layers ($\underline{H}$), used to describe the crystal structure of $Mo_{1-x}B_3$

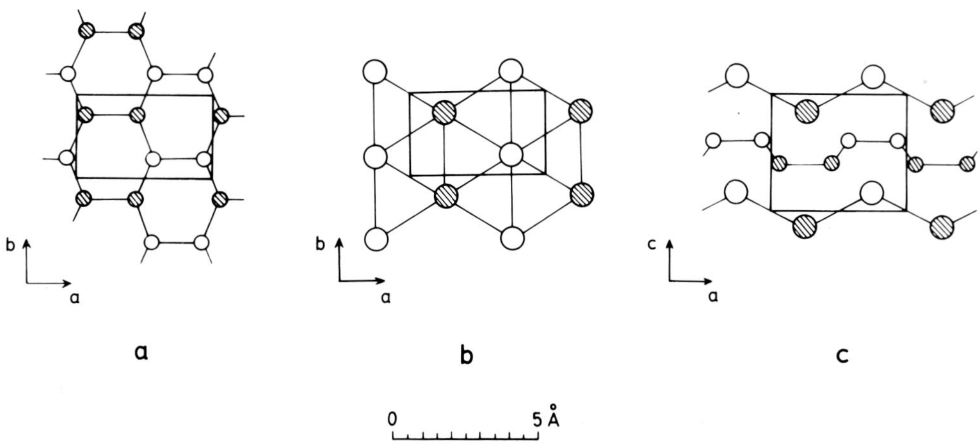

Fig. 4a-c. The crystal structure of RuB_2. (a) A puckered boron layer. (b) A puckered metal layer. (c) The stacking of boron and metal layers

The monoclinic structure of $IrB_{1.35}$ was originally described as a stacking of double layers of metal atoms and puckered boron layers (ARONSSON, 1963). A recent refinement has established the structure of the boron network in more detail (LUNDSTRÖM and TERGENIUS, 1973). The puckered double layers of metal atoms contain isolated boron atoms with trigonal prismatic metal environment (see Fig. 5). These strongly puckered layers form a simple stacking in the c direction with the complicated puckered boron layers shown in Figure 5. There is not much left of the pseudohexagonality in this structure. A variable occupation of the trigonal prisms in the metal double layers is responsible for the homogeneity range of the phase, which has the ideal composition Ir_2B_{3-x}.

The list of representatives (Table 1) would increase considerably if ternary borides were included. In particular, numerous ternary re-

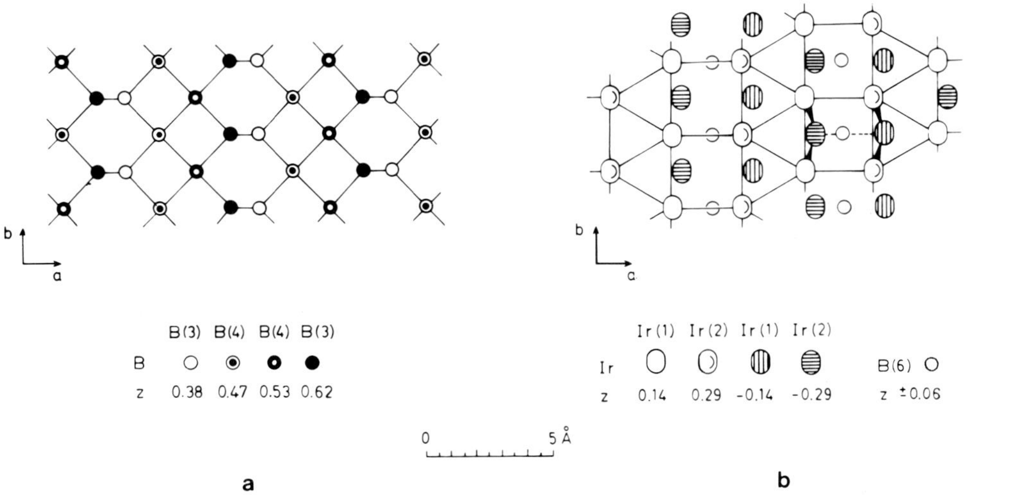

Fig. 5a and b. The crystal structure of Ir_2B_{3-x} consists of a simple stacking of (a) a puckered boron layer and (b) a puckered double layer of metal atoms containing "isolated" boron atoms

presentatives are known for the AlB_2 and ReB_2 type structures (TELEGUS and KUZ´MA, 1971, ROGL et al., 1970a, b).

Several ternary structure types containing two-dimensional boron networks have been described. The orthorhombic $YCrB_4$ structure is built up of boron layers consisting of five- and seven-membered rings (KUZ´MA, 1970). In the space between these layers the large yttrium atoms coordinate the seven-membered boron rings while the smaller chromium atoms coordinate the five-membered rings. More than 30 representatives of this structure type are known (KUZ´MA and SVARICHEVSKAYA, 1972a). The structure of $YCrB_4$ is very closely related to that of ScB_2C_2 (SMITH et al., 1965), which contains layers of boron and carbon atoms forming five- and seven-membered rings. The chromium positions between these layers are, however, empty. The positional parameters for the two structures agree very closely.

Planar boron layers of five-, six- and seven-membered rings are characteristic of the structure of Y_2ReB_6, making the structure an intermediate between the AlB_2-type and the $YCrB_4$-type structure (KUZ´MA and SVARICHEVSKAYA, 1972b). Further ternary borides have been discovered in the Y-B-C system (BAUER and NOWOTNY, 1971). It was shown that the non-metal atoms are arranged in layers, which are built up of squares and eight-membered rings in YB_2C_2 and of squares and seven-membered rings in YB_2C.

c) Structures Based on Boron Chains

The binary structure types, in which the boron atoms form chains with B-B distances of 1.7-1.9 Å, are collected in Table 2. We can distinguish two groups of structures. The first comprises the CrB, MoB, V_5B_6, Ta_3B_4, and V_2B_3 type structures, which are based on double metal layers containing BMe_6 prisms. The second group contains the structure types $Ru_{11}B_8$, o-Ni_4B_{3-x}, m-Ni_4B_3 and FeB, which display more irregular stackings of the prisms.

Table 2. Structure types and representatives with boron chains

Structure type	Representatives
CrB	VB, NbB, TaB, CrB, β-MoB (high temp.), β-WB (high temp.), NiB
V_5B_6	V_5B_6
Ta_3B_4	Ti_3B_4, V_3B_4, Nb_3B_4, Ta_3B_4, Cr_3B_4, Mn_3B_4
V_2B_3	V_2B_3
MoB	α-CrB (low temp.), α-MoB (low temp.), α-WB (low remp.)
FeB	TiB, HfB, MnB, FeB, CoB
o-Ni_4B_{3-x}	o-Ni_4B_{3-x}
m-Ni_4B_3	m-Ni_4B_3
$Ru_{11}B_8$	$Ru_{11}B_8$

The occurrence of boron chains was recognized very early in some of these structures (KIESSLING, 1950). The structures were recently presented in terms of stackings of Me_2B_2 tetrahedra (MATKOVICH, 1974). For the present descriptions, however, we shall use the BMe_6 prisms as the fundamental structural units. The MeB_6 prisms share rectangular faces and form columns with zig-zag chains of boron running along the axis of the column (see projection along the prism axis in Fig. 6). The columns are stacked with the triangular prism faces in common, thus forming layers of prisms, although no boron-boron contacts are formed in the direction of the prism axis. Two types of shift occur in the stacking of the prism layers. The first type is indicated by a change of letter (PM in Fig. 6a) and involves a shift of one half of the prism axis and one half of the edge of the triangular face. The second type is indicated by a dash (MM´) and involves the formation of a common metal layer. In this case the boron atoms from different chains are at bonding distance forming a double chain. Using the notations given above, the structures may be described as follows:

CrB	PMPMPM...
V_5B_6	PP´MPP´M´PP´MPP´M´...
Ta_3B_4	PP´MM´PP´MM´...
V_2B_3	PP´MM´MPP´MM´M...

V_2B_3 thus contains triple chains while V_5B_6 contains double as well as single chains (SPEAR and GILLES, 1969). The MoB-type structure is built up from similar prism layers but every second of the layers is rotated π/2 about a direction perpendicular to the layers. The transposition relation between the CrB and MoB structures has been investigated earlier in detail (BOLLER et al., 1964; HOHNKE and PARTHÉ, 1966). A low-temperature modification of CrB crystallizing in the MoB type structure was recently reported with the cell dimensions a = 2.94 Å and c = 15.72 Å (PAPESCH et al., 1973).

There are some common features between the FeB type structure and the o-Ni_4B_{3-x} type structure (RUNDQVIST, 1959; RUNDQVIST and PRAMATUS, 1967). The structures both contain strings of boron atoms, while in addition o-Ni_4B_{3-x} contains "isolated" boron atoms at the center of trigonal prisms, whose axes are parallel to the main direction of the boron chains in the structure (see Fig. 6b, c).

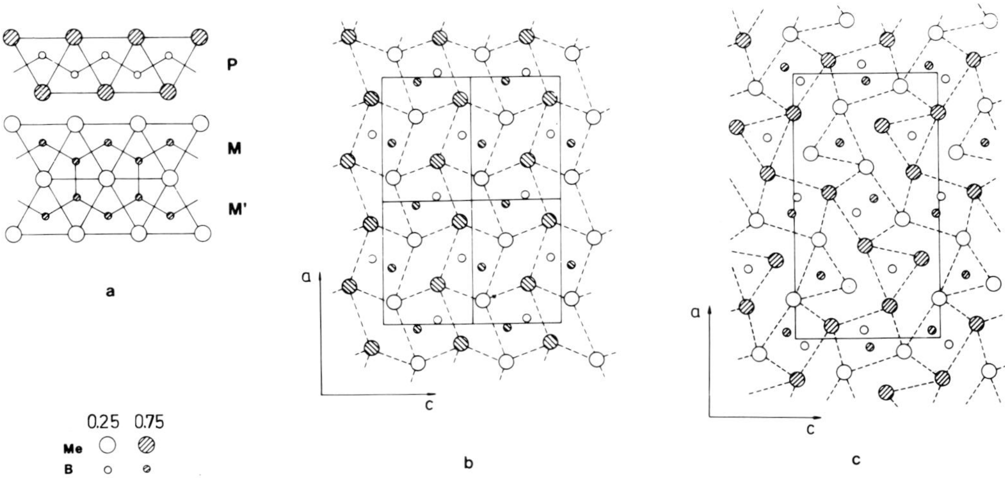

Fig. 6a-c. (a) Prism layers and stacking notation, used to describe the crystal structures of CrB, MoB, V_5B_6, Ta_3B_4 and V_2B_3. (b) The crystal structure of FeB. (c) The crystal structure of $o\text{-}Ni_4B_3$. (b) and (c) from RUNDQVIST, 1959

There are two crystallographically independent boron atoms in $m\text{-}Ni_4B_3$ and they display different metal coordination (RUNDQVIST, 1959). One boron atom has the usual trigonal prismatic environment, while the other is surrounded by eight close metal neighbors situated at the corners of a square (Archimedian) anti-prism. All boron atoms, however, form chains in the [101] direction as illustrated in Fig. 7.

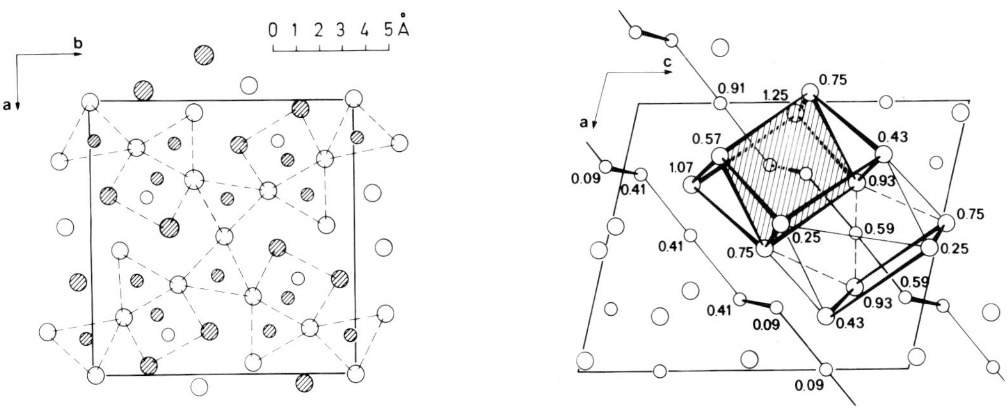

Fig. 7a and b. The crystal structure of (a) $Ru_{11}B_8$, (b) $m\text{-}Ni_4B_3$

In $Ru_{11}B_8$ (ÅSELIUS, 1960) there are triangular prismatically co-ordinated boron atoms, which form branched chains as can be seen from Figure 7. In addition there are "isolated" boron atoms, whose prisms are connected via the triangular faces.

Ternary borides often crystallize in the structure types of Table 2. New ternary representatives were, for instance, discovered for the CrB type structure (KUZ´MA et al., 1970; TELEGUS and KUZ´MA, 1971), the FeB type (ROGL et al., 1970a; ROGL et al., 1972; ROGL and NOWOTNY, 1973a) and the structure types of V_2B_3 and V_5B_6 (CHEPIGA et al., 1972; LUGSCHEIDER et al., 1974). Among the purely ternary structure types involving boron chains, the following should be mentioned: the MoAlB type (JEITSCHKO, 1966; KUZ´MA, 1968), the Mn_2AlB_2 type (BECHER et al., 1966; JEITSCHKO, 1969), the Mo_2BC type (JEITSCHKO et al., 1963) and the UBC type structure (TOTH et al., 1961). Boron chains with a large B-B distance, 1.98 Å, occur in the structure of YBC (BAUER and NOWOTNY, 1971).

d) Structures Based on Boron Pairs

Among the binary transition metal borides there are only three structure types which contain boron pairs. These include the U_3Si_2 structure type with V_3B_2, Nb_3B_2 and Ta_3B_2 as representatives, together with the Cr_5B_3 and $IrB_{0.9}$ (low temp.) structure types with no further known representatives. The U_3Si_2 and Cr_5B_3 structures were earlier classified as belonging to the ´$CuAl_2$´-family in recognition of the similarity between the metal layers in the borides and the layers in the $CuAl_2$ type structure (ARONSSON, 1960a; ARONSSON et al., 1965). We note that the formation of a boron pair requires a common rectangular face between two BMe_6 prisms. Those prisms which share a face, have their prism axes parallel in the U_3Si_2 and Cr_5B_3 structure types but perpendicular in the structure of $IrB_{0.9}$ (ROGL et al., 1971a). One of the boron positions in $IrB_{0.9}$ is only partially occupied.

Ternary borides are known to crystallize in the structure types mentioned. Cr_5PB_2 was, for instance, recently found to crystallize in the Cr_5B_3-type structure (BAURECHT et al., 1971). Of the numerous reports in the literature on ternary phases with the U_3Si_2 structure only a few will be mentioned here, namely those in the Ti-Re-B (KUZ´MA, 1971), Ta-Re-B (KUZ´MA et al., 1971), W-Mn-B and Mo-Mn-B systems (TELEGUS and KUZ´MA, 1971).

There are other ternary borides, which crystallize in new purely ternary structure types characterized by boron pairs or groups of three or four boron atoms. In the hexagonal structure of $ZrCo_3B_4$ with a = 8.418 Å and c = 9.132 Å (VOROSHILOV et al., 1970) the boron pairs arise from BCo_6 prisms sharing a rectangular face, while the zirconium atoms are situated outside the other rectangular faces. The structure has been described as a "superstructure" of the $CaCu_5$-type structure, but it should be noted that the linkage of the BCo_6 prisms is different in the two structures, which leads to the occurrence of boron pairs in $ZrCo_3B_2$ while there are no boron-boron contacts in the corresponding $CaCu_5$-type structure. There is also a modification of $ZrCo_3B_2$ with the $CaCu_5$-type structure, which is described in Section 1e. In W_2CoB_2, the trigonal BW_4Co_2 prisms share a rectangular face consisting of the four tungsten atoms (RIEGER et al., 1966), while in the $CeCo_4B_4$ structure the BCe_2Co_4 prisms are strongly deformed and share a face consisting of two cerium and two cobalt atoms (KUZ´MA and BILONIZHKO, 1971).

In the YCo_2B_2 structure ($ThCr_2Si_2$ type; NIIHARA et al., 1971; ROGL, 1973) the boron atoms have eight close metal neighbors at the corners of an Archimedian anti-prism, like one of the boron atoms in m-Ni_4B_3 (see Sect. 1c.). In YCo_2B_2 two anti-prisms have a common square face, leading to the formation of boron pairs. Additional boron contacts occur in Mo_3NiB_3 (JEDLICKA et al., 1969) and Mo_2IrB_2 (ROGL et al., 1972) in such a way that groups of three boron atoms in the former and four boron atoms in the latter may be regarded as fragments of boron chains, as discussed in the previous Section.

e) Structures Based on Isolated Boron Atoms (Without Boron-boron Contacts)

Structure types with "isolated" boron atoms are collected in Table 3 together with binary transition metal boride representatives. In these structures the boron-boron distances exceed 2.1 Å. The boron atoms in the Mn_4B and Fe_2B ($CuAl_2$) type structures have eight close metal neighbors forming an Archimedian anti-prism. Straight boron chains run through these antiprisms with a boron-boron distance of 2.1 Å or longer. Interatomic distances, based on single crystal refinements, were reported for five of the borides with the $CuAl_2$-type structure (ARONSSON et al., 1968). A ternary $Cr_{1.1}Fe_{0.9}B$ phase crystallizing in the Mn_4B type structure was also investigated, using single crystal X-ray data (BROWN and BEERNTSEN, 1964).

Table 3. Structure types and binary representatives without boron-boron contacts

Structure type	Representatives
Mn_4B	Mn_4B, Cr_2B
Re_3B	Tc_3B, Re_3B
Ni_3B (Fe_3C)	Co_3B, Ni_3B, Pd_3B
Pd_5B_2	Pd_5B_2
Ru_7B_3 (Th_7Fe_3)	Tc_7B_3, Ru_7B_3, Rh_7B_3, Re_7B_3
Fe_2B ($CuAl_2$)	Mn_2B, Fe_2B, Co_2B, Ni_2B, Mo_2B, $Ta_2B_{0.83}$, W_2B
anti-NiAs	PtB, $RhB_{\sim 1.1}$
WC or defect AlB_2	$IrB_{0.9}$ (high temp.), $RuB_{\sim 1.1}$, $OsB_{\sim 1.2}$
NbP or defect α-$ThSi_2$	$IrB_{\sim 1.1}$

Boron atoms with trigonal prismatic metal environment and with all boron-boron distances larger than 2.8 Å are to be found in the structures of Re_3B, Ni_3B, Ru_7B_3 (ARONSSON, 1960a) and Pd_5B_2 (STENBERG, 1961). Further representatives of the cementite-type structure (ROGL and NOWOTNY, 1973a) and the filled-up Re_3B-type structure (KONDRASHEV, 1966; ROGL et al., 1973) were discovered in ternary boride systems.

The anti-NiAs-type structure of PtB and $RhB_{\sim 1.1}$ is the only example of an octahedral metal environment for the boron atom among the binary borides. A tendency towards chain formation is discernible in these structures, the B-B distance being 2.03 Å in PtB. The chains may, however, be incomplete, since the composition $PtB_{0.67}$ has also been reported (WALD and ROSENBERG, 1965).

It is not known if the boron atoms in the structures of $IrB_{0.9}$ (high temp.) (ROGL et al., 1971a), $RuB_{\sim 1.1}$ and $OsB_{\sim 1.2}$ are randomly distributed in the trigonal prismatic holes (defect AlB_2-type structure) or form an ordered arrangement (WC-type structure). Similarly the structure of $IrB_{\sim 1.1}$ might belong to the ordered NbP-type structure or the defect α-$ThSi_2$-type structure with disordered vacancies (LUNDSTRÖM, 1969).

The majority of the purely ternary structure types with "isolated" boron atoms have been collected in Table 4 without any attempt to

Table 4. Some ternary structure types without boron-boron contacts

Structure type	Boron environment	References
$Cr_{23}C_6$ (tau phase)	Square anti-prism	STADELMAIER, 1969
Re_3Al_2B (η´-Mn_3Ni_2Si)	Octahedron	CHABAN and KUZ´MA, 1972
$V_5Ge_3B_x$ ($Mn_5Si_3X_x$)	Octahedron	RIEGER et al., 1965
$RERh_3B_x$ (perovskite)	Octahedron	HOLLECK, 1973
Hf_9Mo_4B (kappa phase)	Trigonal prism	ROGL et al., 1973
$(Re,Fe)_3B(Ti_3P)$	Trigonal prism	GANGLBERGER et al., 1966
$NbFeB(Fe_2P)$	Trigonal prism	KUZ´MA, 1967
$MoCoB$(anti-$PbCl_2$)	Trigonal prism	HASCHKE et al., 1966, JEITSCHKO, 1968
NbCoB	Trigonal prism	KRIPYAKEVICH et al., 1971
$ZrCo_3B_2(CaCu_5)$	Trigonal prism	STADELMAIER and SCHÖBEL, 1969
$Ti_3Co_5B_2$	Trigonal prism	KUZ´MA and YARMOLYUK, 1971

include known representatives. Numerous ternary borides crystallize in the $Cr_{23}C_6$ (tau-phase) type structure (STADELMAIER, 1969) with typical composition close to $Me^{I}_{21}Me^{II}_{2}B_6$ or with approximately equal atomic fraction of the metallic components. An example of the latter composition was recently found in the Cr-Ir-B system (ROGL and NOWOTNY, 1973b). In general the range of homogeneity is fairly small but an extensive range of homogeneity has been reported for the tau phase in the Re-Mn-B system, where the tau phase is claimed to be homogenous from $Re_{4.5}Mn_{18.5}B_6$ to $Re_{14.6}Mn_{8.4}B_6$ (TELEGUS and KUZ´MA, 1969). The structure of the kappa borides is closely related to that of kappa-$Co_3W_9C_4$ (SCHÖNBERG, 1954) but the non-metal content is lower than for the carbides. The composition is close to $Me^{I}_{9}Me^{II}_{4}B$ (ROGL et al., 1971b; ROGL and NOWOTNY, 1973c). In a recent structure determination of Hf_9Mo_4B, it was shown that the boron atoms have a trigonal prismatic metal environment (ROGL et al., 1973).

It is apparent from Table 4 that the octahedral metal environment of the boron atom is more common among the ternary borides than among the binary ones (Table 3). It is interesting to note that full occupation of the octahedral holes is very rare as far as is known at present. The structure of $ZrCo_3B_2$ was shown to be a ternary variant of the $CaCu_5$-type structure with BCo_6 prisms sharing edges (STADELMAIER and SCHÖBEL, 1969). A close relationship was also demonstrated between the $CaCu_5$-type structure and the structures of $CeCo_3B_2$ (KUZ´MA et al., 1969), $CeCo_4B$, $CeCo_{11}B_4$ and $Ce_2Co_7B_3$ (KUZ´MA and BILONIZHKO, 1973). These structures belong to a homologous series of boride structures.

It seems appropriate to conclude the presentation of the crystal chemistry of the transition metal borides with a few general remarks. It was earlier noted (ARONSSON et al., 1965) that the borides cannot be regarded as typical interstitial phases in the sense of HÄGG (1931). The radius ratio exceeds 0.59 except for a few systems, and the crystal structures are, as is evident from the foregoing presentation, much more complex than those of the typical interstitial hydrides, carbides and nitrides. Furthermore, in HÄGG's interstitial phases the non-metal atoms have an octahedral or occasionally tetrahedral metal environment while the trigonal prismatic arrangement is by far the most common among binary as well as ternary borides. In any case a sharp distinction between interstitial phases and other phases is impossible.

A typical feature of the borides is the strong tendency of the boron atoms to form boron-boron bonds, even if the boron content is as low as in $ZrCo_3B_2$. This is in contrast to the behavior of the carbides, for instance, where no carbon-carbon bonds occur in structures with less than 60 at % carbon. As regards the borides it must be remembered that the available structure determinations are only in a minority of cases sufficiently accurate to establish unambiguously the location of the boron atoms. Fractional occupation of these positions has in a few cases been studied as mentioned in the previous Sections. In this context the question of non-stoichiometry among the borides is quite important, as discussed recently (LUNDSTRÖM, 1974).

2. Borides of the IVa, Va, VIa, VIIa and VIII Groups

a) Preparation

α) Introduction

Reactions for preparation of transition metal borides often require very high temperatures. As a consequence it is a difficult task to prepare the homogeneous, single-phase borides of high purity that are desirable for basic research. On the other hand high-temperature powder-metallurgy techniques are attractive for the industrial manufacture of borides. It is not the aim of this Section to give an extensive review of the details of all these methods of synthesis. Rather, after an introductory outline of the various methods of preparation, the most recent developments are reviewed without any attempt to cover the earlier literature. For this, the reader is referred to other sources (KIEFFER and BENESOVSKY, 1963; SAMSONOV et al., 1962; POWELL, 1967). Two recent reviews are also pertinent in this connection, namely by THOMPSON (1970) and by GLASSON and JONES (1969). The former, in particular, contains numerous references to earlier work.

The general methods for the preparation of transition metal borides may be classified according to the following scheme:
(a) Direct synthesis from the elements by reaction sintering, melting or crystallization from melt of another metal. Occasionally a metal hydride is used instead of the pure metal.
(b) Hydrogen reduction of a boron halide and a source of metal such as an oxide, sulfide or halide (the pure metal can also be used). A related method is the high-temperature reaction between boron compounds, such as the halides, alkali tetrahydroborates, diborane or trimethylborane, and a metal halide.
(c) Electrolysis of a melt containing metal oxide and boron oxide or borax, together with a flux of alkali or alkaline-earth halides or fluoroborates.
(d) Reduction of metal oxide by boron (borothermic process), or co-reduction with boron oxide (or other boron compound) by carbon (carbothermic process), or reduction with boron carbide with or without free boron and/or carbon.
(e) Simultaneous reduction of the metal oxide and a source of boron by a strongly electro-positive metal, such as aluminum, magnesium or calcium.
(f) Precipitation of certain borides of group VIII metals from water solution in the vicinity of room temperature.

The reactions (a)-(c) are mainly used on a laboratory scale, although a reaction of type (b), modified to give pure elementary boron, is used for the industrial production of boron fibers. Reactions of type (b) are also used for the vapor deposition of wear-resistant layers of borides on metals. For the industrial production of boride powders the reactions of type (d) are the most attractive ones, although the borothermic process is too wasteful of boron for large-scale production. Reactions of type (e) are of current interest for the preparation of certain boride catalysts for hydrogenation and fuel cells.

Most of the general methods mentioned above have been known for a long time but in the last ten years much work has been devoted to studying them in detail. The main lines of research have been the study of processes for the growth of large single crystals, as well as investigations of commercial production methods and the development of low-temperature methods of synthesis.

β) Direct Combination of the Elements

Although container materials and composition control often present problems, the direct combination of the elements has many advantages in fundamental research, as has been recently discussed by the present author (LUNDSTRÖM, 1969). The elements can be combined by melting and subsequent solidification under a variety of conditions, or by solid-state reaction sintering, or by the crystallization from a melt of another metal ("menstruum technique").

The optimal conditions for the formation of binary platinum metal borides by reaction sintering were recently studied in a series of investigations (SAMSONOV and KOSENKO, 1971; KOSENKO, 1971; SAMSONOV et al., 1971; SAMSONOV et al., 1972; SAMSONOV and KOSENKO, 1973). The sintering temperature necessary for boride formation was found to decrease from Ru to Pd and from Os to Pt. Furthermore, the reaction time was found to increase with increasing boron content of the product.

In the last ten years much work has been done to prepare large single crystals of borides. The fine-grained products generally obtained from the melting and reaction sintering processes may serve as suitable starting materials for single crystal growth. Large single crystals, mainly of diborides but occasionally also of other borides, have been prepared using the chemical transport technique (KLOTZ, 1965; NICKL et al., 1966), the floating-zone technique (HAGGERTY et al., 1969; BROOKES and PACKER, 1968) and the Czochralski (GEORGE et al., 1967) and Verneuil techniques (LYNCH et al., 1966). Large single crystals of transition metal borides can also be prepared from a melt of another metal. The starting materials (metal and boron) react in a metal as solvent, from which the crystals can subsequently be extracted. The more refractory borides of the group IV transition metals were obtained, using copper, lead, tin (JANGG and KIEFFER, 1973), iron (NAKANO et al., 1971) or zinc as solvent metal (OBUKHOV et al., 1968; GURIN et al., 1971).

γ) Reduction of a Metal Compound and Boron Halide with Hydrogen

Several transition metal borides can be prepared by reaction between a boron trichloride-hydrogen mixture and an appropriate source of metal, for instance metal oxides, sulfides or chlorides. It was recently shown for the first time that vanadium, chromium, iron and nickel borides can be prepared in this way from sulfides at lower reaction temperatures than those required for reaction-sintering of the elements (CUEILLERON et al., 1971).

The use of the group IV metal tetrachlorides for a similar reaction has been known for a long time:

$$MeCl_4(g) + 2BCl_3(g) + 5H_2(g) \rightarrow MeB_2(s) + 10HCl(g)$$

Using this vapor-deposition process, the influence of process variables on yield, stoichiometry and microstructure was studied for the titanium, zirconium and hafnium diborides (GEBHARDT and CREE, 1965; PESHEV, 1966). It was shown that fibers of titanium diboride can be prepared by this technique at the very low temperature of 400 °C in an electrical discharge (TAKAHASHI et al., 1971), as well as by thermal decomposition of a mixture of titanium tetrachloride and lithium tetrahydroborate on substrate wires of tungsten and graphite (HOUGH, 1966). Adherent layers of certain metal borides on metal substrate surfaces were obtained by thermal decomposition of metal (Mo, W, Nb, Ta) halides and boron tribromide on a metallic substrate using a solar furnace or induction heating (ARMAS and TROMBE, 1971, 1972; ARMAS, 1973).

δ) Fused-Salt Electrolysis

Electrolysis of a salt-melt containing alkali halides, metal oxides or salts and a boron compound was studied extensively by earlier investigators, notably ANDRIEUX and co-workers (POWELL, 1967; THOMPSON, 1970). The thermodynamic requirements for obtaining single crystals by this technique were investigated (HUGGINS, 1970) as well as the mechanism of formation of tantalum monoboride from a $KCl/K_2TaF_7/KBF_4$ bath (RAMEAU, 1971). As a result it was concluded that primarily elementary tantalum and boron are formed by cathodical reduction of Ta^{5+} and BF_4^- ions. The elements subsequently react to form TaB.

ε) Reduction of a Metal Oxide or Some Other Metal Compound Using Carbon, Boron or Boron Carbide

It was demonstrated that B_2O_2 is formed in the borothermic process for preparation of diborides at temperatures above 1000 °C (BARTON and NICHOLLS, 1966; FAVRE and THÉVENOT, 1971):

$$TiO_2(s) + 4B(s) \rightarrow TiB_2(s) + B_2O_2(g)$$

Addition of carbon to the reaction mixture reduces the oxygen content of the titanium diboride. The sulfide can be used as the source of metal (CUEILLERON et al., 1971). Metal-rich chromium borides could not be obtained with the borothermic process (SEREBRYAKOVA and SAMSONOV, 1967).

Reaction between metal oxides and carbon (carbothermic process), or boron carbide, or both, is of greatest importance for economical large-scale production of borides. The reactions may be exemplified by

$$2MeO_2(s) + B_4C(s) + 3C(s) \rightarrow 2MeB_2(s) + 4CO(g)$$

The product is often contaminated by carbon (MEYER and PASTOR, 1965; SEREBRYAKOVA, 1966) but the process can be modified to give a product of low carbon content (MEERSON and GORBUNOV, 1969; ČECH et al., 1965). Zirconium diboride of reasonable purity can be obtained from a reaction between zircon, boron oxide and carbon in a submerged electric-arc furnace (McMULLEN and McKEE, 1965).

The borothermic preparation of transition metal borides was recently studied extensively by BLIZNAKOV, PESHEV and coworkers (PESHEV et al., 1967; PESHEV and BLIZNAKOV, 1968; PESHEV et al., 1968), who also discussed the relative merits of carbon, boron and boron carbide as reduction materials (PESHEV, 1967; BLIZNAKOV and PESHEV, 1969).

ζ) Reduction by Other Metals

A mixture of a transition metal oxide and boron oxide is reduced by an electro-positive metal such as aluminum, magnesium, or an alkali metal. The purity of the product is fairly low (THOMPSON, 1970), although purities higher than 99.5% were reported using magnesium as the reducing agent (MARKOVSKII and VEKSHINA, 1967). In a variant of this reaction the metal component of the desired boride was used directly as the reducing agent for boron halides (PICHAT, 1967). Titanium diboride was obtained by reaction between boron trifluoride or trichloride and titanium powder. Ferro-boron can be produced from boric acid, aluminum and iron in an arc furnace process (KUMYSH, 1966). The material is, however, also easily made by the carbo-thermic process (ARAVAMUTHAN and SUNDARAM, 1969).

η) Low-Temperature Precipitation of Borides from Solution

A powerful catalyst for fuel cells is obtained when borides are precipitated from a solution of nickel or cobalt acetates by adding a solution of sodium tetrahydroborate (MEIBUHR, 1967). Initially an amorphous product of uncertain composition and constitution is formed. Heat treatment at 300°C-700°C yields crystalline Ni_3B as the main constituent (HOFER et al., 1964; LENFANT et al., 1965; BLIZNAKOV et al., 1972), although the average composition of the precipitate corresponds to Ni_2B. Magnetic and electron diffraction investigations support these observations (ROMANOWSKI, 1967).

b) Properties

α) Introduction

The most conspicous properties of the transition metal borides are the following: high metling points and hardness, good electrical and thermal conductivity, and brittleness. The latter to a certain degree hampers the practical application of the borides. In this Section a few recent developments in our knowledge of these properties are reviewed. Summaries of earlier measurements have been published (ARONSSON, 1960b; STEINITZ, 1960; GREENWOOD et al., 1966). A large number of constants are to be found in SAMSONOV's compilation (SAMSONOV, 1963) and many additional references in a review article from 1970 (THOMPSON, 1970). The reader is referred to these sources for further information. The properties of diborides and of ternary borides are the subjects of two other articles in this book (Chaps. C.XI and C.XII) and therefore they are excluded from this Section.

Only a few of the binary phase diagrams for transition metal-boron systems may be regarded as well established. For certain systems, there has been no phase diagram presented as yet. For other systems, serious discrepancies occur between the phase diagrams presented. Three investigations of the Ni-B system, for instance, have lead to different results as regards the occurrence, mode of formation and melting temperature (a difference of 600°C!) of o-Ni_4B_3 and m-Ni_4B_3 and as regards the occurrence of two boron-rich nickel borides (SCHÖBEL and STADELMAIER, 1965; PORTNOI et al., 1966; SOBOLEV and FEDOROV, 1967). New phase diagrams were also presented for the Mn-B (MARKOVSKII and BEZRUK, 1967), Fe-B (PORTNOI et al., 1969) and Co-B systems (SCHÖBEL and STADELMAIER, 1966), based on microscopic, X-ray diffraction and DTA technique. In the compilation by RUDY one finds the binary systems of the group IV to VI transition metals (RUDY, 1969).

A most valuable aspect of these phase diagrams is that they include information from experimental determinations of the homogeneity ranges of the phases. PORTNOI and his collaborators have also investigated the binary systems Zr-B (PORTNOI et al., 1970), Hf-B (PORTNOI et al., 1971), Ta-B (PORTNOI et al., 1971b), Cr-B (PORTNOI et al., 1969), Mo-B (PORTNOI et al., 1967) and W-B (PORTNOI et al., 1967). The discrepancies between these investigations and those described by RUDY are such that further studies are desirable.

Although there are several borides and boride systems, which need to be characterized more carefully, the majority of measurements of physical properties during the last few years have been carried out on reasonably pure and well-characterized materials, in many cases single crystals.

β) Chemical Properties

Transition metal borides are characterized by chemical inertness in dilute and even in concentrated acids and bases, although there are many conflicting reports on their reactivity (GREENWOOD et al., 1966). There are only a few investigations where the state of division and the purity (phase mixtures have often been used) of the boride have been properly controlled. Frequently, the reaction products have not been characterized at all or very incompletely. Fairly free from objections of this type is however a study of the reactivity of MoB, NiB and some diborides towards aqueous and non-aqueous acids (GREENWOOD et al., 1965). As a result oxidizing acids were found to attack the borides readily, other aqueous acids more slowly. In no case were boranes found among the reaction products from hydrolysis with hydrochloric acid as reported earlier. The reactions of iron and nickel borides with hot aqueous acids have been studied recently (TIMOFEEVA et al., 1970) and support the earlier results. The chemical stability of the borides of group IV to VI metals were studied in alkaline solutions (KUGAI, 1972). In general the decomposition of borides can be effected by fusion with alkali or by heating in a chlorine atmosphere. Fluorine reacts vigorously with borides even at room temperature.

The borides display reasonably good resistance to oxidation in air at high temperature, although the rate of oxidation varies considerably between different borides. Most studies of high temperature oxidation have been carried out on diborides. TiB_2 and ZrB_2, which are the most oxidation-resistant diborides, have been investigated in particular (see Chap. C.XI of this Vol.). In view of the fact that the free energy of formation for most borides is fairly low (80-150 kJ/mole), the oxidation resistance is somewhat surprising. It is surmized, however, that a protective layer of mixed oxides, formed at the surface, is responsible for the slow oxidation rate. An estimation of thermodynamic data for most of the borides was obtained by a combination of theoretical calculations and empirical correlations (KRIKORIAN, 1971; the paper also contains numerous references). Experimental determinations of thermodynamic data were undertaken for nickel borides (OMORI and HASHIMOTO, 1973) and for Mo_2B, W_2B and Re_3B (BAEHREN and VOLLATH, 1969).

Most diborides and certain other borides are stable in the presence of carbon at high temperatures. The cobalt borides react with carbon, forming a ternary tau phase (MARKOVSKII et al., 1971).

An interesting property of borides, especially those of group IV transition metals, is their high resistance to attack by molten metals (for instance iron and aluminum), non-basic slags and salts. The angle of wetting between the molten metal and the boride is of great importance for some applications. It has been reported recently that nickel wets borides of low enthalpy of formation much more efficiently than those of high enthalpy of formation (TUMANOV et al., 1970). A similar observation was earlier made for carbides.

γ) Electrical and Magnetic Properties

Transition metal borides are metallic conductors, with room temperature resistivities in the range 3-200 μohm·cm. The diborides are in general the best conductors but even a boron-rich compound like ZrB_{12} has a resistivity as low as 60 μohm·cm (SAMSONOV, 1963). A positive temperature coefficient of resistivity, characteristic for metals, has also been found for the chromium borides (SEREBRYAKOVA and KOVENSKAYA, 1966), iron borides (KOSTETSKII et al., 1971) and tungsten borides (SATA and URANO, 1970), the resistivity of which were measured from room temperature to about 1000 °C. The temperature-dependence of the electric resistivity of Mn_3B_4 was investigated from -200 °C to 170 °C (NOVOGRUDSKII et al., 1972). The manganese borides, from Mn_4B to MnB_2 (40-85 μohm·cm), have lower electrical resistivity than pure manganese (150 μ·ohm·cm) itself from room temperature up to 400 °C (BEZRUK and MARKOVSKII, 1968). Electrical resistivity and magnetic suceptibility were measured for the monoborides of iron, cobalt and nickel (BUDOZHAPOV et al., 1973).

To the list of known superconducting borides (ARONSSON et al., 1965) the following borides should be added: ZrB_{12} with T_c about 6 K (MATTHIAS et al., 1968; CHU and HILL, 1968), ZrB (probably not a binary compound), HfB and TaB with T_c in the range 3-4 K (SHULISHOVA and SHCHERBAK, 1967).

Most of the borides are paramagnetic. The diborides of group IV to VI metals are, for instance, weakly paramagnetic with the exception of ZrB_2, which is diamagnetic. Ferromagnetic, anti-ferromagnetic or even more complicated magnetic structures occur among the borides of the 3d elements from chromium to nickel. Several investigations have been devoted to this group of borides during the last few years, some of which are reviewed below.

After the pioneering measurements (CADEVILLE and MEYER, 1962; LUNDQVIST et al., 1962) of the temperature-dependence of the susceptibilities of Me_2B and MeB (Me = Cr, Mn, Fe, Co, Ni) phases, numerous reports on the magnetic and electronic properties of these compounds have been published. The magnetic moments of solid solutions of the types $(Me_{1-x}Me'_x)_2B$ and $(Me_{1-x}Me'_x)B$ have been measured and utilized to plot the average moment per transition metal atom versus average atomic number of the transition elements. These plots are very closely related to those of Slater and Pauling for transition metal alloys, although they are shifted towards lower electron concentration as compared to those of the transition metals (CADEVILLE and DANIEL, 1966). In conclusion it was suggested that these borides have a partially filled d band, which is very similar to those of transition metals and alloys, and that a transfer of electrons to this d band has occurred from the transition metal s level or from the boron atoms. The model has gained further support from low-temperature specific-heat measurements (KUENTZLER and MEYER, 1968; KUENTZLER, 1970a, b; 1971). It should be mentioned, however, that objections have been raised to the model presented. The assumption of the close similarity between the d band structures of the boride and the transition metal was questioned in a recent low-temperature specific-heat study (HANSON et al., 1971). Furthermore, a neutron and X-ray diffraction investigation of charge and spin distributions in Fe_2B did not give any support for the transfer of electrons from the boron atoms into the iron 3d band (BROWN and COX, 1971).

A very complex magnetic structure has been discovered for Mn_3B_4. The magnetic susceptibility of Mn_3B_4 powder was measured from liquid nitrogen temperature to about 900°C (HIROTA and YANASE, 1965). An antiferromagnetic (Néel) transition was observed at 392 K and another magnetic anomaly at 226 K. In a subsequent series of papers the properties of the solid solutions of Fe, Ni, Co, Cr, Mo and W in Mn_3B_4 were studied. It was found that the Néel temperature decreases with solute content for Fe (IGA et al., 1966b) as well as for Ni and Co (HIROTA and YANASE, 1966), while the susceptibility at the Néel temperature increases with the iron content but changes slightly with cobalt and nickel content. Chromium-doped Mn_3B_4 displays a ferromagnetic behavior at low temperature (TAWARA et al., 1966). Whether this originates in a claimed ferromagnetic phase Mn_2CrB_4 (a super-structure of Mn_3B_4) or has some other explanation remains to be investigated. Magnetization and NMR studies of the solid solutions of Mo and W in Mn_3B_4 were reported together with Mössbauer studies of iron-doped Mn_3B_4 (IGA and TAWARA, 1968). The experimental data were interpreted in terms of co-existing "ferromagnetic islands in an anti-ferromagnetic sea". Furthermore, it was very hard to reach magnetic saturation of the alloys indicating a complex, helical or canted spin structure for the compounds. The magnetization of pure Mn_3B_4 was recently investigated, using pulsed magnetic fields up to 450 kOe (FAKIDOV and TIMOSHCHUK, 1970; TIMOSHCHUK and FAKIDOV, 1971). In their articles FAKIDOV and TIMOSHCHUK presented a magnetic phase diagram for Mn_3B_4 and, furthermore, discussed a so-called metamagnetic behavior of Mn_3B_4 in strong magnetic fields. A magnetic model for this behavior was already presented earlier (HIROTA and YANASE, 1965).

In the Mn_3B_4 structure the manganese atoms occupy a two-fold (2c) position and a four-fold (4g) position in space group Immm. In the investigations mentioned above, two assumptions regarding the magnetic moments of the manganese atoms were used. It was recently shown using neutron diffraction techniques that manganese atoms at the 2(c) position have $\mu = 2.92\mu_B$ and that those at the 4(g) position have $\mu = 0.44\mu_B$ (NEOV and LEGRAND, 1972). Although the measurements were made on powder and with boron of natural isotopic composition, the results are in good agreement with those from magnetic susceptibility measurements of single crystal Mn_3B_4 (KASAYA et al., 1971).

The anisotropy of the magnetic energy was studied for single crystals of Fe_2B and Co_2B and some solid solutions (IGA et al., 1966a; IGA, 1966; 1970a,b).

Nuclear magnetic resonance and Mössbauer spectroscopy have been used in a few cases to provide information about the electronic configuration of borides. Internal magnetic fields and electric field gradients were studied for Co_2B and Co_3B (SHINOHARA and WATANABE, 1965; KASAYA et al., 1973) as well as a number of monoborides (HIHARA and HIRAHARA, 1965; CREEL and BARNES, 1972). Earlier Mössbauer measurements of iron borides (COOPER et al., 1964; SHINJO et al., 1964) were used to discuss the chemical bonding in these borides. In a recent Mössbauer study of Fe_2B (MURPHY and HERSHKOWITZ, 1973) the easy axis of magnetization was found to change from its room-temperature position in the (001) plane to the [001] direction with increasing temperature, which is in agreement with a study of the magnetocrystalline anisotropy of Fe_2B (IGA et al., 1966a). Other recent studies have been performed on poorly characterized specimens and so are of less value.

c) Applications

The real and potential applications of transition metal borides are based on a unique combination of high melting point, hardness and good electrical and thermal conductivity. The borides are further characterized by a fair corrosion-resistance, chemical inertness in metal melts, good wear-resistance and a thermal shock-resistance much better than that of oxide ceramics. There are, however, also examples of well-established uses, where non of the properties mentioned is of prime importance. Such an example is the use of TiB_2 as a grain-refining agent in aluminum melts (DAVIES et al., 1970). Another example is the use of ferro-boron (mainly iron borides) as a micro-alloying additive in steel to increase the hardenability. There are relatively few well-established applications of borides, where use is made of the characteristic properties mentioned above, although the number of proposed applications is very extensive.

Early expectations that the borides would successfully compete with the refractory carbides as constituents in cemented hard metals have not been fulfilled. It has been difficult to find a useful combination of a refractory boride and a binder alloy with appropriate wetting properties. Cutting tool materials based on Mo_2B with nickel as binder (STEINITZ, 1960) are claimed to be promising, as well as chromium boride with molybdenum as binder phase (MERZ, 1970).

Another method to utilize the good wear-resistance of the borides in spite of their brittleness is to apply a thin boride coating on a metal. Much research is currently devoted to this field and it is only possible to mention a few references in this article. The coating can be obtained by a vapor-deposition process in combination with rapid diffusion of boron into the base metal (boronizing or boriding process, for references see POWELL, 1967), by plasma-spraying of the boride (BENTZ et al., 1968) or by electrodeposition (SCHLAIN et al., 1969). Boride coatings on steel have been produced in a process similar to case-hardening (GALE, 1972), and the properties of such coatings were also recently investigated (MINKEVITCH, 1963).

The combination of high melting point, corrosion and thermal shock-resistance as well as to some extent mechanical strength is desirable for applications in rocket nozzles, combustion chambers and jet engine turbines. One might expect that these requirements could be fulfilled by the properties of uni-directionally solidified eutectic metal-boron systems, a few of which have been explored (CROSSMAN and YUE, 1971; DE SILVA and CHADVICK, 1970). The Ti-TiB eutectic cannot be used since the volume fraction of TiB, and consequently also the strengthening obtainable, is too small (CROSSMAN and YUE, 1971).

At present there is some industrial production of crucibles of titanium diboride and boron nitride for vaporization of metals in high vacuum and at high temperature (THOMPSON, 1974). An improved oxidation resistance as compared to the pure boride is displayed by a composite of zirconium diboride and molybdenum disilicide, which has been proposed for use in components of supersonic aircraft (WHITTMORE, 1968).

It has been proposed to precipitation-harden chromium-base alloys by complex borides. The temperature of formation and stability of the boride precipitates was studied but no information is available as regards the mechanical properties (FILIPPI, 1973).

The resistance of certain borides to attack by molten metals and salts is well known. Since they are also good electrical conductors their use as cathode collector bars for aluminum reduction cells has been suggested and investigated (RANSLEY, 1963). Bars of zirconium diboride have been produced commercially (POWELL, 1967), although high costs have as yet precluded a more general use. General corrosion is slight in this application but it has been found that the sensitivity to intergranular corrosion and brittleness limit the lifetime of the bars (WHITTMORE, 1968). Titanium and zirconium diborides are used as thermocouple protection tubes in steel baths and in aluminum reduction cells (SAMSONOV and KISLYI, 1965; POWELL, 1967; HOFFMANN, 1968). The advantage of a stable protection tube is particularly great in processes like continuous casting. For this application the high thermal conductivity of the borides is also useful. The use of borides as thermocouple material has also been suggested (KISLYI, and SAMSONOV, 1961).

The high neutron-capture cross section of boron has made zirconium diboride attractive for control rods in nuclear energy plants. It has been found, however, that B_4C is a more convenient material for this purpose.

The excellent thermoionic emission power of lanthanum hexaboride has been utilized in the construction of long-life cathodes for electron guns (BROERS, 1967; AHMED and BROERS, 1972; FORD and LICHTMAN, 1973). The lifetime is reported to exceed that of tungsten cathodes by two orders of magnitude.

The preparation of fibers of titanium diboride has been reported (HOUGH, 1966). Use of these fibers in composite materials has been suggested (HOUGH, 1966) but no mechanical data for the composites have been published.

Nickel boride is used as a catalyst in organic chemistry and in fuel cells (MEIBUHR, 1967).

References

AHMED, H., BROERS, A.N.: Lanthanum hexaboride electron emitter. J. Appl. Phys. 43, 2185 (1972)

ANDERSSON, S., CARLSSON, J.O.: The crystal structure of MnB_4. Acta Chem. Scand. 24, 1791 (1970)

ANDERSSON, S., LUNDSTRÖM, T.: The crystal structure of CrB_4. Acta Chem. Scand. 22, 3103 (1968)

ARAVAMUTHAN, V., SUNDARAM, M.: Recent developments in some electrochemical industries. Chem. Age India 20, 281 (1969)

ARMAS, B.: Dépôts de borures de Mo et de W. Rev. Int. Hautes Temp. Réfract. 10, 231 (1973)

ARMAS, B., TROMBE, F.: Dépôts en phase vapeur, sur "front chaud" solaire, de borures de molybdène et tungstène par decomposition thermique de mélanges d'halogénures. C.R. Acad. Sci. (Paris) 272, Ser. C, 286 (1971)

ARMAS, B., TROMBE, F.: Dépôts chimiques, en phase gaseur, de borures de niobium et de tantale. C.R. Acad. Sci. (Paris) 274, Ser. C, 1134 (1972)

ARONSSON, B.: Borides and silicides of the transition metals. Arkiv Kemi 16, 379 (1960a)

ARONSSON, B.: Borides. Part A: Basic Factors. In: Modern Materials (H.H. Hausner, ed.), Vol.II, pp. 143-190. New York and London: Academic Press 1960b

ARONSSON, B.: The crystal structure of RuB_2, OsB_2 and $IrB_{1.35}$ and some general comments on the crystal chemistry of borides in the composition range MeB-MeB_3. Acta Chem. Scand. 17, 2036 (1963)

ARONSSON, B., LUNDSTRÖM, T., ENGSTRÖM, I.: Some aspects of the crystal chemistry of borides, boro-carbides and silicides of the transition metals. In: Anisotropy in Single-crystal Refractory Compounds (F.W. Vahldiek, S.A. Mersol, eds.). New York: Plenum Press 1968
ARONSSON, B., LUNDSTRÖM, T., RUNDQVIST, S.: Borides, silicides and phosphides. London: Methuen 1965
ARONSSON, B., STENBERG, E., ÅSELIUS, J.: Borides of Ruthenium, Osmium and Iridium. Nature 195, 377 (1962)
ÅSELIUS, J.: The crystal structure of $Ru_{11}B_8$. Acta Chem. Scand. 14, 2169 (1960)
BAEHREN, F.D., VOLLATH, D.: Eine Bestimmung der thermodynamischen Daten der Boride Mo_2B, W_2B und Re_3B. Planseeber. Pulvermet. 17, 180 (1969)
BARTON, L., NICHOLLS, D.: The hydrogenation of boron monoxide to diborane and the reactions of boron and boron carbide with titanium and zirconium dioxides. J. Inorg. Nucl. Chem. 28, 1367 (1966)
BAUER, J., NOWOTNY, H.: Der Dreistoff Yttrium-Bor-Kohlenstoff. Monatsh. Chem. 102, 1129 (1971)
BAURECHT, H.E., BOLLER, H., NOWOTNY, H.: Röntgenographische Untersuchungen in den Dreistoffen Cr-P-C, Cr-As-C und Cr-P-B. Monatsh. Chem. 102, 373 (1971)
BECHER, H.J., KORGMANN, K., PEISKER, E.: Über das ternäre Borid Mn_2AlB_2. Z. anorg. allgem. Chem. 344, 140 (1966)
BENTZ, G., PROVOST, G., URBAN, C.: Couches minces de borures et nitrures obtenues par arrosage au plasma. Bull. Soc. Fr. Ceram. 81, 23 (1968)
BEZRUK, E.T., MARKOVSKII, L.Ya.: Some physical properties of manganese borides. Izv. Akad. Nauk SSSR, Neorg. Materialy 4, 447 (1968)
BLIZNAKOV, G., PESHEV, P.: On the preparation and some properties of metal borides. Izv. Khim. Inst. Bulgar. Akad. Nauk II, 391 (1969)
BLIZNAKOV, G., PESHEV, P., GYUROV, G.Ts.: Studies on obtaining nickel borides from solutions. C.R. Acad. Bulgare Sci. 25, 1209 (1972)
BOLLER, H., RIEGER, W., NOWOTNY, H.: Systematische Stapelfehler in der δ-WB-Phase bei Bor-Unterschuss. Monatsh. Chem. 95, 1497 (1964)
BROERS, A.N.: Electron gun using long-life lanthanum hexaboride cathode. J. Appl. Phys. 38, 1991 (1967)
BROOKES, C.A., PACKER, M.E.: The preparation of crystalline refractory carbides and borides by zone melting. In: Special Ceramics 4 (P. Popper, ed.), Brit. Ceram. Res. Assoc. 1968
BROWN, B.E., BEERNTSEN, D.J.: Refinement of an iron chromium boride with the Mn_4B structure. Acta Cryst. 17, 448 (1964)
BROWN, P.J., COX, J.L.: Charge and spin density distributions in the ferromagnetic alloy Fe_2B. Phil. Mag. 23, 705 (1971)
BUDOZHAPOV, V.D., ZELENIN, L.P., CHEMERINSKAYA, L.S., SIDORENKO, F.A. GELD, P.V.: Magnetic susceptibility and electric resistance of iron, cobalt and nickel monoborides. Izv. Akad. Nauk SSSR, Neorgan. Materialy 9, 1447 (1973)
CADEVILLE, M.C., DANIEL, E.: Sur la structure électronique de quelque borures d'éléments de transition. J. Phys. (Paris) 27, 449 (1966)
CADEVILLE, M.C., MEYER, A.J.P.: Points de Curie et moments des borures ferromagnétiques du group du fer de type M_2B, MB, $(M_1,M_2)_2B$ et $(M_1,M_2)B$. C.R. Acad. Sci. (Paris) 255, 3391 (1962)
CĚCH, B., OLEVERIUS, P., SEJBAL, J.: Sintering of zironium boride with activating additions. Powder Met. 8, 142 (1965)
CHABAN, N.F., KUZ´MA, Yu.B.: X-ray diffraction study of Re-Al-B and Re-Si-B ternary systems. Izv. Akad. Nauk SSSR, Neorgan. Materialy 8, 1065 (1972)
CHEPIGA, M.V., KRIVUTSKII, V.P., KUZ´MA, Yu.B.: The system Cr-Ni-B. Izv. Akad. Nauk SSSR, Neorgan. Materialy 8, 1059 (1972)
CHRÉTIEN, A., HELGORSKY, J.: Sur les borures de molybdène et de tungstène MoB_4 et WB_4, composés nouveaux. C.R. Acad. Sci. (Paris) 252 742 (1961)
CHU, C.W., HILL, H.H.: Boron isotope effect in superconducting zirconium dodecaboride. Science 159, 1227 (1968)
COOPER, J.D., GIBB, T.C., GREENWOOD, N.N., PARISH, R.V.: Chemical applications of the Mössbauer effect. I. The bonding in iron borides. Trans. Faraday Soc. 60, 2097 (1964)
CREEL, R.B., BARNES, R.G.: NMR study of transition metal monoborides. J. Chem. Phys. 56, 1549 (1972)

CROSSMAN, F.W., Yue, A.S.: Unidirectionally solidified Ti-TiB and Ti-Ti_5Si_3 eutectic composites. Met. Trans. 2, 1545 (1971)
CUEILLERON, J., LAHET, G., THÉVENOT, F., PÂRIS, R.A.: Nouvelle methode d'obtention de borures metalliques. J. Less Common Metals 24, 317 (1971)
DAVIES, I.G., DENNIS, J.M., HELLAWELL, A.: The nucleation of aluminium grains in alloys of aluminium with titanium and boron. Met. Trans. 1, 275 (1970)
DE SILVA, A.R.T., CHADWICK, G.A.: Fatigue behavior of the unidirectionally solidified fibrous Fe-Fe_2B eutectic alloys. J. Metal Sci. 4, 63 (1970)
FAKIDOV, I.G., TIMOSHCHUK, V.I.: Effect of strong pulsed magnetic fields on the magnetic transitions in Mn_3B_4. Soviet Physics-Solid State 12, 218 (1970). Engl. Transl.
FAVRE, R., THÉVENOT, F.: Préparation et étude des diborures de titane, de zironium et de hafnium. Bull. Soc. Chem. France, Ser. A, 3911 (1971)
FILIPPI, A.M.: Stability of reactive and refractory metal borides in ternary chromium-base alloys. J. Less Common Metals 30, 153 (1973)
FORD, R.R., LICHTMAN, D.: Structure of lanthanum-hexaboride-coated rhenium filaments. J. Appl. Phys. 44, 4378 (1973)
GALE, L.: Case hardening boron can rival WC. Engineer 235, 39 (1972)
GALASSO, F., PINTO, J.: The metal borides in boron fiber cores; identification of MoB_4. Trans. Met. Soc. AIME 242, 754 (1968)
GANGLBERGER, E., NOWOTNY, H., BENESOVSKY, F.: Der Dreistoff Re-Fe-B. Monatsh. Chem. 97, 718 (1966)
GEBHARDT, J.J., CREE, R.F.: Vapor-deposited borides of group IVA metals. J. Amer. Ceram. Soc. 48, 262 (1965)
GEORGE, W.P.R., GOODMAN, C.H.L., STERLING, H.F., WARREN, R.W.: A possible new group of semiconducting compounds. Phys. Stat. Sol. 21, 205 (1967)
GLASSON, D.R., JONES, J.A.: Formation and reactivity of borides. carbides and silicides. J. Appl. Chem. 19, 125 (1969)
GREENWOOD, N.N., PARISH, R.V., THORNTON, P.: Some reactions of metal borides. J. Chem. Soc. p. 545 (1965)
GREENWOOD, N.N., PARISH, R.V., THORNTON, P.: Metal borides. Quart. Rev. 20, 441 (1966)
GURIN, V.N., OBUKHOV, A.P., KORSUKOVA, M.M., TERENT'EVA, S.P., KOZLOVA, I.R., JOFFE, A.F.: Synthese hochschmelzender Verbindungen der Übergangsmetalle aus deren Lösungen in Metallschmelzen. Planseeber. Pulvermet. 19, 86 (1971)
HÄGG, G.: Gesetzmässigkeiten im Kristallbau bei Hydriden, Boriden, Carbiden und Nitriden der Übergangselemente. Z. Phys. Chem. (Leipzig) B12, 33 (1931)
HAGGERTY, J.S., WENCKUS, J.F., LEE, D.W.: Oriented growth and defect structures of single crystal transition metal diborides. In: High Temperature Technology: Proc. of the 3rd Int. Symp. High Temp. Technol. 1967. London: Butterworths 1969
HANSON, B.D., MAHNIG, M., TOTH, L.E.: Low temperature heat capacities of transition metal borides. Z. Naturforschg. A26, 739 (1971)
HASCHKE, H., NOWOTNY, H., BENESOVSKY, F.: Untersuchungen in den Dreistoffen: {Mo,W} - {Fe, Co,Ni} - B. Monatsh. Chem. 97, 1459 (1966)
HIHARA, T., HIRAHARA, E.: Nuclear magnetic resonance in ferromagnetic MnB. J. Phys. Soc. Japan 20, 873 (1965)
HIROTA, H., YANASE, A.: Magnetic properties of Mn_3B_4. J. Phys. Soc. Japan 20, 1596 (1965)
HIROTA, H., YANASE, A.: Magnetic properties of $Mn_{3-x}Ni_xB_4$ and $Mn_{3-x}Co_xB_4$. J. Phys. Soc. Japan 21, 433 (1966)
HOARD, J.L., HUGHES, R.E.: Elemental boron and compounds of high boron content: structure, properties and polymorphism. In: The Chemistry of Boron and its Compounds (E.L. Muetterties, ed.). New York: John Wiley and Sons 1967
HOFER, L.J.E., SCHULTZ, J.F., PANSON, R.D., ANDERSON, R.B.: On the nature of the nickel boride formed by the action of sodium borohydride on nickel salts. Inorg. Chem. 3, 1783 (1964)
HOHNKE, D., PARTHÉ, E.: AB compounds with Sc, Y and rare earth metals. II. FeB and CrB type structures of monosilicides and germanides. Acta Cryst. 20, 572 (1966)
HOLLECK, H.: IV. Int. Conf. on Solid Compounds of Transition Elements, Geneva: April 9-13, 1973
HOUGH, R.L.: Pyrolytic titanium diboride filaments for composites. Amer. Inst. Aeron. Astronaut. J. 4, 107 (1966)

HUGGINS, R.A.: Thermodynamic considerations to the use of electrochemical techniques for the growth of single crystals of borides. Mat. Res. Bull. 5, 391 (1970)
IGA, A.: Magnetocrystalline anisotropy of Co_2B. J. Phys. Soc. Japan 21, 1464 (1966)
IGA, A.: Magnetocrystalline anisotropy in $(Fe_{1-x}Ni_x)_2B$. Japan J. Appl. Phys, 9, 414 (1970a)
IGA, A.: Magnetocrystalline anisotropy in $(Fe_{1-x}Co_x)_2B$. Japan J. Appl. Phys. 9, 415 (1970b)
IGA, A., TAWARA, Y.: Magnetic properties of molybdenum- and tungsten-modified Mn_3B_4. J. Phys. Soc. Japan 24, 28 (1968)
IGA, A., TAWARA, Y., YANASE, A.: Magnetocrystalline anisotropy of Fe_2B. J. Phys. Soc. Japan 21, 404 (1966a)
IGA, A., TAWARA, Y., YANASE, A.: Magnetic properties of Fe-modified Mn_3B_4. J. Phys. Soc. Japan 21, 1640 (1966b)
JANGG, G., KIEFFER, R.: Die Herstellung von Boriden der Übergangsmetalle aus dem Hilfsmetallbad. Monatsh. Chem. 104, 226 (1973)
JEDLICKA, H., BENESOVSKY, F., NOWOTNY, H.: Die Kristallstruktur des W_3CoB_3 und der dazu isotypen Phasen Mo_3CoB_3, Mo_3NiB_3 und W_3NiB_3. Monatsh. Chem. 100, 844 (1969)
JEITSCHKO, W.: The crystal structure of MoAlB. Monatsh. Chem. 97, 1472 (1966)
JEITSCHKO, W.: The crystal structure of MoCoB and related compounds. Acta Cryst. B24, 930 (1968)
JEITSCHKO, W.: The crystal structure of Fe_2AlB_2. Acta Cryst. B25, 163 (1969)
JEITSCHKO, W., NOWOTNY, H., BENESOVSKY, F.: Die Kristallstruktur von Mo_2BC. Monatsh. Chem. 94, 565 (1963)
KASAYA, M., HIHARA, T., KŌI, Y.: Magnetic susceptibility of single crystal Mn_3B_4. J. Phys. Soc. Japan 30, 286 (1971)
KASAYA, M., HIHARA, T., KŌI, Y.: Nuclear magnetic resonance study of Co_2B. J. Phys. Soc. Japan 34, 63 (1973)
KIEFFER, R., BENESOVSKY, F.: Hartstoffe. Vienna: Springer-Verlag 1963
KIESSLING, R.: The crystal structures of molybdenum and tungsten borides. Acta Chem. Scand. 1, 893 (1947)
KIESSLING, R.: The borides of some transition metals. Acta Chem. Scand. 4, 209 (1950)
KISLYI, P.S., SAMSONOV, G.V.: Hochtemperatur-Thermoelemente aus Halbleitern. Planseeber. Pulvermet. 8, 200 (1961)
KLOTZ, H.: Chemischer Transport von Lanthanhexaboride. Naturwiss. 52, 451 (1965)
KONDRASHEV, Yu.D.: The crystal structure of chromium borocarbide. Kristallografiya 11, 559 (1966)
KOSENKO, V.A.: Preparation conditions and properties of rhodium borides. Zh. Prikl. Khim 44, 2099 (1971)
KOSTETSKII, I.I., L'VOV, S.N., KUNITSKII, Yu.A.: Some physical properties of iron borides. Izv. Akad. Nauk SSSR, Neorgan. Materialy 7, 951 (1971)
KRIKORIAN, O.H.: Estimation of heat capacities and other thermodynamic properties of refractory borides. Report UCRL-51043. Springfield, National Technical Information Center 1971
KRIPYAKEVICH, P.I., KUZ'MA, Yu.B., VOROSHILOV, Yu.V.: The crystal structure of NbCoB. Acta Cryst. B27, 257 (1971)
KUENTZLER, R.: Specific heat of nearly ferromagnetic borides $(Co\text{-}Ni)_2B$ and $(Fe\text{-}Co)_2B$. J. Appl. Phys. 41, 908 (1970a)
KUENTZLER, R.: Low temperature specific heat measurements of semi-borides of the first series of transition elements. Phys. Stat. Solid. 41, 291 (1970b)
KUENTZLER, R.: Structure électronique de borures. J. Phys. (Paris), Suppl. de Colloques 32, 634 (1971)
KUENTZLER, R., MEYER, J.-P.: Mesures des chaleurs spécifiques électroniques des borures. C.R. Acad. Sci. (Paris) 266, Ser. B, 755 (1968)
KUGAI, L.N.: Chemical stability of group IV-VI transition metal borides in alkaline solutions. Izv. Akad. Nauk SSSR, Neorgan. Materialy 8, 769 (1972)
KUMYSH, I.S.: Making higher purity ferroboron. Stal' 26, 331 (1966)
KUZ'MA, Yu.B.: The crystal structures of the compounds NbFeB and TaFeB. Dopovidi Akad. Nauk Ukr. RSR p. 939 (1967)
KUZ'MA, Yu.B.: The crystal structures of the compounds NbNiB and TaNiB. Kristallografiya 13, 701 (1968)

KUZ´MA, Yu.B.: Crystal structure of $YCrB_4$ and its analogues. Kristallografiya 15, 372 (1970)
KUZ´MA, Yu.B.: The systems Ti-Co-B and Ti-Re-B. Izv. Akad. Nauk SSSR, Neorgan. Materialy 7, 514 (1971)
KUZ´MA, Yu.B., BILONIZHKO, N.S.: Crystal structure of the compound $CeCo_4B_4$ and its analogues. Kristallografiya 16, 1030 (1971)
KUZ´MA, Yu.B., BILONIZHKO, N.S.: New boride structural types in the homologous series based on the $CaCu_5$ and $CeCo_3B_2$ types. Kristallografiya 18, 710 (1973)
KUZ´MA, Yu.B., KRIPYAKEVICH, P.I., BILONIZHKO, N.S.: Crystal structure of $CeCo_3B_2$ and analogous compounds. Dopovidi Akad. Nauk Ukr. RSR, Ser. A, p. 939 (1969)
KUZ´MA, Yu.B., LAKH, V.I., STADNYK, B.I., CHABAN, N.F.: The system Ta-Re-B. Dopovidi Akad. Nauk Ukr. RSR p. 849 (1971)
KUZ´MA, Yu.B., LAKH, V.I., STADNYK, B.I., KOVALYK, D.A.: The systems Hf-W-B, Hf-Re-B and Nb-Re-B. Porosh. Met. p. 59 (1970)
KUZ´MA, Yu.B., SEREBRYAKOVA, T.I., PLAKHINA, A.M.: The polymorphic transformation of W_2B_5. Zh. Neorgan. Khim 12, 559 (1967)
KUZ´MA, Yu.B., SVARICHEVSKAYA, S.I.: New ternary borides with the $YCrB_4$-type structure. Dopovidi Akad. Nauk Ukr. RSR p. 166 (1972a)
KUZ´MA, Yu.B., SVARICHEVSKAYA, S.I.: The crystal structure of Y_2ReB_6 and its analogues. Kristallografiya 17, 658 (1972b)
KUZ´MA, Yu.B., YARMOLYUK, Ya.P.: The crystal structure of the compound $Ti_3Co_5B_2$. Zh. Strukt. Khim. 12, 458 (1971)
LA PLACA, S., POST, B.: The crystal structure of rhenium diboride. Acta Cryst. 15, 97 (1962)
LENFANT, P., LEGRAS, C., KERVELLA, B.: Sur la structure des catalyseurs "borure de nickel". C.R. Acad. Sci. (Paris) 260, 1636 (1965)
LUGSCHEIDER, E., KNOTEK, O., REIMANN, H.: Der Dreistoffsystem Nickel-Chrom-Bor. Monatsh. Chem. 105, 80 (1974)
LUNDQUIST, N., MYERS, H.P., WESTIN, R.: The paramagnetic properties of the monoborides of V, Cr, Mn, Fe, Co and Ni. Phil. Mag. 7, 1187 (1962)
LUNDSTRÖM, T.: The structure of Ru_2B_3 and $WB_{2.0}$ as determined by single-crystal diffractometry, and some notes on the W-B system. Arkiv Kemi 30, 115 (1968)
LUNDSTRÖM, T.: Preparation and crystal chemistry of some refractory borides and phosphides. Arkiv Kemi 31, 227 (1969)
LUNDSTRÖM, T.: Second International Meeting on Boron Chemistry. Leeds, England 25-29th April, 1974
LUNDSTRÖM, T., ROSENBERG, I.: The crystal structure of the molybdenum boride $Mo_{1-x}B_3$. J. Solid State Chem. 6, 299 (1973)
LUNDSTRÖM, T., TERGENIUS, L.E.: Refinement of the crystal structure of the non-stoichiometric boride $IrB_{\sim 1.35}$. Acta Chem. Scand. 27, 3705 (1973)
LYNCH, C.T., MERSOL, S.A., VAHLDIEK, F.W.: The microstructure of single-crystal titanium diboride. J. Less Common Metals 10, 206 (1966)
MCMULLEN, J.C., MCKEE, W.D., Jr.: Zirconium diboride from zircon. Amer. Ceram. Soc. Bull. 44, 488 (1965)
MARKOVSKII, L.Ya., BEZRUK, E.T.: Phase diagram of the manganese-boron system. Izv. Akad. Nauk SSSR, Neorgan. Materialy 3, 2165 (1967)
MARKOVSKII, L.Ya., BEZRUK, E.T., BERLOVA, G.E.: Stability of nickel and cobalt borides with respect to carbon. Izv. Akad. Nauk SSSR, Neorgan. Materialy 7, 56 (1971)
MARKOVSKII, L.Ya., VERKSHINA, N.V.: A magnesiumthermic method for the preparation of metal borides. Zh. Prikl. Khim. 40, 1824 (1967)
MATKOVICH, V.I.: Structural principles in refractory borides. Pure Appl.Chem. 39, 525 (1974)
MATTHIAS, B.T., GEBALLE, T.H., ANDRES, K., CORENZWIT, E., HULL, G.W., MAITA, J.P.: Superconductivity and antiferromagnetism in boron-rich lattices. Science 159, 530 (1968)
MEERSON, G.A., GORBUNOV, A.E.: Preparation of pure carbon-free borides by a carbothermic process. Izv. Akad. Nauk SSSR, Neorgan. Materialy 5, 2075 (1969)
MEIBUHR, S.G.: Nickel boride hydrogen anodes. Electrochim. Acta 12, 1059 (1967)
MERZ, A.: Eigenschaften und Anwendung hochschmelzender Hartstoffe. Neue Hütte 15, 159 (1970)

MEYER, R., PASTOR, H.: Les borures refractoires: Études particulière des borures de titane et de zirconium: préparation, propriétés et application. Bull. Soc. Franc. Ceram. 66, 59 (1965)

MINKEVITCH, A.N.: Couches de borures obtenues sur les métaux par diffusion. Rev. Met. 60, 807 (1963)

MURPHY, K.A., HERSHKOWITZ, N.: Temperature-dependent hyperfine interactions in Fe_2B. Phys. Rev. B, Solid State 7, 23 (1973)

NAKANO, K., HAYASHI, H., IMURA, T.: Growth of IVa-diborides single crystals in molten iron. Japan J. Appl. Phys. 10, 513 (1971)

NEOV, S., LEGRAND, E.: Neutron diffraction study of the magnetic structure of Mn_3B_4. Phys. Stat. Sol. (b) 49, 589 (1972)

NICKL, J., DUCK, M., PIERITZ, J.: Transport reactions of silicides and borides of transition metals. Angew. Chem. 5, 839 (1966)

NIIHARA, K., SHISHIDO, T., YAJIMA, S.: A ternary rare earth boride with a $ThCr_2Si_2$-type structure. Bull. Chem. Soc. Japan 44, 3214 (1971)

NOVOGRUDSKII, V.N., TIMOSHCHUK, V.I., FAKIDOV, I.G.: Temperature dependence of the electric resistance of Mn_3B_4. Soviet Physics - Solid State 13, 2372 (1972). Engl. Transl.

NOWOTNY, H., HASCHKE, H., BENESOVSKY, F.: Bor-reiche Wolframboride. Monatsh. Chem. 98, 547 (1967)

OBUKHOV, A.P., GURIN, V.N., KOZLOVA, I.R., TERENT'EVA, Z.R., MAZINA, T.I.: Synthesis of refractory compounds of group IV-VI transition metals from molten zinc. Izv. Akad. Nauk SSSR, Neorgan. Materialy 4, 527 (1968)

OMORI, S., HASHIMOTO, Y.: Thermodynamic properties of Ni_3B and Ni_2B. J. Japan. Soc. Powder Met. 20, 80 (1973)

PAPESCH, G., NOWOTNY, H., BENESOVSKY, F.: Untersuchungen in den Systemen: Chrom-Bor-Kohlenstoff, Mangan-Bor-Kohlenstoff und Mangan-Germanium-Kohlenstoff. Monatsh. Chem. 104, 933 (1973)

PESHEV, P.: Preparation of crystalline titanium diboride by vapour deposition process. Izv. Inst. Obsha Neorgan. Khim. Bulgar. Akad. Nauk. 4, 53 (1966)

PESHEV, P.: Étude thermodynamique de certaines réactions de préparation de borures de métaux des groupes II-VI de la classification périodique. Rev. Inst. Hautes Temp. Réfract. 4, 289 (1967)

PESHEV, P., BLIZNAKOV, G.: On the borothermic preparation of titanium, zirconium and hafnium diborides. J. Less Common Met. 14, 23 (1968)

PESHEV, P., BLIZNAKOV, G., LEYAROVSKA, L.: On the preparation of some chromium, molybdenum and tungsten borides. J. Less Common Met. 13, 241 (1967)

PESHEV, P., LEYAROVSKA, L., BLIZNAKOV, G.: On the borothermic preparation of some vanadium, niobium and tantalum borides. J. Less Common Met. 15, 259 (1968)

PICHAT, P.: Préparation de borures au moyen de trifluorure de bore. C.R. Acad. Sci. (Paris) 265, Ser. C, 385 (1967)

PORTNOI, K.I., CHUBAROV, V.M., ROMASHOV, V.M., LEVINSKAYA, M.Kh., SALIBEKOV, S.E.: Phase diagram of the nickel-boron system. Dokl. Akad. Nauk SSSR 169, 1104 (1966)

PORTNOI, K.I., LEVINSKAYA, M.Kh., ROMASHOV, V.M.: Iron-boron phase diagram. Porosh. Met. p. 66 (1969)

PORTNOI, K.I., LEVINSKII, Yu.D., ROMASHOV, V.M., MORDOVIN, O.A., LEVINSKAYA, M.Kh.: Phase diagram of the molybdenum-boron system. Izv. Akad. Nauk SSSR, Metally, p. 171 (1967)

PORTNOI, K.I., ROMASHOV, V.M., BUROBINA, L.N.: Phase diagram of the zirconium-boron system. Porosh. Met. p. 68 (1970)

PORTNOI, K.I., ROMASHOV, V.M., LEVINSKII, Yu.V., ROMANOVICH, I.V.: Phase diagram of the tungsten-boron system. Porosh. Met. p. 75 (1967)

PORTNOI, K.I., ROMASHOV, V.M., ROMANOVICH, I.V.: Chromium-boron phase diagram. Porosh. Met., p. 51 (1969)

PORTNOI, K.I., ROMASHOV, V.M., ROMANOVICH, I.V., LEVENSKII, Yu.B., PROKF'EV, S.A.: Phase diagram of the hafnium-boron system. Izv. Akad. Nauk SSSR, Neorgan. Materialy 7, 1987 (1971)

PORTNOI, K.I., ROMASHOV, V.M., SALIBEKOV, S.E.: Phase diagram of the tantalum-boron system. Porosh. Met. p. 89 (1971b)

POST, B.: Refractory binary borides. In: Boron, Metallo-Boron Compounds and Boranes (R.M. Adams, ed.). New York-London-Sydney: Wiley Interscience 1964

POWELL, C.F.: Borides. In: High-temperature Materials and Technology (I.E. Campbell, E.M. Sherwood, eds.), p. 349. New York: J. Wiley and Sons 1967
RAMAN, A.: On the AlB_2-type phases. Z. Metallk. 58, 179 (1967)
RAMEAU, J.J.: L'étude de la formation du tantale et des borures de tantale par électrolyse ignée. Rev. Int. Hautes Temp. Réfract. 8, 59 (1971)
RANSLEY, C.E.: Refractory carbides and borides for aluminium reduction cells. J. Met. 14, 129 (1963)
RIEGER, W., NOWOTNY, H., BENESOVSKY, F.: Verbindungen mit aufgefülltem Mn_5Si_3-Typ. Monatsh. Chem 96, 98 (1965)
RIEGER, W., NOWOTNY, H., BENESOVSKY, F.: Die Kristallstruktur von W_2CoB_2 und isotypen Phasen. Monatsh. Chem. 97, 378 (1966)
ROGL, P.: Über SE-Metall-Kobaltboride. Monatsh. Chem. 104, 1623 (1973)
ROGL, P., BENESOVSKY, F., NOWOTNY, H.: Über einige Komplexboride mit Platinametallen. Monatsh. Chem. 103, 965 (1972)
ROGL, P., NOWOTNY, H.: Neue Komplexboride. Monatsh. Chem. 104, 943 (1973a)
ROGL, P., NOWOTNY, H.: Neue Tau-Boride. Monatsh. Chem. 104, 1325 (1973b)
ROGL, P., NOWOTNY, H.: Neue Kappa-Phasen. Monatsh. Chem. 104, 1497 (1973c)
ROGL, P., NOWOTNY, H., BENESOVSKY, F.: Komplexboride mit ReB_2-Struktur. Monatsh. Chem. 101, 27 (1970a)
ROGL, P., NOWOTNY, H., BENESOVSKY, F.: Ternäre Komplexboride in den Dreistoffen: {Mo, W} - {Ru,Os} -B und W-Ir-B. Monatsh. Chem. 101, 850 (1970b)
ROGL, P., NOWOTNY, H., BENESOVSKY, F.: Ein Beitrag zur Strukturchemie der Iridiumboride. Monatsh. Chem. 102, 678 (1971a)
ROGL, P., NOWOTNY, H., BENESOVSKY, F.: Komplexboride in den Systemen Hf-Mo-B und Hf-W-B. Monatsh. Chem. 102, 971 (1971b)
ROGL, P., NOWOTNY, H., BENESOVSKY, F.: Neue κ-Boride und verwandte Phasen (Re_3B-Typ, aufgefüllt). Monatsh. Chem. 104, 182 (1973)
ROMANOWSKI, W.: On the nature of catalytically active Ni and Co borides. Rocz. Chem. 41, 423 (1967)
ROMANS, P.A., KRUG, M.P.: Composition and crystallographic data for the highest boride of tungsten. Acta Cryst. 20, 313 (1966)
ROOF, Jr., R.B., KEMPTER, C.P.: New orthorhombic phase in the Ru-B and Os-B systems. J. Chem. Phys. 37, 1473 (1962)
RUDY, E.: Ternary phase equilibria in transition metal-boron-carbon-silicon systems. Part V. Compendium of Phase diagram data. Report AFML-TR-65-2. Springfield: Clearing House for Federal Scientific and Technical Information 1969
RUNDQVIST, S.: An X-ray investigation of the nickel-boron system. Acta Chem. Scand. 13, 1193 (1959)
RUNDQVIST, S., PRAMATUS, S.: Crystal structure refinements of Ni_3B, o-Ni_4B_3 and m-Ni_4B_3. Acta Chem. Scand. 21, 191 (1967)
SAMSONOV, G.V.: Tugoplavkie Soedineniya. Moskva: Metallurgizdat 1963
SAMSONOV, G.V., KISLYI, P.S.: Vysokotemperaturnye nemetallicheskie termopary i nakonechiniki, p. 152. Kiev: Izdatel´stvo "Naukova Dumka" 1965
SAMSONOV, G.V., KOSENKO, V.A.: Investigation of preparation conditions of platinum borides. Porosh. Met. p. 25, 1971
SAMSONOV, G.V., KOSENKO, V.A.: The preparation conditions of ruthenium borides. Zh. Prikl. Khim. 46, 1177 (1973)
SAMSONOV, G.V., KOSENKO, V.A., IVANCHENKO, V.G.: The preparation conditions of palladium borides. Zh. Prikl. Khim. 44, 2304 (1971)
SAMSONOV, G.V., KOSENKO, V.A., RUD´, B.M., SIDOROVA, V.G.: Investigation of preparation and properties of the compound $IrB_{1.1}$. Izv. Akad. Nauk SSSR, Neorgan. Materialy 8, 771 (1972)
SAMSONOV, G.V., MARKOVSKII, L.Ya., ZHIGACH, A.F., VALYASHKO, M.G.: Boron, its Compounds and Alloys. Engl. Transl. Washington: Office of Technical Services, Department of Commerce, 1962
SATA, T., URANO, T.: Reaction between refractory metals and insulating nitrides at high temperature. Yogyo Kyokai Shi 78, 27 (1970)
SCHLAIN, D., McCAWLEY, F.X., WYCHE, C.: Electrodeposition of titanium diboride coatings. J. Electrochem. Soc. 116, 1227 (1969)

SCHÖBEL, J.-D., STADELMAIER, H.H.: Das Zweistoffsystem Nickel-Bor. Z. Metallk. 56, 856 (1965)
SCHÖBEL, J.-D., STADELMAIER, H.H.: Das Zweistoffsystem Kobalt-Bor. Z. Metallk. 57, 323 (1966)
SCHÖNBERG, N.: The structure of the $Co_3W_9C_4$ phase. Acta Met. 2, 837 (1954)
SEREBRYAKOVA, T.I.: Investigation of preparation conditions for cobalt borides. Ukr. Khim. Zh. 32, 1067 (1966)
SEREBRYAKOVA, T.I., KOVENSKAYA, B.A.: Physical properties of chromium borides. Izv. Akad. Nauk SSSR, Neorgan. Materialy 2, 2134 (1966)
SEREBRYAKOVA, T.I., SAMSONOV, G.V.: Investigation of preparation conditions for chromium borides. Zh. Prikl. Khim. 40, 3 (1967)
SHINJO, T., ITOH, F., TAKAKI, H., NAKAMURA, Y., SHIKAZONO, N.: Fe^{57} Mössbauer effect in Fe_2B, FeB and Fe_3C. J. Phys. Soc. Japan 19, 1252 (1964)
SHINOHARA, T., WATANABE, H.: Nuclear magnetic resonance of Co^{59} in ferromagnetic cobalt compounds Co_3B and Co_2B. J. Phys. Soc. Japan 20, 2020 (1965)
SHULISHOVA, O.I., SHCHERBAK, I.A.: Superconductivity of borides of transition and rare-earth metals. Izv. Akad. Nauk SSSR, Neorgan. Materialy 3, 1495 (1967)
SMITH, G.S., JOHNSON, Q., NORDINE, P.C.: The crystal structure of ScB_2C_2. Acta Cryst. 19, 668 (1965)
SOBOLEV, A.S., FEDOROV, T.F.: The phase diagram of the nickel-boron system. Izv. Akad. Nauk SSSR, Neorgan. Materialy 3, 723 (1967)
SPEAR, K.E., GILLES, P.W.: Phase and structure relationships in the vanadium-boron system. High Temp. Sci. 1, 86 (1969)
STADELMAIER, H.H.: Metal-rich metal-metalloid phases. In: Developments in the Structural Chemistry of Alloy Phases (B.C. Giessen, ed.), p. 141. New York-London: Plenum Press 1969
STADELMAIER, H.H., SCHÖBEL, J.-D.: Die Boride $HfCo_3B_2$ und $ZrCo_3B_2$ als ternärer $CaCu_5$-Typ. Monatsh. Chem. 100, 224 (1969)
STEINITZ, R.: Borides. Part B: Fabrication, properties and applications. In: Modern Materials (H.H. Hausner, ed.), Vol. II, pp 191-224. New York-London: Academic Press 1960
STENBERG, E.: The crystal structures of Pd_5B_2,(Mn_5C_2) and Pd_3B. Acta Chem. Scand. 15, 861 (1961)
TAKAHASHI, T., SUGIYAMA, K., SUZUKI, Y.: Fibrous growth of titanium diboride by the discharge method. J. Cryst. Growth 10, 139 (1971)
TAWARA, Y., IGA, A., YANASE, A.: Ferro- and antiferromagnetism in $(Mn_{1-x}Cr_x)_3B_4$. J. Phys. Soc. Japan 21, 476 (1966)
TELEGUS, V.S., KUZ´MA, Yu.B.: X-ray investigation of the rhenium-manganese-boron system. Dopovidi Akad. Nauk Ukr. RSR, Ser. A. p. 945 (1969)
TELEGUS, V.S., KUZ´MA, Yu.B.: Phase equilibriums in the V-Mn-B, Mo-Mn-B and W-Mn-B systems. Porosh. Met. p. 67 (1971)
THOMPSON, R.: The chemistry of metal borides and related compounds. In: Progress in Boron Chemistry (R.J. Brotherton and H. Steinberg, eds.) Vol. 2, p. 173, Oxford: Pergamon Press 1970
THOMPSON, R.: Industrial applications of boron compounds. Pure and Applied Chemistry 39, 547 (1974)
TIMOFEEVA, N.I., LEVINSKAYA, M.Kh., KRAINOVA, Z.I.: Chemical stability of nickel and iron borides. Izv. Akad. Nauk SSSR, Neorg. Materialy 6, 1613 (1970)
TIMOSHCHUK, V.I., FAKIDOV, I.G.: Metamagnetism in Mn_3B_4. Soviet Physics - Solid State 13, 85 (1971). Engl. Transl.
TOTH, L., NOWOTNY, H., BENESOVSKY, F., RUDY, E.: Der Dreistoff U-B-C. Monatsh. Chem. 92, 794 (1961)
TRZEBIATOWSKI, W., RUDZINSKI, J.: The composition and structure of technetium nitride and technetium borides. J. Less Common Metals 6, 244 (1964)
TUMANOV, V.I., GORBUNOV, A.E., KONDRATENKO, T.M.: Wettability and properties of mono- and diborides of group IV and VI metals. Zh. Fiz. Khim. 44, 540 (1970)
VOROSHILOV, Yu.V., KRIPYAKEVICH, P.I., KUZ´MA, Yu.B.: The crystal structure of $ZrCo_3B_2$ and $HfCo_3B_2$. Kristallografiya 15, 934 (1970)
WALD, F., ROSENBERG, A.J.: Constitutional investigations in the boron-platinum system. Trans. Met. Soc. AIME 233, 796 (1965)
WHITTMORE, O.J.: The ultra-refractory ceramics: carbides, nitrides and borides. Trend in Engineering 20, 28 (1968)

X. Single-Crystal Refractory Borides of Transition Metals

V. N. GURIN and V. S. SINELNIKOVA

A number of attractive features make transition metal borides an important subject for scientific and technical study. They possess high melting points and great hardness, as well as resistance to heat and corrosion. Their thermo-physical properties combined with the capability of forming cermets and composites, further recommend them for study.

The use of single-crystal borides appears to be promising in the field of thermoionic cathodes, high-temperature thermocouples, electrodes, corrosion-resistant and abrasive materials, etc. However, it is difficult to point out the difference between the properties of single-crystal borides and those of polycrystalline, sintered and melted specimens because the electrical, mechanical, optical and chemical properties of single-crystal borides have not yet been established. Preparation and further study of single crystals of pure, high purity, alloyed and complex (in solid solution with silicides and carbides) borides may reveal valuable new properties and new areas of application.

1. Specific Features of Boride Single-Crystal Preparation

Compared to other classes of refractory compounds (carbides, nitrides, silicides and oxides), borides of the IV-VI group transition metals exhibit the largest number of noncongruently melting compounds. According to GOLDSCHMIDT (1957), eighteen of the forty-six borides known to date, or 39% (TiB_2, TiB, Ti_2B_5, ZrB, HfB, VB, Nb_3B_2, Nb_3B_4, NbB_2, Ta_2B, Ta_3B_4, Cr_4B, Cr_2B, Mo_2B, Mo_2B_5, MoB_4, WB_4), are noncongruently melting compounds. In addition, the character of melting and the melting points have not yet been established for seven other compounds ($TiB_{1,1}$, HfB_{12}, VB_2, Nb_3B, Nb_2B, Ta_3B, Cr_3B_4). Therefore, direct melting of original compounds cannot be used to produce perfect single-crystals for more than half the boride phases of the IV-VI group transition metals. Three methods which may produce single-crystals of any borides are the gas-phase, solution and solid phase methods.

Another problem in obtaining single-crystal borides is existance of a large number of boron compounds for each transition metal of the IV-VI groups. This creates additional difficulties in the preparation of single-crystals of each individual compound or of its polymorphic modification. Consequently it is necessary to be even more thorough in the selection of techniques for their synthesis and crystallization.

2. Methods of Preparation of Single-Crystal Borides

In the discussions of methods of preparation that follow, we distinguish between the concepts of *method* and the concepts of *technique* (GURIN, 1972). Defining "Method" as consistent with the aggregative state of the original substance (solid state, liquid state or gas state) from which crystals form directly, GURIN (1973) lists four methods of preparing single crystals: solid-phase method, gas-phase method, liquid-phase method and solution method. Each of these methods may have several techniques which can sometimes be combined into analogous groups.

a) Solid-Phase Method

This method consists essentially of grain growth and formation of crystals by solid-phase (including thermo-reactive) diffusion, topochemical solid-phase and heterophase reactions and gas-phase intergrain transport.

The main techniques here are recrystallization and bulk saturation. The solid-phase method can be considered promising in the preparation of single-crystal carbides but it still has no practical application for boride.

Some attempts to produce grain growth and crystals of borides have been reported. KIESSLING (1949) was unable to obtain single crystals of CrB by recrystallization, WATSON and MATCHEN (1967) observed enlargement of boride crystals in the presence of metal oxide and carbon. Such enlargement can be attributed to intercrystalline gas-phase transport with carbon monoxide as the carrier gas. KINOSHITA et al. (1970) studied the growth of ZrB_2 grains by recrystallization. They established a relationship between grains growth in the specimen and density, temperature and heating time. The grains prepared by them increased in size from 14 to 55 μ.

These examples reveal a limited applicability of the solid-phase method for preparation of single-crystal refractory borides. Because the crystals obtained by this method are small and the time required to grow them is long, the solid-phase method can hardly compete with other methods in practical use although its potentialities have not yet been fully explored.

b) Gas-Phase Method

This method involves transporting the substance to be crystallized through the gas-phase to the substrate or a seed with or without a temperature gradient. The main techniques used in the method are: sublimation (at high vapor pressure and congruent vaporization of the substance to be crystallized); all versions of transport (at low vapor pressure of the substance) and chemical reactions in the gas phase (when the substance is to be synthesized prior to its crystallization). The chemical reaction and transport techniques are those most widely used with the borides. PESHEV (1966), PESHEV and NIEMYSKI (1966) and BLIZNAKOV and PESHEV (1969) prepared single crystals, reaching 10mm in size from titanium and boron chlorides using chemical reactions. GEORGE et al. (1967) prepared TiB_2 crystals up to a few mm in size in the same way. NICKL et al. (1966) produced TiB_2 in the form of aggregations of plates about 500 μ in size by the isothermal transport technique.

One of the present authors (SINEL'NIKOVA) prepared single crystals of titanium diboride by deposition from a mixture of boron and titanium chlorides in the presence of hydrogen. The titanium diboride crystals were of hexagonal shape and up to 2 mm in size, with stoichiometric composition. Variation of the B/Ti ratio in the vapor phase from 0.5 to 14 did not effect the chemical composition of titanium diboride.

Single-crystal boron carbide in the form of pyramids, needles and plates up to 4 mm in size, was prepared by MIERZEJEWSKA and NIEMYSKI

(1965) by interaction of carbon and boron chlorides in the presence of hydrogen.

Thus, it can be seen that this method produces single-crystal borides of sufficient size to perform chemical and physical studies. The conditions for the preparation of single-crystal borides of transition metals obtained by the gas-phase method are listed in Table 1.

c) Liquid-Phase Method

In this method the compound to be crystallized is passed through a temperature zone at which it melts. Within this temperature gradient the compound can exist in either solid or liquid form. The method has many modifications differing in technical details. They may be combined in groups of like techniques as was done by GURIN (1973). The most frequently used techniques which apply to borides, are crystal pulling from melt (Czochralski technique), zone melting (the Pfann technique) and flame fusion (Verneuil technique). GEORGE et al. (1967) used the Czochralski technique, and obtained single crystals of TiB_2 and ZrB_2 up to 30 mm. LYNCH et al. (1966), VAHLDIEK (1967) and MERSOL et al. (1968) used the Verneuil technique to prepare single crystals of TiB_2 of up to 13 mm. Zone melting (or, more accurately, the floating-zone technique, which is its vertical version) yields the best results. This technique was used by HAGGERTY et al. (1968), HAGGERTY et al. (1968, 1969) to produce single crystals of ZrB_2 and HfB_2 up to 50 mm in size (with compact polycrystalline specimens of up to 100 mm). These are the best results obtained for preparing single-crystal borides. The present method produces the largest single crystals. The conditions for their preparation are listed in Table 1.

d) The Solution Method

The solution method consists of either spontaneous or controlled crystallization of compounds from metal, salt or hydrothermal solutions. It can be subdivided into a large number of techniques, since there are many solvents where crystallization can be carried out, and a variety of technological versions in which crystallization can be produced. As in the case of the liquid-phase method, all of these can be combined into groups of similar techniques. Among these, only the group involving crystallization from metal solutions (i.e. from solutions in metal melts) was used to produce single crystals of refractory borides. Usually, iron and aluminum were used as the solvent metal. GRUBER (1963) obtained single-crystal borides of titanium, zirconium, niobium, tantalum, molybdenum and tungsten up to 3 mm in size from their solutions in iron. NAKANO et al. (1971) also produced larger crystals of TiB_2, ZrB_2 and HfB_2 from the iron melt. HIGASHI and ATODA (1970) used a solution in aluminium melt to obtain crystals of TiB_2 up to 3 mm in size. GURIN, one of the present authors, and KORSUKOVA (see SINEL'NIKOVA et al. (1976)), also used a solution in aluminium melt. They were the first to prepare crystals of noncongruently melting boride Mo_2B_5 up to 3-4 mm in size and W_2B_5 crystals up to 1.5-2 mm (see Fig. 1). In short, this method also produces single-crystal borides suitable for study. The conditions of their preparation are presented in Table 1[1].

e) A Comparison of Techniques for Preparing Single-Crystal Borides

From the data in Table 1, we may conclude that the largest boride crystals were prepared by the floating-zone technique belonging to the liq-

[1]Some new data on the obtaining of borides single crystals using the solution method are recently reported by NAKANO et al. (1974a) and HIGASHI et al. (1976).

Table 1. Comparison of methods for preparation of refractory boride single crystals

Method	Technique: Reaction etc.	Preparation: T °C	Condition: Gas flow rate, protective atmosphere, matrix solution etc.	Dimensions of single crystal, mm	References
Gas-phase	Chemical reactions	1000÷1400	3.5mol/h, $BCl_3:TiCl_4=2:1$	TiB_2: 5÷10	PESHEV (1966)
	$TiCl_4+2BCl_3+5H_2=TiB_2+10HCl$		H_2: $(TiCl_4+BCl_3)=30:1$		PESHEV and NIEMYSKI (1966)
					BLIZNAKOV and PESHEV (1969)
	-"-"-"	1000÷1500	-	TiB_2: few mm	GEORGE et al. (1967)
	Chemical transport Cl_2, Br_2, J_2	(1100-1050)÷ ÷(950-850)	5-22days; $p=10^{4}÷10^{5}$ torr	TiB_2, ZrB_2, VB_2, CrB, CrB_2: not indicated	NICKL et al. (1966)
	Chemical transport	1020	600h; p=100torr Br_2	TiB_2: ~0.5	NICKL et al. (1973)
	Chemical reactions $TiCl_4+BCl_3+H_2 \rightarrow TiB_2+HCl$	1400÷1700	60 min; B/Ti=0.1;0.5;1; 1.6;2;3;4;14.	TiB_2: 2	The present paper
Liquid-phase	VERNEUIL	-	-	TiB_2: 13; ϕ 7	LYNCH et al. (1966)
					VAHLDIEK (1967)
					MERSOL et al. (1968)
	CZOCHRALSKI	2900÷3040	-	TiB_2, ZrB_2: 30	GEORGE et al. (1967)
	Floating-zone	-	2.5÷20cm/h; Ar, 1÷20atm.	ZrB_2, HfB_2: 50; ϕ 6÷7	HAGGERTY et al. (1968a, 1968b, 1969, 1971)
Solution	in Fe: B(BFe)+M(MFe)→MB+Fe	1400÷2000	1÷100 deg/min	MB (M-Ti, Zr, V, Nb, Ta, Cr, Mo, W): 3	GRUBER (1963)
	"-"-	(1730-1700)÷1400	5deg/h; Ar+(2÷3)%H	TiB_2; ZrB_2, HfB_2: 6÷7	NAKANO et al. (1971)
	in Al: B(BAl)+MAl→MB+(Al)	1450	100h; Ar; HCl	TiB_2: 3	HIGASHI et al. (1970)
	-"-	1100÷1200	15÷20deg/h; Ar; HCl	Mo_2B_5: 4 W_2B_5: 1.5	The present paper

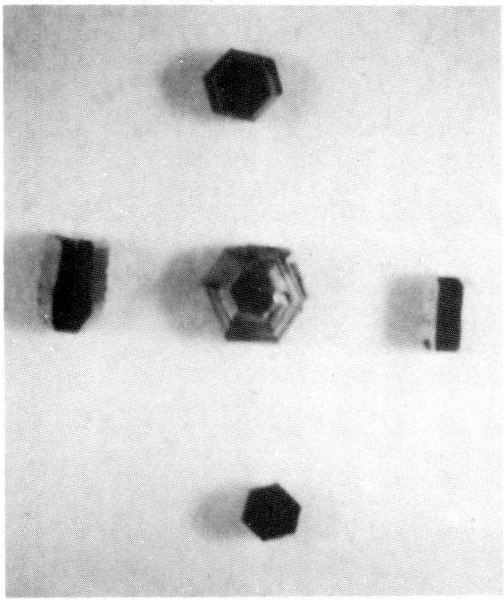

Fig. 1. W_2B_5 Idiomorphic single crystals obtained from solution in molten Al. X 15

uid-phase method. The resulting crystals are nearly one order of magnitude larger than those obtained by the gas-phase and solution methods.

Of all the known experiments, only Korsukova and Gurin, using the solution method, and CLOUGHERTY and POBER (1964) using the liquid-phase method have succeeded in preparing single crystals of noncongruently melting borides Mo_2B_5 and NbB_2.

The crystal preparation temperatures necessary in the liquid-phase method are conditioned by the melting point of borides (at least, the lower limit) and are, therefore, much higher than those needed in the gas-phase and solution methods. In the gas-phase method, the corresponding temperatures are somewhat lower than in the solution method. However, the time required to prepare crystals by the gas-phase method is the longest of the three methods. The solution and, particularly, the liquid-phase method are characterized by shorter production times and are technically simpler. Finally, the solution method is practically the only means of preparing perfect idiomorphic crystals (see Fig. 1). It is possible the properties of such crystals being thoroughly studied could differ from those of other crystalline forms.

Thus, a comparison of the conditions for preparation of single-crystal borides by various methods shows the need for improvements and further developments in each method. Whereas the solid- and liquid-phase methods are more suited for preparation of large single crystals of congruently melting borides, the solution and gas-phase methods should be more practical for obtaining comparatively small crystals as well as films and whiskers of noncongruently melting borides. On the whole, it may be noted that there are very few studies devoted to the development of various methods for obtaining single-crystal refractory borides.

Table 2. Composition and properties of single crystals, grown by different methods

Method, authors, technique, product		Composition, weight %						
		M	B	ΣM+B	Impurities			
Gas phase. (Chem. transport) GEORGE et al. (1967)	TiB_2	-	-	~99 (in starting material)				
Liquid phase.					C	N	O	Fe
(Floating-zone technique)	ZrB_2	80.85	19.01	99.86	0.05	0.004	0.007	0.0001
CHLOUGHERTY and	HfB_2	89.33	10.73	100.06	0.04	0.01	0.02	0.0001
POBER (1964)	NbB_2	81.42	18.47	99.89	0.01	0.005	0.006	0.0001
(CZOCHRALSKI technique)	TiB_2	-	-	~99 (in star.mat.)				
GEORGE et al. (1966)	ZrB_2	-	-					
(Floating-zone technique)	ZrB_2	79.0÷	17.8÷	97.3÷	0.08÷	0.027÷	0.01÷	
HAGGERTY et al. (1968b)		81.1	19.4	99.9	0.0001	0.0001	0.015	
HAGGERTY and LEE (1971)								
(VERNEUIL technique)		68.06÷	30.77÷	99.21÷	0.1÷	0.006÷	0.041÷	0.08÷
VAHLDIEK (1967)	TiB_2	68.44	31.88	99.94	0.05	0.004	0.014	0.001
LYNCH et al. (1966)					Cr:0.015÷0.001; Cu:0.004÷0.002;			
MERSOL et al. (1968)					Si:0.070÷0.001; Al:0.013			
Solution[a]					Al	Fe	Si	
NAKANO et al., (1971,	TiB_2	66.5	30.6	97.1	2.36	0.58	0.70	
1973) (in Fe)	ZrB_2	78.6	18.75	97.35	1.91	0.70	0.17	
	HfB_2	88.1	10.5	98.6	0.35	0.92	0.33	
The present paper	Mo_2B_5	80.2	19.7	99.9	-	-	-	
(in Al)	W_2B_5	89.3	10.5	99.8	-	-	-	

[a]Some new data on the properties (especially on microhardness) of single crystals thus obtained are reported by NAKANO et al. (1973, 1974a,b, 1976).

Properties		
Microhardness, HM kg/mm^2	Electrical Resistivity $\rho\,(25^{\circ}C)$ $\mu\Omega cm$	Density (d). Carrier concentration (electrons/cce), YOUNG's modulus (Y), Shear modulus (G), Elastic modulus ($\bar{Y}$) etc.
		cce.: $(6.3\div 9.0)\times 10^{20}\,e/cm^3$;
-	29÷87	Mobility: 101÷236 cm^2/V_s
		Mean expans. coef: Thermal conduct.:
~1970	5.7	$\alpha_a=7.95\,(^{\circ}C^{-1})\times 10^6$; 0.055 ($\rho^{25}$ =38.8μΩcm) cal/cm.sec.°C
~3700	13.5	α_a=7.49-"- ; -
-	68	α_a=7.32-"- ; 0.040 (ρ^{25} =32μΩcm) -"-
-	114	cce.: $2.7\times 10^{21}\,e/cm^3$; Mobility: 21$cm^2/V_s$
-	13	cce.: $3.5\times 10^{21}\,e/cm^3$; Mobility: 138$cm^2/V_s$
-	-	Critical resolved shear stress: 2.17 kg/mm^2 (2125°C)
2250 [10$\bar{1}$0]÷ ÷3375 (anneal.)	1.0÷1.3 (77 K)	$d^{(77^{\circ}K,\,293^{\circ}K)}$ 4.449÷5.502g/cm^3; Y= =17120÷21400kg/mm^2; $\bar{G}$=6880÷7930kg/mm^2;
2700 [0001] 2800	6.5÷6.8 (293°K)	$\bar{Y}$=18110÷21100; DEBYE T=952÷964°K; Longitudinal velocity $\bar{V}_e$=8020-8550m/sec
		Etch pit densities: Slip planes:
2700÷3900	-	$5\times 10^3\,cm^{-2}$ {0001} <11$\bar{2}$0>
2000÷2600	-	$2.5\times 10^3\,cm^{-2}$ {10$\bar{1}$0} <1$\bar{2}$10>
2000÷2500	-	$5\times 10^4\,cm^{-2}$ {0001}<11$\bar{2}$0> {10$\bar{1}$0}<1$\bar{2}$10>
1960±125 [0001]	17÷50	- -
2400∓200 [0001]	20÷50	- -

3. Properties of Single-Crystal Borides

Our inadequate knowledge of the properties of single-crystal borides is due to the problems encountered in their preparation. This lack of knowledge makes comparison of their properties with those of polycrystalline powders, sintered or pressed materials difficult in many cases. Comparing the properties of boride single crystals obtained by different methods and techniques is not simple either. These difficulties are aggravated if one tries to compare properties of high purity boride single crystals their composition corresponding to definite punkt in the region of homogeneity.

Therefore, Tables 2 and 3 illustrate not what has already been done and studied in the field of single-crystal borides, but rather, what still has to be done. Although probably incomplete, the data presented in the Tables are intended to show not only the main available results reached in the preparation of single-crystal borides by various methods, but the difference in the properties of such single crystals in comparison with polycrystalline borides as well.

a) A Comparison of Composition and Properties of Single-Crystal Borides Obtained by Various Methods

Table 2 contains the data on composition and properties of single-crystal borides whose properties are at least partially known. These crystals were obtained by one or more methods. Note that the large specimens prepared by the liquid-phase method can be studied more thoroughly. Their mechanical properties, in particular, are more readily examined than the small single crystals produced by the gas-phase and solution methods. It is evident that all the methods listed except the solution method are used to obtain single crystals of practically the same compounds. The crystal purity which has been measured is highest for the liquid-phase method and lowest for the solution method. There is too little information on the purity of the gas-phase method. It should be pointed out that the potential of each method for preparation of pure specimens is still not fully explored, so that we can draw no definite conclusion as to which of the methods is the most suitable in this area.

As shown in Table 2, electrical conductivity and microhardness are the only properties which can be used to some extent to compare single crystals obtained by different methods. GEORGE et al. (1967) found that the electrical conductivity of TiB_2, obtained by using the gas-phase and liquid-phase method respectively, differs by a factor of 1.2-4. They started with material of equal purity, but unfortunately, they did not specify the composition of the crystal. These specimens exhibit a still larger difference, a factor of 5-10 (see Table 2) in electron mobility.

The microhardness of ZrB_2 and HfB_2 single crystals prepared by the liquid-phase (the floating-zone technique) and the solution (from solution in iron melt) methods differs strongly. Table 2 lists the extreme values of microhardness of TiB_2, ZrB_2 and HfB_2 which were obtained by NAKANO et al, (1973) at different orientations of indentor with respect to the crystallographic plane of interest. The microhardness of HfB_2 is seen to be much lower than that measured on single crystals prepared by the floating-zone technique. In contrast ZrB_2 exhibits the reverse picture. These facts cannot be explained clearly, since apart from different purity of single crystals produced by different methods, crystallographic orientation of the plane in question and indentor direction are of considerable importance here. Even when the crystallographic orientations coincide, as is the case

Table 3. Comparative properties of single crystals and non-single crystals refractory borides

Authors, material, fabrication, product	Composition ΣM+B weigth %	Properties						
		d, g/cm^3	HM, kg/mm^2	ρ μΩcm	Y, kg/mm^2	σcompr. kg/mm^2	σbend. kg/mm^2	λ cal/cm. sec ^{0}C
MERSOL et al. (1968) Bar TiB_2 a-axis	99.7	4.449÷(77^{0}K, 4.502 293^{0}K)	2700 [0001]	1.0÷1.2 (77^{0}K) 6.6÷6.8 (293^{0}K)	18110	-	-	-
Bar TiB_2 c-axis	-"-	-"-	2350 [10$\bar{1}$0]	1.0÷1.2 (77^{0}K) 6.5÷6.7 (293^{0}K)	20400	-	-	-
Cylinder (Rod) TiB_2 a-axis	-"-	4.502 (77^{0}K,	2650 [0001]	1.3 (77^{0}K) 6.6 (293^{0}K)	18900	-	-	-
Cylinder (Rod) TiB_2 c-axis	-"-	293^{0}K)	2250 [00$\bar{1}$0]	1.2 (77^{0}K) 6.5 (293^{0}K)	21100	-	-	-
Polycrystal cylinder (Rod) TiB_2	-"-	4.330÷4.400 (77^{0}K, 293^{0}K)	2450÷ 2550	2.2÷2.4 (77^{0}K) 8.8÷9.1 (293^{0}K)	-	-	-	-
High pressure, hot-pressed TiB_2	96÷96.4	95% X-ray	~3600 (25^{0}C) ~700 (1000^{0}C)	11.9÷13.4 (25^{0}C)	-	-	-	-
Hot-pressed	95.4÷96.4	-"-	-	11.6÷13.2 (25^{0}C)	-	-	-	0.062 (200^{0}C, 15% por)
CHLOUGHERTY and POBER (1964) Single crystal	99.86	100% X-ray	~1970 (25^{0}C); ~600 (900^{0}C)	5.7 (25^{0}C)	-	-	-	0.055 (ρ^{25}=38.8 μΩcm)
High pressure, hot-pressed ZrB_2	92.9÷98.9	-	~2100 (25^{0}C) ~780 (1000^{0}C)	20.2÷10.2 (25^{0}C)	-	-	-	-
Hot-pressed	92.4÷98.2	-		12.4÷10.3 (25^{0}C)	-	-	-	-
KOTHELNIKOV et al. (1969) Gas-phase TiB_2	-	-	-	-	-	-	36	-
Hot-pressed	-	-	-	14.1÷28.4	54000	165	-	0.058 (20÷150^{0}C (ρ^{25}=28.4μΩcm)
SAMSONOV (1963) Hot-pressed TiB_2	-	4.5	3370±60 (p=30g)	14.4	54000	135	24.5 (1%por)	-
Hot-pressed ZrB_2	-	6.17	2252±22 (p=30g)	16.6	35000	158.7	9.3, (22-24%por)	-

with TiB_2 single crystals, the hardness exhibits a marked difference, apparently because of different indentor directions (see Tables 2 and 3). It is of interest that according to NAKANO et al. (1973) this difference may become as high as 900-1000 kg/mm^2 (~3800-3900 for {11$\bar{2}$0} [0001] and ~2980-3000 for {11$\bar{2}$0} <1$\bar{1}$00>). For ZrB_2 it is about one half of that, while for HfB_2 the difference does not exceed 150-160 kg/mm^2.

These are practically the only possibilities for comparing the properties of single crystals of a compound prepared by different methods to be found in the available works of a general nature. Generally speaking, such a comparison is not rigorous since the purity and degree of perfection of compared single crystals are not the same. For instance, there is no information on dislocation density in the single crystals obtained by the liquid-phase method. Thus, the data indicate the urgent need to conduct a study which would reveal the advantages of each of the methods (and of the techniques involved) used to produce single-crystal refractory borides.

Among interesting results obtained on single crystals, one may note a considerable increase of microhardness after annealing of TiB_2 single crystals prepared by the Verneuil technique (from 2800 to 3375 kg/mm^2), (LYNCH et al., 1966). The authors attribute the increase to dispersion-hardening during annealing due to diffusion into the crystal of the second phase present on its surface. The effect of the second phase of ZrB on the properties of ZrB_2 single crystals obtained by the floating-zone technique is reported by HAGGERTY and LEE (1971). They studied plastic deformation and determined that the slip planes observed in ZrB_2 are typical for hexagonal crystals. The above authors believe that the presence of small needle-like precipitates of the ZrB-phase in ZrB_2 crystals affects the thermomechanical properties of ZrB_2, producing, in particular, a considerable increase of hardness at high temperatures (at 2125°C, the critical resolved shear stress is 2.17 kg/mm^2). These properties are assumed to persist up to 2400°C. The presence of ZrB particles may hinder the motion of dislocations, and thus increase the hardness. On the other hand, such particles reduce the plasticity of ZrB_2. NAKANO et al. (1973) found the major slip planes responsible for the microhardness anisotropy in single-crystal borides prepared by the solution method from iron to be:

(a) for TiB_2 - {0001} <11$\bar{2}$0>
(b) for ZrB_2 - {10$\bar{1}$0} <1$\bar{2}$10>
(c) for HfB_2 - both above.

The authors observed a difference in microhardness between the prismatic and basal planes of single crystals. Indeed, in TiB_2 and HfB_2 the microhardness of prismatic planes is larger than that of basal planes, while in ZrB_2 the reverse is true. In addition, they established that the anisotropy of microhardness, as a function of indentor orientation for TiB_2 and HfB_2, is larger on prismatic than on basal planes, as compared with ZrB_2.

Interesting data were obtained by CLOUGHERTY and POBER (1964) who measured microhardness at various temperatures on single crystals of ZrB_2 and HfB_2 prepared by the floating-zone technique. They revealed a strong decrease in microhardness for these compounds with temperature increases from 25 to 1000°C. For ZrB_2 the microhardness decreases from ~1970 (at 25°C) to ~600 kg/mm^2 (~900°C). In the case of HfB_2, after a drop from ~3700 (25°C) to <800 kg/mm^2 (800°C), the microhardness rises to ~900 kg/mm^2 at 1000°C (in Table 2, the data on the microhardness of single crystals prepared by NAKANO et al.,

1973 and CLOUGHERTY and POBER, 1964 were taken from the graphs). They showed also that the microhardness of HfB_2 and NbB_2 single crystals (and particularly that of non-single crystal materials) decreased strongly with increasing load, so that a sharp drop occurs with the load increasing from 50 to ~300 g while from 300 to 500 g the decrease is insignificant.

b) A Comparison of Properties of Single-Crystal Refractory Borides and Other Materials

The advantages or drawbacks of single-crystal materials become particularly evident when one compares their properties with those of hot-pressed, polycrystalline and other materials. Such a comparison is presented in Table 3 with the data from some of the most representative works. The Table contains the data of MERSOL et al. (1968) indicating also the anisotropy of electrical and mechanical properties of TiB_2. While the anisotropy of electrical properties may practically be considered absent, it is very evident in mechanical properties. Indeed, besides microhardness, other mechanical characteristics also exhibit larger values along the c-axis than in the direction of the a-axis. This is true for both bar and cylindrical (rod-shaped) single-crystal specimens.

As seen from Table 3, the electrical resistivity of single crystals is lower than that of polycrystalline and hot-pressed specimens. For example, ρ of hot-pressed TiB_2 is 2-4 times that of single crystal material. The mechanical properties show the same relationship as the electrical resistivity. This may be attributed partially to higher purity and structural perfection of single crystals.

A comparison of the properties of the materials listed in Table 3 can be performed only on a relative basis since the specimens compared vary in purity and have different characteristics (percentage porosity, crystallographic orientation, etc.). Nevertheless, one may conclude that at the given composition and degree of perfection of single-crystal borides, one can expect the comparison in their properties with those of polycrystalline samples to remain substantially unchanged. Therefore, one of the most important problems in the technology of borides lies in considerable improvement of purity and perfection of the single crystals obtained. Further investigation of their physico-chemical properties is another urgent task.

4. Conclusions

The possibility of broad practical applications of single-crystal refractory borides depends on solving several important problems encountered in their preparation.

First, one may expect a substantial and probably even qualitative change in their properties to result from the following: the alloying of single-crystal borides with various elements and compounds; the preparation of single-crystal boride solid solutions with carbides and silicides; and the preparation of single-crystal borides of semiconductor purity.

Second, the preparation of single-crystals of noncongruently melting borides, particularly compounds rich in metals (e.g. Cr_4B) and boron (HfB_{12}), and a study of their major properties would be of considerable importance.

Third, boride whiskers should be obtained and their characteristics studied.

The solution of these problems can greatly broaden the scope and extent of the use of refractory borides.

References

BLIZNAKOV, G., PESHEV, P.: Preparation and some properties of metal borides. Izv. otdel. chim. nauk Bolg. AN, Vol II, kn. 3, pp. 391-401, Sofija, izd. Bolg. AN, (1969)

CLOUGHERTY, T.V., POBER, R.L.: Physical and mechanical properties of transition metal diborides. Compounds of interest on nuclear reactor technology. Nucl. Metall. 10, 423-443 (1964)

GEORGE, W.P., GOODMAN, C.H.L., STERLING, H.F., WARREN, R.W.: A possible new group of semiconducting compounds. Phys. Stat. Sol. 21, 205-210 (1967)

GOLDSCHMIDT, H.I.: Interstitial Alloys. London: Butterworths 1967

GRUBER, B.A.: Process for the production of metal borides. USA Pat N 3096149 (1963)

GURIN, V.N.: Synthesis methods of transition metals refractory compounds and the trends of their development. Uspechi Chimii 41, 616-647 (1972)

GURIN, V.N.: Preparation methods of single crystals of transition metals refractory compounds. Izv. Akad. Sci. USSR Neorgan. Materialy 9, 1289-1307 (1973)

HAGGERTY, J.S., LEE, D.W.: Plastic deformation of ZrB_2 single crystals. J. Am. Ceram. Soc. 54, 572-576 (1971)

HAGGERTY, J.S., WENCKUS, J.F.: Preparation and characterization of high quality single-crystal refractory metal borides and carbides. Techn. Report AFML-TR-68-228 Sept. (1968a)

HAGGERTY, J.S., O'BRIEN, J.L., WENCKUS, J.F.: Growth and characterization of single-crystal ZrB_2. J. Cryst. Growth 3/4, 291-294 (1968b)

HAGGERTY, J.S., WENCKUS, J.F., LEE, D.W.: Oriented growth and defect structures of single-crystal transition metal diborides. In: High Temperature Technology, pp. 547-555. London: Butterworths 1969

HIGASHI, J., ATODA, T.: Single-crystal growth of titanium diboride in molten aluminium. J. Cryst. Growth 7, 251-253 (1970)

HIGASHI, I., TAKAHASHI, I., ATODA, T.: Crystal growth of borides, and carbides of transition metals from molten aluminium solutions. J. Cryst. Growth 33, 207-211 (1976)

KIESSLING, R.: The binary system chromium-boron. Acta Chem. Scand. 3, 595-602 (1949)

KINOSHITA, M., KOSE, S., HAMANO, J.: Grain growth of zirconium diboride. Bull. Govt. Ind. Res. Inst. Osaka 21, 28-37 (1970)

KOTELNIKOV, R.B., BASHLYKOV, S.N., GALIACBAROV, Z.G., KASHTANOV, A.J.: Especially high melting elements and compounds. Isdatel'stvo Metallurgija, Moskow (1969)

LYNCH, C.T., MERSOL, S.A., VAHLDIEK, F.W.: The microstructure of single-crystal titanium diboride. J. Less-Common Metals 10, 206-219 (1966)

MERSOL, S.A., LYNCH, C.T., VAHLDIEK, F.W.: Defect structure of single-crystal titanium diboride. In: Anisotropy in Single-Crystal Refractory Compounds. (F.W. Vahldiek, S.A. Mersol). Vol II, pp. 41-94. New York: Plenum Press 1968

MIERZEJEWSKA, S., NIEMYSKI, T.: Preparation of crystalline boron carbide by vapour-phase reaction. J. Less-Common Metals 8, 368-374 (1965)

NAKANO, K., HAYASHI, H., IMURA, T.: Growth of IVa-diborides single-crystals in molten iron. Japan J. Appl. Phys. 10, 513-513 (1971)

NAKANO, K., HAYASHI, H., IMURA, T.: Single-crystal growth of IVa-diborides from metal solutions. J. Cryst. Growth 24/25, 679-682 (1974a)

NAKANO, K., IMURA, T., TAKEUCHI, SH.: Hardness anisotropy of single crystals of IVa-diborides. Japan J. Appl. Phys. 12, 186-189 (1973)

NAKANO, K., MATSUBARA, H., IMURA, T.: High temperature hardness of titanium diboride single crystal. Japan J. Appl. Phys. 13, 1005-1006 (1974b)

NAKANO, K., MATSUBARA, H., IMURA, T.: High temperature hardness of IVa-diborides single crystals. J. Less-Common Metals 47, 259-263 (1976)

NICKL, J.J., DUCK, M., PIERITZ, J.: Transport-reaktionen von Siliziden und Boriden der Übergangsmetalle. Z. Angew. Chem. 78, 822-822 (1966)
NICKL, J.J., KOKOUSSAS, I.D., MÜHLRATZER, A.: Untersuchungen über den isothermen chemischen Transport. J. Less-Common Metals 32, 243-273 (1973)
PESHEV, P.: Preparation of crystalline titanium diboride by vapour deposition process. Izvestija in-ta obstsza i neorg. chim 4, 53-67 (1966)
PESHEV, P., NIEMYSKI, T.: Preparation de diborure de titane crystallin au moyen d'une reaction en phase gazeuse. J. Less-Common Metals 10, 133-145 (1966)
SAMSONOV, G.V.: Refractory Compounds. Moskow: Metallurgizdat 1963
VAHLDIEK, F.W.: Electrical resistivity, elastic modulus, and DEBYE temperature of titanium diboride. J. Less-Common Metals 10, 202-209 (1967)
WATSON, C.R., MATCHEN, B.: Method for producing crystalline carbide, boride or silicide. USA Pat N 3348915 (1967)

XI. Properties and Uses of Diborides

J. CASTAING and P. COSTA

Boron gives diborides with a large number of transition metals. Most of them show the hexagonal C32-type structure (ScB_2, TiB_2, VB_2, CrB_2, MnB_2, YB_2, ZrB_2, NbB_2, MoB_2, HfB_2, TaB_2), which appears also for aluminum, magnesium and several rare-earth diborides. Some other diborides (WB_2, ReB_2, RuB_2, IrB_{2-x}) show different structures, based, like the C32 structure, on alternate hexagonal layers of metal and boron (LUNDSTRÖM, 1970). RuB_2 is orthorhombic.

Only the C32-type transition metal diborides have been extensively studied. We shall therefore limit this review to these compounds, with particular emphasis on the physical properties. But the chemical properties and the applications of diborides will also be briefly summarized.

1. Thermodynamic Properties

a) Stability; Homogeneity Range

α) Stability

Transition metal diborides are stable down to room temperature, with the exception of manganese and molybdenum diborides. Manganese diboride decomposes below a temperature of about 1300 °C (it is unstable at 1200 °C and can be readily prepared at 1400 °C) into Mn_3B_4 and MnB_4 (BINDER and POST, 1960; FRUCHART and MICHEL, 1960). Molybdenum diboride is stable between 1600 and 2100 °C only (STEINITZ et al., 1952).

Diborides exist for the stoichiometric composition. The only exception is MoB_2 for which the homogeneity range is shifted towards larger boron contents (Table 1). The width of the homogeneity range of diboride is quite narrow in the case of group IV diborides. It might be wider in the case of group V and VI, mainly at high temperature (RUDY et al., 1963). It should be noticed that the c/a ratio increases with larger boron content except possibly in the case of VB_2; CrB_2 : CASTAING, (1971, p. 48); ZrB_2 : comparison between the values of BERNSTEIN (1963) and GEBHARDT and CREE (1965); NbB_2 : BREWER et al. (1951); MoB_2 and TaB_2 : RUDY et al. (1963); HfB_2 : KRUPKA (1962); UB_2 : HOWLETT (1959) (Table 1).

(Footnotes to Table 1, continued)

[j] MEERSON and SAMSONOV (1954); [k] RUDY et al. (1963); [l] KIESSLING (1949a);
[m] CASTAING (1971); [n] SHAMRAI and FEDOROV (1964); [o] BINDER and POST (1960);
[p] ARONSSON (1960); [q] FRUCHART and MICHEL (1960); [r] NOWOTNY et al. (1960a);
[s] NOWOTNY et al. (1960b); [t] KRUPKA (1962); [u] RUDY et al. (1961);
[v] CLOUGHERTY and BLACKBURN (1963b); [w] KIESSLING (1949b); [x] LUNDING (1961);
[y] MANIELIS et al. (1970); [z] PESHEV et al. (1970); [α] HOWLETT (1959);
[β] SPEAR et al. (1968); [γ] JONES (1954); [δ] HOFMANN and JÄNICHE (1936);
[ε] FELTEN (1956)

Table 1. Crystallographic properties of diborides
a and c : lattice parameters (Å)
Conc. : corresponding value of the concentration (% Boron)
HR and T : homogeneity range (%B) and corresponding temperature (°C)
Lattice parameters for rare-earth diborides are given in BAUER and DEBUIGNE (1973)

	a	c	Conc.	HR	T
ScB_2	3.146[a] 3.148[z]	3.517[a] 3.516[z]	- -		
TiB_2	3.026[b] 3.024[c] 3.028[d,e] - 3.030[f] 3.038[g]	3.213[b] 3.221[c] 3.228[d,e] - 3.229[f] 3.233[g]	- - - - - -	1[h]	-
VB_2	3.006[j] 2.998[d] 2.991[k] 2.996[k] 2.994[β] 2.998[β]	3.056[j] 3.057[d] 3.050[k] 3.054[k] 3.048[β] 3.056[β]	- - B - rich[k] V - rich[k] B - rich[β] V - rich[β]	small[i]	-
CrB_2	2.969[l] 2.972[m]	3.066[l] 3.070[m]	- -	66-70[n]	-
MnB_2	3.007[o] 3.009[p] 3.0081[q]	3.037[o] 3.039[p] 3.0346[q]	- - -		
YB_2	3.298[x] 3.290[y]	3.842[x] 3.825[y]	- -		
ZrB_2	3.167[r] 3.171[g] 3.171[f]	3.529[r] 3.535[g] 3.527[f]	- B - rich[g] 65,3	<1[f]	-
NbB_2	3.100[e] 3.085[e] 3.107[f]	3.264[e] 3.311[e] 3.282[f]	Nb - rich[e] B - rich[e] -	65-70[s]	1.600[s]
MoB_2	3.040[k] 3.026[k]	3.060[k] 3.120[k]	68[k] 75[k]	68-78[k] ≃ 3[k]	T_F[k] 1500[k]
HfB_2	3.142[u] 3.1417[t] 3.1416[t] 3.1460[f]	3.477[u] 3.4742[t] 3.4747[t] 3.4761[f]	 Hf - rich[t] B - rich[t] -	very small[h]	- -
TaB_2	3.078[w] 3.085[f] 3.097[k] 3.065[k]	3.265[w] 3.240[f] 3.225[k] 3.283[k]	- - 66,3[k] 73 quench[k]	64-72[w] 58.9-67.4 64.5-66.9	- 1727[h] 927[v]
UB_2	3.136[e] 3.134[α] 3.129[α]	3.988[e] 3.986[α] 3.989[α]	 U - rich[α] B - rich[α]		
MgB_2	3.0835[γ]	3.521[γ]			
AlB_2	3.01[δ] 3.009[ε]	3.25[δ] 2.262[ε]			

[a] ZHURAVLEV and STEPANOVA (1958); [b] EHRLICH (1949); [c] ZACHARIASEN (1949);
[d] NORTON et al. (1949); [e] BREWER et al. (1951); [f] BERNSTEIN (1963);
[g] GEBHARDT and CREE (1965); [h] KAUFMAN (1963); [i] NOWOTNY et al. (1959a);

Group IV diborides evaporate congruently at high temperature with possibly a slight tendency towards larger boron pressure. This tendency is enhanced for group V diborides. These conclusions are established on the basis of thermodynamical calculations given by KAUFMAN (1963) on TiB_2, ZrB_2, HfB_2, NbB_2 and TaB_2. They are consistent with the results of heat treatments under vacuum on TiB_2, CrB_2 and VB_2 which, above 1500°C, show important boron losses, and the fact that arc-melted chromium diboride ingots include precipitates of lower borides (CASTAING, 1971)

Table 2. Values in kcal/mole of the standard formation enthalpy (ΔH^o_{f298}), melting enthalpy ($\Delta H_{melting}$) and cohesive energy ($E_{coh.}$) computed from the values of ΔH^o_{f298}.

	$-\Delta H^o_{f298}$	$\Delta H_{melting}$ [a]	$E_{coh.}$
MgB_2	22,0[b]	8,97	323,3
AlB_2	-	8,62	-
TiB_2	66,8[c]	24,135	445
VB_2	48,7[d]	20,85	438,2
CrB_2	(33)	16,90	(393)
YB_2	-	16,35	-
ZrB_2	73,8[c]	24,29	484,15
NbB_2	41,9[c]	22,84	480,3
MoB_2	23[e]	18,29	446,1
HfB_2	80,0[c]	25,21	494
TaB_2	50,0[c]	22,84	502,9
UB_2	39,3[f]	18,35	430,3

[a] KRIKORIAN (1971); [b] JANAF, Thermodynamical Tables (1970); [c] SCHICK (1966); [d] SPEAR et al. (1968); [e] TOULOUKIAN (1967); [f] FLOTOW et al. (1969)

β) Mutual Solubility of Diborides

The mutual solubility of transition-metal diborides of the IVth, Vth and VIth column have been extensively studied. Most of them are fully soluble at high temperature (2000-3000°C) and probably also at room temperature. The only exceptions at high temperature are the systems between transition metals with radii which differ by more than approximately 15%, i.e. V and Cr on the one side, Hf and Zr on the other (POST et al., 1954). At low temperature there is also a miscibility gap for $(Cr, Nb)B_2$ and $(Cr, Ta)B_2$ (KUZMA et al., 1969, 1972) and of course a limited solubility of MoB_2 in the other diborides, since it is not a stable compound. Substitution of other transition metals which do not give a C32 type diboride (W, Re) have also been studied (POST et al., 1954; LAKH et al., 1970; ARIEL and BARTA, 1970).

b) Elastic and Thermal Properties

Borides are refractory compounds, with high melting temperatures (T_M, Table 4). T_M is maximum in the first series for TiB_2. Generally speaking the picture of a maximum value of the strength of the oscillators involved in atomic vibrations, as shown by all the elastic and thermal properties summarized below, fits into the frame of a maximum of the cohesive energy (see Table 2) for the IVth column, at least in the case of the first series. For the two others, it might be slightly shifted towards larger electron/atom ratios, with comparable cohesive energies for the IVth and the Vth column.

α) Elastic Properties

There is a fairly large scatter in the published values for the elastic properties (Table 3). For TiB_2 for instance there is a factor 3 between the extreme values of the shear modulus G. In the case of this compound the most reliable values are probably the most recent (WILEY et al., 1969; FRANTSEVICH et al., 1967) and possibly an older determination on a single crystal by GILMAN and ROBERTS (1961). In the latter case, the E_a, E_c, G_a, G_c values given in Table 3 have been calculated from the authors' data on the C_{ij}.

Table 3. Values of the Young modulus (E) and the shear modulus (G) in kbar ($1\ kbar = 10^8\ N/m^2$) for several diborides and values of the Debye temperature (Θ_D, K) deduced from the elastic constant data. In the case of single crystals, values have been given for the a and c directions.

	E		G			
	E_a	E_c	G_a	G_c	Θ_D	
TiB_2	3700[a]	2150[a]	1100[a]	1250[a]		x
	3420[b] to 4118[b]					xx
	3740[c]					xx
	1850[d]	2070[d]	692[d]	785[d]	(568)[d]	x, xxx
	2530[e]		845[e]		(772)[e]	xx, xxx
	4975[f]		2250[f]		790[f]	xx
	5500[f]		2150[g]			
VB_2	3500[g]					xx
CrB_2	(4600)[g]					xx
	2150[c]					xx
ZrB_2	3500[c]					xx
	4700[h]		2050[h]			
	4910[f]		2200[f]		630[f]	
	4300[g]					
HfB_2	5030[f]		2250[f]		477[f]	

Note: Literature values of the Debye temperature have been divided by $\sqrt[3]{3}$ in order to obtain values which correspond to the *acoustic modes only*.
(x) Single crystal. (xx) Porous sample. (xxx) Average value.
[a] (x) GILMAN and ROBERTS (1961); [b] (xx) MANDORF et al. (1961); [c] (xx) NASHPO and SAMSONOV (1957); [d] (x) (xxx) VAHLDICK (1967); [e] (xx) (xxx) IBID VAHLDICK (1967); [f] (xx) WILEY et al. (1969); [g] (xx) FRAUTSEVICH et al. (1967); [h] LANG (1960).

Values of the Debye temperature Θ_D have been given by several authors but using different physical definitions of Θ_D, based on two different velocities v_o and v_m:

$$v_o = v_s\,(2 + 2\,\sigma)^{1/2} \quad \text{(VAHLDIEK, 1967) (rod velocity)}$$
$$v_m = \left[\frac{1}{3}\left(\frac{2}{v_s{}^3} + \frac{1}{v_l{}^3}\right)\right]^{-1/3} \quad \text{(WILEY et al., 1969) (mean sound velocity)} \qquad (1)$$

where v_s and v_l are the shear and the longitudinal velocities, and σ Poisson's ratio.

These values should not then be compared. In our opinion, the second definition should be preferred (ANDERSON, 1963). It leads for TiB_2, ZrB_2 and HfB_2, on the basis of Wiley's results, to values of Θ_D which are close to those which are given below on the basis of low-temperature specific-heat data (1-20 K): 790, 628 and 477 K, against 807, 585 and 499 K respectively; Tables 3 and 4.

β) Thermal Properties at Low Temperature (T<20 K)

The only vibrational excitations which are allowed at low temperature in a solid are *acoustic* modes with characteristic frequencies next to zero. The Debye temperature values are then expected to be very close to those which are derived from the elastic properties. Table 4 summarizes the values of the Debye temperature given by CASTAING et al. (1969), TYAN, TOTH and CHANG (1969), HANSON et al. (1971), CASTAING (1971). The only differences worth noticing between these authors are for VB_2 and CrB_2. In the first case, Hanson's result is the correct one. In the second case, the difference is due to the fact that the αT^3 term of the specific heat includes both phonon and antiferromagnetic-magnon excitations. This effect has been taken into account by Castaing and estimated on the basis of the Néel temperature. Best values (underlined in Table 4) show variations which are consistent with the melting temperatures, through LINDEMANN's formula (1910), with a maximum value of Θ_D for column IV, at least in the first series.

γ) Thermal Properties at Medium and High Temperatures

When temperature is increased the situation is more complex (1) because acoustic modes of higher energy, with a wave number close to the limit of the Brillouin zone are not properly described by Debye's model; (2) because 6N optical modes should be included.

When the crystal - as is the case - includes two types of atoms with $M_1 << M_O$, where M_1 and M_O are the masses of the atoms, the situation is however simplified. The optical branches are very flat and degenerate into an Einstein model with a characteristic frequency $\omega_{Eopt} \simeq \sqrt{2G/M_1}$; besides, the top of the acoustic branch corresponds to a frequency $\omega_{max} \simeq \sqrt{2G/Mo}$ which is much smaller than $\omega_{E,opt}$; for temperatures $T \simeq \tau\omega_{max}/k$ the description of the system may then be reduced to the superposition of an Einstein (or Debye) model representing the acoustic mode, and an Einstein (or Debye) model representing the optical modes with characteristic temperatures Θ_o and Θ_1 verifying $\Theta_1/\Theta_o = (Mo/M_1)^{1/2}$.

This type of analysis has been used by KAUFMAN (1963) to interpret WESTRUM's calorimetry experiments (1963) for TiB_2, ZrB_2, HfB_2, NbB_2 and TaB_2 (Fig. 1). The results are summarized in Table 4. The characteristic temperatures Θ_o and Θ_1, determined for experiments around room temperature, are DEBYE temperatures. It would probably have been more sound, mainly for the heavier diborides (second and third series) to use an Einstein type of analysis. The acoustic temperatures are, as expected, lower than those which were determined at low temperature,

Table 4. Thermal properties of diborides
T_M: Melting temperature (^oC); Θ_D: Debye temperature (K) from low-temperature heat capacity measurements; Θ_{Me} and Θ_B: characteristic temperatures (K) deduced from high-temperature heat capacity measurements; α_A, α_C and Temp. range: thermal expansion along the a and c axis and corresponding temperature range (K); α_M: mean thermal expansion coefficient

	T_M	Θ_D	Θ_{Me}	Θ_B	$\alpha_A x10^6$	$\alpha_C x10^6$	Temp.Rang.	$\alpha_M x10^6$
ScB_2	2250[t]	550[k]			4.4[γ]	4.9[γ]	300-1100[γ]	
TiB_2	2920[a] 2980[b]	820[k] 807[l]	620[z]	1302[z]	7.19[o] 8.00[q] 7.30[p] 6.63[q]	9.77[o] 11.20[q] 10.27[p] 8.65[q]	1100-1800[o] 1000-2300[q] 300-1300[p] 300-1000[q]	8-9.5[r] 4.8[s]
VB_2	2400[a] 2450[c]	810[k] 693[m]						8[s]
CrB_2	1900[a] 2280±20[d] 2200±50[e]	690[k] 570[m]			5[δ]	5[δ]	300-600[δ]	
MnB_2	1968[f]	630[n]						
YB_2	2100[u]				9.4[y]	8.5[y]	300-900[y]	
ZrB_2	3040±50[g] 3000±80[h]	585[l]	440[α]	1277[α]	5.92[o] 8.62[q] 6.60[q]	7.04[o] 7.65[q] 6.78[q]	1100-1800[o] 1000-2300[q] 300-1000[q]	8[r] 6.2[s]
NbB_2	3050[a] 3000[i]	606[k] 617[l]	425[β]	1245[β]	8.69[o] 8.30[q] 5.51[q]	9.42[o] 10.70[q] 9.21[q]	1100-1800[o] 1000-2300[q] 300-1000[q]	7.3[r] 8.2[s]
MoB_2	2100[j]	534[l]			7[δ]	10[δ]	300-900[δ]	
HfB_2	3250±100[a]	499[l]	320[z]	1299[z]	6.64[o] 8.40[q] 6.39[q]	6.89[o] 7.64[q] 6.81[q]	1100-1800[o] 1000-2300[q] 300-1000[q]	7.5[r] 6.6[s]
TaB_2	3200[a] 3100[i]	452[m]	320[z]	1309[z]	8.46[o] 7.82[q] 5.85[q]	10.90[o] 8.40[q] 7.14[q]	1100-1800[o] 1000-2300[q] 300-1000[q]	7.9[r] 8.5[s]
UB_2	2385[v] 2160[w]				9[x]	8[x]	300[x]	

[a] POST et al. (1954); [b] LATVA (1962); [c] NOWOTNY et al. (1959a); [d] MARKOVSKII et al. (1967); [e] KISLYI et al. (1962); [f] MARKOVSKII et al. (1967); [g] GLASER and POST (1953); [h] ISKOLDSKII and BOGORODSKAYA (1957); [i] NOWOTNY et al. (1959b); [j] STEINITZ et al. (1952); [k] CASTAING et al. (1969); CASTAING (1971); [l] Tyan et al. (1969); [m] HANSON et al. (1971); [n] KUENTZLER (1970); [o] BERNSTEIN (1963); [p] SKAAR and CROFT (1973); [q] KEIHN and KEPLIN (1967); [r] POBER and CLOUGHERTY (1963); [s] SAMSONOV et al. (1971c); [t] SAMSONOV (1960); [u] LUNDING (1961); [v] HOWLETT (1959); [w] MATTERSON et al. (1961); [x] BECKMANN (1956); [y] ZHURAVLEV et al (1970); [z] KAUFMAN (1963); [α] WESTRUM and FEICK (1963); [β] WESTRUM and CLAY (1963); [γ] ZHURAVLEV and STEPANOVA (1964); [δ] American Electro. Corp. (1955)

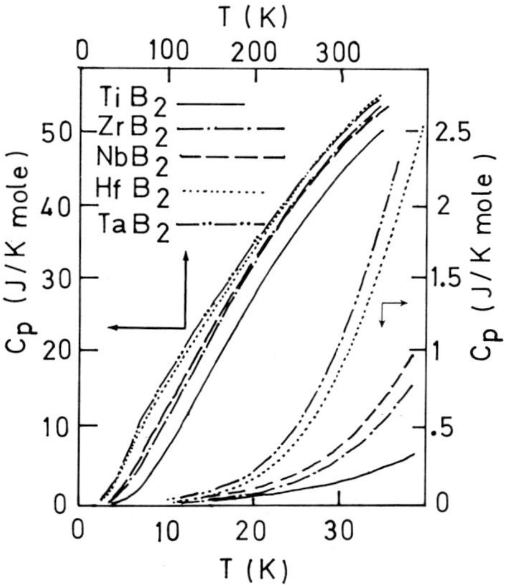

Fig. 1. Heat capacity of refractory metal diborides. (After KAUFMAN, 1963)

or on the basis of the elastic properties: this is due to the fact that at the top of the acoustic branch the density of modes is larger than predicted by Debye's model. We should also notice that Θ_1 is almost a constant, which indicates that for these five compounds the strength of the oscillators is almost the same. This result is consistent (1) with the fact that similar values have been found for the elastic constants of TiB_2, ZrB_2 and HfB_2; (2) with the fact that ZrB_2 and NbB_2, HfB_2 and TaB_2 have the same melting temperatures.

At high temperature, the asymptotic value 9R of the molecular specific heat is reached at approximately 800-1000K (MEZAKI et al., 1962; WALKER et al., 1957; SOUTHERN RESEARCH INSTITUTE, 1962).

δ) Debye Temperature as Deduced from Mean-Square Atomic Displacements

It is possible to reach the value of the mean-square atomic displacements from the attenuation of the X-ray line intensity with increasing temperature, through Debye-Waller's law.

In the case of a monoatomic hexagonal lattice, from a simple evaluation, one can draw the values of μ_a^2 (T) et μ_c^2 (T), mean square displacements along the a and c axis. For a polyatomic lattice, the values of μ_a^2 and μ_c^2 for any atom of the unit cell could in principle be derived from the observation of the attenuation of *all* the lines. In fact the precision is not in general good enough. The work could however be done by MIKSIC (1963) on TiB_2 between 130 and 300°C, which led, as expected, to similar values of the mean square displacements of titanium and boron. On the basis of these results characteristic temperatures could be derived, which were consistent with those determined by KAUFMAN (1963) on the basis of specific heat data at similar temperatures. Besides a strong decrease of the DEBYE temperature when temperature is increased from 600 K to 1600 K has been found by these authors in agreement with the remark made in Section γ.

ε) Thermal Expansion

The thermal expansion of diborides has been determined both by X-ray measurements (BERNSTEIN, 1963; KEIHN and KEPLIN, 1967; SKAAR and CROFT, 1973) or by dilatometry (SAMSONOV et al., 1971c; POBER and CLOUGHERTY, 1963).

The thermal expansion is slightly anisotropic, with a larger dilatation along the c axis. On the opposite, Bernstein has found from the Debye-Waller's factors, larger mean square displacements along a for group IV diborides. Elastic-constant determinations on single crystals are scarce, and lead to divergent results (Table 3). It is then hardly possible to draw any conclusions on the compared strength of the bonding within and between hexagonal planes.

ζ) Thermal Conductivity

All the determinations of the thermal conductivity of diborides have been made on sintered samples and are highly dependent on the state of porosity of the samples. We shall not discuss them here, but limit ourselves to giving two recent references (BRANSCOMB and HUNTER, 1971; SAMSONOV et al., 1971d).

2. Electronic Structure and Related Properties

The physical properties of transition metal diborides which are summarized below, clearly indicate that these compounds are metallic. The crystal structure reveals alternate layers of boron and metal atoms, with lattice parameters which fit with the radii of the metals and the covalent radius of boron. Hardly any anisotropic character occurs however in the physical properties, which is consistent with a value of the c/a ratio close to 1. An electronic structure similar to that of the metals can then be predicted although the electrons of boron should be somehow included in a covalent scheme.

CASTAING et al. (1969) have shown that the d electrons play a dominant part in the bonding properties. On the basis of experimental results and hopping-integral calculations, these authors established a qualitative density-of-states curve (CASTAING, 1971), accounting for 2p and 2s boron states (2x8) and 3d metal states (10 states), with a rather large band (larger than the d band of transition metals) of hybridized (sp) and d states, corresponding to s,p_x and p_y orbitals for boron and d($|m|$ = 1) orbitals for the metal, and another dominantly d ($|m|$ = 2 and $|m|$ = 0) band of smaller width.

The Fermi level for TiB_2 falls at a minimum of the density of states before the incipient part of this second band.

This qualitative analysis has been corroborated by two recent energy band calculations by McALISTER et al. (1972) and by LIU et al. (1975). They both carried out a calculation for CrB_2 using respectively the augmented plane wave (APW) and the KORRINGA-KOHN-ROSTOKER method and applied the rigid band concept to give results for other borides. The first group's results (McALISTER et al., 1972) are in agreement with the fact that the states which are progressively filled in a rigid band scheme when the number of electrons is increased from TiB_2 to CrB_2 are dominantly d(m = 2) states. At lower energy lies a broad band of B(2s,2p) and Ti(3d) character. This description is in agreement with Castaing's qualitative model, predicting for TiB_2 an occupied band of hybrid 2s,2p and 3d states (Fig. 2b, c and d).

The second group (LIU et al., 1975) focused their efforts on energy bands and computed details of the Fermi surface of CrB_2 in order to explain its magnetic properties. Their results (Fig. 2a) seem to be consistent with the former description.

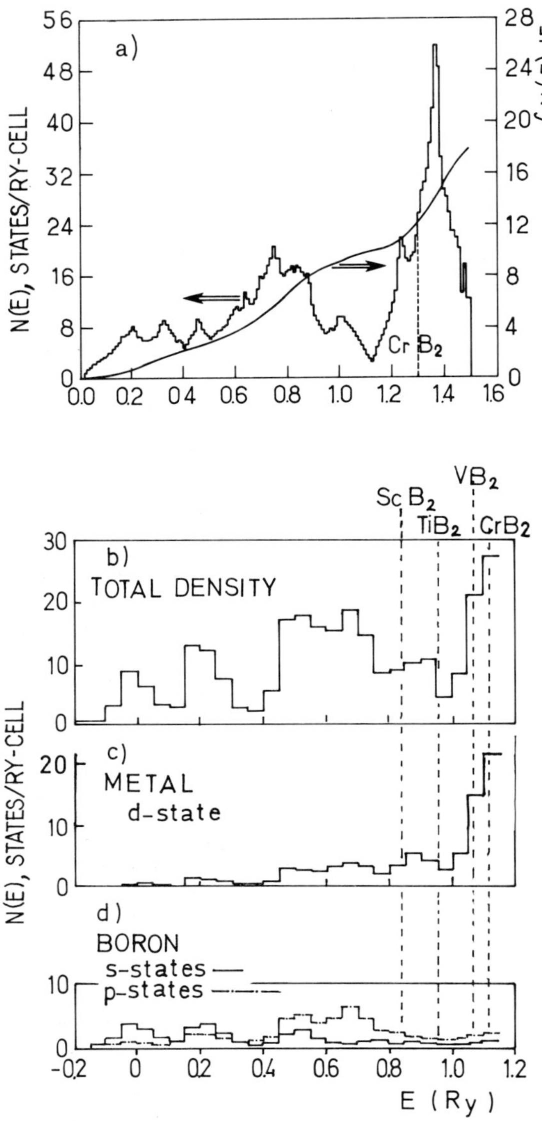

Fig. 2a-d. Computed density of state of diborides; the dashed vertical lines show the position of the Fermi level: (a) density of states and integrated density of states of CrB_2 (after LIU et al., 1975)

(b-d) total, metal and boron density of states (after McALISTER et al., 1972)

Numerous results on properties related to band structure are now available on transition metal diborides. We shall review X-ray emission spectra, low-temperature specific heat, transport properties, magnetic susceptibility and nuclear magnetic resonance, and show that they fit inside the frame of the band model given at the beginning of Section 2.

a) X-ray Emission Spectra

Both metal and boron K emission spectra have been determined consistently by several authors.

Boron K_α spectra (McALISTER et al., 1972; LYAKHOVSKAYA et al., 1970; FRANTSEVICH et al., 1971; HOLLIDAY, 1967; SHASHKINA, 1971) for ScB_2, TiB_2, VB_2, CrB_2, and MnB_2 show a progressive modification corresponding to the emergence of a peak at high energy. This effect is in agreement with the d-character of the density-of-states peak which occurs in the above models for the last borides of the series (VB_2 and onwards). The spectra are spread out towards lower energies. This effect is also in good agreement with band calculations, in particular with McALISTER et al. and should correspond to states with a rather strong (2s) character.

Metal $K_{\beta 5}$ spectra (NEMNONOV and KOLOBOVA, 1966; MENSHIKOV et al., 1963; SHASHKINA, 1971), which yield information on occupied states with (2s-2p) character, show a similar behavior. All these results are consistent with the band models.

A comparison of metal L_α and L_β spectra would have been of particular interest in the case of diborides, since the $|m| = 2$ states have a relatively larger contribution to L_β. If the narrow band which is progressively filled for VB_2 and CrB_2 includes states of this symmetry, an increase of the $\frac{L_\beta}{L_\alpha}$ ratio along the series should be observed. Unfortunately the resolution between the two lines, mainly for TiB_2, does not seem to be reliable; moreover, the inverse trend is observed on pure metals, so that no firm conclusion can be drawn from the experiments of FISHER and BRAUN (1968); FISCHER (1969, 1970); HAGUE (1972).

b) Low-temperature Specific Heat

The electronic specific heat coefficient γ is proportional to the density of states at the Fermi level, if electron-phonon and electron-electron interactions are neglected. A plot of γ against the atomic number maps out the density of states curve if a rigid band model is assumed to be approximately valid. The measured values (Table 5) are slightly smaller than for transition metals, which is consistent with a broader d band. They fit fairly well with the computed values (LIU et al., 1975; McALISTER et al., 1972) and show in particular a minimum for TiB_2. There is however a strong discrepancy in the case of CrB_2, for which magnetic enhancement effects occur, as can be seen by comparison to 4d diborides. These effects will be discussed more extensively in Section 3. Comparison between γ and the theoretical density of states shows besides that electron-phonon enhancement of the specific heat must be fairly small, in agreement with the fact that no diboride is known to be a superconductor.

c) Magnetic Susceptibility and NMR

All diborides except CrB_2 and MnB_2 are non-magnetic, with a temperature-independent susceptibility, after subtraction of a Curie-type contribution of magnetic impurities which are often present in the matrix. Most of the compounds are diamagnetic with the exception of VB_2 (CASTAING et al. 1969; CASTAING, 1971) and ScB_2 (PESHEV et al., 1970; Table 5). This result implies a large unexplained diamagnetic contribution (CASTAING 1971), a negligible STONER enhancement (1938) of the (d) Pauli susceptibility, in contrast to transition metals, and even more surprisingly a very small Van Vleck (d) term. This last result might be the consequence of the Fermi level being situated at the border of two bands of different symmetry.

The analysis of susceptibility may be enlightened by the nuclear magnetic resonance experiments. There is a large amount of work in this field from the earlier experiments on ^{11}B (MALYUCHKOV and POVITSKY, 1962; SILVER and KUSHIDA, 1960, 1963) to more extensive work on the metal nuclei: ScB_2 and YB_2 (CARTER and SWARTZ, 1971; BARNES et al., 1970a), VB_2 (SILVER and KUSHIDA, 1963; JONES et al., 1963) or the boron nucleus (CREEL and BARNES, 1970; BARNES and CREEL, 1969; Table 5).

Table 5. Electronic properties of diborides
γ: electronic specific heat coefficient (mJ/K^2mole)
χ: room temperature molar magnetic susceptibility (10^{-6}emu/mole)
^{11}B K.S. and metal K.S.: boron and metal KNIGHT shift (%)

	ScB_2	TiB_2	$(V\ Ti)B_2$	VB_2	CrB_2	MnB_2	AlB_2
γ	2.2[m]	1.08[m] 1.4[b]	2.97[m]	4.84[m] 4.74[c]	13.6[m]	2.8[a]	
χ	~80[d]	-40[m]	-28[m]	+21[m]	~400[m] 455[e]	4250[f]	
^{11}B K.S.	-0.005 g 0.00(1)[i]	-0.016[h]	-0.01[n]	-0.024[h]	-0.047[h]	-0.05[i]	-0.02[j]
Metal K.S.	+0.07[i] +0.06[g]		-0.38[m]	-0.34[h] -0.31[m]			+0.088[j]
	YB_2	ZrB_2	$Zr_{0.75}Nb_{0.25}B_2$	NbB_2	MoB_2		UB_2
γ		0.93[b]	1.32[c]	2.36[n] 2.33	3.38[b]		9.4[l]
χ		-63[n]		-32[n]	-42[n]		505[l]
^{11}B K.S.	-0.004[i] -0.005[g]	-0.013[h]		-0.008[h]	-0.007[h]		+0.03[k]
Metal K.S.	+0.16[g]						

[a] KUENTZLER (1970); [b] TYAN et al. (1969); [c] HANSON et al. (1971); [d] PESHEV et al. (1970); [e] GUY (1974); [f] CADEVILLE (1966); [g] BARNES et al. (1970a); [h] SILVER and KUSHIDA (1963); [i] CARTER and SWARTZ (1971); [j] KOPP and BARNES (1971); [k] KUZNIETZ and SPOKAS (1970); [l] FLOTOW et al. (1969); [m] CASTAING et al. (1969); [n] CASTAING (1971)

The analysis of the large negative Knight shift of ^{51}V in VB_2 (-0.3%) shows that, in addition to the usual negative contribution of the inner-core d polarization, a large diamagnetic term must be introduced in the partitioning of the local susceptibility (CASTAING, 1971). Boron Knight shifts show small negative values which progressively decrease with increasing atomic number from ScB_2 to CrB_2 (CARTER and SWARTZ, 1971; BARNES et al., 1970a; SILVER and KUSHIDA, 1963). This behavior suggests that inner core polarization effects might also occur on boron (McALISTER et al., 1972). This assumption however is not consistent with the general trend of the density of states in the second series (CASTAING, 1971; SILVER et al., 1963; BARNES et al., 1970a; CARTER and SWARTZ, 1971), and cannot explain the rather strong negative value for AlB_2 (KOPP and BARNES, 1971). All nuclear magnetic resonance spectra are split by quadrupolar effects, due to the non-cubic symmetry of the crystal, the nuclear spin values being greater than 1/2. Many authors have tried to explain the quadrupole effect constants in relation with chemical bonding (SILVER and BRAUN, 1960, 1963). No simple correlation seems to appear, an unknown parameter being the possible occurrence of a non-planar boron network (CARTER and SWARTZ, 1971).

d) Transport Phenomena

The transport properties of diborides are fairly well known. A wide range of measurements on resistivity, thermo-electric power and Hall effect have been performed on these materials. The most valuable have been gathered in Table 6.

Table 6. Transport properties of diborides at room temperature:
ρ: room temperature resistivity (μΩ-cm); $d\rho/dT$: resistivity temperature coefficient (μΩ-cm/ K); S: thermoelectric power (μV/K); R: Hall constant ($10^{-4}cm^3$/Coulomb)

	ρ	$d\rho/dT$	S	R
TiB_2	9[a]; 14.4[c]; 6.6[k];	0.018[f]	-2.9[f]	-24[a]; -19.6[f]
VB_2	38[a]; 19[c]; 22.7[f]	0.048[f]	-5.8[f]	-1.1[a]; -0.82[f]
CrB_2	56[a]; 57[c]; 30[g]		-4.5[g]	-1.1[a]; -1.2[g]
MnB_2	400[b]; 71[e]	0.17[e]	+3.9[e]	
YB_2	39[d]			
ZrB_2	7[a]; 16.6[c]; 9.7[f] 2.3 to 3.4[j]	0.022[f]	-1.2[f]	-20[a]; -19[f]; -12 to -23[j]
NbB_2	12[a]; 34[c]; 25.7[f]	0.026[f]	-3.4[i]	-1[a]; -1.5[f]
MoB_2	30[a]			-0.5[a]
HfB_2	15.8[a]; 12[c]; 10.6[f]	0.034[f]	-1.2[c]; -1[f]	-17[a]; -18[f]
TaB_2	14[a]; 37.4[c]; 32.5[f]	0.039[f]	-2.8[f]	-2.7[a]; 2.1[f]
AlB_2	31 to 77[h]			1.5 to 10[h]

[a] JURETSCHKE and STEINITZ (1958); [b] KASAYA and HIHARA (1970); [c] L'VOV et al. (1962); [d] JOHNSON and DAANE (1963); [e] BEZRUK and MARKOVSKII (1968) (porous sample); [f] SAMSONOV et al. (1971a); [g] SAMSONOV et al. (1971b); [h] SIRTL and WOERNER (1972); [i] KOVENSKAYA (1970); [j] PIPER (1966) (Single crystal); [k] VAHLDIEK (1967) (Single crystal)

The electrical resistivity has been in most cases measured on sintered samples, and it is of the same order as for transition metals. In each transition series the smallest values are obtained for group IV borides: TiB_2 ZrB_2 and HfB_2 (room-temperature resistivity of about 10μΩcm). The resistivity is larger for group V borides, and even more so for group III (39μΩcm for YB_2).

The interpretation of transport properties of systems with complex band structure is always difficult and very seldom leads to satisfactory explanations. In the present case, two types of models could be set forth with little success:
The usual MOTT transition metal model (1964), with a high-mobility, low-density, s-type band, and a low-mobility, high density, d-type band, which leads to a resistivity approximately proportional to the density of states at the Fermi level $n(E_F)$ and a thermoelectric power which has the sign of $\left|\frac{dn(E)}{dE}\right|_{E_F}$, and is proportional to this quantity.
This model is not consistent with some experimental results, namely: (1) The thermoelectric power is negative and large, instead of positive and large as expected, for group V borides; (2) In order to compare resistivity with Mott's theory, one would rather take the values of $\frac{d\rho}{dT}$ than the room-temperature resistivity which includes a

large residual term, which might primarily reflect the concentration of defects of the material. Except for the first series, Table 6 values show small variations within a series and $\frac{d\rho}{dT}$ could then hardly be said to be proportional to the density of states values as given by the electronic specific heat or the Pauli susceptibility.

A model based on the overlap at the Fermi level of two bands with effective masses of opposite signs. This semi-metal type model could be reasonable near the density-of-states minimum. The analysis made in this line by PIPER (1966) for the transport properties of a ZrB_2 single crystal (including magnetoresistance effects) seem satisfactory and consistent, with maybe some restriction on the fact that the number of electrons and holes are supposed to be equal, which, in the rather complex band structure we are dealing with, can hardly be justified. For group III and V borides on the other hand, this model has no validity. It fails to give any reasonable account of the experimental situation: neither the amplitude of the resistivity within a series nor the values and signs of the Hall effect coefficients fit with the predictions of the model.

e) Relation with Lower Boride Properties - Conclusions

All these results give a fairly good scheme of the electronic structure of the transition metal diborides, which is rather similar to that of the transition elements: it is dominated by a rather high density-of-states band deriving mainly from the d states of the metal, with possibly some restrictions in the case of the diborides of group IV, for which the semi-metal description might be partly relevant, in the sense that the Fermi level intersects two bands of different symmetry the first one being almost full and the second almost empty.

The same influence of the d states should be found for lower borides, as shown by the low-temperature heat-capacity experiments by HANSON et al. (1971), TYAN et al. (1969) and KUENTZLER (1970). The main difference between diborides and lower borides is to be found in the superconducting properties: several semiborides are superconductors with rather high transition temperatures: Mo_2B : 5.3 to 5.86 K, W_2B : 3.2 K; Ta_2B : 5.4 K (ENGELHARDT, 1969). The difference might be partly due to the fact that the density of states seems to be higher for lower borides (HANSON et al., 1971, TYAN et al., 1969; KUENTZLER, 1970); but attention should also be paid to phonon spectra anomalies, which might express strong electron-phonon coupling effects similar to those which have been observed in TaC for instance (SMITH and GLÄSER, 1970).

Another interesting feature which can be drawn from the comparison between the properties of various borides of the same metal is the fact that the electronic structure could at least partly be described in terms of local-environment effects: for vanadium borides for instance (BARNES et al., 1970b), the ^{51}V Knight shifts occur with four distinct values correlated with the four types of boron environments of vanadium in these compounds (Fig. 3).

3. Magnetic Diborides

A good proof for the occurrence of magnetic order is the extinction of a nuclear magnetic resonance spectrum when decreasing the temperature. Down to 2 K, the only diborides showing a magnetic transition on the basis of this criterion are chromium diboride and manganese diboride (BARNES and CREEL, 1969). Magnetization measurements allow

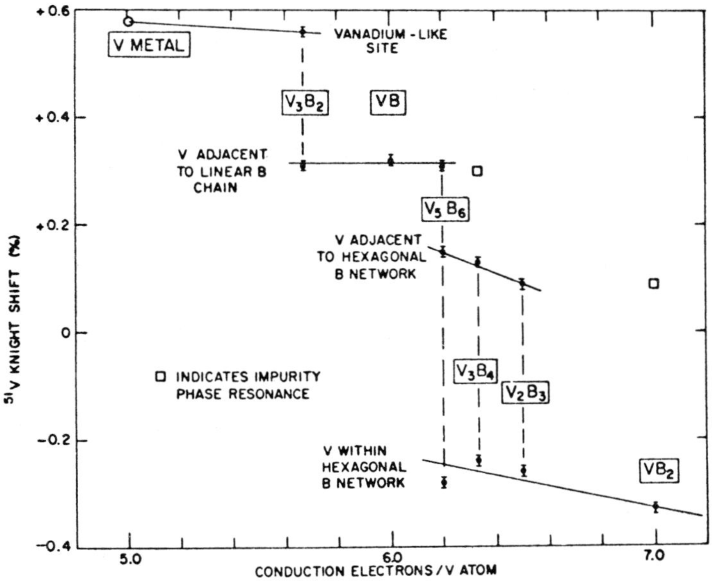

Fig. 3. Knight shifts of the ^{51}V resonances in vanadium borides plotted against the number of conduction electrons per vanadium atom calculated on the basis of the simplified electron transfer theory. (After BARNES et al., 1970b)

the conclusion that CrB_2 is antiferromagnetic with a Néel temperature of 85 K (CASTAING et al., 1969) and MnB_2 is ferromagnetic with a Curie temperature of 140-150 K (BARNES and CREEL, 1969; CADEVILLE, 1966; ANDERSSON et al., 1966). Neutron diffraction experiments have been performed in order to determine the magnetic structure of these compounds. For CrB_2 very faint extra lines appear at 4.2 K (CASTAING, 1971), which are interpreted in a model of itinerant magnetism (nesting of parts of the Fermi surface, LIU et al., 1975), as suggested by experimental results (CASTAING et al., 1972b).

The case of MnB_2 is puzzling: the hyperfine field on the manganese nucleus can be measured by nuclear magnetic resonance up to room temperature (KASAYA and HIHARA, 1969) while it appears to be zero on the boron nucleus at temperatures above 150 K (BARNES and CREEL, 1969). It has been proposed that MnB_2 is an antiferromagnet with a Néel temperature of 760 K, and a magnetic moment on manganese nuclei perpendicular to the c axis. In this structure the boron atoms would occupy sites in the crystal for which the hyperfine field would be zero for symmetry reasons. The weak ferromagnetism observed at low temperature would be due to a slight tilting of the magnetic moment out of the basal plane (KASAYA and HIHARA, 1970). Neutron diffraction experiments are in agreement with this model; the magnetic peak is however still present at 885 K, at variance with the proposed Néel temperature (LEGRAND and NEOV, 1972).

Manganese and chromium diborides seem as a whole to have strongly different behaviors; in MnB_2, manganese atoms apparently show a localized moment (CASTAING 1971, p38), the localized spins interacting both via direct d-d coupling and via s-d exchange with conduction electrons (ROBINSON and ERDÖS, 1972). On the other hand, CrB_2 is an itinerant antiferromagnet, similar to chromium metal. The calculated density of states at the Fermi level for CrB_2 accounts only for about one half of the electronic specific heat. This result is supported by experimental evaluations (CASTAING, 1971; CASTAING et al., 1971a). This enhancement of the specific heat is even stronger for CrB_2-base alloys with VB_2 or MoB_2, the electronic term γT showing a sharp maximum for a concentration which corresponds to the onset of antiferromagnetic ordering at zero temperature (Fig. 4). Simultaneously, low temperature anomalies occur, related to superparamagnetic effects revealed by magnetic susceptibility and nuclear magnetic resonance measurements on $M_{1-x} M'_x B_2$ compounds (M, M' = V, Mn, Cr, Mo; CASTAING, 1971; CASTAING et al., 1971a; CASTAING et al., 1972a; SCHOENBERGER et al., 1971). The increase of the γT term of the specific heat for CrB_2-base alloys has raised the problem of antiferromagnetic spin fluctuations which had not until then received much attention. (MORIYA, 1970; DONIACH, 1968). As shown by CASTAING et al. (1972a) and MORIYA and KATO (1971), such fluctuations could be responsible for the observed specific heat, if flat portions appear on the Fermi surface. This topology has been found in the band calculations by LIU et al. (1975). The situation is however not quite clear since ferromagnetic fluctuations cannot be excluded (CASTAING, 1971; LIU et al., 1975).

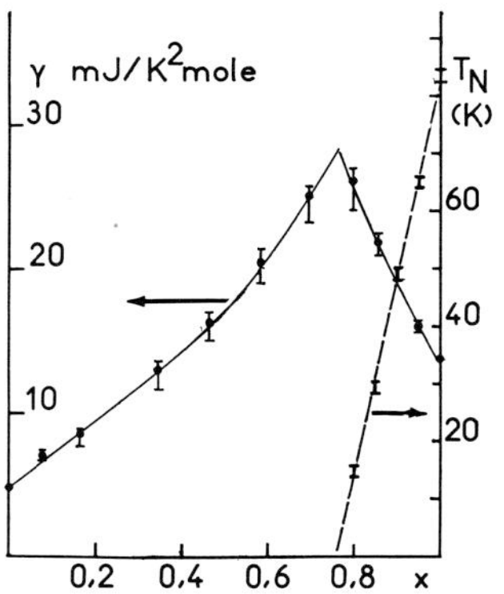

Fig. 4. Variation with chromium content x of the γ coefficient of the linear specific heat term γT and of the Néel temperature of $V_{1-x} Cr_x B_2$ compounds. (After CASTAING et al., 1971b).
Note added in proof: See the specific heat calculation based on spin fluctuations for these alloys by H. HASEGAWA (J. Phys. Soc. Jap. 38, 107, 1975)

4. Plastic Properties

Like most carbides and borides, the transition metal diborides are hard and brittle. Little is known, however, of their actual plastic properties since experiments have been performed either on sintered materials, where the porosity interferes with the mechanical behavior, or on single crystals which were grown from the melt with an excess of metal (for TiB_2) or boron (for ZrB_2), which upon cooling induces the occurrence of fine Widmanstäten platelets or needles of upper (TiB_{2+x}) or lower (ZrB) boride. These platelets produce a hardening of the material. In the latter case, however, experiments have been performed which lead to some knowledge of the intrinsic properties of the material (MERSOL et al., 1968; HAGGERTY and LEE 1971). For two diborides (TiB_2 and ZrB_2) the Burgers vectors of the dislocations have been determined by transmission electron microscopy and have been found in both cases to be parallel to the <11$\bar{2}$0> directions and equal to the a vector of the basal plane, which is the shortest distance between neighboring metal atoms. Two main glide systems are consistent with the Burgers vector: the basal glide (0001), <11$\bar{2}$0> and the prismatic glide {1$\bar{1}$00}, <11$\bar{2}$0>.

At room temperature the slip systems have been deduced from the orientation of slip lines created by micro-hardness indenters. The experiments in both cases show lines primarily consistent with prismatic glide, but basal plane may also operate. Some differences might however occur, probably related to different shapes of the precipitates (needles parallel to the a or c vectors in the case of TiB_2, needles parallel to c and discs in the basal plane in the case of ZrB_2).

At high temperature (2125°C, HAGGERTY and LEE, 1971), ZrB_2 can undergo fairly large permanent deformation before cracking. Slip line traces observed on indentation tests at low temperature show that slip occurs on the basal (0001) plane rather than on the prismatic plane.

The main feature of the stress-strain curve is a large yield drop (Fig. 5), which indicates a small number of mobile dislocations. This effect has been interpreted by HAGGERTY and LEE as a consequence of the dispersion hardening rather than as an intrinsic property comparable to the behavior of germanium single crystals for instance.

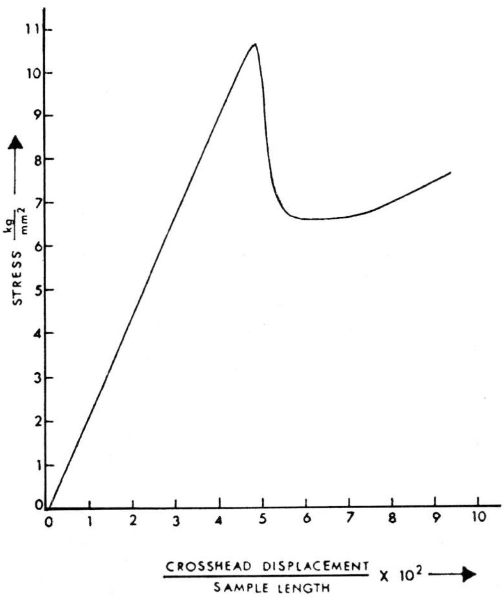

Fig. 5. Stress-crosshead displacement behavior of ZrB_2 single crystal deformed at 2125°C in compression. (After HAGGERTY and LEE, 1971)

Data concerning the hardness of diborides can be found in SAMSONOV (1959), CLOUGHERTY and POBER (1963), VAHLDIEK (1967), NAKANO et al. (1973). The above results show a strong anisotropy of the hardness properties, which might partly be due to the precipitates which are likely to occur upon cooling in single crystals, and a large scatter from one author to the other, with however a trend towards slightly larger values for TiB_2 than for the four other main diborides (ZrB_2, HfB_2, NbB_2 and TaB_2).

5. Chemical Properties and Uses

Several applications of diborides as engineering materials either already exist or can be foreseen as a consequence of (1) their good resistance to chemical agents (2) their low electrical resistivity (3) the fact that they are hard refractory compounds.

This leads to a large potential field of applications, as shown by the great number of patents, provided that oxidation and shock resistance could be improved. Due to their high stability, the most attractive diborides are TiB_2, ZrB_2, HfB_2, NbB_2 and TaB_2.

a) Chemical Properties

The chemical properties of diborides depend greatly upon loosely controlled parameters such as the purity of the materials or the chemical agents, the state of subdivision, adsorption or chemisorption, which can lead to divergent results. In the massive state however, most diborides are chemically inert. Several results have been recently reviewed by THOMSON (1970). GREENWOOD et al. (1966) have summarized the reactivity of diborides with aqueous acids and alkalis. They show *a general inertness to non-oxidizing agents*, but are oxidized by several fused oxides and salts such as lead oxide, or sodium peroxide which reacts violently with VB_2 for instance. In the same way, TiB_2 is stable in the presence of hydrogen chlorine in the 200°-$500^{\circ}C$ range while it reacts with dry chlorine (MEYER and PASTOR, 1965).

In the presence of water vapor, diborides are stable around room temperature (up to $200^{\circ}C$ for TaB_2 and CrB_2; ARABEI et al., 1971).

Higher-temperature properties are of great interest for refractory materials. Titanium, zirconium and niobium diborides are highly resistant to corrosion by molten metals such as boiling zinc, lead, aluminum, copper (THOMSON, 1970; MEYER and PASTOR, 1965). (Nb, Ti)B_2 solid solution or NbB_2-TiN composites also give excellent results (MURATA and MICCIOLI, 1971). This property makes them attractive as materials for electrodes in view of metal refining or aluminum elaboration.

A good corrosion resistance in air is also necessary for engineering materials: the oxidation rate of diborides is slow at lower temperatures, where a protective B_2O_3 glass layer appears on the materials; this has been shown for ZrB_2 by LEBUGLE and MONTEL (1971), GRAHAM et al. (1971), GLEBOV et al. (1972). At temperatures above $1100^{\circ}C$, or in the presence of water vapor, the oxidation rate increases rapidly, the boron oxide layer being destroyed. Additions of silicon carbide or of transition metal silicides improve the oxidation resistance above $1300^{\circ}C$ because of the occurrence of a glassy SiO_2 layer (GRAHAM et al., 1971; TRIPP et al., 1973; PASTOR and MEYER, 1974). In the same way FENTER (1971) has shown that silicon carbide or carbon additions to ZrB_2 and HfB_2 produce materials of higher density, with significantly improved oxidation resistance, thermal stress resistance and mechanical properties for uses in an oxidizing atmosphere at temperatures up to $2200^{\circ}C$.

b) Uses

In their applications, borides have only lately been considered but are leading candidates for use in severe environments. Their chemical stability and inertness makes them suitable as crucible materials for melting metals, or thermocouple protection tubes for measuring temperature in molten metals. This property combined with good electrical and thermal conductivity have made TiB_2 and NbB_2 attractive materials for cathodes or containers in aluminum reduction cells (MEYER and PASTOR, 1965; MURATA and MICCIOLI, 1971; GAIDUCHENKO et al., 1973; POWELL, 1967).

Boride-based composite can also be used as a heating element in corrosive media at high temperature (CARBONNEL and HAMON, 1973; GROPYANOV et al., 1973).

The hardness of borides is so high that they are considered, in spite of their brittleness, as bases for hard-metal work: alloyed with an effective bonding metal, they produce cutting tools comparable to those of carbide-based materials. This property can also be applied to increase the high-temperature wear resistance of a surface by a diboride coating (FUNK et al., 1973).

Borides may also be used to improve the creep behavior of superalloys by fine-particle dispersion hardening (FILIPPI, 1973).

A new field of application was introduced with the aerospace industry: ZrB_2-based composites have been tested in rocket nozzles where they show a good thermal cycling resistance at temperatures up to 2500 °C (FENTER, 1971). Another program has been developed for the use of a diboride as a reradiative nose and leading edge for reusable glide re-entry vehicle, where temperatures are as high as 1600 °C (GANGLER, 1971).

Finally, a quite different field of application lies in the catalytic action of ZrB_2 in the manufacture of diamond crystals (KURATOMI, 1973).

References

AMERICAN ELECTRO. MET. CORP: Report 1.1.54/31.6.55 (AO 71 936) A simplified method for calculating the Debye temperature from elastic constants

ANDERSON, O.L.: J. Phys. Chem. Solids 24, 909 (1963)

ANDERSSON, L., DELBY, B., MYERS, H.P.: Ferromagnetic MnB_2, Solid State Comm. 4, 77 (1966)

ARABEJ, B.G., MYSHKIN, V.A., TIKHONOVA, R.S., FEROLOVA: Corrosion resistance of diborides of chromium, tantalum and their alloys in water in extreme conditions. Zashchita Metallov. 7, 16 (1971)

ARIEL, E. BARTA, J., NIEDZWIEDZ, S.: Tungsten- titanium-boron metastable phase diagram at room temperature. J. Less Common Metals 20, 199 (1970)

ARONSSON, B.: A note on the compositions and crystal structures of MnB_2,Mn_3Si, Mn_5Si_3 and $FeSi_2$. Acta Chem. Scand. 14, 1314 (1960)

BARNES, R.G., CREEL, R.B.: Chromium-like antiferromagnetic behaviour of CrB_2. Phys. Lett. 29A, 203 (1969)

BARNES, R.G., CREEL, R.B., TORGESON, D.R.: Nuclear magnetic resonance in the diborides of scandium and yttrium, J. Chem. Phys. 53, 3762 (1970a)

BARNES, R.G., CREEL, R.B., TORGESON, D.R.: Evidence for localized character of electronic structure in vanadium borides from ^{51}V nuclear magnetic resonance. Solid State Comm. 8, 1411 (1970b)

BAUER, J. and DEBUIGNE, J.: Sur la synthèse des diborures de holmium et de thulium. C.R. Acad. Sc. Paris 277, C 851 (1973)

BECKMAN, G. KIESSLING, R.: Thermal expansion coefficient for uranium boride and β-uranium silicide. Nature 178, 1341 (1956)

BERNSTEIN, H.: X-ray measurements of lattice constants, thermal expansion, and Debye temperature of diborides. RTD TDR 63-4096 p.I30 (1963)

BEZRUK, E.T., MANKOVSKII, L.Y.: Some physical properties of manganese borides. Nieorg. Mat. 4, 447 (1968)

BINDER, I., POST, B.: Manganese diboride, Acta Cryst. 13, 356, (1960)

BRANSCOMB, T.M. HUNTER, O. Jr.: Improved thermal diffusivity method applied to TiB_2, ZrB_2 and HfB_2 from 200 to I300 °C. J. Appl. Phys. 42, 2309 (1971)

BREWER, L., TEMPLETON, D.H., SAWYER, D.L., DAUBEN, C.H.: Refractory borides. J. Am. Ceram. Soc. 34, 173 (1951)

CADEVILLE, M.C.: Propriétés magnétiques des diborures de manganèse et de chrome MnB_2 et CrB_2. J. Phys. Chem. Solids 27, 667 (1966)
CARBONNEL, H., HAMON, L.: Conductive ceramics of high porosity. Ger. Offen. 2 261523 (Cl. C 046) 28 June 1973 - From Chem. Abst. 79, 82840 (1973)
CARTER, G.C., SWARTZ, J.C.: Nuclear magnetic resonance Knight shifts and quadrupole effects in transition metal diborides. J. Phys. Chem. Solids 32, 2415 (1971)
CASTAING, J.: Structure électronique et propriétés magnétiques des diborures de métaux de transition, Thèse Orsay, Publication ONERA 140 (1971)
CASTAING, J., CAPLAIN, R., COSTA, P.: Chaleur spécifique à basse température et susceptibilité magnétique des borures mixtes de molybdène et de chrome. Solid State Comm. 9, 297 (1971a)
CASTAING, J., CAUDRON, R., COSTA, P.: Renforcement du terme de chaleur spécifique électronique à l'apparition de l'ordre antiferromagnétique dans les diborures de métaux de transition. Int. Magn. Conference 1970 J. Phys. 32, C 638 (1971b)
CASTAING, J., CAUDRON, R., TOUPANCE, G., COSTA, P.: Electronic structure of transition metal diborides. Solid State Comm. 7, 1453 (1969)
CASTAING, J., COSTA, P., HERITIER, M., LEDERER, P.: Spin fluctuations in nearly antiferromagnetic $(V, Cr)B_2$. J. Phys. Chem. Solids 33, 533 (1972a)
CASTAING, J., DANAN, J., RIEUX, M.: Calorimetric and resistive investigation of the magnetic properties of CrB_2. Solid State Comm. 10, 563 (1972b)
CLOUGHERTY, E.V., POBER, R.L.: Hot hardness of diborides. RTD-TDR-63-4096 p. 179 (1963a)
CLOUGHERTY, E.V., BLACKBURN, L.D.: Phase boundaries of diborides RTD-TDR-63-4096 p. 293 (1963b)
CREEL, R.B., BARNES, R.G.: ^{11}B spin-lattice relaxation times in transition metal diborides. Phys. Stat. Solidi 41, K27 (1970)
DONIACH, S.: Relation of spin-fluctuation induced mass renormalization to nuclear magnetic relaxation in transition metals and alloys, J. Appl. Phys. 39, 483 (1968)
ENGELHARDT, J.J.: Superconducting isotope effect in molybdenum boride and tungsten boride. Phys. Rev. 179, 452 (1969)
EHRLICH, P.: Binäre Systeme des Titans mit Stickstoff, Kohlenstoff und Bor. Z. Anorg. Chem. 259, 1 (1959)
FELTERN, E.J.: Preparation of AlB_2. J. Am. Chem. Soc. 78, 5977, (1956)
FENTER, J.R.: Refractory diborides as engineering materials . S.A.M.P.E. Quart. 2, 1 (1971)
FILIPPI, A.M.: Stability of reactive and refractory metal borides in ternary chromium-base alloys. J. Less Common Met. 30, 153 (1973)
FISHER, D.W.: Vanadium $L_{II,III}$ X-ray emission and absorption spectra from metal, oxides, nitride, carbide and boride. J. Appl. Phys. 40, 4151, (1969)
FISHER, D.W.: Molecular orbital interpretation from the X-ray L_{II-III} emission and absorption spectra from some titanium and vanadium compounds. J. Appl. Phys. 41, 3561, (1970)
FISHER, D.W., BRAUN, W.L.: Band structure and the titanium $L_{II,III}$ X-ray emission and absorption spectra from pure metal, oxides, nitride, carbide and boride. J. Appl. Phy. 39, 4757 (1968)
FLOTOW, H.E., OSBORNE, D.W., O'HARE, P.A.G., SETTLE, J.T., MRAZEK, F.C., HUBBARD, W.N.: Chromium diboride: preparation, enthalpy of formation at 298.15 K, heat capacity from 1 to 350 K and some derived thermodynamic properties. J. Chem. Phys., 51, 583 (1969)
FRANTSEVICH, I.N., ZHURAKOSKII, E.A. and LYASCHENKO, A.B.: Elastic constants and characteristics of the electronic structure of certain types of refractory compounds obtained by sintering. Izv. Akad. Nauk SSSR - Neorg. Mat. 3, 8, (1967)
FRANTSEVICH, I.N., ZHURAKOSKII, E.A., VASILENKO, N.N.: X-ray emission K-band of boron in transition metal deborides. Soviet Phys. Doklady 15, 970 (1971)
FRUCHART, R., MICHEL., A.: Un nouveau borure de manganèse, MnB_4. C.R. Acad. Sc. Paris, 251, 2953 (1960)
FUNK, R., LUX, B., SCHACHNER, H., TANNENBERGER, H.: Increasing the wear resistance of a surface of a hard metal part. Fr. Demande 2 144 823 (Cl. C 23c) 23 mars 1973 - From Chem. Abst. 79, 107 206 (1973)
GAIDUCHENKO, A.K., KRAT, V.A., OSTROVERKHOV, V.I., KOSHKIN, I.A., DUBOVIK, T.V., KUTSENOK, T.G.: Metalloceramic materials based on titanium boride. USSR 381 702 (Cl.C 22c) 22 mai 1973 - From Chem. Abst. 80, 18518 (1974)

GANGLER, J.J.: N.A.S.A. research on refractory compounds. High Temp., High Press. 3, 487 (1971)
GEBHARDT, J.J., CREE, R.F.: Vapour-deposited borides of group IVa metals. J. Am. Ceram. Soc. 48, 262 (1965)
GILMAN, J.J., ROBERTS, B.W.: Elastic constants of TiC and TiB_2. J. Appl. Phys. 32, 1405 (1961)
GLASER, F.W., POST, B.: System Zr-B. Trans. AIME 197, 1117 (1953)
GLEBOV, L.A., LAVRENKO. V.A., KIRYAKOVA, J.E.: Scale formation in oxygen and behaviour of some high-temperature refractory borides in vacuo. Dopov. Akad. Ukr. S.S.R. Ser. B 34 1014, (1972)
GRAHAM, H.C., DAVIS, H.H., KVERNES, I.A., TRIPP, W.C.: Microstructural features of oxide scales formed on zirconium diboride materials. Materials Science Research Vol 5, p.35 (1971)
GREENWOOD, N.N., PARISH, R.V., THORNTON, P.: Metal borides. Quarterly Rev. 20, 441 (1966)
GROPYANOV, V.M., GROMOVA, E.S., SHUNDA, A.N., TSEBERIN, A.G., KALVERSH, A.C., MOGILENSKII, V.I.: Cermets for heating elements in corrosive media. Ger. Offen. 2 200 665 (Cl.C 046, B 22F) 19 jul 1973 - from Chem. Abst. 79, 96 181 (1973)
GUY, C.N.: The electronic properties of chromium borides. J. Phys. Chem. Solids 37, 1005 (1976)
HAGGERTY, J.S., LEE, D.W.: Plastic deformation of zirconium boride single crystal. J. Am. Ceram. Soc. 54, 572 (1971)
HAGUE, A.: Distributions électroniques des premiers éléments de transition par spectroscopie X. Thesis PARIS VI (1972)
HANSON, B.D., MAHNIG, M., TOTH, L.E.: Low temperature heat capacities of transition metal diborides. Z. Naturforschung 26a, 739 (1971)
HOFMANN, W., JÄNICHE, W.: Das System Aluminium-Bor. Z. Metallk., 28, I (1936)
HOLLIDAY, J.E.: Norelco Reporter 14, 84 (1967)
HOWLETT, B.W.: The uranium-boron system. J. Inst. Met. 88, 91 (1959)
ISKOLDSKII, I.I., BOGORODSKAYA, L.P.: Regarding the possibility of producing metallo-refractory-hard alloys based on borides of Cr, Ti and W. Zh. Prikl. Khim. 30, 177 (1957)
JANAF: Thermodynamical Tables Sup. N°33 (1970)
JONES, M.E., MARSH, R.E.: The preparation and structure of magnesium boride MgB_2. J. Am. Chem. Soc. 76, 1434 (1954)
JONES, W.H., GRAHAM, T.P., BARNES, R.G.: Nuclear magnetic resonance line shapes resulting from the combined effects of nuclear quadrupole and anisotropic shift interactions. Phys. Rev. 132, 1898 (1963)
JOHNSON, R.W., DAANE, A.H.: Electron requirements of bonds in metal borides. J. Chem. Phys. 38, 425 (1963)
JURETSCHKE, H.J., STEINITZ, R.: Hall effect and electrical conductivity of transition metal diborides. J. Phys. Chem. Solids. 4, 118, (1958)
KASAYA, M., HIHARA, T.: Magnetic structure of MnB_2. J. Phys. Soc. Japan 29, 336 (1970)
KASAYA, M., HIHARA, T., KOI, Y.: N.M.R. study in MnB_2. J. Phys. Soc. Japan 26, 1549 (1969)
KAUFMAN, L.: Thermodynamics of stability of diborides. RTD-TDR 63-4096 p. 348 (1963)
KEIHN, F.G., KEPLIN, E.J.: High-temperature thermal expansion of certain group IV and group V diborides. J. Am. Ceram. Soc. 50, 81 (1967)
KIESSLING, R.: Binary system Cr-B, Acta Chem. Scand. 3, 595 (1949a)
KIESSLING, R.: Borides of tantalum. Acta Chem. Scand. 3, 603 (1949b)
KISLYI, P.S., LVOV, S.N., NEMTCHENKO, V.F., SAMSONOV, G.V.: Physical properties of the boride phases of chromium. Porovsh. Met. 6, 50 (1962)
KOPP, J.P., BARNES, R.G.: Nuclear magnetic resonance of ^{11}B and ^{27}Al in aluminum diboride. J. Chem. Phys. 54, 1840 (1971)
KOVENSKAYA, B.A., SEREBRYAKOVA, T.I.: Physical properties of boride phases of niobium. Porovsh. Met. 10, 79, (1970)
KRIKORIAN, O.H.: Estimation of heat capacities and other thermodynamic properties of refractory borides. UCRL Report 51043 (1971)
KRUPKA, M.C.: High temperature vaporization behaviour and thermodynamic properties of the diborides. U.S. At. En. Comm. L.A. 2611 (1962)
KUENTZLER, R.: Chaleurs spécifiques électroniques des borures de manganèse Mn_2B, MnB et MnB_2. C.R. Acad. Sc. Paris 270, B 197 (1970)

KURATOMI, T.: Synthesis of diamond employing calcium carbide as solven and metal borides as catalyst. Japan Kokai 73 00 293 (Cl. 14 El) 05 Janv. 1973 From Chem. Abst. 79, 7466 (1973)

KUZMA, Yu.B., LAKH, V.I., STADNYK, B.I., KOVALYK, D.A.: The systems Hf-W-B, Hf-Re-B and Nb-Re-B. Porovsh. Met. 12, 59 (1970)

KUZMA, Yu.B., TELEGUS, V.S., KOVALYK, D.A.: X-ray investigation of the V-Cr-B., Nb-Cr-B and Mo-Cr-B ternary system. Porovsh. Met. N°5, 80 (1969)

KUZMA, Yu.B., TELEGUS, V.S., MARKO, M.A.: Phase equilibria in the hafnium-niobium-boron and tantalum-chromium-boron systems. Porovsh. Met. N°4, 66 (1972)

KUZNIETZ, M., SPOKAS, J.J.: Nuclear magnetic resonance and relaxation of ^{11}B in Uranium diboride. Bull. Amer. Phys. Soc. 15, 274 (1970)

LANG, S.M.: Properties of high temperature ceramics and cermets; elasticity and density at room temperature. Nat. Bur. Stand. (US) Monograph 6 march 1960

LATVA, J.D.: Selection and fabrication of ceramics and intermetallics. Metal Prog. 82, 139 (1962)

LE BUGLE, A., MONTEL, G.: Detection of two states in zirconium diboride oxydation. C.R. Acad. Sc. Paris Ser. C 273, 546 (1971)

LEGRAND, E., NEOV, S.: Neutron diffraction study of MnB_2. Solid State Comm. 10, 883 (1972)

LINDEMANN, F.A.: Phys. Z. 11, 609 (1910)

LIU, S.H., KOPP, L., ENGLAND, W.B., MYRON, H.W.: Energy bands, electronic properties and magnetic ordering of CrB_2. Phys. Rev. B 11, 3463 (1975)

LUNDIN, C.E.: Rare-earth metals phase diagrams, In: SPEDDING F.H., DAANE A.H. (eds.) The Rare Earths. p. 247 New York: J.Wiley and Sons 1961

LUNDSTRÖM, T.: Crystal chemistry of borides and phosphides. Arkiv för Kemi 31 237, (1970)

L'VOV, S.N., NEMCHENKO, V.F.: Possible application of the single zone representation to the diborides of the group IV transition metals. Izv. Vyssh. Ucheb. Zaved. Fiz. II, 59 (1968)

L'VOV, S.N., NEMCHENKO, V.F., SAMSONOV, G.V.: Influence of the non-metal atoms on the electrical properties of the refractory compounds of the transition metals. Porovsh. Mel. 4, 3 (1962)

LYAKHOVSKAYA, I.I., ZIMKINA, T.M., FOMICHEV, V.A.: Boron K spectra from transition metal diborides and LaB_6, BaB_6 and AsB compounds. Soviet Phys. Solid. State 12, 138, (1970)

Mc ALISTER, A.I., CUTHILL, J.R., WILLIAMS, M.L., DOBBYN, R.C.: Electronic structure of diborides of the 3d metals. In: Proc. of the International Symposium on X-ray Spectra and Electronic Structure of Matter. Munich 1972

MALYUCHKOV, O.T., POVITSKY, V.A.: Nuclear magnetic resonance study of transition metal borides and pure boron. Porovsh. Mel. 4, 26 (1962)

MANIELIS, R.M., TIELIOKOVA, T.M., GRICHINA, L.N.: Structure and various properties of yttrium diboride. J. Neorg. Mater. 6 1184 (1970)

MANDORF, V., HARTWIG, J., SELDIN, E.J.: Tech. Mem. TMC -49 Fostoria Dev. Lab (1961)

MARKOVSKII, L.YA., BEZRUK, E.T.: Phase diagram of the Mn-B system. Izv. Akad. Nauk SSSR. Neorg. Mat. 3, 2165 (1967)

MARKOVSKII, L.YA., VEKSHINA, N.V., SHTRIKHMAN, R.A.: Micromethod for the determination of the fision temperature of highly refractory materials. Ogneupory 22, 42 (1957)

MATTERSON, K.J., JONES, H.J., MOORE, N.C.: 4th Plansee Seminar. (1961)

MEERSON, G.A., SAMSONOV, G.V.: Preparation of the borides of high melting metals by the vacuum thermal process. Zh. Prikl. Khim. 27, 1115 (1954)

MENSHIKOV, A.Z., NEMNONOV, S.A.: Influence of the chemical bonds on the valence state of chromium in different compounds. Izv. Akad. Nauk SSSR Fiz. Ser. 27, 402 (1963). See also: PETROV, E.V. and TSVETKOV, V.P.: X-ray K absorption spectra of the Zirconium phases ZrB , ZrC and ZrN. Akad. Nauk Ukr. SSR. Met. Fiz. 40, 98 (1972)

MERSOL, S.A., LYNCH, C.T., VAHLDIEK, F.W.: Defect structure of single-crystal titanium diboride. In: Anisotropy in Single-Crystal Refractory Compounds. New York: Plenum Press 2-41, 1968

MEYER, R., PASTOR, H.: Preparation, propriétés et applications des diborures de titane et de zirconium. Bull. Soc. Franc. Ceramique 66, 59 (1965)

MEZAKI, R., TILLEUX, E.W., BARNES, D.W., MARGRAVE, J.L.: Int. Conf. Thermodynamics of Nuclear Materials Vienna May 1962
MIKSIC, M.G.: Thermal vibrations and bonding effects in TiB_2, Polytechnic Institute of Brooklyn Report NR-032-414 (1963)
MORIYA, T.: Spin fluctuations in nearly antiferromagnetic metals. Phys. Rev. Letters 24, 1433 (1970)
MORIYA, T., KATO, T.: Note on the spin fluctuations in nearly ferro- and antiferromagnetic metals. J. Phys. Soc. Japan 31, 1016 (1971)
MOTT, N.F.: Electrons in transition metals. Adv. Phys. 13, 325 (1964)
MURATA, Y., MICCIOLI, B.R.: Inhibition of grain growth in niobium diboride. Amer. Ceram. Soc. Bull. 50, 182 (1971)
NAKANO, K., INEURA, T., TAKEUCHI, S.: Hardness anisotropy of single crystals of IVa-diborides. Jap. J. Appl. Phys. 12, 186 (1973)
NEMNONOV, S.A., KOLOBOVA, K.M.: The X-ray spectra, electron structure and properties of titanium compounds. Fiz. Met. Metalloved. 22, 36 (1966)
NESHPOR, V.S., SAMSONOV, G.V.: The problem of brittleness of metallic compounds. Fiz. Met. Metalloved 4, 181 (1957)
NORTON, J.T., BLUMENTHAL, H., SINDEBAND, S.J.: Structure of diborides of Ti, Zr, Cb, Ta and V. Trans. AIME, 185, 749 (1949)
NOWOTNY, H., BENESOWSKY, F., KIEFFER, R.: Beitrag zum Aufbau des System Vanadin-Bor. Z. Metallk. 50, 258 (1959a)
NOWOTNY, H., BENESOVSKY, F., KIEFFER. R.: Beitrag zum Aufbau der Systeme Niob-Bor und Tantal-Bor. Z. Metallk. 50, 417 (1959b)
NOWOTNY, H., BENESKOVSKY, F., RUDY, E., WITTMANN, A.: Aufbau und Zunderverhalten von Niob-Bor-Silicium Legierungen. Monatsh. Chem. 91, 975 (1960b)
NOWOTNY, H., RUDY, E., BENESOVSKY, F.: Untersuchungen in den Systemen Zirkonium-Bor- Kohlenstoff und Zirkonium-Bor- Stickstoff. Monatsh. Chem. 91, 963 (1960a)
PASTOR, H., MEYER, R.: Effet d'additions de siliciures metalliques aux borures de titane et de zirconium sur leur comportement en frittage et leur résistence à l'oxydation à haute température. Rev. Int. Htes Temp. et Refract. 11, 41 (1974)
PESHEV, P., ETOURNEAU, J., NASLAIN, R.: Sur le système bore -scandium. Mat. Res. Bull. 5, 319 (1970)
PIPER, J.: Galvanomagnetic effects in single-crystal ZrB_2. J. Phys. Chem. Solids 27, 1907 (1966)
POBER, R.L., CLOUGHERTY, E.V.: Coefficient of linear thermal expansion of diborides. RTD-TDR 63-4096 Part I Report (1963)
POST, B., GLASER, F.W., MOSKOWITZ, D.: Transition metal diborides. Acta Met. 2, 20, (1954)
POWELL, C.F.: Ch. 10 - p.349, High Temperature Materials and Technology, I.E. CAMPBELL, I.E., SHERWOOD, E.M. eds. sponsored by the Electroch. Soc. New-York: John Wiley and Sons, 1967
ROBINSON, J.M., ERDOS, P.: s-d exchange model of weak ferromagnetism with an application to MnB_2. Phys. Rev. B 6, 3337 (1972)
RUDY, E., BENESOVSKY, F.: The systems Hf-B-N and Zr-B-N. Monatsh. Chem 92, 415 (1961)
RUDY, E., BENESOVSKY, F., TOTH, L.: Investigation of the ternary systems of the group V and VI metals with B and C. Z. Metallk. 54, 345 (1963)
SAMSONOV, G.V.: Visnyk Akad. Ukrain. SSR, 6, 27 (1959)
SAMSONOV, G.V.: Preparation and properties of scandium diboride, Dok. Akd. Nauk SSSR 133, 1344 (1960)
SAMSONOV, G.V., KOVENSKAYA, B.A., SEREBRYAKOVA, T.I.: Some physical characteristics of diborides of group IV and V transition metals. Izv. Vissh. Ucheb. Zaved. Fiz. 14, 19 (1971a)
SAMSONOV, G.V., KOVENSKAYA, B.A., SEREBRYAKOVA, T.I.: On chemical bond in transition metals. Dop. Akad. Ukrains. SSR, A(12), 1127 (1971b)
SAMSONOV, G.V., KOVENSKAYA, B.A., SEREBRYAKOVA, T.I.: Some physical characteristics of the diborides of transition metals of group IV and V. Izv. Vuzov Fizika I, 19 (1971c)
SAMSONOV, G.V., KOVENSKAYA, B.A., SEREBRYAKOVA, T.I., TEL'NIKOV, E.YA.: Thermal expansion of the diborides of group IV and V transition metals. Tepl. Vys. Temp. 9, 195 (1971d)

SCHICK, H.L.: Thermodynamics of Certain Refractory Compound. New York: Academic Press 1966

SCHOENBERGER, R.J., BARNES, R.G.: N.M.R. study of the antiferromagnetic transition in $V_xCr_{I-x}B_2$. AIP Conf. Proc.; Magnetism and magnetic materials, 5, 392 (1971)

SHAMRAI, F.I., FEDOROV, T.F.: Phase diagram of the system Cr-B (1964) Quoted by F.A. SHUNK, Constitution of Binary Alloys. New York: McGraw Hill 1969

SHASHKINA, T.B.: X-ray emission spectra of manganese diborides. Phys. Stat. Sol. (b) 44, 571 (1971)

SILVER, A.H., BRAY, P.J.: N.M.R. study of bonding in some solid boron compounds. J. Chem. Phys. 32, 288 (1960)

SILVER, A.H., KUSHIDA, T.: Nuclear magnetic resonance in transition-metal diborides. J. Chem. Phys. 38, 865 (1963)

SIRTL, E., WOERNER, L.M.: Preparation and properties of aluminum diboride single crystals. J. Cryst. Growth 16, 215 (1972)

SKAAR, E.C., CROFT, W.J.: Thermal expansion of TiB_2. J. Am. Ceram. Soc. 56, 45 (1973)

SMITH, H.G., GLÄSER, W.: Phonon spectra in TaC and HfC. Phys. Rev. letters 25, 1611 (1970)

SOUTHERN RESEARCH INSTITUTE: A.S.C. Tech. Rep. 62-675 (1962)

SPEAR, K.E., SCHÄFER, H., GILLES, P.W.: Thermodynamics of the vanadium diborides. In: High Temperature Technology (Asilomar 1967) London: Butterworths 1968

STEINITZ, R., BINDER, I., MOSKOWITZ, D.: System molybdenum-boron and some properties of the molybdenum borides. Trans. AIME, 983 (1952)

STONER, E.C.: Collective electron ferromagnetism. Proc. Roy. Soc. (London) A165, 372 (1938)

THOMPSON, R.: Chemistry of metal borides and related compounds. Progress in boron chemistry Vol. II, p. 173, BROTHERTON R.J. and STEINBERG H. (eds.) New York: Pergamon Press 1970

TOULOUKIAN, Y.S.: Thermphysical Properties of High Temperature Solid Materials. New York: MacMillan 1967

TRIPP, W.C., DAVIS, H.H., GRAHAM, H.G.: Effect of silicon carbide addition on the oxidation of zirconium diboride. Amer. Ceram. Soc. Bull. 52, 612 (1972)

TYAN, Y.S., TOTH, L.E., CHANG, Y.A.: Low temperature specific heat study of the electron transfer theory in refractory metal diborides. J. Phys. Chem. Solids 30, 785 (1969)

VAHLDIEK, F.W.: Electrical resistivity, elastic modulus and Debye temperature of titanium diboride. J. Less Common Metals 12, 202 (1967)

WALKER, B.E., EWING, C.T., MILLER, R.R.: Heat capacity of titanium diboride. J. Phys. Chem. 61, 1682 (1957)

WESTRUM, E.F. Jr., CLAY, G.A.: $NbB_{I.963}$: the heat capacity and thermodymic properties from 5 to 350 K. J. Phys. Chem. 67, 2385 (1963)

WESTRUM, E.F. Jr., FEICK, G.: Zirconium diboride: heat capacity and thermodynamic properties from 5 to 350 K. J. Chem. Eng. Data 8, 193 (1963)

WESTRUM, E.F. Jr.: Heat capacity of diborides. RTD-TDR 63-4096 Report (1963)

WILEY, D.E., MANNING, W.R., HUNTER, J.R.: Elastic properties of polycrystalline TiB_2, ZrB_2 and HfB_2 from room temperature to 1300 K. J. Less Common Metals 18, 149 (1969)

ZACHARIASEN, W.H.: Crystal chemistry of the 5f series elements: VIII crystal structure studies of Uranium silicides and of $CeSi_2$, $NpSi_2$ and $PuSi_2$. Acta Cryst. 2, 94 (1949)

ZHURAVLEV, N.N., STEPANOVA, A.A.: Structure of ScB_2, Kristallografiya 3, 83 (1958)

ZHURAVLEV, N.N., BIELONSOVA, I.A., MANIELIS, R.M., OIELOUSOVA, N.A.: X-ray diffraction determination of the thermal expansion coefficient of lanthanum and yttrium borides. Kristallografiya, 15, 836 (1970)

ZHURAVLEV, N.N., STEPANOVA, A.A.: X-ray determination of the thermal dilatation coefficient of ScB_2. Porovsh. Met. 6, 83 (1964)

XII. Ternary Metal Borides

H. NOWOTNY and P. ROGL

1. Introduction

A classification of ternary metal borides has recently been given (NOWOTNY, 1972). In the meantime quite a few well-defined ternary metal borides have been detected and characterized, particularly by Russian authors.

The classification of the ternary borides can best be made in the sense as for binary metal borides following the scheme (see Appendix pxxx) established by KIESSLING (1950). The mode of classification widely reflects the size of the boron aggregates in dependence on the ratio metal/boron. Whether a true boron-boron aggregate does exist or not, may be decided by means of the boron-boron distances. In very few cases only a simple scheme of binding can be developed even for borides where the bond type within the boron-boron aggregate is unambiguously known. Thus the views about the nature of the chemical bonding are quite contradictory. In general the opinion has been expressed that a boron → metal transfer takes place for boron-rich borides, while the opposite should occur for metal-rich borides.

2. Preparation

Most of the material described in the following Chapter has been prepared by direct reaction of the pure components (metal 1, metal 2, and boron). In order to obtain well-defined ternary metal borides it is important to maintain an inert atmosphere or vacuum during sintering, melting (preferentially arc melting) and heat treatment, and to avoid any contamination of the material by the equipment. Because of the high temperature often involved in the procedure, the pick-up of oxygen, nitrogen, carbon and other elements easily takes place and more complex phases can be stabilized. Compounds such as $(Cr,Ir)_3B_2$ are obviously stable only in the presence of some oxygen or nitrogen (ROGL and NOWOTNY, 1974a). The ternary boride MoAlB also forms by reduction of Mo- and B-oxides by aluminum. However, the various methods using oxides and a reduction agent seem to be rather limited for the preparation of ternary borides with the possible exception of metal carbo-borides. Boriding of binary alloys by thermal decomposition of boron halides and similar processes (pyrolysis) should be applicable to prepare ternary metal borides.

3. Classification of Ternary Metal Borides

As has recently been mentioned, only a small portion of this vast area has been investigated so far and very few systematic studies have been carried out on physical and chemical properties of ternary metal borides.

According to a survey by GOLDSCHMIDT (1967) on ternary compounds within the combinations: transition element - transition element -

boron as well as transition element - B group element - boron, the term interboride systems has been introduced for boride solid solutions, which will not be discussed here and "double boride" of independent structure as the well-defined ternary metal borides are mainly characterized on the basis of structural chemical investigations. As can be expected, complete or extended solid solutions occur between isotypic binary metal borides which is particularly true for the same or neighboring group number e.g. $(Nb,Ta)_3B_2$ or $(V,Mo)B_2$ (KUZ'MA et al., 1972).

There are however quite a few examples where a high temperature binary metal boride can be stabilized to lower temperature by metal/metal substitution. Thus looking at an isothermal section the boride behaves like a ternary metal boride as is shown for e.g. $Hf_{0.2}W_{0.8}B$ (KUZ'MA et al., 1970a), $Mo_{0.7}Fe_{0.1}B$ or $W_{0.7}Fe_{0.1}B$ (HASCHKE et al., 1966) all having CrB type structure. A large number of ternary metal borides exhibits crystal structures which derive by ordering of a binary metal boride structure, such as Mo_2FeB_4 which possesses the ordered Cr_3B_4 type structure etc.

From the general features of metal borides, given in Chapter C IX one can immediately recognize that metal-rich borides are in structure closely linked to so-called interstitial compounds (or interstitial alloys), where mainly isolated nonmetal atoms and subsequently boron atoms occur within metal atom polyhedra. In contrast because of the strong tendency in forming boron-boron-aggregates, metal borides of medium or high boron content are characterized by all kinds of building groups, such as B_2 pairs, B_3,B_4,B_6 groups, various chains, networks etc. Thus boron-rich metal borides compare to some extent to graphite compounds and the relationship, as we will see later on, clearly shows up within the group of so-called mixed interstitial compounds, metal carboborides.

a) Ternary Metal Borides with High Ratio Metal/Boron

α) Topological Packed Structures

From a structural chemical viewpoint metal-rich borides quite often display interstitial character and this is especially true for high ratios M+M'/B. That is why relatively small additions of boron to binary alloys may sometimes stabilize well-known elemental or binary compound types just by filling the voids. Thus there was a μ-phase $Hf_{0.45}Al_{0.50}B_{0.05}$ described (RIEGER et al., 1964a); furthermore a partially filled β-Mn type structure occurs with $Mo_{\sim 4.5}Re_{\sim 15.5}B_{\sim 2.2}$ (KUZ'MA et al., 1970b) and $W_{6.7}Re_{13.3}B_{1.1}$ (KUZ'MA et al., 1968). Nevertheless there are other metal-rich ternary borides such as Re_3Al_2B (CHABAN and KUZ'MA, 1972) which represent an ordered Ti_2Ni (Mn_3Ni_2Si)-type structure. According to these findings boron acts like silicon and occupies a metal site. A cubic Laves phase occurs with $(Ni_{0.63}Be_{0.37})(Be_{1.40}B_{0.60})$. Boron obviously stabilizes the ternary compound inasfar as no nickel beryllide exists of this or a similar structure (BECHER and NEIDHARD, 1966a).

β) K-borides, $(M_9M'_4B)$

Very recently a number of so-called K-borides have been detected having most likely a formula $M_9M'_4B$ such as Hf_9Mo_4B. The crystal structure of these compounds strongly resembles that of K-carbides

---

[1]The metal atoms are eventually in a partial ordering.

although the mode of filling differs. The icosahedra are filled by metal atoms, the triangular metal prism by boron atoms, while the metal octahedra are empty (ROGL et al., 1973a). Because of the complex mode of filling somewhat different compositions for the K-borides have earlier been proposed. Representatives are: $Zr_9\{Mo,W^2,Re\}_4B$, $Hf_9\{W,Re,Os\}_4B$ and solid solutions $(Hf,Al)_9(Mo,Al)_4B$ (ROGL and NOWOTNY, 1973b).

τ-borides, $(M,M')_{23}B_6$. Numerous complex borides having $Cr_{23}C_6$ or ordered $Cr_{23}C_6$ type structure have been collected by STADELMAIER (1969). These compounds which are called τ-phases, also link transition elements and B group elements. More recently new representatives of τ-borides have been detected, as is shown in Table 1 (ROGL and NOWOTNY, 1973c). It is of interest to notice that besides $Ce_2Ni_{21}B_6$ obviously no τ-phases exist with other rare earth elements (La,Pr,Nd, Sm,Eu) (KUZ'MA and BILONISHKO, 1971). Similarly no rare earth cobalt borides of this type have been found so far. The τ-borides Hf-{Co,Ni}-B have also been confirmed by KUZ'MA (1972) who reported a homogenous domain of $Hf_{2-4.3}Ni_{21-18.7}B_6$ (800°C quench). Similarly the τ-boride $Re_{12}Ni_{11}B_6$ was corroborated which extends at 800°C over a range corresponding to 40-45 a/o Re and 20 a/o B (CHEPIGA, 1972). A more detailed inspection leads to the assumption that the ordered type corresponds to a formula $M_{21}M'_2B_6$ while compositions having a ratio of $M/M' \approx 1-2$ display a random metal distribution, occasionally a partial ordering may occur.

Table 1. Novel ternary τ-borides; τ-borides already described in literature, see STADELMAIER (1969)

$Ce_2Ni_{21}B_6$	(KUZ'MA and BILONISHKO, 1971)
$Cr_{\sim 13}Ir_{\sim 10}B_6$	(ROGL and NOWOTNY, 1973c)
$Mn_{\sim 16}Ir_{\sim 7}B_6$	(ROGL and NOWOTNY, 1973c)
$Fe_{10-15.4}Ir_{13-7.6}B_6$	(ROGL and NOWOTNY, 1973c)
$Co_{\sim 15}Ir_{\sim 8}B_6$	(ROGL and NOWOTNY, 1973c)

It is of interest to note that no binary metal boride of this structure type exists. For a fictitious "$Fe_{23}B_6$" a lattice parameter ($a \approx 10.7$ Å) has been obtained by extrapolation of the respective solid solutions $(Re,Fe)_{23}B_6$, $(Fe,Ir)_{23}B_6$ and $Fe_{23}(C,B)_6$.

τ-carboborides (Borocarbides). A new τ-carboboride was found with $Co_{11}B_2C$ (MARKOVSKII et al., 1971). The large extent of substitution for τ-carboborides has been shown by BORLERA and PRADELLI (1971) who demonstrated the formation of $(Cr,Fe)_{23}(C,B)_6$. For the complete solid solution $Fe_{23}(C_{0.75}B_{0.25})_6$-$Cr_{23}(C_{0.25}B_{0.25})_6$ a relatively sharp minimum of the lattice parameter (a = 10.55 Å) (Fig. 1) was found at 30 mol % Cr-carboboride. An ordering of the Cr- and Fe-atoms is to take place according to the above-mentioned authors. If ordering is ignored, which is true for high temperatures, the τ-phase $Fe_{23}(C_{0.5}B_{0.5})_6$ stems finally from the series: $Cr_{23}C_6 \rightarrow Cr_{23}(C,B)_6 \rightarrow (Cr,Fe)_{23}(C,B)_6 \rightarrow Fe_{23}(C,B)_6$. The competition of metal antisquare prism for $Fe_{23}(C,B)_6$ and the trigonal metal prism for $Fe_3(C,B)$ with cementite structure is obvious.

---
[2] $(Zr,W)_{13}B$ metastable.

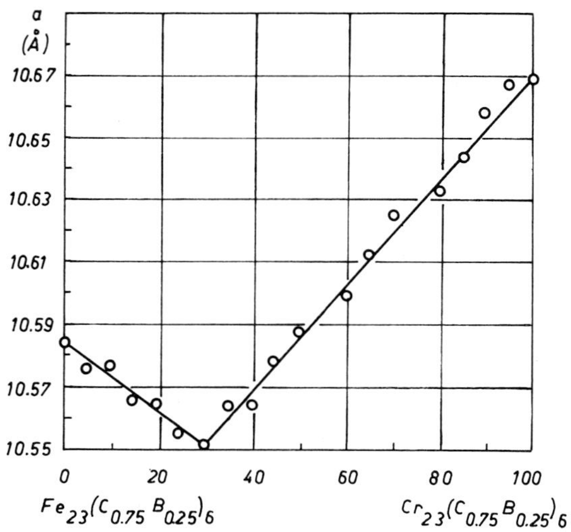

Fig. 1. Lattice parameter a of solid solutions: $Fe_{23}(C_{0.75}B_{0.25})_6$ - $Cr_{23}(C_{0.75}B_{0.25})_6$. (From BORLERA and PRADELLI, 1971)

γ) Perovskite Borides, (MM'_3B, $M_3M'B$)

Ternary metal borides having a fairly high metal/boron ratio are encountered with perovskite borides. There is a strong relationship with the former boride class which can also be attributed to a complex perovskite structure, e.g. $(Co_{13})(Co_8)Hf_2B_6$ displaying two clusters (Co_{13}) and (Co_8) instead of single atoms (GOODENOUGH and LONGO, 1970). Only recently HOLLECK (1973) has discovered a series of ternary borides of formula MRh_3B_{1-x} with M = Sc,Y,La,Ce,Pr,Nd,Sm,Eu,Gd,Tb,Dy, Ho,Er,Tm,Yb,Lu,Th and U. This category of compounds strongly supports the idea of similar stabilizing of complex metallic phases by either boron or carbon or another nonmetal element. The stabilizing mode can also be recognized from the perovskite boride $Ni_3InB_{0.5}$. The particular case represents the generation of boride structures by filling parent lattices, in other words formation of interstitial compounds.

δ) $M_5M'_3B_x$ and $M_5(X,B)_3$-borides

This class of ternary borides also exhibits the role of boron as a stabilizer, which can be recognized from the fact that many of the compounds develop from the M_5X_3 binaries, thus again high metal/boron ratios result. There are however, borides such as $Ta_5Al_3B_x$ or $Nb_5Ge_3B_x$ ($x < 1$) having the filled-up Mn_5Si_3 type structure which are obviously ternary borides. Some more representatives have recently been found, such as $Nd_5Si_3B_{1.5}$ (Mn_5Si_3 type structure), $La_5Ge_3B_x$, $\{Gd,Tb,Dy\}_5$-$\{Si,Ge\}_3B$ (MAYER and FELNER, 1974).

A number of silico-borides or phosphorous borides are structurally related. These compounds crystallize either with an ordered Cr_5B_3 or W_5Si_3 type structure (ARONSSON et al., 1965).

More recent examples have been found with Cr_5PB_2 (ordered Cr_5B_3 type structure), $Cr_{\sim 4.5}P_2B$ (BAURECHT et al., 1971) and $Ni_{\sim 4.6}Si_2B$ (ordered W_5Si_3 type) (URAZ and RUNDQVIST, 1970). As the phosphorous-borides can partially be considered as pseudosilicides a smaller metal/nonmetal ratio may be more characteristic for stability.

It is of interest to notice that the Cr_5B_3 type structure belongs to the $CuAl_2$-type family, as is also the case for the U_3Si_2-type structure. In fact the B-B pair formation in Cr_5PB_2 resembles that of borides having the ordered U_3Si_2 type structure (see below). In this context the boride $Cr_{1.8}W_{3.2}B_3$ should be mentioned, which possibly derives from Cr_5B_3.

ε) *Ternary Borides of Formula (M,M')$_3$B and M$_3$(X,B)*

Having a relatively high metal/boron ratio, ternary borides of the above formula compete with respect to structure type of Ti_3P versus Fe_3C, as was recently shown for iron containing ternary borides (ROGL and NOWOTNY, 1973d). The tendency for the formation of the cementite structure is well known from the existence of Co_3B, Ni_3B and Pd_3B. As one can see from Table 2, the somewhat larger number of outer electrons per atom e/a appears to favor the Fe_3C type structure prior to the Ti_3P type structure. This does however not hold for $Fe_3(P_{0.52-0.06}B_{0.48-0.94})$ crystallizing with Ti_3P type structure (RUNDQVIST, 1962). On the other hand a larger ratio r_B/r_M indicates higher stability of the latter named type. But $(Fe,Si)_3(B,Si)$ was found to be of the Fe_3C type structure (CHABAN and KUZ'MA, 1970).

Carboborides e.g. with cementite type $(Cr,Fe)_3(B,C)$ and $(Cr,Fe)_7(B,C)_3$ with Cr_7C_3-type structure derive from Fe_3C and Cr_7C_3; however no corresponding binary borides of iron and chromium respectively do exist.

The evaluation of the stability of a ternary boride (and this is a very general problem) cannot be made by analogy. It depends on the mutual affinity of the binaries as well as on the specific interaction. One could imagine that the corresponding iron-tungsten boride does not exist simply because tungsten itself is extremely stable, forming an equilibria: W + Fe_2B.

Table 2. Ternary borides of formula $(Fe,M')_3B$; M' = Mo,W,Re,Ru,Os,Rh,Ir,Pd,Pt

Phase		e/a	r_B/r_M[a]	Structure type	
$Fe_{2.6}Mo_{0.4}B$		6.55	0.69	Ti_3P	
$(Fe,W)_3B$		-	-	-	not observed
$(Fe,Tc)_3B$		-	-	-	not observed
$Fe_{2.6}Re_{0.4}B$	(GANGLBERGER, 1966)	6.65	0.69	Ti_3P	
$Fe_{2.6}Ru_{0.4}B$	(GANGLBERGER, 1966)	6.75	0.69	Ti_3P	
$Fe_3(P_{0.52-0.06}B_{0.48-0.94})$		-	-	Ti_3P	
$(Fe,Os)_3B$		-	-	-	not observed
$(Re,Co)_3B$	(KRYPYAKEVICH et al., 1972)			Ti_3P	
$Fe_{2.2}Rh_{0.8}B$		6.95	0.69	Fe_3C	
$Fe_{2.2}Ir_{0.8}B$		6.95	0.68	Fe_3C	
$(Fe,Pt)_3B$		-	-	-	not observed

[a] r_M = average metal radius

Besides ternary metal borides which obey the interstitial principle recognizable from the parallelism to the corresponding ternary metal carbides, nitrides and so forth, there is also another class which can be assigned substitutional type of metal borides because of their crystal structure created at least in a formal way by replacing small metal atoms by boron atoms. The mode can clearly be shown in the series of rare earth cobalt borides.

ζ) *Complex Borides Having a Structure Type Related to $CaCu_5$*

Metal-rich borides sometimes clearly display the metal-like character of boron as can be seen from a formal substitution of small metal atoms by boron atoms. This is the case e.g. for $CeCo_3B_2$ crystallizing with an arrangement which is close to that of $CeCo_5$ ($CaCu_5$-type structure). $CeCo_3B_2$ corresponds to an ordered $CaCu_5$-type inasfar as the 2c) and 3g) positions are occupied by two boron atoms and three small metal atoms. The following compounds have been found to be of this type: {Sc,Y,Ce,Sm,Gd,Tb,Dy,Ho,Er,Tm,Yb,Lu}Co_3B_2 (KUZ'MA et al., 1969a). Generally one observes a linear dependence of the average cell dimensions with the trivalent R.E.ion radii for almost all R.E. containing compounds (ROGL, 1973).

STADELMAIER and SCHÖBEL (1969) derived the same structure type for $ZrCo_3B_2$ and $HfCo_3B_2$. The first named boride was earlier reported as "$ZrCo_3B$". VOROSHILOV et al. (1970) have later shown that a superstructure according to $a\sqrt{3}$ and 3c occurs with these compounds.

As related to close-packed phases (volume substitution $(R_M/R_{M'})^3 \approx 3$) the coordination numbers are high, 14 and 20 for Zr, 14 and 13 for Co, and nine neighbors for boron. One can still recognize the trigonal metal prism (6 Co) and additional two Zr-atoms and one boron atom in a plane between. That means the boron atoms are shifted from the $CaCu_5$-positions forming B-B-pairs at a distance of 1.70 Å. Evidently no analogous Hf-Ni-boride exists.

Derivative structures which form a regular series of B/M'-substitution have been found by KUZ'MA and BILONISHKO (1973). Almost all rare earth cobalt borides (as well as the corresponding Y compounds) do exist: $RECo_4B$, $RE_3Co_{11}B_4$; $RE_2Co_7B_3$ and the already mentioned $RECo_3B_2$ according to a ratio B/B+Co = 3/15, 4/15, 4.5/15, 6/15 (Fig. 2). It should be emphasized, that in contrast to $ZrCo_3B_2$ there is no B-B-pair formation. The crystal structure of the borides {Y,La,Ce,Pr,Sm,Gd, Tb,Dy,Ho,Er,Yb} Ni_4B also derives from a two units arrangement of the $CaCu_5$-type structure (NIIHARA et al., 1973). More recently U{Co,Ni}$_4$B and U{Fe,Co,Ni}$_3B_2$ phases were detected, also belonging to $CaCu_5$-type derivatives (VALYOVKA and KUZ'MA, 1974a).

η) *The Ternary Boride $Ti_3Co_5B_2$*

The structure type was explained by KUZ'MA and YARMOLYUK (1971) in a topochemical way as a sum of the $CeCo_3B_2$ array + 2MM' body-centered cells forming centered cuboctahedra. Subsequently there is no B-B pair formation, only isolated boron atoms occur.

ν) *Ternary Metal Borides of Composition MM'B and $(M,M')_2B$*

Trigonal prismatic [M_6B] and anti-square [M_8B] groups are mainly encountered. Thus isolated boron atoms occur. Closely related are the ordered Fe_2P-type structure and ordered anti-$PbCl_2$-type structure and subsequently their representatives, as is shown in Figure 3. The latter compounds such as MoCoB or ReCoB are named E-phases (JEITSCHKO, 1968). Ni_6Si_2B also belongs to the Fe_2P-type structure; this compound may however be considered as a boron-stabilized silicide rather than a boride (RUNDQVIST and JELLINEK, 1959).

Related but more complex, the crystal structure of NbCoB and TaCoB was also found, Figure 3 (KRYPYAKEVICH et al., 1971). The array is an intermediate between the E-phase type and NbFeB (KUZ'MA, 1967).

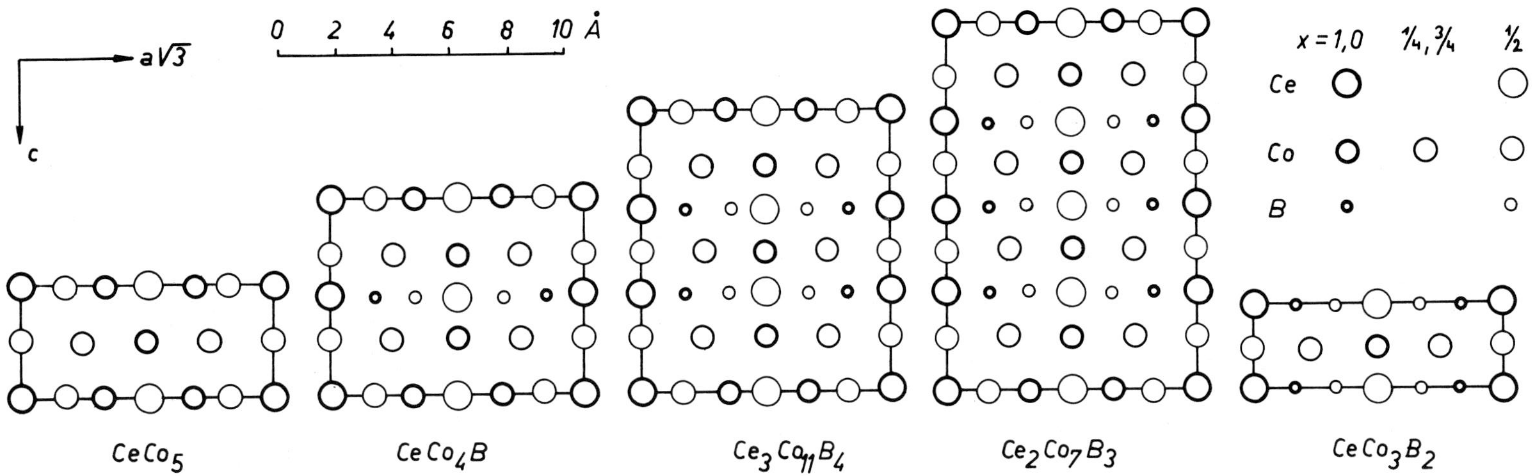

Fig. 2. Generation and stacking mode of borides with $CaCu_5$-type structure derivatives. (From KUZ'MA and BILONISHKO, 1973)

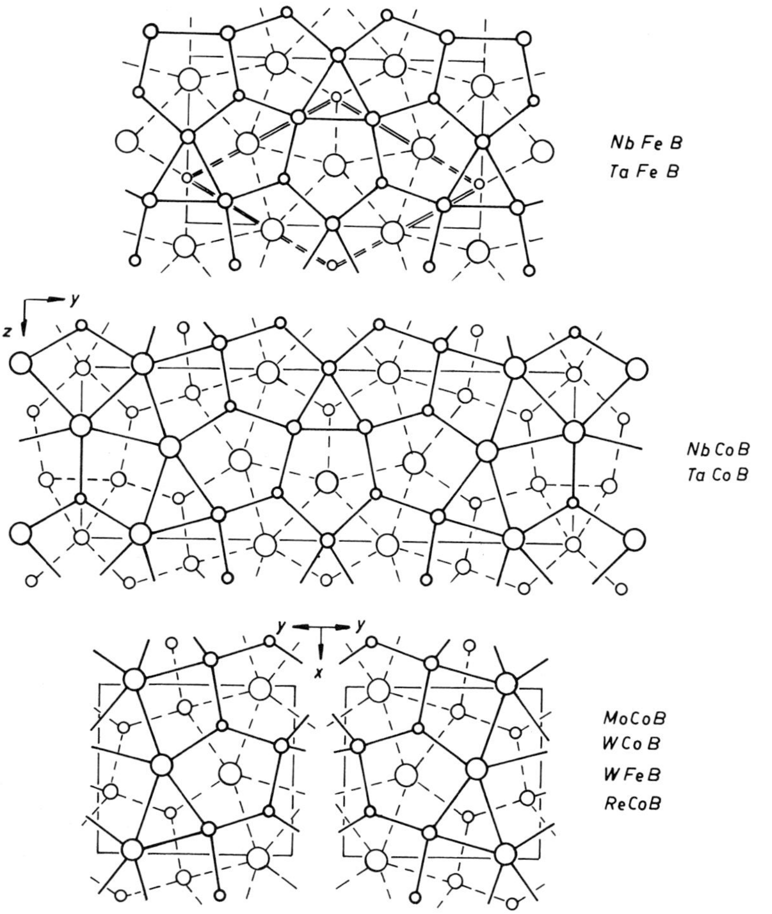

Fig. 3. Comparison and relationship of Fe_2P, Ni_2Si and NbCoB structure-types. (From KRYPYAKEVICH et al., 1971; KUZ'MA et al., 1971)

The trigonal prismatic building group occurs furthermore within the compounds NbNiB, TaNiB (KUZ'MA, 1968) and MoAlB (JEITSCHKO, 1966). This class of compounds already forms a branch to monoboride types inasfar as B-B-chain formation plays a role in the chemical bonding. A similar case will be discussed with the Mn_2AlB_2-type structure (BECHER et al., 1966b) or with the isotypic Fe_2AlB_2 (JEITSCHKO, 1969).

ι) Ternary Borides Having $CuAl_2$ and $CuMg_2$-Type Structures

The archimedian antiprism occurs with $V_{0.95-0.63}Re_{1.05-1.37}B$; there is no binary boride M_2B for vanadium or rhenium either (KUZ'MA and KOVALYK, 1969). On the other hand the isotypic borides $V_{0.15-0.20}Cr_{1.85-1.80}B$ and $Mo_{0.24-0.39}Cr_{1.76-1.61}B$ (KUZ'MA et al., 1969b) are raising the problem of whether they derive from binaries or not, inasfar as Mo_2B certainly exists, while Cr_2B with $CuAl_2$-type structure is still questionable. It might therefore be that "Cr_2B" with $CuAl_2$-type structure is some sort of a stabilized (or impurity) form. It is interesting enough to mention that $(Cr,W)_2B$ extends almost to the Cr-B-side. Thus, very likely the only stable Cr_2B crystallizes with the Mg_2Cu-type structure (or like Mn_4B, having voids). Ternary borides of this type are recorded for $V_{0.51-0.36}Re_{1.49-1.64}B$ and $Re_{1.5}Mn_{0.5}B$ (TELEGUS and KUZ'MA, 1969a) although the latter compound seems to derive from the binary Mn_4B by an additional filling of the voids, as is the case for $(Cr,Mn)_2B_{0.5-1}$ (TELEGUS and KUZ'MA, 1969b).

κ) $(Re,Co)_7B_4$

The structure type determined by KRYPYAKEVICH et al. (1972) belongs also to the $CuAl_2$ structure family, according to the sequence:

$$Fe_2B \rightarrow (Re,Co)_7B_4 \rightarrow Cr_5B_3 \rightarrow Mo_2FeB_2. \quad (1)$$

$(Re,Co)_7B_4$ subsequently consists of a formal combination of the Fe_2B and the Mo_2FeB_2 array along the c-axis according to the following sum: $(M,M')_7B_4 = 2M_2B + M_2M'B_2$.

λ) MM'_2B_2 Borides with Ordered $BaAl_4$ ($ThCr_2Si_2$)-Type Structure

Representatives are {Y,La,Nd,Pr,Sm,Gd,Tb,Dy,Ho}Co_2B_2 (NIIHARA et al., 1971, 1973). Because of the participation of large atoms there is no boron-boron pair formation, although there is a remarkable similarity to the Mo_2CoB_2-type structure. The antisquare building element RE_4Co_4 with isolated boron atoms still dominates. The structural competition was discussed in detail by ROGL (1973).

b) Ternary Borides of Formula $M_2M'B_2$ and $(M,M')_3B_2$

There are two major structure types, the ordered U_3Si_2- and the Mo_2CoB_2-type structure, which both display a similar building element namely B-B pairs in trigonal prisms. Representatives are shown in Table 3. The close relation between both the arrangements can be seen from the existence of low and high temperature modifications for W_2FeB_2. The first reported boride of U_3Si_2 type has been found as an isolate from Incoloy (BEATTIE, Jr, 1958). In some cases the ordering is obviously only partial. Topochemically the U_3Si_2-type structure can be considered as a sum of MB_2 elements (AlB_2 type) + MM' bcc cell.

Table 3.

U_3Si_2-type		W_2CoB_2-type	
Ti_2ReB_2	(KUZ'MA, 1971)	Mo_2CoB_2	(RIEGER et al., 1966)
$Nb_{1.5}Re_{1.5}B_2$	(KUZ'MA et al., 1970a)	Mo_2NiB_2	(RIEGER et al., 1966)
$Ta_{1.5}Re_{1.5}B_2$	(KUZ'MA et al., 1971a)	W_2FeB_2 (l)	(RIEGER et al., 1966)
$Mo_{2-1.5}Cr_{1-1.5}B_2$		W_2CoB_2	(RIEGER et al., 1966)
$(Mo,Mn)_3B_2$	(TELEGUS and KUZ'MA, 1971)	W_2NiB_2	(RIEGER et al., 1966)
$(W,Mn)_3B_2$	(TELEGUS and KUZ'MA, 1971)		
Mo_2FeB_2	(RIEGER et al., 1964b)		
$W_{1.75}Fe_{1.25}B_2$ (h)			

α) Borides of Formula $M_3M'B_3$ and $M_2M'B_2X_n$

In borides of formula $\{W,Mo\}_3\{Co,Ni\}B_3$, B-B-B-groups occur, that is to say, the tendency for forming larger boron-boron aggregates (towards B-B-chains etc.) already starts. This can be seen from the topochemical mode of W_3CoB_3 which is a sum of W_2CoB_2 + WB (JEDLICKA et al., 1969).

The same is true, in the formation of larger boron-boron aggregates, for the Mo_2IrB_2-type structure (ROGL et al., 1972). V_2OsB_2, V_2IrB_2, Mo_2OsB_2, Mo_2IrB_2, $(Mo,Ru)_3B_2$ and $(W,Ru)_3B_2$ are very likely of the same type. However these borides may contain some oxygen or nitrogen. It should be mentioned that B_3- and B_4-groups have not been observed with binary metal borides as yet. The Mo_2IrB_2 structure type seems to be an ordered Y_3Co_2-type structure which recently was determined by MOREAU et al. (1975).

Rare earth Co_4B_4 (rare earth = Y,Ce,Gd,Tb,Dy,Ho,Er,Tm,Lu). Although the ratio (M+M')/B is close to that of monoborides (boron-boron chains) the presence of the large rare earth atoms still paralyzes the boron-boron chain formation, only B-B pairs occur (1.72 Å) (KUZ'MA and BILONISHKO, 1972). The structure has been explained by means of irregular polyhedra of CN = 24, 12 and 8. There are however some unusually large Ce-B distances accounted for.

β) Ternary Borides with Boron-boron Chain Formation

As already mentioned boron-boron chains occur with the MoAlB- and the Mn_2AlB_2-type structure (Fig. 4). Cr_2AlB_2 and Co_2AlB_2 are according to CHABAN and KUZ'MA (1973b) of the same type (Mn_2AlB_2).

Most typical are the metal monoborides which occur with the interrelated CrB-, FeB- and MoB-type structures. With respect to ternary borides a fairly systematic transformation by shift from one type to another takes place by metal/metal substitution. For instance $(V_{0.4-0.6}Ru_{0.6-0.4})B$, $(Cr_{0.6-0.7}Re_{0.4-0.3})B$ (TELEGUS et al., 1969), $(Cr_{0.65-0.4}Ru_{0.35-0.6})B$, $(Mo_{0.5}Ru_{0.5})B$, $(W_{0.5}Ru_{0.5})B$, $(Cr_{0.67}Os_{0.33})B$ and $(Ru_{0.25-0.6}Ni_{0.75-0.4})B$ crystallize with FeB-type structure (ROGL and NOWOTNY, 1973d). As already mentioned there are quite a few metal-metal borides having CrB-type structure which behave at a certain temperature like ternary borides. In general, there is no indication of metal ordering in this class of monoboride-structures. Transformation of the CrB type structure into the MoB type was shown for $V_{0.4}Re_{0.6}B$ and $Cr_{0.25}Re_{0.75}B_{1-x}$.

An interesting ternary boride was found with $W_2Ir_3B_{\sim 5}$ ($Mo_{2.5}Ir_{2.5}B_5$) displaying boron-boron chains (B-B = 1.75 Å) which come fairly close together (2.00 Å). In such a way, the tendency for the generation a two-dimensional network shows up (ROGL et al., 1972).

γ) Ternary Borides Having V_5B_6-, Cr_3B_4-, Cr_3AlB_4- and V_2B_3-type Structures

The corresponding compounds display besides single chains, also double and triple chains. Characteristic metal-metal borides are listed in Table 4.

With respect to a fixed ratio (M+M')/B, one observes that large metal atoms such as rare earth, as already emphasized, seem to decrease the tendency for the formation of boron-boron aggregates, while e.g. aluminum favors boron-boron aggregation (see Al-containing ternary borides).

c) Ternary Borides with Network Structures

The same regularity is met with compounds corresponding to the diboride ratio. While $(Mo,Ru)B_2$, $(W,Ru)B_2$, $(Mo,Os)B_2$, $(W,Os)B_2$, $(Mo,Ir)B_2$, $(W,Ir)B_2$ possess a typical diboride structure with puckered boron-

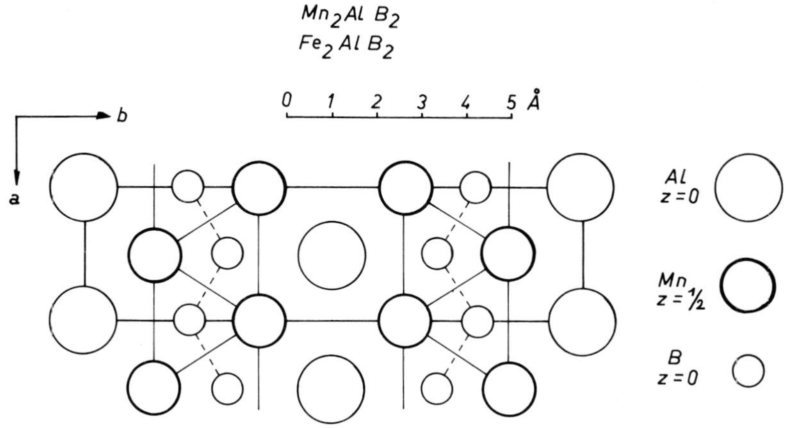

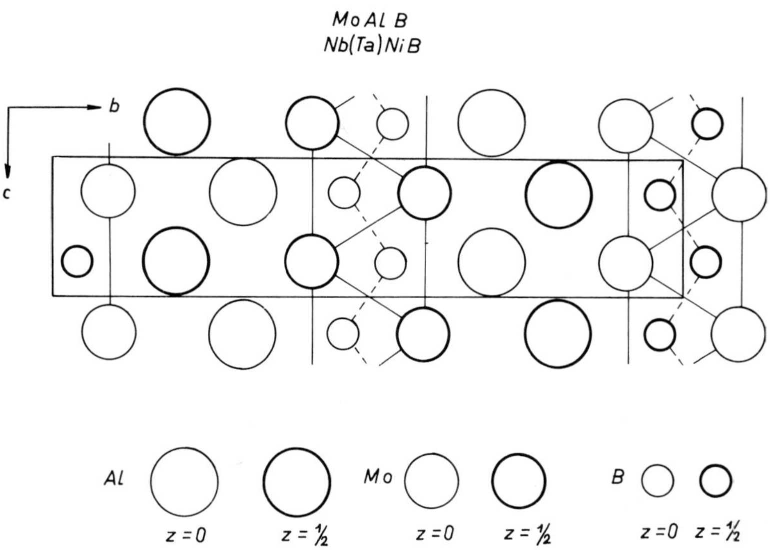

Fig. 4. Boron chain formation in aluminum-containing borides MoAlB and Mn_2AlB_2. (From JEITSCHKO, 1966; BECHER et al., 1966b)

boron nets (ROGL et al., 1970), the boron atoms in $CeCr_2B_6$ form only a loose framework consisting of 14-membered rings (KUZ'MA and SVARICHEVSKAYA, 1972a; KUZ'MA et al., 1973). Some distances B-B = 2.11 Å are however almost too large for bonding. It is not surprising that the structural element of W_2CoB_2 also occurs (pair formation).

Y_2ReB_6 can be regarded as a complete network structure, consisting of 5-, 6- and 7-membered rings; nevertheless there are some boron-boron distances (1.96, 1.87 Å) which indicate a somewhat weaker B-B-bonding. The interesting array, Fig. 5, was worked out by KUZ'MA and SVARICHEVSKAYA (1972b). Numerous representatives {Gd,Tb,Dy,Ho,Er,Tm,Lu}$_2ReB_6$, have been detected by these authors. The ternary borides are obviously limited by the size of the rare earth element.

Table 4.

Single and double chains	$Cr_2Ni_3B_6$ (CHEPIGA et al., 1972)		
Double chains	Mo_2FeB_4 $MoFe_2B_4$ (GLADYSHEVSKY et al., 1966)	W_2FeB_4 $Mo_{1.3}Ru_{1.7}B_4$	Cr_3AlB_4 (KUZ'MA et al., 1972a)
Triple chains	$VCoB_3$ (KUZ'MA and STARODUB, 1973) $Cr_{1.5}Ni_{0.5}B_3$ (CHEPIGA et al., 1972)		

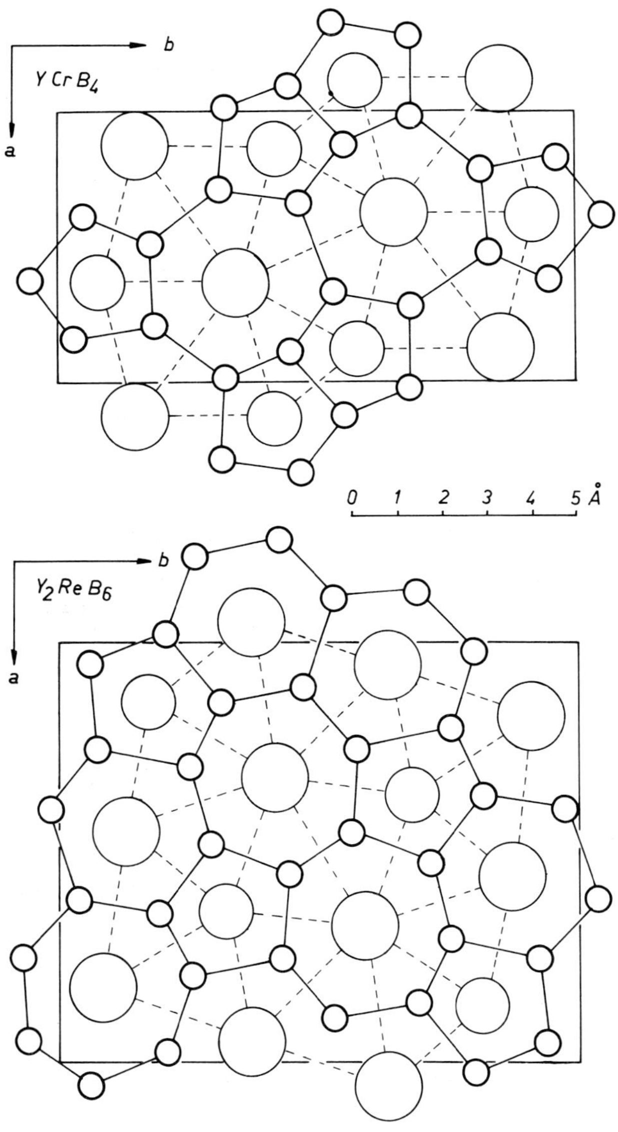

Fig. 5. Boride-structures of $YCrB_4$ and Y_2ReB_6 with boron nets consisting of 5-, 6- and 7 membered rings. (From KUZ'MA, 1970a, 1970b; KUZ'MA and SVARICHEVSKAYA, 1972b)

$YCrB_4$. Similar series {Ce,Gd,Tb,Dy,Ho,Er,Tm,Lu}CrB_4, {Y,Gd,Tb,Dy,Ho,Er}MoB_4, {Y,Gd,Tb,Dy,Ho,Er}WB_4, {Y,Gd,Tb,Dy,Ho,Er}ReB_4 etc. do exist for another diboride arrangement. In contrast to the above array the network is built up by five- and seven-membered rings (KUZ'MA, 1970a, KUZ'MA, 1970b, KUZ'MA and SVARICHEVSKAYA, 1972c).

$ThMoB_4$. Boron networks built up by five- and seven-membered rings also occur for Th{V,Mo,W,Re}B_4. The crystal structure was recently determined (ROGL and NOWOTNY, 1974b). U{Mo,W,Re}B_4 are also isotypic with $ThMoB_4$-type structure (ROGL and NOWOTNY, 1975).

For higher boron/metal ratios it was shown that the substitution of Mo and W by noble metals easily stabilizes the Mo_2B_5-type structure of W_2B_5 to lower temperatures e.g. $W_{4/3}Pd_{2/3}B_5$ and the same is true for the $Mo_{2-x}B_{\sim 9}$($Mo_{1-x}B_3$) type structure e.g. $MoRhB_{\sim 9}$ or $W_{4/3}Ni_{2/3}B_{\sim 9}$ (NOWOTNY et al., 1967). As stated by LUNDSTRÖM and ROSENBERG (1973) there is only a two-dimensional network present, while other proposals assume a corrugated network.

d) Boron-Rich Ternary Borides

These mainly derive from the elementary boron structures and three-dimensional boron networks dominate the structure type (MATKOVICH and ECONOMY, 1970) (HOARD and HUGHES, 1970).

Thus the boron-rich ternary borides display characteristic building groups such as (B_{12}) icosahedra linked by different atoms.

$AlBe_{0.8}B_{24.2}$($Al(Be_{0.8}B_{0.2})B_{24}$) As was shown by KROGMANN and BECHER (1972), the Be-Al-boride is a derivative of the tetragonal boron I, which is itself very likely stabilized by carbon or nitrogen. The Al-atoms in $AlBe_{0.8}B_{24.2}$ occupy the tetrahedral voids formed by four B_{12} icosahedra and it was assumed that the Be-atoms are substituting boron atoms in the cluster.

$MgAlB_{14}$. A complete determination of the crystal structure by MATKOVICH and ECONOMY (1970) reveals the icosahedral building groups B_{12} which like spheres form a close packed array. The 8 linking metal atoms and 8 extra boron atoms act like fillers (occupying the interstities) for 4 B_{12} clusters. As shown by these authors for the boron modifications and boron derivative structures, binary and ternary, the external bond of a boron atom of the B_{12} icosahedron can vary from 2 to 12.

α) Carboborides

AlC_4B_{24}($Al_{2.1}C_8B_{51}$). According to WILL (1969) and PEROTTA et al. (1967, 1969) four icosahedra occupy the 4c position and are linked by carbon bridges $(B_{12})_3$-C-B-C-$(B_{12})_3$. The aluminum atoms are randomly distributed in the channels but attached to the boron atoms in the bridge. To overcome the electron deficiency in a framework of (B_{12}) or $(B_{12})_4$ groups some metal atoms fill the array, in agreement with the bonding theory in boron rich borides by LONGUET-HIGGINS and ROBERTS (1954, 1955).

AlC_4B_{40}. Again B_{12} clusters form the building groups, but the structure type derives from the rhombohedral cell ($B_{12}C_3$) according to $B_{13.34}C_{1.34}Al_{0.33}$. Only the linking between the B_{12} clusters (icosahedra) is a peculiar one, inasfar as besides stretched C-B-C and C-B-B-chains, C-Al-B-bridges, forming an angle, also occur (NEIDHARD et al., 1970) (PEROTTA et al., 1969).

e) Carbon-Containing Metal Borides

Out of the so-called mixed interstitial compounds, borocarbides or carboborides are of interest. Typical carboborides are e.g. ScB_2C_2, GdB_2C_2, UBC, Cr_3BC (Cr_7BC_4) (KONDRASHEV, 1966), Mo_2BC.

While for YBC (BAUER and NOWOTNY, 1971), DyBC (BAUER and DEBUIGNE, 1972), UBC (TOTH et al., 1961) and Mo_2BC (JEITSCHKO et al., 1963) the usual structural elements are encountered, B-B-zigzag chains and trigonal metal prisms (carbon in an octahedral surrounding), new building groups show up with ScB_2C_2 (SMITH et al., 1965), YB_2C_2 and YB_2C (BAUER and NOWOTNY, 1971).

Boron-carbon networks occur, consisting of five- and seven-memebered rings within ScB_2C_2, four- and seven-membered rings within YB_2C and four- and eight-membered rings within YC_2B_2. The latter-named compound (and crystal structure), corresponds to GdC_2B_2 and in general more to the carboborides LnC_2B_2 (Ln = La,Pr,Nd,Gd,Tb,Dy,Ho,Er,Yb) described by SMITH and GILLES (1967). Some R.E. carboborides of formula R.E.C_2B_2 were also described by BEZRUK and MARKOVSKII (1972). While TmB_2C_2 and LuB_2C_2 have been characterized as to be isotypic to GdC_2B_2 or more precisely to YC_2B_2, no specific data were given for TbB_2C_2 and YbB_2C_2, but there is no doubt that the latter compounds also belong to the same type. This is also true for CeB_2C_2, SmB_2C_2, EuB_2C_2, TmB_2C_2 (FISHEL and EICK, 1969) and LuB_2C_2 (NORDINE et al., 1964). The array of YB_2C and DyB_2C can be regarded as a topochemical sum of the diboride + Y(Dy)B_2C_2 (Fig. 6). There were some more carboborides of rare earth and thorium (TOTH et al., 1961a) found, which are however not sufficiently characterized.

4. Ternary Metal Borides Which are not Completely Characterized

Various ternary metal borides such as MgNiB (KUZ'MA, 1974) have also been described without detailed structural chemical data. No complete compilation will be given in here.
Six compounds were found by KUZ'MA et al. (1973a) within the La-Ni-boron system. Lattice parameters and space group were reported for $LaNi_5B$, $LaNi_4B$, $LaNi_3B$, $LaNi_{12}B_6$ and $LaNi_2B_2$. Besides these ternary borides a compound $LaNi_{12}B_2$ was detected. Furthermore borides such as $Ce_3Fe_{16}B$, Ce_2FeB_3, $CeFe_2B_2$ (approximative composition) have been reported (BILONISHKO and KUZ'MA, 1972). A more detailed description for $CeCo_{12}B_6$ ($YCo_{12}B_6$) was presented by NIIHARA and YAJIMA (1972); a resemblance to $LaNi_{12}B_6$ is obvious. Some more Ce-Co-borides seem to exist (BILONISHKO and KUZ'MA, 1974). Besides the already known τ-phase a number of Ce-Ni-borides also exist: $CeNi_5B$, $CeNi_4B$, $CeNi_{12}B_6$, $CeNi_2B_3$. Space group and cell parameters have partially been determined. Various ternary borides have partially been characterized within the Mn-Al-B, Co-Al-B and Ni-Al-B systems, such as Mn_5AlB_{11}, $Co_3Al_2B_5$, α,β-Ni_8AlB_{11} and Ni_5AlB_4 (CHABAN and KUZ'MA, 1973b, 1973c). From a survey investigation of {Fe,Ni}-Ga-B it turns out, that no ternary boride forms for iron, only the τ-boride for cobalt, but three more Ni-Ga-borides besides the τ-boride exist (CHABAN and KUZ'MA, 1973a).

5. Properties and Applications of Ternary Metal Borides

There are only a few investigations of physical and chemical properties known. Measurements have mainly been made on complete and terminal solid solutions of binary metal borides and may reflect in some cases on the behavior of ternary borides as well.

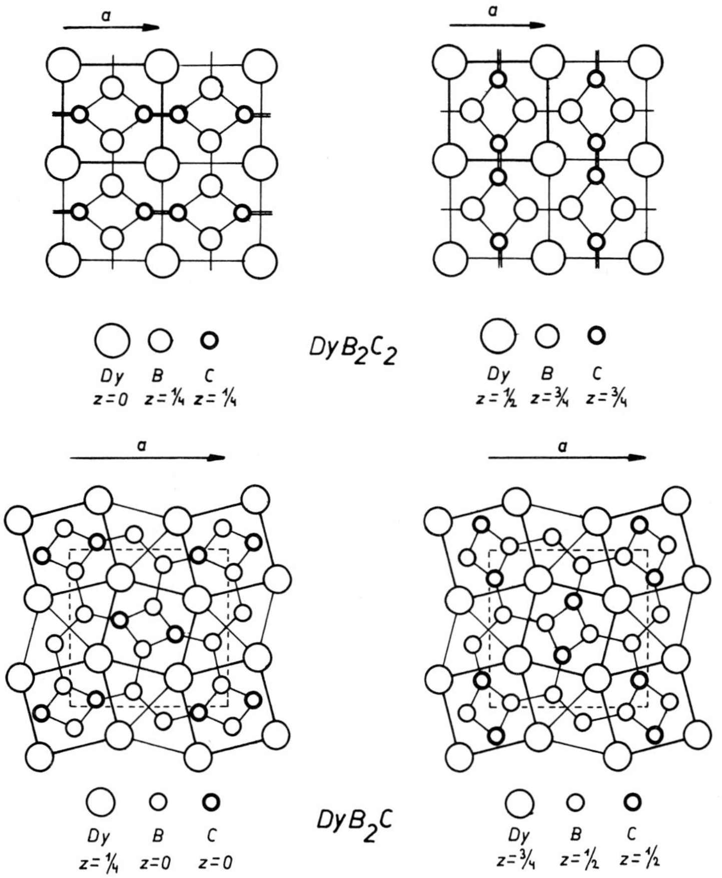

Fig. 6. Crystal structures of DyB_2C_2 and DyB_2C. (From BAUER and DEBUIGNE, 1972) {Tb,Ho,Er,Tm}B_2C. (From BAUER and DEBUIGNE, 1975)

a) Thermal Properties

For the K-boride Hf_9Mo_4B a mp of 1880°C was found (ROGL et al., 1971). A melting point of 1800°C for Mo_2NiB_2 has been reported by KOLOMITZEV and MOSKALEVA (1966), however a different cell (tetragonal) was assumed. TmB_2C_2 and LuB_2C_2 melt congruently at ca. 2050°C.

Heat-capacity measurements have been carried out at low temperatures for a number of transition metal borides of type $(M,M')_mB_n$ (KUENTZLER, 1970; HANSON et al., 1971). There are however no indications for the presence of singularities with respect to a composition which could be considered as a ternary boride. The main conclusion was that the electronic behavior in $(M,M')_2B$-borides compares well with that of M-M'-alloys. According to both sets of e/a (borides and alloys) an electron transfer is assumed to take place from the boron atoms to the 3d-band of the transition element atoms.

b) Electrical and Magnetic Properties

The electrical resistivity for Mo_2NiB_2 was found to be 41.5 μΩcm. While this boride exhibits normal metallic behavior, $YMoB_4$ e.g. on

the other hand displays decreasing electrical resistivity with increasing temperature (VALOVKA and KUZ'MA, 1974b).
It is of interest to note that in some boride solid solutions various properties may change in a characteristic way; the strong increase of transition temperature of superconductivity in transition metal borides is remarkable. While MoB_2 and NbB_2 are no superconductors, $Zr_{0.13}Mo_{0.87}B_2$ (T_c~10 K) or $Sc_{0.1}Nb_{1.9}B_2$ (T_c~6.6 K) display fairly high T_c values. COOPER et al. (1970) consider a ratio e/a between 3.5 and 4 as a favorable one for the occurrence of superconductivity. The borocarbide Mo_2BC has also a fairly high T_c value (~4 K) particularly in view of its irregular structure type (TOTH and ZBASNIK, 1968). Very recently it was shown that a number of Mo_2IrB_2-type phases are superconducting such as $Mo_{1.5}Ru_{1.5}B_2$ (T_c = 3.72-3.60 K) (VANDENBERG et al., 1975).

Fe_5PB_2 has a CURIE point of 628 K (BLANC et al., 1967); $Fe_3P_{0.18}B_{0.82}$ of 800 K (FRUCHART et al., 1962). According to NIIHARA and YAJIMA (1972) the compound formulated as $CeCo_{12}B_6$ displays ferromagnetic behavior below 140 K. The paramagnetism above this temperature (Curie-point) obeys the Curie-Weiss law. It was found that $TbCo_2B_2$ and $DyCo_2B_2$ are ferromagnetic at low temperature, while $CeCo_3B_2$, $LaCo_2B_2$, $SmCo_2B_2$ and $GdCo_2B_2$ behave antiferromagnetic (SOBCZAK, 1975). Some of the τ-borides, such as $(Fe_{1-x}Co_x)_{20}Al_3B_6$ etc. have been well characterized by magnetic measurements. A survey of these data is presented by GOODENOUGH and LONGO (1970). A more complex τ-boride $Co_{51-9}Fe_{9-18}$-$Ti_{0.02-1}Al_{6.5-11.2}B_{17-25.9}$ has been proposed for application of magnetic head assemblies (HIROTA, 1965a, 1965b, 1967, 1968).

Ternary solid solutions such as $(M,M')_mB_n$ also allow clarifying the detailed mechanism of magnetic properties. As was shown e.g. by IGA et al. (1966) for $(Mn,Fe)_3B_4$ a strengthening of the interlayer antiferromagnetic coupling of Mn_3B_4 takes place by substituting Mn-atoms by some Fe-atoms. $Mn_{3-x}\{Fe,Co,Ni\}_xB_4$ ($x<0.5$) materials have been proposed for switching systems, because of the strong increase of the magnetization at -24 °C (YANASE et al., 1969). In contrast to the Fe etc. additions to Mn_3B_4, it was shown by IGA and TAWARA (1968) that Cr, Mo and W exert selective influence. From magnetic measurements and lattice parameter dependence it was concluded that a ferromagnetic Mn_2CrB_4 should exist. Furthermore a preferential substitution of Mo and W atoms at 2c) metal sites was detected.

Magnetization measurements on solid solutions of binary metal borides give a hint of the electronic state in metal borides in general. If the metal atoms occupy only one crystallographic site, such as Fe in FeB or Fe_2B, the saturation magnetization is believed to be a sensitive test for the donor model whether boron transfers electrons into the 3d-band of the transition element atom or not. The variation of two different interstitial atoms such as boron and carbon should than be informative about such a donor hypothesis because of the difference of electrons of the donor atoms. Unfortunately sufficient C/B substitution is possible only in Fe_3C e.g. where different Fe-sites exist. According to BERNAS et al. (1967) an increase of the average moment of iron of 0.5 μ_B per boron atom takes place. Mößbauer effect investigations on $Fe_3(C,B)$ were interpreted by BERNAS et al. (1967) as compatible with the donor model under the assumption of a strong hybridisation of spdγ orbitals (s and p orbital stemming from the nonmetal atoms). It is stated furthermore that the nonmetal atoms (boron and carbon) exert a rather short range influence on the Fe atoms. There are no doubt still many shortcomings in the theory of the electronic state.

$(Fe,Ni)_2B$-borides having some amount of carbon were studied by Mößbauer spectroscopy (FLECHON et al., 1971). No particular conclusion can be made, however, inasfar as a detailed characterization of the phases present is missing. Mößbauer measurements were also made for Mo_2FeB_2 (U_3Si_2-type structure and W_2FeB_2 (W_2CoB_2-type structure), (DE JOUNG and BARNES, 1973).

Similarly, magnetocrystalline anisotropy investigations of $(Fe,Co)_2B$ and $(Fe,Ni)_2B$ solid solutions do not reveal any peculiarity (IGA, 1970).

c) Mechanical Properties

Compounds such as $\{Fe,Co,Ni\}TaB_2$ and "$\{Fe,Co,Ni\}_5TaB_2$" are mentioned in literature as well as some microhardness values (LAVENDEL, 1961) for the first named compounds (~1700; ~1550; ~1650 kp/mm^2). There are however no specific data about these phases. On the other hand a more recent investigation (KUZ'MA et al., 1971) revealed the occurrence of phases having the following compositions: TaNiB (see p. 420) and $TaNiB_2$, found in the Ta-Ni-B system at 950°C, TaFeB (see p. 420), $Ta_3Fe_3B_4$ and $TaFeB_3$ for the corresponding Fe-containing system at 950°C. Microhardness measurements were made for Mo_2NiB_2 by KOLOMITZEV and MOSKALEVA (1966); a value of 1514 kp/mm^2 was reported. Microhardness measurements were also made for TmB_2C_2 and LuB_2C_2 by BEZRUK and MARKOVSKII (1972), a value of 1550 ± 50 kp/mm^2 has been given.

Occasionally ternary metal borides are listed such as '$ZrCrB_2$', 'Zr_2MoB_4' which could not be found in the corresponding ternary systems, although e.g. improvement of the mechanical properties of compact Zr-diboride bodies by addition of Cr,Mo,W,Re has been observed (MAYER and PASTOR, 1969).

Appendix. Classification of ternary borides

Structure type	Possible formula	Representative of binary borides	Representative of ternary borides	Note	Structural elements
filled β-Mn	$(M,M')_5B_{1-x}$ $M_3M'_2B_{1-x}$	--	$Mo_{\sim4.5}Re_{\sim15.5}B_{\sim2.2}$	--	metal frame-work
$Cr_{23}C_6$	$(M,M')_{23}B_6$	--	$Ce_2Ni_{21}B_6$	ord.	-"-
			$Fe_{10-15.5}Ir_{13-7.5}B_6$	p.ord.	
K-boride related to $W_9Co_3C_3$	$M_9M'_4B$	--	Hf_9Mo_4B	ord.	-"-
Mn_3Ni_2Si (Ti_2Ni)	$(M_3B)M'_2$	--	Re_3Al_2B	ord.	-"-
filled Cu_3Au	$M_3M'B_{1-x}$	--	$Ni_3InB_{0.5}$	ord.	-"-
W_6Fe_7 (μ)	$M_6M'_7B_x$	--	$Hf_{0.45}Al_{0.5}B_{0.05}$	ord.	-"-
$MgCu_2$	$M(MB)_2$	--	$[Ni,Be][Be,B]_2$	--	-"-
filled Mn_5Si_3	$M_5M'_3B_{1-x}$	--	$Ta_5Ge_3B_{1-x}$	ord.	isolated boron atoms
W_5Si_3	$M_5(M'B)_3$	--	$Cr_{\sim4.5}P_2B$	--	-"-
$Ti_3Co_5B_2$	$(M,M')_4B$	--	$Ti_3Co_5B_2$	ord.	-"-
Re_3B	M_3B	Re_3B	--		-"-
filled Re_3B	$M_3B_xC_y$	--	$Cr_3B_{0.2}C_{0.8}$	p.ord.	-"-
Ti_3P	$(M,M')_3B$	--	$Mo_{0.4}Fe_{2.6}B$	--	-"-
Fe_3C	-"-	Co_3B	$Fe_{2.2}Ir_{0.8}B$	--	-"-
Mn_5C_2	M_5B_2	Pd_5B_2	--	--	-"-
Fe_2P	MM'B	--	NbFeB	ord.	-"-
Ni_2Si	-"-	--	MoCoB	ord.	-"-
NbCoB	-"-	--	TaCoB	ord.	-"-
$CuAl_2$	$(M,M')_2B$	Ta_2B	$V_{0.95-0.63}Re_{1.05-1.37}B$	--	-"-
Mg_2Cu	-"-	Cr_2B $Mn_2B_{0.5}$	$Re_{1.5}Mn_{0.5}B$	--	-"-
$(BaAl_2)Al_2$	$(M,M')_3B_2$	--	$LaCo_2B_2$	ord.	-"-

Appendix (continued)

Structure type	Possible formula	Representative of binary borides	Representative of ternary borides	Note	Structural elements
$(Re,Co)_7B_4$	$(M,M')_7B_4$	--	$(Re,Co)_7B_4$	--	isolated boron atoms
Th_7Fe_3	M_7B_3	Ru_7B_3	--		-"-
$CaCu_5$ and family	$M(M'B)_5$	--	$CeCo_3B_2$ etc.	ord.	-"-
NiAs	MB	$RhB_{1.1}$	--		-"-
$ThSi_2$	MB_{2-x}	$IrB_{1.1}$	--	boron deficiency	-"-
WC	MB_{1-x}	$IrB_{1-x}(h)$	--		-"-
$IrB_{1-x}(l)$	-"-	$IrB_{1-x}(l)$	--		-"-
Cr_5B_3	$(M,M')_5B_3$	Cr_5B_3	$(Cr_{1.8}W_{3.2}B_3)$	--	boron pairs + isol. B-atoms
U_3Si_2	$(M,M')_3B_2$	V_3B_2	Mo_2FeB_2	ord.	boron pairs
			$W_{1.75}Fe_{1.25}B_2$	p.ord.	-"-
W_2CoB_2	-"-	--	W_2NiB_2	ord.	-"-
$HfCo_3B_2$ related to $CaCu_5$	$M(M'_3B_2)$	--	$ZrCo_3B_2$	ord.	-"-
$CeCo_4B_4$	$(M,M')_5B_4$	--	$CeCo_4B_4$	ord.	-"-
W_3CoB_3	$(M,M')_4B_3$	--	W_3NiB_3	ord.	B_3-groups
Mo_2IrB_2	$(M,M')_3B_2$	--	Mo_2IrB_2	ord.	B_4-groups
FeB	MB	CoB	$Cr_{0.67}Os_{0.33}B$	--	boron-chains
MoB	-"-	MoB(l)	(Re_3CrB_{4-x})	--	-"-
CrB	-"-	MoB(h)	$(Hf_{0.2}W_{0.8}B)$	--	-"-
MoAlB	$(M,M')_2B$	--	TaNiB	ord.	-"-
m-Ni_4B_3	M_4B_3	m-Ni_4B_3	--		-"-
Mn_2AlB_2	$(M,M')_3B_2$	--	Fe_2AlB_2	ord.	-"-
Mo_2BC	M_2BC	--	Mo_2BC	ord.	-"-
o-Ni_4B_{3-x}	M_4B_{3-x}	o-Ni_4B_{3-x}	--		boron chains + isol. B-atoms
$Ru_{11}B_8$	$M_{11}B_8$	$Ru_{11}B_8$	--		
YBC	MBC	--	GdBC	ord.	branched chains
UBC	-"-	--	UBC	ord.	-"-

Appendix (continued)

Structure type	Possible formula	Representative of binary borides	Representative of ternary borides	Note	Structural elements
$W_2Ir_3B_{6-x}$	$(M,M')_5B_6$	--	$W_2Ir_3B_5$	ord.	two dimensional networks starting from double chain formation
			$Mo_{2.5}Ir_{2.5}B_5$	p.ord.	
Cr_3AlB_4	$(M,M')_4B_4$	--	Cr_3AlB_4	ord.	boron double chains
Ta_3B_4	$(M,M')_3B_4$	V_3B_4	$MoFe_2B_4$	ord.	-"-
V_5B_6	$(M,M')_5B_6$	V_5B_6	$Cr_2Ni_3B_6$	--	boron double chains + boron chains
V_2B_3	$(M,M')_2B_3$	V_2B_3	$VCoB_3$	ord.	boron triple chains
			$Cr_{1.5}Ni_{0.5}B_3$	--	
$CeCr_2B_6$	$(M,M')B_2$	--	$CeCr_2B_6$	ord.	starting network formation
$IrB_{1.35}$	MB_{2-x}	$IrB_{1.35}$	--		two dimens. boron networks
$YCrB_4$	$(M,M')B_2$	--	$CeCrB_4$	ord.	boron (or B-C) networks
ScB_2C_2 related to $YCrB_4$	MB_2C_2	--	ScB_2C_2	ord.	-"-
YB_2C_2	-"-	--	GdB_2C_2	ord.	-"-
YB_2C	MB_2C	--	DyB_2C	ord.	-"-
Y_2ReB_6	$(M,M')B_2$	--	Y_2ReB_6	ord.	-"-
AlB_2	-"-	YB_2	--		-"-
ReB_2	-"-	ReB_2	$(W,Ir)B_2$	--	-"-
RuB_2	-"-	OsB_2	--		-"-
W_2B_5	$(M,M')_2B_{5-x}$	W_2B_5	--		-"-
		Ru_2B_3	--		network + isol. B-atoms
Mo_2B_5	-"-	W_2B_5	$W_{1.37}Pd_{0.63}B_5$	--	

Appendix (continued)

Structure type	Possible formula	Representative of binary borides	Representative of ternary borides	Note	Structural elements
$Mo_{0.8}B_3$ ($Mo_2B_{\sim 9}$)	$(M,M')_{0.8}B_3$	$W_{0.8}B_3$	--		network+
Sm_2B_5	M_2B_5	Sm_2B_5	--		boron frame-work
YB_4	MB_4	ThB_4	--		-"-
MgB_4	-"-	MgB_4	--		-"-
CrB_4	-"-	CrB_4	--		-"-
MnB_4	-"-	MnB_4	--		-"-
CaB_6	MB_6	YB_6	--		-"-
UB_{12}	MB_{12}	ZrB_{12}	--		-"-
$MgAlB_{14}$	$(M,M')_2B_{14}$	--	$MgAlB_{14}$		-"-
$ZnB_{\sim 22}$	$MB_{\sim 20-40}$	$ZnB_{\sim 22}$	--		-"-
$REB_{\sim 100}$	$MB_{\sim 100}$	$YB_{\sim 66}$	--		-"-
.					.
.					.
.					.
B					boron

ord. = ordered p.ord. = partially ordered

Note Added in Proof

In the meantime some more ternary borides have been detected and characterized:

$Li_3Ni_{20}B_6$ ($Cr_{23}C_6$-type), $Li_5Ni_{33}B_{16}$, LiNiB	W. JUNG, Naturwissenschaften 63, 246 (1976)
$MgNi_3B_2$ (distortion-derivative of $CeCo_3B_2$) Na_4CaB	H.H. STADELMAIER and C.P. DOSHI, paper presented at the V. Internat. Conference on Solid Compounds of Transition Elements, Uppsala, Sweden, June 21-25 (1976)
Y_3ReB_7 (Y_3ReB_7-type)	Y. KUZ'MA, S. MIKHALENKO, Dopov. Akad. Nauk. RSR, Ser. A. 11, 1029 (1976)
$(Nb,Ta)_5Ge_2B$ (Cr_5B_3-type)	M.A. MARKO, Y. KUZ'MA, E.I. GLADYSHEVSKII, Dopov. Akad. Nauk RSR, Ser A, 6, 555 (1976)
$Mo_{1.7}Ge_{0.3}B$ (Mg_2Cu-type)	M.A. MARKO, P.S. SAAK, Y. KUZ'MA Neorgan. Mater. 12(7), 1307 (1976)
$Hf_3Ir_5B_2$ ($Ti_3Co_5B_2$-type) $(Ta,Nb)_{2-3}Ru_{21-20}B_6$ ($Cr_{23}C_6$-type) $(Nb,Ta)NiB_2$ ($NbCoB_2$-type)	P. ROGL, H. NOWOTNY to be published
$(Os,V)B_2$ (ReB_2-type) $Os(V,Mo,W)_2B_2$ ($IrMo_2B_2$-type)	P. ROGL, E. RUDY to be published

References

ARONSSON, B., LUNDSTRÖM, T., RUNDQVIST, S.: Borides, Silicides and Phosphides London: Methuen 1965

BAUER, J., DEBUIGNE, J.: Dysprosium borocarbides, C.R. Acad. Sci. (Paris) Ser. C. 274, 1271 (1972)

BAUER, J., DEBUIGNE, J.: J. Inorg. Nucl. Chem. 37, 2473 (1975)

BAUER, J., NOWOTNY, H.: Der Dreistoff Yttrium-Bor-Kohlenstoff, Mh. Chem. 102, 1129 (1971)

BAURECHT, H.E., BOLLER, H., NOWOTNY, H.: Röntgenographische Untersuchungen in den Dreistoffen Cr-P-C, Cr-As-C und Cr-P-B, Mh. Chem. 102, 373 (1971)

BEATTIE, H.J. jr.: The crystal structure of a M_3B_2-type double boride, Acta Cryst. 11, 607 (1958)

BECHER, H.J., NEIDHARD, H.: Untersuchungen im System Nickel-Beryllium-Bor und Kobalt-Beryllium-Bor, Z. anorg. allg. Chem. 344, 125 (1966a)

BECHER, H.J., KROGMANN, K., PEISKER, E.: Über das ternäre Borid Mn_2AlB_2, Z. anorg. allg. Chem. 344, 140 (1966b)

BERNAS, H., CAMPBELL, I.A., FRUCHART, R.: Elektronenaustausch und Mössbauer-Effekt von Zwischengitterverbindungen auf Eisenbasis J. Phys. Chem. Solids 28, 17 (1967)

BERNAS, H., CAMPBELL, I.A., FRUCHART, R., KISER, S., SENATEUR, J.P.: Coll. Int. CNRS 157, 381 (1967)

BEZRUK, E.T., MARKOVSKII, L.Y.: Thulium, lutetium, terbium and ytterbium borocarbides and some of their properties, Zh. Prikl. Khim. 45, 1 (1972)

BILONISHKO, N.S., KUZ'MA, Y.B.: Cerium-iron-boron system, Neorgan. Mater. 8(1), 183 (1972)

BILONISHKO, N.S., KUZ'MA, Y.B.: Cerium-cobalt-boron system, Neorgan. Mater. 10(2), 265 (1974)

BLANC, A.M., FRUCHART, E., FRUCHART, R.: Magnetic and crystallographic study of $(Fe_{1-x}Cr_x)_3P$ solid solutions and of the ferromagnetic Fe_5B_2P phase, Ann. Chim. (Paris) 2, 251 (1967)

BORLERA, M.L., PRADELLI, G.: Cubic borocarbides of the type $(Fe,Cr)_{23}(C,B)_6$, Metall. Italiana 63(3), 107 (1971)

CHABAN, N.F., KUZ'MA, Y.B.: Phase equilibriums in manganese-silicon-boron and iron-silicon-boron systems, Neorgan. Mater. 6(5), 1007 (1970)

CHABAN, N.F., KUZ'MA, Y.B.: X-ray diffraction study of rhenium-aluminum-boron and rhenium-silicon-boron ternary systems, Neorgan. Mater. 8(6), 1065 (1972)

CHABAN, N.F., KUZ'MA, Y.B.: Ternary systems composed of gallium, boron, and iron, cobalt, or nickel, Dopov. Akad. Nauk, Ukr. RSR. Ser. A, 35(6), 550 (1973a)

CHABAN, N.F., KUZ'MA, Y.B.: Ternary systems Cr-Al-B and Mn-Al-B, Neorgan. Mater. 9(7), 1908 (1973b)

CHABAN, N.F., KUZ'MA, Y.B.: Isothermal cross sections of the (cobalt, nickel)-(aluminum, silicon)-boron system, Neorgan. Mater. 9(12), 2136 (1973c)

CHEPIGA, M.V.: The rhenium-nickel-boron system, Visnik, Ser. Khim. Vip. 14, 27 (1972)

CHEPIGA, M.V., KRIVUTSKII, V.P., KUZ'MA, Y.B.: Chromium-nickel-boron System, Neorgan. Mater. 8(6), 1059 (1972)

COOPER, A.S., CORENZWIT, E., LONGINOTTI, L.D., MATTHIAS, B.T., ZACHARIASEN, W.H.: Superconductivity: the transition temperature peak below four electrons per atom, Proc. Nat. Acad. Sci. U.S 67, 313 (1970)

FISHEL, N.A., EICK, H.A.: Some new lanthanide diboridedicarbides J. Inorg. Nucl., Chem. 31, 891 (1969)

FLECHON, J., VARNIER, Ch., LE CÄER, G., GENIN, J.M.: Structure of chemical iron-nickel-boron deposits studied by the Mössbauer Effect, C.R. Acad. Sci. (Paris) Ser. B, 272, 437 (1971)

FRUCHART, E., TRIQUET, A.M., FRUCHART, R., MICHEL, A.: Two new ferromagnetic compounds in the ternary system Fe-P-B, C.R. Acad. Sci. (Paris) 255, 931 (1962)

GANGLBERGER, E.: Herstellung von komplexen Phasen des Typus: Übergangselement-(B-Element)-Nichtmetall und deren Kristallchemie, Thesis, Univ. Vienna, 1966

GLADYSHEVSKY, E.I., FEDOROV, T.F., KUZ'MA, Y.B., SKOLOZDRA, R.W.: The molybdenum-iron-boron system, Poroshkov. Met. 6(4), 55 (1966)

GOLDSCHMIDT, H.J.: Interstitial Alloys London: Butterworths 1967
GOODENOUGH, J.B., LONGO, J.M.: In: Landolt-Börnstein, Numerical Data etc., Berlin-Heidelberg-New York: Springer 1970
HANSON, B.D., MAHING, M., TOTH, L.E.: Low temperature heat capacities of transition metal borides, Z. Naturforschg. 26a, 739 (1971)
HASCHKE, H., NOWOTNY, H., BENESOVSKY, F.: Untersuchungen in den Dreistoffen: {Mo,W}-{Fe,Co,Ni}- B, Mh. Chem. 97, 1459 (1966)
HIROTA, H.: Soft magnetic alloy, Germ. P. 1 291 903 (1965a)
HIROTA, H.: Ferromagnetic materials, USP 3 403 996 (1965b); Ferromagnetic materials, Fr. P. 1511 196 (1967); Ferromagnetic materials, Brit. P. 1163 517 (1968)
HOARD, J.L., HUGHES, R.E.: In: Muetterties, E.L. (ed.) The Chemistry of Boron and its Compounds.New York , London, Sydney: Wiley and Sons 1970
HOLLECK, H.: Paper at the 4th. Int. Conf. Solid Compounds of Transition Elements Geneva, April 1973
IGA, A.: Magnetocrystalline anisotropy in $(Fe_{1-x}Ni_x)_2B$, System Japan J. Appl. Phys., 9(4), 414 (1970)
IGA, A.: Magnetocrystalline anisotropy in $(Fe_{1-x}Co_x)_2B$, System Japan J. Appl. Phys. 9(4), 415 (1970)
IGA, A., TAWARA, Y.: Magnetic properties of molybdenum- and wolfram-modified Mn_3B_4, J. Phys. Soc. Japan 24, 28 (1968)
IGA, A., TAWARA, Y., YANASE, A.: Magnetic properties of Fe-modified Mn_3B_4, J. Phys. Soc. Japan 21, 1640 (1966)
JEDLICKA, H., BENESOVSKY, F., NOWOTNY, H.: Die Kristallstruktur des W_3CoB_3 und der dazu isotypen Phasen Mo_3CoB_3, Mo_3NiB_3 und W_3NiB_3, Mh. Chem. 100, 844 (1969)
JEITSCHKO, W.: Die Kristallstruktur von MoAlB, Mh.Chem. 97, 1472 (1966)
JEITSCHKO, W.: The crystal structure of MoCoB and related compounds, Acta Cryst. 24, 930 (1968)
JEITSCHKO, W.: The crystal structure of Fe_2AlB_2, Acta Cryst. B25, 163 (1969)
JEITSCHKO, W., NOWOTNY, H., BENESOVSKY, F.: Die Kristallstruktur von Mo_2BC, Mh. Chem. 94, 565 (1963)
DE JOUNG, D.B., BARNES, R.G.: ^{56}Fe Mössbauer study of some $M_2(M')B_2$ borides J. Phys. Chem. Solids, 34, 139 (1973)
KIESSLING, R.: The borides of some transition elements, Acta Chem. Scand. 4, 209 (1950)
KOLOMITZEV, P.T., MOSKALEVA, N.V.: Phase composition and some properties of alloys of the molybdenum-nickel-boron system, Poroshkov. Met. 6(8), 86 (1966)
KONDRASHEV, Y.D.: Crystalline structure of chromium borocarbide, Kristallografiya 11, 559 (1966)
KROGMANN, K., BECHER, H.J.: Strukturbestimmungen eines aluminium- und beryllium-haltigen tetragonalen Bors der Formel $AlBe_{0.8}B_{24.2}$, Z. anorg. allg. Chem. 392, 197 (1972)
KRYPYAKEVICH, P.I., KUZ'MA, Y.B., CHEPIGA, M.V.: Crystal structure of a $(Re,Co)_7B_4$ compound, Dopov. Akad. Nauk. Ukr. RSR, Ser. A. 34(9), 856 (1972)
KRYPYAKEVICH, P.I., KUZ'MA, Y.B., VOROSHILOV, Y.V., SHOEMAKER, Cl. B., SHOEMAKER, D.B.: The crystal structure of NbCoB, Acta Cryst.27, 257 (1971)
KUENTZLER, R.: Specific heat of nearly ferromagnetic borides $(Co-Ni)_2B$ and (Fe-Co)B, J. Appl. Phys. 41(3), 908 (1970)
KUZ'MA, Y.B.: Crystal structures of NbFeB compounds, Dopov. Akad. Nauk. Ukr. RSR, Ser. A, 29(10), 939 (1967)
KUZ'MA, Y.B.: Crystalline structures of NbNiB and TaNiB compounds, Kristallografiya 13(4), 701 (1968)
KUZ'MA, Y.B.: Crystal structure of the compound $YCrB_4$ and its analogs, Soviet Phys. Crystall. 15(2), 312 (1970a)
KUZ'MA, Y.B.: New ternary compounds with $YCrB_4$-type structure, Dopov. Akad. Nauk. Ukr. RSR. Ser. A. 32(8), 756 (1970b)
KUZ'MA, Y.B.: Titanium-cobalt-boron and titanium-rhenium-boron systems, Neorgan. Mater. 7(3), 514 (1971)
KUZ'MA, Y.B.: Phase equilibria in the ternary hafnium-cobalt-boron, hafnium-nickel-boron and niobium-manganese-boron systems, Visnik, Ser. Khim. Vip. 14, 20 (1972)
KUZ'MA, Y.B.: Phase equilibria in the system magnesium-nickel-boron, Visnick, Ser. Khim. 16, 58 (1974)

KUZ'MA, Y.B., BILONISHKO, N.S.: Cerium-nickel-boron system, Neorgan. Mater. 7(4), 620 (1971)

KUZ'MA, Y.B., BILONISHKO, N.S.: Crystal structure of the compound $CeCo_4B_4$ and its analogs, Soviet. Phys. Crystall. 16, 897 (1972)

KUZ'MA, Y.B., BILONISHKO, N.S.: New structural types of borides of a homologous series on the basis of calcium-copper ($CaCu_5$) and cerium cobalt boride ($CeCo_3B_2$) types, Kristallografiya 18(4), 710 (1973)

KUZ'MA, Y.B., BILONISHKO, N.S., NIMKOVICH, E.M.: Lanthanum-nickel-boron system Dopov. Akad. Nauk. Ukr. RSR, Ser. A. 35(10), 939 (1973a)

KUZ'MA, Y.B., KOVALYK, D.A.: Vanadium-rhenium-boron system, Neorgan. Mater. 5(10), 1687 (1969)

KUZ'MA, Y.B., KRYPYAKEVICH, P.I., BILONISHKO, N.S.: Crystal structure of $CeCo_3B_2$ and analogous compounds, Dopov. Akad. Nauk. Ukr. RSR. Ser. A. 31(10), 939 (1969a)

KUZ'MA, Y.B., KRYPYAKEVICH, P.I., CHABAN, N.F.: Crystal structure of Cr_3AlB_4, Dopov. Akad. Nauk. Ukr. RSR. Ser. A. 34(12), 1118 (1972a)

KUZ'MA, Y.B., LAKH, V.I., STADNYK, B.I., CHABAN, N.F.: Tantalum-rhenium-boron system, Dopov. Akad. Nauk. Ukr. RSR. Ser. A. 33(9), 849 (1971a)

KUZ'MA, Y.B., LAKH, V.I., STADNYK, B.I., KOVALYK, D.A.: Hafnium-tungsten-boron, Hafnium-rhenium-boron, and niobium-rhenium-boron systems, Poroshkov. Met. 10(12), 59 (1970a)

KUZ'MA, Y.B., LAKH, V.I., STADNYK, B.I., VOROSHILOV, Y.V.: Phase equilibriums in zirconium-rhenium-boron and tungsten-rhenium-boron systems, Poroshkov. Met. 8(6), 47 (1968)

KUZ'MA, Y.B., LAKH, V.I., TELEGUS, V.S., STADNYK, B.I., BORBYLEVICH, N.N.: Molybdenum-rhenium-boron ternary system, Izv. Vyssh. Ucheb. Zaved. Tsved. Met. 13(3), 160 (1970b)

KUZ'MA, Y.B., RUDAKEVICH, O.M.: Crystal structure of the compounds cobalt tantalum boride (TaCoB) and cobalt niobium diboride ($NbCoB_2$), Dopov. Akad. Nauk. Ukr. RSR. Ser. A., 35(2), 164 (1973)

KUZ'MA, Y.B., STARODUB, P.K.: Vanadium-iron-boron, vanadium-cobalt-boron and vanadium-nickel-boron systems, Neorgan. Mater. 9(3), 376 (1973)

KUZ'MA, Y.B., SVARICHEVSKAYA, S.I.: Crystal structure of the $CeCr_2B_6$ compound, Kristallografiya 17(5), 939 (1972a)

KUZ'MA, Y.B., SVARICHEVSKAYA, S.I.: Crystal structure of the Y_2ReB_6 compound and its analogs, Kristallografiya 17(3), 658 (1972b)

KUZ'MA, Y.B., SVARICHEVSKAYA, S.I.: New borides with a structure of the $YCrB_4$ type, Dopov. Akad. Nauk. Ukr. RSR. Ser. A. 34, 166 (1972c)

KUZ'MA, Y.B., YARMOLYUK, Y.P.: Crystal structure of $Ti_3Co_5B_2$, Zh. Strukt. Khim. 12, 458 (1971)

KUZ'MA, Y.B., MARKO, M.A., PETROVSKAJA, M.V.: X-ray investigations of the systems V-{Nb,Ta}-B, {Nb,Ta}-B, Nb-Ta-B and V-{Mo,W}-B, Visnik. Ser. Khim., Vip. 13, 3 (1972)

KUZ'MA, Y.B., SOBOLEV, A.S., FEDOROV, T.F.: Phase equilibrium in the tantalum-iron-boron and tantalum-nickel-boron ternary system, Poroshkov. Met. 11(5), 81 (1971)

KUZ'MA, Y.B., SVARICHEVSKAYA, S.I., FOMENKO, V.N.: Lanthanum-chromium-boron and cerium-chromium-boron systems, Neorgan. Mater. 9(9), 1542 (1973)

KUZ'MA, Y.B., TELEGUS, V.S., KOVALYK, D.A.: X-ray diffraction study of vanadium-chromium-boron, niobium-chromium-boron, and molybdenum-chromium-boron systems, Poroshkov. Met. 9(5), 79 (1969b)

LAVENDEL, H.W.: Alloys of tantalum diboride with iron, cobalt and nickel, Planseeber. Pulvermet. 9, 80 (1961)

LONGUET-HIGGINS, H.C., ROBERTS, V.M. de: The electronic structure of the borides MB_6. The electronic structure of an icosahedron of boron atoms, Proc. Roy. Soc. (L) A 224, 336 (1954), A 230, 110 (1955)

LUNDSTRÖM, T., ROSENBERG, I.: Crystal structure of the molybdenum boride ($Mo_{1-x}B_3$). J. Solid State Chem. 6(2), 299 (1973)

MARKOVSKII, L.Y., BEZRUK, E.T., BERLOVA, G.E.: Stability of nickel and cobalt borides with respect to carbon, Neorgan. Mater. 7(1), 56 (1971)

MATKOVICH, V.I., ECONOMY, J.: Structure of $MgAlB_{14}$ and a brief critique of structural relationship in higher borides, Acta Cryst. 26, 616 (1970), Electron. Technol. 3(1-2), 159 (1970)

MATSUSHITA: Feromagnetic materials mainly consisting of cobalt, iron and boron, Fr. P. 1 554 425 (1968)

MAYER, I., FELNER, I.: Nowotny phase of M_5X_3 type rare earth silicides and germanides with boron, J. Less-Common Metals 37, 171 (1974)

MAYER, R., PASTOR, H.: Etude du frittage et de la compression à chaud des diborures de titane et de zirconium avec addition de mêtaux de transition, et de quelques propiêtés des alliages obtenus, Planseeber. Pulvermet. 17, 111 (1969)

MOREAU, J.M., PARTHÉ, E., PACCARD, D.: The orthorhombic structure of Y_3Co_2, a shift structure variation of the monoclinic Dy_3Ni_2 Type, Acta Cryst. B31, 747 (1975)

NEIDHARD, H., MATTES, R., BECHER, H.J.: Zur Darstellung und Struktur eines aluminiumhaltigen Borcarbids, Acta Cryst. B26, 315 (1970)

NIIHARA, K., KATAYAMA, Y., YAJIMA, S.: The YNi_4B compound in yttrium-nickel-boron system, Chem. Lett. (Chem. Soc. Japan), 46, 613 (1973)

NIIHARA, K., YAJIMA, S.: New ternary compound in rare earth-cobalt-boron system, Chem. Lett. (Chem. Soc. Japan), 875 (1972)

NIIHARA, K., SHISHIDO, T., YAJIMA, S.: A ternary rare earth boride with a $ThCr_2Si_2$-type structure, Bull. Chem. Soc. Japan 44, 3214 (1971)

NIIHARA, K., SHISHIDO, T., YAJIMA, S.: The crystal data of ternary rare earth borides, RCo_2B_2, Bull. Chem. Soc. Japan 46, 1137 (1973)

NORDINE, P.C., SMITH, G.S., JOHNSON, Q.: preparation and cell parameters of LuB_2C_2, Lawrence Radiation Laboratory Rep. No. UCRL-12205, Livermore, California (1964)

NOWOTNY, H.: In: Roberts, L.E.J. MTP Int Review of Science, London: Butterworths 1972

NOWOTNY, H., HASCHKE, H., BENESOVSKY, F.: Bor-reiche Wolframboride, Mh.Chem. 98, 547 (1967)

PERROTTA, A.J., TOWNES, W.D., POTENZA, J.A.: Crystal structure of $C_8Al_{2.1}B_{51}$, Acta Cryst. B 25(7), 1223 (1969)

PERROTTA, A.J., TOWNES, W.D., POTENZA, J.A.: Crystal structure of $C_8Al_{2.1}B_{51}$ (AlB_{10}). Nat. Bur. Stand. (U.S.) Spec. Publ. 1967 No. 301 283

RIEGER, W., NOWOTNY, H., BENESOVSKY, F.: Untersuchungen in Systemen: Übergangsmetall (T)-Bor-Aluminium, Mh. Chem. 95, 1417 (1964a)

RIEGER, W., NOWOTNY, H., BENESOVSKY, F.: Die Kristallstruktur von Mo_2FeB_2, Mh.Chem. 95, 1502 (1964b)

RIEGER, W., NOWOTNY, H., BENESOVSKY, F.: Die Kristallstruktur von W_2CoB_2 und isotypen Phasen, Mh. Chem. 97, 378 (1966)

ROGL, P.: Über SE-Metall-Kobaltboride, Mh. Chem. 104, 1623 (1973)

ROGL, P., BENESOVSKY, F., NOWOTNY, H.: Über einige Komplexboride mit Platinmetallen, Mh. Chem. 103, 965 (1972)

ROGL, P., NOWOTNY, H.: Neue K-(Kappa-)Phasen, Mh. Chem. 104, 1497 (1973b)

ROGL, P., NOWOTNY, H.: Neue τ-Boride, Mh. Chem. 104, 1325 (1973c)

ROGL, P., NOWOTNY, H.: Neue Komplexboride, Mh. Chem. 104, 943 (1973d)

ROGL, P., NOWOTNY, H.: Neue Phasen mit Mo_2IrB_2-Typ, Rev. Chim. Min. 11, 547 (1974a)

ROGL, P., NOWOTNY, H.: Ternäre Komplexboride mit $ThMoB_4$-Typ, Mh. Chem. 105, 1082 (1974b)

ROGL, P., NOWOTNY, H.: Uran-haltige Komplexboride, Mh. Chem. 106, 381 (1975)

ROGL, P., NOWOTNY, H., BENESOVSKY, F.: Neue K-Boride und verwandte Phasen (Re_3B-Typ aufgefüllt), Mh. Chem. 104, 182 (1973a)

ROGL, P., NOWOTNY, H., BENESOVSKY, F.: Komplexboride mit ReB_2-Struktur; Ternäre Komplexboride in den Dreistoffen: {Mo,W}-{Ru,Os}-B und W-Ir-B, Mh. Chem. 101, 27, 850 (1970)

ROGL, P., NOWOTNY, H., BENESOVSKY, F.: Komplexboride in den Systemen Hf-Mo-B und Hf-W-B, Mh. Chem. 102, 971 (1971)

RUNDQVIST, S.: X-Ray Investigations of the ternary system Fe-P-B; some features of the systems Cr-P-B, Mn-P-B, Co-P-B and Ni-P-B, Acta Chem. Scand. 16, 1 (1962)

RUNDQVIST, S., JELLINEK, F.: The structures of Ni_6Si_2B, Fe_2P and some related phases, Acta Chem. Scand. 13, 425 (1959)
SMITH, G.S., JOHNSON, Q., NORDINE, P.C.: The crystal structure of ScB_2C_2, Acta Cryst. 19, 668 (1965)
SMITH, P.K., GILLES, P.W.: High temperature rare earth-boron-carbon studies III. J. Inorg. Nucl. Chem. 29, 375 (1967)
SOBCZAK, R.: private communication 1975
STADELMAIER, H.H.: In: Giessen, B.C. Developments in the structural chemistry of alloy phases. p. 141, New York-London: Plenum Press 1969
STADELMAIER, H.H., SCHÖBEL, J.: Die Boride $HfCo_3B_2$ und $ZrCo_3B_2$ als ternärer $CaCu_5$-Typ, Mh. Chem. 100, 224 (1969)
TELEGUS, V.S., KUZ'MA, Y.B.: X-ray investigation of a rhenium-manganese-boron system, Dopov. Akad. Nauk. Ukr. RSR, Ser. A. 10, 951 (1969a)
TELEGUS, V.S., KUZ'MA, Y.B.: X-ray structural study of the chromium-manganese-boron system, Visnik, Ser. Khim. 11, 21 (1969b)
TELEGUS, V.S., KUZ'MA, Y.B.: Phase equilibrium in vanadium-manganese-boron, molybdenum-manganese-boron, and tungsten-manganese-boron systems, Poroshkov. Met. 11(1), 67 (1971)
TELEGUS, V.S., KUZ'MA, Y.B., STEFANISHINA, C.K.: Chromium-rhenium-boron ternary systems, Poroshkov. Met. 9(2), 64 (1969)
TOTH, L.E., BENESOVSKY, F., NOWOTNY, H., RUDY, E.: Der Dreistoff: Thorium-Bor-Kohlenstoff, Mh. Chem. 92, 956 (1961a)
TOTH, L.E., NOWOTNY, H., BENESOVSKY, F., RUDY, E.: Der Dreistoff: Uran-Bor-Kohlenstoff, Mh. Chem. 92, 794 (1961)
TOTH, L.E., ZBASNIK, J.: Low-temperature heat capacities of superconducting molybdenum carbides, Acta. Met. 16, 1177 (1968)
URAZ, A.A., RUNDQVIST, S.: A Ternary W_5Si_3-type phase in the Ni-Si-B system, Acta Chem. Scand. 24, 1843 (1970)
VALOVKA, I.P., KUZ'MA, Y.B.: New ternary borides with structures $CeCo_3B_2$ and $CeCo_4B$ type, Dopov. Akad. Nauk. Ukr. RSR. Ser. A, 36(11), 1029 (1974a)
VALOVKA, Y.P., KUZ'MA, Y.B.: The temperature dependence of the resistivity of some ternary borides. Visnik, Ser. Khim 15, 10 (1974b)
VANDENBERG, J.M., MATTHIAS, B.T.: CORENZWIT, E., BARZ, H., Superconductivity of some binary and ternary transition-metal borides, Mat. Res. Bull. 10, 889 (1975)
VOROSHILOV, Y.V., KRYPYAKEVICH, P.I., KUZ'MA, Y.B.: Crystal structure of $ZrCo_2B_3$ and $HfCo_3B_2$, Kristallografiya 15, 934 (1970)
WILL, G.: Die Kristallstruktur von C_4AlB_{24}, Acta Cryst. B 25(7), 1219 (1969)
YANASE, A., TAWARA, Y., IGA, A.: Ferromagnetic material. USP 3424687 (1969)

XIII. Rare Earth-Boron Phase Equilibria

K. E. SPEAR

1. Introduction

Rare earth (RE) borides are fascinating high-temperature materials with extremely interesting chemical and structural properties. In spite of this, detailed studies of RE-B systems are rare. As a result, many fundamental as well as potentially useful properties of these materials have not been examined. One reason for this is that a solid coherent base of phase equilibria information has been lacking, but is needed for preparing and understanding the behavior of these materials during studies of chemical, physical, and/or mechanical properties.

Very recently, SPEAR (1976) compiled and analyzed the phase data available for binary RE-B systems and predicted unmeasured properties needed for the construction of temperature-composition phase diagrams. All the information was graphically depicted in terms of such diagrams for the seventeen RE-B systems (Sc, Y, La, and Ce through Lu). Prior to this, phase diagrams had been published for only seven of these systems: La-B (JOHNSON and DAANE, 1961), Ce-B (STECHER et al., 1965), Sm-B (SOLOVYEV and SPEAR, 1972), Eu-B (SAVITSKII et al., 1971), Gd-B (MANELIS et al., 1966), Y-B (LUNDIN, 1961), and Dy-B (SPEAR, 1974). Diagrams from work still in progress had also been reported by SPEAR and PETSINGER (1972) for the Gd-B system, and by IMPERATO and SPEAR (1972) for the Er-B system.

The brief review presented here on phase equilibria in binary rare earth boride systems summarizes the more detailed information given previously by SPEAR (1976). Earlier reviews by POST (1964), SMITH (1964), SAMSONOV (1965), and HOARD and HUGHES (1967) provided excellent summaries of rare earth boride properties, but little phase information was available.

The current phase equilibria knowledge is reviewed in four sections. Compound formation is discussed first, the limited information on homogeneity ranges second, vaporization behavior third, and binary temperature-composition phase diagrams fourth. Many features of the given diagrams are estimates (the entire Pm-B diagram is estimated), but such diagrams provide the most concise summaries of reported data. In addition, the diagrams provide researchers with reasonable estimates of phase equilibria properties which are yet to be measured.

2. Binary Phases Reported

Formulas ranging from RE_2B to REB_{100} have been reported for rare earth boride phases. The well-established binary compounds are designated in Table 1. Five structure types include most of the established rare earth borides. Very recent data and some reported compounds not included in Table 1 are briefly discussed below.

Table 1. Established rare earth boride phases[a]

Rare earth metal	Structure type					
	AlB_2 (hex)	ThB_4 (tetr)	CaB_6 (cub)	UB_{12} (fcc)	YB_{66} (fcc)	Other
Sc	X	-	-	-	-	ScB_{12}
Y	X	X	X	X	X	
La	-	X	X	-	-	
Ce	-	X	X	-	-	
Pr	-	X	X	-	-	PrB_x $(x < 4)$
Nd	-	X	X	-	X	Nd_2B_5
Pm[b]		(+)	(+)		(+)	(Pm_2B_5)
Sm	-	X	X	-	X	Sm_2B_5
Eu	-	-	X	-	-	
Gd	X	X	X	-	X	Gd_2B_5
Tb	X	X	X	X	X	
Dy	X	X	X	X	X	
Ho	X	X	X	X	X	
Er	X	X	-	X	X	
Tm	X	X	-	X	X	
Yb	X	X	X	X	X	
Lu	X	X	-	X	X	

[a]Symbols designate the following: X = established phases,
- = not observed, and
(+) = predicted to form.

[b]System not studied.

The most metal-rich rare-earth boride reported is La_2B by MARKOVSKII and VEKSHINA (1965), but its existence has not been confirmed. They best prepared this phase by sintering mixtures of metal and boron at 800°C. Earlier studies by JOHNSON and DAANE (1961) included searching for a compound between La and LaB_4 which might be stable only at high or low temperatures, but they found no evidence for such a compound.

The existence of a number of AlB_2-type rare earth boride phases is now well established. POST (1964) listed diborides of Gd, Tb, Dy, Ho, and Er from unpublished work of LA PLACA. Recent studies have confirmed the existence of these diborides: Gd (SPEAR and PETSINGER, 1972), Dy (SPEAR, 1974), Er (IMPERATO and SPEAR, 1972), Tb and Ho (SPEAR, unpublished research), and Yb (BAUER, 1974). However, GdB_2 is unstable below about 1280°C (SPEAR and PETSINGER, 1972). CASTELLANO (1972) reported the growth of single crystals of ErB_2 and TmB_2. Diborides of Sc (ZHURAVLEV and STEPANOVA, 1958; PRZYBYLSKA et al., 1963; PESHEV et al., 1970), Y (LUNDIN, 1961) and Lu (PRZYBYLSKA et al., 1963) have also been reported.

A new type of metal-rich boride with a formula Sm_2B_5 was recently discovered independently by SOLOVYEV and SPEAR (1972), and by SCHMIDT et al. (1973). The latter authors derived the formula from a single crystal structure analysis of this monoclinic phase. The isomorphous gadolinium compound was also prepared by SPEAR and PETSINGER (1972), but attempts to prepare the Dy (SPEAR, 1974), Tb, Ho (SPEAR, unpublished research) and Er (IMPERATO and SPEAR, 1972) analogs of this phase have failed. Large platelets of new metal-rich boride phases of Pr and Nd were produced by FISK et al. (1972) in attempts to grow tetraboride crystals. SCHMIDT (private communication, 1974) reports Nd_2B_5 is isomorphous with Sm_2B_5, but that PrB_x ($x < 4$) has a different low symmetry structure.

SAVITSKII et al. (1971) made extensive attempts to prepare EuB_4, but always observed the condensed metal and hexaboride phases. FISK et al. (1972) attempted to prepare EuB_4 by precipitation from molten aluminum, but were also unsuccessful. The ScB_4 phase has been reported (SAMSONOV, 1965), but probably does not exist as is evidenced by the inability of PRZYBYLSKA et al. (1963) and PESHEV et al. (1970) to prepare this phase.

The Sc, Er, Tm, and Lu hexaborides appear to be unstable with respect to the other condensed phases in their respective systems. MAR (1973) attempted to prepare ErB_6 by techniques previously applied to this system, and was equally unsuccessful. However, by slowly cooling a molten sample from 2300°C in a boron nitride crucible, he obtained significant amounts of ErB_6. Later he reported (NICHOLS et al., 1973) these ErB_6 crystallites were $(Er_{0.8}Ca_{0.2})B_6$. Although HoB_6 is also difficult to prepare, its existence as a stable phase in the Ho-B system does not appear to be in doubt.

The scandium dodecarboride phase is not listed in the same column as the other dodecarborides in Table 1 since it has a different structure (MATKOVICH et al., 1965; PESHEV et al., 1970).

The formula now used for the most boron-rich binary boride phase is REB_{66}, based on the crystallographic formula of YB_{66} determined by RICHARDS and KASPER (1969), and supported by NASLAIN et al. (1971) by a single crystal analysis on ThB_{66}. SPEAR and SOLOVYEV (1972) and SCHWETZ et al. (1972) simultaneously determined which rare earth metals form the REB_{66} phase. The rare earths Y, Sm, and Gd through Lu form REB_{66} phases, while La, Ce, and Pr do not. Attempts to prepare the Sc (CARLSSON and LUNDSTRÖM, 1970) and Eu (SCHWETZ et al., 1972) REB_{66} phases were also unsuccessful. The biggest uncertainty is with NdB_{66}. SPEAR and SOLOVYEV (1972) prepared this phase, while SCHWETZ et al. (1972) obtained a mixture of NdB_6 plus boron.

The existence or lack of existence of a particular phase type is correlated with the ionic radii of the rare earth metals in Table 2. This table was recently discussed by SPEAR (1976), and is similar to a table given by HOARD and HUGHES (1967) in which they correlated the formation of a particular phase type with metallic radii. The divalent radii of Eu and Yb are used along with the trivalent radii of the other metals. Ytterbium is also listed as a +3 ion since lattice constant correlations indicate trivalent ion character in all but the hexaboride phase (SPEAR, 1976). Physical property results indicate the valence states listed in the table are those exhibited by the RE elements in their borides. The newly-discovered Sm_2B_5 type phases are not included in the table since so little is known about them.

The trends are perfectly clear in Table 2, and are emphasized even more by the order of arrangement of the phase types. The dodecaborides favor smaller radii metals, and the hexaborides favor larger radii metals. The predictions indicated in the table for the formation of PmB_{66}, PmB_4, and PmB_6 require no further explanations.

Table 2. Correlation of rare earth ionic radii with the existence of the various phase types

Element	Ionic Charge	Ionic[a] Radius (Å)	Phase Types[b]				
			UB_{12}	AlB_2	YB_{66}	ThB_4	CaB_6
Eu	+2	1.15					X
La	+3	1.13				X	X
Ce	+3	1.08				X	X
Pr	+3	1.06				X	X
Nd	+3	1.05			X	X	X
Pm	+3	1.04			(+)	(+)	(+)
Sm	+3	1.03			X	X	X
Gd	+3	1.00		X	X	X	X
Yb[c]	+2	0.99	X	X	X	X	X
Tb	+3	0.98	X	X	X	X	X
Y	+3	0.97	X	X	X	X	X
Dy	+3	0.96	X	X	X	X	X
Ho	+3	0.95	X	X	X	X	X
Er	+3	0.93	X	X	X	X	X
Tm	+3	0.92	X	X	X	X	
Yb[c]	+3	0.91	X	X	X	X	
Lu	+3	0.90	X	X	X	X	
Sc	+3	0.81	X	X			

[a]The ionic radii were taken from DICKINSON (1970), except for La, Y and Dy. He lists values of La (1.12), Y (0.93), and Dy (0.97).
[b]X = established phases; (+) = predicted phases (see Table 1).
[c]Ytterbium is listed twice in the table, both as a +2 and a +3 ion.

3. Homogeneity Ranges

Composition widths of boride phases and the mechanisms by which the crystal structures can accommodate changes in boron-to-metal ratios have not been determined for most rare earth borides. In no case has a single phase composition width of a rare earth boride been determined as a function of temperature. Most of the available data are for the hexaboride phases, the most studied phase-type of all rare earth boride phases.

Data for hexaborides indicate that they are essentially stoichiometric at their metal-rich phase boundary, and become boron-rich by the formation of metal vacancies. Studies of ThB_6, SmB_6, and EuB_6 have been the most extensive. ETOURNEAU et al. (1971) prepared thorium boride samples with B:Th molar ratios ranging from 6 to 24, and measured the compositional dependence of the ThB_6 lattice constant, of both X-ray and neutron diffraction intensities, and of the density. The results were all consistent and indicated that composition changes produce essentially no change in the boron sublattice, but the occupancy of the metal sites decreases from about 100 percent to 78 percent with increasing boron content.

NIIHARA (1971) performed similar studies on SmB_6 by making measurements of lattice constant, X-ray diffraction intensities, density and resistivity as a function of composition. Again a metal vacancy defect model was indicated for boron-rich samples. Samarium hexaboride was found to be stable between compositions of SmB_6 and $Sm_{0.68}B_6$. YAJIMA and NIIHARA (1971) report studies which gave similar results for the hexaborides of Ce, Pr, Gd, and Tb.

SCHWETZ and LIPP (1973) studied the homogeneity range of EuB_6 samples prepared by vacuum-sintering, hot-pressing, and arc-melting. They determined composition limits of EuB_6 and $Eu_{0.90}B_6$, and obtained a lattice constant versus composition curve. Both wet-chemical analyses and microprobe analyses were used to measure compositions.

JOHNSON and DAANE (1961) found evidence for a homogeneity range in LaB_6 from density measurements as a function of composition. The results indicated a metal vacancy model for the defect structure of the boron-rich samples, which is in agreement with other hexaboride studies. However, they found no change in lattice constant with composition, which is in contrast to the results of other hexaborides.

Evidence for homogeneity ranges in hexaborides has also been obtained by others in terms of lattice constant changes with composition (CeB_6 - STECHER et al. (1965); SmB_6 - SOLOVYEV and SPEAR (1972); EuB_6 - FELTEN et al. (1958); GdB_6 - SPEAR and PETSINGER (1972); DyB_6 - SPEAR (1974)).

As the boron content of the hexaborides increases, the lattice constant of SmB_6 decreases, the LaB_6 value remains constant, and the other values increase. The mechanism explaining these changes with composition is still not clear, but SPEAR (1976) has given a possible explanation in terms of the electronic state of the metals in these compounds.

Very limited studies have been performed on the hexagonal rare earth diborides, and nothing has been reported with regard to homogeneity ranges or defect structures. However, transition metal diborides such as NbB_2 and TaB_2 exhibit composition ranges that extend on both sides of the stoichiometric composition, as POST (1964) discusses in some detail. Compounds with the new type of Sm_2B_5 boride structure have not been studied with regard to possible solid solution ranges.

EICK and MULFORD (1969) found a decrease of about 1% in the lattice constants of AmB_4 as the composition was changed from Am-rich to B-rich. FELTEN et al. (1958) report that LaB_4 appears to have a wide range of homogeneity, but provide no evidence for the apparent width. JOHNSON and DAANE (1961) report that the homogeneity range of LaB_4 is very narrow. In studies of DyB_4, SPEAR (1974) found no significant difference in lattice constants of Dy-rich and B-rich samples which had been annealed at 1800°C. Lattice parameter measurements by EICK and GILLES (1959) on several tetraborides gave no evidence of solid solution ranges.

The only variation from stoichiometry in REB_{12} phases reported is for YB_{12}. JOHNSON and DAANE (1963) found they could prepare this phase only with a nonstoichiometric composition of $Y_{0.92}B_{12}$. Their density measurements indicated the formation of a metal vacancy defect structure. The lattice constant of DyB_{12} was measured by SPEAR (1974) for both Dy-rich and B-rich compositions after annealing at 1800°C. The values were practically identical, and in agreement with previously reported values, which supports a narrow homogeneity range for DyB_{12}.

The only data available related to homogeneity ranges in REB_{66} phases are lattice constant values compiled by SPEAR (1976) for RE-rich, single-phase, and B-rich samples. The DyB_{66} phase has been studied most completely, and serves as an example. SPEAR (1974) measured respective Dy-rich and B-rich lattice constants of 23.466 and 23.419 Å for 1800°C annealed samples; single-phase values of 23.441 Å (SCHWETZ et al., 1972) and 23.422 Å (SPEAR and SOLOVYEV, 1972) lend support to the accuracy of the phase boundary values. These observations indicate a measurable homogeneity range exists in DyB_{66}. The single crystal studies on REB_{66} phases (RICHARDS and KASPER, 1969; NASLAIN et al., 1971) indicate a metal site occupancy in the structure of 50%. Changes in this percent occupancy provide the most reasonable explanation for a homogeneity range in REB_{66} phases.

4. Vaporization Behavior

The possible importance of vaporization effects in rare earth boride experiments is not readily apparent from the temperature-composition phase diagrams shown in the next section. The presence of a vapor phase on these diagrams signifies the sum of the partial pressures of metal and boron species is 1 atmosphere (atm), an extremely high pressure when considered in terms of mass losses by evaporization. Vaporization rates may still be appreciable at lower total pressures of metal and boron species, even if an inert gas is maintained at 1 atm.

The preferential loss of either metal or boron during the vaporization of RE-B phases can cause significant composition changes in condensed samples. The rates at which composition changes occur depend on the relative and absolute partial pressures of the gaseous species (and thus temperature and composition), and on the experimental arrangement. Mass spectrometric experiments on rare earth borides by SMITH (1964), STURGEON and EICK (1965) McGRATH (1972), and others have shown the only major gaseous species to be the gaseous elements.

Most rare earth boride systems contain a congruently vaporizing binary phase. The phases that are known or predicted to vaporize congruently are depicted in Figure 1. SMITH (1964) determined the vaporization

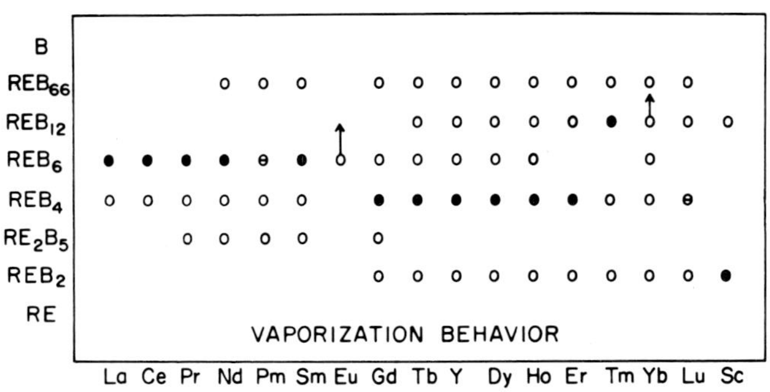

Fig. 1. Graphical representation of the vaporization behavior of RE-B systems. *Circles:* existing phases in each system. *Filled circles:* congruently vaporizing phase. *Line through a circle:* the phase is predicted to vaporize congruently. *Arrows:* signify composition changes observed in solid samples during the course of evaporation

behavior of many RE-B systems, and reviewed previously reported information. McGRATH (1972) and ETOURNEAU et al. (1972a, 1972b) have recently extended this work. The REB_6 phases of La, Ce, Pr, Nd, Sm, and probably Pm vaporize congruently, while the REB_4 phases of Gd, Tb, Y, Dy, Ho, Er, and probably Lu vaporize congruently. The only rare earth diboride to vaporize congruently is ScB_2 (PESHEV et al., 1970), and the only rare earth dodecaboride exhibiting congruent vaporization is TmB_{12} (McGRATH, 1972). In all cases except for TmB_{12}, these congruently vaporizing phases also melt congruently and are the highest melting phases in their respective binary systems. The TmB_{12} phase probably melts peritectically to give a higher melting TmB_4 phase plus a boron-rich liquid, but experimental phase data are not available for this system.

The two metals Eu and Yb have the lowest boiling points (highest vapor pressures) of the rare earth metals, and their borides tend to vaporize metal preferentially. McGRATH (1972) presents inconclusive evidence that indicates EuB_6 vaporizes congruently. SAVITSKII et al. (1971) found EuB_6 decomposes at high temperatures to give europium vapor and a boron-rich liquid. The preferential vaporization of metal from EuB_6 to leave a solid residue of β-rhombohedral boron was observed by ETOURNEAU et al. (1972a) at lower temperatures. SMITH (1964) and McGRATH (1972) both found ytterbium borides richer in metal than YbB_{66} preferentially lost metal by vaporization, but it is not known if the same is true for YbB_{66}, or if this latter phase vaporizes congruently.

5. Binary Temperature-Composition Phase Diagrams

Binary phase diagrams for the seventeen RE-B systems are graphically given in this section. The text and references accompanying the diagrams are limited. The recent extensive review by SPEAR (1976) provides most available details, and all but one of the diagrams given here are identical to those given in the earlier review.

The phase diagrams of systems that have many similarities are discussed together. However, before discussing specific systems, a few comments are given regarding melting data, solid solutions, and the general labeling of these diagrams.

The melting temperatures and melting behavior of rare earth boride phases have not been measured in most cases; more than half of the melting temperatures given in the diagrams were estimated. The maximum uncertainty in these values is estimated to be no larger than ± 200°, and in many cases the uncertainty is probably much less. The predictions were based on (1) published lower limits for melting temperatures, (2) apparent and surmised trends in published values, and (3) logical reasoning based on phase relationships (see SPEAR, 1976, for details and references).

In the present diagrams, estimated melting temperatures are placed in parentheses, and estimated phase boundaries are indicated by dashed lines. The rare earth metal melting points, boiling points, and solid-state transformations are values from HULTGREN et al. (1973). The presence of boron is assumed to have no effect on these transformations. The melting point of elemental boron has been recently given as 2042°C by MAR (1972), 2074°C by KIMPEL and MOSS (1968), and 2092°C by HOLCOMBE et al. (1973). The higher values are preferred.

Only qualitative indications of homogeneity ranges for the binary compounds are shown on the diagrams since so little information is available, as was discussed in Section 3. Most of the phases are shown as "line compounds." Information on the solid solubility of boron in the RE metals, or of the metals in boron is sparse, but available data indicate these solubilities are very small.

Each phase diagram given is for a constant pressure of 1 atm. Regions of the diagrams that involve a vapor phase, and other regions whose identity may be confusing are labeled with respect to the phases present. A vapor phase is designated by V, a liquid phase by L, and a solid phase by its chemical formula.

a) La-B and Ce-B Systems

The La-B phase diagram shown in Figure 2 is a revision of the one given by JOHNSON and DAANE (1961). Revisions include new melting point data for LaB_6 and a modification of the boron-rich region of the diagram.

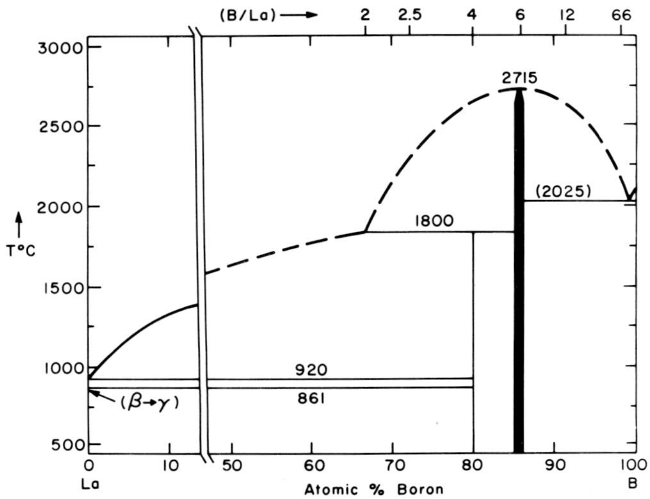

Fig. 2

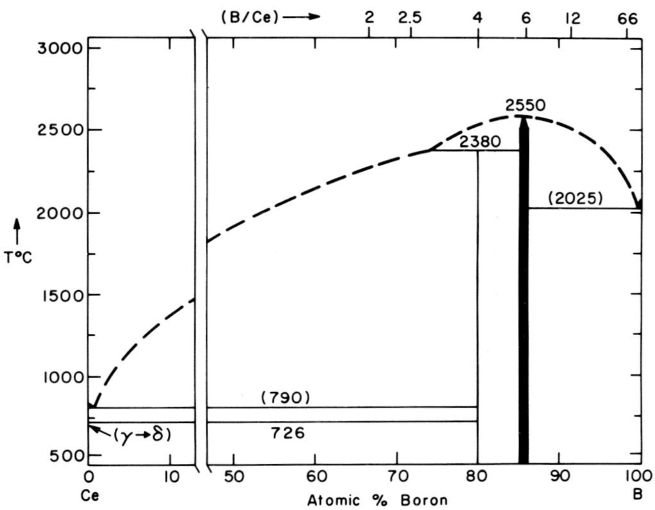

Fig. 3

A different type of modification of the boron-rich region of the diagram was given by SPEAR (1976). STECHER et al. (1965) published a partial diagram for the similar Ce-B system. They show a melting temperature of about 2200°C for a sample with 94 atomic % B, while the given diagram in Figure 3 has an estimated CeB_6-B eutectic temperature of 2025°C.

b) Pr-B, Nd-B, Pm-B, and Sm-B Systems

The phase diagrams for the Pr-B, Nd-B, Pm-B, and Sm-B systems are shown in Figures 4, 5, 6, and 7 respectively, and are believed to be very similar. However, the REB_{66} compound is found in all but the Pr-B system. The completely estimated diagram for the Pm-B system is assumed to be almost identical to that for the Nd-B and Sm-B systems.

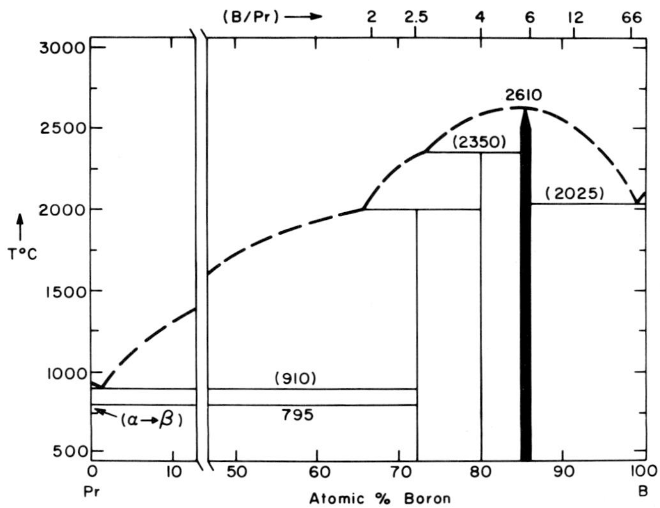

Fig. 4

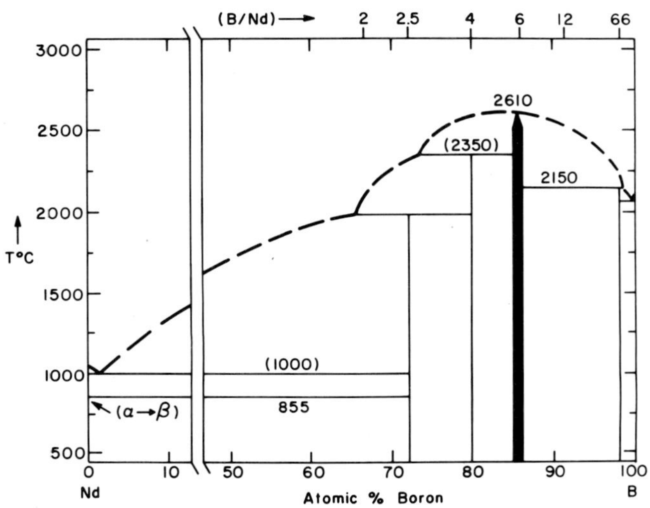

Fig. 5

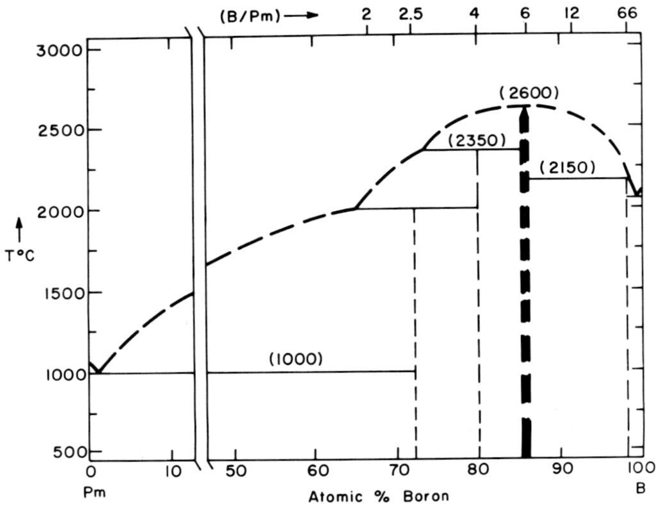

Fig. 6

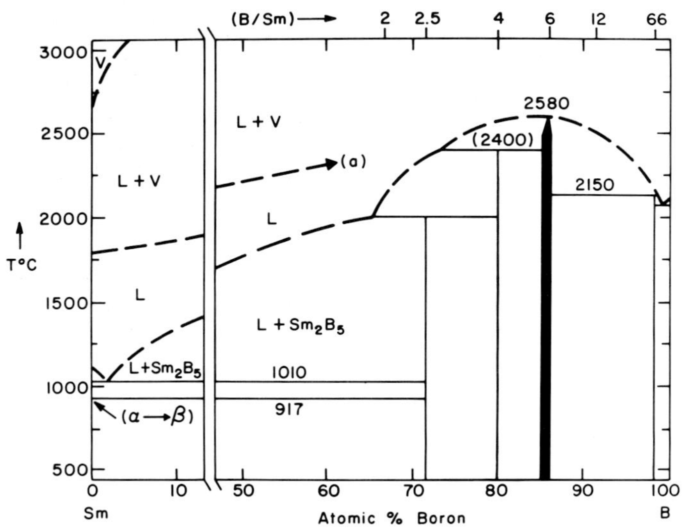

Fig. 7

The dashed boundary curve in the Sm-B diagram separating the metal-rich liquid from the two-phase liquid-vapor region was positioned by assuming thermodynamic ideality for Sm-metal in the liquid. However, this dashed boundary curve must increase it slope considerably as the boron content increases. Since SmB_6 vaporizes congruently, the Sm pressure over this phase must be smaller than that of boron over SmB_6, which in turn must be smaller than the pressure of pure boron. At the melting temperature of SmB_6, the vapor pressure of elemental boron is only about 1 torr according to HULTGREN et al. (1973).

c) Eu-B System

The Eu-B diagram shown in Figure 8 is the same general type of diagram as given by SAVITSKII et al. (1971). However, these authors give no information about the pressure of the vapor phase in their system,

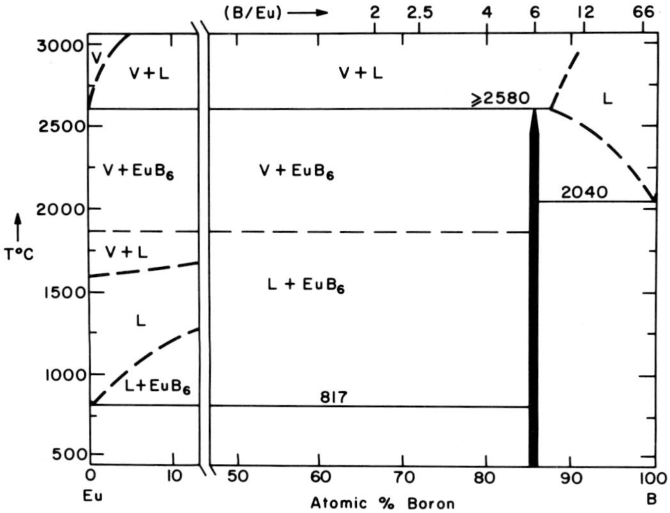

Fig. 8

while the present diagram is a 1 atm constant pressure diagram. The very metal-rich region of the diagram given by SAVITSKII et al. (1971) does not appear to have been estimated very well. No two-phase vapor-liquid region is evident, and the single-phase liquid region shown to exist between 826 and 1439°C contains almost no boron. According to HULTGREN et al. (1973), the boiling point of Eu-metal is 1597°C. The lower temperature limit for the two-phase region containing 1 atm vapor and EuB_6 must therefore be higher than 1597°C, and not the 1439°C value given by SAVITSKII et al. (1971).

d) Gd-B System

A slight modification of the tentative Gd-B phase diagram reported by SPEAR and PETSINGER (1972) is shown in Figure 9 and is based on unpublished data of BLANKS and SPEAR (1974). The changes pertain to

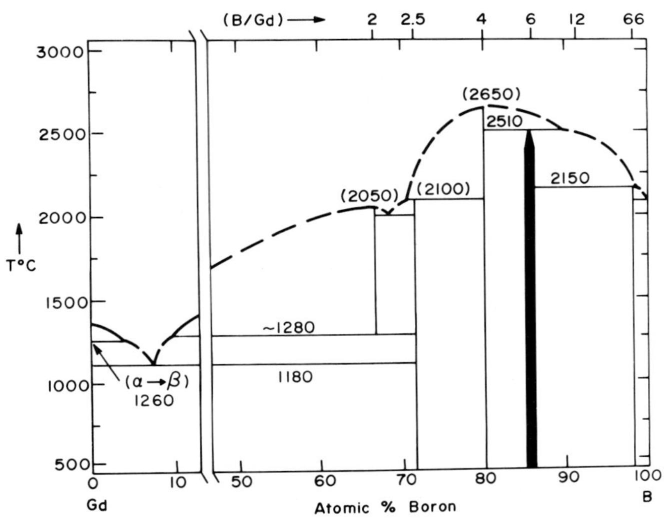

Fig. 9

equilibria involving Gd-metal. An earlier diagram reported by MANELIS et al. (1966) contains only the two phases GdB_4 and GdB_6 instead of the five intermediate phases that exist in this system. SPEAR and PETSINGER (1972) found that GdB_2 dissociates to Gd plus Gd_2B_5 at temperatures close to 1280°C and lower, but this decomposition temperature is not yet known precisely. BLANKS and SPEAR (1974) showed that Gd and Gd_2B_5 form a eutectic at 1180°C.

e) Tb-B and Y-B Systems

The Tb-B diagram shown in Figure 10 is largely estimated, and is assumed to be very similar to the diagram for the Y-B system shown in Figure 11. The Y-B diagram is that given by LUNDIN (1961), except for revisions reported by OLIVER and BROWER (1971) for the boron-rich region of the system. Not shown on the diagram in Figure 11 is the Y-Yb_2 eutectic composition of 25.5 atomic percent boron given by LUNDIN (1961).

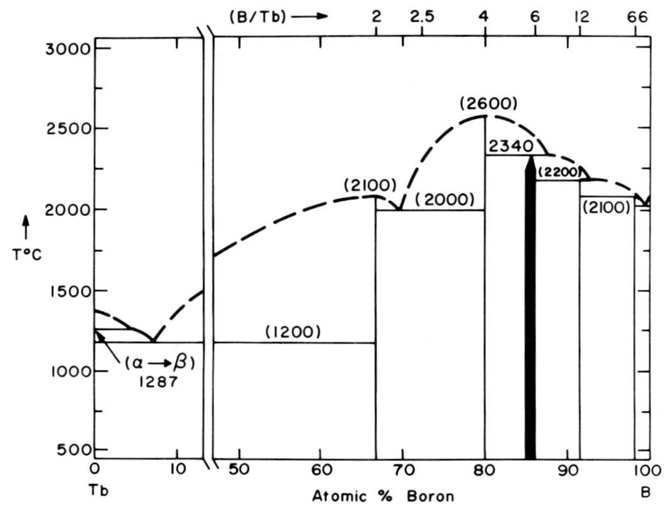

Fig. 10

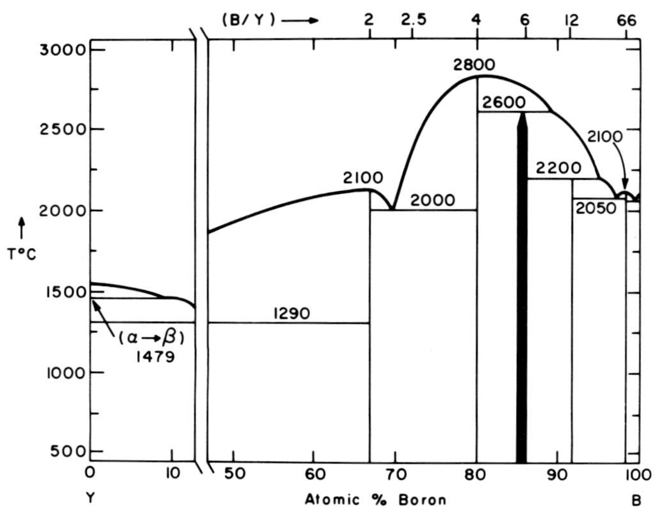

Fig. 11

f) Dy-B and Ho-B System

The Dy-B diagram shown in Figure 12 is that given by SPEAR (1974), except for inclusions of the Dy-metal transformation at 1384°C and the presence of a 1 atm vapor phase. The Ho-B system shown by the diagram in Figure 13 is expected to be very similar in behavior to the Dy-B system.

g) Er-B, Tm-B and Lu-B Systems

The phase information on the Er-B system by IMPERATO and SPEAR (1972; unpublished research, 1974) was used to produce the diagram shown in Figure 14. The Tm-B and Lu-B diagrams shown in Figures 15 and 16 respectively are similar to the Er-B diagram. As with the Sm-B system, the slope of the dashed boundary curve that separates the metal-rich

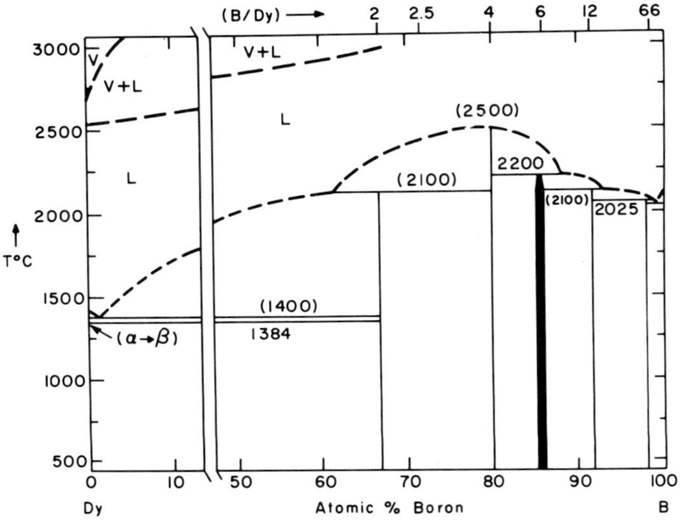

Fig. 12

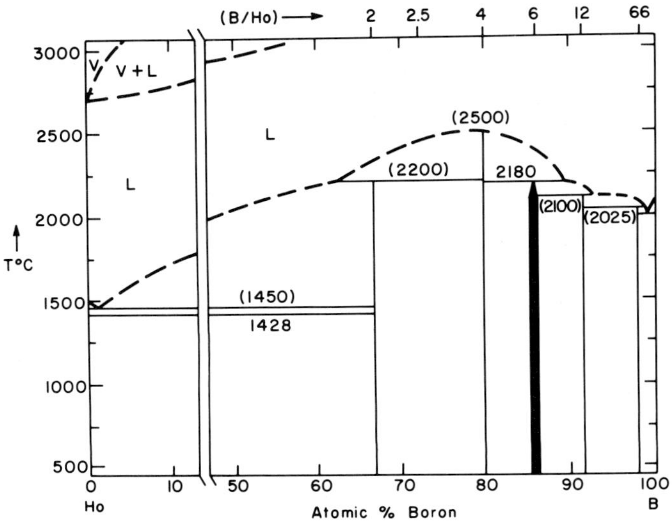

Fig. 13

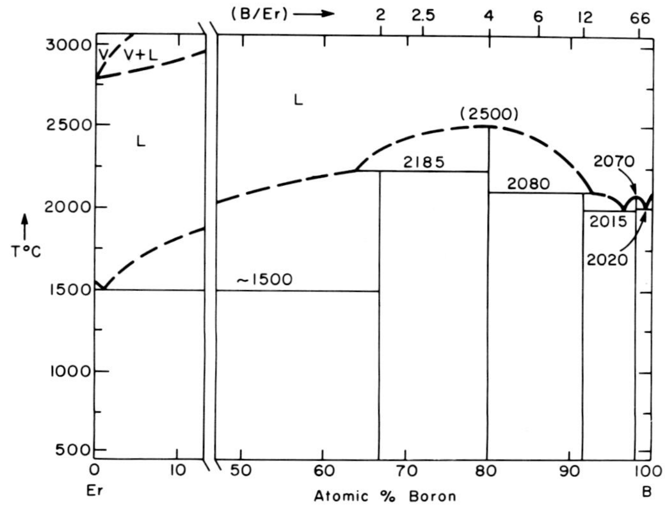

Fig. 14

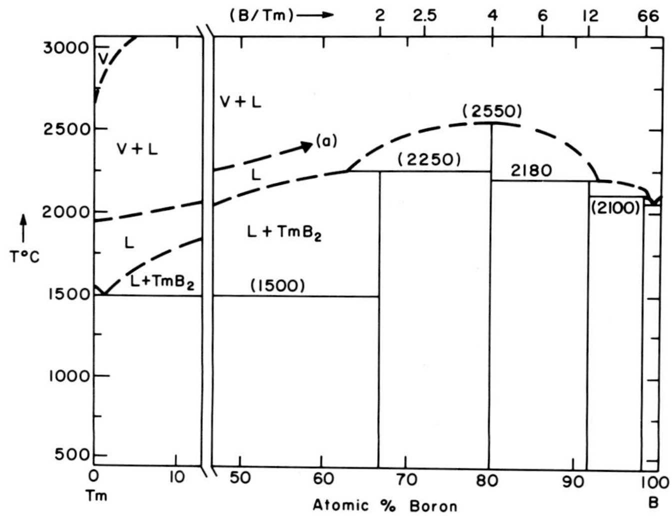

Fig. 15

liquid from the two-phase liquid-vapor region in the Tm-B system probably increases considerably with increasing boron content.

h) Yb-B System

The low boiling point (high vapor pressure) exhibited by Yb-metal causes the 1 atm vapor phase to play a dominate role in the Yb-B phase diagram shown in Figure 17. The given diagram is of the general type to be expected, but the temperatures at which the binary borides dissociate to a 1 atm vapor and a more boron-rich condensed phase are unknown.

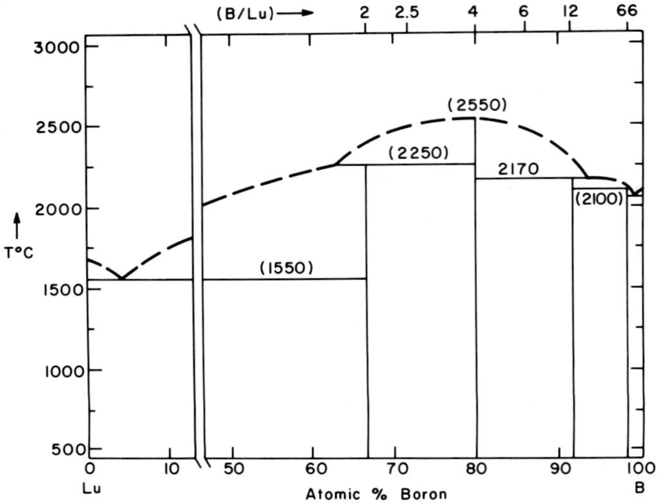

Fig. 16

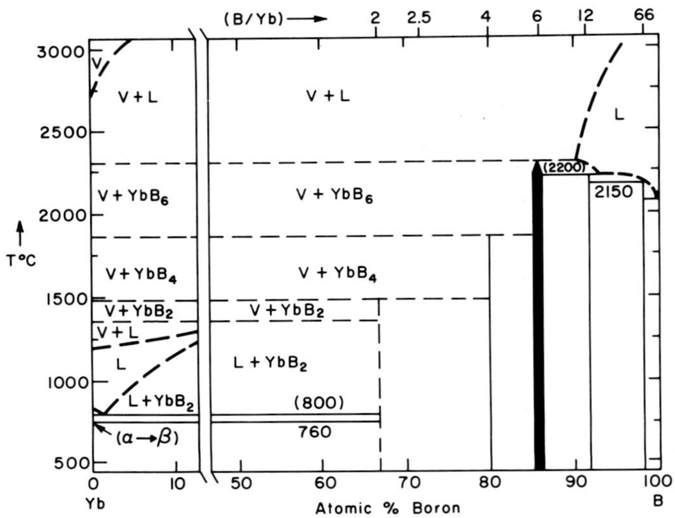

Fig. 17

i) Sc-B System

The Sc-B diagram is shown in Figure 18. It is unique in that the other RE-B systems that form the REB_2 and REB_{12} phases also exhibit a higher melting REB_4, a REB_{66}, and in some cases a REB_6 phase.

6. Concluding Remarks

The phase diagrams given in the preceding section should not be considered "established", but only as a means of summarizing our current knowledge of the equilibria in these systems. Isolated experimental observations on RE-B systems that would ordinarily be lost in laboratory notebooks can now be checked against the best current phase diagram, and perhaps used to improve the accuracy of the diagram.

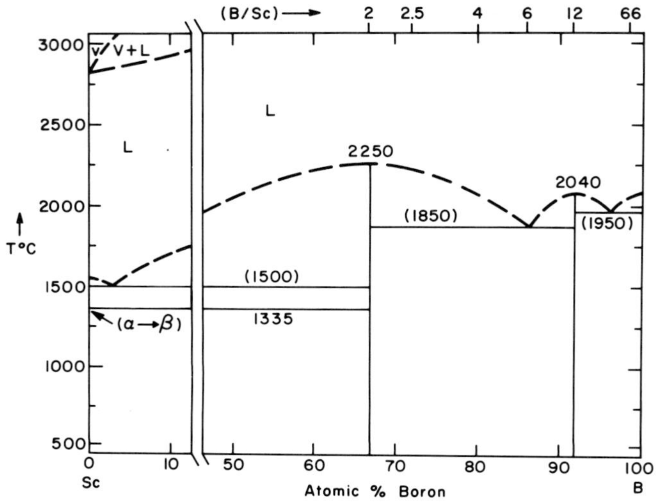

Fig. 18

In addition to the obvious need for experimental data on properties discussed in this review, other related types of chemical and thermodynamic information are also badly needed for RE-B systems. Experimental thermodynamic data are still sparse. Most phases are unstudied with regard to low temperature heat capacity, high temperature heat content, and the corresponding absolute entropy values as a function of temperature. Enthalpy or free energy of formation data are known for only a few of the phases, and the uncertainties in most of these values are large. The partial pressures of the vapor species over RE-B systems also need to be measured for most systems.

A significant number of ternary systems that include a rare earth metal, boron, and a third element have been partially studied. The most frequently reported systems include RE-B-C and RE-B-M, where M is a transition element. A systematic analysis of these phase data would provide very useful information on the relative stabilities of the binary RE-B phases.

Our understanding of the solid-state and high-temperature chemistry of rare earth boride systems could be greatly enhanced by measurements of the above-mentioned chemical and thermodynamic properties, and by cross-checking these properties for internal consistency.

Acknowledgments. The author gratefully acknowledges financial support from the National Science Foundation (Grant GH-34515). He also wishes to acknowledge the scientists who provided manuscripts of their unpublished research, and Academic Press, Inc., for permission to publish the phase diagrams given by SPEAR (1976).

References

BAUER, J.: Sur le diborure d'ytterbium. C.R. Acad. Sci. (Paris) Ser. C. 279, 501-504 (1974)

CARLSSON, J.-O., LUNDSTRÖM, T.: The solution hardening of β-rhombohedral boron. J. Less Common Metals 22, 317-320 (1970)

CASTELLANO, R.N.: Crystal growth of TmB_2 and ErB_2. Mat. Res. Bull. 7 (4), 261-265 (1972)

DICKINSON, S.K.: Ionic, covalent, and metallic radii of the chemical elements. AFCRL-70-0727 (December 1970)

EICK, H.A., GILLES, P.W.: Precise lattice parameters of selected rare earth tetra- and hexa-borides. J. Amer. Chem. Soc. 81, 5030-5032 (1959)

EICK, H.A., MULFORD, R.N.R.: Americium and neptunium borides. J. Inorg. Nucl. Chem. 31, 371-375 (1969)

ETOURNEAU, J., MERCURIO, J.P., NASLAIN, R.: Etude comparative de la stabilitê thermique des hexaborures de terres rares. C.R. Acad. Sci. (Paris) 275, 273-276 (1972a)

ETOURNEAU, J., MERCURIO, J.P., NASLAIN, R., HAGENMULLER, P.: Etude comparative de la stabilitê thermique des tetraborures de terres rares. C.R. Acad. Sci. (Paris) 274, 1688-1691 (1972b)

ETOURNEAU, J., NASLAIN, R., LA PLACA, S.: L'Hexaborure de thorium nonstoechiometrique $Th_{1-x}B_6$. J. Less Common Metals 24, 183-194 (1971)

FELTEN, E.J., BINDER, I., POST, B.: Europium hexaboride and lanthanum tetraboride. J. Amer. Chem. Soc. 80, 3479 (1958)

FISK, Z., COOPER, A.S., SCHMIDT, P.H., CASTELLANO, R.N.: Preparation and lattice parameters of the rare earth tetraborides. Mat. Res. Bull. 7 (4), 285-288 (1972)

HOARD, J.L., HUGHES, R.E.: Elementary boron and compounds of high boron content: Structure, properties, and polymorphism. In: The Chemistry of Boron and Its Compounds (E.L. MUETTERTIES, ed.), pp. 25-154 New York: Wiley and Sons, 1967

HOLCOMBE, C.E., Jr., SMITH, D.D., LORE, J.D., DUERKSEN, W.K., CARPENTER, D.A.: Physical-chemical properties of β-rhombohedral boron. High Temp. Sci. 5 (5), 349-357 (1973)

HULTGREN, R., DESAI, P.D., HAWKINS, D.T., GLEISER, M., KELLEY, K.K., WAGMAN, D.D.: Selected Values of the Thermodynamic Properties of the Elements. Metals Park, Ohio: American Society for Metals 1973

IMPERATO, E.G., SPEAR, K.E.: Investigations on the erbium-boron system. Presented at the 74th Ann. Meet. Amer. Ceram. Soc. (1972)

JOHNSON, R.W., DAANE, A.H.: The lanthanum-boron system. J. Phys. Chem. 65, 909-915 (1961)

JOHNSON, R.W., DAANE, A.H.: Electron requirements of bonds in metal borides. J. Chem. Phys. 38 (2), 425-432 (1963)

KIMPEL, R.F., MOSS, R.G.: Melting point of 98.9 to 99.6% pure boron. J. Chem. Eng. Data 13 (2), 231-234 (1968)

LUNDIN, C.E.: Rare earth metal phase diagrams. In: The Rare Earths (F.H. SPEDDING and A.H. DAANE, eds.), pp. 224-385. New York: Wiley and Sons 1961

MANELIS, R.M., MEERSON, G.A., ZHURAVLEV, N.N., TELYUKOVA, T.M., STEPANOVA, A.A., GRAMM, N.V.: Vacuum-thermal preparation of yttrium and gadolinium borides, and some of their properties. Sov. Powd. Met. 11 (47), 904-909 (1966)

MAR, R.W.: High-temperature thermal analysis of high boron alloys using automatic optical pyrometry. Thermochimica Acta 4, 367-376 (1972)

MAR, R.W.: The nature of erbium hexaboride: 1. Conditions for its formation. J. Amer. Ceram. Soc. 56 (5), 275-278 (1973)

MARKOVSKII, L.Y., VEKSHINA, N.V.: A new boride phase with highmetal content in the lanthanum-boride system. Russ. J. Appl. Chem. 38 (9), 1908-1911 (1965)

MATKOVICH, V.I., ECONOMY, J., GIESE, R.F., Jr., BARRETT, R.: The structure of metallic dodecaborides. Acta Cryst. 19, 1056-1057 (1965)

McGRATH, L.G.: Vaporization thermodynamics of the rare earth borides. New Mexico State Univ.; Las Cruces: Ph. D. Thesis 1972

NASLAIN, R., ETOURNEAU, J., KASPER, J.S.: Structure cristalline de la phase du système bore-thorium. Un nouveau borure à structure type YB_{66}. J. Solid State Chem. 3 (1), 101-111 (1971)

NICHOLS, M.C., MAR, R.W., JOHNSON, Q.: Questions concerning the existence of erbium hexaboride: The crystal structure of $(Er_{0.8}Ca_{0.2})B_6$. J. Less Common Metals 33, 317-320 (1973)

NIIHARA, K.: The preparation and nonstoichiometry of samarium hexaboride. Bull. Chem. Soc. Japan 44 (4), 963-967 (1971)

OLIVER, D.W., BROWER, G.G.: Growth of single crystal YB_{66} from the melt. J. Cryst. Growth 11, 185-190 (1971)

PESHEV, P., ETOURNEAU, J., NASLAIN, R.: Sur le système bore-scandium. Mat. Res. Bull. 5, 319-328 (1970)
POST, B.: Refractory binary borides. In: Boron, Metallo-Boron Compounds and Boranes (R.M. ADAMS, ed.), pp. 301-371. New York: Interscience Publishers 1964
PRZYBYLSKA, M., REDDOCH, A.H., RITTER, G.J.: The preparation and structure of lutetium diboride, scandium dodecaboride and lutetium antimonide. J. Amer. Chem. Soc. 85, 407-411 (1963)
RICHARDS, S.M., KASPER, J.S.: The crystal structure of YB_{66}. Acta Cryst. B25, 237-251 (1969)
SAMSONOV, G.V.: High-temperature Compounds of Rare Earth Metals with Nonmetals. New York: Consultants Bureau 1965
SAVITSKII, E.M., ARABEI, B.G., BAKARINOVA, V.I., SALIBEKOV, S.E., RAMASHOV, V.M., TIMOFEEVA, N.I.: Reactions in the system europium-boron. Inorg. Mater. 7 (4), 539-541 (1971)
SCHMIDT, P.H., COOPER, A.S., LA PLACA, S.J.: Personal communication (1973)
SCHWETZ, K., ETTMAYER, P., KIEFFER, R., LIPP, A.: Über die Hektoboridphasen der Lanthaniden und Aktiniden. J. Less Common Metals 26, 99-104 (1972)
SCHWETZ, K., LIPP, A.: Über die Existenzbreite der Europium-Hexaboridphase. J. Less Common Metals 33, 295-303 (1973)
SMITH, P.K.: Lanthanide boride systems and properties. University of Kansas: Ph. D. Thesis 1964. (Also USAEC Report No. COO-1140-103, 1964)
SOLOVYEV, G.I., SPEAR, K.E.: Phase behavior in the Sm-B system. J. Amer. Ceram. Soc. 55 (9), 475-479 (1972)
SPEAR, K.E.: Phase Equilibria and Related Properties of the Dysprosium-Boron System. Proc. Internat. Sympos. on Boron and Borides, Tbilisi, Georgia, USSR, Oct. 9-11 (1972). Acadamy of Sciences USSR, Moscow (1974), pp. 207-215 (in Russian)
SPEAR, K.E.: Phase behavior and related properties of rare-earth borides. In: Phase Diagrams: Materials Science and Technology (A.M. ALPER, ed.) New York: Academic Press 1976, pp. 91-159
SPEAR, K.E., PETSINGER, D.W.: Phase behavior in the gadolinium-boron system. Presented at the 74th Ann. Meet. Amer. Ceram. Soc. (1972)
SPEAR, K.E., SOLOVYEV, G.I.: High boron content rare-earth borides. In: Solid State Chemistry (R.S. ROTH and S.J. SCHNEIDER, eds.), pp. 597-604. Proc. 5th Mat. Res. Symp., NSF Spec. Publ. 364 (1972)
STECHER, P., BENESOVSKY, F., NOWOTNY, H.: Untersuchungen über das Legierungsverhalten von Boriden in den Dreistoffen Cer-Thorium(Uran)-Bor. Planseeber. Pulvermet. 13, 37-46 (1965)
STURGEON, G.D., EICK, H.A.: Samarium borides. The tetraboride-hexaboride conversion. J. Phys. Chem. 69 (11), 3705-3708 (1965)
YAJIMA, S., NIIHARA, K.: The nonstoichiometry of rare earth hexaborides. Preprint of paper 9th Rare Earth Res. Conf. Blacksburg, Virginia, 1971
ZHURAVLEV, N.N., STEPANOVA, A.A.: Determination of the structure of ScB_2 by X-ray diffraction. Sov. Phys.-Cryst. 3, 76-77 (1958)

XIV. Metallic Borides: Preparation of Solid Bodies - Sintering Methods and Properties of Solid Bodies

H. PASTOR

Powder metallurgy can proceed in two fundamentally different ways for the fabrication of products from powders of refractory borides. One way - conventional sintering - includes molding or compacting of products from the powder (compression in steel die, isostatic compression, injection molding, rolling, slip casting etc.) with subsequent sintering of the compacts. The other way - hot pressing - combines the pressing and sintering processes and is discussed in Chapter C.XIII. For both ways, the raw boride material must previously be milled to a sufficiently fine particle size.

Taking into account the three main steps of the conventional sintering process (milling, molding or compacting and sintering) we shall review successively the sintering of pure refractory borides and boride-metal systems. The properties (mainly density and mechanical properties) of the sintered products will be reported in relation to sintering parameters.

Sintering and hot pressing of refractory borides have been reviewed in some papers (SAMSONOV and PORTNOI, 1961; PASTOR, 1969).

1. Sintering of Pure Borides

a) Comminution of Powders

In order to obtain fully dense products, fine powders must be used, that is with a mean particle size ranging from 0.1 to 10 µm, preferably 0.1 to 5 µm. This fineness can be achieved by milling the raw boride.

PASTOR (unpublished) has investigated the comminution of TiB_2 powders in a classical ball mill with the following characteristics:

Diameter of vessel	10 cm
Volume of vessel	1150 cm^3
Speed of rotation	90 rpm
Balls/barrel volume ratio	0.50
Powder/barrel volume ratio	0.30
Diameter of the balls: steel	10-12 mm
WC-Co	10-15 mm

Cyclohexane was added to give complete submersion of the balls.

Starting with a powder of 6.6 µm average particle size, it was found that, within the milling time range of 6-24 h mean diameter d, BET specific surface and pollution (wt% pollutant content) obey the general law:

$$\log(\text{property}) = A \log(\text{time in h}) + B$$

with the following values for the constants:

	Steel balls			WC-Co balls		
	d (μm)	S (m^2/g)	Fe (wt%)	d (μm)	S (m^2/g)	W (wt%)
A	- 0.286	0.104	0.846	- 0.552	0.610	1.011
B	0.938	- 0.127	- 0.846	1.063	- 0.611	- 0.583

It can be seen that pollution is quite proportional to milling time (particularly for WC). The optimal milling time was determined at 20 h, permitting a decrease in the mean particle size from 6.6 to 2 μm, the iron contamination being of 2.2 wt%. Iron can easily be removed by chemical washing with 10N HCl aqueous solution (MEYER and PASTOR, 1965). Milling with WC-Co balls must be avoided owing to the high tungsten contamination (7 wt%); furthermore tungsten carbide cannot be removed by any known means.

Kinetics of comminution fo TiB_2 and ZrB_2 powders in a planetary centrifugal mill have been extensively studied by DOBROVOL'SKII et al. (1967) and MATSERA et al. (1973a, b). Industrial powders with particles less than 56 μm in size were comminuted in a planetary mill with the following technical specifications:

Number of milling vessels.	4
Volume of each milling vessel	113 cm^3
Speed of arm	415 rpm
Speed of ring	914 rpm
Acceleration of vessel centers	27 m/sec^2
Electric motor rating	1.7 kw

The milling elements were balls of ball-bearing steel (1% C; 1.5% Cr). For DOBROVOL'SKII et al. (1967) the optimal grinding time for ZrB_2 powder is 5 min, giving an average particle size of 1.6 μm with very low contamination (0.05 wt% Fe). The operational parameters were: ball/vessel volume ratio = 0.80, powder/vessel volume ratio = 0.25. MATSERA et al. (1973a, b) have determined the following operational parameters for TiB_2 and ZrB_2 powders:

Ball/vessel volume ratio	0.50
Powder/vessel volume ratio	0.30
Acetone/vessel volume ratio	0.10-0.20
Oleic acid (surfactant)/charge volume ratio	0.10-0.15
Ball diameter	10-11 cm
Grinding time	30 min

Under these optimal conditions the planetary mill yield powders of less than 1 μm of mean particle size (specific surface: 43-46 m^2/cm^3), iron contamination being 4.2 wt% for TiB_2 and 3.1 wt% for ZrB_2. Comminutability of materials generally increases with increasing Young's modulus, shear modulus, microhardness and brittleness, just as it does in vibratory milling.

The kinetics of comminution of ZrB_2 powder in a vibratory mill have been investigated by D'YAKONOVA and KISLYI (1972). The vibratory mill operated at a frequency of 1500 vibrations/min and an amplitude of 3.5 mm, with a ball to powder weight ratio of 1:1. Effective comminution occurs during the first 2-3 h. As a result of pickup of mill material, contamination (3 % by weight for 2 h milling time) increases rapidly with increasing milling time and is always greater than in a ball mill.

According to SMITH (1970), starting with a TiB_2 powder of 3-4 μm mean particle size, one pass through a vibratory mill (frequency: 2800 cycles/min; amplitude: 2.5 mm) produces an average size reduction of 10%; two passes bring a further 5-8% reduction. Milling time and contamination are not given.

Ultrasonic comminution of refractory compound powders (including TiB_2) has been investigated by MATSERA et al. (1971). The technical specifications of the ultrasonic bath were as follow:

Floor surface area	64 cm^2
Effective volume	700 cm^2
Vibration frequency	19.4-20.0 kHz

Powders have initially less than 40 μm particle size and are comminuted in alcohol with an excess nitrogen pressure of 4 atm. For TiB_2, the mean particle diameter is found to be 2.5, 1.85, 1.82, 1.58, 1.36, 1.30 μm respectively for 0, 2, 5, 10, 15 and 20 min milling time. The intensity of the process reaches a maximum during the first 2-5 min and consequently there is virtually not enough time for contamination.

b) Compacting or Molding of Powders

α) Pressing

Conventional Pressing. Pressing is usually carried out within a hard metal die between two steel or hard metal punches. In order to confer enough mechanical strength to the blank to permit further manipulations without risk, removable organic binders are mixed in the powder. The following binders are commonly used, dissolved in a convenient volatile solvent: camphor, natural or synthetic waxes, latex, methyl polymethacrylate, polyvinyl alcohol, carboxymethylcellulose, ammonium alginate. Some of them also act as a lubricant, so minimizing the wear of the die.

The compressibility of boride powders can be investigated by conventional measurement of the dimensions and weights of the blanks and (or) the measurement of the stroke of the punches with a cathetometer, or by the electrical conductivity method (owing to the metal-type conductivity of borides).

According to BABICH et al. (1960) (see also SAMSONOV et al., 1960), the process of densification of TiB_2, CrB_2 and $(Ti, Cr)B_2$ powders is well described by the following equations:

for TiB_2 : $\log P = -11.07 \log \beta + 3.02$

for CrB_2 : $\log P = -10.48 \log \beta + 3.25$

for $(0.8\ Ti, 0.2\ Cr)B_2$: $\log P = -11.29 \log \beta + 3.24$

where P = compacting pressure, in the range 0.5-7 $tons/cm^2$

β = relative volume = material to blank density ratio.

Spring-back of these materials is very important and its relationship to pressure is associated with their brittleness and lack of ductility. Stratification phenomenon occurs in the pressure range above 3 or 4 $tons/cm^2$. Holding under pressure for a short while has practically no effect on the density of the blanks. The technological ideal pressure may be taken as 3 $tons/cm^2$. Two-stage compacting is also advisable at this pressure, the compact being rubbed through a sieve (105 μm) after the first compacting.

SAMSONOV and NESHPOR (1955) have used the electrical conductivity method to investigate the compressibility of refractory compound powders (including TiB_2 and ZrB_2) in the pressure range 190-1700 kg/cm^2. The variation of the specific electrical conductivity χ with the relative density θ (apparent density of the blank/density of the powder material) can be satisfactorily expressed by the equation:

$$\chi = K.\ e^{-\frac{A}{\theta}}$$

where K and A are constants. This equation is valid for $0.35 < \theta < 0.80$. The constant A, called the brittleness factor, characterizes the tendency to brittle disintegration of the material and is found to be 2.66 for TiB_2 and 3.27 for ZrB_2.

STRUK (1966) uses the same method to investigate the compressibility of TiB_2, ZrB_2, CrB_2 and W_2B_5 powders in the pressure range 5-75 kN/cm^2 (~ 500-7,500 kg/cm^2). Compressibility is well described by two equations of the same type as above, each having its coefficients A_1 and K_1 or A_2 and K_2. On a semilogarithmic graph (log χ versus $\frac{1}{\theta}$) the point of intersection of the two representative straight lines indicates a change in the rate of increase of electrical conductivity which slows down when pressure increases (consequently when $\frac{1}{\theta}$ decreases). It is assumed that, as pressure increases, densification is achieved at first mainly by improved particle stacking and brittle fracture of interlocking surface features, leading to a partial spheroidization of the particle shape. Above a certain pressure, particle rupture with formation of a large number of fine grains takes place, thus decreasing the rate of growth of electrical conductivity by increasing the contact electrical resistance. By analogy with the preceding findings, the constant A_1 characterizes the brittle fracture of interlocking surface features and the constant A_2 the brittle rupture of the material itself. The values of constants A and K are reported in the following Table.

Refractory boride	TiB_2	ZrB_2	CrB_2	W_2B_5
Pressure range (kN/cm^2)	5-40	5-35	5-50	5-40
K_1	4990	6080	4990	8350
A_1	5.21	4.99	3.33	3.22
Pressure range (kN/cm^2)	40-70	35-75	50-70	40-70
K_2	2220	2180	2240	2360
A_2	2.38	2.30	1.30	2.00

The borides may be arranged in the following order of increasing brittleness (constant A_2): CrB_2, W_2B_5, ZrB_2 and TiB_2. It must be pointed out that the addition of plasticizer (a solution of synthetic rubber in gasoline) has practically no effect on densification.

Isostatic Pressing. BUMM and LIEPELT (1972) have experimented with isostatic pressing on several metal, oxide, carbide, nitride and boride powders (including CrB, CrB_2 and TiB_2) up to 15,000 atm. The particle size of the boride powders was less than 8 μm. The highest increase rate of relative density is observed in the pressure range below 6000 atm. At 15,000 atm, the highest relative density (75% of theoretical density) is observed for CrB and the lowest (68%) for TiB_2.

β) Extrusion - Die Pressing

The basic principles of the extrusion-die production of pipes and rods from powders of refractory compounds have been developed by KISLYI and SAMSONOV (1962). Although the process is described in detail only for molybdenum disilicide, it can be transposed to borides. The flowsheet of the production of parts is as follows:

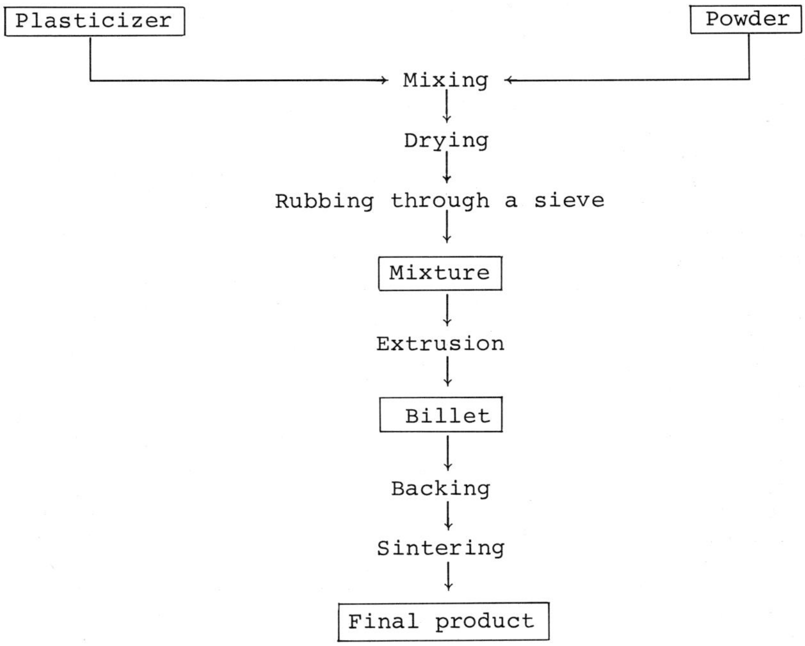

The following plasticizers were tested: starch, polyvinyl alcohol, rubber, paraffin, bakelite, clay and water glass. The last two must be avoided for borides with which they react. Cracking of starch and bakelite left too much free carbon in the sintered product.

γ) Hot Casting Under Pressure

Hot casting under pressure is widely used for the manufacture of complex-shaped parts in the ceramic industry. This method has been applied by MEDVEDEV et al. (1971) for the shaping of parts of lanthanum hexaboride. A slip was prepared by mixing, in a variable-speed propeller-type mixer, the LaB_6 powder with a molten binder (at 80-90°C) composed of 85 wt% paraffin and 15 wt% beeswax. The optimum binder content for a low viscosity is 10.5 wt%. Casting was carried out under a pressure of 5-8 atm. The castings reproduced accurately the mold shape and had high strength and good surface finish. The binder was removed in two steps. The first step was carried out by slow heating to 120, 140, 160 and finally 180°C, the parts being embedded in calcined alumina powder; with one h heating plateau at each temperature, 70-80% of the binder was removed. Complete removal of the 30-20% remaining binder occurred during sintering.

δ) Slip Casting

Slip casting is a common practice in the ceramic industry. The material to be cast is milled to an average particle size of a few microns. A slip is made by mixing the finely divided material with a liquid suspending medium.

The slip is then poured into a suitable mold (e.g. a plaster of Paris mold). The liquid in the slip is drawn into the mold by capillary forces and the solids are deposited in coherent form.

According to a patent by REDDY et al. (1966), fluid slips stable for 24 h are obtained with powdered nitrides, carbides and borides of tungsten, molybdenum, chromium, titanium zirconium, aluminium, and also zircon, zirconia, molybdenum disilicide, in xylene, kerosine, toluene, benzene, naphta, acetone or alcohol, if 5-15% by weight polymeric cyclopentadiene is mixed with the liquid. The slip, containing 11-63 vol% of solids of 1.3 μm average particle size can be dried conventionally in absorbent plaster molds to give strong machinable castings, with high density and minimum shrinkage. The slip characteristics for borides TiB_2, ZrB_2 and CrB_2 in a xylene suspending media are as follows:

Boride	wt% cyclopentadiene in xylene	Solids		Sintered density[a] (% of theoretical)
		Content (vol.%)	Average particle size (μm)	
TiB_2	5	44.5	2.5	92-94
ZrB_2	7	50.9	-	89
CrB_2	7	53.8	2.2	-

[a]Sintering temperature: 1700-2200°C
Holding time at sintering temperature: 1-3 h

Slip casting of zirconium diboride has been studied by DOBROVOL'SKII et al. (1966), testing aqueous solutions of sodium alginate, polyvinyl alcohol, starch, sodium polymethacrylate and carboxymethylcellulose. A 3 wt% aqueous solution of carboxymethylcellulose proved to be the best dispersing media. The best casting properties of the slip were obtained at a solid to liquid ratio of 78 : 22 and are not improved by any change in the pH. The slip manifested a tendency to thixotropic thickening, the viscosity after 3 h at rest being almost twice as great as the initial value. Sintering (2200°C for 40 min in hydrogen stream) takes place with a large but uniform shrinkage and provides parts with a porosity of about 10%.

AUGUSTINIK and DROZDETSKAYA (1967) recommend a 1.3 wt% aqueous solution of sodium carboxymethylcellulose as the best suspending medium for slip casting of zirconium diboride (mean particle size: 0.19 μm).

c) Sintering

Sintering, activated sintering and reaction-sintering have been tested for borides in order not only to obtain maximum density of sintered bodies, but also to prepare special porous parts with controlled porosity.

The green compacts are very often presintered in a vacuum or in a stream of neutral gas (Ar, H_2) at a temperature in the range 800-1400°C. Presintering is intended to ensure the complete removal of the organic binder in order to avoid later pollution of the sintering furnace by pyrolysis of the by-products. It also facilitates machining or finishing of the parts, which would be very difficult and expensive after final sintering because of the hardness of borides.

Sintering is usually carried out in a vacuum in a resistance furnace fitted up with a graphite, tantalum or tungsten resistor, or in a high-frequency furnace with a graphite susceptor. In some cases, sintering is also carried out in a stream of neutral gas (Ar, H_2). Graphite plates are commonly used as supports, but it must be remembered that some borides, by contact with graphite, form eutectics melting in the temperature range 2200-2500°C. In order to limit interaction between the graphite plate and the parts being sintered, it is advisable to sprinkle the plate with the boride powder. The best, but not the cheapest way is evidently to use sintered supports of the same material as that being sintered, although some sticking is to be expected.

Owing to the high sintering temperatures employed, and grain growth, high losses of material (by volatilization of boron or boride) are often observed. In order to limit these losses, the part to be sintered can be embedded in powder of the same boride. This expedient has been used by MEERSON et al. (1966) for the sintering of YB_6 and LaB_6 parts.

As we shall see below, sintering of pure refractory borides requires high sintering temperatures (higher than 0.7 times the absolute melting temperature) and rarely leads to fully dense parts. These facts motivate the study of activated sintering. The methods of activation can be classified in two groups:

1. Activation by a small addition (a few percent) of selected metals, mainly transition metals: e.g. fine powders of Fe, Ni, Co, Cr, Pt, Pd, etc. or sometimes chlorides of these metals when they are easily reducible to metal by hydrogen during presintering. The mechanism of activation by metals is not yet clearly understood (surface tension, surface diffusion, etc.) and seems to be related to the electronic structure of the metal addition.

2. Activation by powders in a special state of activity, for instance very fine (submicronic) powders.

Reaction-sintering may be related to this last group of activated sintering. It consists in preparing and sintering the boride in a single operation by compacting reactants able to generate the pure boride (and gaseous or volatile by-products) at the sintering temperature. In the case of direct synthesis (metal + boron) fully dense pure boride parts can be anticipated. But in the case of reactants (e.g. metal oxide + boron carbide) generating gases, this gas formation is often a barrier to densification and leads to porous parts.

Sometimes fully dense bodies are not required. For instance, filters are prepared by sintering porous materials. Porous sintering can be achieved by: use of coarse grain or granulated powders, forming without pressure (e.g. loosely molded powder), use of a pore generating agent (generally a removable organic material in powder form).

α) Sintering of Borides of Rare Earth Metals

Table 1 summarizes the sintering characteristics of borides of rare earth metals.

Table 1. Sintering characteristics of borides of rare earth metals

Boride	Binder nature and content (wt%)	Compacting pressure (kbar)	Sintering temperature (°C)	Holding time (min)	Sintering atmosphere or vacuum V (Pressure in torr)	Relative density (%)	Remarks	References
YB_6	Aq. soln. 10% PVA[a]	4-5	2327	30	V : $1-3\times10^{-5}$	71	Weight loss: 10.5%	MEERSON et al. (1966)
LaB_6	-	-	1375	15	-	-	-	LAFFERTY (1951)
LaB_6	Aq. soln. 10% PVA	4-5	2407	30	V : $1-3\times10^{-5}$	82	Weight loss: 15%	MEERSON et al. (1966)
LaB_6	Aq. soln. 2% PVA	6-7	2300-2350	60	V : $5\text{-}10\times10^{-3}$	99.2	LaB_6 powder packing	SHLYUKO et al. (1967)
LaB_6	Glycerin + alcohol	8	2000	120	V : 10^{-5}	81	Weight loss: 9%	MEERSON and MAMEDOV (1967)
LaB_6	Glycerin + alcohol	8	2000	120	Ar : 798-836	89.5	5%	KUDINTSEVA et al. (1968)
LaB_6	Glycerin + alcohol	8	2000	120	H2 : 798-836	100	6%	
LaB_6	85 paraffin + 25 beeswax	Hot casting: P = 5-8 atm.	2300-2350	60	V : $5-10\times10^{-3}$	95-97	LaB_6 powder packing	MEDVEDEV et al. (1971)
LaB_6	Aq. soln. 3% PVA	1.5-2.5	2350-2400	60-90	V : -	100	weight loss and grain growth	KUZNETSOVA et al. (1972)
EuB_6	none	3.35	2227-2727	960	V : -	63	weight loss: 40%	HOYT and CHORNE (1960)

[a]PVA: polyvinyl alcohol

The recent developments in electronic technology have created a demand for shaped parts of rare earth metal hexaborides and particularly of lanthanum hexaboride.

MEERSON et al. (1966) have investigated the sintering of two batches of lanthanum hexaboride, one referred to as LaB_6 (N°1) (with 44.5% of grains less than 2.8 µm in diameter, and 42.8% larger than 6.4 µm), produced by borothermal reduction of lanthanum oxide La_2O_3; the other referred to as LaB_6 (N°2) (with 64.6% of grains less than 2.8 µm in diameter, and 23.4% larger than 6.4 µm), obtained by borocarbothermal reduction of La_2O_3 (boron carbide as reductant) which contains 0.10% carbon. Sintering is carried out in a vacuum, and it is found that by increasing the sintering temperature, the relative density of LaB_6 samples increases and reaches a maximum value of 0.82 and 0.77 for LaB_6 (N°1) and LaB_6 (N°2) respectively at 2407°C. Consequently, the maximal bending strength (850-900 kg/cm^2) is also obtained at 2407°C. Intensive shrinkage (10 to 12% linear shrinkage) occurs within the temperature range 2207-2407°C. Weight loss by evaporation of boride can be decreased by embedding the parts in LaB_6 powder, as shown below:

Type of Sintering	wt% loss of LaB_6 at temperature (°C)				
	1997	2197	2297	2407	2447
Without embedding	4.1	5.7	8.5	13.4	15.1
With embedding	2.0	2.5	3.0	4.2	4.7

Intensive grain growth (from 70 to 250-300 µm) is observed within the temperature interval 2397-2427°C, but it should be mentioned that these temperatures comprise ~ 96% of the absolute melting point. Thus sintering of LaB_6 at temperatures close to its melting point favors collective recrystallization and consequently leads to the highest relative density, strength and linear shrinkage of the samples. At temperatures above 2427°C these characteristics drop back because of intensive evaporation. It is established that borocarbothermal LaB_6 sintered under identical conditions to analogous samples of borothermal LaB_6 exhibits somewhat less shrinkage, lower density and strength.

In further work, MEERSON et al. (1968) add a third batch of LaB_6 powder (LaB_6-N°1, fine) obtained by planetary centrifugal milling of LaB_6 (N°1) and characterized by most particles of 1 µm in diameter. A compressibility study shows that LaB_6 (N°1, fine) powder has the best pressability. The highest green density (70-75%) for the three powders investigated occurs for 8 tons/cm^2 pressure. Sintering has been studied in vacuum (10^{-5}torr) or in argon or hydrogen (1.05-1.10 atm) in a furnace fitted with a tungsten heating element, and led to the following results for LaB_6 (N°1) batch (green porosity: 30%):

Sintering conditions (2000°C, 2h)	Residual porosity after sintering (%)			Wt loss of specimens after sintering (%)		
	In vacuum	In Ar	In H_2	In vacuum	In Ar	In H_2
Without embedding	24	14	0	20	9	-
Embedding in pure LaB_6	22	12.5	0	15	7	8
Embedding in technical LaB_6	19	10.5	0	9	5	6

A density close to the theoretical value is achieved by sintering in argon at 2000-2450°C or in hydrogen at 1800-2200°C. Hydrogen has thus an activating effect on sintering, which is explained by the formation of gaseous boranes or $B_x O_y H_z$ species thus setting up a certain concentration of free vaporized lanthanum and creating the requisite conditions for gaseous transfer, stimulating in turn shrinkage and grain growth. In fact, the greatest grain growth took place during activated sintering in a hydrogen atmosphere: for instance, after 2 h sintering at 2200°C the 1 μm powder formed grains 20-25 μm in diameter in vacuum, about 60 μm in argon, and about 150 μm in hydrogen. It is demonstrated that the use of industrial LaB_6 hampers densification during sintering as a result of the evaporation of impurities.

The thermoemissive properties of the previous sintered LaB_6 specimens have been investigated by KUDINTSEVA et al. (1968) as a function of composition, impurity content and porosity. It was found that excess boron in LaB_6 increases the work funtion, while a decrease in the content of other rare earth elements reduces the work function. Addition of 10 wt% tungsten also lowers the work function by approximately 0.06 eV but seems of no effect on densification. An increase in specimen porosity from 0 to 20% reduces the work function by 0.2 eV. The work function of specially pure, stoichiometric and nonporous LaB_6 is found to be 2.76 eV.

According to SHLYUKO and CHERNYAK (1967) the best packing material composition in order to limit vaporization of LaB_6 compacts is 30% LaB_6 granules and 70% LaB_6 powder. This composition secures adequate transport of LaB_6 through the gaseous phase and prevents the packing material from sticking to specimens during sintering.

KUZNETSOVA et al. (1972) were able to produce large-sized parts of LaB_6 (up to 100 mm in diameter and 3 mm thick) by sintering at 2350-2400°C, but were also confronted with the evaporation phenomena, which they tried to avoid by lowering the sintering temperature with the use of activators. The effect of additions on the sintering behavior of LaB_6 at 2000°C is illustrated as follows:

Composition	Shrinkage	Composition	Shrinkage
LaB_6	1	LaB_6	1
LaB_6 + 1 % Ir	6	LaB_6 + 1 % Ni	7
LaB_6 + 2 % Ir	7	LaB_6 + 2 % Ni	7
LaB_6 + 3 % Ir	7.5	LaB_{66}+ 3 % Ni	9
LaB_6 + 4 % Ir	8	LaB_6 + 4 % Ni	9
LaB_6 + 5 % Ir	10	LaB_6 + 5 % Ni	10

The optimum amount of nickel is 1.5-2.0 wt% and leads to specimens with 85-87% relative density by sintering at temperatures as low as 2000°C. Activation is explained by the formation of liquid nickel borides and free lanthanum. However, the presence of nickel boride in the specimen will increase the rate of evaporation under operating conditions.

Vacuum sintering of yttrium hexaboride has been studied by MEERSON et al. (1966). The maximum relative density (71% of theoretical) is obtained at 2327°C. The bending strength increases monotonically up to 380-400 kg/cm^2 at 2377°C.

β) Sintering of Borides of Actinide Metals

MATTERSON and JONES (1961) have studied the preparation and properties of UB_2, UB_4 and ThB_4 with a view to their possible interest as nuclear fuels. The powders (passing through a 100-mesh sieve) were prepared by milling the resulting products of direct synthesis (metal + boron, in argon) and were contaminated with oxide (1 to 5%) and higher borides (1 to 5%). Hydrostatically pressed samples were induction-heated in vacuum, in a graphite crucible. At 2000°C and a pressure less than 10^{-5} torr all the borides lost boron, producing the lower boride and/or metal. Densification could only be achieved by hot pressing.

γ) Sintering of Borides of Group IV B Metals: Ti, Zr, Hf

Table 2 summarizes the sintering characteristics of borides of group IV B metals. It is not easy to compare the various results given in the literature, because whereas the sintering parameters (compacting pressure, sintering temperature, holding time at sintering temperature, sintering atmosphere) are carefully mentioned, the authors generally do not take the precaution of mentioning grain size (mean grain size or particle size distribution) and impurity content of their initial powders, both of which obviously influence sintering. However it can be seen that parts of pure borides with less than 10% porosity can be obtained from pure powders with mean particle size of a few microns by sintering compacts (pressed under 2 and 4 kbar) at 2200-2300°C for less than one hour. As will be seen later, fully dense parts can only be obtained by activating the sintering process by metal additions.

Sintering Characteristics of Pure TiB_2 and ZrB_2

Titanium Diboride. According to SAMSONOV and KOVAL'CHENKO (1961) the sintering process of titanium diboride is divided into three stages: (1) a slight densification at temperatures up to 1900°C; (2) practically no shrinkage between 1900 and 2100°C; (3) a further densification step, by plastic flow, above 2100°C, where intensive grain growth occurs. Increasing holding time at these temperatures above 60 min has practically no effect on densification.

These results are confirmed in great part by COBLE and HOBBS (1963) for whom the sintering features of TiB_2 are as follows: (1) a short densification step where the density increases linearly with the logarithm of increasing time; (2) densification approaches a limit depending on the sintering temperature and resulting from the equilibrium between the rates of densification and evaporation of the diboride; (3) the densification limit does not depend on the particle size of the initial powder; (4) formation of pores bigger than the initial ones. These features result from the fact that while mass transfer occurs by evaporation-condensation and volume diffusion processes at practically the same rate, the activation energy for the first process is less than that for the second.

KISLYI and ZAVERUKHA (1970) have investigated the characteristics of the sintering process of TiB_2 (particle size less than 1 μm; iron content: 0.7% by weight) in a vacuum of 10^{-4} torr, by measuring shrinkage, electrical resistivity and weight of the samples as a function of temperature and time. It is found that at low temperatures (up to 1100°C) an increase of contact surfaces between boride particles occurs by a surface diffusion process. At temperatures above 1800°C the evaporation-condensation mechanism contributes to mass transfer.

KISLYI et al. (1972a) have also investigated the behavior during sintering of specimens of TiB_2 powders with 120, 40, 6.5, 1.8 and 1.1 μm mean

Table 2. Sintering characteristics of diborides of group IV metals (TiB_2, ZrB_2, HfB_2)

Boride	Mean particle size (µm)	Binder nature and content (wt%)	Compacting pressure (kbar)	Sintering temperature (°C)	Holding time (min)	Sintering atmosphere or vacuum V: (P in torr)	Relative density (% theo.)	Remarks	References
TiB_2	2 - 3	$FeCl_3$ + alcohol: 1% Fe	3	1800	60	V:10^{-1}	74.5	Increasing holding time above 60 min has practically no effect on densification	BABICH et al. (1960)
TiB_2	2 - 3	$FeCl_3$ + alcohol: 1% Fe	3	2000	60	V:10^{-1}	78.4		SAMSONOV and PORTNOI (1961)
TiB_2	2 - 3	$FeCl_3$ + alcohol: 1% Fe	3	2200	60	V:10^{-1}	81.6		SAMSONOV and KOVAL'CHENKO (1961)
TiB_2	2 - 3	$FeCl_3$ + alcohol: 1% Fe	3	2300	60	V:10^{-1}	85.7		
TiB_2	2 - 3	$FeCl_3$ + alcohol: 1% Fe	3	2400	60	V:10^{-1}	90.9		
TiB_2	<250	Polyester: 7%	1	1900	360	H_2: 760	57.5	-	PECHINEY (1960a)
TiB_2	-	none	1	1900	360	V: 10^{-1}	-	Reaction-sintering	PECHINEY (1960b)
TiB_2	-	-	-	-	-	-	93-97	-	MANDORF et al. (1963)
TiB_2	5 - 20	Starch + water (300g/l):15-20%	none	2200-2250	15	H_2: 760	92-95	Extrusion-die pressing	SAMSONOV and PORTNOI (1961)
TiB_2	-	Aq.soln. 10% PVA[a]	10.1	2200	45	V: 10^{-4}	81	Weight loss: a few %	COBLE and HOBBS (1973)
TiB_2	5	Paraffin wax: 2%	0.28	2300	120	H_2 or He: 760	80	-	FARRIOR, 1965
TiB_2	-	Epoxy resin + acetone: 3%	1.5	1850	180	Ar: 760	85.7	-	
TiB_2	-	Epoxy resin + acetone: 3%	1.5	2000	180	Ar: 760	84.6	-	} REIDL and HARTWIG (1965)
TiB_2	-	Epoxy resin + acetone: 3%	1.5	2130	180	Ar: 760	85.6	-	
TiB_2	4.2	Camphor (2%) + ether	5	2000	60	V: 5.10^{-2}	69.0	Fe content: 0.35%	MEYER and PASTOR (1965)
TiB_2	4.2	Camphor (2%) + ether	5	2000	60	V: 5.10^{-2}	92.4	Fe content: 0.58%	

TiB_2	-177,+149	Aq.soln. 5% PVA	1	2100	60	H_2: 760	71	-	PRSHEDROMIRS-KAYA and SLEPTSOV(1966)
TiB_2	-177,+149	Aq.soln. 5% PVA	1	2250	60	H_2: 760	75	-	
TiB_2	2.5	Xylene/CPD[b] (85/15): 16%	none	2000	120	Ar: 760	92-94	Slip casting	REDDY et al. (1966)
TiB_2	-	Aq.soln. 1% PVA	0.5	2150-2200	30	-	77.5	- 74, + 44 } Pore size	SLEPTSOV et al. (1969)
TiB_2	-	Aq.soln. 1% PVA	0.5	2150-2200	30	-	73.0	-420, +177 } Pore size	
TiB_2	-	none	12	1900	720	Ar: 760	90	Isostatic pressing	FITZER (1973)
TiB_2	-	none	12	2200	10	Ar: 760	90	Loss of B above 2200°C	
ZrB_2	-	Starch + water (300g/l):15-20%	none	2300	15	H_2: 760	94-97	Extrusion-die pressing	SAMSONOV and PORTNOI (1961)
ZrB_2	-	Aq.soln. 5% PVA	none	2050-2100	60-80	H_2: 760	88-94	Extrusion-die pressing. 2% C inbihit sintering	KISLYI (1962)
ZrB_2	-	Aq.soln.10% PVA	10.1	2200	45	V: 10^{-4}	86	weight loss: 3.1%	COBLE and HOBBS (1973)
ZrB_2	-75, +44	Rubber (5%) or PVA (3%)	0.6	2000	30	-	63	+ 3% Ni or Co	SLEPTSOV et al. (1965)
ZrB_2	5	Paraffin wax: 2%	0.28	2200	60-120	H_2 or He: 760	80	-	FARRIOR (1965)
ZrB_2	<20	Rubber + benzene	1.5(?)	2000	60	V: 10^{-2}	78.2	-	CECH et al. (1965)
ZrB_2	3	-	-	2200-2250	40	H_2: 760	85.8	-	KISLYI and KUZENKOVA (1965)
ZrB_2	5.2	Aq.soln. 5% PVA	1	2100	240	Ar: 760	90	Powder containing ZrB_{12}	KUZENKOVA and KISLYI (1966a)
ZrB_2	5.2	Aq.soln. 5% PVA	1	2100	240	Ar: 760	69	Powder without ZrB_{12}	KUZENKOVA and KISLYI (1966b)
ZrB_2	-177,+149	Aq.soln. 5% PVA	1	2000	60	H_2: 760	61	-	PRSHEDROMIRS-KAYA and SLEPTSOV(1966)
ZrB_2	-177,+149	Aq.soln. 5% PVA	1	2150	60	H_2: 760	66	-	
ZrB_2	3	Xylene/CPD (93/7)	none	2000	120	Ar: 760	89	Slip casting	REDDY et al. (1966)

Table 2. Continued

ZrB_2	<40	Aq.soln. (3%) CMC[c]: 22%	none	2200	40	H_2: 760	88	Slip casting 0.5-0.7% Fe content	DOBROVOLSKII et al. (1966)
ZrB_2	< 7	Rubber + benzene (Fe content 0.08%)	0.5-1	1900	120	V: 5.10^{-4}	80.7	Without additives	MEERSON and GORBUNOV (1968)
ZrB_2	< 7	Rubber + benzene (Fe content 0.08%)	0.5-1	2100	120	V: 5.10^{-4}	83.0	Without additives	MEERSON and GORBUNOV (1968)
ZrB_2	< 7	Rubber + benzene (Fe content 0.08%)	0.5-1	1800	120	V: 5.10^{-4}	85.3	with ZrO_2 + B additions	MEERSON and GORBUNOV (1968)
ZrB_2	< 7	Rubber + benzene (Fe content 0.08%)	0.5-1	1900	120	V: 5.10^{-4}	98.8	with ZrO_2 + B additions	MEERSON and GORBUNOV (1968)
ZrB_2	< 7	Rubber + benzene (Fe content 0.08%)	0.5-1	2100	120	V: 5.10^{-4}	99.5	with ZrO_2 + B additions	MEERSON and GORBUNOV (1968)
ZrB_2	see Remarks	-	-	2200	60	V: 10^{-4}-10^{-5}	59.8	S:1.4m^2/g Specific surface of the powder	GROPYANOV (1969)
ZrB_2	see Remarks	-	-	2200	60	V: 10^{-4}-10^{-5}	64.8	S:1.9m^2/g Specific surface of the powder	GROPYANOV (1969)
ZrB_2	see Remarks	-	-	2200	60	V: 10^{-4}-10^{-5}	68.2	S:2.3m^2/g Specific surface of the powder	GROPYANOV (1969)
ZrB_2	see Remarks	-	-	2200	60	V: 10^{-4}-10^{-5}	72.7	S:4.0m^2/g Specific surface of the powder	GROPYANOV (1969)
ZrB_2	1	Starch: 2%	1	2200	90	Ar: 760	82-85	-	KUZENKOVA et al. (1971) KISLYI et al. (1971) KISLYI et al. (1972b) KISLYI et al. (1973)
HfB_2	5	Paraffin wax: 2%	0.28	2200	60-120	H_2 or Hc: 760	80	-	FARRIOR (1965)

[a]PVA: polyvinyl alcohol [b]CPD: cyclopentadiene [c]CMC: carboxymethylcellulose

particle size, under conditions of dynamic heating at a constant rate (10°C/min) and also during isothermal sintering at temperatures of 1800, 2000, 2100 and 2200°C. Powders (mixed with binder: a 3 wt% solution of rubber in benzine) were pressed into specimens with a porosity of 36-38%. Sintering was carried out in argon. Shrinkage ($\frac{dl}{l_o}$) begins above 1500°C and is greater, the smaller the mean particle size (d) of the initial powder, as expressed by the equation:

$$\frac{dl}{l_o} = \frac{K_1}{d^{0.3}}$$ (where K_1 is a constant).

Activation energy for mass transfer by grain boundary diffusion during the first stage of sintering is 74 ± 6 Kcal/mole. The shrinkage rate in the final stage of sintering is described by an equation of diffusion creep with an activation energy of mass transfer (volume diffusion) of 185 ± 15 Kcal/mol.

Zirconium Diboride. KISLYI (1962) points out that sintering of zirconium diboride parts in conversion natural gas (75-76% H_2, 22-23% CO, 1-2% CO_2, 0.5% H_2O) is inhibited at temperatures up to 1000°C by the separation of free carbon from the gas phase on the grain boundaries. Powders containing more than 2% free carbon are not sintered at 2050-2100°C and melt at higher temperatures (2250°C). Sintering must be preceded by grinding the powder to 2-10 µm mean particle size.

According to KUZENKOVA and KISLYI (1966b) the shrinkage of compacted bodies of ZrB_2 powders is practically the same in the direction of application of the compacting force and in the radial direction. At temperatures up to 1700°C there is a small shrinkage, depending linearly upon holding time, which results only from the ordering of grain boundaries. At sintering temperatures of 1800-2200°C and for holding times of up to 30 min a strong shrinkage occurs by slip processes taking place at grain boundaries under the influence of surface tension. Simultaneously a slight grain coarsening takes place. At high temperatures and holding time longer than 30 min, shrinkage results from viscous flow and the energy of activation for grain growth is 28.7 ± 9.6 Kcal/mol (KUZENKOVA and KISLYI, 1966a), in agreement with the energy of activation for the diffusion of boron in ZrB_2 along vacancies at boron sublattice sites. It is also found that zirconium dodecaboride, if present, activates shrinkage by accelerating diffusion processes along grain boundaries.

According to HOFMANN and ADAM (1970) there is a linear dependence between the grain size of ZrB_2 powder and the apparent density after sintering, the correlation coefficient being - 0.84. For equal grain size the apparent density increases linearly with increasing boron content in ZrB_2.

Activated Sintering

Sintering can be activated either by using finer powders or by adding a few percent of transition metals.

According to GROPYANOV (1969), within the 1850-2110°C temperature range the diffusion coefficient of defects in the ZrB_2 crystal lattice decreases and the activation anergy of diffusion increases with an increase (by vibromilling) in the surface area of the original ZrB_2 powders. This is explained by the fact that grinding must reduce the defects concentration in the mass of the material, which thus becomes stronger. With a reduction in the defects concentration, the activation energy of diffusion increases and the diffusion coefficient falls to a level characterizing the diffusion of single vacancies.

Activation of the sintering of ZrB_2 powder (containing 0.7 wt% free carbon) takes place by adding zirconia and boron (MEERSON and GORBUNOV, 1968). It is assumed that carbon reduces zirconia, producing highly active zirconium atoms which combine with boron atoms to form fine and highly active ZrB_2 nuclei; this process begins on the surface of the boride grains and spreads throughout the crystals.

COBLE and HOBBS (1963), ČECH et al. (1965) and RASSKAZOV (1970) have studied the influence of small additions of metals on the sintering behavior of titanium and zirconium diborides. Their sintering features and results are summarized in Table 3.

From COBLE and HOBBS'S (1963) results it can be pointed out that cobalt, nickel and especially chromium additions activate the sintering of titanium diboride. However the most noticeable effect occurs for samples with up to 10 wt% chromium, which must be preferably considered as a pseudobinary TiB_2-Cr alloy.

Some additions, especially cobalt, niobium, chromium and particularly rhenium, accelerate the sintering process of zirconium diboride (ČECH et al., 1965). This influence is associated with an appreciable contraction of the boride crystal lattice, indicating a gradual substitution of the added metal atoms for the zirconium atoms.

The sintering kinetics curves worked out by RASSKAZOV (1970) show that, at the relatively low sintering temperatures investigated (1650 to 2000°C), molybdenum and tungsten additions (up to 5 wt%) seem to inhibit the sintering of zirconium diboride. A slight increase in relative density is observed for the sample containing 10% Mo, which can be considered a pseudobinary alloy. According to RASSKAZOV, sintering depends on activated volume diffusion, on the formation of a solid solution, and on surface diffusion. On the other hand, small amounts (0.25-1 wt%) of boron activate the sintering process, a relative density of 80-85% being achieved already at 1800-1900°C.

Reaction-sintering as explained in a French patent (PECHINEY, 1960b) is unable to give dense bodies.

Production of Intentionally Porous Bodies

In many fields of engineering there is a need for materials retaining their strength at high temperatures for long periods and having, at the same time, high oxidation, erosion and thermal shock resistance. Refractory carbides, nitrides and borides are such materials and, among them, porous sintered borides are suitable for the filtration of some metallic melts, hot gases and corrosive media. This has inspired a number of papers on the manufacture of boride filters.

SLEPTSOV et al. (1965), PRSHEDROMIRSKAYA and SLEPTSOV (1966) and SLEPTSOV et al. (1969) have investigated the technological factors for the preparation of porous TiB_2 and ZrB_2 bodies by sintering freely-poured spheroidized powders or by reaction-sintering, thus determining optimum conditions for the desired porosity. The amounts of open and closed pores are determined, as well as pore size distribution. The permeability of such filters is studied.

SAMSONOV and MATSERA (1972, 1973) have investigated the manufacture and properties of porous fiber materials made from refractory compounds (including TiB_2 and ZrB_2). They have studied the extrusion of fibers (150 μm or less in diameter) from plasticized powder mixtures (using synthetic rubber solution in benzene, starch or paraffin wax as plasticizers). The optimum mixture compositions and extrusion conditions to

Table 3. Activated sintering of TiB_2 and ZrB_2

Boride	Nature and content (Wt%) of activating addition	Nature and content (Wt%) of binder	Compacting pressure (kbar)	Sintering temperature (°C)	Holding time (min)	Atmos-phere	Relative density (%)	References
	none				45		81.0	
	1 Co				8		87.5	
	5 Co	aqueous 10% solution of polyvinyl alcohol			8		84.9	
	1 Ni				30	Vacuum 10-4 torr	89.0	COBLE and HOBBS (1963)
TiB_2	10 Ni		1.7 - 10.1	2200	30		83.5	
	1 Cr				15		78.5	
	10 Cr				15		95.5	
	5 Si				10		82.0	
	none			2000		V	78.2	
	1 Fe			2200		V	89.2	
	1 Co			2000		Ar	99.2	
	1 Ni			2000		Ar	98.2	
	0.5 Ni + 0.5 Fe			2000		Ar	96.6	
	0.5 Ni + 0.5 Co			2000		Ar	98.4	
	1 Ti	Solution of synthetic rubber in benzene		2200		V	67.7	CECH et al. (1965)
ZrB_2	1 Zr		1.5	2200	60	Ar	98.0	
	1 Nb			2200		Ar	99.0	
	1 Cr			2000		Ar	99.7	
	1 Re			2000		Ar	97.4	
	1 Re			2200		Ar	100	
	0.5 Ti + 0.5 Cr			2200		Ar	100	
	0.5 Re + 0.5 Cr			2200		V	88.7	
	1 Mo			1800			73.5	
	2 Mo			1800			73.0	
	3 Mo			1800			71.5	
	4 Mo			1800			70.0	
	5 Mo	Alcoholic 2% solution of polyvinyl-butyryl		1800			69.5	
ZrB_2	10 Mo		1.5	1800	60	-	75.5	RASSKAZOV (1970)
	1 W			1900			71.5	
	2 W			1900			67.8	
	3 W			1900			66.5	
	4 W			1900			65.5	
	5 W			1900			64.3	

ensure continuous filament discharge are thus determined. Sintering conditions securing the maximum possible strength of the sintered fibers are determined, and this strength can be increased using activators (3 wt% Fe). Porous materials are produced from these sintered fibers. Shrinkage during sintering of these materials is determined primarily by the initial porosity of the fibers, no change in inter-fiber porosity (about 45%) being observed.

δ) Sintering of Borides of Group V B Metals: V, Nb, Ta

Fully dense bodies can hardly be obtained by conventional sintering of vanadium, niobium or tantalum diborides, as can be seen in Table 4 which summarizes the sintering characteristics of these borides.

ELLIOTT and LEVENDEL (1967) have investigated the initial stage of sintering of TaB_2 in the temperature range 1200-1500°C under vacuum. Mass transport occurs either by volume or grain boundary diffusion; a differentiation between the two processes could not be made. The activation energy for the sintering process is found to be 90 Kcal/mole. Excess boron additions (up to 5 wt%) seem to inhibit sintering of TaB_2.

ε) Sintering of Borides of Group VI B Metals: Cr, Mo, W

From Table 5 it can be seen that available literature provides practically no valuable information on the sintering of group VI B metal borides.

2. Sintering of Boride-metal Composites

This review has been limited to boride-metal (MB_x-M') pseudo-binary systems containing more than 50 volume % of boride. Infiltrated borides have also been excluded.

a) Rare Earth Metal Boride-metal Systems

Table 6 summarizes the sintering characteristics of rare earth metal boride-metal systems.

According to SAMSONOV and KONDRATOV (1969) the specific electrical resistance of YB_6-Re cerments (with 1 to 95 wt% YB_6) sintered at 2000°C for 5 h increases with YB_6 content, reaching a maximum for Re-50 % YB_6. The lowest work function is obtained for Re-20 % YB_6. Rhenium reacts with YB_6; the resulting alloys have the following phase composition:

Re - 1 % YB_6 : Re(Y), Re_3B
Re - 20 % YB_6 : Re(Y)B_3, Y(Re)B_4
Re - 50 % YB_6 : Re(Y)7B_3, Y(Re)B_4, ReB_3
Re - 70 % YB_6 : YB_6, Y(Re)B_4
Re - 95 % YB_6 : YB_6, Y(Re)B_4.

In the same sintering conditions tungsten reacts also with YB_6 (KUDINTSEVA et al., 1970) giving highly porous parts with the following phase compositions:

W - 1 % YB_6 : W(Y), W_2B
W - 5 % YB_6 : W(Y), W_2B
W - 15 % YB_6 : W(Y), W_2B, Y(W)B_4
W - 25 % YB_6 : -Wb, Y(W)B_4, W_2B
W - 50 % YB_6 : W_2B_5, Y(W)B_4
W - 70 % YB_6 : W_2B_5, Y(W)B_4, YB_6

Table 4. Sintering characteristics of diborides of group V B metals (VB_2, NbB_2, TaB_2)

Boride	Binder nature and content (wt%)	Mean particle size (µm)	Compacting pressure (kbar)	Sintering temperature (°C)	Holding time (min)	Sintering atmosphere or vacuum V (Pressure in torr)	Relative density (%)	Remarks	References
VB_2	Paraffin: 2	5	0.28	2000	120	H_2 or He: 760	> 80	-	FARRIOR (1965)
NbB_2	Aq.soln.10% PVA[a]	-	10.1	2420	2	V: 10^{-4}	76	-	COBLE and HOBBS (1963)
NbB_2	Aq.soln.10% PVA[a]	-	10.1	2650	14	V: 10^{-4}	73.4	Weight loss: 3%	COBLE and HOBBS (1963)
NbB_2	Paraffin: 2	5	0.28	2300	120	H_2 or He: 760	> 80	-	FARRIOR (1965)
TaB_2	Paraffin: 2	5	0.28	2300	120	H_2 or He: 760	> 80	-	FARRIOR (1965)
TaB_2	none	4	2.11	2100	30	V: 10^{-5}	94	-	ELLIOTT and LAVENDEL (1967)

[a]PVA: polyvinyl alcohol

Table 5. Sintering characteristics of borides of group VI B metals

Boride	Binder nature and content (wt%)	Mean particle size (µm)	Compacting pressure (kbar)	Sintering temperature (°C)	Holding time (min)	Sintering atmosphere or vacuum V (Pressure in torr)	Relative density (%)	Remarks	References
CrB	-	< 53	5	1600-1700	30	-	-	-	TAKAGI et al. (1956)
CrB_2	Alcohol soln.of $FeCl_3$ (1% Fe)	2-3	3	2000	60	V: 10^{-1}	76	CrB_2 melts at 2100°C	BABICH et al. (1960)
CrB_2	Paraffin: 2	5	0.28	1900	10	H_2 or He: 760	> 80	-	FARRIOR (1965)
CrB_2	Xylene + 7% CPD: 46.2 vol.%	2.2	none	1700	60-180	Ar: 760	-	Slip casting	REDDY et al. (1966)
MoB_2	Paraffin: 2	5	0.28	1900	60	H_2 or He: 760	> 80	-	FARRIOR (1965)

Table 6. Sintering characteristics of rare earth metal boride-metal systems

Boride	Metal (wt%)	Particle size Boride (μm)	Particle size Metal (μm)	Binder (wt%)	Compacting pressure (kbar)	Sintering temperature (°C)	Holding time (min)	Sintering atmosphere (P in (torr)	Density (g/cm³)	Porosity (%)	Volume change (%)	Resistivity[a] (μΩ.cm)	Work[b] function (eV)	References
YB_6	5 Re	< 43	< 43	Paraffin: 1%	10	2000	300	$V:10^{-3}$	3.84		-13.2	79.5	3.64	SAMSONOV and KONDRATOV (1969)
YB_6	30 Re	< 43	< 43	Paraffin: 1%	10	2000	300	$V:10^{-3}$	4.1		- 7.1	81.5	3.85	
YB_6	50 Re	< 43	< 43	Paraffin: 1%	10	2000	300	$V:10^{-3}$	4.75		- 0.46	109.3	3.78	
YB_6	80 Re	< 43	< 43	Paraffin: 1%	10	2000	300	$V:10^{-3}$	10.8		+ 6.17	100.8	3.47	
YB_6	30 W	< 43	< 43	Paraffin: 1%	10	2000	300	$V:10^{-3}$	4.65	37.8		41.7	-	KUDINTSEVA et al. (1970)
YB_6	50 W	< 43	< 43	Paraffin: 1%	10	2000	300	$V:10^{-3}$	6.54	40.5		39.4	3.85	
YB_6	75 W	< 43	< 43	Paraffin: 1%	10	2000	300	$V:10^{-3}$	9.79	45.4		144.5	-	

[a]Resistivity: specific electrical resistance at room temperature (μΩ.cm)

[b]Work function at 1500°C (eV)

The highest specific electrical resistance is obtained for the W-25% YB_6 cermet.

b) Group IV B Metal Boride-Metal Systems

Tables 7 and 8 summarize the sintering characteristics of TiB_2-metal and ZrB_2-metal systems respectively. The systems investigated are mainly those with chromium, molybdenum, tungsten, rhenium, iron, cobalt and nickel additions. Pressed compacts of 70 TiB_2-30 Zr composition sintered for half an hour at 2000°C show practically no shrinkage (NELSON et al., 1952). No example of sintering of (TiB_2 or ZrB_2) - (V, Nb or Ta) systems can be found.

According to MEYER and PASTOR (1969), chromium, molybdenum, tungsten and rhenium react to a certain extent with titanium and zirconium diboride during sintering as follows:

$$TiB_2 \text{ (or } ZrB_2) + Cr \xrightarrow{2000^\circ C} \underset{\text{cubic}}{TiB \text{ (or } ZrB)} + \underset{\text{orthorhombic}}{CrB}$$

$$TiB_2 \text{ (or } ZrB_2) + Mo \xrightarrow{2050^\circ C} \underset{\text{cubic}}{TiB \text{ (or } ZrB)} + \underset{\text{orthorhombic}}{\beta - MoB}$$

$$TiB_2 \text{ (or } ZrB_2) + W \xrightarrow{2100-2400^\circ C} \underset{\text{cubic}}{TiB \text{ (or } ZrB)} + \underset{\text{orthorhombic}}{\beta - WB}$$

$$3TiB_2 \text{ (or} 3ZrB_2) + 7Re \xrightarrow{2200^\circ C} \underset{\text{cubic}}{3TiB \text{ (or} 3ZrB)} + \underset{\text{hexagonal}}{Re_7B_3}$$

The sintered specimens are rather porous and very brittle. The highest densification is achieved with additions of molybdenum or rhenium for TiB_2 and with additions of molybdenum for ZrB_2; low contents of chromium (~ 7.5% Cr) also seem favorable for both borides. However, in practice, none of these four refractory metal additions was able to improve the high-temperature (1200°C) resistance to oxidation in air of these borides.

α) Titanium Diboride-Metal Systems

TiB_2-(Cr, Mo or W) Systems

Sintering of 70 TiB_2-30 Cr samples at 1927°C for half an hour in argon gives porous parts, displaying however a good resistance to oxidation in air at 1093°C (weight gain of 0.4 mg/cm^2/h; NELSON et al., 1952), in good agreement with the observations of MEYER and PASTOR (1969) on the 77.5 TiB_2-22.5 Cr composition.

According to TANGERMANN'S (1963) results it seems preferable to sinter TiB_2-Cr samples in a hydrogen flow; the sintering temperature can thus be lowered to 1650°C and dense but quite brittle parts with high hardness are obtained.

The sintering of TiB_2-Cr samples at 2000°C for two hours in argon gives parts with porosity (6.4, 12 and 22%) which increases with the chromium content (7.5, 15 and 22.5 wt% respectively), thus lowering the bending strength. A high weight loss (5.2, 9.2 and 15.2 % respectively) is observed and attributed to volatilization of chromium (MEYER and PASTOR, 1969). In comparison, the porosity (6.7-7.3) of TiB_2-Mo samples sintered in vacuum (10^{-3} torr) at 2050°C for two h is rather low and practically independent of the molybdenum content.

Table 7. Sintering characteristics of TiB_2-metal systems

Boride	Metal	Particle size		Binder	Compacting pressure	Sintering temperature	Holding time
		Boride	Metal				
	(wt%)	(µm)	(µm)	(wt%)	(kbar)	(°C)	(min)
TiB_2[a]	30 Zr	< 43	< 43	none	3.45	1982	30
TiB_2	30 Cr	< 43	< 43	none	3.45	1927	30
TiB_2	30 Fe	< 43	< 43	none	3.45	1788	30
TiB_2	30 Co	< 43	< 43	none	3.45	1799	30
TiB_2	30 Ni	< 43	< 43	none	3.45	1760	30
TiB_2	50 Fe	< 90	< 90	-	-	1530	-
TiB_2	20 Co	< 90	< 90	-	-	1530	-
TiB_2	20 Ni	< 90	< 90	-	-	1530	-
TiB_2	16 Mo	-	-	-	-	-	-
TiB_2	20 Mo	< 40	-	-	2.35	2150	120
TiB_2	4.60 Fe	0.29	1.7	-	0.6	2000	45
TiB_2	8.22 Fe	0.29	1.7	-	0.6	1850	45
TiB_2	14.90 Fe	0.29	1.7	-	0.6	1820	45
TiB_2	18.93 Fe	0.29	1.7	-	0.6	1800	45
TiB_2	24.5 Fe	0.29	1.7	-	0.6	1780	45
TiB_2	29.89 Fe	0.29	1.7	-	0.6	1720	45
TiB_2	38.42 Fe	0.29	1.7	-	0.6	1700	45
TiB_2	48.00 Fe	0.29	1.7	-	0.6	1700	45
TiB_2	15 Cr	< 1	-	Polystyrene + benzene	1.5	1650	60
TiB_2	21 Fe	< 1	-	"	1.5	1600-1650	60
TiB_2	23 Ni	< 1	-	"	1.5	1600-1650	60
TiB_2	10 Fe Si	< 1	-	"	1.5	1600	60
TiB_2	21 (Fe+3.8 B)	< 1	-	"	1.5	1600-1650	60
TiB_2	23 (Co + 4 B)	< 1	-	"	1.5	1600-1650	60
TiB_2	19Co+1.8B+0.2Cr	< 1	-	"	1.5	1800	60
TiB_2	1 Co+4 Cr+0.2 B	< 1	-	"	1.5	1800	60
TiB_2	23 (Ni + 4 B)	< 1	-	"	1.5	1600-1650	60
TiB_2	10 (Cu + 2 B)	-	-	-	0.78-3.94	1200	120
TiB_2	7.5 Cr	4-7	< 43	Camphor	6	2000	120
TiB_2	15.0 Cr	4-7	< 43	Camphor	6	2000	120
TiB_2	22.5 Cr	4-7	< 43	Camphor	6	2000	120
TiB_2	7.5 Mo	4-7	< 43	Camphor	6	2050	120
TiB_2	15.0 Mo	4-7	< 43	Camphor	6	2050	120
TiB_2	22.5 Mo	4-7	< 43	Camphor	6	2050	120
TiB_2	7.5 W	4-7	< 43	Camphor	6	2400	120
TiB_2	15.0 W	4-7	< 43	Camphor	6	2400	120
TiB_2	22.5 W	4-7	< 43	Camphor	6	2400	120
TiB_2	7.5 Re	4-7	< 43	Camphor	6	2200	120
TiB_2	15.0 Re	4-7	< 43	Camphor	6	2200	120
TiB_2	3,6,9,12 vol.% Mo, Ni, Co	1-10	-	-	0.77	~ melting point metal	30-120
TiB_2	6 Mo	< 1	< 1	-	-	2200	0
TiB_2	6 Mo	< 1	< 1	-	-	2200	30
TiB_2	6 Mo	< 1	< 1	-	-	2200	60
TiB_2	6 Mo	< 1	< 1	-	-	2200	120
TiB_2	10 NiAl	10	10	none	4	1800	180

[a] TiB_2 which contained 4.17% free carbon was previously made to react with Ti (TiB_2/Ti =
[b] Porosity calculated assuming that theoretical density of the cermet is additive (this
π: porosity (%) - Δ 1/1: shrinkage (%) - Δ W: weight loss (%)

Sintering atmosphere (P in torr)	Density (g/cm^3)	π (%)	$\Delta l/l$ (%)	ΔW (%)	HV or R_A	σ_b (hbar)	References
Ar: 760	3.66	-	0	-	-	-	NELSON et al. (1952)
Ar: 760	4.39	-	7.37	-	-	-	
Ar: 760	4.60	-	8.11	-	-	-	
Ar: 760	5.00	-	11.45	-	-	-	
Ar: 760	4.88	-	9.68	-	-	-	
H_2: 760	6.05	-	-	-	-	23.2	ISKOLDSKY and BOGORODSKAYA (1957)
H_2: 760	5.22	-	-	-	-	34	
H_2: 760	5.48	-	-	-	-	-	
-	-	21	-	-	-	6-9	SAMSONOV (1958)
-	-	-	-	-	-	-	JAYNE (1961)
Ar: 760	4.7		20.0		HV -	-	FUNKE et al. (1961)
Ar: 760	4.6		17.0		1798	45	
Ar: 760	4.8		16.9		1713	52	
Ar: 760	4.95		15.6		1675	64	
Ar: 760	4.85		-		-	-	
Ar: 760	4.90		17.0		1338	40	
Ar: 760	5.23		17.5		1367	51	
Ar: 760	5.34		21.4		823	55	
H_2: 760	5.99	-	20	-	R_A:92.8	37	TANGERMANN (1963)
H_2: 760	5.9	-	20	-	90.7	49	
H_2: 760	5.5	-	20	-	91.8	36	
H_2: 760	5.25	-	23	-	89.0	40	
H_2: 760	6.3	-	25	-	92.9	50	
H_2: 760	6.3	-	25	-	92,0	26	
H_2: 760	4.4	-	10	-	87.6	16.5	
H_2: 760	5.6	-	25	-	94.4	45.8	
H_2: 760	6.4	-	25	-	92.5	29.5	
H_2: 760	-	-	-	-	-	-	U.S. BORAX (1965)
Ar: 760	4.36	6.4[b]	-	5.23	-	25.0	MEYER and PASTOR (1969)
Ar: 760	4.21	12.0	-	9.2	-	22.8	
Ar: 760	3.84	22.0	-	15.2	-	10.6	
V: 10^{-3}	4.41	6.7	-	1.5	-	25.8	
V: 10^{-3}	4.60	6.8	-	1.2	-	22.4	
V: 10^{-3}	4.79	7.3	-	1.3	-	18.6	
V: 10^{-3}	4.12	14.0	-	5.6	-	20.0	
V: 10^{-3}	4.30	15.7	-	2.0	-	22.0	
V: 10^{-3}	4.64	15.0	-	1.3	-	-	
V: 10^{-3}	4.57	4.8	-	1.3	-	21.5	
V: 10^{-3}	4.78	6.6	-	1.3	-	19,6	
-	low	high	-	-	-	-	PENDSE and PRATT (1972)
V	-	18	12	-	-	-	ZAVERUKHA (1972)
V	-	11	15	-	-	-	
V	-	10	16	-	-	-	
V	-	8	17	-	-	-	
Ar: 760	-	18	-	-	-	-	BROUSSAUD et al. (1974)

60/10) at 1871°C to form TiC.
implies that there is no interaction between TiB_2 and the metal).
HV: VICKERS Hardness; R_A: ROCKWELL A Hardness; σ_b: bending strength (hbar)

Table 8. Sintering characteristics of ZrB_2-metal systems

Boride	Metal (wt%)	Particle size: Boride (μm)	Particle size: Metal (μm)	Binder (wt%)	Compacting pressure (kbar)	Sintering temperature (^oC)	Holding time (min)
ZrB_2	20 Cr	3	< 40	none	3.45	1802	2-3[a]
ZrB_2	20 Cr	< 40	< 40	none	3.45	1802	2-3[a]
ZrB_2	15 Co	3	< 53	none	3.45	1682	2-3[a]
ZrB_2	20 Co	< 40	< 53	none	3.45	1760	2-3[a]
ZrB_2	20 Co	3	< 53	none	3.45	1566	2-3[a]
ZrB_2	20 Ni	< 40	9-14	none	3.45	1704	2-3[a]
ZrB_2	20 Ni	3	9-14	none	3.45	1704	2-3[a]
ZrB_2	20Co+4Zr	< 40	< 53	none	3.45	1760	2-3[a]
ZrB_2	10cO+10Cr	3	< 53	none	3.45	1802	2-3[a]
ZrB_2	6 Fe	150	3.8	Tetrahydronaphtalene:2%	0.7	-[b]	-[b]
ZrB_2	12 Fe	150	3.8	Tetrahydronaphtalene:2%	0.7	-	-
ZrB_2	18 Fe	150	3.8	Tetrahydronaphtalene:2%	0.7	-	-
ZrB_2	6 Co	150	5.6	Tetrahydronaphtalene:2%	0.7	-	-
ZrB_2	12 Co	150	5.6	Tetrahydronaphtalene:2%	0.7	-	-
ZrB_2	18 Co	150	5.6	Tetrahydronaphtalene:2%	0.7	-	-
ZrB_2	6 Ni	150	16.0	Tetrahydronaphtalene:2%	0.7	-	-
ZrB_2	11 Ni	150	16.0	Tetrahydronaphtalene:2%	0.7	-	-
ZrB_2	16 Ni	150	16.0	Tetrahydronaphtalene:2%	0.7	-	-
ZrB_2	12 Cr	18	8.5	Tetrahydronaphtalene:2%	0.7	-[b]	-[b]
ZrB_2	17 Cr	18	8.5	Tetrahydronaphtalene:2%	0.7	-	-
ZrB_2	22 Cr	18	8.5	Tetrahydronaphtalene:2%	0.7	-	-
ZrB_2	12 Mo	18	11	Tetrahydronaphtalene:2%	0.7	-	-
ZrB_2	18 Mo	18	11	Tetrahydronaphtalene:2%	0.7	-	-
ZrB_2	22 Mo	18	11	Tetrahydronaphtalene:2%	0.7	-	-
ZrB_2	13 W	18	6.5	Tetrahydronaphtalene:2%	0.7	-	-
ZrB_2	17 W	18	6.5	Tetrahydronaphtalene:2%	0.7	-	-
ZrB_2	24 W	18	6.5	Tetrahydronaphtalene:2%	0.7	-	-
ZrB_2	0 Mo	38	38	-	-	2200-2250	40
ZrB_2	5 Mo	38	38	-	-	2200-2250	40
ZrB_2	10 Mo	38	38	-	-	2200-2250	40
ZrB_2	20 Mo	38	38	-	-	2200-2250	40
ZrB_2	40 Mo	38	38	-	-	2200-2250	40
ZrB_2	5 Mo	-	-	-	-	2200	150-180
ZrB_2	10 Mo	-	-	-	-	2200	150-180
ZrB_2	15 Mo	-	-	-	-	2200	150-180

Sintering atmosphere (P in torr)	Density (g/cm^3)	π (%)	Δ 1/1 (%)	Δ W (%)	HV or R_A	σ_b (hbar) σ_c	References
Ar: 760	6.38	-	-	-	-	23.5	McTAGGART et al. (1954)
Ar: 760	6.43	-	-	-	-	15.0	
Ar: 760	5.63	-	-	-	-	8.7	
Ar: 760	6.00	-	11.3	-	-	14.4 σ_b	
Ar: 760	5.48	-	-	-	-	6.2	
Ar: 760	6.24	-	16.0	-	-	15.5	
Ar: 760	5.71	-	10.8	-	-	11.7	
Ar: 760	5.82	-	-	-	-	-	
Ar: 760	5.95	-	-	-	-	8.1	
-	6.18	0.5	-	-	R_A:86	23.8	TYRRELL and HOUCK (1963)
-	6.18	1.8	-	-	88.5	39.9	
-	6.02	4.5	-	-	81	16.1	
-	6.12	1.8	-	-	85	33.6 σ_c	
-	6.32	2.3	-	-	88	47.6	
-	6.48	2.8	-	-	90.5	36.4	
-	6.02	3.8	-	-	88.5	28.7	
-	6.22	2.2	-	-	89	43.4	
-	6.46	1.5	-	-	91.5	35.7	
-	5.89	4.7	-	-	R_A:91	σ_c 17.5	TYRRELL and HOUCK (1964)
-	6.03	3.7	-	-	92	28.0	
-	6.00	5.4	-	-	92	43.4	
-	5.96	6.7	-	-	89	15.4	
-	6.27	4.6	-	-	83	σ_c 19.6	
-	6.42	5.0	-	-	92	14.0	
-	6.36	4.9	-	-	86	24.5	
-	6.66	4.4	-	-	88	8.8	
-	6.84	6.8	-	-	92	7.0	
H_2: 760	5.20	14.2	10.5	-	-	-	KISLYI and KUZENKOVA (1965) KISLYI and KUZENKOVA (1966a)
H_2: 760	5.89	5.3	13.4	-	-	-	
H_2: 760	5.92	6.8	15.7	-	-	-	
H_2: 760	6.23	8.7	11.9	-	-	-	
H_2: 760	5.20	31.9	4.8	-	-	-	
Ar: 760	-	5.9	-	-	-	-	KISLYI and KUZENKOVA (1966b)
Ar: 760	-	7.2	-	-	-	-	
Ar: 760	-	10.7	-	-	-	-	

Table 8. Continued

Boride	Metal (wt%)	Particle size Boride (μm)	Particle size Metal (μm)	Binder (wt%)	Compacting pressure (kbar)	Sintering temperature (^{o}C)	Holding time (min)
ZrB_2	7.5 Cr	4-7	< 43	Camphor	6	2000	120
ZrB_2	15.0 Cr	4-7	< 43	Camphor	6	2000	120
ZrB_2	22.5 Cr	4-7	< 43	Camphor	6	2000	120
ZrB_2	7.5 Mo	4-7	< 43	Camphor	6	2050	120
ZrB_2	15.0 Mo	4-7	< 43	Camphor	6	2050	120
ZrB_2	22.5 Mo	4-7	< 43	Camphor	6	2050	120
ZrB_2	7.5 W	4-7	< 43	Camphor	6	2100	120
ZrB_2	15.0 W	4-7	< 43	Camphor	6	2100	120
ZrB_2	22.5 W	4-7	< 43	Camphor	6	2100	120
ZrB_2	7.5 Re	4-7	< 43	Camphor	6	2200	120
ZrB_2	15.0 Re	4-7	< 43	Camphor	6	2200	120
ZrB_2	0 W	1	1	Starch: 2%	1	2200	90
ZrB_2	5 W	1	1	Starch: 2%	1	2200	90
ZrB_2	10 W	1	1	Starch: 2%	1	2200	90
ZrB_2	20 W	1	1	Starch: 2%	1	2200	90
ZrB_2	30 W	1	1	Starch: 2%	1	2200	90
ZrB_2	40 W	1	1	Starch: 2%	1	2200	90
ZrB_2	50 W	1	1	Starch: 2%	1	2200	90
ZrB_2	3,6,9,12 vol. % Mo,Ni,Co	1-10	-	-	2.35	2150	120

[a]Sintering temperature was reached in 5 to 6 hours.

[b]Resistance sintering

[c]Porosity calculated assuming that theoretical density of the cermet is additive

π: porosity (%) - Δ 1/1: shrinkage (%) - Δ W: weight loss (%) -

σ_b: bending strenght (hbar)

σ_c: compressive strength (hbar)

Sintering atmos- phere (P in torr)	Density (g/cm^3)	π (%)	Δ 1/1 (%)	Δ W (%)	HV or R_A	σ_b (hbar) σ_c	References
Ar: 760	5.75	6.7^c	-	~ 1	-	σ_b -	MEYER and PASTOR (1969)
Ar: 760	5.60	10.0	-	5	-	13.9	
Ar: 760	5.17	18.0	-	-	-	-	
V: 10^{-3}	6.09	3.0	-	1.2	-	21.4	
V: 10^{-3}	6.23	4.0	-	0.8	-	-	
V: 10^{-3}	6.43	3.5	-	0.6	-	-	
V: 10^{-3}	4.95	23.0	-	0.7	-	-	
V: 10^{-3}	4.82	29.0	-	0.8	-	20.0	
V: 10^{-3}	5.23	27.5	-	0.7	-	22.2	
V: 10^{-3}	5.54	14.0	-	0.2	-	-	
V: 10^{-3}	5.68	16.7	-	0.3	-	-	
Ar: 760	6.21	15-18	-	-	-	σ_b 15.0	KUZENKOVA et al. (1971) KISLYI et al. (1971) KISLYI et al. (1972b) KISLYI and KUZENKOVA (1973)
Ar: 760	6.53	8-10	-	-	-	17.5	
Ar: 760	6.72	10-12	-	-	-	18.5	
Ar: 760	7.05	10-12	-	-	-	20.0	
Ar: 760	7.53	10-12	-	-	-	16.0	
Ar: 760	8.60	12-16	-	-	-	12.5	
Ar: 760	9.40	16-20	-	-	-	7.0	
-	low	high	-	-	-	-	PENDSE and PRATT (1972)

(this implies that there is no interaction between ZrB_2 and the metal)

HV: VICKERS Hardness; R_A: ROCKWELL A Hardness

Sintering in a vacuum of 94 TiB_2-6 Mo samples has been studied by ZAVERUKHA (1972), by measuring their change of resistivity as a function of temperature and holding time at sintering temperature. It is found that shrinkage starts above 1100°C, but the rate of shrinkage increases sharply only above 1400°C. During isothermal treatment the highest rate of shrinkage is observed within the first 20 min, after which it decreases and remains practically constant. The porosity of 94 TiB_2-6 Mo samples sintered at 2200°C decreases (18, 11, 10 and 8 %) with increasing holding time at sintering temperature (0, 30, 60 and 120 min).

According to KISLYI and KUZENKOVA (1973) the shrinkage of the 94 TiB_2-6 Mo composition starts only at 1600°C.

The sintered 80 TiB_2-20 Mo composition is recommended as a hard cutting material for use in high-speed machining (JAYNE, 1961).

Sintering of TiB_2-W compositions (7.5 to 22.5 wt% W) at 2400°C for 2 h in a vacuum gives parts with no less than 15 % porosity. These porous samples have a poorer resistance to oxidation in air at 1200°C than pure TiB_2 (MEYER and PASTOR, 1969).

TiB_2-(Fe, Co or Ni) Systems

Phase diagrams of the TiB_2-(Fe, Co or Ni) systems have been determined by sintering pressed mixtures of powders (SAMSONOV, 1958). The three pseudobinary systems display a simple eutectic diagram with the following characteristics:

Metal	Mole % TiB_2 at eutectic point	Eutectic temperature (°C)	Mol % TiB_2 solubility at eutectic temperature
Fe	1.5 - 2	1250	$<$ 0.5
Co	~ 4	1220	~ 0.5
Ni	2 - 3	1180	0.5 - 1

The conditions for the production of virtually non-porous TiB_2-Fe cermets by sintering in argon have been determined by FUNKE et al. (1961). Hardness of these alloys decreases with increasing iron content. Tests of the cutting properties of these TiB_2-Fe cermets showed that they have a maximum stability at a cutting speed of 90 m/min. The best cutting properties were found for the alloy with 15% Fe which had a higher stability than the classical 79WC-15TiC-6Co "hard metal" cutting tool (see YUDKOVSKII et al., 1964). FUNKE and YUDKOVSKII (1963) have also studied the oxidation of TiB_2-(Fe, Co or Ni) cermets in air in the temperature range 500-1000°C.

According to TANGERMANN (1963), TiB_2 is completely wetted by nickel and not wetted at all by iron or cobalt (contact angle greater than 90°); nevertheless iron permits full densification. The volatilization of iron during sintering is limited by operating in a gas flow (1 atm); hydrogen, as a sintering atmosphere, has a better influence on densification than argon, probably because of the reduction of the superfecial oxides of the metal constituant. Utilization of (Fe, Co or Ni)-B alloys as binder results in a higher shrinkage leading to complete densification.

FeSi (TANGERMANN, 1963) or NiAl (BROUSSAUD et al., 1974) are poor binders for TiB_2.

TiB_2-(Co or Ni, Cr, B) cermet compositions have been patented for cutting-tool utilization (M.P. GmbH, 1963).

TiB_2-35% stainless steel or nickel, with a Vickers hardness of 1200-1400, has been patented for watch case fabrication (C.I.M.E., 1971).

β) Zirconium Diboride-Metal Systems

ZrB_2-(Cr, Mo or W) Systems

Chromium does not appear to be good binder for zirconium diboride, and seems to favor grain growth (McTAGGART et al., 1954). ZrB_2-Cr cermets are brittle and rather porous, even when sintered at 2000°C (MEYER and PASTOR, 1969). However the addition of 22% Cr yields the highest compressive strength and resistance to mechanical shock, and the addition of 17% Cr yields the highest oxidation resistance in air at 1200°C (TYRRELL and HOUCK, 1964).

In contrast to the conclusions of TYRRELL and HOUCK (1964), molybdenum appears as the best refractory metal binder for ZrB_2, giving cermets with rather low porosity (MEYER and PASTOR, 1969) but low mechanical strength and a poor oxidation resistance. According to KOVAL'CHENKO and SAMSONOV (1959) the shrinkage of ZrB_2-Mo samples (from 0.5 to 90 mol % Mo) starts above 1800°C. These results are supported by the investigations of KISLYI and KUZENKOVA (1965, 1973), who were able to produce gas-proof protective thermocouple sheaths (porosity: 5.3 %) by sintering 95 ZrB_2-5 Mo mol % compositions for 40 min at 2200°C in hydrogen flow; under these conditions, a liquid eutectic composition spreads along the boundaries of the two components, promoting the viscoplastic flow process and favoring the sealing of pores and defects. During isothermal holding in the temperature range 1800-2200°C, very intensive shrinkage is observed in the initial 20-30 min; then, this shrinkage slows down and can be described as a viscous flow resulting from diffusional processes. The energy of activation of the densification process for ZrB_2-Mo cermets (367 kJ/mol for the 5% mol Mo, for instance) is less than the energy of activation of densification of pure ZrB_2 (678 kJ/mol); thus molybdenum activates diffusion processes during sintering (KISLYI and KUZENKOVA, 1966a). ZrB_2-Mo alloys containing less than 15 mol % Mo are solid solutions of molybdenum in ZrB_2, the micro-hardness of which decreases with increasing molybdenum content. If zirconium dodecaboride is present in the initial diboride powder, the alloys formed contain ZrB (KISLYI and KUZENKOVA, 1966b).

Resistance sintering of ZrB_2-W compositions in conditions not well defined yields slightly porous specimens (TYRRELL and HOUCK, 1964). However, by sintering at 2100°C for two h in a vacuum it was not possible to get parts with less than 23% porosity (MEYER and PASTOR, 1969). According to KUZENKOVA et al. (1971), the highest densification (8-10% porosity) is obtained for the ZrB_2-5 wt% W composition by sintering for 1½ h in argon (25°C/min) at a temperature above 1400°C. ZrB_2-W specimens increase appreciably in volume as compared with pure ZrB_2 samples. Shrinkage occurs on subsequent heating only above 1800-1900°C. In the case of a rapid heating rate (700-750°C/min) up to the isothermal holding temperature of 2200°C, at first shrinkage is more intense than that of specimens slowly heated up to this temperature, but subsequently a slight swelling occurs. In both cases swelling is attributed to the formation of new phases (solid solution of tungsten in ZrB_2, and β-WB stabilized by zirconium). The electrical resistivity of sintered ZrB_2-W cermets (in the range 0-50 wt% W) decreases with increasing tungsten content. A maximum is observed for compressive strength (47 Hbar) and bending strength (19 Hbar) for the cermets with 40 wt% and 20 wt% W respectively (KUZENKOVA et al., 1971; KISLYI et

Table 9. Sintering characteristics of Group VI B metal boride-metal systems

Boride	Metal (wt%)	Particle size: Boride (μm)	Particle size: Metal (μm)	Binder (wt%)	Compacting pressure (kbar)	Sintering temperature (°C)	Holding time (min)
Cr_2B	40 Cr/Mo (8/2)	-	-	-	-	1530-1600	-
Cr_2B	50 Cr/Mo	-	-	-	-	1530-1600	-
Cr_2B	60 Cr/Mo	-	-	-	-	1530-1600	-
Cr_2B	15 Fe	-	-	-	0.77	1593	30
Cr_2B	20 Fe	-	-	-	0.77	1593	30
Cr_2B	15 Co	-	-	-	0.77	1593	15
Cr_2B	15 Ni	-	-	-	0.77	1593	5
Cr_2B	5 Ni + 10 Fe	-	-	-	0.77	1593	15
Cr_2B	5 Ni + 15 Co	-	-	-	0.77	1593	10
Cr_2B	13.8 Fe	0.5-4	-	Polystyrene	1.5	1250	120
Cr_2B	13.0 Co	0.5-4	-	Polystyrene	1.5	1220	120
Cr_2B	13.0 Ni	0.5-4	-	Polystyrene	1.5	1220	120
Cr_2B	20.0 Cr	0.5-4	-	Polystyrene	1.5	1190	120
CrB	20 Fe	90	90	-	-	1530	-
CrB	20 Fe	90	90	-	-	1575	-
CrB	50 Fe	90	90	-	-	1575	-
CrB	10 Co	90	90	-	-	1530	-
CrB	20 Co	90	90	-	-	1530	-
CrB	10 Ni	90	90	-	-	1490-1530	-
CrB	10 Ni				2.1	1450	100
CrB	15 Ni				1.06	1400	100
CrB	15 Ni				0.42	1450	100
CrB	20 Ni				2.1	1300	100
CrB	20 Ni				0.42	1300	100
CrB	20 Ni				2.1	1350	100
CrB	20 Ni				2.1	1400	100
CrB	20 Ni				2.1	1450	100
CrB	20 Ni				2.1	1500	100
CrB	20 Ni				2.1	1550	100
CrB	4.6 Fe	0.5-2	-	Polystyrene	1.5	1270	120
CrB	12.7 Fe	0.5-2	-	Polystyrene	1.5	1350	120
CrB	5 Co	0.5-2	-	Polystyrene	1.5	1300	120
CrB	5 Ni	0.5-2	-	Polystyrene	1.5	1300	120
CrB	14 Ni	0.5-2	-	Polystyrene	1.5	1300	120
CrB_2	10 Ni	-	-	-	-	1500	-
CrB_2	20 Ni	-	-	-	-	1500	-
WB	30 Co	< 37	1	Camphor: 1%	1	1300	60
WB	30 Ni	< 37	3.1	Camphor: 1%	1	1200	60
W_2B_5	1 Ni	< 40	-	Rubber + benzene	1	1400	-
W_2B_5	3 Ni	< 40	-	Rubber + benzene	1	1400	-
W_2B_5	6 Ni	< 40	-	Rubber + benzene	1	1400	-
W_2B_5	10 Ni	< 40	-	Rubber + benzene	1	1400	-
W_2B_5	20 Ni	< 40	-	Rubber + benzene	1	1200	-
W_2B_5	30 Ni	< 40	-	Rubber + benzene	1	1200	-

π: porosity (%) - Δ 1/1: shrinkage (%) - Δ W: weight loss (%)

HV: VICKERS Hardness; R_A: ROCKWELL A Hardness - σ_b: bending strength (hbar)
σ_t: tensile strength (hbar)

Sintering atmosphere (P in torr)	Density (g/cm^3)	π (%)	Δ 1/1 (%)	Δ W (%)	HV or R_A	σ_b, σ_t (hbar)	References
H_2: 760	6.91	-	-	-	R_A: 88	-	AEMC (1955) SECEMAEU (1957) AEMC (1958) M.P. GmbH (1959)
H_2: 760	-	-	-	-	86	-	
H_2: 760	7.05	-	-	-	82	σ_b 25.9	
-	4.36	-	-	-	-	σ_t: 14	EVANS (1956)
-	5.6	-	-	-	-	-	
-	5.86	-	-	-	-	10.4	
-	5.70	-	-	-	-	30.8	
-	5.43	-	-	-	-	12.4	
-	6.13	-	-	-	-	22.0	
H_2: 760	6.73	-	16	-	R_A: -	σ_b 17	TANGERMANN (1963)
H_2: 760	6.90	-	19	-	-	35	
H_2: 760	6.90	-	17.5	-	86.4	35.5	
H_2: 760	6.18	-	25	-	84	-	
H_2: 760	6.06	-	-	-	-	σ_b 16.5	ISKOLDSKY and BOGORODSKAYA (1957)
H_2: 760	6.17	-	-	-	-	-	
H_2: 760	5.93	-	-	-	-	-	
H_2: 760	4.6	-	-	-	-	17.1	
H_2: 760	6.23	-	-	-	-	-	
H_2: 760	4.6-4.77	-	-	-	-	-	
	6.18	1.0			HV20:771		MORDUCH (1962)
	6.31	0.5			734		
	6.28	1.0			830		
	6.26	2.9			852		
	6.27	2.8			1090		
	6.31	2.2			777		
	6.44	0.1			682		
	6.39	0.9			710		
	6.15	4.6			804		
	6.05	6.1			939		
H_2: 760	6.08	-	15	-	R_A: -	σ_b 32	TANGERMANN (1963)
H_2: 760	5.56	-	18	-	91.6	15	
H_2: 760	5.70	-	15.5	-	-	25	
H_2: 760	5.97	-	17	-	78	31	
H_2: 760	5.63	-	10	-	77	23	
-	-	-	-	-	HV:1500	σ_t: 10.5	C.I.M.E. (1971)
-	-	-	-	-	1400	8.5	
V: 10^{-2}	13.0	-	-	-	R_A: 90	σ_b 190	KANEKO et al. (1973)
V: 10^{-2}	13.0	-	-	-	85	225	
V: 10^{-3}	-	8	-	-	-	-	KIPARISOV et al. (1970)
V: 10^{-3}	-	7	-	-	-	-	
V: 10^{-3}	-	2	-	-	-	-	
V: 10^{-3}	-	1	-	-	-	-	
V: 10^{-3}	-	0	-	-	-	-	
V: 10^{-3}	-	6	-	-	-	-	

al., 1971; KISLYI and KUZENKOVA, 1973). KISLYI et al. (1972b) have developed a technique for the manufacture of protective thermocouple sheaths from ZrB_2-W cermets.

ZrB_2-(Fe, Co or Ni) Systems

According to McTAGGART et al. (1954) ZrB_2-20 wt% Fe, Ni or Co, ZrB_2-10% Co-10% Cr compositions sintered in argon at temperatures equal to or above the melting points of the respective metals consist of a continuous ZrB_2 skeleton filled in with the binding metal. In all cases the boride underwent a considerable grain growth (the most important being observed with cobalt) which was related to the fluidity of the binder metal.

ZrB_2-(Fe, Co or Ni) cermets are brittle and have rather poor oxidation resistance. Cobalt (18%) provides the least brittle bond, and nickel the hardest (16%) and most resistant to oxidation (11%). Iron reacts with the formation of tetragonal Fe_2B thus precluding the use of iron as binder (TYRRELL and HOUCK, 1963).

A 76 ZrB_2-3 Ni-20 Cr-1 B (wt%) cermet has been patented as cutting-tool material (M.P. GmbH, 1957), while a 70 ZrB_2-20 Ni-10 Co (wt%) cermet is recommended as an unscratchable watch case material (C.I.M.E., 1971).

Group VI B Metal Boride-metal Systems

Table 9 summarizes the sintering characteristics of group VI B metal boride-metal systems. The systems investigated up to now concern only the borides Cr_2B, CrB, CrB_2, WB and W_2B_5 with mainly iron-group metal binders.

A great deal of work has been devoted to the study and development of Cr_2B-(80 Cr-20 Mo) alloys (trade-name: Borolite IV) made either by hot pressing or sintering (A.E.M.C., 1955; S.E.C.E.M.A.E.U., 1957; A.E.M.C., 1958; M.P. GmbH, 1959). These cermets have relatively good bending strength and hardness, a good electrical conductivity, a high resistance to oxidation in air at 1000°C (weight gain: 0.010 $mg/cm^2/h$), but a rather low impact strength.

Sintered specimens of the Cr_2B-rich Cr_2B-(Fe, Co, Ni) systems have been proposed for the manufacture of bodies resistant to molten zinc at 600°C (EVANS, 1956).

According to TANGERMANN (1963), Cr_2B and CrB both react with iron, cobalt and nickel, forming double borides; thus, densely sintered specimens of the Cr_2B-(Fe, Co, Ni) and CrB-(Fe, Co, Ni) systems are very brittle. These results confirm the previous work of ISKOLDSKY and BOGORODSKAYA (1957) on the CrB-(Fe, Co, Ni) systems.

MORDUCH (1962) has studied in particular the sintering of CrB-Ni alloys. He points out the formation of brittle complex borides by reaction of the boride with the binder. The maximum density for compacts containing 20 wt% Ni and sintered for 100 min is obtained at 1400°C. Below and above this temperature the density falls. Changes in compacting pressure do not markedly affect the density of the sintered parts.

A CrB_2-20 (or 10) wt% Ni cermet sintered at 1500°C, with a Vickers hardness of 1400-1500 but only 8.5-10.5 hbar bending strength, has been proposed as an unscratchable material for watch-cases (C.I.M.E., 1971).

According to KIPARISOV et al. (1970) the ditungsten pentaboride W_2B_5 is fairly wetted by molten nickel, thus enabling the sintering of W_2B_5-Ni cermets (1 to 30% Ni) to take place at a relatively low temperature (1200-1400°C) and resulting in quite dense specimens.

The shrinkage of WB-30 Co and WC-30 Ni compacts increases sharply above 1000° and 900°C respectively and stops at 1280° and 1230°C respectively. In both materials, unknown phases could not be identified (complex borides).

3. Conclusion

Solid boride bodies can be prepared by a variety of sintering methods. However, in the present state of research, only titanium and zirconium borides have been intensively studied, mainly because of the potential of their outstanding physical and mechanical properties. Unfortunately, studies of sintering are somewhat rare and incomplete and its mechanisms are not well elucidated. Conventional sintering of pure borides generally leads to porous bodies with consequently low mechanical strength. Densification can be improved by activating the sintering processes by adding small quantities of metals. These activation mechanisms are not yet well understood.

With regard to binders, investigators have as yet not discovered a plastic and non-reactive metal binder for borides. This is certainly the most important problem facing research in the near future, since the industrial development of borides in such fields as cutting tools, and applications requiring high corrosion, heat and/or oxidation resistance, depends on finding a suitable binder to compensate the disadvantages of excessive brittleness of borides.

References

A.E.M.C. (American Electro Metal Corp.): Cemented Borides. Summary Progress Report. PB 121346. Wash. D.C.: O.T.S. 1955

A.E.M.C. (American Electro Metal Corp.): Cemented Metallic Borides. Final Summary Report. PB 139415. Wash. D.C.: O.T.S. 1958

AUGUSTINIK, A.I., DROZDETSKAYA, G.V.: Conditions for making manufactured objects from metal-like refractories by dross casting. Izvest. V.U.Z., Khim. Khim. Tekhnol. 10, 556-558 (1967)

BABICH, B.N., PORTNOI, K.I., SAMSONOV, G.V.: Compacting and sintering of boride powders. Metalloved Term. Obrabot. Metallov, 1, 31-35 (1960). Translation in English: HB-5301; H. BRUTCHER, Altadena, California; or AD 273601, Washington, D.C.: O.T.S.

BROUSSAUD, D., ACCARY, A., PASTOR, H., MEYER, R.: Préparation et propriétés des cermets TiB_2-NiAl. High Temp. High Pressures 6, 455-461 (1974)

BROUSSAUD, D., PASTOR, H., MEYER, R., ACCARY, A.: On the behavior of titanium diboride towards nickel-aluminium Ni-Al. In: Modern Developments in Powder Metallurgy. Vol. 8. (H.H. HAUSNER, W.E. SMITH, eds.) pp. 589-601. Princeton, N.J.: M.P.I.F., A.P.M.I. (1974)

BUMM, H., LIEPELT, H.: Results of isostatic pressing of metallic and non-metallic powders up to 15000 atmospheres. Z. Werkstofftechnik 3, 364-368 (1972)

ČECH, B., OLIVIERUS, P., SEJBAL, J.: Sintering of zirconium boride with activating additions. Powder Met. 8, (N°15), 142-151 (1965)

C.I.M.E. (Compagnie Industrielle des Métaux Electroniques): Boitiers de montre en matériau extra-dur. Fr. 2067099, (Aug. 20, 1971)

COBLE, R.L., HOBBS, H.A.: Sintering. In: Investigation of Boride Compounds for Very High Temperature Applications. (L. KAUFMAN, E.V. CLOUGHERTY, eds.), Report AD 428006, pp. 82-120. Springfield, Va.: N.T.I.S. 1973

DOBROVOL'SKII, A.G., DOBROVOL'SKII, G.G., LYUDVINSKAYA, T.A., POPICHENKO, E.Y.: Slip casting of zirconium boride. Izvest. Akad. Nauk S.S.S.R., Neorg. Mater. 2, 864-869 (1966) Translation in English: Inorg. Mater. 2, 735-739 (1966)

DOBROVOL'SKII, A.G., NIKITIN, Y.I., SOKHIN, M.S., TUTAKOV, O.V.: Preparation of powders on a planetary mill. Poroshk. Metal 6, 1-7 (1967). Translation in English: Soviet Powder Met. 6, 429-433 (1967)

D'YAKONOVA, L.V., KISLYI, P.S.: Comminution characteristics of zirconium diboride. Poroshk. Metal. (1), 1-5 (1972). Translation in English: Soviet Powder Met. (1), 1-4 (1972)

ELLIOTT, A.G., LAVENDEL, H.W.: Sintering of TaB_2. In: Sintering and Related Phenomena. (G.C. KUCZYNSKI, N.A. HOOTON, C.F. GIBBON, eds.) pp. 565-579. New York: Gordon and Breach, 1967

EVANS, R.M.: Boride parts resist molten Zinc. Materials and Methods 43, 132-133 (1956)

FARRIOR, G.M.: An investigation of the physical properties of several transition metal diborides and of binary diboride solid solutions. Dissert. Graduate Council Univ. Tennessee in partial fulfillment of the requirements for the degree of Dr. of Philos. 158 pp. (1965)

FITZER, E.: Classical sintering, reaction-sintering, and hotpressing of high-melting AB_2-hard alloys. Arch. Eisenhüttenw. 44, 703-709 (1973)

FUNKE, V.F., YUDKOVSKII, S.I., SAMSONOV, G.V.: Alloys in the system B-Ti-Fe. Zhur. Prikladn, Khim. 34, 1013-1020 (1961). Translation in English: Russ. J. Appl. Chem. 34, 973-978 (1961)

FUNKE, V.F., YUDKOVSKII, S.I.: High-temperature oxidation of the alloys of titanium boride with iron group metals. Zhur. Fiz. Khim. 37, 1557-1562 (1963). Translation in English: Russ. J. Phys. Chem. 37, 835-837 (1963)

GROPYANOV, V.M.: Effect of dispersion on powder sintering. Ogneupory 34, 46-52 (1969). Translation in English: Refractories 34, 769-775 (1969)

HOFMANN, J., ADAM, J.: Influence of some characteristics of powdered zirconium diboride on its sinterability (in Czech.) Pokroky Praskove Met. (1), 3-12 (1970)

HOYT, E.W., CHORNE, J.: Preparation of self-bonded borides. Vallecitos Atomic Laboratory, General Electric Co., Pleasanton, Calif., CEAP-3332. Wash. D.C.: O.T.S. 1960

ISKOLDSKY, I.I., BOGORODSKAYA, L.P.: The possibility of production of metalloceramic (sintered) hard alloys based on chromium, titanium and tungsten borides. Zh. Priklad. Khim. 30, 177-185 (1957). Translation in English: Russian J. Appl. Chem. 30, 181-188 (1957)

JAYNE, T.D.: (to Rand Development Corp., Cleveland, Ohio). Tool material. US 2996793, (Aug. 22, 1961)

KANEKO, A., NISHIMURA, T., ISHIBASHI, O., TOMONO, S.: The sintering process of WB-Co and WB-Ni. Nippon Tungsten Review 6, 80-86 (1973)

KIPARISOV, S.S., NIKIFOROV, O.A., KUZ'MINA, Y.K.: Possible sintering of tungsten borides with nickel. Izvest. V.U.Z., Tsvetn. Met. 13, 93-95 (1970)

KISLYI, P.S.: Sintering of blanks obtained by pressing powders of refractory compounds. Poroshkov. Met. (5), 68-78 (1962). Translation in English: Soviet Powder Met. (5), 363-370 (1962)

KISLYI, P.S., KUZENKOVA, M.A.: Gas-impermeable protective thermocouple sheaths from zirconium boride. Poroshk. Met. (1), 32-36 (1965). Translation in English: Soviet Powder Met. (1), 23-26 (1965)

KISLYI, P.S., KUZENKOVA, M.A.: Regularities of sintering of zirconium diboride-molybdenum alloys. Poroshk. Met. (5), 16-23 (1966a). Translation in English: Soviet Powder Met. (5), 360-365 (1966a)

KISLYI, P.S., KUZENKOVA, M.A.: Some properties of ZrB_2-Mo alloys. Izvest. Akad. Nauk S.S.S.R., Neorg. Mater. 2, 2139-2144 (1966b). Translation in English: Inorg. Mater. 2, 1850-1854 (1966b)

KISLYI, P.S., KUZENKOVA, M.A.: Cermets and alloys of refractory compounds with molybdenum and tungsten. In: Trudy Vsesoyuz. Nauch.-Tekhn. Konf. po Metallokeram. Mater. i Izdeliyam. (ed., N.V. MANUKYAN et al.), pp. 226-230. Erevan, 1973

KISLYI, P.S., KUZENKOVA, M.A., STRUK, L.I., MORGUNOVSKAYA, B.A., ZVAZHENKO, V.K., BOL'SHECHENKO, A.G., OSTROVERKHOV, V.I.: Protective sheaths of thermocouples for automatic thermal control over metallurgical processes. Poroshk. Met., (12), 49-56 (1972). Translation in English: Soviet Powder Met. (12), 987-992 (1972b)

KISLYI, P.S., KUZENKOVA, M.A., ZAVERUKHA, O.V.: On the sintering process of zirconium diboride with tungsten. Phys. Sintering 3, 29-44 (1971)

KISLYI, P.S., KUZENKOVA, M.A., ZAVERUKHA, O.V.: The effect of the particle size of powders on the sintering kinetics of titanium diboride. Phys. Sintering, 4, 107-118 (1972a)

KISLYI, P.S., SAMSONOV, G.V.: Basic principles of the extrusion-die production of pipes and rods from powders of highmelting compounds. Poroshkov. Met. (3), 31-48 (1962). Translation in English: Soviet Powder Met. (3), 164-177 (1962)

KISLYI, P.S., ZAVERUKHA, O.V.: Vacuum sintering of titanium diboride. Poroshk. Met. (7), 32-35 (1970). Translation in English: Soviet Powder Met. (7), 549-551 (1970)

KOVAL'CHENKO, M.S., SAMSONOV, G.V.: Study of alloys of zirconium boride with molybdenum. Voprosy Poroshkovoi Met. i Prochnosti Materialov, 7, 18-24 (1959)

KUDINTSEVA, G.A., KONDRATOV, I.Y., KUZNETSOVA, G.M.: Preparation and certain properties of cermet alloys of tungsten with yttrium hexaboride. Izvest. Akad. Nauk S.S.S.R., Neorg. Mater. 6, 1335-1336 (1970). Translation in English: Inorg. Mater. 6, 1172-1173 (1970)

KUDINTSEVA, G.A., KUZNETSOVA, G.M., NAMEDOV, F.G., MEERSON, G.A., TSAREV, B.M.: Production of sintered cathodes from pure lanthanum hexaboride and determination of their emission properties. Izvest. Akad. Nauk S.S.S.R., Neorg. Mater. 4, 49-53 (1968). Translation in English: Inorg. Mater. 4, 38-42 (1968)

KUZENKOVA, M.A., KISLYI, P.S.: Grain growth in zirconium diboride during sintering. Poroshk. Met. (1), 12-16 (1966a). Translation in English: Soviet Powder Met., (1), 10-13 (1966a)

KUZENKOVA, M.A., KISLYI, P.S.: The mechanism of shrinkage of zirconium diboride during sintering. Poroshkov. Met. (2), 46-55 (1966b). Translation in English: Soviet Powder Met. (2), 114-121 (1966b)

KUZENKOVA, M.A., KISLYI, P.S., ZAVERUKHA, O.V., KUZ'MA, Y.B.: Structure and properties of zirconium diboride - tungsten alloys. Poroshk. Met. (11), 32-37 (1971) Translation in English: Soviet Powder Met. (11), 879-883 (1971)

KUZNETSOVA, G.M., KUDINTSEVA, G.A., LIKHACHEVA, T.K., SUCHKOV, L.K.: Production of large-sized lanthanum hexaboride parts. Poroshk. Met. (10), 46-51 (1972). Translation in English: Soviet Powder Met. (10), 810-814 (1972)

LAFFERTY, J.M.: Boride cathodes. J. Appl. Phys. 22, 299-309 (1951)

MANDORF, V., HARTWIG, J., SELDIN, E.J.: High temperature properties of titanum diboride. In: High Temperature Materials. (G.M. AUTL, W.F. BARCLAY, H.P. MUNGER, eds.), Vol. II, pp. 455-467. New York: Wiley Interscience 1963

MATSERA, V.E., PUGIN, V.S., DOBROVOL'SKII, A.G., NIKITIN, Y.I., POGORELYI, B.V., STRASHINSKAYA, L.V., GORNOSTAEV, V.F.: Ultrasonic comminution of refractory compound powders. Poroshk. Met. (12), 11-16 (1971). Translation in English: Soviet Powder Met. (12), 948-952 (1971)

MATSERA, V.E., PUGIN, V.S., DOBROVOL'SKII, A.G., GORNOSTAEV, V.F., STRASHINSKAYA, L.V., DERENOVSKAYA, N.A.: Comminution of powders in the planetary centrifucal mill. I. Determination of the optimum comminution conditions. Poroshk. Met. (6), 11-15 (1973a). Translation in English: Soviet Powder Met. (6), 447-450 (1973a)

MATSERA, V.E., PUGIN, V.S., STRASHINSKAYA, L.V., DOBROVOL'SKII, A.G., NIKITIN, Y.I., POGORELYI, B.V.: Comminution of powders in the planetary centrifugal mill. II. Some milling characteristics. Poroshkov. Met. (7), 11-17 (1973b). Translation in English: Soviet Powder Met. (7), 525-530 (1973b).

MATTERSON, K.J., JONES, H.: A study of the tetraborides of uranium and thorium. Trans. Brit. Ceram. Soc, 60, 475-493 (1961)

McTAGGART, G.D., WILLMORE, T.A., BENNETT, D.G.: Metal bonded zirconium diboride. Washinton, D.C.: O.T.S. 1954

MEDVEDEV, O.G., TRUNOV, G.V., CHERNYAK, L.V., SHLYUKO, V.Y.: Production of parts of complex shape in lanthanum hexaboride. Poroshk. Met. (3), 101-102 (1971). Translation in English: Soviet Powder Met. (3), 250-251 (1971).

MEERSON, G.A., GORBUNOV, A.F.: Activated sintering of zirconium boride. Izvest. Akad. Nauk S.S.S.R., Neorg. Mater. 4, 320-324 (1968). Translation in English: Inorg. Mater. 4, 267-270 (1968)

MEERSON, G.A., MAMEDOV, F.G.: Investigation of sintering conditions for lanthanum hexaboride. Izvest. Akad. Nauk S.S.S.R., Neorg. Mater., 3, 942-949 (1967). Translation in English: Inorg. Mater. 3, 844-850 (1967)

MEERSON, G.A., MANELIS, R.M., TELYUKOVA, T.M.: Peculiarities of the production of objects from lanthanum and yttrium hexaborides by sintering under vacuum. Izvest. Akad. Nauk S.S.S.R., Neorg. Mater. 2, 291-298 (1966). Translation in English: Inorg. Mater. 2, 250-255 (1966)

MEYER, R., PASTOR, H.: Les borures réfractaires. Étude particulière des borures de titane et de zirconium: préparation, propriétés et applications. Bull. Soc. Fr. Céramique (N°66), 59-80 (1965)

MEYER, R., PASTOR, H.: Frittage et compression à chaud des diborures de titane et de zirconium. Planseeber. Pulvermet. 17, 111-122 (1969)

MORDUCH, D.W.: An investigation into the suitability of using borides of the transition elements for cutting tool and drawing die purpose. Thesis for Diploma of Imperial College of Science and Technology, London 1962

M.P. GmbH (Metallwerk Plansee GmbH): Bodies of chromium boride and chromium molybdenum and their production. Brit. 809555 (Feb. 25, 1959)

NELSON, J.A., WILLMORE, T.A., BENNETT, D.G.: Metal bonded titanium diboride. Report PB 143651, Washington, D.C.: O.T.S. 1952

PASTOR, H.: Le frittage et la compression à chaud des composés réfractaires non oxydes: borure, carbures, nitrures et siliciures. L'industrie Céramique (N°615), 89-104 (1969)

PECHINEY: Réfractaires, Fr. 1215 862, (April 21, 1960a)

PECHINEY: Réfractaires spéciaux, Fr. 1216 522, (April 26, 1960b)

PENDSE, D.R., PRATT, P.L.: A study of the physical and mechanical properties of cemented borides. In: Special Ceramics. Vol. 5. (P. POPPER, ed.), pp. 135-146. Stoke-on-Trent: The British Ceramic Research Association, 1972

PRSHEDROMIRSKAYA, E.M., SKEPTSOV, V.M.: Some regular features in the sintering of porous parts from high-melting-point compounds. Poroshk. Met. (4), 11-15 (1966). Translation in English: Soviet Powder Met. (4), 269-272 (1966)

RASSKAZOV, N.I.: Sintering of zirconium diboride in the presence of molybdenum, tungsten and boron. In: Third International Powder Metallurgy Conference, Karlovy Vary, Czechoslovakia, Vol. 2, pp. 228-238 (1970)

REDDY, R.L., MONTGOMERY, L.C., GRULKE, C.A.: Slip casting composition. Brit. Pat. 1052 590 (Dec. 30, 1966)

REIDL, R.W., HARTWIG, J.: Titanium diboride articles and methods for making same. US 3215 545 (Nov. 2, 1965)

SAMSONOV, G.V.: Nature of interaction between titanium boride and the iron group metals. Metalloved. Term. Obrabot. Metal. (1), 35-38 (1958). Translation in English: HB 4122, H. BRUTCHER, Altadena, California

SAMSONOV, G.V., KONDRATOV, I.Y.: Cermet alloys of rhenium with yttrium hexaboride. Poroshk. Met. (8), 87-91 (1969). Translation in English: Soviet Powder Met. (8), 673-676 (1969)

SAMSONOV, G.V., KOVAL'CHENKO, M.S.: Some features of the sintering of powders of refractory compounds. Poroshk. Met. (1), 2029 (1961). Translation in English: AEC-tr-5604. Washington, D.C.: O.T.S.

SAMSONOV, G.V., MARKOVSKII, L.Y., ZHIGACH, A.F., VALYASHKO, M.G.: Bor, ego soedineniya i splavy. Kiev.: Akad. Nauk. Ukrain. S.S.R., 1960. Translation in English: Boron, its Compounds and Alloys. AEC-tr-5032, Washington, D.C.: O.T.S. 1962

SAMSONOV, G.V., MATSERA, V.E.: Manufacture and properties of powder metallurgical refractury compound fibers and porous fiber materials. Poroshk. Met. (9), 42-47 (1972). Translation in English: Soviet Powder Met. (9), 719-723 (1972)

SAMSONOV, G.V., MATSERA, V.E.: Pore-rich fiber materials made from high-melting compounds. Powder Met. Int. 5, 133-134; 136 (1973)

SAMSONOV, G.V., NESHPOR, V.S.: Compressibility of powders of borides, carbides and nitrides of refractory metals. Doklady Akad. Nauk S.S.S.R. 104, 405-408 (1955). Translation in English: HB-3661, H. BRUTCHER, Altadena, California

SAMSONOV, G.V., PORTNOI, K.I.: Splavy na osnove tugoplavkikh soedineniy. Moscow: Gosudarst. Nauchno-Tekhn. Izdat. Oborongiz, 1961. Translation in English: Alloys Based on High Melting Compounds. AD 283 859. N.T.I.S., Springfield, Va.: N.T.I.S. 1962

S.E.C.E.M.A.E.U. (Société d'Electrochimie, d'Electrométallurgie et des Aciéries Electriques d'Ugine): Procédé de fabrication d'un mélange de borure de chrome et d'alliage chrome-molybdène et produit obtenu par ce procédé. French Pat. 1147 790 (Nov. 29, 1957)

SHLYUKO, V.Y., CHERNYAK, L.V.: Investigation of the sintering conditions of parts made from lanthanum hexaboride. Poroshk. Met. (12), 41-45 (1967). Translation in English: Soviet Powder Met. (12). 969-972 (1967)
SLEPTSOV, V.M., PRSHEDROMIRSKAYA, E.M., KUKOTA, Y.P.: Porous materials from carbides and borides of the transition metals. Poroshk. Met. (10), 85-90 (1965). Translation in English: Soviet Powder Met. (10), 849-853 (1965)
SLEPTSOV, V.M., PRSHEDROMIRSKAYA, E. M., VITRYANYUK, V.K.: Porous materials from refractory compounds. Rev. Int. Hautes Temper. et Refract. 6, 131-135 (1969)
SMITH, E.A.: Comminuting very hard and soft solids. Metals and Materials 4, 426-428 (1970)
STRUK, L.I.: Study of the densification of refractory compound powders by the electrical conductivity method. Poroshk. Met. (3), 14-20 (1966). Translation in English: Soviet Powder Met. (3), 184-188 (1966)
TAKAGI, R., TAMURA, K., YOSHIDA, S.: Some properties of sintered CrB. Nippon Kinzoku Gakkaishi 20, 375-377 (1956)
TANGERMANN, I.: On the behavior of heavy-metal borides toward binder metals. Neue Hütte 8, 359-365 (1963). Translation in English: HB-6007, H. BRUTCHER, Altadena, California
TYRRELL, M.E., HOUCK, C.W.: Metallic binders for zirconium diboride: iron, cobalt and nickel. US Bureau of Mines, RI-6262 (1963)
TYRRELL, M.E., HOUCK, C.W.: Metallic binders for zirconium diboride: chromium, molybdenum and tungsten. US Bureau of Mines, RI-6549 (1964)
U.S. BORAX (United States Borax and Chemical Corporation): Improvements in or relating to abrasion-resistant materials. Brit. 994 435 (June 10, 1965)
YUDKOVSKII, S.I., EIKHMANS, E.F., GUSEVA, A.N., FUNKE, V.F., ROMANOV, K.F., SMIRNOV, F.F.: Cutting and physical mechanical properties of alloys with a titanium boride base. Sbornik Tr. vsesoyuz. nauchn.-issled. Inst. Tverd. Splavov 5, 130-141 (1964)
ZAVERUKHA, O.V.: Sintering titanium diboride - molybdenum compacts. In: Tekhnologiya polucheniya novikh materialov, pp. 93-98. Kiev: Akad. Nauk. Ukr. S.S.R., Institut Problem Materialovedeniya, 1972. Translation in English: Technology of Producing New Materials. JPRS-59873, pp. 53-58; N.T.I.S., Springfield, Va.: N.T.I.S. 1973

XV. Magnetic Properties of Borides

K. H. J. BUSCHOW

1. Introduction

In many borides the boron atoms appear to be arranged in such a manner as to form as many boron bonds with each other as possible. Thus layer networks of boron atoms can be discerned in the rare earth borides of the type RB_2 and in the compounds of the types RB_4, RB_6 and RB_{12}; it is even possible to speak of three-dimensional networks. The presence of these boron bonds characterizes the borides as a separate class of compounds. It is generally accepted that the presence of the boron bonds is the reason that the borides are electron-deficient, i.e. that the conduction electron concentration in these compounds is considerably reduced if compared with the formal number of valence electrons available. The metallic properties of the borides would be accounted for accordingly by the excess of valence electrons not required by the boron bonds. (For a description of the crystal structures of the borides and the various models dealing with the corresponding electronic structures we refer to other chapters of this volume). Depending on the crystal structure and the nature of the constituent elements the electron deficiency referred to would vary over a wide range and be responsible for the fact that the boron intermetallic compounds comprise superconductors and conductors as well as semiconductors. In as far as the origin of the magnetic moments in the borides and their coupling are determined by the conduction electron concentration, an intimate connection between the magnetic and electrical properties can be expected.

In this review we will first discuss the magnetic properties of transition metal borides, in which the magnetism is due to the 3d electrons residing in narrow energy bands, and where fairly high ordering temperatures can be expected. Secondly we will discuss the magnetic properties of rare earth borides, in which the magnetic moments can be considered as strongly localized, and an indirect exchange mechanism leads to relatively low magnetic ordering temperatures. Finally we will discuss the few examples of actinide borides in which a situation intermediate between those of the rare earth and transition metal borides is expected.

2. Transition Metal Borides

The transition metal borides can be classified as refractory hard metallic materials. They generally display remarkable chemical stability, while in several cases their electrical properties are superior to those of the parent metals.

The magnetic properties of the transition metal borides vary appreciably, both with the nature of the transition metal component and with composition. Some of the borides are Pauli-paramagnetic, some of them are strongly ferromagnetic with ordering temperatures well above room temperature, but several of them are also diamagnetic.

This section will be restricted to borides of the transition elements Ni, Co, Fe and Mn; their properties are the most interesting and give rise in most cases to a well-defined magnetic moment. The available experimental material can be regarded as sufficient for this group to illustrate the main features of the physics of the 3d electron magnetism in transition metal borides.

The magnetic properties as determined by bulk magnetic measurements are summarized in Table 1. None of the three nickel borides gives rise to the formation of a magnetic moment at the nickel atoms. LUNDQUIST et al. (1962) reports a very weak temperature-independent paramagnetism for NiB with $\chi = 4 \times 10^{-8}$ emu, while KOTETSKII and L'VOV (1972) found that this compound can be classified rather as diamagnetic. Paramagnetism was also reported by these authors for Ni_2B and Ni_3B, which agrees with the results of the investigation by CADEVILLE and MEYER (1962).

Table 1. Crystallographic and magnetic data for boride compounds of the transition metals Ni, Co, Fe and Mn

Compound	Crystal structure	Ordering temperature (K)	Saturation moment (μ_B)	References
NiB	o.rh.(CrB)	diamagnetic		KOSTETSKII and L'VOV (1972)
Ni_2B	tetr.(Al_2Cu)	paramagnetic		KOSTETSKII and L'VOV (1972)
Ni_3B	o.rh.	paramagnetic		KOSTETSKII and L'VOV (1972)
CoB	o.rh.(FeB)	diamagnetic		CADEVILLE and MEYER (1962)
Co_2B	tetr.(Al_2Cu)	$T_C = 429$	0.76	FRUCHART (1963)
Co_3B	o.rh. (Fe_3C)	$T_C = 747$	1.12	FRUCHART (1963)
FeB	o.rh.(FeB)	$T_C = 598$	1.12	LUNDQUIST et al. (1962)
Fe_2B	tetr.(Al_2Cu)	$T_C = 1013$	1.92	CADEVILLE and MEYER (1962)
MnB	o.rh.(FeB)	$T_C = 573$	1.92	LUNDQUIST et al. (1962)
MnB_2	hex.(AlB_2)	$T_C = 157$	0.25	CADEVILLE (1966)
Mn_2B	tetr.(Al_2Cu)	paramagnetic		MIRJASOV and PARSANOV (1959)
Mn_3B_4	o.rh.(Ta_3B_4)	$T_N = 392$	--	HIROTA and YANASE (1965)
Mn_4B	monocl.(Mn_4B)	paramagnetic		MIRJASOV and PARSANOV (1959)

Of the three cobalt borides only the first (Table 1) does not give rise to the formation of a cobalt moment. In an early investigation CoB was reported to be ferromagnetic below $T_c = 477$ K (LUNDQUIST and MYERS, 1961) but CADEVILLE and MEYER (1962) showed that CoB is actually diamagnetic (with $\chi = -0.3 \cdot 10^{-6}$ emu/g at 290 K) and decomposes at temperatures higher than 200°C. CREEL and BARNES (1972) indirectly confirm the results of CADEVILLE and MEYER in the sense that their NMR data on CoB are incompatible with magnetic ordering taking place above liquid nitrogen temperature. The compounds Co_2B and Co_3B are both ferromagnetic. Both the saturation moment and the Curie temperature are seen to increase with cobalt concentration. The magnetic data given for Co_2B in Table 1 are in good agreement with those reported by CADEVILLE and MEYER (1962).

Iron forms only two borides, and both compounds display strong ferromagnetism. The results obtained for FeB by LUNDQUIST (1962) are in good agreement with the data reported for this compound by CADEVILLE and DANIEL (1966).

Only in the first three manganese boron compounds listed in Table 1 do the Mn atoms give rise to a magnetic moment. The magnetic ordering temperatures increase with increasing Mn content. The ordering is of the ferromagnetic type for MnB_2 and MnB but of the antiferromagnetic type for Mn_3B_4. The results reported for MnB by LUNDQUIST (1962) are in good agreement with the data of CADEVILLE and DANIEL (1966). Regarding the magnetic properties of the compound MnB_2, the data reported by various authors are at variance: LUNDQUIST and MYERS (1961) report that according to their results the compound MnB is the only ferromagnetic phase above liquid-nitrogen temperature. In their study of the magnetization and the Curie temperature of various manganese boron alloys *versus* composition, the phase MnB_2 has probably been overlooked: in view of its low saturation and Curie temperature (see Table 1) it can readily be obscured in the various magnetization *versus* temperature curves by the presence of small quantities of the phases Mn_3B_4 and MnB. Better agreement with regard to MnB_2 exists between the results of CADEVILLE (1966) and those of ANDERSON et al. (1966). The latter authors report T_c = 143 K and a Bohr magneton number of 0.19 μ_B per Mn atom. Their results were obtained by measurements on a sample of MnB_2 and corrected for the magnetic contribution of small quantities (1.7%) of the phase Mn_3B_4. The two compounds with the highest Mn concentration appear not to give rise to magnetic ordering (MIRJASOV and PARSANOV, 1959).

Apart from the binary compounds listed in Table 1, numerous pseudobinary compounds have been investigated. This applies especially to the compounds of the types $M_{1-x}M'_xB$ and $(M_{1-x}M'_x)_2B$, where M and M' represent elements of the first series of transition metals. The variations with composition of the saturation moment per M atom and of the Curie temperature are shown in Figures 1 and 2, where they are compared with the data for the corresponding transition metal alloys $M_{1-x}M'_x$ (CADEVILLE and DANIEL, 1966).

Series of this type have proved very useful for understanding the 3d electron magnetism of the transition metal compounds. LUNDQUIST and MYERS (1961) have already pointed out the great similarity in behavior between the monoborides and the pure metals. The monoborides in fact show magnetic properties similar to those of the pure metals but with the number of outer electrons approximately 1.2 larger than in pure metals. The additional outer electrons are provided by the boron atoms. Guided by the results of his magnetic measurements and the considerations of KIESSLING (1950) concerning bond formation between the chains of boron atoms in the monoborides, LUNDQUIST (1962) proposed an energy band scheme for the monoborides as shown in Figure 3. In addition to a broad, shallow conduction band and a narrow 3d band (i.e. the two-band model proposed for the pure transition elements by MOTT) a narrow covalent bonding band and a nonbonding band can be discerned in Figure 3. These are the result of the 2s, 2p hypridization of the 2s and 2p boron electron wave functions required by the crystal symmetry. Each of these two latter bands can contain two electrons. The non-bonding band lies above the Fermi level whereas the bonding band lies below the Fermi level in such a manner as to make some emptying of this band

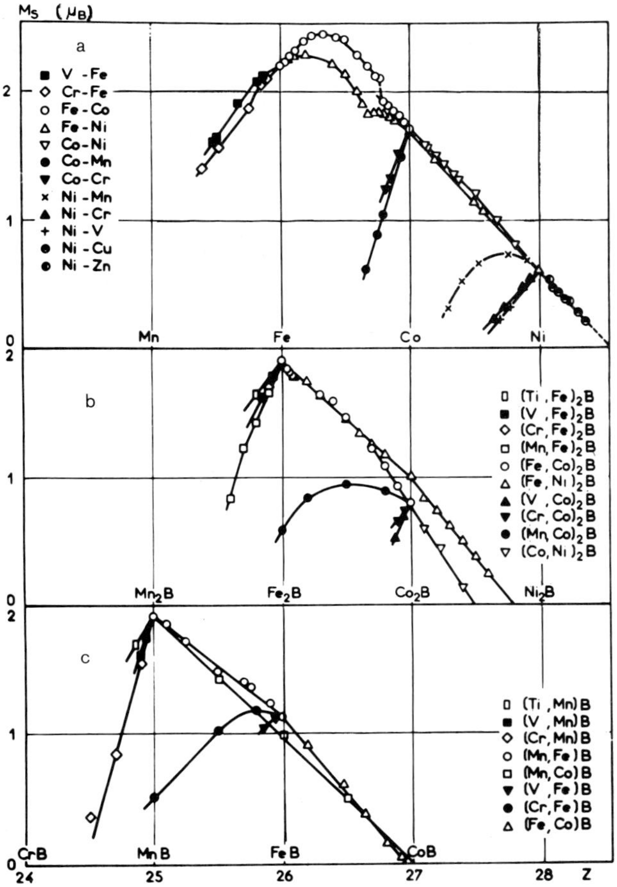

Fig. 1a - c. Saturation moments per transition metal atom as a function of mean atomic number:
(a) transition metal alloys
(b) type M_2B compounds
(c) type MB compounds

possible. In this way about two of the three valence electrons of the B atoms are used for bonds between the boron atoms, and a minimum of one electron contributes to the 3d and/or conduction band. Figure 3 is possibly subject to the following changes: As seen from Table 1, both NiB and CoB are diamagnetic. It is therefore more appropriate to shift the Fermi level of CoB towards some higher energy, corresponding to a situation where the 3d band is filled completely. Furthermore it is not strictly necessary to have as strong a splitting of the bonding and antibonding 2s, 2p bands as assumed in Figure 3. There might even be a moderate overlap of these two bands. In Figure 3 no account has yet been taken of the exchange splitting of the 3d band. The linear variation of the saturation moments in progressing from MnB to CoB has been interpreted by LUNDQUIST and MYERS (1961) to mean that there is only one unsaturated 3d half-band in MnB and FeB. The rapid decrease in magnetization in the series $Mn_{1-x}Cr_xB$ was attributed to a rapid development of unsaturation in both of the 3d half-bands.

CADEVILLE and DANIEL (1966) showed that the similarity in magnetic behavior between transition metal alloys and transition metal borides is not restricted to the MB series but holds equally well for the M_2B series. Inspection of Figure 1 shows that there is a step-by-step shift to the left corresponding to a shift in atomic number of 0.8 - 0.9 of

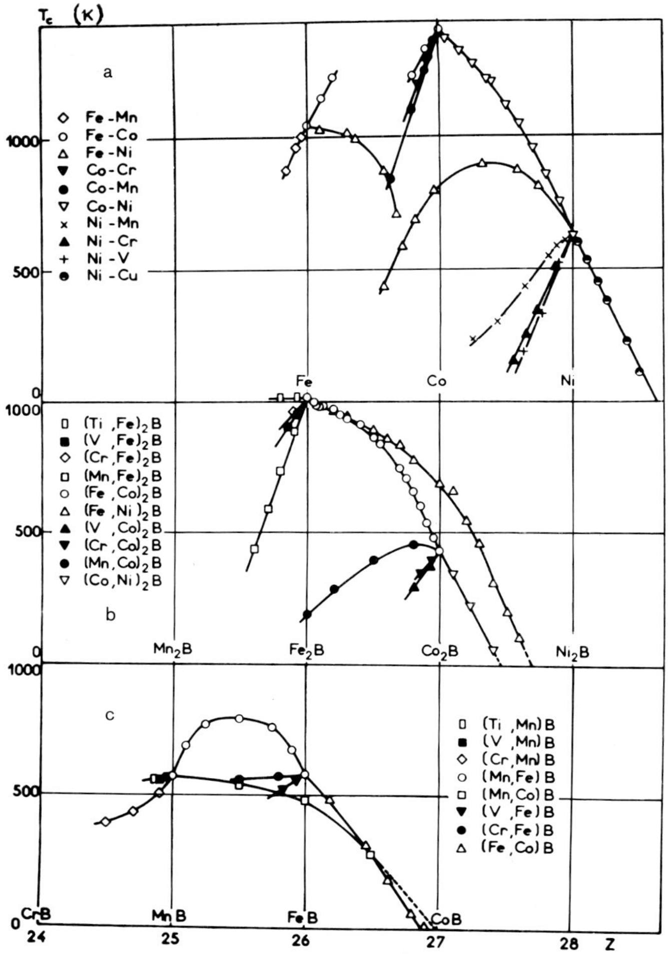

Fig. 2a - c. Ferromagnetic CURIE temperatures as a function of mean atomic number (a) transition metal alloys (b) type M_2B compounds (c) type MB compounds

a unit in the case of the M_2B compounds and 1.7 - 1.8 units in the case of MB compounds. Roughly speaking, this means that each boron atom in both types of compounds contributes 1.8 electrons to the 3d band. CADEVILLE and DANIEL also show that the tight binding approximation can formally describe the moment variations presented in Figure 1 as far as negative slopes dM_s/dZ are concerned. With only one half-band unsaturated $dM_s/dZ = -1$ is expected and actually observed experimentally. This approximation fails, however, to explain the positive dM_s/dZ slopes in Figure 1. In the corresponding compounds the overall density of states very likely comprises high local contributions combined with small effective exchange interactions between the 3d-electrons. This gives rise to a large residual resistivity and a marked moment reduction. The correlation between these two quantities was well established by CADEVILLE and DANIEL (see Figs. 1 and 4), who explained these data in terms of the Friedel virtual bound state concept.

In conclusion we would mention that an analysis of the magnetic properties of the compounds MnB_2 and CrB_2 (CADEVILLE, 1966) showed that in this case, too, the boron atoms contribute about 1.8 (actually 1.7) electrons to the transition metals' 3d band. So it seems that this

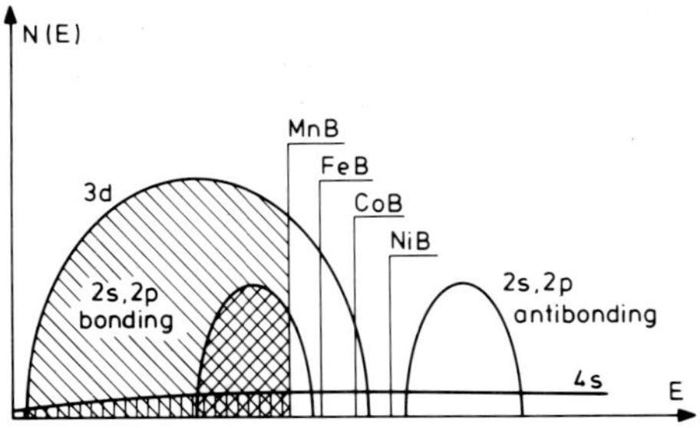

Fig. 3. Energy band scheme for the monoborides of Mn, Fe, Co and Ni. *Vertical lines:* Fermi surfaces suggested for the various substances

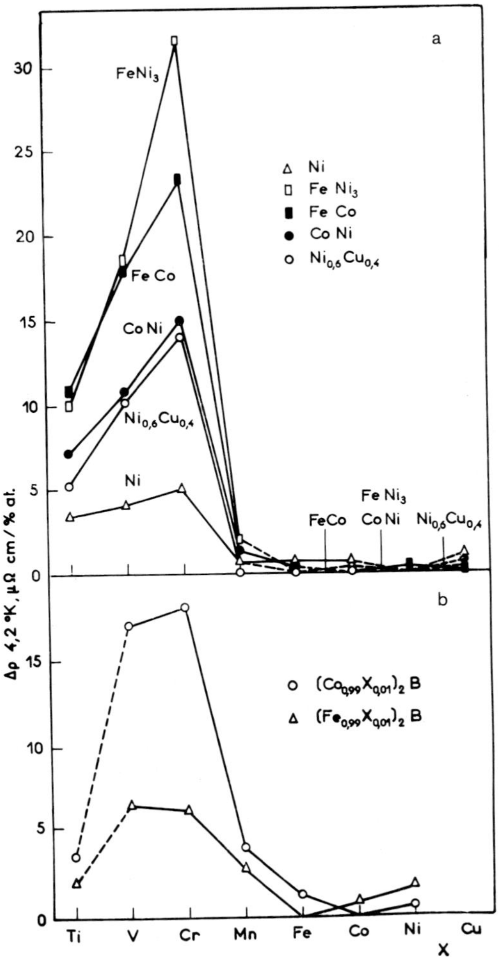

Fig. 4a and b. Residual resistivity for alloys and compounds containing 1 at % of an impurity element *(on horizontal scale)* (a) transition metal alloys (b) type M_2B compounds

electron contribution - 1.8 electron per boron atom - is comparatively independent of the large differences in composition and/or crystal structure encountered with the compounds M_2B, MB and MB_2. In this respect it is interesting to mention a suggestion by TYAN et al. (1969) that it is not necessary to have an actual electron transfer between the boron and transition metal atoms. The presence of the boron atoms could also stabilize the 3d bands with respect to the s conduction band and in this way effectuate s ↔ d transfer. CREEL and BARNES (1972) showed, however, that the measured ^{11}B quadrupole couplings and asymmetry parameters obtained from their monoboride NMR spectra (TiB, VB, CrB, CoB, NiB, MoB and WB) are quite small and consistent with the transfer of electrons away from the boron atoms into the transition metals' 3d band. Experimental evidence of charge transfer in M_2B-type compounds by means of high-resolution bremsstrahlung was obtained recently by KOHM and MERZ (1974).

3. Rare Earth Borides

The following types of rare earth borides have been reported (GSCHNEIDNER, 1961): RB_2, RB_x, RB_4, RB_6 and RB_{12}. More recently BUCHER et al. (unpublished results) observed also compounds of the type R_2B_5 for R = Pr, Nd, Sm and Gd. The exact stoichiometry of the RB_x compounds (x = 3 to 4) has not yet been established and we will therefore disregard these compounds in the present review. The magnetic data for the compounds RB_4, RB_6 and RB_{12} reported below are taken from the results of investigations which have already been reported in the literature, while the data for the compounds RB_2, R_2B_5 and partially also RB_4 are derived from results not yet published.

a) RB_2 Compounds

The magnetic properties of the diboride compounds RB_2 were determined on polycrystalline material obtained by fusing stoichiometric proportions of the starting materials (rare earths 99.9 percent purity, boron 99.99 percent purity) in an arc furnace. X-ray diffraction showed that the samples were single-phase in the cases of R = Tb, Dy, Ho and Er. The X-ray diagrams were indexed on the basis of the hexagonal AlB_2-type unit cell. The lattice constants are given in Table 2. They are in good agreement with data in literature. It proved impossible to prepare AlB_2 type compounds with rare earth elements lighter than R = Tb.

Table 2. Crystallographic and magnetic data for some rare earth diborides (AlB_2 structure)

Compound	Lattice constants a(Å)	Lattice constants c(Å)	Ordering temperature (K)	As. Curie temperature (K)	$\mu_{eff}(\mu_B)$	$\mu_o(\mu_B)$
TbB_2	3.290	3.878	T_c = 151	+151	9.96	4.0
DyB_2	3.287	3.845	T_c = 55	+ 33	10.7	5.4
HoB_2	3.281	3.813	T_c = 15	+ 25	10.2	7.5
ErB_2	3.271	3.782	T_c = 16	+ 9	9.47	5.1

The magnetic properties of the diborides with R = Tb, Dy, Ho and Er are summarized in Table 2. Above the magnetic ordering temperature (T_c) the magnetic susceptibility displays Curie-Weiss behavior with effective moments (μ_{eff}) close to the corresponding values of $g\sqrt{J(J+1)}$. Except in the case of TbB_2 we observed considerable differences between T_c and the asymptotic Curie temperatures (θ_p) derived from extrapolations of the χ^{-1} *versus* T plots. This suggests that the magnetic ordering phenomena are more complicated than simple ferromagnetic ordering. This follows also from the shape of the magnetization *versus* tempera-

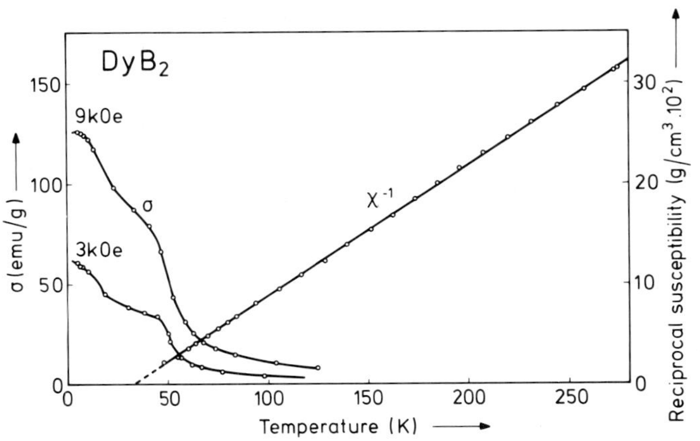

Fig. 5. Temperature dependence of the magnetization *(left-hand scale)* and reciprocal susceptibility *(right-hand scale)* of the compound DyB_2

ture curves of TbB_2 and DyB_2 (Fig. 5), which seem to involve two steps. Studies of the field dependence of the magnetization at 4.2 K suggest that relatively strong magneto-crystalline anisotropies are involved. For all the compounds investigated saturation was still not reached at 18 kOe, the ultimate field strength applied in this investigation. This anisotropy is very probably induced by crystal fields. The magnetic moments (μ_o) attained with 18 kOe are listed in Table 2; they fall appreciably below the corresponding values of $gJ\mu_B$.

b) R_2B_5 Compounds

Rare earth borides of the R_2B_5 with R = Pr, Nd, Sm and Gd type were observed by BUCHER et al. (unpublished observations) to crystallize with monoclinic symmetry and to display silvery aspects. The Néel temperatures of these compounds determined by these authors are 15.1 K, 32.7 K, 23.5 K and 50 K respectively. The Néel temperature of the last compound, namely Gd_2B_5 could be determined less accurately due to some traces of Gd metal still present in the sample. Paramagnetic Curie temperatures were determined for Pr_2B_5 (θ_p = 14 K) and Nd_2B_5 (θ_p = 14 K).

c) RB_4 Compounds

The rare earth tetraborides are tetragonal (UB_4-type). The magnetic properties of these compounds, as obtained by BUSCHOW and CREIJGHTON (1972) are summarized in Table 3. The lattice constants and the effective moment of the compound TmB_4 are taken from the results of PADERNO and POKRZYWNICKI (1967) and PADERNO and SAMSONOV (1961) respectively, while the critical temperatures of this compound are those given by BUCHER et al. These authors also determined the magnetic properties of most of the other RB_4 compounds. Their results were obtained from material grown from Al flux. They are given in brackets in Table 3. Several of the heavy rare earth compounds have also been investigated by PADERNO and POKRZYWNICKI (1967) as well as by FISK et al. (1971).

The compounds LuB_4 (BUCHER et al.) as well as LaB_4 and YB_4 (BUSCHOW and CREIJGHTON, 1972) are diamagnetic. The diamagnetic character of these compounds shows that the charge transfer from R to B has led to a situation in which the 5d(4d) electron states of the R atoms are empty. The diamagnetic nature of these compounds could moreover indicate that the RB_4 compounds in which R is trivalent are semiconductors rather than conductors. However, in view of the electrical measurements performed on some of the RB_4 compounds by JOHNSON and DAANE (1963) as well as by FISK et al. (1971), this possibility can be excluded.

Table 3. Crystallographic and magnetic data for some rare earth tetraborides (UB_4 structure)

Compound	Lattice constants a (Å)	c (Å)	Ordering temperature (K)	As. Curie temperature (K)	$\mu_{eff}(\mu_B)$	$g[J(J+1)]^{\frac{1}{2}}$
CeB_4	7.208	4.092	...	...	...	2.54
PrB_4	7.246	4.121	T_C = 25 (24.2)	5 (+27)	3.53	3.58
NdB_4	7.225	4.103	T_N = 7 (6.85)	-12	3.61	3.62
SmB_4	7.178	4.072	T_N = 57 (31)	...	...	...
GdB_4	7.145	4.050	T_N = 42 (45.3)	-66	7.98	7.94
TbB_4	7.122	4.032	T_N = 43 (44.7)	-43 (-22)	9.30	9.72
DyB_4	7.102	4.018	T_N = 21 (20.7)	-22 (-11)	10.26	10.63
HoB_4	7.086	4.009	T_N = 7 (7.3)	-14 (- 7)	10.50	10.60
ErB_4	7.070	3.996	T_N = 13 (15.0)	-10 (+15)	9.52	9.60
TmB_4	7.05	3.99	T_N = 11.8	3	7.6	7.68
YbB_4	7.059	3.992	T_N = 10	...	...	4.54

With the exception of SmB_4 and YbB_4 there exists reasonable agreement regarding the magnetic ordering temperatures. For the compounds DyB_4 and HoB_4 BUCHER et al. report a second magnetic transition at 13.1 K and 5.05 K respectively. One might wonder whether these are the true magnetic ordering temperatures and whether the observed transition temperatures near 21 K and 7 K for DyB_4 and HoB_4 respectively are due to the presence of hexaborides. However, the X-ray investigation performed by BUCHER et al. as well as that by BUSCHOW and CREIJGHTON (1972) show the presence of DyB_4 and HoB_4 only. The agreement between the magnetic ordering temperatures of SmB_4 and YbB_4 is less satisfactory. It is possible that for samarium and ytterbium the valence state in RB_4 is ill-defined so that small deviations from stoichiometry as well as the presence of lattice faults may have a large influence. In that case different conditions of preparation may lead to different results. BUCHER et al. did not observe magnetic ordering in YbB_4 at temperatures higher than 1.3 K. It is unlikely that the ordering temperature reported for YbB_4 by BUSCHOW and CREIJGHTON (1972) is due to traces of YbB_6, because the latter compounds does now show magnetic ordering (see below). Note that for YbB_4 a deviating valency follows also from an analysis of the lattice constants and from the absence of Curie Weiss behavior at any temperature.

A study of the lattice constants in Table 3 shows that Ce in CeB_4 also has an anomalous valency. BUCHER et al. observed a temperature-independent magnetic susceptibility ($\chi_M = 540.10^{-6}$ cm^3/mole) in the range 300 K down to about 50 K, from which it can be derived that Ce in CeB_4 is tetravalent rather than trivalent. The value of the temperature-independent portion of the magnetic susceptibility of the sample studied by BUSCHOW and CREIJGHTON is somewhat smaller ($\chi_M = 420.10^{-6}$ cm^3/mole). It is possible that also in this case there is a large effect due to differences in preparation conditions.

There exists a wide discrepancy regarding the paramagnetic Curie temperatures reported in the various investigations. The greatest difference is reported for TmB_4 : θ_p = +3 K (BUCHER et al.) compared to θ_p = -23 K (PADERNO and POKRZYWNICKI, 1967). It is difficult to trace the origin of these discrepancies. First, there may be a large (crystal field-induced) magnetic anisotropy in these compounds which gives rise to different values of the susceptibility when measured parallel or perpendicular to the c-direction. A large spread in θ_p values may be introduced if the various investigations refer to measurements involving a prevalence of different crystallographic directions. Second, one cannot exclude discrepancies arising from differences in stoichiometry and/or lattice faults in view of the large variety of conditions of preparation.

From the results of BUCHER et al. and those of BUSCHOW and CREIJGHTON it seems well established that PrB_4 shows a ferromagnetic transition, (Fig. 6) while for the other compounds it is of the antiferromagnetic type.

A change in the sign of θ_p and a change from antiferromagnetism to ferromagnetism are most unusual for magnetic compounds in which the coupling between the magnetic moments proceeds in an indirect way by means of spin polarization of the conduction electrons. In the RKKY approach the expression for the asymptotic Curie temperature reads as follows (DeGENNES, 1962):

$$\theta_p = -3\pi Z^2 (k_B E_F)^{-1} (g-1)^2 J(J+1) |J(o)|^2 \sum_j F(2k_F R_{ij}) \quad (1)$$

$$F(x) = (x \cos x - \sin x)/x^4 \quad (2)$$

where Z is the mean number of conduction electrons, J(o) the effective s-f exchange integral, E_F the Fermi energy and k_F the Fermi momentum. The function $\sum_j F(2k_F R_{ij})$ takes account of the non-uniform spin polarization of the conduction electrons by the rare earth moment. The summation extends over all rare earth atom shells with a rare earth atom at the origin. Due to the non-uniform character (see Eq. (2)) of the conduction electron spin polarization, the contributions of the various shells may have a difference in sign and to a large extent cancel each other out. It can be seen from Eq. (1) that the sign of the function $\sum_j F(2k_F R_{ij})$ determines the sign of θ_p. In the free electron model $k_F \propto R_{ij}^{-1}$, so that the quantity $\sum_j F(2k_F R_{ij})$ is the same for all compounds within a series of isostructural compounds, which means that Eq. (1) does not allow a sign reversal of θ_p within such a series.

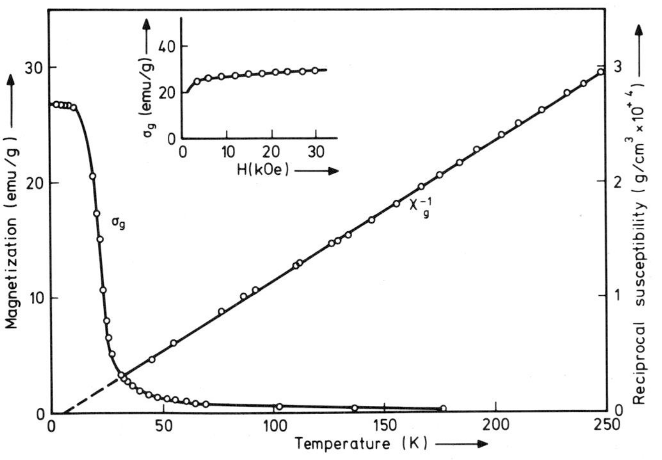

Fig. 6. Magnetization (σ_g) and reciprocal susceptibility (χ_g^{-1}) of the compound PrB_4 as a function of the temperature. *Inset:* field dependence of the magnetization at 4.2°K

KAPLAN and LYONS (1963) and LEVY (1969) considered the influence of the orbital angular momentum on the exchange interaction. A more sophisticated expression for θ_p, which also takes into account the spin orbit splitting of the conduction electrons band states, has been given by LEVY (1969). In contradistinction to Eq.(1), it proves possible with LEVY's formalism to account for a sign reversal of θ_p within a series of isostructural compounds. BUSCHOW and CREIJGHTON (1972) have applied this to the tetraboride compounds and used it to explain the ferromagnetic nature of the compound PrB_4.

It is unfortunate that LEVY's expression for θ_p contains many adjustable parameters which are not easily checked independently by other experiments. The explanation put forward by BUSCHOW and CREIJGHTON therefore loses something of its strength. The hyperfine field at a boron site in RB_4 depends on the conduction electron spin polarization at this site, induced by the polarizing influence of the various rare earth neighbor shells. The expression for the hyperfine field in LEVY's formalism therefore contains similar parameters to the expression for θ_p. Nuclear magnetic resonance at the boron sites would thus be useful as an independent check. Investigations performed by CREIJGHTON et al. (1973) showed, however, that the interpretation of the ^{11}B NMR spectra (which involve a superposition of the spectra of the three crystallographically independent B sites in RB_4) is a difficult task in itself. It proved less suitable for checking the proposed orbital contribution to the exchange interaction. A decomposition of the NMR satellite spectrum of ^{11}B in NdB_4 is shown in Figure 7.

In view of the uncertainty involved in the above explanation of the sign change of θ_p within the isostructural series RB_4 it seems reasonable not to exclude other explanations. One possibility, for instance, is a relatively large contribution of 5d electrons to the indirect coupling of the moments. Recent band structure calculations (HARMON and FREEMAN, 1974) on gadolinium metal indicate that the major part

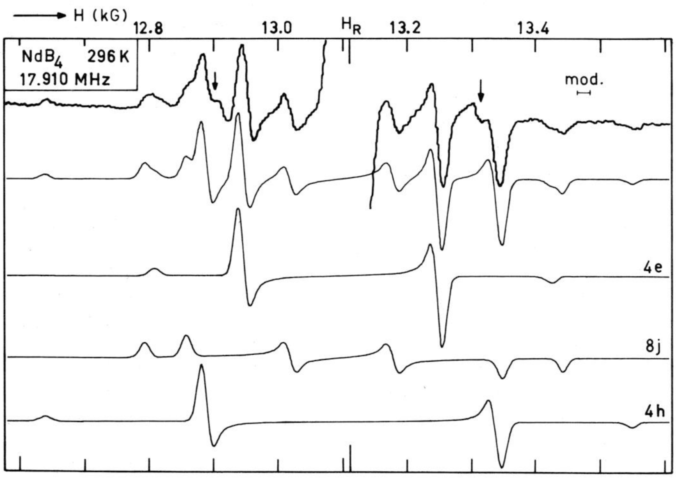

Fig. 7. Measured *(upper curve)* and simulated *(lower curves)* derivative satellite spectra of ^{11}B in polycrystalline NdB_4 at 296 K and 17.910 MHz. The low-field ($H < H_R$) and high-field parts of the spectrum were each obtained from 92 scans of 10 min (2×15.3 h). The modulation was 20 G peak to peak. Two arrows: positions of the 90^0 satellite peaks of ^{11}B in NdB_6

of the conduction electron density in the asymptotic region has a 5d rather than a 6s character. If this is more or less also the case in the present compounds, the moment coupling in RB_4 can be considered as consisting of two contributions involving conduction electrons with 6s and 5d character respectively. While the former reside in broad energy bands, the 5d electrons occupy relatively narrow bands and this makes the exchange coupling proceeding via the 5d electrons more sensitive to lattice contraction or expansion. If the two combinations are of different sign, their relative importance will change in the progress from compounds with large lattice constants to those with low lattice constants and thereby produce a change in sign of the total exchange coupling. The possibility of a two-band model can be checked by EPR measurements (DAVIDOV et al., 1973). The presence of d electrons, which are always unbottlenecked, requires a large g-shift and a large temperature dependence of the line width. The results of FISK et al. (1971) obtained on GdB_4 point indeed in this direction, but further studies will be required in order to check the amount of bottlenecking of the s electrons. On the other hand it should be noted that an explanation in terms of a changing d-electron contribution seems less likely in view of the diamagnetic character of the RB_4 compounds in which R represents La, Y and Lu.

d) RB_6 Compounds

The series of hexaborides forms by far the most interesting series of boride compounds. It comprises an almost unequalled diversification in magnetic and electronic properties, as it contains compounds which are diamagnetic, Pauli-paramagnetic, antiferromagnetic or ferromagnetic as well as compounds which are normal conductors, superconductors or semiconductors.

Most of the elctrical and magnetic properties of the RB_6 compounds can be regarded as very sensitive to the number of valence electrons available to the conduction band. A considerable dependence of these properties on the exact composition of the RB_6 phase may therefore be expected. Recently SCHWETZ and LIPP (1973) showed that for EuB_6 a large homogeneity region exists, extending even to $Eu_{0.9}B_6$. SCHWETZ and LIPP also discuss the presence of homogeneity regions in most of the other hexaborides and argue that this need not always be accompanied by easily detectable shifts in the lattice constants. It is therefore very likely that the discrepancies emerging from the numerous magnetic and electrical investigations to be reported below have their origin partly in differences in composition dependent on the conditions of preparation.

The structure of the hexaborides (CaB_6 type) is relatively simple. It can be visualized as a CsCl structure in which R atoms reside on the Cs sites while the Cl sites are occupied by small octahedra of B atoms. These B octahedra are oriented in such a way that they form a three-dimensional network. A survey of the lattice constants and the magnetic properties is given in Table 4. The lattice constants were taken from GSCHNEIDNER (1961). The ordering temperatures are those of MATTHIAS et al. (1968). Where these differ appreciably from other reported values the latter are indicated in brackets (but see below). The most exhaustive investigation into the paramagnetic regime of these compounds was carried out by PADERNO et al. (1967). Their results are contained in the fourth column of Table. 4.

The first compound, LaB_6, was reported to be a Pauli paramagnet (PADERNO et al., 1967), becoming superconducting (although incomplete) below 5.7 K (MATTHIAS et al., 1968). This, however, was queried by ETOURNEAU et al. (1970), who showed that LaB_6 is diamagnetic rather than Pauli-paramag-

Table 4. Crystallographic and magnetic data for some rare earth hexaborides (CaB_6 structure)

Compound	Lattice constant (A)	Ordering temperature (K)	Asymp. Curie temperature (K)	$\mu_{eff}(\mu_B)$	$g[J(J+1)]^{\frac{1}{2}}$
LaB_6	4.154	diamagnetic $\chi_g = -0.3 \cdot 10^{-6}$			
CeB_6	4.139	$T_N = 3(23)$	- 76	2.49	2.54
PrB_6	4.130	$T_N = 7$	- 68	3.59	3.58
NdB_6	4.125	$T_N = 8.6$	- 42	3.54	3.62
SmB_6	4.130	not observed	no Curie-Weiss behavior		...
EuB_6	4.178	$T_c = 8.5$	+ 9	8.1	7.94
GdB_6	4.111	$T_N = 18(13.5)$	- 55	8.01	7.94
TbB_6	4.102	$T_N = 23$	- 35	9.43	9.72
DyB_6	4.098	$T_N = 21.5$	- 21	10.63	10.63
HoB_6	4.097	$T_N = 9$	--	--	10.60
YbB_6	4.143	not observed	-161	--	4.54

netic (above 77 K) and that above 1.3 K LaB_6 behaves as a normal conductor. The break in the temperature dependence of the specific heat near 6 K was ascribed by these authors to the impurity content and/or stacking faults which could be varied by changing the conditions of preparation.

Cerium hexaboride is an antiferromagnet with a Néel temperature below 3 K (GEBALLE et al., 1968). The ordering temperature has been determined more accurately by LEE and BELL (1972): $T_N = 2.31$ K. The magnetic susceptibility obeys a Curie-Weiss law with a free-ion slope down to about 150 K but rises above the extrapolated Curie-Weiss value below this temperature. This suggests the presence of strong crystalline electric fields which, for cubic symmetry, split the $J = 5/2$ state of Ce^{3+} into a quartet state (Γ_8) and a doublet state (Γ_7). GEBALLE et al. (1968) found a saturation moment of 0.56 μ_B per Ce at 1.3 K, which indicates that the Γ_7 state is the lower. NICKERSON and WHITE (1969) tried to analyze the experimental $\chi^{-1}(T)$ curve in terms of crystal field effects, but found that these effects alone are unable to account for the marked decrease of $\chi^{-1}(T)$ below 150 K. In their model they assume an indirect exchange interaction between the localized 4f electrons (s_f), proceeding via the conduction electrons (s), i.e. involving the Hamiltonian $J_{sf}\vec{s}_f \cdot \vec{s}$. In view of the difference between the angular probability distributions of the cubic wave functions of the states Γ_8 and Γ_7, NICKERSON and WHITE argue that in CeB_6 the s-f exchange integral J_{sf} should depend also on the orbital state of the 4f electron and hence be different for Γ_8 and Γ_7: for the Γ_8 state the probability distribution protrudes into the space between the boron tetrahedra and therefore the s-f exchange interaction can be expected to be relatively strong. As the temperature decreases, the Γ_8 state becomes more and more depopulated, leaving only the weakly coupled Γ_7 state. Magnetic ordering involves primarily the Γ_7 state, whereas the value of the asymptotic Curie temperature (θ_p) depends to a relatively large extent on Γ_8. This would make the large difference between $T_N = 2.31$ K and $\theta_p = -76$ K understandable. The low-temperature specific heat of CeB_6 was recently investigated by LEE and BELL (1972), who interpreted the large specific heat above T_N = 2.31 K with the aid of a decorated Ising model. Their analysis gives independent support for NICKERSON and WHITE's proposal that the exchange interaction involving the excited state (Γ_8) is substantially stronger than that involving the ground state (Γ_7). The specific heat data and an extension of the susceptibility measurements to temperatures below 77 K permitted a more accurate determination of the overall crystal field splitting (Δ) in CeB_6. By fitting their data LEE and BELL found Δ/k_B = 70 K, compared with the value Δ/k_B = 320 K proposed by NICKERSON and WHITE. The exchange interaction involving the Γ_8 level is estimated by LEE and BELL to be 28 times as high as that involving the Γ_7 ground state and these authors point to the possibility that CeB_6 barely misses ordering at some temperature higher than T_N = 2.31 K and that fluctuations and shortrange order may be very important. They show that, in terms of the decorated Ising model, short-range order can be induced by exchange involving an excited crystal field state, as that state becomes thermally populated.

In view of the results outlined above, a spin-disorder resistivity which increases markedly as the temperature rises is expected. That the situation is more complicated than this is indicated by the resistivity data of FISK et al. (1969), who report a Kondo-like minimum in the temperature dependence of the resistivity of CeB_6.

Praseodymium hexaboride orders antiferromagnetically at 7 K (MATTHIAS et al., 1968). The ordering temperature was determined more accurately by specific heat measurements. GEBALLE et al. report a very sharp heat capacity peak at 6.92 K. These authors observed a Curie-Weiss type behavior of $\chi(T)$ between 15 K and room temperature, corresponding to an effective moment per Pr atom of 3.56 μ_B and asymptotic Curie temperature of -44 K. This latter value is appreciably lower than the result of PADERNO et al. (1967), who report θ_p = -68 K, obtained by extrapolation from liquid-nitrogen temperature. The nature of the magnetic ordering phenomenon was studied in detail by LEE et al. (1970) using specific heat and resistivity measurements. LEE et al., guided by the results of NICKERSON and WHITE (1969) on CeB_6, assume that the nature of the crystal field leads to a splitting of the J = 4 state with a non-magnetic singlet state lying lowest. They showed that the extreme sharpness of the specific heat anomaly at T_N (see Fig. 8) is due to a first-order transition which for a singlet ground state system can be expected within the Wang-Cooper model (WANG and COOPER, 1969). By comparing the magnetic specific heat and the temperature derivative of the resistivity, they confirmed that appreciable short-range order above T_N is absent in PrB_6. This contrasts more or less with the view of LEE and BELL (1972) who qualitatively interpret the large specific heat above T_N as being due to short-range order induced by a relatively strong exchange interaction in a thermally populated excited crystal field state.

Neodymium hexaboride orders antiferromagnetically. The values reported for the ordering temperature are T_N = 8.4 K (WESTRUM et al., 1966) and T_N = 8.6 K (MATTHIAS et al., 1968). Heat capcity measurements on NdB_6 performed by WESTRUM et al. showed a cooperative λ-type anomaly near 7.5 K and a Schottky anomaly with a maximum near 50 K. These authors analyzed the heat capacity data in terms of crystal field splitting of the ten-fold degenerate ground state of the J = 9/2 state into a low-lying doublet state and two excited quartet states. The best fit to their data was obtained with energy separations to the excited quartet levels of 115 and 320 K respectively. The proposed level scheme is consistent with the level schemes mentioned above for the isostructural compounds CeB_6 and PrB_6.

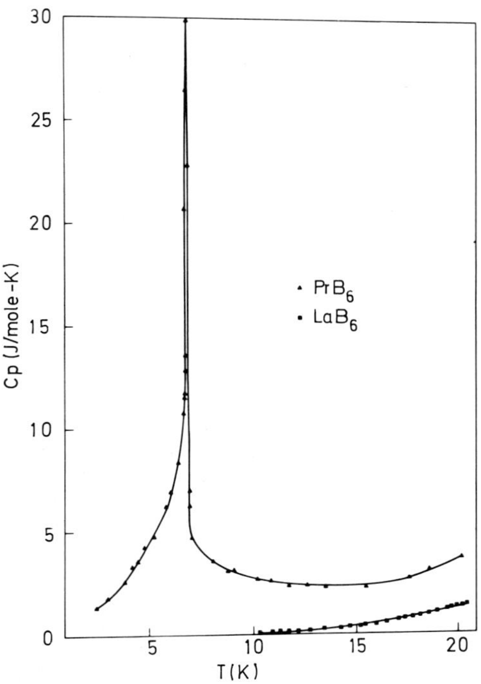

Fig. 8. Specific heat of the compounds PrB_6 and LaB_6. The specific heat of LaB_6 represents the approximate contributions due to lattice vibrations and conduction electrons in PrB_6

Samarium hexaboride does not show magnetic ordering down to 0.35 K. Compared with the magnetic properties of the other RB_6 compound this behavior is quite remarkable and has been interpreted by MENTH, et al. (1969) as being due to the Sm ions having their non-magnetic divalent configuration at low temperatures. At higher temperatures the Sm ions change from the divalent to the trivalent configuration and as a consequence SmB_6 changes from a semiconductor to a metal. Figure 9 shows the temperature dependence of the resistance of SmB_6 as reported by MENTH et al.. It reflects the exponential dependence of the carrier concentration due to the thermally ionized Sm^{2+} ions. As further

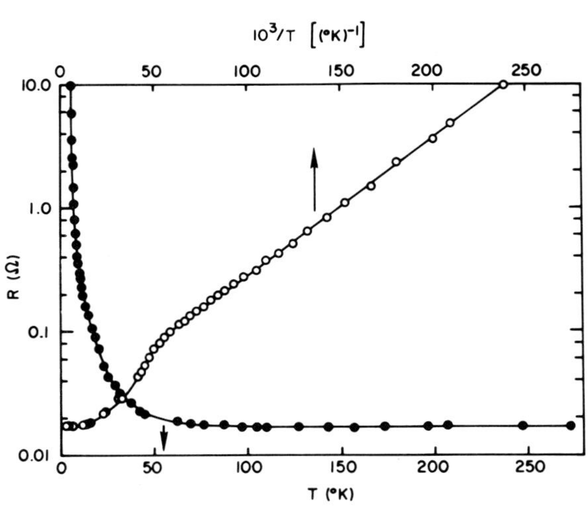

Fig. 9. Temperature dependence of the electrical resistivity of SmB_6. *Black dots:* resistivity versus T; *open circles:* resistivity versus $10^3/T$

evidence for the valence change of the Sm ions, MENTH et al. mention that the temperature dependence of the size of the SmB_6 unit cell fails to show normal thermal expansion but rather passes through a minimum near 150 K. The temperature dependence of the reciprocal molar susceptibility of SmB_6 obtained by these authors is shown in Figure 10. The behavior below 10 K is ascribed to rare earth impurities present in the Sm starting material. At low temperatures the susceptibility is determined by divalent Sm and a temperature-independent van Vleck-type susceptibility is expected. At higher temperatures the excited trivalent state becomes thermally populated and eventually the magnetic properties will be governed by the magnetism of the Sm^{3+} ions. Even if the activation energy between the divalent state and the trivalent state were known and the Sm^{3+} ion concentration could be estimated, a quantitative interpretation of the susceptibility *versus* temperature curve of SmB_6 seems almost impossible without additional experimental data. The reason for this is the low g value and the small energy separation between the ground multiplet (J = 5/2) and higher multiplets, which can lead to an appreciable crystal field or exchange field induced mixing of the excited levels into the ground state multiplet. Apart from the activation energy mentioned, at least three more parameters are required for a quantitative description of the temperature dependence of the susceptibility (De WIJN et al., 1973). The situation seems still more complicated, however. COHEN et al. (1970) have investigated the temperature dependence of the isomer shift accompanying the Sm ion valence change. According to these authors, some of the Sm ions may be present in a new electronic configuration ($4f^s5d^x6s^y$) rather than as trivalent ions. MAPLE and WOHLLEBEN (1971) explained the magnetic behavior of SmB_6 in terms of the Friedel-Anderson theory as modified by Hirst. They propose a model in which the two valency levels are tied to the Fermi level, which stabilizes an intermediate valence. There exists a fluctuation frequency between the two configurations which is much too high for Mössbauer experiments to be able distinguish between them.

Europium hexaboride is a semiconductor in close agreement with europium being divalent. GEBALLE et al. (1968) showed that EuB_6, in contrast to all other RB_6 compounds giving rise to magnetic ordering, becomes ferromagnetic rather than antiferromagnetic upon cooling. The same ordering temperature is observed for EuB_6 samples differing in stoichiometry and electrical resistivity. On the grounds of this insensitivity of the ordering temperature to conductivity, GEBALLE et al. suggested

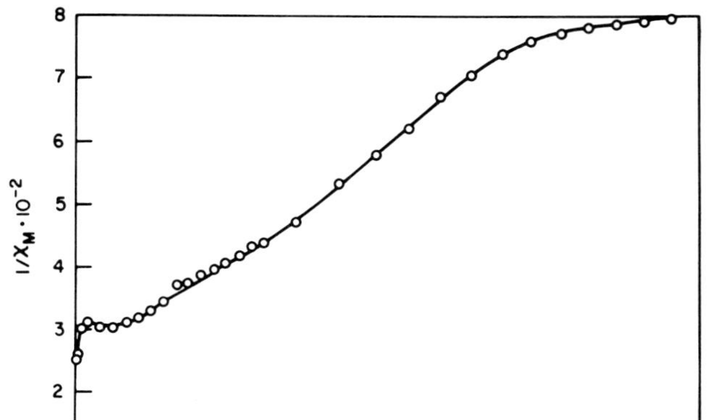

Fig. 10. Reciprocal molar susceptibility of SmB_6 as a function of temperature

that the ferromagnetism of EuB_6 is due to a super-exchange mechanism rather than to the involvement of conduction electrons.

A different explanation for the ferromagnetism of EuB_6 has been put forward by FISK (1971), who suggested that instead of the normal RKKY mechanism responsible for the antiferromagnetic exchange in, for instance, GdB_6, the Bloembergen-Rowland interaction should be applied in the case of EuB_6. The Bloembergen-Rowland interaction is a modification of the RKKY interaction and applies to semiconductors with a valence band whose width is much smaller than that of band gap to conduction band. Using the valence for the squared s-f exchange integral J_{sf}^2 extracted from the experimental data available for GdB_6, FISK was able to account for both the magnitude and the sign of θ_p for EuB_6. WOOD (1971), however, has queried whether such a procedure is justified, since J_{sf}^2 represents different quantities in the case of a semiconductor and a conductor. He suggested that, for a highly degenerate semiconductor such as EuB_6, the indirect exchange interaction may be considered as consisting of three contributions: (1) the normal RKKY interaction involving the conduction electrons, (2) the Bloembergen-Rowland interaction calculated as if conduction electrons were absent, and (3) a correction term for the Bloembergen-Rowland interaction, taking into account the unavailability of the lowest state in the conduction band. WOOD discussed the prevalence of the three contributions for a varying conduction electron concentration due to different preparative conditions and showed that the details of the indirect exchange interaction are apparently quite complicated.

NMR measurements on EuB_6 and a number of other hexaborides were performed by McNIFF and SHAPIRO (1963), who observed a negative Knight shift at the B^{11} sites which was even larger than in GdB_6. These authors tried to obtain an indication of the electron density variation effect in EuB_6 compared to GdB_6 by making magnetic effects the same in both compounds and substracting the GdB_6 results from the EuB_6 data. Even if the RKKY approach held without any reservation, this approach does not seem justified in view of the large difference expected between the Fermi wave vectors of the two compounds.

Gadolinium hexaboride orders antiferromagnetically at low temperatures. Two different values have been reported for the Néel temperature: T_N = 18 K obtained by GEBALLE et al. (1968) compared to T_N = 13.5 K and T_N = 14 K found by COLES et al. (1962) by means of electrical resistivity measurements and susceptibility measurements, respectively. COLES et al. additionally furnished experimental evidence that the exchange interaction between the localized 4f moments in the RB_6 compounds which experiment and theory indicate to be metallic in character proceeds via the conduction electrons. They showed that the temperature dependence of the electrical resistivity has indeed the spin disorder contribution expected on this basis. Above the Néel temperature they also observed electron spin resonance signals with the characteristics to be expected of a Gd^{3+} ion embedded in a metallic matrix. The exchange interaction in GdB_6 as well as in EuB_6 and various compounds of the type $Gd_xLa_{1-x}B_6$ was studied in greater detail by means of EPR by SPERLICH and JANNECK (1972) and SPERLICH et al. (1973). The strength of the exchange coupling between Gd moments, estimated from the exchange-narrowed contribution to the line width, was found to be in good agreement with the strength of the exchange coupling derived from susceptibility data.

The hexaborides of terbium, dysprosium and holmium are all antiferromagnets. The Néel temperatures of TbB_6 and DyB_6 given by MATTHIAS et al. (1968) are in good agreement with the values obtained by FISK et al. (1971). Inspection of Table 4 shows that it is apparently only in the case of DyB_6 that $\theta_p/T_N \sim -1$ as expected for magnetic ordering with a NaCl-type magnetic unit cell. For TbB_6 and GdB_6 the quantity θ_p is appreciably higher in absolute value than T_N and this might be indicative of a different type of magnetic ordering in these compounds (FISK et al., 1971). According to FISK et al., this apparent deviation from the RKKY model may be the result of a splitting of the conduction electron Brillouin zone, which can significantly modify both the conduction electron energy spectrum and the RKKY interaction. This splitting is greater or smaller according to the spin moment of the rare earth element.

There is some doubt about the existence of hexaborides for the rare earth elements erbium and thulium (GEBALLE et al., 1968; GSCHNEIDNER, 1961). NICHOLS et al. (1973) have shown that ErB_6 is actually metastable at room temperature. No magnetic data for ErB_6 and TmB_6 have been reported in literature.

The magnetic properties of ytterbium hexaboride were studied in detail by ETOURNEAU et al. (1970), who showed that Yb in YbB_6 is almost exclusively in the divalent state. The weak temperature-dependent paramagnetism with $\chi^{-1}(T)$ following Curie-Weiss behavior in the range 77 to 400 K could, according to these authors, be explained by the presence of about 2 percent of trivalent Yb or by the presence of other paramagnetic impurities. There is as yet no explanation for the rather high value of θ_p in YbB_6. It is known, however, that a changing valence of Yb atoms can give rise to an apparently high value of θ_p (HAVINGA et al., 1973).

Yttrium hexaboride has been reported to become superconducting at temperatures below 6.5 - 7.1 K (MATTHIAS et al., 1968). The same authors also determined the coefficient of the electronic contribution to the specific heat and found a value of about 7×10^{-4} cal.mole^{-1}deg^{-2}. The sign and magnitude of the B^{11} Knight shift observed in YB_6 by GOSSARD and JACCARINO (1962) and by McNIFF and SHAPIRO (1963) seem to be at variance. FISK et al. (1969) investigated the depression of the superconducting transition temperature YB_6 by rare earth impurities. This depression was found to be roughly proportional to the de Gennes factor $(g-1)^2J(J+1)$ except in the case of Eu, Yb, Tm and Ce. The different effect of Eu and Yb impurities reflects the divalent character of these elements, while the extremely large effect of Ce is interpreted by FISK et al. as being indicative of a virtual 4f level. The large effect of Tm is not understood.

In conclusion, it might be said that the family of rare earth hexaborides constitutes a series of compounds with markedly varying physical properties, but that the magnetic interactions in these substances are fairly complex and only partly understood at the moment. There is no doubt that the main interaction in compounds in which the magnetic rare earth component is trivalent is of the RKKY type. We would again refer in this connection to the results obtained for rare earth impurities in YB_6 by FISK et al. (1969) and the Knight shift results by GOSSARD and JACCARINO (1962). It is remarkable, however, that the values observed for the asymptotic Curie temperature of the RB_6 compounds under discussion do not follow the trend of the de Gennes factor $(g-1)^2J(J+1)$. This trend is followed approximately for GdB_6, TbB_6 and DyB_6 but is deviated from quite considerably for the group of RB_6 compounds with the light rare earth elements. It is not unlikely that, depending on the rare earth component, a varying degree of participation of 5d electrons in the indirect exchange coupling is involved. In this respect it is interesting to note that the nuclear quadrupole interaction detected by ^{11}B NMR (GOSSARD and JACCARINO, 1962) in RB_6 compounds shows a fractional increase which varies far more markedly over this series (about 20%) than the inverse cube of the interatomic spacing (about 1%). An interaction involving d electrons has also been suggested by FISK (1969) to explain the results he obtained on Ce- and Pr-doped YB_6. On the other hand, the EPR results obtained by SPERLICH et al. (1973) on compounds of the $Gd_xLa_{1-x}B_6$ series do not suggest any appreciable s-d exchange interaction.

e) RB_{12} Compounds

Rare earth dodecaborides do not seem to exist for the light rare earth elements and the first few of the heavy rare earth elements. Magnetic data have appeared in literature only for the RB_{12} compounds with R representing Ho, Er, Tm, Yb and Lu. According to MATTHIAS et al. (1968), the compounds HoB_{12}, ErB_{12} and TmB_{12} order antiferromagnetically at 6.5 K, 6.5 K and 4.2 K repsectively. LuB_{12} is reported to become superconducting below 0.48 K. The compound YbB_{12} was investigated by FISK et al. (1969) and reported to have a highly unusual resistance behavior. These authors found that Yb is trivalent in YbB_{12} and suggest that YbB_{12} is the virtual 4f-hole analog of CeB_6. Their investigation was, however, greatly hampered by the difficulty of obtaining YbB_{12} as a single-phase compound.

4. Actinide Boron Compounds

Magnetic investigations on actinide boron compounds were carried out only on a relatively small number of compounds, owing to the radioactive hazard involved with most of the actinide elements. This review will therefore present only the results reported for Th-B compounds and U-B compounds.

Thorium forms only two compounds with boron, namely ThB_4 (UB_4 structure type) and ThB_6 (CaB_6 structure type). Susceptibility measurements on these compounds were performed by AUSKERN and ARONSON (1968). ThB_4 was found to be a Pauli paramagnet with $\chi(300\ K) = 0.04 \cdot 10^{-6} cm^3/g$, in good agreement with the results obtained by BUSCHOW and CREIJGHTON (1972). The compound ThB_6 is diamagnetic, with $\chi(300\ K) = -0.08 \times 10^{-6}$ cm^3/g, and superconducting with $T_c = 0.74$ K (MATTHIAS et al., 1968).

Uranium forms three boron compounds: UB_2 (AlB_2 structure type), UB_4 (UB_4 structure type) and UB_{12} (UB_{12} structure type). Magnetic measurements on these compounds were performed by TROĆ et al. (1971). The temperature dependence of the susceptibility of the three uranium borides is shown in Figure 11. UB_2 exhibits a temperature-independent susceptibility over the complete temperature range investigated. The susceptibility of UB_{12} decreases only slightly with temperature with an almost constant slope over the entire range. UB_4 has a susceptibility maximum of about 150 but shows Curie-Weiss behavior above 450 K with $\mu_{eff} = 3.4\ \mu_B$ per U atom and an asymptotic Curie temperature of -680 K. TROĆ et al. have interpreted the magnetic data obtained for the uranium borides on the basis of band theory rather than on that of localized 5f moments.

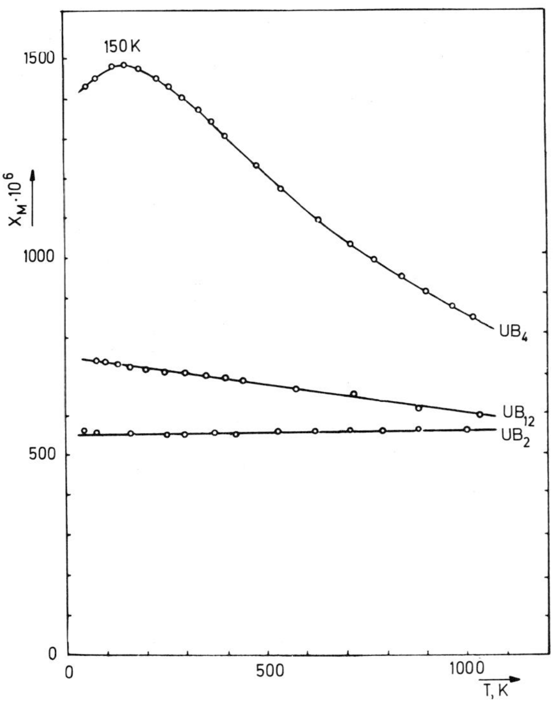

Fig. 11. Temperature dependence of the molar susceptibility of the compounds UB_2, UB_4 and UB_{12}

References

ANDERSON, L., DELBY, B., MYERS, H.P.: Ferromagnetic MnB_2. Solid State Comm. 4, 77-78 (1966)

AUSKERN, A.B., ARONSON, S.: Electrical properties of thorium borides. J. Chem. Phys. 49, 172-276 (1968)

BARBARA, B., BECLE, C., NGUYEN, N.N., SIAUD, E.: Magnetic properties and magnetic structure of TbGa and ErGa compounds. Proc. 3rd Rare Earth Conference, Inst. Phys. London 1971

BUCHER, E., SCHMIDT, P.H., CASTELLANO, R.N., DERNIER, P.D., COOPER, A.S. (unpublished results)

BUSCHOW, K.H.J., CREIJGHTON, J.H.N.: Magnetic properties of rare earth tetraborides. J. Chem. Phys. 57, 3910 1972

CADEVILLE, M.C.: Proprietes magnetiques des diborures de manganese et de chrome MnB_2 et CrB_2. J. Phys. Chem. Sol. 27, 667-670 (1966)

CADEVILLE, M.C., DANIEL, E.: Sur la structure électronique de quelques borures d'éléments de transition. J. Phys. (Paris) 27, 449-457 (1966)

CADEVILLE, M.C., MEYER, A.: Points de Curie et moments des borures ferromagnêtiques du groups du fer de type M_2B, MB, $(M_1, M_2)_2B$ et $(M_1, M_2)B$. C.R. Acad. Sci. (Paris) 255, 3390-3391 (1962)

CASTELLANO, R.N.: Crystal growth of TmB_2 et ErB_2. Mat. Res. Bull. 7, 261-266 (1972)

COHEN, R.I., EIBSCHÜTZ, M., WEST, K.W., BUEHLER, E.: Electronic configuration of SmB_6. J. Appl. Phys. 41, 898-899 (1970)

COLES, B.R., COLE, T., LAMBE, T., LAURANE, N.: Electrical resistivity and paramagnetic resonance in gadolinium hexaboride. Proc. Phys. Soc. (London) 79, 84-86 (1962)

COLES, B.R., GRIFFITH, D.: Antiferromagnetic behavior of GdB_6. Proc. Roy. Soc. London A77, 213-215 (1961)

CREEL, R.B., BARNES, R.G.: NMR study of transition metal monoborides: I. ^{11}B Nuclear electric quadrupole interaction parameter. J. Chem. Phys. 56, 1549-1562 (1972)

CREIJGHTON, J.H.N., LOCHER, P.R., BUSCHOW, K.H.J.: Nuclear Magnetic resonance of ^{11}B at the three boron sites in rare-earth tetraborides. Phys. Rev. B11, 4829-4843 (1973)
DAVIDOV, D., CHELKOWSKI, A., RETTORI, C., ORBACH, R., MAPLE, M.B.: Electron spin resonance and superconductivity in $Gd_xLa_{1-x}Al_2$ intermetallic compounds. Phys. Rev. B7, 1029-1038 (1973)
DeGENNES, P.G.: Interaction indirectes entre couches 4f dans les métaux de terres rares. J. Phys. rad. 23, 510-524 (1962)
DeWIJN, H.W., van DIEPEN, A.M., BUSCHOW, K.H.J.: Effect of crystal fields on the magnetic properties of Sm intermetallic compounds. Phys. Rev. B7, 524-533 (1973)
ETOURNEAU, J., MERCURIO, J.P., NASLAIN, R., HAGENMULLER, P.: Structure electronique de quelques hexaborures de type CaB_6. J. Solid State Chem. 2, 333-342 (1970)
FISK, Z.: On the ferromagnetism of EuB_6. Phys. Letters 34A, 261-262 (1971)
FISK, Z.: Magnetic susceptibility of Ce and Pr in SrB_6 and YB_6. Phys. Letters 30A, 152-153 (1969)
FISK, Z., MATTHIAS, B.T., CORENZWIT, E.: Rare earth impurities in YB_6 and ZrB_{12}. Proc. Nat. Acad. Sci. U.S. 64, 1151-1153 (1969)
FISK, Z., TAYLOR, R.H., COLES, B.R.: Anomalous magnetic behavior of gadolinium borides. J. Phys. C4, L292-L295 (1971)
FRUCHART, R.: Moments magnetiques du borures de cobalt Co_3B, Co_2B et du borure de fer FeB. C.R. Acad. Sci. (Paris) 256, 3304-3305 (1963)
GEBALLE, T.H., MATTHIAS, B.T., ANDRES, K., MAITA, J.P., COOPER, A.S., CORENZWIT, E.: Magnetic ordering in the rare-earth hexaborides. Science 160, 1443-1444 (1968)
GOSSARD, A.C., JACCARINO, V.: Boron nuclear magnetic resonance in rare-earth intermetallic compounds. Proc. Phys. Soc. 80, 877-881 (1962)
GSCHNEIDER, K.A.: Rare Earth Alloys, New York: Van Nostrand Co., 1961
HARMON, B.N., FREEMAN, A.J.: Augmented-plane-wave calculation of indirect exchange matrix elements for gadolinium, Phys. Rev. (B) 10, 4849-4855 (1974)
HAVINGA, E.E., BUSCHOW, K.H.J., van DAAL H.J.: The ambivalence of Yb in $YbAl_2$ and $YbAl_3$. Solid State Comm. 13, 621-727 (1973)
HIROTA, H., YANASE, A.: Magnetic properties of Mn_3B_4. J. Phys. Soc. Japan 20, 1596-1599 (1965)
JOHNSON, R.W., DAANE, A.H.: Electron requirements of bonds in metal borides. J. Chem. Phys. 38, 425-432 (1963)
KAPLAN, T.A., LYONS, D.H.: Theory of exchange interaction in rare-earth metals. Phys. Rev. 129, 2072-286 (1963)
KIESSLING, R.: The borides of manganese Acta Chem. Scand. 4, 146-150 (1950)
KOHM, A., MERZ, H.: Charge transfer in some transition metal semiborides. Phys. Stat. Sol. (b) 61, 142-152 (1973).
KOTETSKII, I.I., L'VOV, S.N.: Fiz. Metal Metalloved 33, 773-780 (1972)
LEE, K.N., BACHMANN, R., GEBALLE, T.H., MAITA, J.P.: Magnetic ordering in PrB_6. Phys. Rev. B11, 4580-4585 (1970)
LEE, K.N., BELL, B.: Exchange interaction and fluctuations in CeB_6. Phys. Rev. B6, 1032-1040 (1972)
LEVY, P.M.: Indirect exchange via spin orbit coupled states. Solid State Comm. 7, 1813-1818 (1969)
LUNDQUIST, N., MYERS, H.P.: The saturation magnetization of the monoborides of Mn, Fe and Co. Ark. Fys. 20, 463-471 (1961)
LUNDQUIST, N., MYERS, H.P., WESTIN, R.: The paramagnetic properties of the monoborides of V, Cr, Mn, Fe, Co and Ni. Phil. Mag. 7, 1187-1195 (1962)
LUNDQUIST, N., MYERS, H.P., WESTIN, R.: The paramagnetic properties of the monoborides of V, Cr, Mn, Fe, Co and N. Phil. Mag. 7, 1187-1195 (1972)
MAPLE, M.B., WOHLLEBEN, D.: Nonmagnetic 4f shell in the high pressure phase of SmS. Phys. Rev. Lett. 27, 511-515 (1971)
MATTHIAS, B.T., GEBALLE, T.H., ANDRES, K., CORENZWIT, G., HULL, G.W., MAITA, J.P.: Superconductivity and antiferromagnetism in boron rich lattices. Science 159, 530-531 (1968)
McNIFF, E.J., SHAPIRO, S.: Nuclear magnetic resonance in metal hexaborides. J. Phys. Chem. Sol. 24, 939-945 (1963)
MENTH, A., BUEHLER, E., GEBALLE, T.H.: Magnetic and semiconducting properties of SmB_6. Phys. Rev. Lett. 22, 295-297 (1969)

KIRJASOV, N.Z., PERSANOV, A.P.: Izv. Akad. Nauk. SSSR. Ser. Fiz. 23, 285-288 (1959)

NICHOLS, M.C., MAR, R.W., JOHNSON, Q.: Questions concerning the existence of erbium hexaboride; the crystal structure of $Er_{0.8}Ca_{0.2}B_6$. J. Less Common Metals 33, 317-320 (1973)

NICKERSON, J.C., WHITE, R.M.: Theory of the susceptibility of CeB_6. J. Appl. Phys. 40, 1011-1012 (1969)

PADERNO, Yu.B., POKRZYWNICKI, P.: Magnetic properties of some heavy rare earth tetraborides. Phys. Stat. Sol. 24, K11-K12 (1967)

PADERNO, Yu. B., POKRZYWNICKI, A., STALINSKI, B.: Magnetic properties of some rare earth hexaborides. Phys. Stat. Sol. 24, K73 (1967)

PADERNO, Yu.B., SAMSONOV, G.V., Zhur. Strukt. Khim. 2, 213-214 (1961)

SCHWETZ, K., LIPP, A.: Über die Existenzbreite der Europiumhexaboridphase. J. Less Common Metals 33, 295-303 (1973)

SPERLICH, G., JANNECK, K.H.: Exchange interaction in GdB_6 and EuB_6 by electron spin resonance. Int. J. Magnetism 3, 157-159 (1972)

SPERLICH, G., JANNECK, K.H., BUSCHOW, K.H.J.: Exchange narrowing of the ESR spectra of metallic $Gd_xLa_{1-x}B_6$ (x = 1 to 0.01) Phys. Stat. Sol. (b) 57, 701-706 (1973)

TROC, R., TRZEBIATOWSKI, W., PIPREK, K.: Magnetic properties of uranium borides and of the uranium beryllide UB_{13}. Bull. Acad. Pol. Sci. Sér. sci. chim., 19, 427-432 (1971)

TYAN, Y.S., TOTH. L.E., CHANG, Y.A.: Low temperature specific heat study of the electron transfer theory in refractory metal borides. J. Phys. Chem. Sol. 30, 785-792 (1969)

WANG, Y.L., COOPER, B.R.: Magnetic ordering in the metals with singlet crystal field ground state II. Phys. Rev. 185, 696-712 (1969)

WESTRUM, E.F., CLEVER, H.L., ANDREWS, J.T.S., FEICK, G.: Thermodynamics of lanthanium and neodymium hexaborides. In: Rare Earth Research L. Eyring, ed., Vol. III, pp. 597-605, New York: Gordon and Breach 1966

WOOD, V.E.: Indirect exchange interaction in EuB_6. Phys. Lett. 37A, 357-358 (1971)

D. Special Applications

I. Chemical Vapor Deposition of Boron Filament

V. KRUKONIS

During the decade, 1960-1970, boron filament progressed from its position of a $5000 per pound laboratory curiosity to its present status of a production level reinforcement fiber exhibiting superior properties for advanced composite applications. Boron filament is currently manufactured in quantities of about 30,000 lbs/year for primary use in military aircraft structures. Boron composites are being used extensively on the U.S. military aircraft, F-14 and F-15, and are being studied as a potential reinforcement material for commercial aircraft such as the DC-10, Boeing 737, and European A300 Airbus. From 1974 on, several thousand pounds of boron composites will be used on the Space Shuttle Orbiter structure, and, finally, boron composites are finding increasing application in stiffening golf shafts, tennis rackets, and bicycle frames.

Several excellent reviews of the technological achievements in boron filament research and development activity have been written previously WAWNER (1967), GALASSO (1969), LINE and HENDERSON (1969), and the reviews cover work up to the 1968 time period. This Section will present more recent work, but before discussing the advances in boron filament technology which have occurred during the present decade, a brief background of the process and the state of technological development that existed in 1968 will be presented.

1. Background

The chemical vapor deposition of boron from a mixture of boron trichloride and hydrogen onto a heated wire substrate was first reported by WEINTRAUB (1911), but it was work at Texaco Experiments, Incorporated, that demonstrated that a continuous high-strength, high-modulus, low-density fiber reinforcement material could result from this vapor deposition process (TALLEY, 1959). Boron filament was formed which exhibited a tensile strength of 500 Ksi, a modulus of 58 million psi, and a density of 2.6 g/cm^3. Because of the potential importance of such a material as a reinforcement in composites for advanced aircraft applications, the United States Air Force Materials Laboratory funded further research and process development effort at Texaco and also at a number of other laboratories. During the early and mid 1960s United Aircraft Research Laboratory, AVCO Systems Division, General Electric Research Laboratory, and Battelle Memorial Institute, among others, carried out deposition process development and filament fracture analysis studies. By the late 1960s, however, only two companies in the United States were still involved with boron filament research and production, namely AVCO Systems Division and Hamilton-Standard, a division of United Aircraft Corporation. In Europe at this same period, Wacker-Chemie, École Nationales Supérieure des Mines, and Philips Laboratories were carrying out boron filament research.

a) Boron Deposition Process

Boron filament is produced via chemical vapor deposition on a continuously moving tungsten substrate, and a schematic diagram of a basic deposition unit is shown in Figure 1. The reactor module consists of a glass deposition tube which is fitted with gas inlet and outlet ports, two mercury-filled electrodes, a variable-DC power supply connected to the two electrodes, a tungsten let-off system, and a boron filament take-up unit. The tungsten substrate, typically 0.0005" in diameter, is drawn through the reactor and is heated to about 1300°C by means of the variable power supply. (Prior to entering the deposition reactor shown schematically in Figure 1, the tungsten substrate is first passed through a short "cleaning" stage, not shown in the figure, in which the substrate is heated to candescence in a hydrogen atmosphere in order to remove residual lubricants which are used in the production of the tungsten.) A stoichiometric mixture of boron trichloride and hydrogen is introduced at the top of the reactor, and at the 1300°C level a mantle of boron is deposited via the reaction

$$BCl_3 + 3/2H_2 \longrightarrow B + 3HCl \qquad (1)$$

The exhaust gases are removed through the outlet port at the bottom of the reactor.

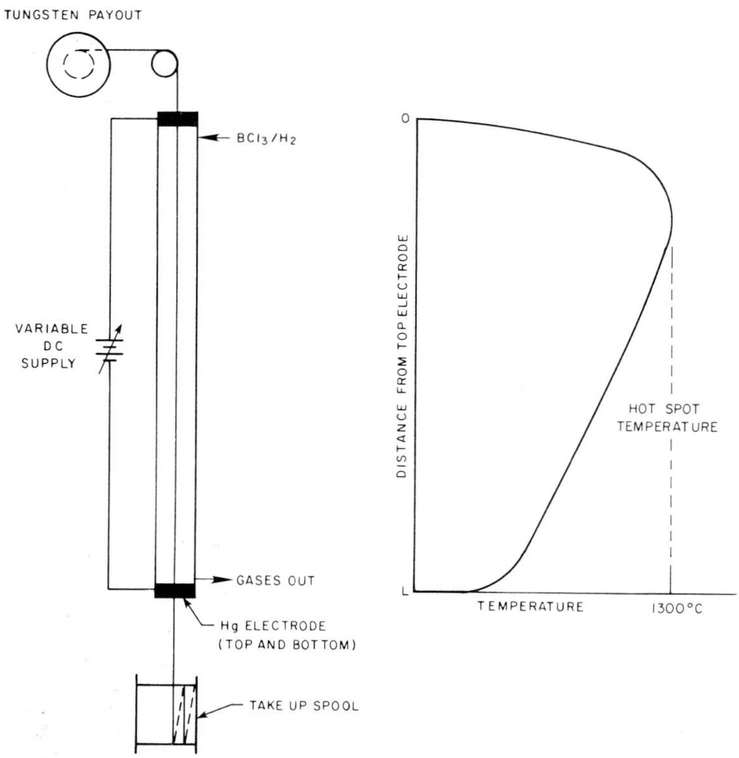

Fig. 1. Schematic diagram of boron filament reactor and its temperature profile

Several chemical reactions occur simultaneously during the formation of the filament; for example, boron is deposited from the gas phase via the halide-hydrogen reaction, other by-products such as B_2H_6, $HBCl_2$, and B_2H_5Cl are formed in the gas phase in addition to the HCl indicated by the stoichiometric relation above, and the tungsten core reacts with the boron mantle to form a number of tungsten boride forms. The boride phases which are formed are δ-WB, W_2B_5, and WB_4, species which have been identified by Debye-Scherrer techniques CHRETIEN and HELGORSKY (1961), GALASSO and PATON (1966), WAWNER and SATTERFIELD (1967a). Equilibrium effects due to the buildup of HCl in the gases can cause a decrease in deposition rate GRUBER (1970) and because of the equilibrium effects, the gases are metered at such a rate that during the residence time in the reactor only a few percent of the boron trichloride is converted to boron. It is interesting to point out here that the boron halide reduction reaction occurs with a standard free energy change of about +15 kcal/gmol B, i.e. for the stoichiometric equation above, $\Delta F^{o}_{1300^{o}C}$ = +15 kcal (JANAF TABLES); thus, the reaction for the formation of boron filament exhibits probably the largest ΔF^{o} (or analogously, the lowest equilibrium yield) of any large scale production process. Because of the low conversion per pass of the boron trichloride, a production plant for manufacturing boron filament contains condensation and purification equipment for separating the HCL and recycling the unreacted boron trichloride.

The reaction profile, i.e. the buildup of boron and the interaction between the boron and the tungsten core as the filament passes through the reactor, has been determined by detailed examinations of filament "tails", i.e. that length of filament which remains in the reactor when the power supply and the take-up unit are de-energized simultaneously; analysis of the tail portion, which progresses continuously from bare tungsten at the entrance of the reactor to fully formed filament at the bottom, provides a quantitative picture of reaction phenomena that occur on the moving filament.

Photomicrographs of cross-sections of filament progressively formed at three positions in the reactor are shown in Figure 2. The top photograph is a polished cross-section of the form which is present about one inch below the entrance electrode. A thin outer layer is barely visible on the periphery of the tungsten and has been identified as a zone of δ-WB; the remainder of the core consists of unreacted tungsten. The next photograph shows a cross-section of filament formed about 2 inches below the electrode. The cross-section (which has been treated with Murakami's etch to accent the phases) shows two outer layers in addition to unreacted tungsten; X-ray analysis indicates the presence of three boride phases, namely δ-WB, W_2B_5 and WB_4. The individual boride phases, W_2B_5 and WB_4, cannot be distinguished optically in the outer layer, but they have always been found in admixture even in fully-formed boron filament. Previous studies have concluded that the phase, WB_4, forms only when the boride layer is under a high pressure compression that results during the reaction of a boron mantle with the core WAWNER (1967), LINE and HENDERSON (1969); this conclusion was probably based on the work of CHRETIEN and HELGORSKY who reacted boron and tungsten to WB_4 at high pressure. However, the photomicrograph which shows no presence of a boron mantle, combined with the X-ray identification of WB_4 challenges the previous hypothesis that WB_4 is formed only during reaction of boron and a boride under the high compressive stresses of the boron mantle BERNEBERG and KRUKONIS (1969). A few inches lower, as shown in the last photograph in Figure 2, a boron mantle is formed as well as a larger layer of W_2B_5 and WB_4 also evident in the figure. Farther along in the deposition process, the boron layer forms progressively thicker, the δ-WB phase disappears, and the tungsten core reacts completely to a mixture of only W_2B_5 and WB_4.

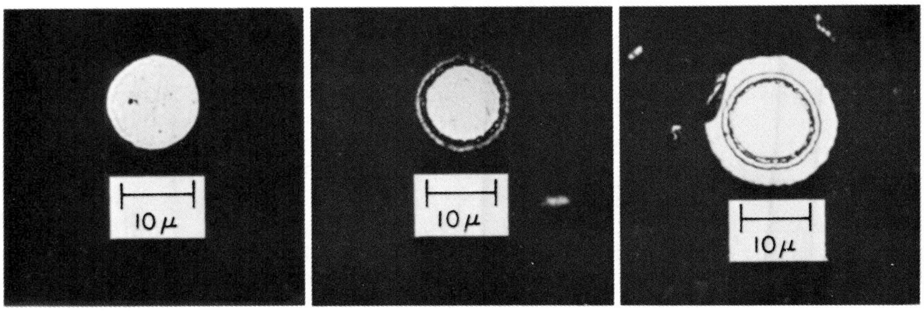

1" BELOW ELECTRODE 2" BELOW ELECTRODE 4" BELOW ELECTRODE

Fig. 2. Cross sections of boron filament formed at various positions in reactor

Figure 1 discussed previously, also shows the temperature profile of a reactor in which filament is heated with a single DC power supply. As the filament increases in diameter, its resistance decreases; with a constant current through the filament the temperature at the reactor outlet is 200-300°C lower than at the entrance. The temperature distribution shown in Figure 1, in particular, the sharp rise after the substrate enters the reactor followed by the monotonic decrease, is also influenced by the simultaneous formation of borides phases and the disappeareance of high conductivity tungsten; the highest temperature along the profile is called the deposition or "hot spot" temperature.

Figure 3 shows photomicrographs of the surface of 4-mil boron filament and a polished cross-section of 4-mil boron. An axially-aligned "ear of corn" or "kernel" structure is evident on the surface, and the core of fully formed 4-mil boron consists of W_2B_5 and WB_4 (but, as stated earlier, the individual phases cannot be distinguished in the core.)

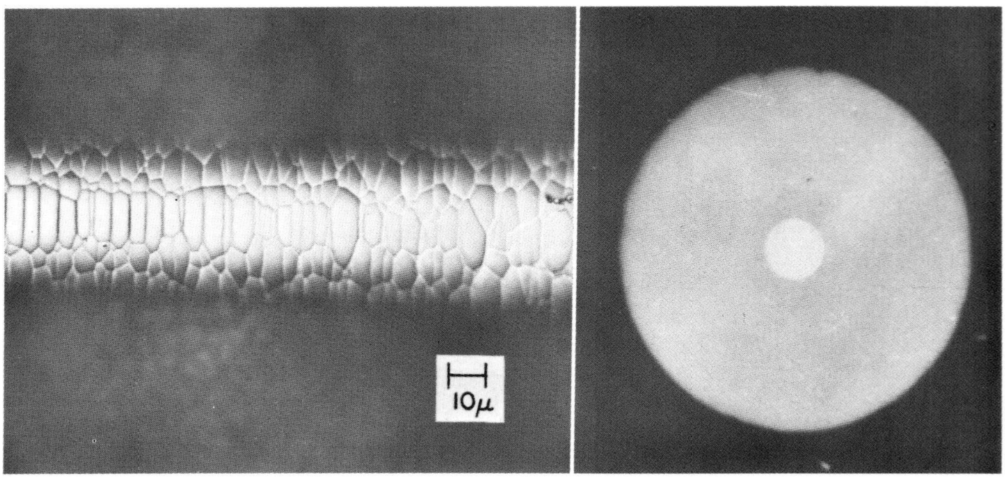

Fig. 3. Surface and cross section of 4 mil boron

b) Flaws in Boron Filament

The influence of flaws on the strength of boron filament was recognized in early process-development studies, and substantial effort was directed to establishing the causes and cures of these flaws. Typical boron filament exhibits an average tensile strength of about 500-550 Ksi, if all production parameters are held at their optimum values. If, however, the deposition temperature is varied over wide limits, if the tungsten substrate is not adequately cleaned before entering the deposition reactor, or if impurities are present in the gas phase, anomalous growths and stress concentrations can be created in the filament which can reduce the tensile strength to as low as 100 Ksi. Very strong filament usually fractures into many pieces during tensile testing, but when a low strenght filament is tensile tested, the two pieces of the fracture frequently "save", i.e. both pieces remain within the gauge length of the tensile machine; optical examination of the saved fracture faces has been found to be an effective technique to delineate the various causes of fracture WAWNER and SATTERFIELD (1967b).

Figure 4 is a photomicrograph of a filament fracture face which shows an anomaly associated with an imperfection on the tungsten substrate, and the anomaly is visible as a dark spot at about 12 o'clock on the core. The imperfection shown in Figure 4 resulted in a stress concentration near the core which reduced the filament tensile strength to 240 Ksi. Figure 5 is a photograph of another type of anomalous growth, namely a crystalline flaw. These anomalous boron growths form when the deposition temperature is higher than the normal 1300°C, and these crystalline deposits can result in tensile strengths as low as 100 Ksi.

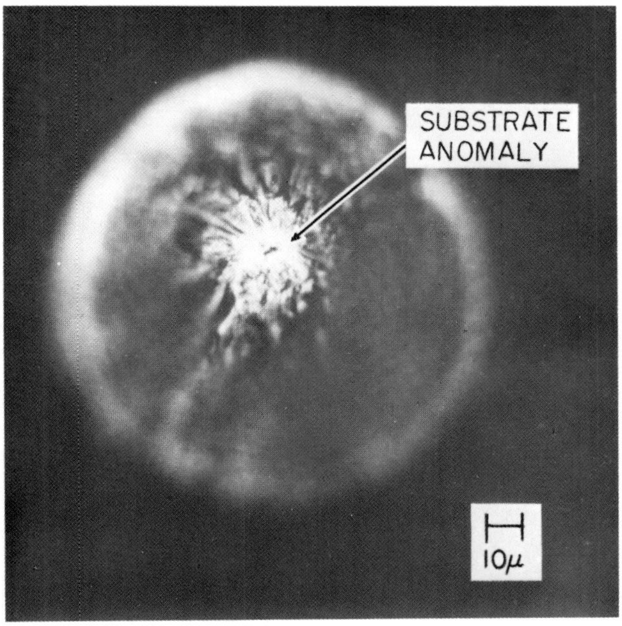

Fig. 4. Anomaly on tungsten substrate which produced low strength fracture

It has been experimentally determined that there are internal stresses in the boron mantle which are tensile at the tungsten-boron interfaces and which gradually progress to compressive stresses at the surface WAWNER (1967) LINE and HENDERSON (1969). In addition to this normal

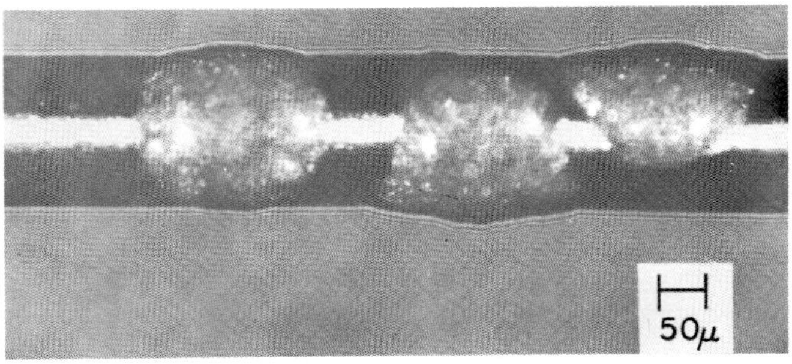

Fig. 5. Polycrystalline growths on surface of boron filament

stress distribution in high strength filament there is also present in boron produced at deposition temperatures below about 1250°C an aggravated internal stress situation which results in very low strength filament. The low strength fracture site has been identified by micro-optical examination of filament fracture faces, and the particular internal stress concentration in the interior of the filament which causes the low strength manifests itself as a crack. A photomicrograph of the crack is shown in Figure 6, and this particular failure mode has been named the "crack tip" WAWNER (1967). The previous authors who considered internal stresses concluded that the crack pre-exists in boron, and furthermore they concluded that even high-strength boron contains a small radial crack. It will be shown later, when more recent work is discussed, that the crack is not a pre-existing one, neither in low-nor in high-strength boron, but instead it will be shown that it appears only when the filament is fractured.

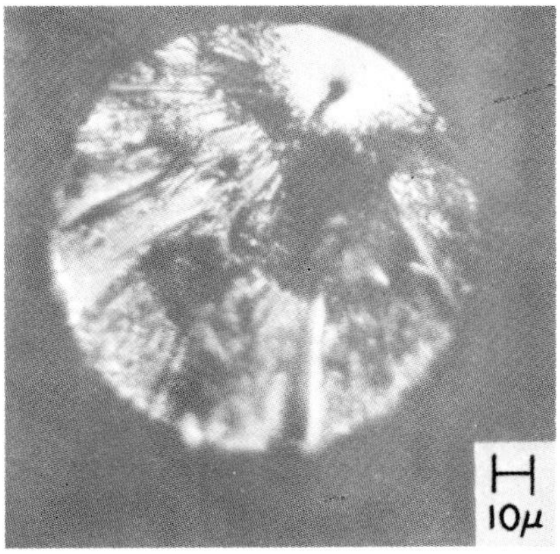

Fig. 6. Filament fracture face exhibiting the crack tip flaw

During the process development work of the early 60s the previous flaw types were identified as the predominant causes of low strength boron. Other flaws, for example, "spikes", have appeared intermittently, but they have not been studied in detail. An example of a group of spikes on a boron filament fracture end is shown in Figure 7, the name deriving from the sharp points at the end of each spurious growth. Because of the presence of the sharp points, it was reasoned that the spike was a result of a vapor-liquid-solid (VLS) growth mechanism caused by some impurity on the surface of the tungsten substrate which alloyed with the boron to form a low temperature eutectic solution WAGNER and ELLIS (1965). To test the VLS hypotheses, deposition tests were carried out on gold-plated tungsten, and a photograph of

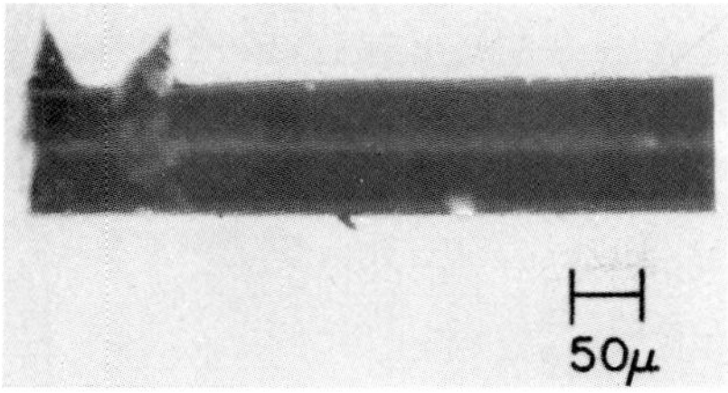

Fig. 7. Spikes on boron which result in low tensile strength

the resultant boron filament produced is shown in Figure 8; a profusion of spikes along the length of the filament is evident and ostensibly verifies the VLS mechanism KRUKONIS (1969).

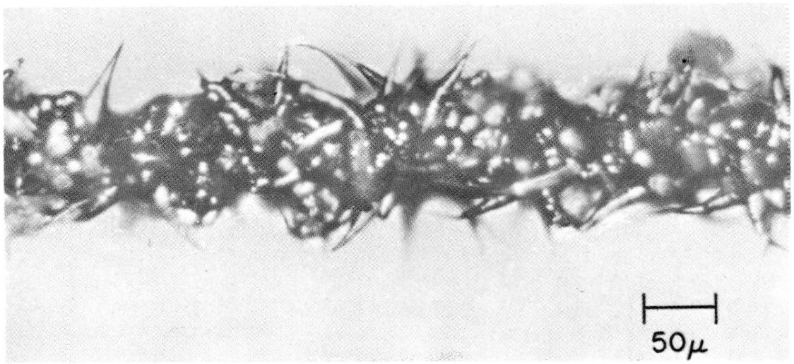

Fig. 8. V-L-S growths on boron deposited on gold-plated tungsten

For completeness in this discussion, a photomicrograph of a "normal" fracture face of boron which failed at a high strength of 535 Ksi is shown in Figure 9. The genesis of the fracture of high strength filament is somewhat more difficult to delineate from micro-optical analysis, but the fracture, which initiates at the core, is caused by stress concentrations related to the inherent tensile-compressive stress distributions previously discussed; the particular fracture nucleation site in Figure 9 is visible as a very small shiny area at the filament core.

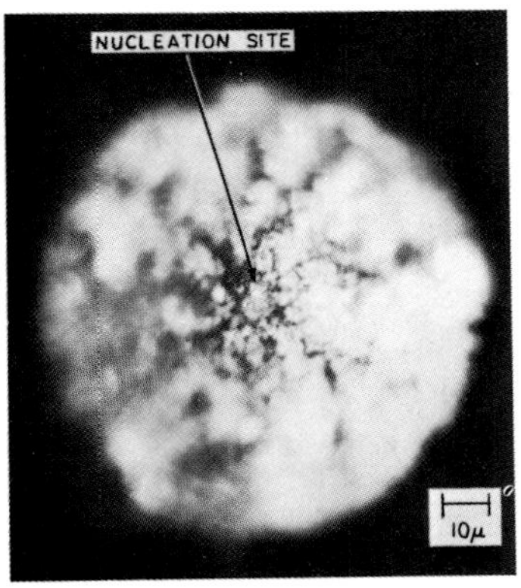

Fig. 9. Fracture face of high strength boron

The understanding of flaws and their removal to as great an extent as possible was the predominant factor in raising the strength of boron and in increasing confidence in the reinforcement fiber for advanced aircraft applications.

2. Boron Filament Research and Process Development After 1968

The major portion of research and new process development carried out since 1968 has been directed to achieving lower-cost boron filament. By mid-1968 the price of boron filament had dropped from the several thousands of dollars per pound cost, representative of the early 1960s, to about $300/lb, and by that time the filament had proven itself as a reinforcement fiber for advanced composites; increasing uses of the material motivated scale-up in production by both manufactureres, and a large emphasis was placed on lowering the $300/lb cost even further.

In 1968, the tungsten substrate contributed about $65 to a pound of 4-mil boron filament, the largest single factor in the $300/lb figure. Since the substrate represented almost 25% of the cost, methods to reduce the contribution of the substrate were investigated, and research was directed to two major categories, namely the formation of larger diameter boron and the deposition of boron on potentially less expensive carbon substrate.

a) Large Diameter Boron Filament

In early 1969, an improved 6-mil diameter boron was introduced on the market as a result of the economic factors of tungsten cost mentioned previously. Because of the shorter length per pound of filament, namely about 70,000 ft for 4-mil filament versus about 31,000 ft for 6-mil, the $65/lb tungsten contribution was decreased to only $30 in a pound of 6-mil boron. (Subsequently, in late 1969, 5.6-mil boron became the standard large diameter size, the rationale for the choice of 5.6-mil being one of convenience, i.e. the areal ratio is 2.)

Deposition of 5.6-mil boron is carried out in the same reactors as those used for 4-mil boron, the larger diameter filament being produced by reducing the drawing rate through the deposition chamber. In a 6-ft long, single-stage reactor, 4-mil filament is produced at about 15 ft/min, while 5.6-mil boron is produced at about 6 1/2 ft/min. At the same weight production rate, i.e. in pounds of filament per unit time, the "equivalent" drawing rates for 4-mil and 5.6-mil are 15 and 7 1/2 ft/min filament, respectively, (ignoring second-order density differences between the two). The slight decrease in weight-rate of 5.6-mil boron relative to 4-mil is caused by a lower average filament temperature. As discussed previously in reference to the temperature profile in Figure 1, the temperature of 4-mil filament is low at the reactor exit because of resistance considerations; the resistance (and thus, temperature) of 5.6-mil boron is even lower at the exit than that of 4-mil boron resulting in a reduced drawing rate; however, the production rate of 5.6-mil boron cannot be increased by using a higher current because of the crystalline flaw limitations described previously, i.e. at deposition temperatures above 1300 C^{o} low strength polycrystalline growths appear on the filament as shown by the photomicrograph of Figure 5.

In 1969 research and development activity in boron/aluminum (B/Al) composites was increasing in the U.S., and in a number of studies it was found that 4-mil boron suffered from low transverse strength in an aluminum matrix, frequently failing at below 15 Ksi KREIDER et al. (1971). Splitting of the filament was found to be responsible for the low transverse strength, and Figure 10 is a scanning electron micrograph (SEM) of a fractured, cross-plied, B/Al composite which illustrates the splitting phenomenon TOTH (1974). Subsequently a solution to the splitting problem of 4-mil filament in B/Al composites was found in the larger diameter boron; workers at United Aircraft Research Laboratory demonstrated that B/Al composites produced with 5.6-mil boron filament failed in the matrix at transverse strengths over 20-30 Ksi and was not limited by filament splitting KREIDER et al (1971). Thus, 5.6-mil boron became the standard metal matrix reinforcement filament, not only for the considerations of decreased tungsten cost mentioned previously, but for transverse strength considerations as well.

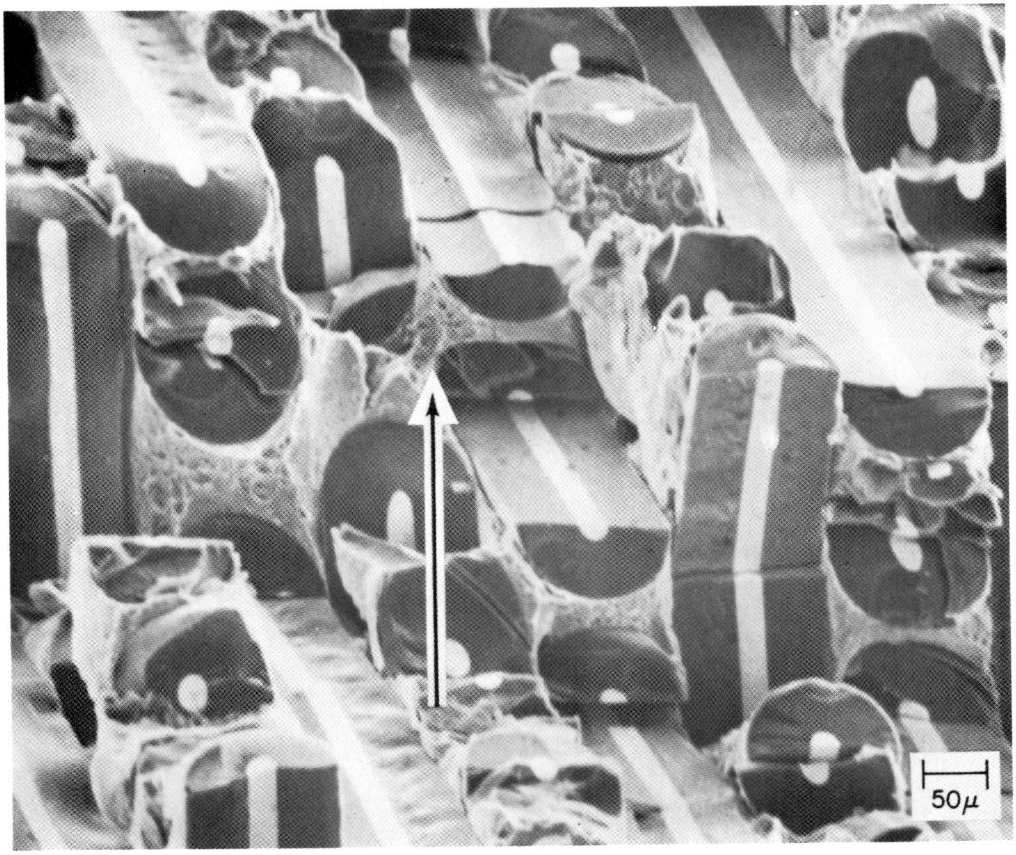

Fig. 10. SEM of fractured B/Al composite showing split filament

In a brief aside here and relative to the previously discussed hypothesis of the pre-existing crack, the breaking load which fractured the composite shown in Figure 10 was applied perpendicular to the 90° filaments as indicated by the arrow, and it can be seen that all the 90° filaments have been split in a plane essentially perpendicular to the applied load. If a pre-existing crack were present in the filament, this uniformity in the splitting plane would be absent, and, instead, splitting would probably occur in random planes. Other work investigating this splitting phenomenon has been carried out at AVCO and also suggests that the crack is not a pre-existing one. For example, Figure 11 is a photomicrograph of the fracture face of a cleaved boron filament; cleavage was effected by scratching the boron surface with 30μ diamond paste and subsequently bending the filament until a surface-initiated fracture occurred. In the figure a radial crack is evident, and the fracture nucleation site is also evident as a chipped surface as indicated. The radial crack in Figure 11 is located essentially 180° from the site of the fracture, and furthermore, in every cleaved filament examinde it was always found to be present 180° from the nucleation site KRUKONIS (1969). If the radial crack pre-existed in boron, such consistency in its relation to the cleavage point would be highly improbable. (The reason for the consistent 180° relation, however, has not yet been explained.)

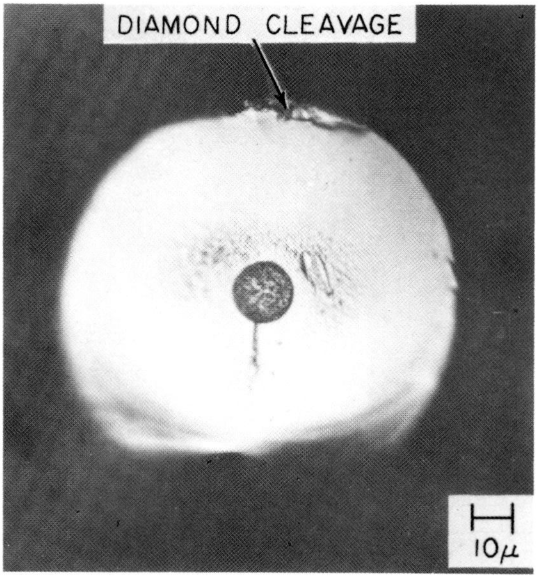

Fig. 11. Photomicrograph of diamond cleaved boron filament showing radial crack

The formation of filament of diameter larger than about 5.6 (or 6) mils presented some initial difficulty; for example, 8-mil diameter boron (which contains only about 1/4 of the tungsten and, thus, 1/4 of the tungsten cost of 4-mil filament) could be produced in a single-stage reactor at a drawing rate of only 1/2 ft/min (which on a weight basis is "equivalent" to a 4-mil drawing rate of only 2ft/min). The increased diameter (and hence decrease resistance) relative to both 4 and 5.6-mil boron caused the temperature of 8-mil boron to fall to below 800°C at the exit electrode, and the lower average filament temperature was responsible for the low drawing rate. In addition to the obvious economic disadvantage of such a very low production throughput, the tensile strength of the 8-mil diameter boron produced in a single-stage DC reactor was also found to be very low, frequently less than 300 Ksi, but optical examination of the fracture faces was not successful in delineating the stress concentrations or anomalies which were responsible for the low strength. With subsequent changes in the 8-mil deposition process, however, the low strength fractures of 8-mil boron were eliminated.

As was stated earlier, the current in the filament cannot be increased above a certain level because of the crystalline boron limitation (which is the same whether the diameter is 4, 5.6, or 8-mils), and thus solely DC heating with high current in a single-stage reactor could not be used to produce 8-mil boron at high rates. Other means of heating the filament in order to increase the drawing speed were tested, and Very High Frequency (VHF) heating was developed as a reliable method to heat the large diameter portion of the filament DE BOLT (1969), DE BOLT et al. (1971). A schematic diagram of a VHF-augmented reactor, i.e. one using both DC and VHF power supplies is shown in Figure 12. DC power is introduced through the mercury electrodes, and VHF power, at about 63 MHz, is introduced at three positions in parallel along the filament length via coupling loops, each a tuned section about 1/8 wavelength (2 ft)long. The control of the power input within a given section is accomplished by adjusting the shape of the coupling loop wire and by tuning each loop closer to or farther from resonance using a vacuum capacitor; with this tuning circuitry, it is possible to adjust the temperature profile within the localized areas as shown in Figure 12. Because of the ability to couple VHF power into any selected portion, and specifically into the large diameter section, of the filament, the deposition rate of 8-mil boron could be increased to about 5 ft/min while simultaneously maintaining the hot spot temperature at 1300°C. (The 5 ft/min drawing rate for 8-mil boron corresponds, on a weight basis, to the equivalent of about a 20 ft/min drawing rate for 4-mil boron.)

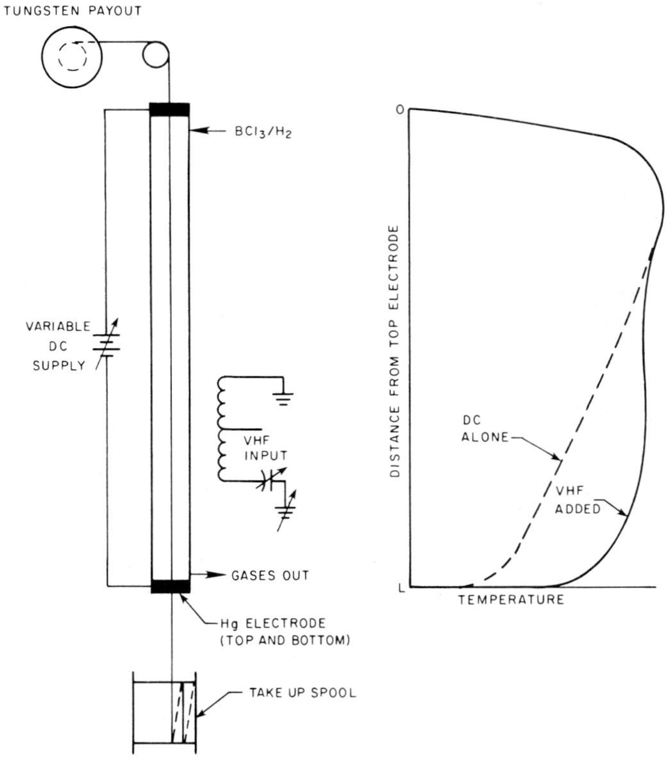

Fig. 12. Schematic diagram of VHF-augmented reactor and its temperature profile

Figure 13 shows the surfaces of 8-mil boron produced in single-stage DC reactor and in a VHF-augmented reactor, respectively. The morphology of the DC-produced 8-mil filament shown in Figure 13 exhibits very small scale nodules which are characteristic of boron which has deposited at low temperature. The temperature at the exit of the VHF-augmented 8-mil reactor is about 1000 F, similar to that existing in a single-stage, 4-mil reactor, and the surface of the VHF-augmented 8-mil boron shown in Figure 13 exhibits an ear of corn structure similar to that of the 4-mil boron shown in Figure 3 although there is also some finer scale kernel structure evident on the surface of the 8-mil boron. (The magnification values for the respective 8-mil boron surfaces shown in Figure 13 are correct, incidentally, and the optical effects are cause by differences in light reflection phenomena between the two surface structures.) As a demonstration of the contribution of 0.5-mil tungsten to boron filament, cross-sections of 4-, 5.6-, and 8-mil boron are shown in Figure 14.

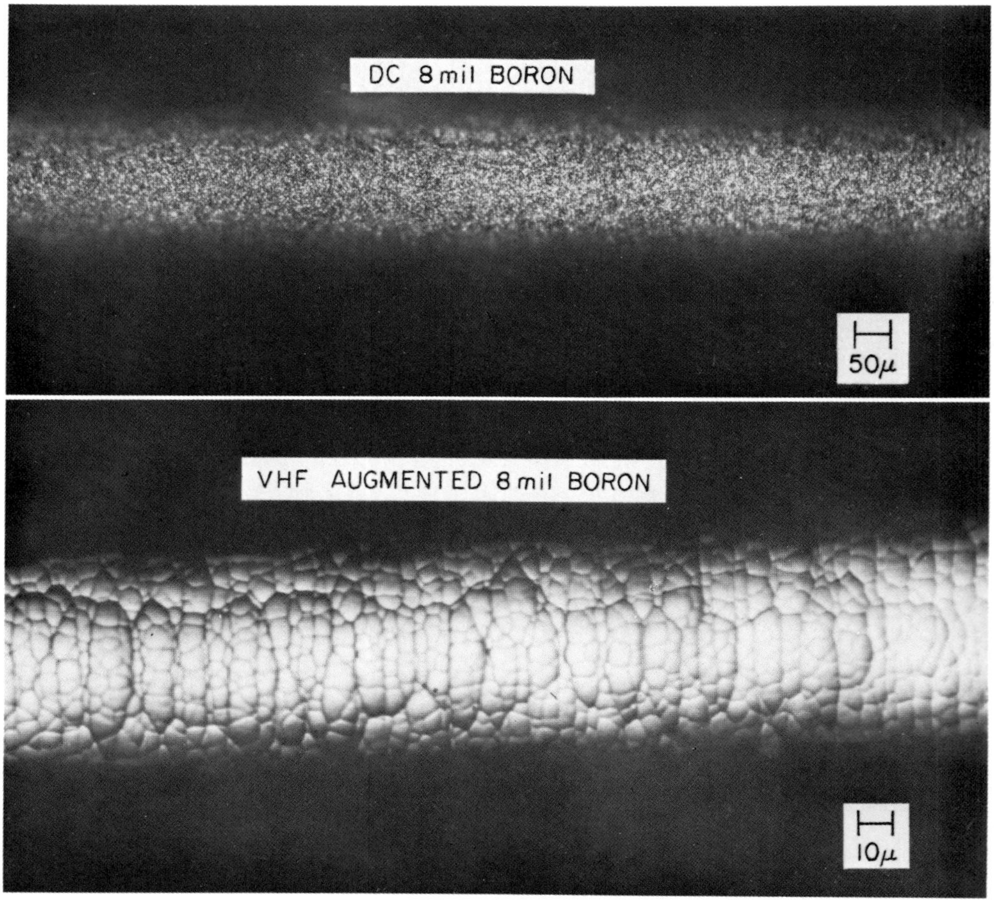

Fig. 13. Comparison of surfaces of 8 mil boron filament

The current production of 8-mil boron is only about 200 lbs/yr compared to the 30,000 lbs/year level for 4-mil boron, but as the cost

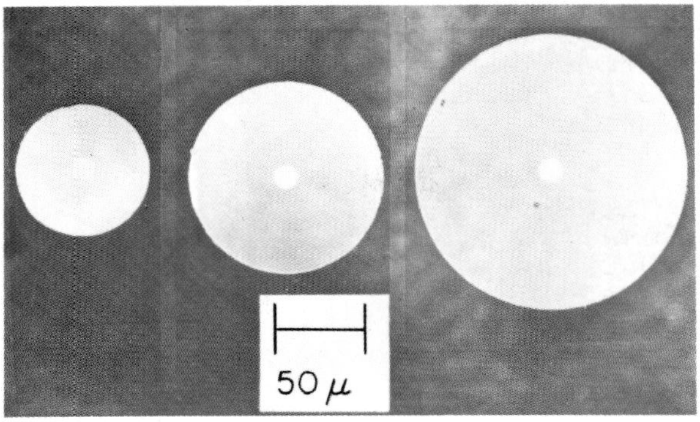

Fig. 14. Comparison of cross sections of 4, 5.6, and 8 mil boron filament

advantages and enhanced mechanical properties continue to be demonstrated, its use will probably increase in both organic and metal matrix applications. One of the major potential applications is for production of B/Al fan blades for advanced turbo-jet engines, and initial work has shown that B/Al composites made with 8-mil boron exhibit superior impact properties KREIDER et al. (1971).

b) Boron Deposited on a Carbon Substrate

As was stated earlier, tungsten contributed about $65 to a pound of 4-mil boron filament in the 1968 period, and in addition to the effort carried out by AVCO and Hamilton-Standard on large diameter boron, research was initiated at AVCO with funding from the Air Force Materials Laboratory directed to another method of decreasing substrate cost, that is the deposition of boron on a carbon monofilament substrate, a material which could be spun from coal-tar pitch and which was projected to be of potentially much lower cost than tungsten.

During the research and process development studies on boron on carbon filament (B/C) new information was obtained concerning fundamental processes in operation during the vapor deposition of boron. For the initial B/C research, deposition studies were carried out in a "static" reactor, i.e. one in which the substrate is held stationary, power impressed on the filament between the two electrodes, and, after the deposition is completed, the power de-energized. During tests with the carbon monofilament substrate, an unusual phenomenon was observed a few seconds after deposition was started, namely very bright zones appeared at multiple closely space positions along the entire length of the filament. These bright spots resembled a string of Italian Christmas tree lights, and the term "light bulb effect" or "light bulbing" to describe the phenomenon appeared in the literature DE BOLT et al. (1971).

In order to determine the cause of the bright spots which occurred during deposition, filament which exhibited the "light bulb" phenomenon was examined metallographically, and a photomicrograph of axially cross-sectioned B/C filament is shown in Figure 15; the figure shows that the substrate has experienced multiple fractures a few mils apart, and it can also be seen that the boron mantle is slightly thicker at each fracture point. Based upon the nature of the cross-section of Figure 15 and combined with the observation of the bright spots that occurred during deposition, an explanation of the light bulb effect was developed. It was proposed that after the boron mantle has built up to some level, the substrate fractured (albeit for reasons not then understood), and the current path then

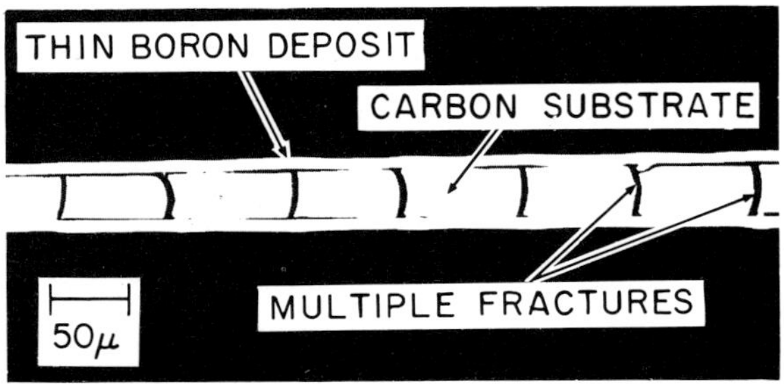

Fig. 15. Axial cross section of boron on carbon showing substrate fracture caused by elongation

shifted from one of parallel conduction in both the carbon and boron to one of conduction in just the boron mantle. The decreased area for current conduction, and thus the increased resistance, caused high temperature zones at the fracture points, these high temperature points being the "ligh bulbs" which appeared along the filament.

In continuing work the cause of substrate fracture was subsequently related to a change in dimensions of the filament, i.e. an elongation or expansion of the amorphous boron which occurs *after* the boron is deposited on the carbon substrate DIEFENDORF and MEHALSO (1971). Boron elongation, which manifested itself as an increase in length of the boron filament during deposition, was first determined experimentally by monitoring the position of a weight attached to the substrate; as the deposition proceeded in the static reactor, it was found that the filament progressively extended. Figure 16 gives the results of boron elongation measurements; other data have shown that the filament undergoes as much as 5 percent elongation during the deposition process. Based upon a simple calculation it can be concluded that the dimensional changes occurring in the boron mantle are primarily plastic in nature; for example, if the strain were elastic, a 5% elongation on the approximately 60 million psi modulus boron mantle would stress portions of the boron to 3 million psi, a value far beyond any tensile strength reported for the filament. Although the stresses in the boron mantle are partially accommodated by plastic rearrangement, the substrate, because of integral bonding of the mantle to the carbon, experiences a strain impressed upon it by the elongating boron which results in the multiple fractures shown in Figure 15; it is pertinent to emphasize again that only the substrate and not the boron, fractures because of the elongation strain.

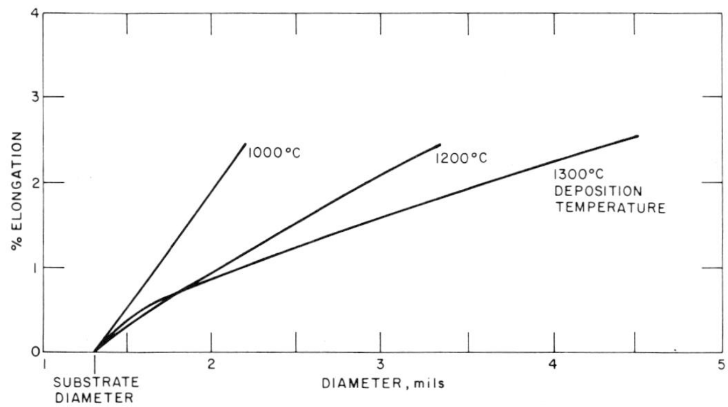

Fig. 16. Elongation during deposition of boron on carbon

The phenomenon of boron elongation which caused light bulbing and substrate fracture problems limited the development of a large scale process to produce boron on carbon for almost two years. For example, in tests using continuous production reactors, the substrate fracture phenomenon caused run stoppages after a few minutes of operation, and moreover, much of the B-C filament that was produced was of low strength. On occasion, however, tensile strengths as high as 700 Ksi were achieved which pointed out that B/C filament could be a viable material if the light bulb problem could be overcome DIEFENDORF and MEHALSO (1972).

The detrimental effects of elongation on filament properties and on reactor operation were subsequently circumvented. It was found that by first depositing a thin layer of pyrolytic graphite (PG) on the carbon substrate, the fracture during boron deposition could be eliminated DE BOLT (1970), DE BOLT (1972), BASCHE et al. (1972), and, hence, boron on carbon filament became a potential new, high strength, lower cost reinforcement fiber. Polished cross-sections of 4- and 5.6-mil B/C are shown in Figure 17, and the carbon substrate, typically 1.3-mils in diameter, the PG layer of about 0.1-mil in thickness, and the boron mantle are distinguishable. A larger magnification scanning electron micrograph of 4-mil B/C is shown in Figure 18, and it can be discerned that there is no bonding between the carbon substrate and the PG layer nor between the PG and the boron. The separation between layers arises because of the expansion phenomenon which occurs probably in both the PG and in the boron during the deposition processes; a similar expansion or rearrangement phenomenon may be occurring during the stress recrystallization of pyrolytic graphite BLACKMAN and UBBELOHDE (1962). Since the expansion is probably two or three-dimensional, radial separations between layers result as shown in Figure 18. It is this separation which is ostensibly responsible for eliminating substrate fracture, i.e. since no bond exists between the boron and the carbon substrate, the boron expansion/elongation phenomenon, which to be sure, is still in operation during the deposition, does not impress a strain on the carbon. In Figure 18, a narrow reaction or diffusion layer is also evident exterior to the PG layer. The material is relatively insoluble in nitric acid; it contains boron and carbon, but B_4C has not been identified in the layer DE BOLT, et al. (1972).

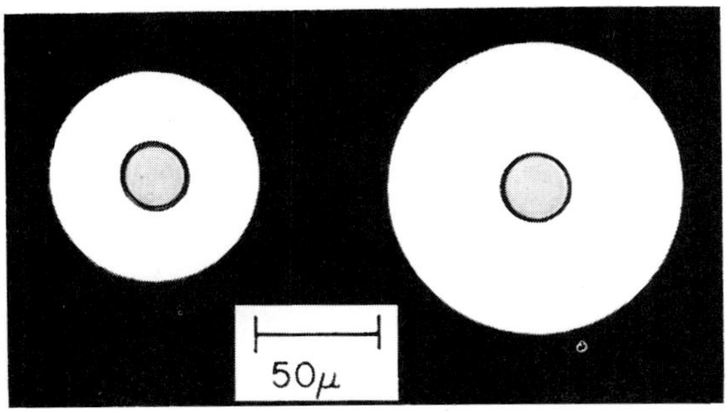

Fig. 17. Polished cross sections of 4 mil and 5.6 mil boron on carbon

A photomicrograph of the surface of carbon monofilament is shown in Figure 19 and is compared to the surface of tungsten substrate; it can be seen that the surface of carbon is almost optically smooth, whereas the tungsten possesses axial die grooves, Analogously, Figure 20 shows a comparison of the surface structure of boron on carbon filament and

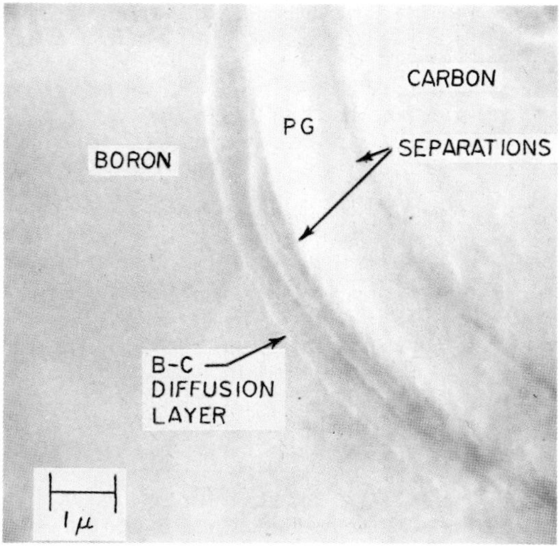

Fig. 18. SEM of the interfaces of boron, PG, and carbon substrate

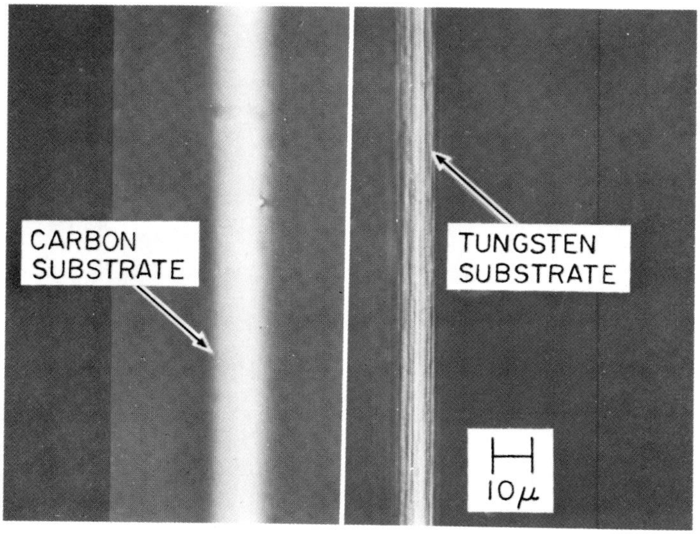

Fig. 19. Comparison of the surfaces of carbon and tungsten substrates

the surface of boron on tungsten. The axially-aligned "kernels of corn" are again evident on the boron on tungsten filament in Figure 20 (as they were in the higher magnification photomicrograph shown in Figure 3); however, because the carbon monofilament substrate shown in Figure 19 possesses no die grooves or other distinguishable surface features (except for an occasional isolated defect seen on the carbon in Figure 19), the boron "kernels" on B/C filament nucleate and grow randomly and exhibit spatial symmetry rather than the axial symmetry seen on boron on tungsten.

A photomicrograph of a high strength B/C fracture is shown in Figure 21. This particular filament fractured at 596 Ksi, and the figure

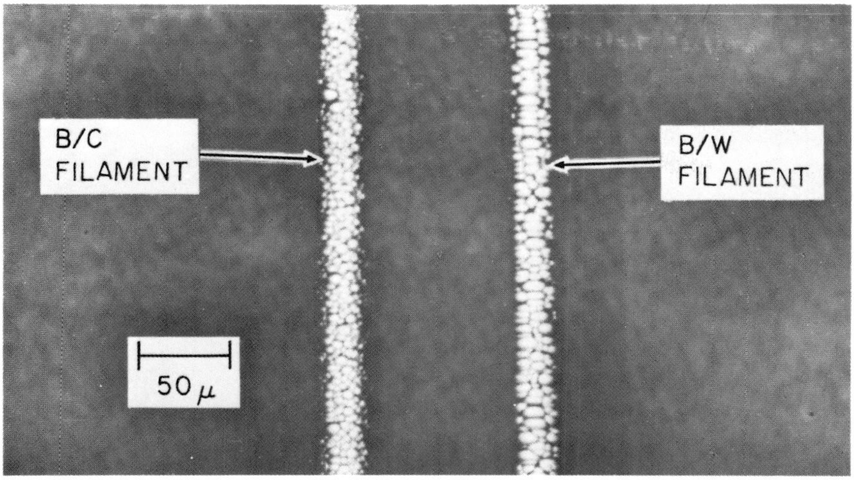

Fig. 20. Comparison of the surfaces of boron filament produced on carbon and on tungsten substrates

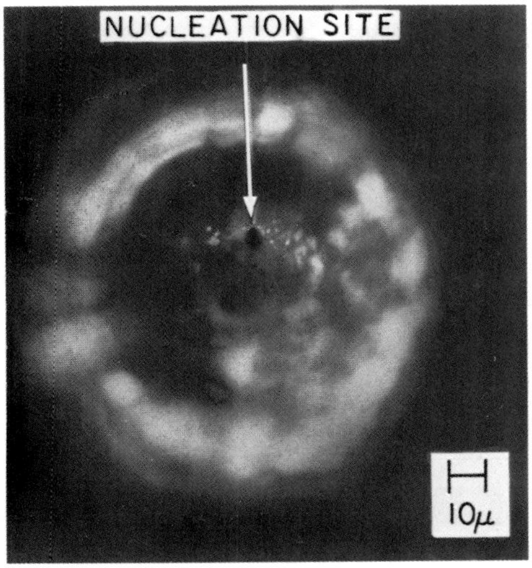

Fig. 21. Fracture face of high strength boron on carbon

shows that the fracture nucleation occurred at the core-boron interface, just as was seen in Figure 9. Analogously, B/C filament, just like boron on tungsten filament, was found to possess an internal stress distribution varying from tensile stresses at the boron-PG interface to compressive in the outer layers DE BOLT et al. (1972), and in addition, both low and high strength B/C filament was found to exhibit the "radial crack" phenomenon similar to that shown previously in Figure 11. A diamond cleaved B/C filament face is shown in Figure 22 and the crack is evident; however, as the case in cleaved boron on tungsten, the crack was always found to be oriented 180° from the fracture nucleation site, suggesting that no pre-existing radial crack is present in B/C filament either KRUKONIS (1972). The authors mentioned earlier WAWNER (1967), LINE and HENDERSON (1969) concluded that internal stresses and the radial crak were caused by

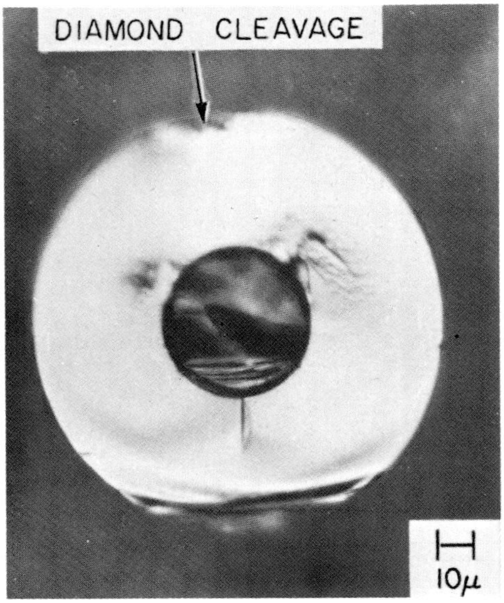

Fig. 22. Diamond cleaved boron on carbon filament showing radial crack

volume changes occurring during reaction of boron and the tungsten substrate. Since at the deposition temperature of 1300°C there is virtually no reaction between boron and carbon, the presence of the radial crack in Figure 22 cannot be related to chemico-volumetric changes in the core, and instead, this author suggests that the radial crack in both B/C and B/W filament results from boron elongation which is the predominate cause of the tensile-compressive stress distribution. Again, in repetition, the radial crack is not pre-existing in the interior of the filament, but forms (somehow) during fracture or cleavage.

To date a fundamental understanding of the origin and nature of boron elongation has still not been achieved; however, the deposition of the thin layer of pyrolytic graphite was completely successful in eliminating the light bulb problem At present about 100 lbs/year of both 4 and 5.6 mil B/C is being evaluated by aircraft companies for applications as a potentially lower cost boron filament.

3. Coated Boron Filaments

Besides their current primary use in epoxy composites, boron filaments are also being tested extensively in aluminum. Briefly, B/Al composites are produced by diffusion bonding boron filament and aluminum foils at a temperature of about 965°F for 30 minutes using a consolidation pressure of about 5000 psi. Diffusion bonding at 965°F requires relatively large manufacturing equipment for the formation of sheets or plates; for example, in order to satisfy the very high 5000 psi requirement, a bonding press with a 1000 ton capacity can process an area of only 3ft^2. Bonding at higher temperatures at which the aluminum matrix can flow more readily and therefore can bond at lower pressures is being investigated as a means of circumventing the large equipment requirements. However, at bonding temperatures above approx. 1000°F, the boron filament experiences a strength degradation because of reactions between the filament and the aluminum matrix; a B/Al composite (50 v/o) bonded at a temperature of 1050°F exhibits a strength of only 100 Ksi compared to about 230 Ksi for standard diffusion bonded B/Al material; therefore, although low pressure, high temperature B/Al consolidation is an attractive manufacturing alternative to high pressure diffusion bonding, the problem of boron-matrix reaction presents itself, and in order to circumvent the degradation problem which occurs at high temperature, various diffusion barriers have been deposited on boron to preclude boron-aluminum reactions during B/Al composite manufacture; to date, the coated filament forms which have been found to exhibit reasonable good high temperature stability in aluminum are those which have been coated with silicon carbide, with boron nitride, or with boron carbide.

a) Silicon Carbide-Coated Boron

Silicon carbide-coated boron, developed in the mid-60s HOUGH (1969) STARKS et al. (1966), and currently manufactured by Hamilton-Standard under the trade name Borsic, is the coated boron form which has received the major portion of experimental effort in high temperature B/Al consolidation. The filament has a silicon carbide (SiC) layer about 0.1-mil thick and is produced by vapor depositing the SiC layer onto the boron using a mixture of hydrogen and methyldichlorosilane BASCHE and KUNTZ (1971). The high temperature tensile and ageing properties of Borsic have been described in other works VELTRI and GALASSO (1968), BASCHE et al. (1968); for example, the SiC-coated boron does not degrade after heating in air to 1000oF whereas boron drops from an average 500 Ksi to below 200 Ksi. The SiC-coated filament has been tested extensively in aluminum composites, and it has exhibited enhanced properties over boron filament in high temperature, low pressure bonding KEBLER and TUCKER (1972).

b) Boron Nitride-Coated Boron

Nitride-coated boron was first studied at Lockheed MSD CAMAHORT (1968); the material is produced by first oxidizing a very thin layer of the boron filament surface at about 1000°C and subsequently converting the boron oxide to boron nitride by reaction with ammonia at about 1100°C. This layer of boron nitride formed, is substantially thinner than the 0.1-mil layer of SiC on Borsic filament discussed previously, and it is typically less than 0.5μ thick; an example of the thin coating is shown in Figure 23, an SEM photograph of a fractured nitrided filament. Results of tests with nitrided 4-mil boron at Lockheed CAMAHORT (1968) and with nitrided 5.6-mil boron at AVCO, CAMAHORT and KRUKONIS (1972) showed that the material maintains enhanced properties relative to boron in high temperature B/Al consolidation. For example, B/Al composites fabricated at 1150°F using virgin boron degrade to a strength below 100 Ksi, whereas nitrided boron composites fabricated at the same temperature have exhibited strengths as high as 200 Ksi. An Air Force Materials Laboratory funded study of oxidizing and nitriding parameters is currently in progress to optimize nitrided boron production conditions KRUKONIS et al. (1974); B/Al composite strengths as high as 294 Ksi have been obtained and point out the potential of nitrided boron as a filament form capable of use in high temperature consolidation. On the other hand, although high longitudinal tensile strengths have been measured, the transverse properties of B/Al composites fabricated with nitrided boron have been limited to a maximum of about 15 Ksi because of debonding between the matrix and the nitride coating, and the study mentioned previously is also directed to circumventing the debonding phenomenon.

Besides providing a coating which is resistant to attack by aluminum in high temperature bonding, the nitriding process was found to increase the tensile strength of boron filament. Figure 24 is a histogram of nitrided 5.6-mil filament and is superposed on a histo-

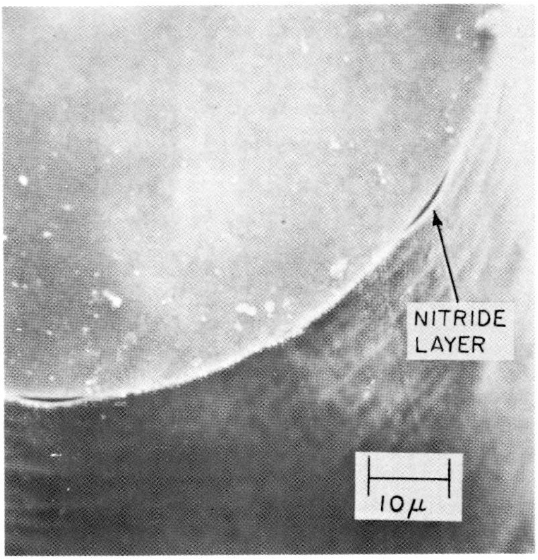

Fig. 23. SEM of fractured nitrided boron showing thin nitride layer

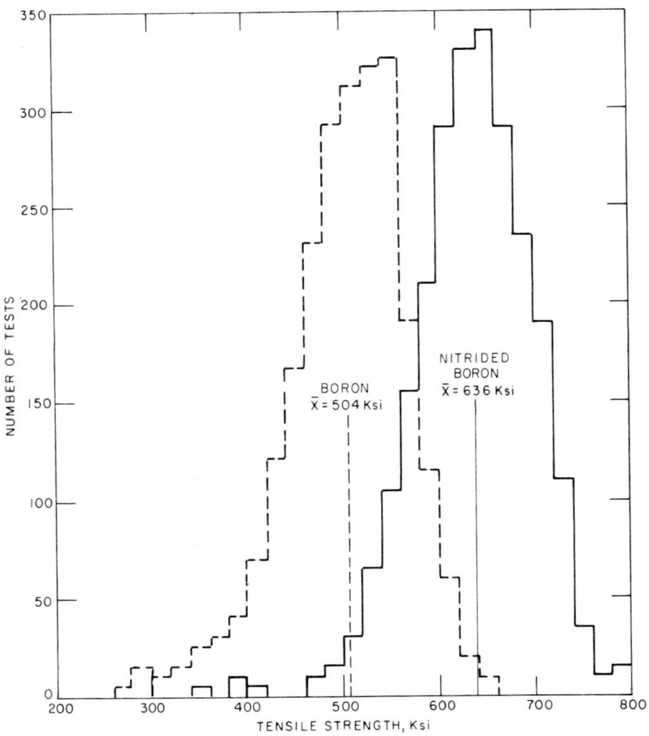

Fig. 24. Comparison of tensile strength histograms of boron and nitrided boron filament

gram of production grade 5.6-mil boron. Although the strength increase mechanism is not yet completely understood, it is probably related to some annealing and rearrangement of the internal stresses in the filament and is not related specifically to the formation of boron nitride on the surface. For example, a similar increase in strength during other secondary treatments of boron has also been reported; the deposition of pyrolytic graphite on boron HOUGH (1971) the deposition of a B_4C layer on boron KRUKONIS (1973), the heating of boron to 1300°C in a hydrogen atmosphere DE BOLT (1972), and a yet undisclosed low temperature boron treatment process being investigated by Northrup Corporation BROWN (1974) all result in boron filament with enhanced tensile strength.

Interestingly, the flexural or "loop" strength of boron is also increased by the nitriding step. Typical flaw-free, virgin 5.6-mil boron exhibits a loop strength of about 1.2 million psi, i.e. it can be bent to a radius of about 1/8" before breaking, its surface stress approximated by the formula,

$$\sigma = \frac{dE}{D} \tag{2}$$

where (in consistent units)

σ is the surface stress
d is the filament diameter
D is the filament loop diameter
E is the modulus of boron

Nitride 5.6-mil boron, on the other hand, can be looped to a smaller radius yielding a surface stress of about 1.4 million psi. The increase in the tensile strength combined with the simultaneous increase in the surface flexural strength indicates that more than just a rearrangement and annealing of internal stresses in boron is occurring during the nitriding process. For example, if only a re-distribution of the internal stresses, say, a decrease in the tensile stresses at the core-boron interface, were responsible for the increase in tensile strength of the nitrided filament, the loop strength would have to decrease based upon a simple static force analysis; yet, the loop strength has been found to be higher, and this phenomenon, like boron elongation and like the "radial crack" described previously, has not yet been explained.

4. Other Boron Forms

Very recent work has been directed to other boron filamentary forms, e.g. to boron ribbon with a large aspect (width to thickness) ratio. The work was motivated by both technical and economic factors. NASA/Langley studies have shown that a large aspect ratio ribbon can result in high transverse strength composites FOYE (1966), and NASA funded two programs, one at United Aircraft Research Laboratory and another at AVCO Systems Division, to develop a process to produce boron ribbon. A carbon ribbon substrate was used by both groups of researchers, but boron elongation precluded the achievement of high strength boron ribbon. The expansion phenomenon, which previously had manifested itself as primarily an axial elongation on (round) B/C, resulted in a two or three dimensional expansion during the formation of ribbon B/C. In addition to the light-bulbing phenomenon caused by the axial strain and fracture of the carbon substrate, a transverse fracture in the carbon substrate occurred; an example of the fracture is shown in Figure 25, which is a photomicrograph of a cross-section of a deposit of boron on carbon ribbon substrate DE BOLT et al. (1975). Even by first depositing pyrolytic graphite on the ribbon, the technique which circumvented the problem for (round) B/C filament, the expansion/light-bulbing problems could not be overcome, and the strongest boron ribbon produced exhibited a tensile strength of only about 80 Ksi BASCHE

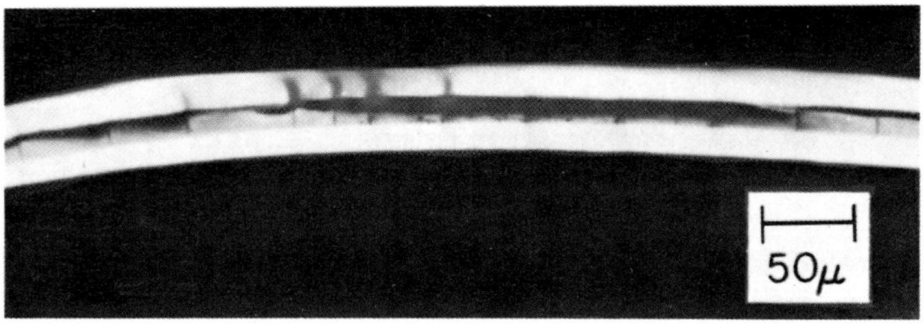

Fig. 25. Cross section of boron ribbon showing transverse fracture in the ribbon substrate

and JACOB (1973). During subsequent work at AVCO it was found that silicon carbide could be deposited on carbon with no expansion problems, and it was shown that a ribbon of silicon carbide on carbon about 75-mils wide x 4-mils thick could be produced with a tensile strength of over 200 Ksi DE BOLT and KRUKONIS (1973); a cross-section of SiC on carbon ribbon is shown in Figure 26.

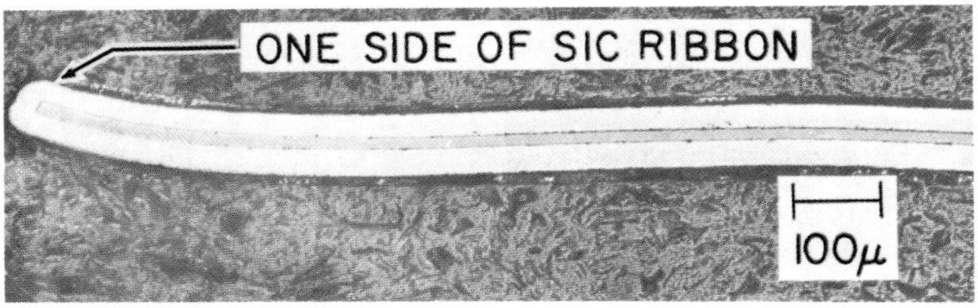

Fig. 26. Cross section of silicon carbide ribbon

5. Closing Remarks

After the high strength and high stiffness of boron was first demonstrated in 1959, process and applications development at a large number of companies rapidly carried the material from the laboratory to large-scale production, and within only five years after the filament was first produced, it became the base-line structural material for the horizontal stabilizer of an advanced U.S. aircraft, the F-14 Tomcat. Furthermore, the superior properties of boron in advanced composite structural applications motivated General Bernard SCHRIEVER, Commander, Air Force Systems Command and Director of Project Forecast to proclaim, "We are on the threshold of the greatest single material advance that has been made in the last 3000 years" SCHRIEVER (1964).

The mid-1970s find this Chemical Vapor Deposition process expanded to a production level of 30,000 lbs/year with levels of well over 100,000 lb projected by 1980. The uses of boron in advanced military aircraft structures are being extended to applications in high performance turbo-jet engines, commercial aircraft structures, and to other commercial uses where its high strength and high stiffness can be utilized to provide low weight. Sporting and leisure products such as golf shafts, tennis reckets, and bicycle frames are receiving increasingly greater development effort to achieve lower weight, higher stiffness product forms. In addition, the future will see large applications of boron composites in high speed, light weight, mass transportation vehicles, in advanced stationary gas turbines, and in other energy production, storage, and saving devices.

Acknowledgments. The author would like to acknowledge the work of two co-researchers at AVCO, Dr. Harold E. DeBolt, and Mr. Thomas W. Henze, who became engrossed in boron filament technology long ago and who have continued the progress in developing new material forms, increasing boron properties, and reducing production costs, very frequently working under the disadvantage of not being able to pursue in depth such important areas as understanding the nature of boron elongation, the nature of the strength increase occurring during nitriding, the effect of pyrolytic graphite in eliminating light bulbs, or so many other still ill-understood phenomena.

Funding by the United States Air Force Materials Laboratory in almost all facets of boron filament research has played an important part in advancing the state of development ot its present stage, and the entire boron industry must acknowledge this funding as well as the technical discussions and critical program reviews with AFML personnel. During the '60's and '70's George Peterson, Robert Neff, Frank Fecheck, and Robert Toma shot among many, many others of AFML, monitored, questioned, criticized, and coerced to force rapid development progress on boron filament, and it is to a large extent this "guidance" which carried boron filament from the development phase to large-scale production and application in a short time.

The author would also like to acknowledge the stimulating discussions with personnel from the competition, United Aircraft Corporation, who, although they are formidable adversaries encountered in the pursuit of programs and funding for boron filament advancement, are, above all, scientist who strive to understand and circumvent many of the still illunderstood phenomena referred to above.

Finally, the author and his co-researchers acknowledge the work of Dr. F.E. Wawner of the University of Virginia, who in the mid '60s was instrumental in delineating many of the causes of boron fracture and who, for the past five years, has been involved, or perhaps more quantitatively stated, inextricable entwined with the advancement of filament technology at AVCO.

References

BASCHE, M., FANTI, R., GALASSO, F.: Preparation and properties of silicon carbide-coated boron filaments. Fiber Sci. Technol. 1, 19-21 (1968)

BASCHE, M., FANTI, R., GALASSO, F., KUNTZ, U., SCHILLE, R.: Method for producing boron-carbon fibers. U.S. Pat. 3, 679, 475 (1972)

BASCHE, M., JACOB, B.: Development of a Process for Producing Ribbon shaped Boron Filaments, CR-132356. Printed by United Airways Corp. East Hartford, Conn. 1973

BASCHE, M., KUNTZ, U.E.: U.S. Pat. 3, 622, 369 (1971)

BERNEBERG, P.L., KRUKONIS, V.J.: Unpubl. Indt. Res Developm (IRAD-ZOO1), (1969)

BLACKMAN, L.C.F., UBBELOHDE, A.R.: Stress recrystallization of graphite. Proc. Roy. Soc. 266, 20-35 (1962)

BROWN, G.G.: Personal communication of work being carried out at Northrup Research Lab. (1974)

CAMAHORT, J.L.: Protective coating by surface nitridation of boron filament. Jour. Comp. Mater. 2, 104-111 (1968)

CAMAHORT, J.L., KRUKONIS, V.J.: Unpubl. Ind. Res. Develop. (IRAD-Y243), (1972)

CHRETIEN, A., HELCORSKY, J.: Sur les borones de molybdone et de tungstene MoB_4 et WB_4, Composes Nouveaux. C.R. Acad. Sci. (Paris) 252, 742-744 (1961)

DEBOLT, H.E.: Unpubl. Ind. Res. Develop. (IRAD-Z001), (1969)
DEBOLT, H.E.: Unpubl. Ind. Res. Develop. (IRAD-Z610), (1970)
DEBOLT, H.E.: Unpubl. Ind. Res. Develop. (IRAD-Y246), (1972)
DEBOLT, H.E., DIEFENDORF, R.J., GRUBER, P.E., JOO, L.A., KRUKONIS, V.J., MCKEE, J.A., MEHALSO, R.M.: Lower cost high strength boron filament, AFML-TR-70-287. Printed by AVCO Corporation, Lowell, Mass. (1971)
DEBOLT, H.E., KRUKONIS, V.J.: Development of a process for producing ribbon shaped filaments, NASA CR-132319. Printed by AVCO Corporation, Mass. (1973)
DEBOLT, H.E., KRUKONIS, V.J., MC KEE, J.A., PRESCOTT, R., WAWNER, F.E.: Development and demonstration of a low cost boron filament formation process, AFML-TR-72-271. Printed by AVCO Corporation, Lowell, Mass.(1972)
DIEFENDORF, R.J., MEHALSO, R.M.: Carbon monofilament as a substrate for CVD filament. Adv. Mat. Composites and Carbon, Amer. Cer. Soc., 51-58 (1971)
DIEFENDORF, R.J., MEHALSO, R.M.: Chemical vapor deposition of boron on a carbon monofilament substrate. Proc. 3rd Int. Conf. Chem. Vapor Deposition. Salt Lake City, 552-560 (1972)
FOYE, R.L.: An evaluation of various engineering estimates of the transverse properties of unidirection composites. SAMPE J. 10, G31-G42 (1966)
GALASSO, F.S.: High Modulus Fibers and Composites. New York-London-Paris: Gordon and Breach (1969)
GALASSO, F.S., PATON, A.: The tungsten borides in boron fibers. Trans. AIME 236, 1751-1752 (1966)
GRUBER, P.E.: On the kinetics of chemical vapor deposition. 2nd Int. Conf. Chem. Vapor Deposition, Los Angeles (May 1970)
HOUGH, R.L.: U.S. Pat. 3, 437, 511 (1969)
HOUGH, R.L.: High strength, continuous large diameter composite fibers. 2nd ASTM Conf. Composite Materials, Anaheim (1971)
JANAF: Thermochemical Tables, PB-168370, U.S. Dep. Commerce (1965)
KEBLER, R.W., TUCKER, R.C.: Master alloy B/Al tape for low pressure consolidation, IR-363(III). Printed by Union Carbide Corporation, Indianapolis, Ind. (1972)
KREIDER, K.G., DARDI, L., PREWO, K.: Metal matrix composite technology, AFML-TR-71-204. Printed by United Aircraft Corporation, East Hartford, Conn. (1971)
KRUKONIS, V.J.: Unpubl. Ind. Res. Develop. (IRAD-Z001), (1969)
KRUKONIS, V.J.: Unpubl. Ind. Res. Develop. (IRAD-Y246), (1972)
KRUKONIS, V.J.: Unpubl. Tensile strength of B_4C-coated boron filament produced by Soc. Nat. Poudres et Explosifs (1973)
KRUKONIS, V.J., HAUZE, A.W., WAWNER, F.E.: Exploratory development and evaluation of low cost boron aluminum composites, AVSD-0242-74-CR. AVCO Corporation, Lowell, Mass. (1974)
LINE, L.E., JR., HENDERSON, U.V.: Boron filament and other reinforcements produced by chemical vapor plating. In: Handbook of Fiberglass and Advanced Plastics Composites pp 201-236. (G. LUBIN, ed.), New York-London (1969)
SCHRIEVER, B.A.: A Look at the Future. 26th Conf. Aviation Writers' Assoc. Bal Harbour (May 1964)
STARKS, D.F., HOUGH, R.L., GOLF, L.C.: Surface finishes for boron filaments. AIAA Journ. 4, 1818-1821 (1966)
TALLEY, C.P.: Mechanical properties of glassy boron. J. Appl. Phys. 30, 1114-1115 (1959)
TOTH, I.: Original SEM photograph supplied by TRW (1974)
VELTRI, R., GALASSO, F.: Tensile strength of boron filament coated with silicon carbide and of uncoated boron filament at elevated temperatures. Nature 220, 781-782 (1968)
WAGNER, R.S., ELLIS, W.C.: The vapor-liquid-solid mechanism of crystal growth - Its applications to silicon. Trans AIME 233, 1053-1064 (1965)
WAWNER, F.E., Jr.: Boron filaments. In: Modern Composite Materials pp 244-269. (L. BROUTMAN and R. KROCK, eds) Reading: Addison Wesley (1967)
WAWNER, F.E., SATTERFIELD, D.B.: Observations on the tensile strength of amorphous boron. Appl. Phys. Letters 11, 192-194 (1967a)
WAWNER, F.E., SATTERFIELD, D.B.: Strength, uniformity and fracture characteristics in boron filaments. In: Boron Filament Process Development, AFML-TR-67-120, Vol. IA, Appendix C, pp C1-C43. Printed by Texaco Experiment Incorporated, Richmond, Va. (1967b)
WEINTRAUB, E.: On the properties and preparation of the element boron. J. Ind. Eng. Chem. 3, 299-301 (1911)

II. Boron Carbide Fibers from Carbon Fibers

W. D. SMITH

1. Introduction

There are many properties of boron carbide which make it attractive in a fibrous form. These include high strength, high modulus of elasticity, low density, chemical stability and good oxidation resistance. However, the process of fiberizing such a refractory, hard material by the usual methods, is virtually impossible. Thus, other methods are needed to overcome these problems. Starting with a boron-coated tungsten wire, HIGGINS et al. (1969) deposited B_4C from a mixture of BCl_3, CH_4 and H_2 at 1100-1200°C. HOUGH and GOLF (1966) used organoboranes to deposit B_4C on tungsten and HAMLING (1968) prepared partially converted B_4C fibers by impregnation of rayon with a soluble boron compound followed by heating. GATTI et al. (1966) formed B_4C whiskers by vaporization of B_4C powder followed by condensation in a temperature gradient. None of these methods, however, prepared fibers that were completely B_4C and continuous. It will be shown that multifilament carbon yarn can be totally (or partially) converted to continuous B_4C yarn by reaction with BCl_3 and H_2 under controlled conditions at high temperatures.

Figure 1 shows a polished cross section of B_4C fibers that are almost totally converted to B_4C. The small cores visible in some fibers are unreacted carbon. Figure 2 shows some partially converted B_4C fibers with larger carbon cores.

The general process of B_4C fiber preparation can be characterized as chemical conversion of a precursor fiber. Briefly, this involves conversion of an easily fiberized material into one that would be difficult to fiberize. This process has been discussed by ECONOMY et al. (1973) for several metal carbide fiber systems and will be expanded upon here for the specific case of B_4C fibers.

2. B_4C Fiber Preparation

Boron carbide fibers were prepared by the reaction of carbon yarn with BCl_3 and H_2 at high temperatures according to the following equation:

$$4\ BCl_3 + 6\ H_2 + C\ \text{fibers} \longrightarrow B_4C\ \text{fibers} + 12\ HCl \quad (1)$$

This reaction occurs in two steps in the temperature range of 1600-1900°C.

$$2\ BCl_3 + 3\ H_2 \rightarrow 2\ B + 6\ HCl \quad (1a)$$

$$4\ B + C \rightarrow B_4C \quad (1b)$$

The reactor consisted of a carbon resistance tube furnace which acted as the reaction vessel and heat source. The gases BCl_3, H_2 and diluent argon entered the furnace at one end, reacted in the hot zone and reaction products exited at the other end. The carbon yarn could

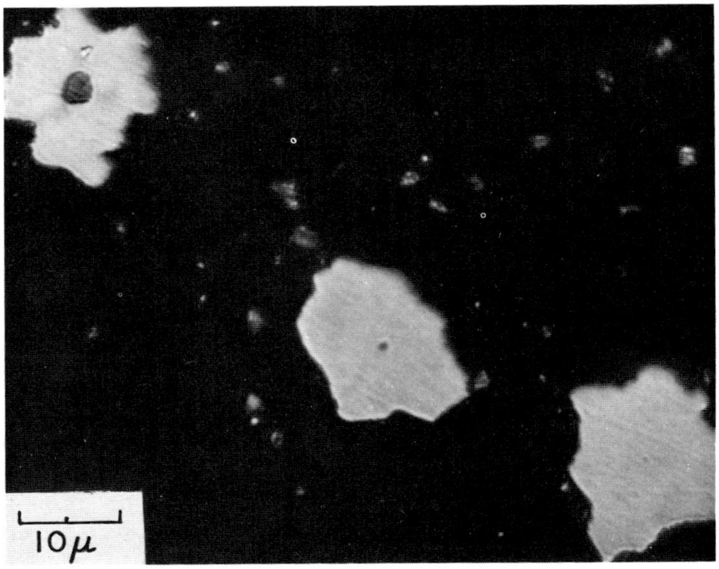

Fig. 1. Cross section of B_4C fibers with little or no carbon cores

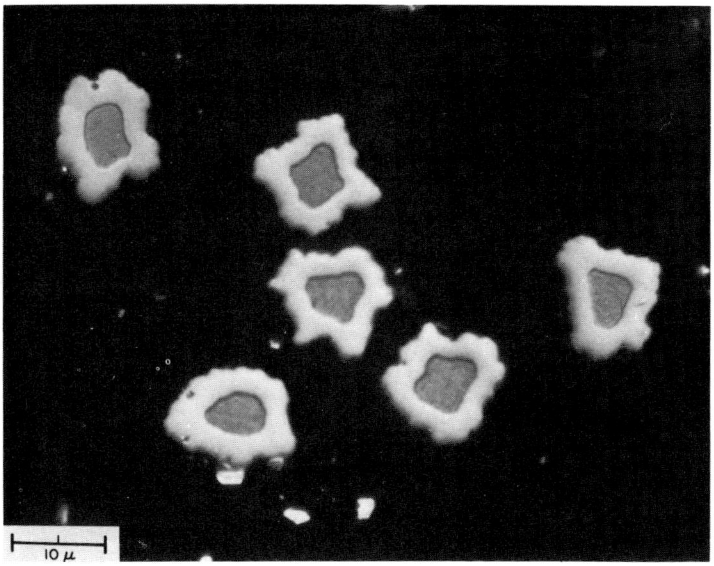

Fig. 2. Cross section of B_4C fibers with moderate sized carbon cores (partially converted)

either be held stationary in the furnace or passed through to produce either short pieces or continuous lengths of B_4C yarn. Reaction temperatures were about 1600-1900°C for reaction times of 1 to 10 min. Multifilament carbon yarn was chosen as the precursor instead of monofilament for several reasons. (1) Multifilament is readily available and fairly low in cost, (2) it has good strength and can be conveniently handled, (3) fiber diameters of the carbon yarn are very small with about a 10 micron average and (4) a greater output of B_4C yarns could be achieved. Monofilament carbon on the other hand, is (1) less readily available and (2) much weaker and more difficult to handle.

Of the two steps in the reaction, the reaction of boron and carbon [Eq. (1b)] is the rate-limiting step. This is determined by the fact that under isothermal conditions the reacted yarn will change from flexible and unfused to brittle and fused as the flow rates of BCl_3 and H_2 are both increased. This brittle and fused yarn is caused by an excessive boron deposition which cannot be reacted quickly enough with the carbon fiber. The excess boron then bridges or fuses fibers together at contact points to yield a brittle structure. To determine the reactivity of boron with carbon yarn at high temperature, the reaction was carried out in two separate steps. First boron was deposited on carbon yarn comprised of 720 10μ-diameter carbon fibers at a temperature (1000°C) below which boron and carbon would react. The amount of boron deposited was in excess of that needed to completely react the carbon to B_4C. The coated carbon yarns were then heated at 1800°C in argon for time periods from 0.1-6.7 min. After heat treatment, unreacted boron was removed by treatment with HNO_3 at 100°C and the samples weighed. X-ray diffraction patterns were then taken to verify that only B_4C was formed and that no boron remained. Weight gain was taken to be the amount of boron reacted to form boron carbide. Figure 3 shows the results of these experiments. The ordinate axis could be expressed as either carbon yarn reacted (w/o of the total present) or boron reacted (w/o of the total boron needed to form all B_4C). In either case, the reaction shows a very rapid initial rate which then gradually slows as the reaction progresses. This can be explained in the following manner. When the boron-coated carbon yarn is initially heated, the boron and carbon are in direct contact with each other and react very rapidly. However, as a reaction layer of B_4C forms and becomes thicker, the boron and carbon must diffuse through this layer to react. Since B and C are essentially in excess, these data specify the minimum time necessary to reach any level of conversion desired. The rate of boron deposition then must not exceed the rate at which it reacts with the carbon fiber or excess boron will deposit and fused fibers will result. To control the amount of boron deposited in reaction step Eq. (1a), it is necessary only to control the flow rates of BCl_3 and H_2 for any constant temperature and time condition.

To verify that reaction step Eq. (1a) totally described the reduction of BCl_3 by H_2 at 1800°C, the exhaust gases were analyzed for the possible presence of chloroboranes by the gas chomatographic method of MYERS and PUTNAM (1962). Using a n-hexadecane column, the only gases found were HCl and BCl_3. H_2 was not found since BCl_3 was in excess. No chloroborane compounds were found even though the method could detect the presence of B_2H_5Cl and $BHCl_2$. These results are in agreement with those of ATTWOOD and SHELTON (1970) who state that complete reduction of BCl_3 by H_2 to B is favored at these high temperatures.

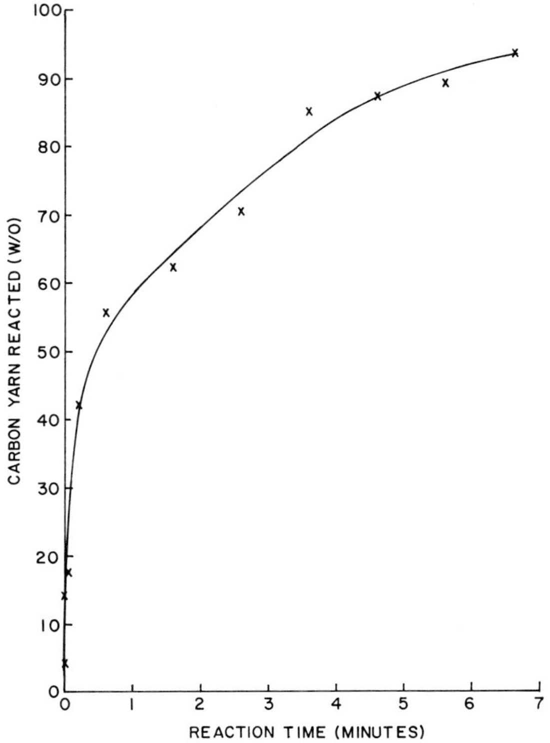

Fig. 3. Reactivity of boron with carbon yarn at 1800°C

3. High Strength Fibers

Fiber strength is a parameter that is highly dependent on porosity, internal and surface flaws, crystallite size, uniformity and many other factors. For B_4C fibers, one factor was found to be fiber curling during reaction. Curling would cause low strength, since when a tensile load was applied to these curled high modulus fibers, the fibers would break at the curled sections before the fiber straightened. This curling is a consequence of the volume change of the fibers during the conversion reaction. A simple calculation for the molar volume change of carbon fiber at a density of 1.42 g/cm^3 to boron carbide fiber at 2.51 g/cm^3 shows an increase in volume of 160%. This results in an increase in both fiber diameter and length. Fiber diameter measurements show an increase from about 10μ to over 12μ.

To improve B_4C fiber strength, fiber curling was essentially eliminated by employing a high temperature treatment under tension. It was found that when B_4C fibers were heated under tension near their melting point, the fibers straightened and room temperature tensile strength increased. The results of high temperature and tension treatments on curled B_4C yarns are shown in Table 1. B_4C yarns comprised of 720 filaments were treated for time periods of five min. The room-temperature filament strengths show an improvement factor of from around two to more than four. This treatment does cause some grain growth which may decrease strength if longer times and higher temperatures are used. A change is also noticeable in the physical appearance of the yarn. Figure 4 shows a B_4C yarn before treatment and Figure 5 shows this yarn after treatment. Clearly the fibers have been straightened. This straightening most likely occurs by annealing which relieves the

Table 1. Improvement of B_4C fiber strength by high temperature treatment under tension

B_4C sample no.	Initial fiber tensile strength (psi x 10^{-3})	Treatment Temp. ($^{\circ}C$)	Treatment Load (g)	Final fiber tensile strength (psi x 10^{-3})
1	44	2150	27	113
2	91	2150	100	154
3	43	2200	27	136
4	128	2200	27	200
5	42	2200	53	162
6	70	2200	53	176

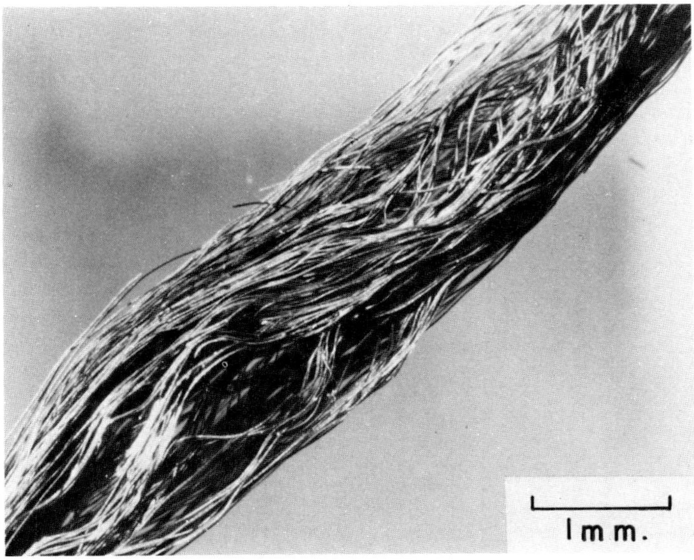

Fig. 4. B_4C yarn with curled filaments

internal stresses in the fibers. Fibers are not broken in this process, indicating that the load tends only to remove the curling and not to apply a stress greater than the strength of the fibers.

4. Properties of B_4C Fibers

a) Mechanical

Boron carbide fibers in the form of a 1-ply untwisted yarn (tow) are flexible, easily handled and strong. The fibers are flexible enough to be wound around a 0.5 inch mandrel without breakage and can be processed with most filament and yarn-handling equipment. A listing of the average mechanical properties of the fibers is given in Table 2. Tensile strength has been improved to over 300,000 psi by better con-

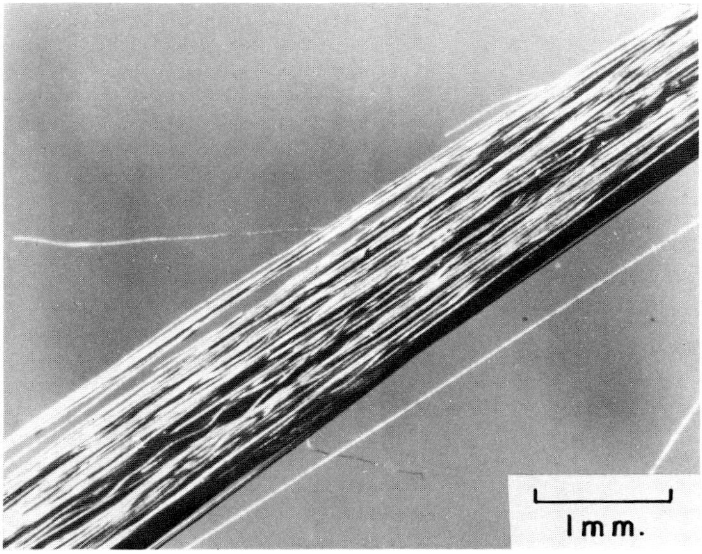

Fig. 5. B_4C yarn with straightened fibers after high temperature treatment under tension

Table 2. Properties of B_4C fibers

Tensile strength (psi x 10^{-3})	300 - 350
Modulus of elasticity (psi x 10^{6})	40 - 60
Density (g/cc)	2.20 - 2.35
Fiber diameter (μ)	11 - 12
Form	1 ply, 720 filament yarn

trol of the high temperature-tension treatment step. The range of values given for the modulus, density and fiber diameter indicate the dependence of these properties on the amount of B_4C phase present. In general, these properties are directly proportional to the amount of the B_4C phase, while tensile strength is dependent on this and many other variables as stated previously. Thus, it is possible to have high strength even though modulus, density or fiber diameter may be varied.

b) Chemical

B_4C fibers are fairly inert towards most acids, bases and solvents. There is no reaction with acids such as concentrated and dilute HCl, HNO_3 and H_2SO_4. There was a slight surface change on the fibers when heated in 90% H_2O_2 at 100°C for 3h. Also boiling 2N NaOH reacted only slightly with the B_4C fibers.

The fibers are resistant to such gases as air and chlorine. Oxidation in air does not begin until the temperature reaches 750°-800°C. They are quite stable toward chlorine with no reaction after 15 min exposure at 700°C. With longer times of 30 and 90 min the fibers showed weight losses of 10 and 30%.

Stability toward molten metals was also very good. No reaction was found after immersion in either molten aluminum or magnesium after 20 min at 900°C.

5. Behavior at High Temperatures

In order to convert a carbon fiber completely to B_4C, a diffusion mechanism was proposed by which boron diffuses inward and carbon diffuses outward, This is given support by the fact that fibers which are totally B_4C can be prepared by the reaction of boron with carbon fibers. These fibers reach theoretical density of 2.51 g/cc, indicating no porosity.

Additional evidence of mobility in boron carbide fibers having unreacted carbon cores can be seen in cross sections of fibers given post treatments. If the fibers thus formed, as shown in Figure 2, are heated at about 2270°C for a time period of 30 min, a change in the fiber cross section occurs. Initially, there are two distinct and separate phases - carbon and boron carbon. Upon heating, the boundary tends to disappear and both carbon and boron carbide phases start to diffuse into one another. Figure 6 shows the fiber cross sections after heat treatment. The highly reflective phase is B_4C while the dark spots are graphite. X-ray diffraction patterns taken before and after heat treatment show no difference in lattice parameter of the boron carbide phase. However, the week diffuse carbon line has been replaced by well-crystallized graphite. This further suggests that graphitization has been catalyzed to some extent by the presence of boron carbide.

6. X-ray Diffraction Analysis

X-ray diffraction patterns of B_4C fibers were taken with a 114.6 mm Debye-Scherrer camera using filtered copper radiation. The samples were single yarn bundles with the fibers glued together. Exposure time was about three h at 40 KV and 21 ma. Both partially and totally converted fibers were measured and the results are shown in Table 3. There seems to be no significant difference among these patterns, even for the sample heated to 2300°C. Lines due to carbon or graphite have been deleted.

The lattice parameters were calculated using a least-squares method and extrapolation to 180° 2 θ. Table 4 compares published lattice parameters with those of B_4C fibers. Average values for B_4C fibers are $a = 5.596 \pm .003$Å, $c = 12.07 \pm .03$Å and $c/a = 2.157$.

7. B_4C Fiber Reinforced Composites

The principal application of B_4C fibers is for the reinforcement of organic and inorganic matrix material. The high strength, high modulus and low density of B_4C fibers make them ideal for high performance and aerospace applications.

B_4C fibers with properties shown in Table 2 were incorporated into both epoxy and carbon matrix composites. Epoxy composites were prepared by cutting the fibers to mold length, adding resin, and pressing to dimension under heat and pressure. A cross section of B_4C fibers in

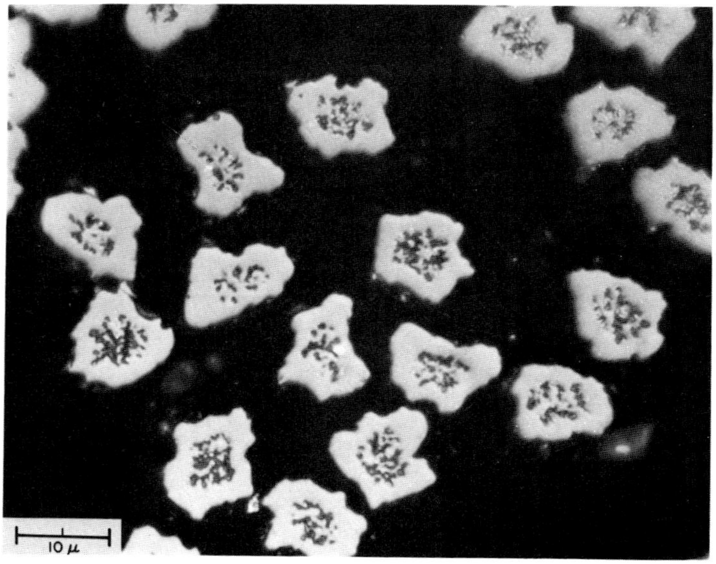

Fig. 6. Cross section of partially converted B_4C fibers after treatment at 2270 C for 30 min

an epoxy matrix is shown in Figure 7. Carbon matrix composite fabrication was similar except that three reimpregnation-carbonization cycles were added to reduce porosity. The epoxy composites have flexural strengths indicating about a 75% translation of fiber properties to composites' properties. Flexural modulus is good and another important property, interlaminar shear strength, is quite high even without surface treatment (see Table 5). Carbon matrix composites had lower flexural strengths because of porosity in the matrix, but had similar modulus and interlaminar shear values. The high interlaminer shear strengths of B_4C fiber-C matrix composites is quite significant since most carbon fiber-C matrix composites have very low values. Mc LOUGHLIN (1970) reported values of 2000-4500 psi for unidirectional carbon-carbon composites with a fiber content of 75-80% by weight. B_4C-C composites routinely have interlaminar shear strengths of 6000 psi. The reason for this difference is not completely understood but seems to be due to the rougher surface of the B_4C fibers which may increase bonding.

8. Summary

B_4C continuous filament yarn with good physical and chemical properties can be prepared from a precursor carbon yarn by reaction with BCl_3-H_2 mixtures at temperatures of 1600-1900°C. Conversion from carbon to B_4C occurs by way of a two-step reaction, (1) reduction of BCl_3 to boron and (2) reaction of boron with the carbon yarn. The second step was found to be rate-limiting since the reaction of B and C is slowed appreciably by formation of a B_4C reaction layer. Filament curling was identified as the reason for low strength in the yarn thus formed, but this could be eliminated by a high-temperature treatment

Table 3. X-ray diffraction patterns of B_4C fibers

hkl	B_4C fibers (fully converted) d (A°)	B_4C fibers (with carbon cores) d (A°)	Heat-treated[a] B_4C fibers (with carbon cores) d (A°)
101	4.49	4.49	4.49
003	4.02	4.00	4.02
012	3.78	3.80	3.78
110	2.80	2.80	2.79
104	2.56	2.56	2.56
021	2.38	2.37	2.37
211	1.810	1.812	1.805
205	1.710	1.710	1.705
303	1.494	1.499	1.496
125	1.460		
108	1.441	1.445	
220	1.399	1.397	1.378
131	1.335	1.336	1.336
223	1.321	1.318	1.318
132	1.311		
306	1.256	1.256	1.254
042	1.187	1.188	1.187

[a]1h at 2300°C in argon

Table 4. Comparison of B_4C lattice parameters

	a (A°)	c (A°)	c/a
ALLEN (1953)	5.61	12.07	2.151
YAKEL (1973)	5.6008	12.0721	2.1554
B_4C Fibers			
fully converted	5.596	12.08	2.159
carbon core	5.595	12.07	2.157
carbon core +H.T.[a]	5.596	12.06	2.155

[a]Heat-treated 1h at 2300°C in argon

under tension to yield high-strength fibers. B_4C fibers were shown to have an average tensile strength of 300,000 psi and modulus of elasticity of 50 x 10^6 psi at a density of 2.35 g/cc. The fibers are almost inert to hot acids and bases and quite resistant to chlorine up to about 700°C and air up to almost 800°C. Mobility in the B_4C-C system was demonstrated by heat treatment of partially converted B_4C fibers. X-ray diffraction analysis shows B_4C fibers to have about the same lattice parameters as bulk even in the presence of excess carbon and after heat treatment near their melting point.

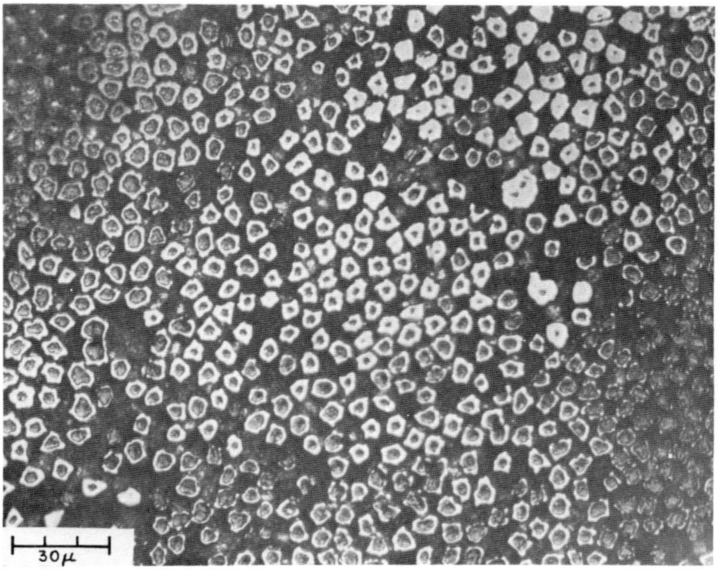

Fig. 7. Cross section of B_4C fibers in an epoxy matrix composite

Table 5. Mechanical properties of B_4C fiber unidirectional composites

	Epoxy matrix	Carbon matrix
Flexural strength (psi x 10^{-3})	120	53
Flexural modulus (psi x 10^{-6})	23	20
Interlaminar shear strength (psi)	7500	6000
Density (g/cc)	1.8	1.6
Fiber volume (%)	55	55

B_4C fibers were incorporated into epoxy and carbon matrix composites and showed good reinforcing properties with especially high interlaminar shear strengths.

References

ALLEN, R.D.: The solid solution series, boron-boron carbide. J. Amer. Chem. Soc. 75, 3582-3583 (1953)

ATTWOOD, B., SHELTON, R.A.J.: Thermodynamic properties of $BHCl_2$ from the equilibration of BCl_3-H_2 mixtures. J. Less Common Metals 20, 131-134 (1970)

ECONOMY, J., SMITH, W.D. and LIN, R.Y.: Preparation of refractory fibers by chemical conversion method. J. Appl. Polym. Sci., Appl. Polym. Symp. 21, 131-141 (1973)

GATTI, A., MANCUSO, C., FEINGOLD, E. and JAKAS, R.: Study of the growth parameters involved in synthesizing boron carbide filaments. NASA Accession No. N66-28634, Rept. No. NASA-Cr-75673 (1966)

HAMLING, B.H.: Metal carbide fibers. U.S. Pat. 3,403,008 (1968)

HIGGINS, J.B., GATTI, A. and GEBHARDT, J.J.: Preparation and properties of boron carbide continuous filaments. J. Electrochem. Soc. 116, 137-143 (1969)

HOUGH, R.L., GOLF, L.C.: Boron carbide filaments from organoboranes. Soc. Aerosp. Mater. Process Eng., Nat. Symp. Exhib., 10th, San Diego, Calif. (1966)

McLOUGHLIN, J.R.: New high temperature carbon fibre composite. Nature 227, 701 (1970)

MYERS, H.W., PUTMAN, R.F.: Determination of chloroboranes, diborane (6), and hydrogen chloride by gas chromatograph. Anal. Chem. 34, 664-668 (1962)

YAKEL, H.L.: Lattice expansions of two boron carbides between 12 and 940 °C. J. Appl. Cryst. 6, 471-472 (1973)

III. Boron Nitride Fibers

J. ECONOMY and R. LIN

1. Introduction

The preparation of boron nitride was first reported by Balmain (BALMAIN, 1842) in the middle of the nineteenth century. It remained as a laboratory curiosity until the mid part of the current century. Hot pressing techniques were devised permitting commercial development of molded samples (TAYLOR, 1957, 1958, 1959). Interest was especially heightened because of its unusual combination of properties which included excellent electrical insulating character, thermal stability, corrosion resistance and mechanical properties. A second advance in the commercial use of BN occurred with the successful preparation of boron nitride fiber (ECONOMY and ANDERSON, 1966a). The availability of boron nitride in a fiber form has greatly increased the scope of its application. In this paper the preparation, structure, properties and application of boron nitride fiber are reviewed.

2. Preparation of Boron Nitride Fiber

Boron nitride fiber is prepared by reacting boric oxide fiber with ammonia at elevated temperatures (ECONOMY and ANDERSON, 1969; ECONOMY et al., 1969). This preparation is an excellent example of a more general synthetic route pioneered by Carborundum scientists for making refractory fibers, known as the chemical conversion of a precursor fiber (CCPF) (ECONOMY et al., 1973). The unique feature of this method lies in the fact that fiber forming is achieved by chemically converting a precursor fiber which is either commercially available or can be easily prepared. The preparation of boron nitride fiber consists of three key steps:

1. Fiberization of B_2O_3 melt
2. Nitriding of B_2O_3 fiber
3. Stabilization of BN (graphitization)

Fiberization of boric oxide is easily accomplished due to the excellent rheological behavior of the boric oxide melt. Nitriding of boric oxide fiber with ammonia occurs via a series of solid-gas reactions. High temperature treatment is carried out to remove trace oxides and stabilize the product.

The overall reaction can be expressed by the following chemical equation:

$$B_2O_3 + 2\ NH_3 \rightarrow 2\ BN + 3\ H_2O \qquad (1)$$

The actual reaction mechanism is very complex and involves a series of heterogeneous reactions which in many ways have a formal resemblance to the charring and graphitization of organic materials. By monitoring the weight changes during the reaction and chemical analysis of the by-products and the fiber, Economy, Anderson and Matkovich (ECONOMY et al., 1969) were able to propose a reaction mechanism for the boron nitride formation.

$$nB_2O_3 \text{ (fiber)} \xrightarrow[200^\circ C]{NH_3} (B_2O_3)_n \cdot NH_3 \quad (2)$$

$$(B_2O_3)_n \cdot NH_3 \xrightarrow[350^\circ C]{NH_3} (BN)_x(B_2O_3)_y \cdot (NH_3)_z + H_2O \quad (3)$$

$$(BN)_x(B_2O_3)_y \cdot (NH_3)_z \xrightarrow{1500^\circ C} BN \text{ (fiber)} + B_2O_3 + H_2O + NH_3 \quad (4)$$

Reactions in Eqs. (2) and (3) are the rate-determining steps of the process. Ammonia diffuses into the fiber forming the addition complex indicated in Eq. (2). At temperatures in excess of 350°C the reaction with ammonia changes to a substitution type giving off water Eq. (3). It should be noted that the substitution reaction in Eq. (3) leads directly to compositions which can be interpreted as mixtures of cyclic BN and boroxine rings. Further nitriding during Eq. (3) leads to coalescence of the cyclic rings into a highly disordered three-dimensional layer-type BN structure. The reaction at this stage is greatly facilitated by the presence of a microporous structure which permits ready diffusion of NH_3 and H_2O into and out of the fiber. As the reaction progresses, the nitrogen content of the fiber increases. The final conversion step is a fast process which can be completed in a matter of seconds. The diffusion of ammonia through the microporous structure of the fiber is strongly influenced by the diamèter of the fibers. The larger the fiber diameter, the longer the time required to reach a certain degree of conversion. The effect of fiber diameter on the rate of nitriding is illustrated in Table 1.

Table 1. Effect of fiber diameter on degree of nitriding at 700°C

Time at 700°C h.	Fiber diameter microns	Nitrogen content %
0.5	6.3	44
	10.8	42
	17.5	34
	34.0	22
2.0	6.3	49
	10.8	47
	17.5	41
	34.0	30
14.0	6.3	51
	10.8	50
17.	17.5	45
	34.0	32

It appears likely that the size of the diffusing species can also influence chemical reactivity and rate of diffusion. To test this effect a series of experiments was carried out where the partially nitrided fibers were reacted with amines with different alkyl group substituents.

$$\text{BN fiber (partially nitrided)} + \underset{(400\ cc/min)}{\text{amine}} \xrightarrow[10\ h\ at\ 700^\circ C]{to\ 700^\circ C\ at\ 25^\circ C/h} \text{nitrided fiber} \quad (5)$$

The carbon content was used as a measure of the rate of reaction. Results are summarized in Table 2. The difference in carbon content in the final product suggests a size effect on reactivity.

Table 2. Effect of amine structure on the rate of nitriding

Amine	Carbon content in nitrided fiber %
CH_3NH_2	23.7
$(CH_3)_2NH$	11.8
$(CH_3)_3N$	10.6
$CH_3CH_2NH_2$	8.1

3. Structure of Boron Nitride Fibers

Boron nitride has a crystal structure similar to that of graphite (PEASE, 1952), (ECONOMY and ANDERSON, 1967); but there is a distinct difference in the nature of the layer packing. In boron nitride the hexagonal rings of atoms are packed directly on top of each other, whereas in graphite, half the atoms lie between the centers of the hexagonal rings of adjacent layers (Fig. 1). The distance between layers is about 3.33 A., while the B-N and C-C bond distances within the rings are 1.45 and 1.41 A., respectively. Another crystallographic form of boron nitride, called turbostratic, has also been reported (THOMAS, J. et al., 1963). The turbostratic structure in both BN and carbon can best be described as consiting of two-dimensional layers of hexagonal rings which are stacked at regular intervals; however, there is a complete lack of orientation from layer to layer. Neither turbostratic boron nitride nor carbon possesses a three-dimensional lattice structure, therefore, this system is distinguished by the absence of any hkl reflections.

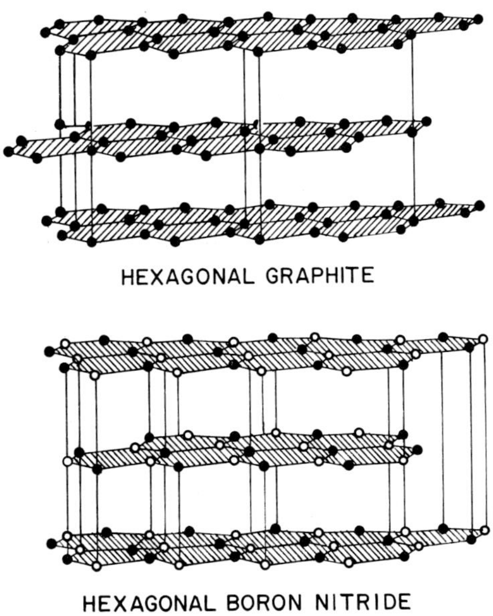

Fig. 1. Hexagonal structure of graphite and boron nitride

Preliminary analysis of the boron nitride fibers by X-ray indicated a diffraction pattern similar to that reported for turbostratic boron nitride. (See Fig. 2.) The 002 peak corresponds to the distance between layers while the 10 peak describes the distance across the planar ring. From this type of pattern the crystallite size can be calculated from the degree of line-broadening of the diffuse lines. There is an inconsistency in interpreting the structure of the fibers as turbostratic since their properties differ considerably from those reported for turbostratic boron nitride. For example, turbostratic boron nitride reportedly dissolves in water with evolution of heat. It has also been noted that it dissociates to its elements on heating at temperatures in excess of 2000°C. The fibers on the other hand are quite stable to refluxing water and withstand temperatures in excess of 2500°C under an inert atmosphere.

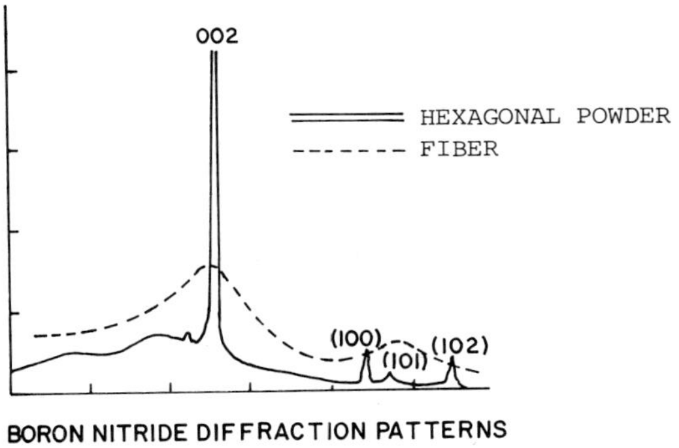

Fig. 2. X-ray diffraction patterns of boron nitride

Electron diffraction studies were also carried out and it was found that in addition to the turbostratic phase a three-dimensional crystal structure was present in the fibers which showed well-defined hkl reflections (Table 3). However, a number of lines reported for the hexagonal form were missing and in particular the prominent 102 line which appears at 1.817 Å. This could not be considered conclusive evidence for the presence of a three-dimensional crystal structure since electron diffraction does not give as representative diffraction patterns as X-rays. By increasing exposure time X-ray diffraction also showed the presence of weak hkl lines. Again the 102 line of the hexagonal form was missing, suggesting the presence of a different crystallographic form of boron nitride. On the basis of the available information, it appears, therefore, that boron nitride fibers consist of a major phase of turbostratic boron nitride and a three-dimensional ordered phase which is not the well-known hexagonal form.

It was found that with relatively fine-grained structures of less than 100 Å, only the turbostratic phase could be detected. Presumably particle size effects would wash out the possible presence of this new phase. The existence of the new phase could easily be detected in fibers with grain sizes in excess of 150 Å. An X-ray pattern of such fiber is shown in Figure 3 where the presence of the (112) reflection with d = 1.18 Å suggests a three-dimensional phase. This particular line is common to both the hexagonal and rhombohedral forms of boron nitride. However, the absence of the (102) line at 1.82 Å clearly rules out the hexagonal phase while the absence of lines at about 1.99 Å (101) and (102) argues against the rhombohedral phase. This reflection as well as one observed at d = 1.00 Å can be interpreted in terms of a small primitive unit cell with a = 1.445 Å and

Table 3. Comparison of X-ray and electron diffraction interlayer spacings for boron nitride fibers

hkl	Hexagonal BN (d) X-ray, Å	BN fibers (d) electron diffraction, Å.
002	3.33	3.33
100	2.17	2.17
101	2.06	--
102	1.817	--
004	1.667	--
103	1.552	--
104	1.322	--
110	1.253	1.26
112	1.173	1.178
200	1.086	1.083

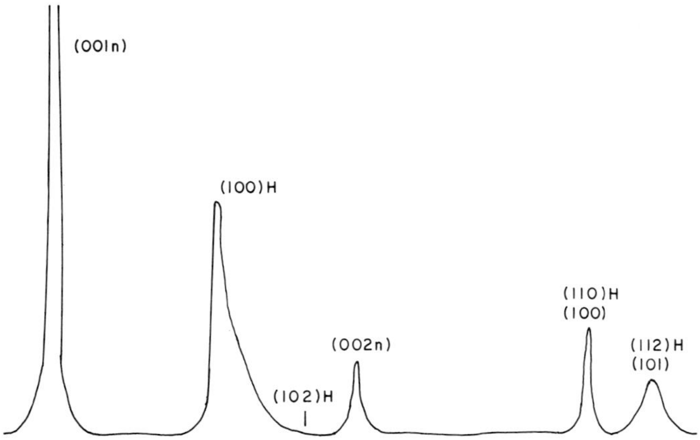

Fig. 3. X-ray diffraction pattern of boron nitride fibers with grain size in excess of 150 Å

c = 3.33 Å. The relationship of this small primitive hexagonal cell to regular hexagonal boron nitride is shown in Figure 4. It is characterized by random occupancy of the atomic positions and the unit cell content is 1/3 BN formula weight. This phase appears to be very stable since thermal cycling up to 2500°C does not lead to any recrystallization. If anything, the turbostratic material may convert to the primitive hexagonal structure. Some insight to the unusual stability of this phase can be gained by examining more closely the various possible packing sequences in BN. In Figure 5 some of the possible packing sequences of the hexagonal layers of BN are indicated. Of these only the hexagonal with rings packed directly above each other and the rhombohedral with a three-layer repeat unit have been observed. The other stacking sequences which are indicated can be obtained by translation of neighboring layers. Conversion of the rhombohedral to the hexagonal form seems unlikely, since both translation and a 60° rotation of alternate layers is required. This is not the case in graphite where the hexagonal and rhombohedral phases are in equilibrium. Hence in BN, it should be recognized that two geometrical isomers exist. In fact it seems reasonable to conjecture that the primitive hexagonal structure is derived from random stacking of these geometrical isomers.

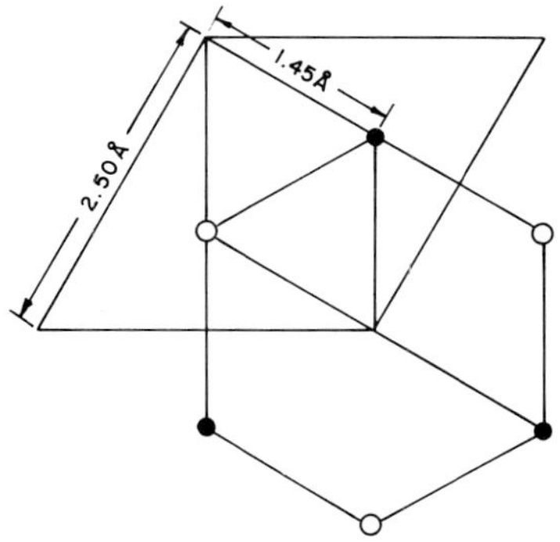

Fig. 4. Relationship between the unit cell of regularly stacked BN layers (2.50Å) and that resulting from random stacking (1.45 Å)

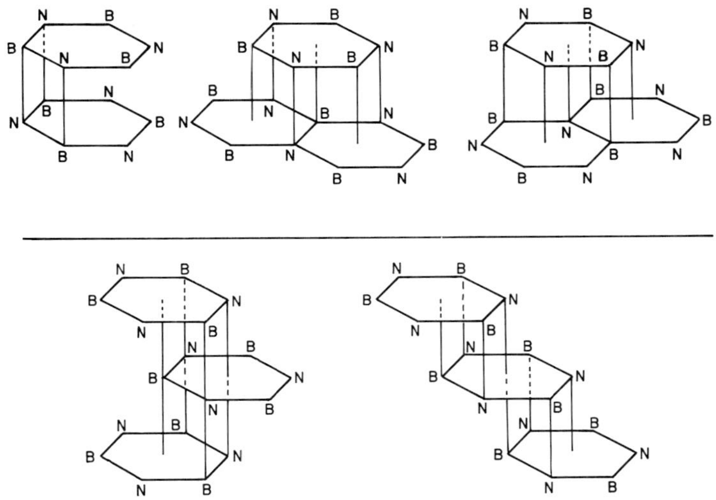

Fig. 5. Various ways of stacking of boron nitrogen layers

Since the turbostratic structure represents the major phase in the fibers, a number of questions remain to be answered as to the reported properties of this phase and those of the fibers. From studies of the fibers it appears that at least some of the properties are determined by the size of the crystallites and differences in the interlayer spacing. Reports by earlier workers (THOMAS et al., 1963), which indicated poor water resistance and thermal stability for turbostratic boron nitride, were based on the use of samples with crystallites of 15-40 Å. and an interlayer spacing of 3.56 Å. In this study it has been found that the crystal dimensions of the turbostratic boron nitride in the fibers can range from dimensions of 15-40 Å. up to several thousand angstroms. It has also been found that the interlayer spacing can approach a value of 3.33 Å. which is identical to that of the hexagonal form. It therefore appears that the only factor which distinguished the turbostratic form from the hexagonal form is the absence of any orientation in the c direction. Otherwise the size and shape of the crystallites may be similar and in fact the various forms may exist within the same crystallite.

4. Properties of Boron Nitride Fibers

a) Correlation of Structure and Properties

The unusual properties of the fibers can be related directly to the nature of the BN structure (ECONOMY 1967). For example, thermal degradation reactions require cleavage of many bonds across the planar network. This is statistically unlikely. Perhaps even more important is the absence of any grain growth or phase changes up to at least 2500°C. Most refractory fibers at temperatures of 700 to 1000°C suffer sharp losses in properties due to grain instability. As shown in Table 4 there is practically no change in the crystallite dimensions of L_a and L_c during thermal cycling, although there is a small reduction in the d_{002} interlayer spacing.

b) Thermal and Oxidative Stabilities

Under inert or reducing conditions, boron nitride fibers are stable up to very high temperatures (ECONOMY et al., 1966b). Initially a small weight loss is noted at elevated temperatures, but, as can be noted in Table 5, the fibers do not display any additional weight loss during prolonged heating at 2100°C. The observed thermal resistance is in contradiction to a report that boron nitride at 1900°C has a vapor pressure of 150 mm. From other data on boron nitride, it can be predicted that the fibers should be stable to 3000°C under inert conditions.

Table 4. Thermal cycling produces no significant change in BN fibers

Conditions		Crystal dimensions, Å L_a	L_c	d_{002}
untreated fibers		72	22	3.50
2100°C	2-1/2 h	61	24	3.45
2100°C	9-1/2 h	66	27	3.44
2400°C	20 min	74	23	3.45
2500°C	20 min	73	34	3.41

Table 5. Thermal stability of boron nitride fibers in nitrogen

Temp. (°C)	Time (h)	Weight loss (%)
1900	0.25	--
2100	2.5	3.53
2100	9.5	3.70
2400	0.33	7.3
2500	0.33	7.4

The extreme thermal stability of both graphite and boron nitride fibers depends on the planar network of hexagonal rings which are packed in such a way that multiple cleavage of bonds across the planar network is necessary for thermal decomposition.

Boron nitride fibers are superior to graphite fibers in oxidation resistance, as shown in Figure 6. Graphite fibers oxidize completely in the temperature range of 600-700°C, while boron nitride fibers

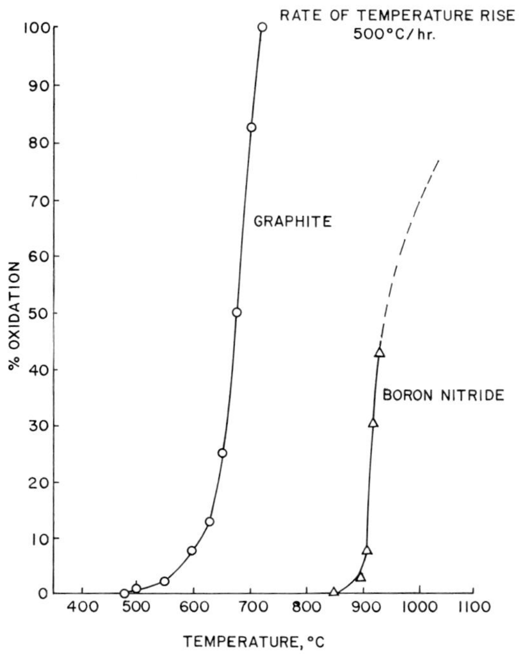

Fig. 6. Rising temperature oxidation of boron nitride and graphite fibers

begin to oxidize appreciable at 900°C. Oxidation studies carried out under isothermal conditions indicate incipient oxidation at 855°C, as shown in Figure 7. At 840°C no weight change could be detected after several hours. Oxidation of boron nitride results in a boric oxide coating which reportedly acts as a glaze and affords protection up to 1300°C.

A number of glass fibers, such as quartz, Refrasil , S-glass, and Fiberfrax ceramic fiber display good thermal stabilities in air in the temperature range of 800-1200°C; however, embrittlement of the fiber due to devitrification takes place at temperatures of over 1000°C. Applications where glass fibers are flexed require the presence of a sizing to protect the surface of the fiber. This drastically limits high-temperature applications, since there are no sizings available for use at temperatures in excess of 300°C.

c) Mechanical Properties

Boron nitride fibers with tensile strengths as high as 210,000 lb/in^2 have been ovserved using one inch gauge lengths and a cross-head movement of 0.2 in/min; however, in a typical sample the strength values usually range from 50,000 to 125,000 lb/in^2. The elastic modulus was found to vary over a range of 4 x 10^6 to 12 x 10^6 lb/in^2, with a break elongation of about 2% to 3%. These values are summarized and compared against commercially available glass fibers and carbon fibers in Table 6. The size of the crystallites in the boron nitride fibers plays an important role in determining mechanical properties. Presumably, for optimum properties, a polycrystalline fiber should possess small crystallites with strong grain boundaries, so that the energy involved in crack initiation and propagation will be dispersed by the numerous crystal boundaries present at all angles in the body. Crystallite size, as measured from X-ray line broadening in boron nitride fibers, is very small with dimensions of 50-150 Å. The fine grain size can also

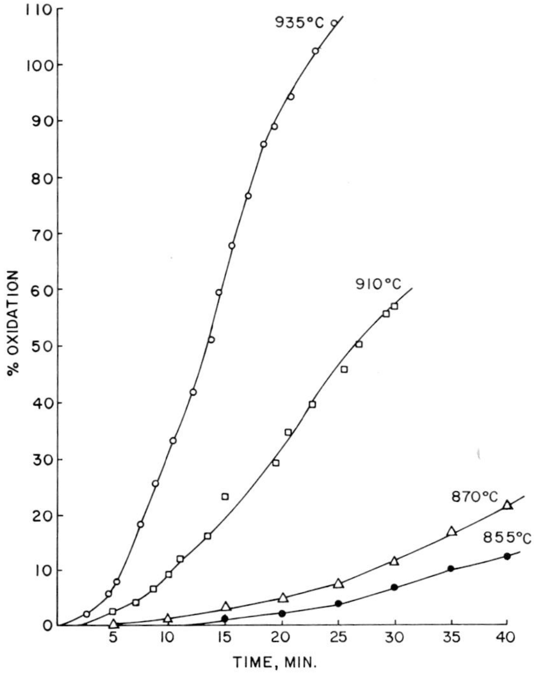

Fig. 7. Isothermal oxidation of boron nitride fibers

Table 6. Comparison of strength properties of boron nitride fibers

	Tensile strength lb/in^2	Elastic modulus 10^6, lb/in^2	Elongation at break %
E-glass	200-220,000	10.5	3.4
Carbon	50-100,000	4-9	2-3
Boron nitride	50-210,000	4-12	2.3

be noted from an electron micrograph of a fiber cross section shown in Figure 8.

Thermal cycling of boron nitride fibers up to 2200°C under an inert atmosphere does not affect the room temperature strength. Preliminary tensile tests at 900°C indicate that the fibers retain their strength at elevated temperatures. Data reported on solid pieces of boron nitride show some loss in properties in the temperature range of 400-1400°C followed by an increase at higher temperatures. The decrease in strength of solid boron nitride is associated with the presence of the boron oxide impurities.

More recently ECONOMY and LIN (1972) reported the feasibility of producing boron nitride fiber with a modulus exceeding 50 x 10^{0} psi. The increase in stiffness was accomplished by orienting the crystallites along the direction of tension, a technique related to that used in the production of high modulus graphite fiber.

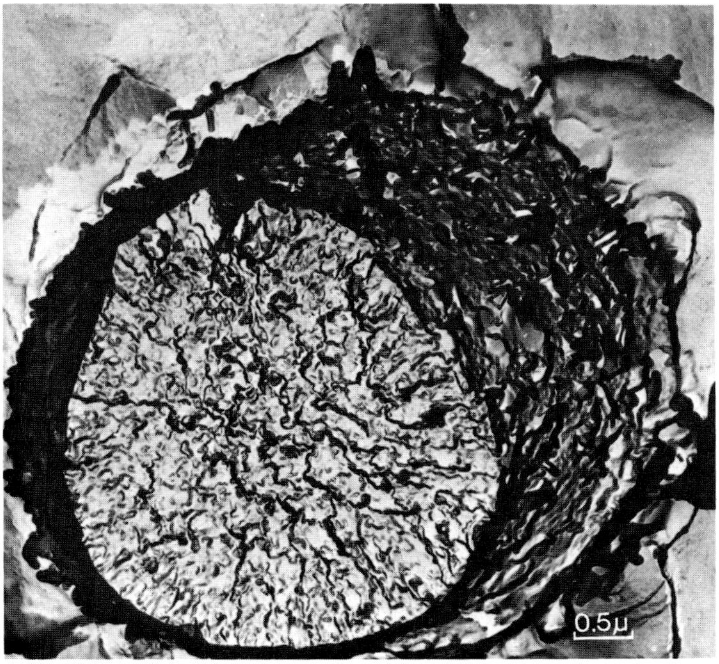

Fig. 8. Electron micrograph of a boron nitride fiber cross section

d) Corrosion Resistance

Boron nitride fibers appear to be extremely corrosion resistant (ECONOMY and ANDERSON 1966b). As can be seen in Table 7, the fibers are more resistant to refluxing acid and base than either carbon fibers or glass fibers. The effect of hydrofluoric acid at room temperature is also shown in this Table. Chemical analysis of the boron nitride fibers before and after refluxing in the acid or alkali indicated that the composition was unaltered.

Table 7. Comparative corrosion resistance of boron nitride fibers

Reagent	Temp. ^{o}C	Time h.	Boron nitride fibers % wt. loss	Carbon fibers % wt. loss	Pyrex fibers % wt. loss
5% H_2SO_4	100	3	0.4	2.4	0.8
5% NaOH	100	3	0.1	2.3	51.2
		(0.002)			100
48% HF	25 -	(3)		+5.6	
		(300)	0.0		

The effect of water on the boron nitride fibers was investigated under a variety of conditions as shown in Table 8. The fibers are essentially inert to water at room temperature; however, under extended reflux conditions, weight losses of 0.0% to 0.1% may be observed. On occasion, higher weight losses have been noted, but the integrity of the fibers in all cases is retained and this minor weight loss is probably associated with the presence of trace amounts of poorly crystallized boron nitride. At higher temperatures of 775° and in an atmosphere of steam, a weight loss of 5% is observed after 1 h.

Table 8. Resistance of boron nitride fibers to water

Reagent	Temp. °C	Time	Weight loss %
H_2O	25	2 weeks	0.0
	25	8 weeks	0.0
	100	75 h	0.1
Steam	600	1 h	0.0
	700	1 h	0.8
	775	1 h	5.0

Corrosion data for solid boron nitride indicate that the fibers should resist chlorine up to 700°C and will be inert to melts such as silicon, copper, cryolite, and sodium chloride. The boron nitride fibers were not attacked by either molten cast iron or aluminum.

Boron nitride fiber has been actively evaluated as a separator in high energy battery system (GAY, et al., 1973; NELSON, 1974; NELSON et al., 1974). In this application the corrosion resistance and the thermal stability of the boron nitride fiber are utilized. According to the test for Li/S battery system, the BN separator performed without failure in the molten LiF-LiCl-KF at 400°C in excess of 1000 h. This new finding will have a strong technological significance in the area of both material development and energy conservation.

e) Electrical Properties

The electrical resistivity of boron nitride fibers has been studied up to temperatures of 800°C. Results showed that the fibers possess exceptionally good electrical resistivity over a wide range of temperatures varying from 10^{14} ohm-cm at 25°C to 10^{10} ohm-cm at 800°C. In Table 9 a comparison between high temperature electrical resistivity of boron nitride, aluminum oxide, and beryllia is given. Boron nitride displays a fairly high thermal conductivity comparable to stainless steel. In Table 10 a comparison is made between the thermal conductivity of hot-pressed BN, Al_2O_3, and BeO. BN retains its thermal conductivity as the temperature increases, whereas Al_2O_3 and BeO show a considerable drop with increasing temperature. The combination of high electrical resistivity with good thermal conductivity is very important for high-temperature electrical insulating material. It is noteworthy that boron nitride also displays a very low dielectric loss and a low dielectric constant.

The high electrical resistivity of BN is rather surprising since graphite, which possesses the same number and arrangement of electrons, is an electrical conductor. Because of the shortened bond length be-

Table 9. Electrical resistivity

Temp °C	BN fibers	Resistivity, Al_2O_3	ohm-cm. BeO
25	10^{14}	10^{14}	
500	--	6.3×10^{10}	
800	10^{10}		1.8×10^{1}
1000	--	2×10^{6}	

Table 10. Thermal conductivity

Temp °C	Thermal conductivity, cgs Hot pressed BN	Al_2O_3	BeO
100	--	0.04-0.069	0.521
300	0.0687	--	--
400	--	0.02-0.031	0.16-0.22
700	0.0646	--	--
800	--	0.015-0.017	0.062-0.11
1000	0.0637	0.014-0.015	0.045-0.068

tween B and N of 1.45 Å, it has been suggested that the structure possesses aromatic character similar to that of graphite. Certainly the requirements for aromaticity are met, such as availability of three pairs of electrons for resonance and a planar ring. Yet, the high electrical resistivity of BN which would be associated with the absence of freely resonating electrons has prompted many workers to describe the structure in terms of single bonds only (see Fig. 9). Thus, the single bond model would involve only trigonal bonds while the ionic double bond model would depend on the nitrogen donating its pair of electrons to the boron. Supporting evidence for the resonating structure comes from NMR studies where coupling constants of B^{11} isotope show a donation of 0.45 electrons to the boron from the nitrogen (SILVER and BRAY, 1960). This indicates about 45% double bond and 55% single bond character. Data in support of the single bond concept are obtained from the very low diamagnetic susceptibility moments of -0.4×10^{6} cgs electromagnetic units, indicating the absence of large orbits of resonating electrons.

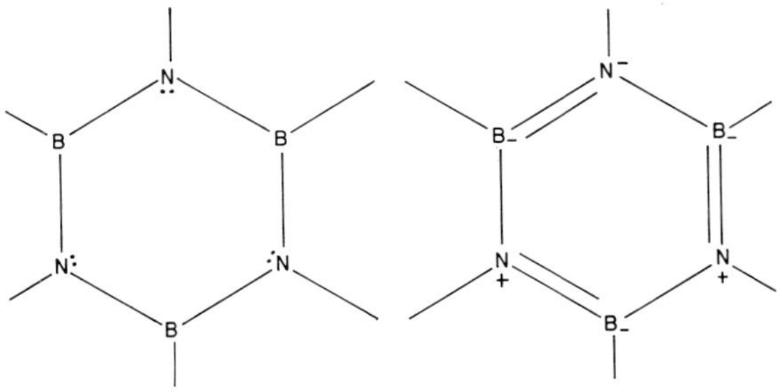

Fig. 9. Possible boron nitride structure

Certainly the existence of a partial ionic double bond in BN with resonating character appears justified from the shortened bond distances and NMR data; however, the assumption that such a bond should provide freely resonating electrons for conduction does not necessarily follow. In graphite where the environment is essentially neutral, the π electrons are delocalized, i.e. not attached to any atom and hence are free to move under the influece of an electrical field. In BN the ionic network which results from a partial electron transfer from nitrogen to boron may be sufficient to create potential wells, thus localizing the electrons so that they are incapable of moving when placed in an electric field. Retention of good electrical resistivity up to 2000-3000°C suggests a high degree of stability for this ionic double bond. The absence of delocalization of electron was obtained from a careful examination of the three dimensional electron density map obtained from X-ray analysis of a single crystal of BN (ECONOMY and MATKOVICH, 1972).

References

BALMAIN, W.H.: J. Prakt. Chem. 27, 422 (1842)

ECONOMY, J.: Boron nitride fibers. Res. Development. 18, No. 6, 8 (1967)

ECONOMY, J., ANDERSON, R.V.: A new route to boron nitride, J. Inorg. Chem. 5, 989 (1966a)

ECONOMY, J., ANDERSON, R.V.: Properties and uses of boron nitride fibers. Textile Res. J. 36, No. 11, 994 (1966b)

ECONOMY, J., ANDERSON, R.V.: Boron nitride fibers. J. Polymer Sci. Part C, No. 19, 283-297 (1967)

ECONOMY, J., ANDERSON, R.V.: U.S. Pat. 3,429, 722 (1969)

ECONOMY, J., ANDERSON, R.V., MATKOVICH, V.I.: Preparation of BN fibers. Appl. Polymer Symp. 9, 377 (1969)

ECONOMY, J., LIN, R.Y.: U.S. Pat. 3, 688, 059 (1972)

ECONOMY, J., MATKOVICH, V.I.: Stacking of hexagonal BN layers. Int. Symp. Boron. TBILISI, USSR (1972)

ECONOMY, J., SMITH, W.D., LIN, R.Y.: Preparation of refractory fiber by chemical conversion method. Appl. Polymer Symp. No. 21, 131-141 (1973)

GAY, E.C., STEUNENBERG, R.K., BATTLES, J.E., CAIRNS, E.J.: 8th Intersoc. Energy Convers. Eng. Conf. Proc. 96, (Aug. 13-16, 1973)

NELSON, P.A.: The lithium/sulfur battery. Paper Great Lakes Power Club, Winter Meeting (Feb. 15, 1974)

NELSON, P.A., CHILENSKAS, A.A., WALSH, W.J., GAY, E.C., YAO, N.P., SHIMOTAKE, H.: ANL High energy batteries for electrical vehicles. 3rd Int. Elec. Vehicle Symp. (Feb 19-20, 1974)

PEASE, R.S.: Acta Cryst. 5, 356 (1952)

SILVER, A.H., BRAY, P.J.: J. Chem. Phys. 32, 288 (1960)

TAYLOR, K.M.: U.S. Pat. 2, 808, 314 (1957)

TAYLOR, K.M.: U.S. Pat. 2, 855, 316 (1958)

TAYLOR, K.M.: U.S. Pat. 2, 888, 325 (1959)

THOMAS, J., WESTON, N.E., O'CONNOR, T.E.: J. Amer. Chem. Soc. 84, 4619 (1963)

IV. Reinforcing Properties of ALB_2 Flakes

L. C. WOHRER and A. A. WOSILAIT

1. Introduction

Current high-strength, high-stiffness refractory fibers are being used increasingly in structural hardware replacing heavier metals such as aluminum and steel. Yet the one-dimensional nature of fibers poses distinct limitations on their use in components requiring truly orthotropic behavior. Certainly the greatest potential for developing reinforced composites with optimum mechanical properties in two dimensions would be in the use of refractory fibers or flakes. For instance, to achieve planar reinforcement with fibers or whiskers requires the use of multiple crossplies which results in approximately 60% reduction in mechanical properties in any direction. With flakes it is possible to achieve planar reinforcement directly since the flakes display planar isotropic character. That this is possible has been demonstrated with glass flake-reinforced composites where a Young's modulus of 6×10^6 psi and a flexural strength of 50×10^3 psi is observed in planar directions (OLEESKY and MOHR, 1964). Flake-reinforced metals have been produced by unidirectional solidification of binary eutectic alloys such as $CuAl_2$ in Al. The tensile strength of such a composite was 43,000 at a 45% volume concentration of $CuAl_2$ flakes (HERTZBERG, 1965). The shortcomings with this system lie in the high density of the flake material, inadequate control of the flake concentration and problems in obtaining useful sizes. There have been some attempts to utilize SiC flakes in resinoid composites; however, poor translation of properties was observed primarily because of the low aspect ratio. Thus, a definitive answer as to the potential of single crystal flakes in composites was not possible.

AlB_2 flakes were selected for this study since they display high strength and modulus exceeding 500×10^3 psi and 60×10^6 psi, respectively, as do other refractory diborides. Equally as important, they possessed a relatively low density of 2.95 g/m^3.

AlB_2 flakes were under investigation at The Carborundum R & D Laboratories for several years. Much of this effort has been directed at establishing conditions for crystallization of AlB_{12} from molten aluminum to optimize size and minimize impurities such as AlB_{12} and other reaction products (MATKOVICH et al., 1964). A preliminary evaluation of the AlB_2 flakes in resinoid composites produced a major improvement in mechanical properties over the best results obtained with fibers or whiskers (ECONOMY et al., 1969: WOHRER et al., 1966). Thus the potention usefulness of flakes in lightweight structures of extreme stiffness for aerospace and aircraft applications has been recognized.

This paper presents the work conducted to purify and classify the crude flakes according to aspect ratio, evaluate the effect of aspect ratio, volume loading, matrix material, void content, direction of load and machining upon composite performance. The study is strictly experimental, as micromechanical treatments of discontinuous reinforcements in elastic-plastic matrixes can be found elsewhere (PIGGOT, 1966; RILEY, 1968; PADAWER and BEECHER, 1970; PIGGOT, 1973; PIGOTT, 1973).

Binary flake composites displaying a flexural strength and moduli of 109 x 10^3 psi and 44 x 10^6 psi, respectively, have been obtained. In terms of planar-specific flexural strength and modulus these values amount to 1.2 x 10^6 ins and 500 x 10^6 ins.

Ternary composites containing flakes between plies of graphite fibers have displayed several previously unexpected properties. Most noticeable has been the excellent ability to absorb energy under impact testing and the notably good retention of strength and stiffness after crude machining. In general, the stress-strain properties of the ternary composites were well between those of binary AlB_2 composites on the one end and unidirectional and orthotropic binary fiber composites on the other end. The stiffness of the ternary composites, though less than that of binary flake composites, still compared favorably with that of binary graphite fiber composites. Equally important, the procedure of making ternary composites is suitable for the fabrication of large structures involving vacuum bag techniques. It is also evident that further improvements, particularly in strength of composites, can be expected when platelets of greater perfection and aspect ratio become available.

2. Experimental

a) Source of AlB_2 Flakes

The AlB_2 flakes used in this study were provided by The Carborundum Company in the form of aluminum ingots containing 5 to 25% flakes. These flakes had been crystallized in molten aluminum under conditions defined in the Al-B phase diagram below.

b) Screening and Elutriation

A procedure consisting of repeated screening and aqueous elutriation was developed to purify and separate flakes into fractions of the following diameter ranges:

Diameter (μ)
a) 1410 to 2380
b) 707 to 1410
c) 420 to 707

Screening was carried out under water to prevent flake damage. Occasionally formed clusters of flakes could be dispersed by adding acetone. Screening was repeated three times to obtain a satisfactory diameter classification. Elutriation of the crude flakes was accomplished by using a four-foot-long glass tube with an ID of 55mm containing a 22mm OD inner tube to serve as an annulus. The water entered at the botton tangent to the diameter which provides a spiraling upward flow. For elutriation the flow rate of water was varied starting with the slowest. It was increased step-wise after practically no flakes left the column at the top.

c) Diameter and Thickness Distribution of Flakes

Flake characterization was accomplished as follows:
Distribution and Means. Maximum and minimum diameters of 100 flakes are measured to the nearest 100 μ using a microscope. The thickness of an equal number of flakes is measured from a polished cross section

of a composite using a magnification of 600 x or 1000 x depending on the thickness of the flakes.

Means: 100 to 150 flakes are weighed on a micro balance and the total area is determined by a digital scanning microscope. The diameter of the flakes is obtained from the area to give the maximum length, from one corner of a hexagon to the opposite corner. The thickness of the flakes is calculated using the formula:

$$\frac{\text{Weight}}{\text{Area x Spec Grav}} = \text{thickness.}$$

The means obtained by both techniques were averaged. The narrowest thickness and diameter distributions were obtained using a low flow rate corresponding to a Reynold's number of 200. The broadest distributions were displayed by flakes elutriated at a high flow rate corresponding to a Reynold's number of 800. Figure 1 denotes a typical distribution for flakes possessing an aspect ratio, i.e. diameter/thickness of 274.

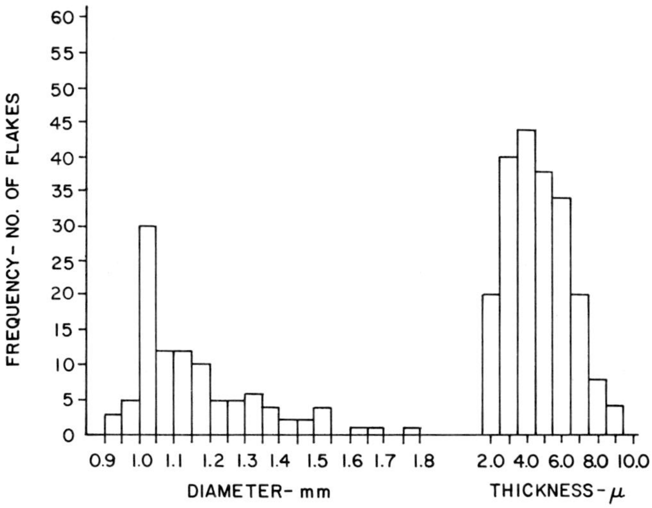

Fig. 1. Diameter and thickness distribution of AlB_2 flakes D/T~274

d) Flake Composites

Composite Preparation and Testing. Standard fiber reinforced composite prepregging and compression molding techniques were adapted to the preparation of flake reinforced composites. ERLA 4617 epoxy (Union Carbide Corp.) was selected as a matrix due to its low viscosity at room temperature, good prepregging characteristics, and good mechanical properties.

The resin is mixed with m-phenylene diamine, i.e. m-PDA (27 parts of m-PDA per 100 parts of ERLA 4617) heated to 65-70°C until the m-PDA is completely dissolved. The mixture is then heated with agitation at 70°C for 4h and cooled to room temperature. A weighed amount of AlB flakes is mixed with the resin and spread over a 5" x 5" area onto a poly-propylene screen so that the flakes assume alignment while most of the excess resin is removed. The mixture is then heated for 45 to

60 min in a vacuum oven at 85-100°C in order to obtain a 40 mil thick prepreg that can be handled at room temperature. The pregreg is cut to fit the mold and laid inside the mold cavity. The mold is then heated to 100°C for 3h and excess resin is pressed out of the cavity while the resin is slowly advancing. The pressure is applied slowly so that the flakes can still realign and distribute themselves uniformly in the mold during the initial resin removal. At the end of 3h, the mold is closed to the desired thickness, while the pressure on the composite is increased to about 500 psi. It is important that the final closing of the mold concurs with the gelling of the resin if void-free composites are to be obtained. The mold is then heated to 120°C for 4h and finally to 160°C for 16h. Then it is cooled to room temperature and the composite is removed.

The composites were molded into strips of 5 x 0.5 x 0.08 ins of panels of 6 x 3 x 0.06 ins from which specimens were cut. They were tested for flexural strength, flexural modulus, sonic modulus and interlaminar shear after cutting the samples to the following sizes:

Flexural strength:	1.25" span, 3 pt. loading, 0.02" head speed, span-to-depth ratio ranging from 16:1 to 20:1
Flexural modulus:	2.5 x 0.5 x 0.08 in or 1.5 x 0.5 x 0.062 in
Sonic modulus:	5 x 0.5 x 0.08 in
Interlaminar shear:	0.60 x 0.2 x 0.08 in, span-to-depth ratio 6:1

Ternary composites were made in a similar way except that plies of flakes were alternatingly layed up with plies of graphite fibers to obtain a balanced structure.

3. Composite Properties

a) Composite Properties vs. Flake Aspect Ratio

Composites were prepared using 15 different groups of flakes, varying in diameter, thickness and aspect ration with ERLA 4617 as a matrix. Two test specimens were made from each group for flexural strength and modulus, and sonic modulus tests. The flake dimensions and test results are listed in Table 1 in the order of declining flexural strength. The values for v/o AlB_2, voids, and densities were calculated from constituent densities.

A flexural strength and modulus approaching 109 ksi and 37 x 10^6 psi, respectively, were exhibited by specimens containing the largest flakes. These flakes possessed an aspect ratio in the range of 290-340. A lower strength and stiffness was shown by composites containing lower aspect ratio flakes. The reinforcing action of flakes was diminished to that of a common filler when the aspect ratio was below 60. The values for flexural moduli were generally in very good agreement with the corresponding sonic moduli.

The average composite flexural strengths (adjusted to 60 v/o) are once more listed in Table 2, now grouped according to flow rates and screen sizes used in flake classification. As can be seen in this Table, drastic increases in the reinforcing power of flakes occur at constant flow rates with increasing flake diameter, and at decreasing flow rates when the flake diameter is constant. A few inconsistencies are apparent: for instance, composites from group 5 should possess higher flexural strength (about 95 ksi would be expected instead of the 83.9 ksi).

Table 1. Properties of AlB_2 composites vs. aspect ratio

Flake group No.	Average flake Diam. (μ)	thickness	Aspect ratio	v/o AlB_2	Voids (%)	Flexural strength (psi X 10^3)	Flexural modulus (psi X 10^6)	Sonic modulus (psi X 10^6)	Density (g/cc)
1	2269	6.65	341	60.4	0.87	109.0	33.3	----	2.28
				61.2	4.42	86.6	29.4	----	2.26
2	1260	4.6	274	60.5	0.1	93.3	31.8	33.4	2.28
				62.5	0.1	95.8	32.5	34.8	2.33
3	2154	5.8	370	62.5	0	89.3	35.6	34.8	2.32
				62.5	0	87.0	32.8	33.1	2.30
4	2298	7.9	291	63.3	0	89.9	37.6	35.8	2.33
				63.3	0	86.4	36.4	34.2	2.32
5	1208	3.9	309	58.4	4.3	83.2	31.8	26.4	2.22
				58.4	4.3	79.9	29.7	26.5	2.17
6	754	3.5	215	59.8	3.0	79.2	31.9	32.6	2.24
				59.8	3.0	75.4	30.9	----	----
7	803	3.5	229	57.2	3.5	70.9	29.3	27.5	2.19
				57.2	3.5	72.0	28.9	26.9	2.16
8	1360	6.6	206	55.9	1.8	65.1	25.3	33.7	2.18
				55.9	1.8	63.4	26.6	----	2.18
9	1251	4.8	260	60.5	7.8	52.2	24.5	25.9	2.16
				60.5	7.8	53.9	23.9	25.1	2.17
10	677	4.3	157	51.6	8.8	53.2	20.6	21.0	2.03
				51.6	8.8	56.0	22.5	21.2	2.04
11	1412	9.6	147	58.7	4.9	52.8	27.8	27.7	2.20
				58.7	4.9	54.4	28.5	25.6	2.20
12	610	6.2	98	54.5	8.1	51.5	22.2	25.6	2.09
				55.6	14.8	42.5	17.2	21.3	2.03
13	512	7.0	73	53.36	8.5	36.5	15.9	15.6	2.04
				53.36	8.5	35.7	15.8	15.8	2.06
14	615	12.4	50	49.7	9.4	27.9	11.3	11.5	1.97
				49.7	9.4	24.4	11.1	11.6	1.98
15	593	9.7	61	50.0	10.5	21.8	8.8	9.0	1.09
				50.0	10.5	20.3	8.3	8.3	1.97

The presence of 4.3% voids together with slightly more contaminants may be responsible for the lower value. Even more pronounced is the strength reduction of composites from group 9. A flexural strength of about 80 ksi was expected instead of the actual 52.7 ksi. The 7.8% voids, and perhaps the presence of few large particulate contaminants may have caused the low strength. On the other hand, composites from group 12 possessed higher than expected strength in spite of high void content. Examination of polished composite cross sections revealed in this case that the flakes were, in fact, thinner than expected and of higher aspect ratio. Presumably, these flakes were initially clustered and required a higher flow rate for elutriation. Composites from groups 6 and 7 exhibited identical strength as was expected since flake dimensions were very similar. The composites from groups 10 and 13, however, possessed much different strength although they should have been equal. An examination of flake size

Table 2. Composite flexural strength (psi x 10^{-3}) vs. flow rate and flake size (Adj. to 60 v/o AlB_2)

Flow rate (Re≠)	Flakes collected on screens (opening in μ) 1410-2380	707-1410	420-707	250-420
200	109 (1)	83.9 (5)	74.8 (6) 74.9 (7)	40.7 (13) 63.5 (10)
350	---	90.8 (2)	---	---
400	84.6 (3)	52.7 (9)	---	---
550	---	69.0 (8)	---	---
800	83.5 (4)	54.0 (11)	51.0 (12)	25.1 (15) 31.6 (14)

Figure in parenthesis denotes flake group from Table 1

distributions indicates the presence of coarser flakes in group 10 in spite of identical elutriation conditions. Differences in the structure (geometric shape) and wettability (air bubbles on the surface during elutriation) are potential reasons for this discrepancy. Thus, further refinement of experimental conditions may eliminate these discrepancies.

The dependence of flexural strength upon aspect ratio is shown in Figure 2. The volume loading of flakes was adjusted to 60 v/o for direct comparison. Figure 3 shows the dependence of the flexural and sonic moduli upon aspect ratio also adjusted to 60 v/o flake loading. The strong influence of flake aspect ratio on composite modulus is strikingly evident.

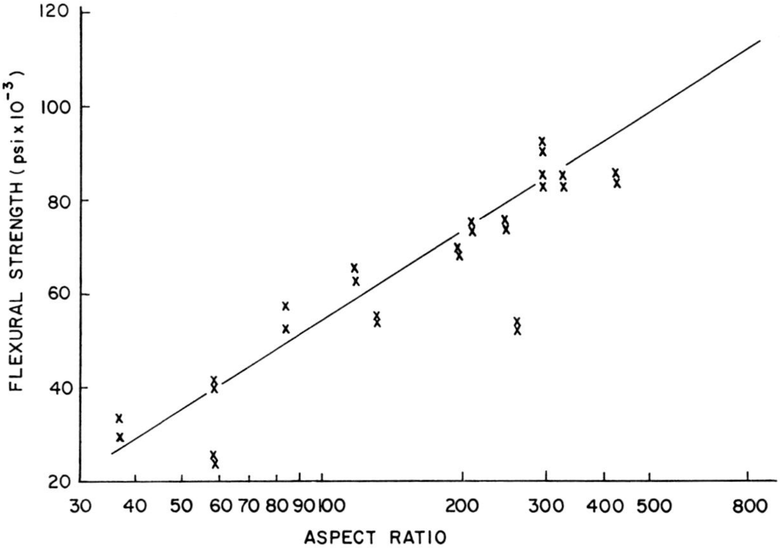

Fig. 2. Flexural strength of AlB composites vs. aspect ratios adjusted to 60 v/o AlB_2

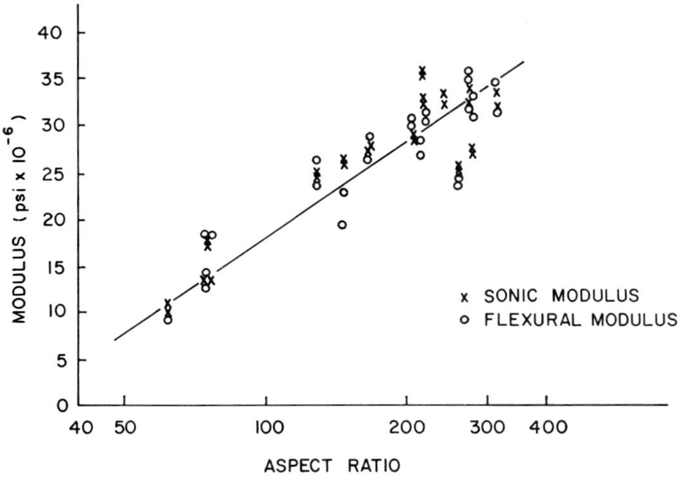

Fig. 3. Composite flexural modulus vs. aspect ratio of flakes (adjusted to 60 v/o AlB_2)

Figure 4 shows the effect of aspect ratio on tensile and compressive strength of flake composites. The interlaminar shear strength was very little affected by the flake's aspect ratio. The failure stresses ranged from 6800 to 8300 psi when tested using a short beam with a span-to-depth ratio of 6:1. However, the actual interlaminar shear strength may be higher since all specimens displayed a tensile failure mode as opposed to a stretched-out stress curve indicative for a shear failure mode. There is also no clear trend between impact strength and aspect ratio, as can be seen in Table 3. The impact strength of un-notched specimens ranged from 1.6 ft lb/in to 28.8 ft lb/in. Low values were realized when the specimen broke into two compact pieces. High values materialized when the specimen shattered (delaminated). Notched specimens displayed always a low impact strength of about 1.0 ft lb/in notch.

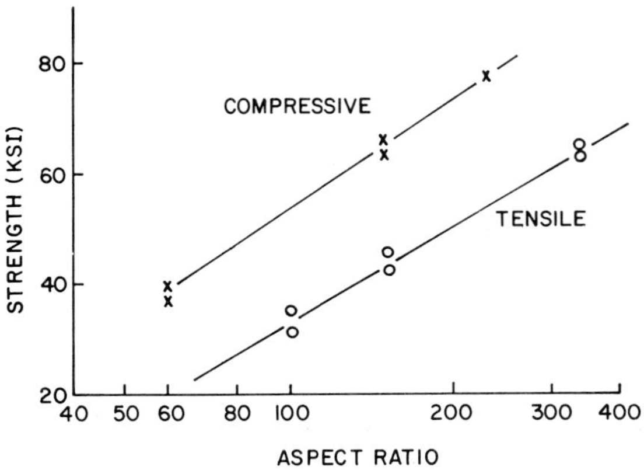

Fig. 4. Tensile and compressive strength of AlB_2 composites vs. aspect ratio adjusted for 60 v/o AlB_2

Table 3. Impact strength of AlB_2 composites

Flake aspect ratio	AlB_2 volume loading (%)	Void content (%)	Impact strength[a] (ft lb/in) un-notched
59	59.0	3.8	1.66
59	59.4	1.9	1.93
100	59.6	3.4	9.46
100	59.0	2.3	8.19
100	60.1	2.4	14.0
100	60.2	2.1	16.8
150	60.8	8.0	3.7
150	60.3	2.4	3.7
150	60.1	1.4	3.0
150	60.1	1.5	8.0
200	60.4	1.6	4.0
200	60.4	3.6	4.0
229	60.5	0.7	5.6
229	60.7	1.2	4.7
250	60.2	0.8	5.4
250	60.1	1.9	28.8

[a]In plane of flakes

b) Composite Properties vs. AlB_2 Volume Loading

Composites were prepared using three groups of flakes (ELLIOTT 1965; RILEY, 1968; PADAWER and BEECHER, 1970) in volume loadings ranging from 23 to 67%. (For flake dimensions see Table 1.) The flexural strengths and moduli are shown in Figures 5 and 6. They increase linearily with increasing flake content. The scatter of the test data is presumably due to varying void content, small irregularities in orientation and distribution of flakes in the composite, excessive stress concentration around sharp edges of thick flakes and particulate contaminants.

c) Effect of Matrix

To date, two epoxy resins, ERLA 4617 and ERL-2256 (Union Carbide Cor.) were evaluated using flakes of the group 12 (Table 1). The test data are summarized in Table 4, adjusted for a flake loading of 50 v/o. Although the ERLA 4617 composites possessed a much higher void content, their flexural strength and moduli were still greater by 34% and 26%, respectively. Presumably, the higher modulus of the ERLA 4617 produces a more favorable ^{E}Flake/^{G}Matrix ratio thus sharply improving the load transfer from flake to matrix.

d) Micro Structure

Photomicrographs of polished composite cross-sections reveal interesting details in the microstructure of flake composites. Figure 7 depicts the structure of a near perfect composite specimen that failed at a flexural stress of 109,000 psi. Perhaps one of the most serious factors responsible for the scatter in composite strength, particularly the comparatively large difference between flexural and tensile strengths, may be found in particulate inclusions. Typical inclusions, as shown in Figure 8, present sites of high stress concentration, extreme localized pressure during molding which may cause fracture of adjacent flakes. resulting in loss of composite strength.

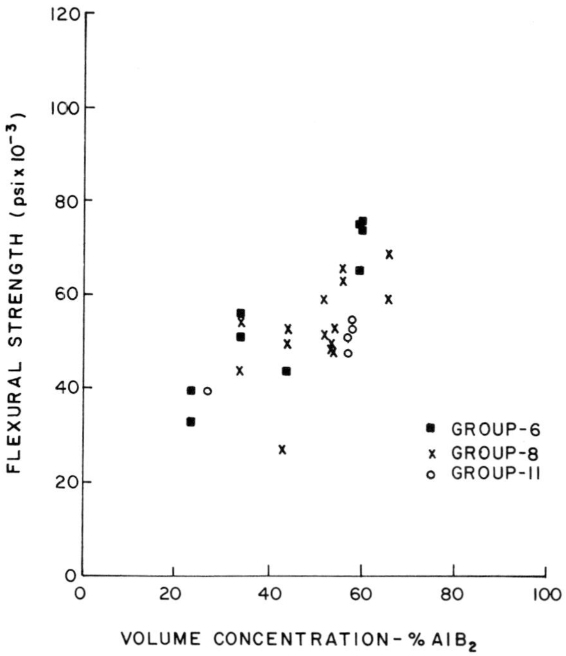

Fig. 5. Flexural strength (psi x 10^{-3})

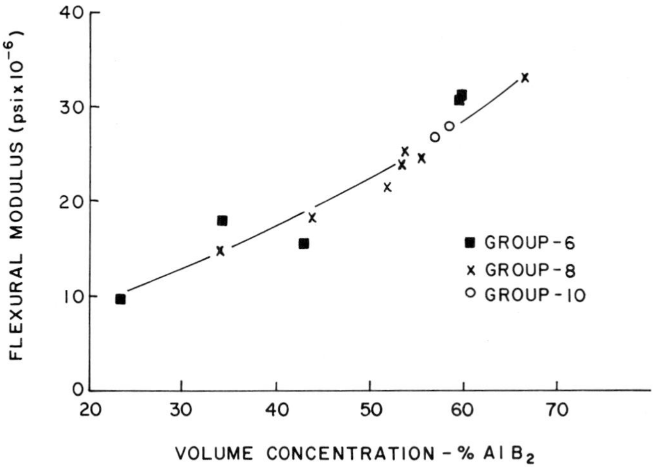

Fig. 6. Flexural modulus vs. flakes concentration

In comparison, the response of flexural strength to variations in void content was surprisingly flat as can be seen in Figure 9.

Microscopic examination of polished cross-sections in the length and transverse directions of the molded specimen showed no differences. Consequently, mechanical composite properties were assumed to be isotropic in the plane of the flakes. This assumption was verified by cutting and testing specimens from panels at angles of 0°, 45° and 90°. Statistical treatment of the test data thus obtained indicated no differences in tensile, flexural strengths and moduli due to orientation.

Table 4. Comparison of ERL 2256 and ERLA 4617 composites - AlB_2 adjusted to 50 v/o group 12 flakes

Specimen no.	ERL 2256 Voids (%)	ERL 2256 Flexural strength (psi X 10^{-3})	Flexural modulus (psi X 10^{-6})	ERLA 4617 Voids (%)	ERLA 4617 Flexural strength (psi X 10^{-3})	Flexural modulus (psi X 10^{-6})
1	1.1	37.9	14.7	6.9	55.0	18.4
2	1.1	41.5	16.2	6.9	58.0	17.5
3	3.1	43.4	20.0	4.8	45.9	19.8
4	3.1	39.9	14.1	2.4	49.1	19.7
5	1.5	40.6	13.7	0.7	49.6	18.1
6	1.5	30.2	13.4	1.6	51.3	18.4
7	1.1	17.9	13.4	1.3	45.0	17.4
8	1.1	42.5	12.5	6.5	41.3	16.7
9	3.6	46.0	14.4	7.2	46.5	19.3
10	3.6	27.2	13.6	6.1	36.9	16.2
11	0.7	21.6	12.2	4.9	47.5	19.3
12	0.7	33.4	12.4	4.5	44.3	16.5
13	4.4	33.8	15.4	5.3	42.8	16.4
Averages	2.0	35.1	14.3	4.6	47.2	18.0
Property of resin matrix		16.0	0.551	0	19.2	0.815

Examintaion of fracture surfaces produced evidence of excellent bonding between flake and matrix. As shown in Figure 10, failure occurred due to shear fracture in the matrix leaving the flakes covered with a thin resin layer.

e) Effect of Machining and Holes on Composite Strength

Careful machining of flake composites, i.e. cutting specimens under moderate pressure with a water-cooled diamond blade, did not affect composite properties adversely. This became apparent by comparing properties of specimens molded to size with those cut from a panel. In fact, the cut specimens displayed about 20 higher flexural strengths. The lower strength of the molded specimens is caused by disorientation of flakes at the edges. The difference diminishes with increasing size of the test specimen.

Small losses in composite tensile and flexural strength were encountered when holes were drilled through the specimens using a conventional high speed drill of 1/8th in diameter so that the cross sectional area was reduced by 1/4th. The strength losses ranged from 7 to 21% but were mostly around 10%. The tensile, flexural and sonic moduli, however, were not affected adversely.

f) Ternary Composites

The term "Ternary Composites" refers to composites containing two types of reinforcements in a common matrix such as AlB_2 flakes and

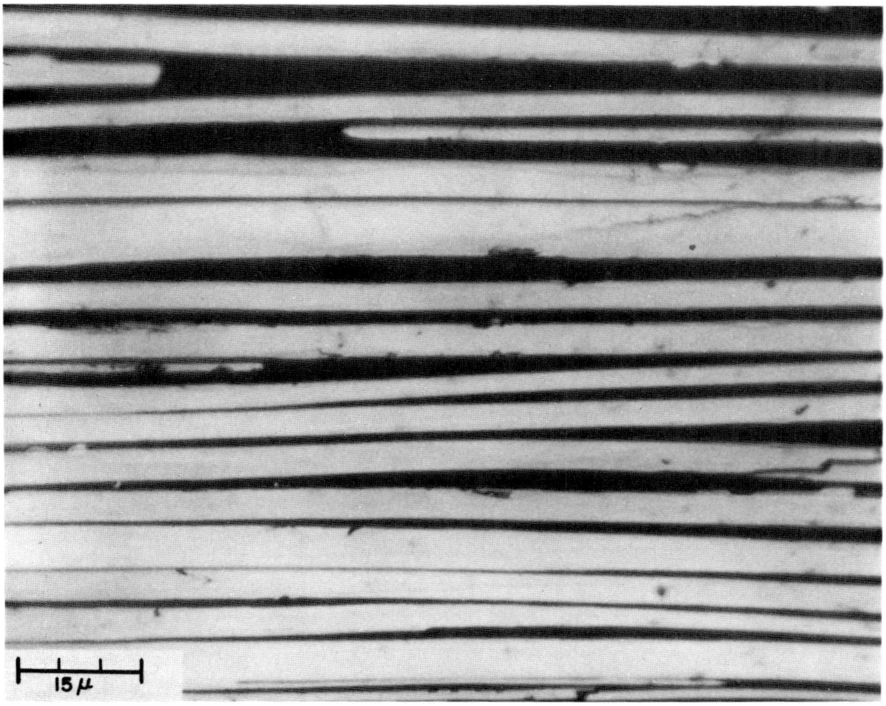

Fig. 7. Polished cross-section of composite with 11-5 flakes

graphite fibers, while the term "Binary Composites" refers to those containing only one type of reinforcement in a matrix such as flakes or fibers. The addition of fibers in the composite increases the possible design variation substantially. Fibers between flake plies can be used in one direction. They can also be applied in multiple directions to approach orthotropic behaviour. In addition, composites can be layed up with plies of flakes or of fibers on both surfaces, greatly altering properties such as flexural strength and modulus, particularly if the number of plies in the composite is small. Another factor influencing composite performance is the difference in modulus between flakes and fiber. For optimum reinforcing power the moduli of both reinforcing components should be as closely matched as possible so that the load can be distributed uniformly.

A series of ternary composites using Hercules Type A-S and HMS graphite fibers were prepared and evaluated. The composites consisted of three plies unidirectional graphite prepreg and two plies of flake prepregs interspaced. Figure 11 denotes a polished cross-section of these composites.

The mechanical properties of those composites are listed in Table 5, in comparison with binary AlB_2 flake composites and binary orthotropic graphite fiber composites. The data are average values of at least two tests. They indicate that all ternary composites combine high strength and stiffness in the fiber direction when compared to the binary flake and the orthotropic fiber composites. The transverse strength, of course, is lower but the transverse moduli compare, again favorably, to those of the orthotropic fiber composites. The values

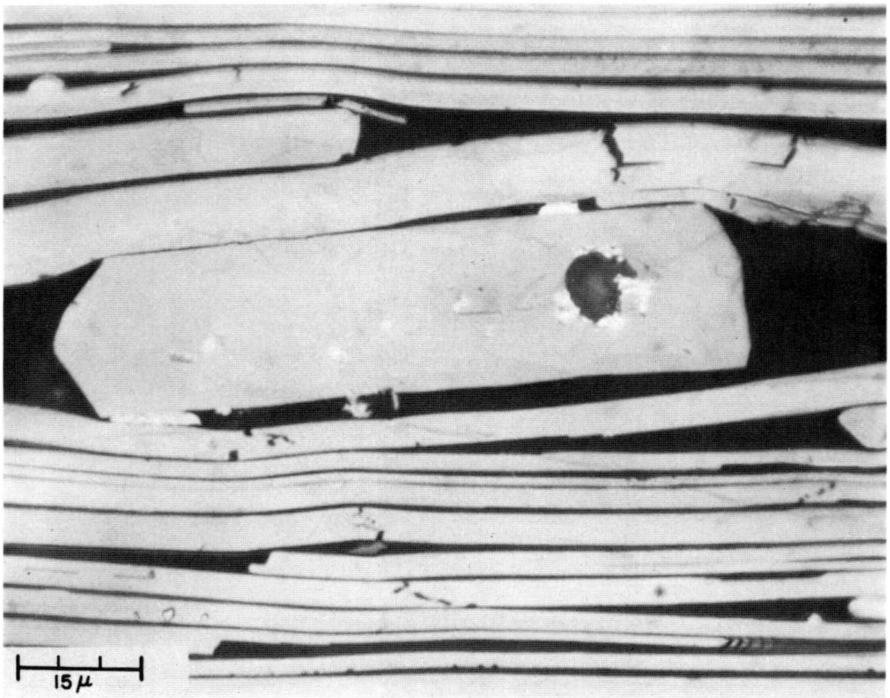

Fig. 8. Polished cross-section of composite with 11-5 flakes

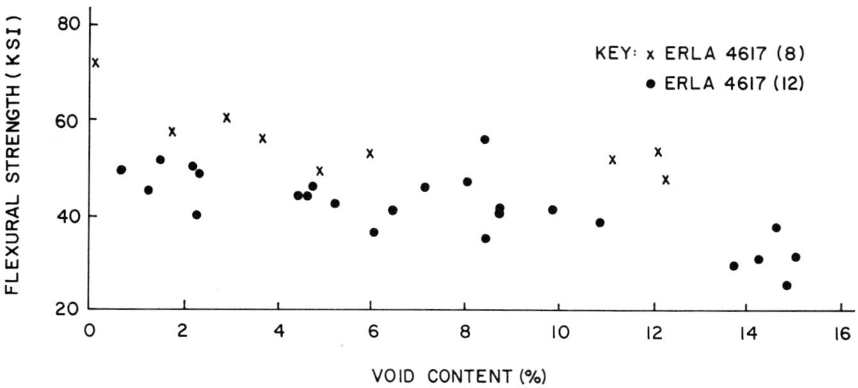

Fig. 9. Influence of voids on flexural strength of composites adjusted for 50 v/o AlB_2

Fig. 10. Fracture surface showing resin cover on flakes

for interlaminar shear strength indicate the minimum expectation since all specimens displayed a tensile failure mode. The stress-strain curves of all ternary composites, particularly when tested under tension, are characterized by the presence of a knee occuring at about 50 to 60 percent of the failure stress as can be seen in Figure 12 which is caused by delamination between the stiffer flake and the fiber layers. Of particular interest is the unexpectedly high impact strength exceeding those of binary flake and fiber composites considerably. Thus, the composites are capable of absorbing a large amount of energy resulting in a frequently observed complete delamination of the structure upon impact.

4. Summary

An effective technique of repeated wet screening and elutriation was developed to purify and classify flakes according to aspect ratio, i.e. diameter/thickness, and minimize the presence of particulate contaminants. The distribution of diameters and thicknesses depended considerably on the flake perfection. Twinning and large surface steps broadened the distribution.

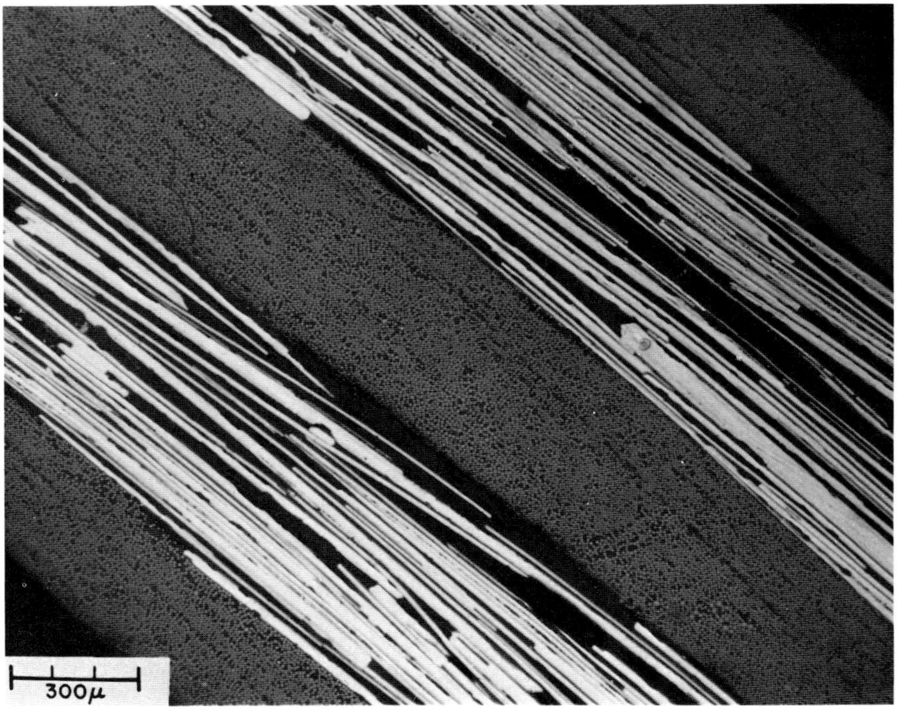

Fig. 11. Ternary composite

Flakes could be readily aligned and compacted into the organic matrix after optimizing prepreg and compression molding procedures. The aspect ratio of flakes was found to be the most important parameter affecting the composite strength. The modulus was also strongly dependent on aspect ratio. Flexural strength and moduli of 109 x 10^3 and 37 x 10^6 psi, respectively, were achieved by a composite containing 60 v/o flakes with an aspect ratio of about 250. Composite strengths and moduli increased rapidly with increasing volume loading of flakes. Composites containing up to 67 v/o in an epoxy matrix were prepared, although the optimum loading (at least for larger flakes) appeared to be 55 to 60 v/o. The tensile strength and modulus approached 70 x 10^3 psi and 45 to 10^6 psi, respectively. Composite properties were also affected by the modulus of the matrix. Voids in composites caused only a modest loss of strength and stiffness but particulate includions appeared to affect composite properties more adversely. Isotropic composite behavior in the plane of the flakes was proven by testing specimens in directions of 0-45-90°. Isotropy was observed regardless of flake dimensions. Both binary and ternary flake composites displayed excellent machining characteristics, i.e. an almost 100% retention of stiffness and strength after machining.

Table 5. Summary of properties of binary AlB and ternary AlB graphite composites matrix: ERLA 4617

Flake volume	60	17	---	33	---	29	25	0		
Flake aspect ratic	239	239	---	239	---	239	239	-		
Fiber volume	0	25-Type HMS	---	25-Type HMS	---	29-Type A-S	33-Type A-S	57-Type HMS 0 - 45 - 90°		
Direction	0-90°	0°	90°	0°	90°	0°	90°	0°	45°	90°
Tensile strength (10^3 psi)	67.2	80	30	90	19	123	23	44	--	23
Tensile modulus (10^6 psi)	44.3	20	11	33	15	21	15	10.3	--	12.9
Flexural strength (10^3 psi)	109	--	--	--	--	--	--	---	--	----
Flexural modulus (10^6 psi)	35.8	--	--	--	--	--	--	---	--	----
Flexural strength (10^3 psi)	81.9	85	25	95	25	115	28	111	51	50
Flexural modulus (10^6 psi)	40.7	20	15	27	10	22	10	20	7	5
Compressive strength (10^3 psi)	78	--	--	45	22	45	13	---	--	----
Compressive modulus (10^6 psi)	40	--	--	--	--	--	--	---	--	----
Interl. shear strength[a] (10^3 psi))	8.0	7.5	6.0	8.0	7.1	8.1	7.3	7.0	6.0	6.0
Shear modulus (10^6 psi)	13.6	---	---	---	---	---	---	---	---	---
Impact strength (ft lb/in)	29	---	---	32	---	44	---	14	---	---

[a]Tensile failure mode

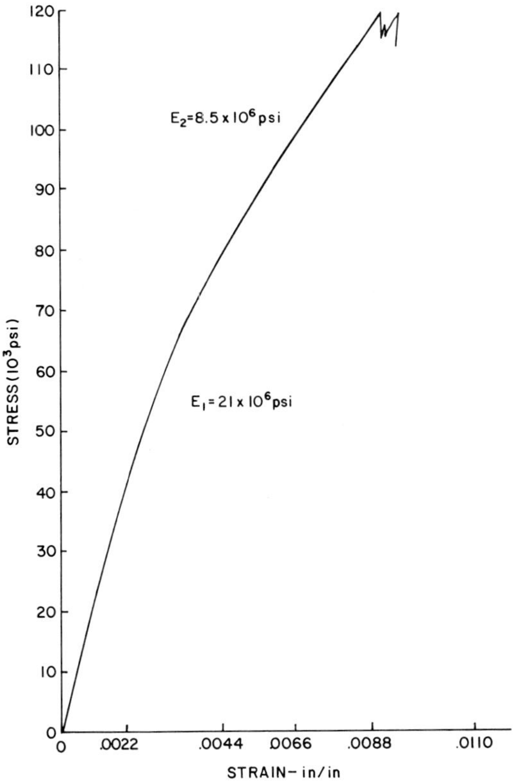

Fig. 12. Stress-strain curve of ternary composites

References

ECONOMY, J., MATKOVICH, V.I., WOHRER, L.: Single Crystal Flake Reinforced Composites, Sampe J. Jan. (1969)

ELLIOTT, R.P.: Constitution of Binary Alloys, 1st Supplement, p. 26 New York: McGraw Hill (1965)

HERTZBERG, R.W.: Potential of Unidirectionally Solidified Eutectic Alloys as Reinforced Composites, Fiber Composite Materials, ASM, p. 77-87 (1965)

MATKOVICH, V.I., ECONOMY, J., GIESE, R.F.: Presence of Carbon in Aluminum Borides, J. Amer. Chem. Soc. 86, 2337 (1964)

OLEESKY, S., MOHR, G.: Handbook of Reinforced Plastics, Reinhold Publishing Corp. p. 148-149 (1964)

PADAWER, G.E., BEECHER, N.: Strength and Sitffness of Planar Reinforced Plastic Resins, Polymer Eng. Sci. 10, 185, (1970)

PIGGOTT, M.R.: Theory of Fiber Strengthening, Acta Met. 14, 1429 (1966)

PIGGOTT, M.R.: Steel Disc Reinforced Polycarbonate, J. Mat. Sci. 8, 1373-1382 (1973)

PIGGOTT, M.R.: The Effect of Aspect Ratio on Toughness in Composites, J. Mat. Sci. 9, 494-502 (1974)

RILEY, V.R.: Fibre/Fibre Interaction, J. Comp. Materials 2, 436 (1968)

WOHRER, L., FRECHETTE, F., ECONOMY, J.: Advances in Reinforced Plastics, Machine Design, 138-140, Dec. 22 (1966)

WRIGHT, E.H., WILLEY, L.A.: Aluminum Binary Equilibrium Diagrams, Alcoa Res. Lab. Tech. Paper Nr. 15 (1960)

V. Amorphous Boron Films

K. MOORJANI and C. FELDMAN

1. Introduction

Amorphous solids have received considerable attention in the past few years. Their novel features contain much exciting physics covering the entire range from the most fundamental studies to device characteristics. An extended study on amorphous boron films has been in progress at the Applied Physics Laboratory and the results, along with those of other researchers, are presented and discussed in this paper. The contents of the article are restricted to amorphous films as opposed to the bulk form of amorphous boron. However, a comparison between the two forms is carried out wherever it is necessary and possible.

Due to the complex nature of the properties of amorphous semiconductors, much insight can be gained by comparing results on the amorphous form with those on the corresponding crystalline phase. This has certainly proved valuable for silicon and germanium, and an attempt in this direction is described here for boron. The problem, however, is more complicated for boron; the properties of crystalline boron are not as well established as those of silicon and germanium. Producing relatively defect-free crystals of boron is difficult. Moreover, the attempts to produce relatively pure samples of crystalline boron have not been entirely successful. Most of the controversy in the existing literature can perhaps be ascribed to these factors and to the complex structure of crystalline boron which has eluded meaningful theoretical investigation.

It is often remarked that the transport properties of crystalline boron are similar to those of amorphous semiconductors. Whether this behavior is an inherent property of crystalline boron, or whether the similarity arises due to the presence of a large number of defects and impurities in crystalline boron, remains to be resolved.

It should be pointed out that no precise definition of "amorphous" phase exists, as yet. From studies on amorphous forms of silicon and germanium, it is known that substrate temperature, deposition rate, deposition time, vacuum level, etc., are crucial parameters in determining the degree of disorder present in the samples; but such a detailed study of amorphous boron has not yet been carried out. It should therefore be emphasized that, though samples described here are amorphous, their degree of non-crystallinity or "amorphicity" is not well established.

2. Preparation Techniques

Boron thin films have been prepared by vacuum deposition, sputtering, rolling liquid and pyrolytic decomposition. Of the four, vacuum deposition has proven to be the most satisfactory technique for producing amorphous films. Pyrolytic chemical deposition techniques (ARMINGTON et al., 1964) require hot substrates ($> 1000^{\circ}C$) and result in single crystal or polycrystalline rather than amorphous layers. Liquid rolled films (TAVADZE et al., 1972) also exhibit polycrystallinity with appreciable distortion of the crystal lattice. The sputtering technique (ROSE, 1967), used in order to avoid the high source temperatures required during vacuum deposition, suffers from a lack of efficient cathode configurations and slow deposition rates. In the vacuum deposition process, specialized joule heating of boron filaments has been used (McELLIGOTT and ROBERTS, 1966), however the most satisfactory method is direct electron bombardment of boron. Two basic types of electron bombardment techniques have been used: the O'BRYAN (1934) arrangement using electron-emitting filaments coaxial with a boron rod or crucible (HILL, 1956); MUGGLETON and HOWE, 1961; VAN AUDENHOVE et al., 1963; ARNISON and GILMOUV, 1966) and the focused electron beam usually located at some distance from the boron (FELDMAN et al., 1972); PALETNIK et al., 1972).

Because of the high melting point of boron, the design of the source and its crucible is very important. In crucible design, surface area of the melt as well as temperature determines the deposition rate. In order to avoid contamination from crucible material, techniques consist of allowing the outside walls of the boron to form its own crucible (VAN AUDENHOVE et al., 1963) or to melt the boron in tantalum or molybdenum crucibles, maintaining the interface between the crucible and the melt at low temperatures (MUGGLETON and HOWE, 1961; FELDMAN et al., 1972). Tungsten crucibles react with boron at high temperatures and are therefore unsuitable (TUCKER, 1966). Uncooled crucibles of carbon have been shown to lead to carbon contamination of the boron deposit (FELDMAN et al., 1972). The water-cooled molybdenum crucible and the electron gun used by the authors are illustrated in Figure 1.

There is no agreement in the literature on the values of the melting point (NEWKIRK, 1964) or of the vapor pressure at the melting point of boron. Fortunately, the vapor pressure at the melting point is sufficiently large to assure a high rate of deposition. BURNS et al. (1967) measure a vapor pressure of about 2×10^{-2} Torr at a melting point of $2130 \pm 40^{\circ}C$. At this vapor pressure, the rate of mass lost from the source is 78.3 $\mu g/cm^2$-sec. SEARCY and MYERS (1957) at about the same temperature ($2132^{\circ}C$) obtained a vapor pressure of 1.2×10^{-2} Torr, but did not observe melting. BURNS et al. (1967) have shown that the evaporation coefficient of boron is approximately unity and that most of the material leaving the melt consists of single atomic B (or B^+) rather than B_2 or B_3 (TUCKER, 1966). Ion beam sputtering used for analyses described below indicates, however, that in sputtering some clusters of B_2, $B_3 \ldots B_8$ are ejected from the solid (FELDMAN et al., 1972).

Both the boron vapor and the radiant energy from the hot source strike the substrate at a rate inversely proportional to the square of the source to substrate distance. A compromise is usually maintained among the deposition rate (which should be fast to maintain purity) the requirement for uniformity of deposit, and the temperature rise of the substrates. Substrate temperatures between $100^{\circ}C$ and $300^{\circ}C$ (ARNISON and GILMOUV, 1966); FELDMAN et al., (1972) are usually maintained at distances from the source of approximately 13 cm (ARNISON and GILMOUV, 1966) to 35 cm (VAN AUDENHOVE et al., 1963) with deposition rates of 430 Å/sec to 3.6 Å/sec, (note 1000 Å = 23 $\mu g/cm^2$).

Substrates usually consist of borosilicate glass or fused silica; sometimes other materials are used for special purposes. The substrates must be cleaned by the best available techniques and outgassed carefully. In order to obtain pure samples, care must be exer-

Fig. 1. Bell jar configurations showing a molybdenum crucible and electron gun

cized in using only low vapor pressure materials in the vacuum chamber and in obtaining the lowest vacuum pressure possible (FELDMAN et al., 1972). For the films deposited in this laboratory, deposition rates of about 20 Å/sec are obtained at a distance of 20 cm from a molybdenum crucible about 3.5 cm in diameter. This deposition rate indicates a source mass loss of about half that quoted above and is more in agreement with the lower vapor pressure of SEARCY and MEYERS (1957). However there is uncertainty in the actual melt diameter since there is a temperature gradient near the cool crucible walls.

In some of the samples discussed here, formed at the Applied Physics Laboratory, carbon was introduced into the boron film; this was carried out by admitting methane or acetylene into the vacuum chamber through a calibrated leak valve. An approximate correlation could then be made between the partial pressure of the gas introduced and the amount of carbon measured in the film by sputter-ion source mass spectrometry. Typical conditions used in depositing the films studied here are listed in Table 1.

3. Structure and Impurity Analysis

The amorphous nature of vacuum deposited boron samples has been established by electron microscopy (FELDMAN et al., 1965). A recent

Table 1. Typical deposition conditions

	Sample B93	Sample B98
Crucible	Molybdenum liner	Molybdenum liner
Bell jar	Stainless steel	Stainless steel
Substrate	Fused silica	Fused silica
Source-substrate distance	19 cm	19 cm
Substrate temperature	$300^{\circ}C$	$300^{\circ}C$
Electron beam power	1.2 kW	1.1 kW
Deposition time	7.5 min	5 min
Thickness	9400 Å	7100 Å
Deposition rate	1250 Å/min	1420 Å/min
Vacuum before	6×15^{-9} Torr	4×10^{-9} Torr
Vacuum during deposition	4×10^{-5} Torr pressure of C_2H_2	5×10^{-7} Torr

confirmation of the amorphous structure has been provided by PALETNIK et al. (1972). Their work indicates that the films obtained at substrate temperatures T_s lower than $550^{\circ}C$ are amorphous, but at higher values of T_s a heterogeneous mixture of amorphous and tetragonal structure is observed. The concentration of tetragonal boron increases with increasing T_s and reaches 100% at $T_s > 700^{\circ}C$. However, the annealing of amorphous films at elevated temperatures leads to crystallization in the β modification.

KATADA (1966) has discussed the radial distribution function obtained from an electron diffraction study of evaporated amorphous boron films. He concludes that the basic 12 atom icosahedral unit common to all modifications of crystalline boron exists also in amorphous boron films and that the boron-boron bond lengths have approximately the same values in the two phases. However, the arrangement of icosahedra is found to be random in the amorphous phase. Similar studies on bulk amorphous boron, conducted by BADZIAN (1970), corroborate the above conclusions and indicate that amorphous films of boron and bulk amorphous boron possess the same structure.

When conducting research on materials, it is imperative that an impurity analysis be carried out. The lack of analysis in the past has made progress in understanding thin film phenomena difficult. Fortunately, a relatively new instrument, the sputter-ion source mass spectrometer, sometimes termed "secondary ion mass spectrometer" (SIMS), is sufficiently developed and available so that impurities in films, including gaseous species, may be adequately determined (EVANS, 1972). The particular spectrometer at APL uses a high energy noble gas ion beam (usually 10 keV argon) which bombards the sample surface, ejecting from the film: boron ions, boron ion clusters (B_2^+, $B_3^+ \ldots B_8^+$), ion complexec BC^+), and impurity ions. The ejected ions are then analyzed by a mass spectrometer. A typical spectrum of a sample containing carbon is illustrated in Figure 2. The instrument has a provision for examining a narrow energy window (50 eV) of the ejected particles at energies between 0 eV to 500 eV. This provision frequently allows one

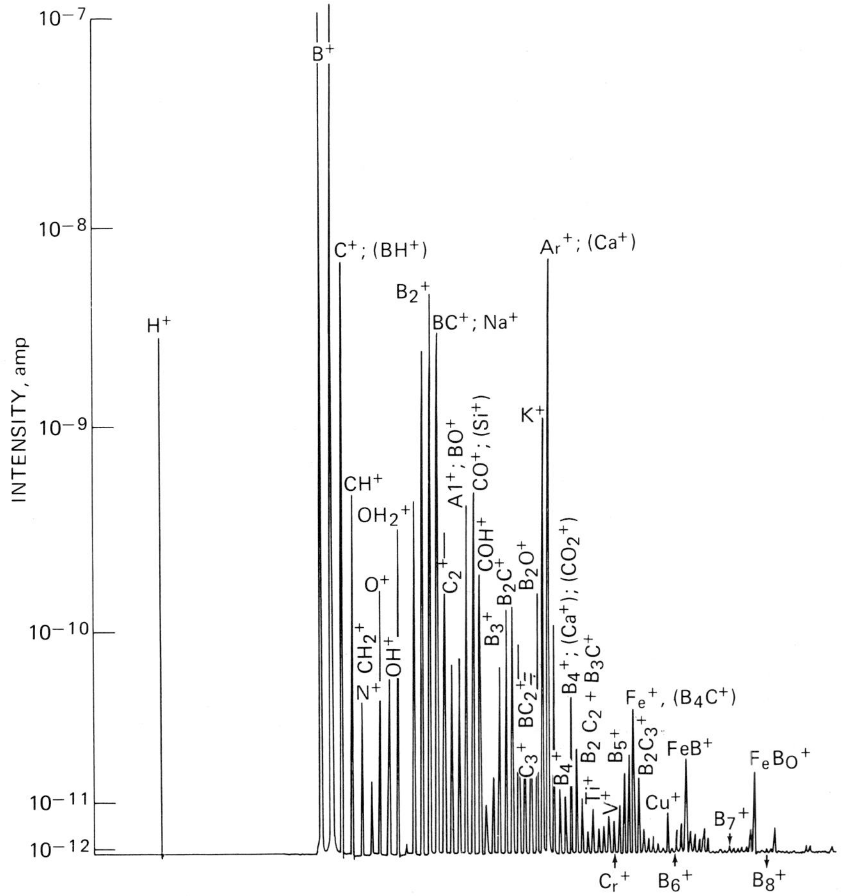

Fig. 2. Polyatomic mass spectrum of a boron

to separate complexes from single atoms since energies of the complex species fall off more rapidly than single atomic species. In separating carbon from BH, for example, an energy window at 200 eV, where the carbon species dominate over the BH species, was used. Complexes such as BH, CH, also aid in determining impurity content. A listing of the impurities observed in typical films as well as in bulk materials is given in Table 2.

Ion cluster distributions were analyzed in the belief that they reflect or represent short-range order or nearest-neighbor interactions (FELDMAN and SATKIEWICZ, 1973). These ion clusters are groups of atoms ejected from the surface in one piece, and a study of their distribution shows sensitivity to impurity content and state of the film. Although the use of cluster spectra in analyzing short-range structure has not been demonstrated unambiguously, comparison of graphite and diamond polyatomic spectra does show persistent differences in cluster ion population (HERZOG et al., 1973). Details of the cluster spectra as well as impurity analysis have been described by FELDMAN et al. (1972).

Table 2. Impurities in boron samples (ppm-atomic)

Species	Bulk boron	B93B(+C)	B98B
H	20	≤ 20	≤ 20
Li	ND	14	16
C	140	(9.6%)	1000
N	300	7	150
O	430	370	1300
Na	ND	12	6
Mg	ND	31	~ 2
Al	28 (BO^+)	15	60
Si	170 (CO^+)	120	120
Cl	ND	ND	~ 2
K	0.4	8	~ 0.4
Cu	7 (+Ar^+)	20	20
Ti	ND	~ 0.4	~ 0.7
V	ND	~ 4	~ 14
Cr	ND	9	10
Fe	ND	61	210
Cu	ND	~ 66	ND
Ba	0.4	ND	ND

4. Optical Properties

The analysis of the optical constants of crystalline solids has proved to be a worthwhile tool in the study of electronic band structure. It is therefore hoped that disorder-induced changes in the band structure can be usefully investigated by this technique. However, as recently discussed for the case of amorphous silicon films (MOORJANI et al., 1974), some intrinsic difficulties exist in obtaining the optical constants of films from transmission and refelction measurements. These difficulties are briefly summarized in this Section, and within these limitations, new data on amorphous boron are presented, analyzed, and compared with earlier work in the literature.

The optical transmission spectra of amorphous boron films were obtained with a double-beam recording spectrophotometer using a SPEX Model 1400 dual.grating monochromator. Tests on transmission standards from the National Bureau of Standards indicate that the absolute accuracy of the measured transmission coefficient is approximately ±0.005, except for regions of very low transmission where it is better than ±10%.

The real and imaginary parts of the refractive index are obtained by measuring the transmission coefficient T on two simultaneously deposited films which are essentially identical in all physical respects except for thickness. The method for the computer analysis of the data, taking into account the interference effects in the film and multiple reflections in the substrate, has been described elsewhere (MOORJANI and FELDMAN, 1970a). The measured values of T are varied by the experimental error $\Delta T \simeq \pm 0.005$ to obtain the corresponding errors in the optical constants. Typical results of such an analysis on a set of films are shown in Figure 3. From the equation relating T to various parameters, it is easy to show that the error $\Delta\alpha$ in the absorption coefficient α is given by

$$\Delta\alpha \approx -\frac{1}{2d}\frac{\Delta T}{T^2}e^{-\alpha d}, \qquad (1)$$

where d is the thickness of the film. The errors shown in Figure 3 are compatible with those computed from Eq. (1).

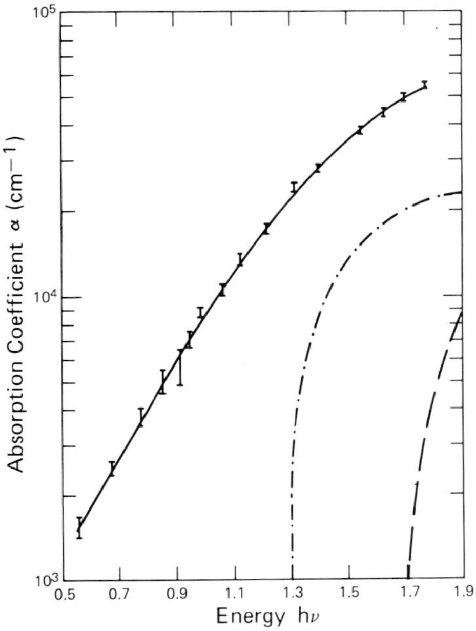

Fig. 3. Spectral dependence of absorption coefficient ——— present data on amorphous boron. -•-•- data on crystalline boron (KIERZEK-PERCOLD et al. 1970). ----- data on crystalline boron (WERHEIT 1970)

Due to the thicknesses used in the present analysis (0.33μm-0.55μm), according to Eq. (1) large errors exist for values of $\alpha \lesssim 10^3$/cm. Much thicker films are needed to obtain reasonably accurate values of α in this region. However, for values of $\alpha \gtrsim 10^5$/cm (in the energy range 2.0 eV-6.0 eV) the error reduces rapidly, and earlier results (MOORJANI and FELDMAN, 1968, 1970a) should still be valid. In the above energy range KIERZEK-PERCOLD et al. (1967, 1970) have reported structure in the absorption coefficient α for β-rhombohedral boron. This structure arises from the optical transitions at critical points in the Brillouin zone and disappears in amorphous boron due to the lack of long-range order (MOORJANI and FELDMAN, 1968, 1970a).

From Figure 3 it is observed that, in the energy range shown, the absorption coefficient for amorphous boron lies much higher than in crystalline boron and that the absorption edge is considerably broadened and shifted toward lower energies. The value of the refractive index at 2μm is found to be in the range 3.25-3.35, which is somewhat higher than the reported value of 3.1 (WERHEIT, 1970) for crystalline boron. It should be noted that the optical features observed for amorphous boron are similar to those observed for amorphous silicon and germanium. Based on studies of silicon (MOORJANI et al., 1974), it is expected that as the degree of disorder in amorphous films is decreased by increasing the deposition temperature, for example, the absorption edge and the refractive index will approach those of the crystalline phase.

It is customary to analyze the spectral dependence of α in order to arrive at the value of the band-gap parameter. However such a procedure is not strictly valid since it neglects the spectral dependence of the refractive index. Therefore we have analyzed the spectral dependence of ε_2, the imaginary part of the dielectric constant. As shown in Figure 4, in the energy range 0.9 eV-1.7 eV, ε_2 can be well fitted to the expression for non-direct transitions,

$$(h\nu)^2 \varepsilon_2 \propto (h\nu - E_g)^2, \quad (2)$$

with the "band-gap" parameter E_g = 0.71 eV. This value differs slightly from the ones reported previously: 0.56 by WERHEIT and BINNENBRUCK (1972), and 0.62 by FELDMAN et al. (1972). The present value (0.71 eV) should be more reliable, since the spectral dependence of the refractive index has been included in the analysis.

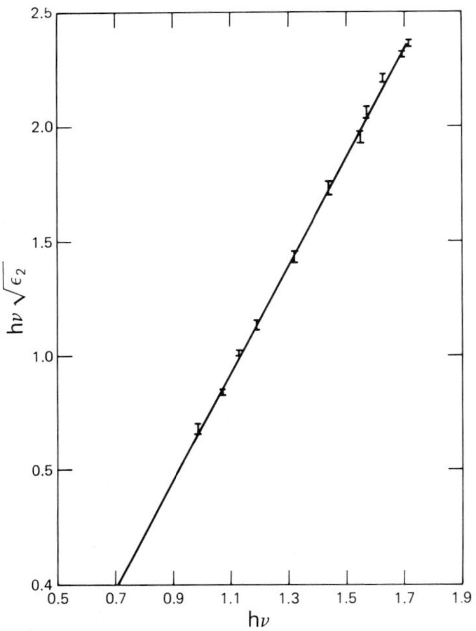

Fig. 4. $h\nu\sqrt{\varepsilon_2}$ versus energy $h\nu$ for amorphous boron

For the lower values of α, in the limited energy range 0.55 eV-0.9 eV, the data is more uncertain but can be fitted to the exponential $e^{h\nu/E_a}$ with $E_a \cong 0.21$ eV, which is in good agreement with the value of 0.19 eV reported by BEREZIN et al. (1972).

In terms of the recently proposed models of the density of states in disordered semiconductors (MOTT and DAVIS, 1971), E_g corresponds to the "gap" between band-like states for non-direct optical transitions, while E_a is an activation energy for transitions between states in the exponentially decaying tails to the density of states which are induced by disorder. Such models have been quite successfully applied to amorphous semiconducting alloys but their validity for monoatomic amorphous semiconductors is not entirely clear.

5. Electrical Properties

The electrical properties of amorphous films on substrates may be measured in planar or sandwich type configurations. In order to avoid contamination of boron, electrodes are frequently deposited in a separate chamber. The required configurations are obtained through the use of etched metal masks. In Figure 5, a four-probe planar configuration and sandwich configuration on fused silica are illustrated. The four-probe configuration serves to eliminate contact errors, while the sandwich configuration indicates whether the resistivity is influenced by the film surface. The films studied in this laboratory exhibited p-type conductivity; no signficant difference was observed between the resistivities measured in the two configurations. Typical resistivity values of films and bulk materials are given in Table 3. The reported resistivity of bulk crystalline material (Table 3; also see Fig. 15 in WERHEIT, 1970) is considerably higher than that of amorphous thin films. The addition of carbon into the films greatly increases their resistivity. This is the opposite to the observations in the crystalline material where the addition of carbon decreases the resistivity (GEIST et al., 1970). Compositions with carbon content almost up to that of boron carbide (B_4C) were examined and the resistivity continued to increase. The resistivity of bulk crystalline boron carbide is considerably less than that of boron. An analysis of the films shows that the increase in resistivity with carbon is due neither to nitrogen, oxygen nor hydrogen in the film. The increase of resistivity in the thin amorphous films cannot be ascribed to the thinness of the films since the mean free paths of the charge carriers in boron (SOSNOWSKI, 1970) are believed to be much smaller than the film thickness studied.

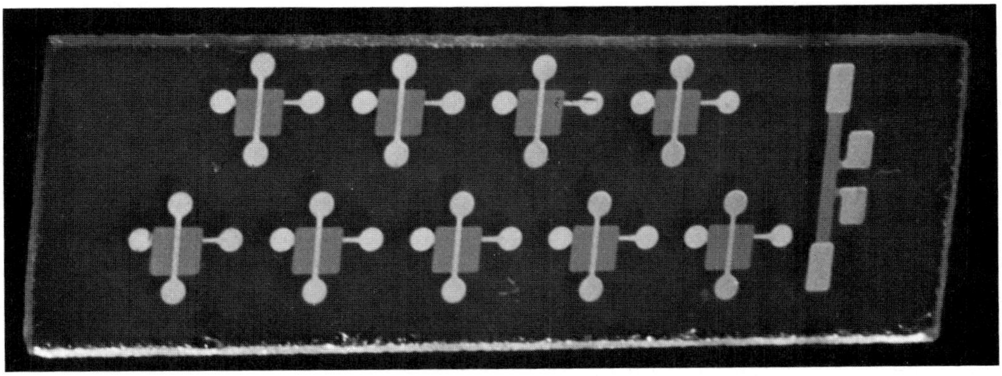

Fig. 5. Sample configuration for electrical measurements

A plot of carbon content versus logarithm of resistivity in the samples, beginning with the purest sample, yields a smoothly increasing curve. This smooth variation of resistivity with carbon content indicates that phase separation or precipitation of carbon to grain boundaries is unlikely. The position of the carbon in the amorphous network may account for the disparity in resistivity between crystalline and non-crystalline species.

Table 3. Boron resistivities

		ρ (Ωcm)
Film	B98 (0.1%C)	7.50 x 10^2
	B91B (7.2%C)	6.1 x 10^3
	B93 (9.6%C)	3.6 x 10^4
Bulk crystal type	Boron (~3%C)	3 x 10^1 [a]
	Boron (99.999%)	3 x 10^5 [a]
	Boron (99.9995%) (~0.02%C) [c]	5 x 10^6 [b]

[a] GEIST et al. (1970)
[b] MAJNI and PRUDENZIATI (1971)
[c] Sputter-ion source measurements on sample kindly sent by M. PRUDENZIATI

To throw further light on the effect of carbon in amorphous boron layers, electron spin resonance (ESR) measurements were carried out on samples doped with various amounts of carbon (FELDMAN et al., 1975). While the carbon concentration was increased by more than an order of magnitude (from 0.4% to 12%), the spin concentration remained relatively constant, indicating no equivalence between density of free carriers and spin concentration. This is in contrast to the results of GEIST et al. (1970) on β-rhombohedral boron, but in agreement with the recent work of GEWINNER et al. (1973). A slight decrease in spin density with hydrogen content was observed in amorphous boron (FELDMAN et al., 1975).

The temperature dependence of resistivity in the range 300°K-1000°K has been reported (FELDMAN et al., 1972). Above 700°K, both doped and undoped films exhibit an activation energy of approximately 0.65 eV. This value, which represents the separation between the band edge and the Fermi level, is in excellent agreement with the value of 0.68 ± 0.03 eV reported for crystalline boron by various authors (see WERHEIT, 1970). At lower temperatures, two activation energies in the ranges 0.28 eV-0.42 eV and 0.14 eV-0.15 eV have been observed in amorphous films. These values are independent of the amount of carbon in the films.The lower value is in fair agreement with the one found by GOLIKOVA et al. (1968) for β-rhombohedral boron. However, they found variation in activation energy with the number of current carriers. The higher activation energy (0.28 eV-0.42 eV) should be compared with the value 0.23 eV-0.36 eV observed by PRUDENZIATI et al. (1972). The reason for this equivalency and the non-variation of activation energy with carbon content is not entirely clear. Even the purest films and bulk material contain carbon as well as other impurities. Determination of levels and temperature behavior have not yet been carried out. The low value of resistivity of purer films compared to the value for crystalline boron and impure films seems to indicate the presence of compensating impurities in impure films and perhaps even in crystalline boron (GOLIKOVA et al., 1970). Carbon in amorphous boron films may, in fact, behave as a compensating impurity (FELDMAN et al., 1976).

6. High Field Effects

In bulk boron deviations from ohmic behavior (WEINTRAUBE, 1909) and breakdown effects (LYLE, 1918) have been known for a long time. Various studies on non-linear effects have been reported in the literature (DIETZ and HELMBERGER, 1965; KLEIN, 1968; PRUDENZIATI et al., 1973). Non-linear effects in thin vacuum deposited amorphous films have been described by FELDMAN (1964) and an abrupt transition from high to low resistivity with negative resistance (switching) was reported by FELDMAN and GUTIERREZ (1968). Thin amorphous films and crystalline as well as amorphous phases of bulk boron exhibit similar high field effects. This is perhaps due to the fact that both the amorphous films and the bulk materials are semiconductors possessing low mobilities. Because of the high temperature required to bring about crystallization in thin films, the electrical properties of polycrystalline films have not been studied.

The typical current-voltage effects observed in amorphous boron films are illustrated in Figure 6. The low field region, A-B, appears ohmic, while in the B-C portion, strong deviations from ohmic behavior are observed. In this non-linear region, the relationship between current and voltage has been reported to be space charge dependent (FELDMAN and GUTIERREZ, 1968),

$$I \propto V^2, \tag{3}$$

or Frenkel-Poole dependent (MOORJANI and FELDMAN, 1970b),

$$I = I_o \exp (\beta V^{\frac{1}{2}}), \tag{4}$$

where $\beta = (e^3/\varepsilon d)^{\frac{1}{2}}/kT$ depends on the dielectric constant ε, temperature T and the thickness of the film d.

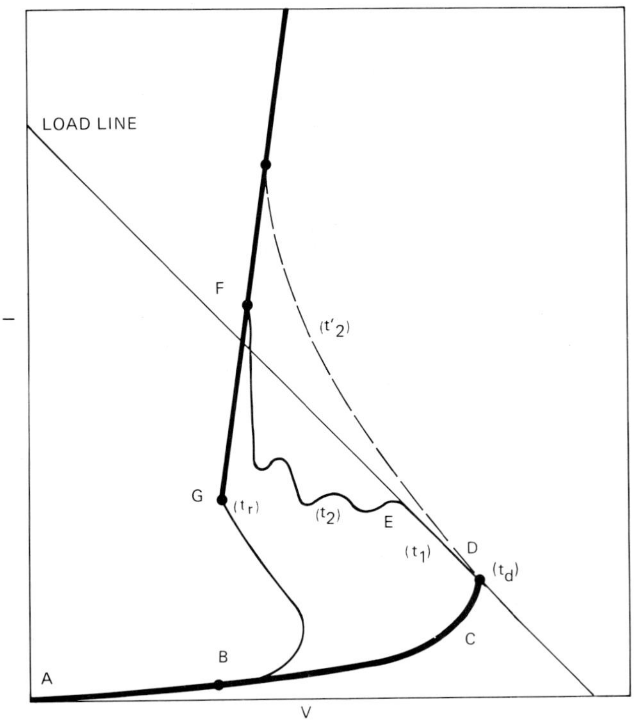

Fig. 6. Current-voltage characteristics defining switching parameters

Some controversy exists over the constant β in Eq. (4), depending on whether the Frenkel-Poole or the Schottky effect dominates. The Frenkel-Poole effect is expected from traps in the material, while the Schottky effect would be due to energy barriers at the electrode. Feldman et al. (1975) pointed out that β in reality is β/γ where γ goes from 1 to 2, depending on the amount of compensation. Thus a knowledge of the Fermi level in the system is required in order to distinguish between the two cases. Eq. (4) is dominant at liquid nitrogen temperatures in pure films; at higher temperatures Eq. (3) seems to hold equally well. In β-rhombohedral boron, space charge limited currents have been observed by PRUDENZIATI et al. (1973), while BEREZIN et al. (1972) have reported the existence of the Frenkel-Poole effect.

The region C-D, (Fig. 6) has a stronger dependence on voltage and follows a relationship (FELDMAN and GUTIERREZ, 1968; MOORJANI and FELDMAN, 1970b),

$$I \propto V^n . \qquad (5)$$

This sudden current increase C-D is observed in films but not in the crystalline material (PRUDENZIATI et al., 1973).

At the point D on the switching curve there is generally a time delay t_d before the sample switches to point F. During continuous cycling, separate delay times have been observed from D to E and from E to F. However in single-pulse switching only one switching time is observed, t_p (CHARLES and FELDMAN, 1975). On decreasing the voltage the current falls to point G where it again irreversibly switches back to the original trace with a characteristic recovery time, t_r. Note the portions F-G as well as A-D in Figure 6 are reversible in that current follows decreasing or increasing voltage changes. Typical times associated with the switching process are shown in Table 4. During the transition D-F a narrow conducting filament has been observed to form (FELDMAN and MOORJANI, 1970) which is always approximately in the center of the electrode area (CHARLES and FELDMAN, 1975).

Various theories have been postulated for the observed switching effect in boron. The early researchers, as well as recent ones, believe the effect may be completely thermal in nature. The negative temperature coefficient of resistance of boron causes the resistance to decrease as the sample warms and, as it warms, more current flows, and so on, causing thermal runaway. Recent measurements by PRUDENZIATI et al. (1973) in crystalline boron have shown that this may be the case at high temperatures, but at low temperature the effect may be due to a double injection phenomenon. Neither the thermal nor the double injection explanation seems tenable in thin films in view of the recent measurements on the time constants (CHARLES and FELDMAN, 1975). An electrothermal mechanism has been postulated (FELDMAN and CHARLES, 1974) in which the sample interior initially heats up by a thermal process but the region next to the electrodes remains relatively cool and non-conducting. At sufficiently high fields electron multiplication or tunneling occurs between the electrodes and the edge of the central hot region. This distance is less than the sample thickness. The high field brings about the rapid current increase characterized by Eq. (3). A highly conducting path is thus created directly across the sample and the capacitor C formed by the sample electrodes may now discharge through the film. This discharge forms the conducting filamentary path. The recovery time and delay time are thus electrothermally controlled, while the switching times represent RC time constants where R represents the resistance of the load and the electrode. PRUDENZIATI (personal communication) has observed radiation during switching at energies

Table 4. Typical switching parameters (Sample B93)

		$300^{\circ}C$	$77^{\circ}K$
Switching threshold conditions	V_{th} (V)	12.0	12.5
	I_{th} (mA)	3.0	0.11
	P_{th} (mW)	36.0	1.38
Preswitching parameters	R (kΩ)	14.0	115.0
	C (nF)	2.8	2.5
Switching delay time	t_d (μs)	33.0	35.5
Switching times	t_1 (μs)	2.5	2.3
	t (μs)	7.3	4.4
Recovery time	t_r (μs)	63.7	50.9
Single-pulse switching time	t_p (μs)	5.95	5.25
Calculated RC time constant	t_p (μs)	6.2	5.4

approximately equal to the band gap in thin film amorphous samples made in this laboratory as well as in crystalline samples. This radiation indicates that a recombination process within the hot thermally generated filament is taking place.

7. Summary

The preparation and properties of amorphous boron films have been reviewed in the present article. Vacuum deposition provides the most satisfactory production technique; and it is shown that, with appropriate care, relatively pure films can be obtained.

The structural studies of amorphous boron films indicate that the basic 12-atom icosahedral unit of crystalline boron is retained, with approximately the same values of bond lengths, in the amorphous phase. The icosahedra, however, have random orientations in the amorphous state. The presence of essentially the same short-range order in the two phases accounts for amorphous boron exhibiting the semiconducting properties of crystalline boron. But the properties are modified due to the absence of long-range order in the amorphous phase.

The spectral dependence of the absorption coefficient of β-rhombohedral boron exhibits structure connected with band to band transitions at the critical points in the Brillouin zone. This structure disappears in amorphous boron due to the lack of long range order. In comparison to the fundamental absorption edge of crystalline boron, the edge in amorphous boron is broadened and shifted towards lower energies. For low values ($\lesssim$ 10 /cm) of absorption coefficient α, inherent difficulties exist in the determination of optical constants from transmission measurements on thin films. For higher values of α, however, the data can be fitted to the expression for non-direct optical transitions with the "band gap" parameter equal to 0.71 eV.

The effects of disorder on conductivity are more difficult to delineate mainly due to the presence of a large number of impurities in both crystalline and amorphous boron. The two phases exhibit essentially the same thermal activation energies as determined from the temperature dependence of resistivity in the range $300^{\circ}K-1000^{\circ}K$. The high temperature activation energy of 0.65 eV for both doped and undoped boron films is in good agreement with the values of 0.68 ± 0.03 reported for β-rhombohedral boron by various researchers.

The reduction of impurities (mainly carbon) leads to an enormous drop in resistivity, indicating the presence of compensating impurities in impure samples. However, the reduced impurity content does not lead to changes in the thermal activation energies. This would indicate that even the most pure films contain the impurities responsible for the levels associated with the activation energies but the densities of these levels may be changed.

In contrast to crystalline boron, doping amorphous films with carbon increases the resistivity, and no connection is observed between the number of free current carriers and the spin concentration as determined by electron spin resonance measurements.

The crystalline as well as amorphous boron exhibits non-linear current-voltage characteristics and the accompanying switching phenomenon. Various switching parameters for amorphous boron films have been determined and a brief explanation of the phenomenon is presented.

In conclusion, it should be noted that the work reviewed here is a mere beginning towards the understanding of amorphous boron films. As stated in the introduction, further work is needed to delineate the effects of various deposition parameters on the structure and properties of boron films. Measurements on purer samples are essential in order to separate the effects of disorder from those of impurities. The relationship between the optical and the thermal "band-gap" parameter for amorphous boron is not entirely clear and indeed the nature of the conduction process in the amorphous phase is not quite understood. Hopefully the present review will provide an impetus to further research.

Acknowledgments. This work was supported in part by Naval Ordnance Systems Command, Contract N00017-72-C-4401, Task A13B. The assistance and contributions of N.A. Blum, H.K. Charles, Jr. and F.G. Satkiewicz to various aspects of the present paper are gratefully acknowledged.

References

ARMINGTON, A.F., POTTER, W.D., TANNER, L.E.: Single crystal boron films on silicon. J. Appl. Phys. 35, 730-731 (1964)

ARNISON, G.T.J., GILMOUV, R.J.: The preparation of thin selfsupporting boron films using a heated substrate. Nucl. Inst. Meth. 45, 178 (1966)

BADZIAN, A.R.: Radial distribution function of non-crystalline boron. In: Boron, Preparation and Properties (T. Niemyski, ed.) Vol. III, pp. 143-146. Warsaw: Pol. Sci. Publ. 1970

BEREZIN, A.A., ZAITSEV, V.K., KAZANIN, M.M., TKALENKO, E.N.: Electrical conductivity of β-rhombohedral boron in a strong electric field. In: Boron, Preparation, Structure and Properties (F.N. Tavadze, ed.) Vol. IV, pp. 91-97, Moscow: Sci. Publ. House 1974

BURNS, R.P., JASON, A.J., INGHRAM, M.G.: Evaporation coefficient of boron. J. Chem. Phys. 46, 394-395 (1967)

CHARLES, H.K., JR., FELDMAN, C.: Switching times in amorphous boron, boron plus carbon, and silicon thin films. J. Appl. Phys. 46, 819-830 (1975)
DIETZ, W., HELMBERGER, H.: Boron semiconductor devices. In: Boron, Preparation, Properties and Applications (G.K. GAULÉ, ed.) Vol. 2, pp. 301-316. New York: Plenum Press 1965
EVANS, C.A.: Secondary ion mass analysis: a technique for three-dimensional characterization. Analytical Chem. 44, 67A-80A (1972)
FELDMAN, C.: Electrical conduction in amorphous boron. Nature 203, 964-965 (1964)
FELDMAN, C., CHARLES, H.K., JR.: Electrothermal model of switching in amorphous boron and silicon thin films. Solid State Commun. 15, 551-554 (1974)
FELDMAN, C., CHARLES, H.K., JR., SATKIEWICZ, F.G., BOHANDY, J.: Electrical properties of carbon-doped amorphous boron films. Proc. 5th Int. Symp. Boron and Borides. J. Less-Common Metals 47, 141-145 (1976)
FELDMAN, C., GUTIERREZ, W.A.: Switching and negative resistance in amorphous boron layers. J. Appl. Phys. 39, 2474-2476 (1968)
FELDMAN, C., MOORJANI, K.: Observation of filament formation in amorphous films during switching. Thin Solid Films 5, R1-R4 (1970)
FELDMAN, C., MOORJANI, K., BLUM, N.A.: Mass spectrometry, optical absorption and electrical properties of amorphous boron films. In: Boron, Preparation, Structure and Properties (F.N. Tavadze, ed.) Vol. IV pp. 130-139, Moscow: Sci. Publ. House 1974
FELDMAN, C., ORDWAY, F., ZIMMERMAN, W., MOORJANI, K.: In: Boron, Preparation, Properties and Applications, (G.K. Gaulé, ed.) Vol. 2, pp. 235-260. New York: Plenum Press 1965
FELDMAN, C., SATKIEWICZ, F.G.: Mass spectra analyses of impurities and ion clusters in amorphous and crystalline silicon films. J. Electrochem. Soc. 120, 1111-1116 (1973)
GEIST, D., MAYER, J., PEUSSNER, H.: Electrical conductivity and movable carrier electron paramagnetic resonance. In: Boron, Preparation and Properties (T. Niemyski, ed.) Vol. 3, pp. 207-214. Warsaw: Pol. Publ. 1970
GEWINNER, G., KUBLER, L., KOULMANN, J.J., JAEGLÉ, A.: EPR investigations of paramagnetic centers in β-rhombohedral boron. Phys. Stat. Sol. (b) 59, 395-402 (1973)
GOLIKOVA, O.A., ZHUBANOV, M.Zh., KLIMASHIN, G.M.: Electrical properties of boron. Soviet Physics Semiconduct. 2, 451-454 (1968)
GOLIKOVA, O.A., ZHUBANOV, M.Zh., KLIMASHIN, G.M.: Electrical properties of boron. In: Boron, Preparation and Properties (T. Niemyski, ed.) Vol. III, pp. 269-274. Warsaw: Pol. Sci. Publ. 1970
HERZOG, R.F.K., POSCHENRIEDER, W.P., SATKIEWICZ, F.G.: Observation of clusters in a sputtering ion source. Radiation Effects 18, 199-205 (1973)
HILL, H.A.: Evaporation of boron. Rev. Sci. Inst. 27, 1086 (1956)
KATADA, K.: An electron diffraction study of evaporated boron films. Japan. J. Appl. Phys. 5, 582-587 (1966)
KIERZEK-PERCOLD, E., KOLODZIEJCZAK, J., MICHALAK, A., PRACKA, I.: Optical constants of β-rhombohedral boron in the region 1 eV - 6 eV. In: Boron, Preparation and Properties (T. Niemyski, ed.) Vol. III, pp. 219-226, Warsaw: Pol. Sci. Publ. 1970
KIERZEK-PERCOLD, E., KOLODZIEJCZAK, J., PRACKA, I.: Optical constants of β-rhombohedral boron in the region 1.2 to 6.2 eV. Phys. Stat. Sol. 22, K147-K150 (1967)
KLEIN, W.: Semiconductor properties of boron in the electrical breakdown range. J. Appl. Phys. 39, 5797-5798 (1968)
LYLE, F.W.: The breakdown effect in boron conductors. Phys. Rev. 11, 253-260 (1918)
MAJNI, G., PRUDENZIATI, M.: Thermoelectric power in crystalline boron. Phys. Stat. Sol. 5, K129-K132 (1971)
McELLIGOTT, P.E., ROBERTS, R.W.: Preparation and chracterization of evaporated boron films. J. Appl. Phys. 37, 1992-1994 (1966)
MOORJANI, K., BLUM, N.A., FELDMAN, C.: Comments on the determination and analysis of optical constans of thin silicon films: In: Amorphous and Liquid Semiconductors (J. Stuke and W. Brenig, eds.) pp. 563-566, London: Taylor and Francis Ltd. 1974
MOORJANI, K., FELDMAN, C.: Optical transitions in amorphous boron films. Solid State Commun. 6, 473-475 (1968)
MOORJANI, K., FELDMAN, C.: Optical constants of amorphous boron. In: Boron, Preparation and Properties (T. Niemyski, ed.) Vol. III, pp. 265-268. Warsaw: Pol. Sci. Publ. 1970a

MOORJANI, K., FELDMAN, C.: Electrical conduction in amorphous boron and silicon.
MOTT, N.F., DAVIS, E.A.: Electronic Processes in Non-Crystalline Materials. Oxford: Clarendon Press 1971
MUGGLETON, A.H.F., HOWE, F.A.: The preparation of thin self supporting boron films. Nucl. Inst. Meth. 13, 211-214 (1961)
NEWKIRK, A.E.: Elemental boron. In: Boron, Metallo-Boron Compounds and Boranes (R.M. Adams, ed.) pp. 233-299. New York London Sidney: Wiley Interscience Pub. 1964
O'BRYAN, H.M.: Evaporation techniques for highly refractory substances. Rev. Sci. Inst. 5, 125-126 (1934)
PALETNIK, L.S., NECHITAILO, A.A., KOZ'MA, A.A., FUKS, M.YA., TSOMAYA, K.P.: The structure of boron films condensed in high vacuum. In: Boron, Preparation, Structure and Properties (F.N. Tavadze, ed.) vol. IV, pp. 139-142, Moscow: Sci. Publ. House 1974
PRUDENZIATI, M., LANZI, A., MAJNI, G.M., MATAVASI, G.: Switching effect in β-rhombohedral boron. Phys. Stat. Sol. (a) 18, 651-659 (1973)
PRUDENZIATI, M., MAJNI, G., ALBERIGI QUARANTA, A.: Determination of trapping centers in Beta-rhombohedral boron. J. Phys. Chem. Solids 33, 245-254 (1972)
ROSE, A.: Sputtered boron films on silicon surface-barrier detectors. Nucl. Inst. Meth. 52, 166-170 (1967)
SEARCY, A.W., MEYERS, C.E.: The heat of sublimation of boron and the gaseous species of the boron-boric oxide system. J. Phys. Chem. 61, 957-960 (1957)
SOSNOWSKI, L.: The present state of research on boron. In: Boron, Preparation and Properties (T. Niemyski, ed.) Vol. III, pp. 15-19. Warsaw: Pol. Sci. Publ. 1970
TAVADZE, G.F., KHANTADZE, D.V., TSAGAREISHVILI, G.V., TAVADZE, F.N., ANTADZE, M.E., NIKAIDZE, Sh.G., AVLOKHASHVILI, DZH. A.: Liquid-rolled boron films. In: Boron, Preparation, Structure and Properties (F.N. Tavadze, ed.) Vol. IV, pp. 145-146, Moscow: Sci. Publ. House 1974
TUCKER, C.W.: Interaction of boron with tungsten single crystal substrates. Surface Science 5, 179-186 (1966)
VAN AUDENHOVE, J., ESCHBACH, H.L., MORET, H.: Deposition by electron bombardment and weighing under vacuum of thin high purity boron layers. Nucl. Inst. Meth. 24, 465-470 (1963)
WEINTRAUBE, E.: Preparation and properties of pure boron. Trans. Amer. Electrochem. Soc., 165-185 (1909)
WERHEIT, H.,: Die Halbleitereigenschaften des Bors. Festkörperprobleme X, 189-225 (1970)
WERHEIT, H., BINNENBRUCK, H.: Comparative investigation of the optical absorption and the photoeffects in β-rhombohedral boron. In: Boron, Preparation, Structure and Properties (F.N. Tavadze, ed.) Vol. IV, pp. 110-115, Moscow: Sci. Publ. House 1974

VI. Boride Coatings

A. P. EPIK

Boride coatings on metallic materials are of great technological importance primarily because of their hardness and wear resistance. At the same time they possess, as a rule, higher oxidation stability in gaseous and liquid reactive media than corresponding pure metals and alloys. Therefore, investigation of the preparation techniques of determination of properties and the exploration of possible areas for application of boride coatings is of practical interest. Some papers have appeared over the years dealing with particular problems of boride coatings, but the number of reviews summarizing available information is rather small.

In the present paper an attempt is made to analyze the present state in development of boride coatings. However, because of space limitations only the most important main principles are emphasized. More detailed descriptions of one or another particular problem have been reported by LYAKHOVICH and VOROSHNIN (1964), MINKEVICH (1965), GLUKHOV (1970), SORKIN (1972), SAMSONOV and EPIK (1973)

1. Methods of Preparation of Boride Coatings

Among several methods of preparing boride coating the method of diffusion saturation is most prominent. This method has been most thoroughly studied, continuosly developed and improved and now finds growing practical applications. Other methods, such as chemical deposition from gaseous phase, spraying, building up, electric-spark alloying etc. have been insufficiently studied and, due to their specific character, have limited practical use. Consequently, additional attention will be given to the method of diffusion saturation in subsequent chapters.

a) Diffusion Saturation

Diffusion saturation of metallic survace with boron can be carried out in a wide temperature range (550-1200°C) from liquid, gaseous and solid media. In this case phase composition of diffusion boride coatings, their structure, thickness, residual stresses and structure and composition of the transition zone (under the boride coating) would be mainly determined by the following factors:

1. composition of the saturating medium
2. saturation temperature and time
3. composition and structure of the saturating material.

As a rule, *Liquid-Phase Boriding* is performed in borax-based melts, either with or without electrolysis by means of reducing agents such as B_4C, SiC, silico-calcium etc. The corresponding processes are usually referred to as electrolysis or liquid boriding. The most detailed study of mechanisms associated with the process of electrolysis and liquid boriding of iron alloys was made by LYAKHOVICH et al. (1974). The ideas available on the mechanism of deposition of active boron atoms or ions on saturated metallic surface were briefly reviewed by SAMSONOV and

EPIK (1973). Therefore, without going deper into specific problems, we shall mention that by varying compositions of boriding liquid media and saturation procedures, one can control the rate of layer growth, phase composition, structure and consequently, properties.

All liquid-phase processes of boriding which are characterized by a high rate of grwoth of the coating, nondeficiency and low cost of starting materials, have essential limitations. Borided parts undergo thermal shock on immersion into and on extraction from the melt. Surface melt residues must be washed out and subsequently purified; parts of complex forms or large dimensions are difficult or impossible to boride, etc. Therefore, the use of these processes for a wide range of articles and on an industrial scale appear difficult.

Gaseous Boriding, particularly by the powder method, does not suffer from the limitations mentioned above and is in our opinion the most promising process for industrial use. The same viewpoint is expressed in a number of the latest reviews by other authors: MATUSCHKA (1973), KONOWROCKA and TACIKOWSKI (1973), KRZYMINSKI and KUNST (1973). "Pure" gaseous boriding, using mixtures of boron halogenides or boron hydride with hydrogen are unlikely to find applications on an industrial scale in the near future since these mixtures are not only toxic and liable to explode but are also expensive and require complex equipment.

Among various boron-containing materials for boriding, the most suitable one is commercial boron carbide (EPIK et al. 1970). It was shown, furthermore, that oxygen (DERKACH et al., 1971) and halogen (SOSNOVSKY et al. 1972) can activate the boriding process. Thus, boron can be transfered in the form of suboxides (B_2O_2, BO) or boron subhalogenides (BF, BF_2 etc.) which subsequently decompose on metallic surface, and form active boron atoms which diffuse into it. The advantage of power boriding is that commercial boron carbide is easy to procure and is comparatively inexpensive. It can also be used repeatedly. It does not require special technical equipment and highly qualified personnel gives reliable reproducible results and can yield complex machine parts. Utilization of haloid activators, particularly those containing fluorine, can produce rates of diffusion saturation approaching those of liquid-phase processes. The temperatures of both gaseous and liquid-phase boriding are between 900 and 1100°C for the majority of metals and alloys. The time of exposure varies from one to six hours. The thickness of the boride coating formed is from 25 to 250 μ (thicker coatings are not practical).

b) Chemical Precipitation from Gaseous Phase

Reduction of halogenides (usually chlorides) of boron on a heated surface is the most commonly used method for chemical precipitation of borides by the gaseous phase. One of the first studies on the process of gaseous-phase precipitation of borides was reported by BECKER (1933). He showed that to produce good quality coatings the optimal temperature of substrate and concentration (partial pressure) of components of the vapor-gaseous reaction mixture for each system must be experimentally selected. POWELL (1970) reviewed the chemical precipitation of refractory compounds, and presented optimal conditions for gaseous-phase deposition of borides (Table 1).

As mentioned in the above work (POWELL, 1970), it is desirable to have in boride precipitation both the metal and the boron compound to contain the same halogen. Thus, for example, VCl_4 and BCl_3 halogenides used for vanadium boride precipitation interact at room temperature and must be mixed immediately in a nearby reaction chamber. When using compounds of the same halogen, however, vapors of both

Table 1. Conditions for a gaseous-phase deposition of borides (with hydrogen as a reducing agent)

Metal boride[a]	Sample temperature °C	Metal compound	Vaporization temperature °C	Boron compound	Vaporization temperature °C
Ti	1100-1300	$TiCl_4$	20	BBr_3	20
Ti	1100-1300	$TiCl_4$	50	BCl_3	-22
Zr	1700-2500	$ZrCl_4$	300-350	BBr_3	20
Hf	1900-2700	$HfCl_4$	300-350	BBr_3	20
V	900-1300	VCl_4	20	BBr_3	20
V	900-1300	VCl_4	75	BCl_3	-22
Ta	1300-1700	$TaCl_5$	170-190	BBr_3	20
W	1300-1700	WCl_6	250-350	BBr_3	20
W	800-1200	WCl_6	235	BCl_3	-22

[a]Exact compositions are not determined

halogenides can be mixed at the temperature determined by properties of the less volatile substance and subsequently introduced into the reaction chamber making precipitation rapid and continuous. Accordingly, the use of commercial BCl_3 with metal chlorides is recommended. However, it is practically impossible to produce high quality coatings from niobium, tantalum, molybdenum and tungsten borides using this method since pure metals begin to precipitate at the temperatures lower than those necessary to form borides. In this case, the resulting coatings contain a considerable amount of pure metal or powder-like precipitate poorly bonded with the sample surface.

Boride deposition can also be carried out by thermal decomposition of corresponding boron hydrides at comparatively low temperatures (200-400°C). Deposition takes place best at reduced pressure or by using a neutral (reducing) gas-carrier. As drawbacks of this method one should point out the tendency to excessive boron deposition, partial decomposition of boron hydrides at vaporization temperatures, flammability of some boron hydrides. [e.g. $Zr(BH_4)_4$ and $HF(BH_4)_4$] in dry air and general difficulties in the boron hydrides preparation in comparison with chlorides.

In summary, the method of chemical gaseous-phase deposition, despite its versatility and universality, has a number of essential limitations among which are relatively complex equipment, the necessity of strict technological sequence, the difficulty of producing coatings sufficiently uniform in thickness, composition and structure on large parts or articles with complex profile and the danger of explosion and toxicity of the vapor-gaseous mixtures used. These limitations are a serious hindrance to practical application of this method on a commercial scale.

c) Spraying and Building-Up

Borides or boride-based alloys can be deposited on a metallic surface as protective coatings by spraying or by building-up methods. However, these methods are not yet widely used because the technology is still in development and the process is difficult to control. A study of the process of plasma spraying of TiB_2, ZrB_2, HfB_2, NbB_2, TaB_2, CrB, CrB_2, MoB and Mo_2B_5 and properties of resulting coatings was reported by BENTZ et al. (1968) and LECRIVAIN and PROVOST (1968).

Powder spraying with a grain size of 10-55 was carried out both in the air and in a hermetically sealed chamber with protective atmosphere of argon or nitrogen. Different operating conditions and types of plasma burners were used. It was found that upon spraying in a hermetically sealed chamber with protective atmosphere appreciable quality coatings are produced and the initial composition of the spraying powders is practically preserved. The coefficient of power utilization for various borides is from 30-70% and relative density can reach values of 80-90%. On spraying in air, the porosity of coatings increases and their composition changes appreciably, mainly at the expense of oxide formation. Subsequent annealing of coating at 500-2000°C can significantly change their properties. The properties' change include residual stress density, microhardness, mechanical strength, thermal and electrical conductivity, etc. Clearly, future studies of the process of plasma spraying, investigation of coating properties and determination of their fields aof application would be encouraged by the development of a technique for large-scale production of a flexible cord. The core of this cord would consist of a power of the spraying material with a small amout of a binder, enclosed in a thin plastic casing made of organic substance. According to CLEMENT (1968) the advantages of this method are chemical purity of coatings, stability of the process, small material losses, full automation, high uniformity of coatings, etc. Flexible cord can be used successfully to spray self-fluxing materials of the Ni-Cr-B-Si type and other alloys containing appreciable amounts of boron and borides. In his review MECKELBURG (1970) pointed out that borides sprayed by the plasma method can be used in nuclear reactors and also as wear and heat-resistant materials in other technical fields. Some pure borides and alloys containing borides can be deposited using the flame spraying and detonation spraying methods, but information on these coatings is rather limited and of preliminary nature.

Although pure borides, as a rule, are not used for building-up, alloys based on them are finding more and more practical application. The components of these alloys are usually: TiB_2, CrB_2, $(Ti, Cr)B_2$ and W_2B_5 borides, sometimes combined with carbides. Thus, according to ORESHKIN et al. (1972), one of the most promising building-up compounds based on borides has the composition (wt%): 80% $(Ti, Cr)B_2$ + 15% W_2B_5 + 5% B_4C. Comparative tests show that steam shovel teeth produced by building-up method in this composition exhibited 8-10 times less wear than the heat-treated ones of 35 HML steel. This and other building-up alloys based on borides can be recommended for strengthening cutting tools in digging and highway engineering machines, and for other parts exposed to extensive abrasion and erosion.

As mentioned above, at present, diffusion boride coatings on metallic materials are best developed and are finding an ever wider application. Use of these coatings in machine and tool applications is discussed in the following sections.

2. Boride Coatings on Steels and Cast Irons

a) The Technology of Boriding

Boriding of cast irons and structural and tool steels is most often carried out in liquid with or without electrolysis and power-like media. The composition of some media and saturation procedures used on a commercial scale are given in Table 2.

Table 2. Saturating media and boriding procedures

No.	Composition of saturating medium	Recommended temperature of saturation °C	Time necessary to produce layer of 100-200μ h	Presence of borides in a layer FeB	Fe_2B
1.	Powder of commercial boron carbide, particle size 80-160	900-1000	2-4	Yes	Yes
2.	The same powder plus 1-2% fluorine activator (NaF, KBF_4, NH_4F and others)	900-1000	2-3	Yes	Yes
3.	Borax of "4-th" type (electrolysis at current density 0.2-0-25 A/cm^2)	900-950	1-2	Yes	Yes
4.	70-80% borax plus 20-30% reducing agent (silicomanganese, silico-calcium, ferromanganese ferrosilicium)	950-1000	2-3	No	Yes

It should be mentioned that choice of boride layer depth and corresponding saturation procedures are determined primarily be the type of boriding material and the purpose of preparation. Experience shows that sufficiently high quality is obtained on carbon steels, cast irons and low-alloy steels with layers 150-200μ thick, and on high-alloy steels with layers up to 100μ. Data from the literature make it possible to compare the activity of different saturating media (Table 3). Kinetics of boriding of some tool steels is shown in Table 4. Boriding of parts can be followed with heat treatment if necessary. In this case, after unpacking, the parts are heated in protective medium (salt bath), and thermal treatment (quenching, tempering) is performed according to the method specific to the type of steel. Steels having a low critical rate of quenching, such as HVG, H12Fl, H12M and others, reach a great hardness of base (up to 58-60 HRC) after rapid cooling of containers in running cold water. Subsequently, they need only be tempered. Good results are obtained by isothermal quenching of borided steels.

b) Phase Composition, Structure and Properties of Coatings

Phase composition, structure and properties of diffusion boride coatings are determined by the activity of the saturating medium, temperature and time of boriding and methods of subsequent therman treatment. Variation in any of these given components can cause a wide variation in the final properties of coatings. Naturally they also depend to some extent on the chemical composition of initial boriding material.

Table 3. Thickness of boride layer at saturation in different media

No.	Method and Procedure of Boriding	Layer Thickness Armko-Fe	Steel 45	Steel 48
1.	Electrolysis in Borax ($950^{\circ}C$, 4h)	250	220	200
2.	Liquid composition 70% borax + 30% B_4C ($950^{\circ}C$, 4h)	200	150	120
3.	M mixture, containing 84 B_4C + 16 borax (950°, 6h)	170	130	110
4.	Liquid, composition 79% borax + 15% NaCl + 6% B (powder) (950°, 4h)	240	210	170
5.	P powder, commercial B_4C ($950^{\circ}C$, 4h)	220	160	140
6.	M mixture, 98% B_4C + 1.5% KBF_4 to 0.5% was (950°, 3h)	230	210	190

All compositions given in wt.%.

Table 4. Dependence of boride layer depth on temperature and duration of boriding in powder of boron carbide

Type of Steel	Temperature $^{\circ}C$	Depth of Boride Layer, μ 1	2	4	6
U8 U10	850	15	30	55	70
	900	20	40	80	115
U12	950	30	60	150	200
	1000	50	100	165	225
HVG	850	10	20	50	60
HNV	900	15	35	65	100
HC	950	25	50	95	150
HF	1000	30	95	180	220
H12M	850	8	15	20	25
H12F1	900	12	20	35	45
	950	15	25	50	65
H12	1000	28	50	100	125
	1050	50	80	135	170
4H5MFS	850	8	15	20	30
4H5V2FS	900	12	25	45	65
3H2V8	950	25	40	75	90
4H3M3F	1000	40	75	120	160
	1050	60	100	160	215

Indicated groups of steel types are borided at approximately equal rate

Boride coatings usually contain Fe_2B internal phase and FeB external phase. The FeB phase possesses an orthorhombic lattice, density of 6.7 g/cm^3 and a high level of hardness (180-2400 kg/mm^2), the Fe_2B phase has a tetragonal body-centered lattice, density of 7.32 g/cm^3 and a lower degree of hardness (1400-1700 kg/mm^2).

In alloyed steels these phases contain a small amount of alloying element atoms which replace iron atoms in the lattice. High-carbon steels and cast irons can form carboboride phases which correspond approximately to the formulas $Fe_3(B,C)$ and $Fe_{23}(B,C)_3$. The carbon to boron ratio in these phases can vary. In alloyed steels carboborides of the $Me_{23}(C,B)_6$-type can be formed. All these phases are usually distributed in the transition zone under a continuous boride layer. The structure and phase composition of the transition zone itself depend primarily on the chemical composition of the boriding alloy but also on the boriding method and subsequent thermal treatment. The base of the transition zone is usually a solid solution of boron and other elements in α-iron.

Boride layers have a jagged structure caused by the great difference in the rate of boron diffusion along the various crystallographic directions of the boride phases. The points are directed perpendicularly to the saturated surface, for example, the diffusion front. An especially clear jagged structure is seen on iron low-carbon and low-alloy steels. The size of point protrusion decreases with the increasing content of alloying elements, the internal layer boundary possessing a more uniform character. Typical structures of boride layers on some types of tool steels are given in Figure 1.

Boride coatings possessing a high degree of hardness and increased high-temperature strength show good resistance to different types of wear, in particular, to abrasion, heat and erosion. Stability of borides against corrosion in solutions of a number of acids, alkalis, and salts is considerably higher than corresponding metals. It is, therefore, expedient to use them when a coating must combine a high degree of hardness and increased heat and corrosion resistance. The influence of boride coatings on mechanical properties has not yet been adequately studied. In the work of KARPENKO et al. (1971), data are given on the influence of boriding on some steel properties.

A generalized conclusion can be drawn that boride coatings increase ultimate strength and yield point by 10-20%, raise cyclic strength up to 25% and corrosion fatigue resistance (up to 200%) but decrease all plasticity characteristics. However, the data obtained should be considered as preliminary, since the scope of investigations has been comparatively small and has not included all the factors influencing the characteristics of coatings.

c) Examples of Practical Application

Boride diffusion coatings find a wide range of application for many types of structural parts, instruments and equipment. In numerous publications, partially summarized by LYAKHOVICH and VOROSHNIN (1967), GLUKHOV (1970), SORKIN (1972), examples are given illustrating the effective application of boriding methods to increase the service life of articles for various purposes. The list of parts whose stability can be sharply increased using diffusion boride coatings includes bushings and track pins for vehicles, bushings of mud pumps, chain bolts in scooping machines, clutches, push cams, clipping and conveying grips, draw plates, mold plates for brick pressing, mold matrices for pressing plastic materials, metallic and ceramic powders, dies for cold and hot deformation of metals, dies for casting nonferrous alloys, etc.

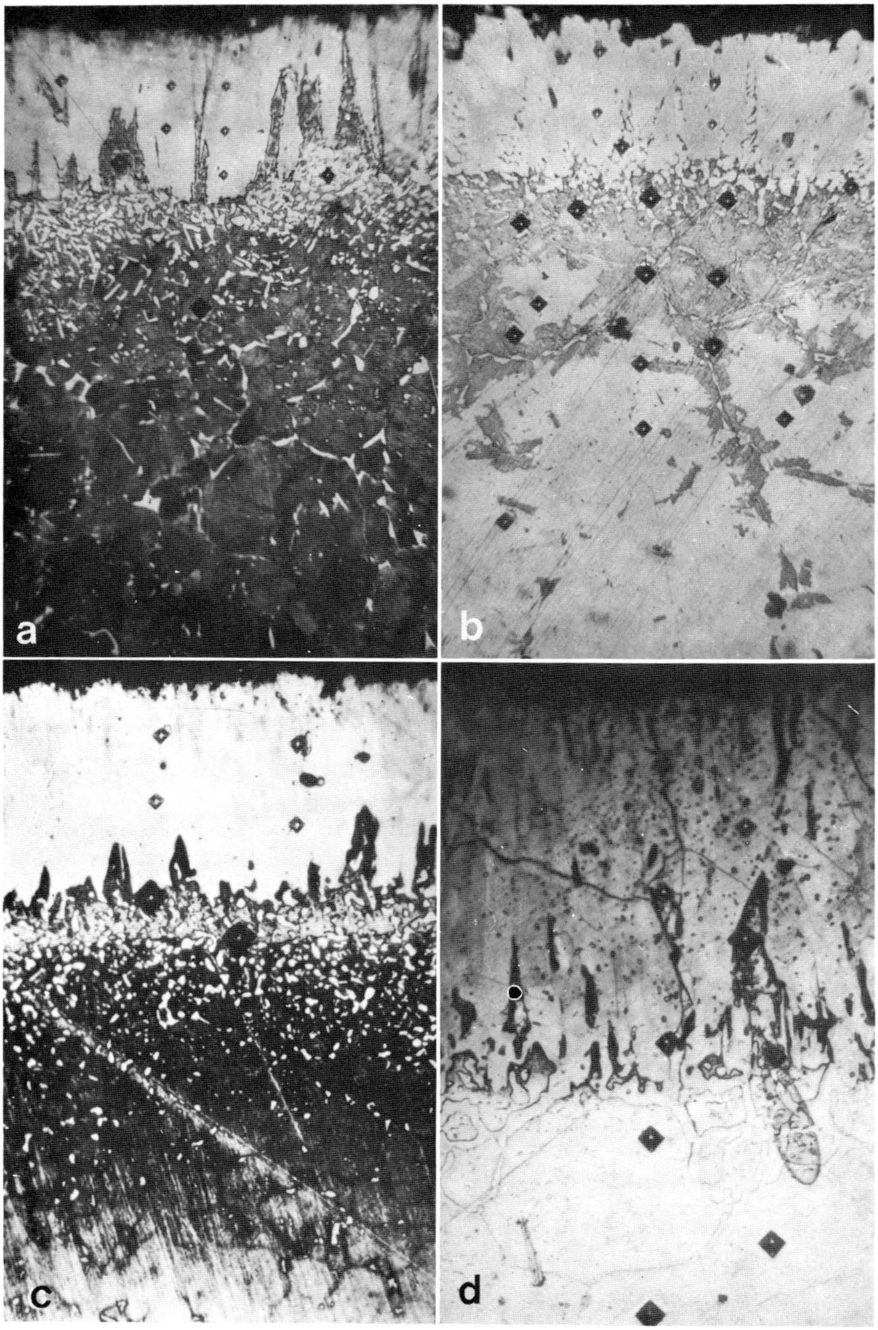

Fig. 1a-e. Typical structures of diffusion boride coatings on tool steels. (a) 9HS (b) HVG (c) 4H5V2FS (d) 3H3M3F (e) 8H4V2S2MF

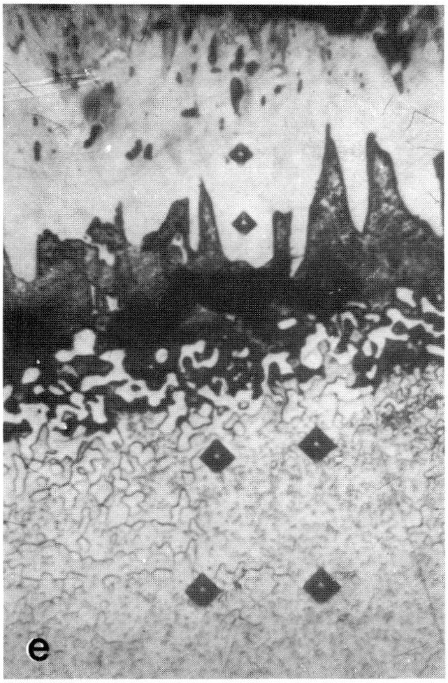

Fig. 1e

Table 5 shows the results of production tests of parts borided in powder mixtures based on boron carbide.

d) Future Prospects of Boride Coating

The perfection of different boriding techniques, improvement of coating quality and the extension of their fields of application is being investigated in various fields. Latest publications indicate the following basic directions.

1. Development of compositions which permit boriding at comparatively low temperatures (550-700 °C) satisfy requirements for a practical application rate, provides methods of decreasing deformation of the borided articles, reducing residual stresses in a layer, strengthening its binding with the base and decreasing brittleness of coatings.

2. Development of powder compositions with sufficiently high activity which do not need hermetically sealed containers (fusible blocks). STEINECK (1973) reported the development of this type of granulator based on boron carbide with fluorine-containing activators and amorphous carbon.

3. Development of boriding compositions and saturating methods which insure formation of a (Fe_2B) boride layer of the required depth. Such a composition for liquid boriding, containing borax, sodium chloride and silico-manganese, was proposed by LYAKHOVICH et al. (1974).

4. Development of technologically feasible methods for producing alloyed boride coatings (alloying with silicon, aluminum, manganese, chromium and others).

5. Develop more intensified boriding processes. In this area PASECHNIK and KOROTKOV (1967) have shown that using reverse current of variable

Table 5. Test results of borided parts

Name of tools	Type of tool steel	Average stability of parts (h)	Average stability of borided parts (h)	Service life increase (times)
Tools for cold deformation of metals				
Matrices and dies for bending and stretching of steel	H12M U8			
Bending of stainless steel	8HF	400-500	4000-5000	10
Stretching of stainless steel		100-150	1200-1500	10-11
Bending of carbon steel		1000-1500	10000-12000	8-10
Stretching of carbon steel		300-500	2500-4000	8-10
Caulk dies	H12M	3000	12000-15000	4-5
Dies for cold pressing	H12M	1000-1500	8000	5-8
Press molds for pressing ceramics and metal ceramics	U8 HVG,H12M	10000 10000	250000 60000-80000	25 6-8
Bolt dies	H12M	2000	16000-19000	8-10
Tools for hot deformation of metals				
Molds for casting non-ferrous alloys	5HNV 3H2V8	2000	5000-6000	2.5-3
Dies for swaging	5HNM EI-958	1200-1500	3500-4500	2.5-3
Broachings for hot deformation of metals	35HGS 38HMUA	260-350	760-1000	3-4

polarity while boriding of cast irons by electrolysis, one can increase the saturation rate by 1.5-2 times, other conditions being equal. The possibility of intensifying the process of diffusion layer growth particularly in boriding carbon steel with thermal cycling in the region of critical points ($800 \pm 100^{\circ}C$) was demonstrated by PANICH and LEPESHEVA (1974).

In conclusion, it should be pointed out that in many cases diffusion boriding is technically and economically more effective than other methods of chemical-thermal treatment. Thus, it appears more promising for future development and practical applications.

3. Boride Coatings on Refractory Metals and Alloys

Boriding of refractory metals of subgroups (IV-VIA) of the Periodic Table, and alloys based on them, has been less investigated than iron alloys. Only recently have investigations of both boriding techniques and properties of boride coating on refractory metals been undertaken. They can be of independent significance primarily as wear-resistant and anti-frictional materials. Furthermore, borides are usually considerably more corrosion and scale-resistant than corresponding metals. Data are available to show that boride coatings are "a barrier," inhibiting diffusion of carbon, silicon, aluminum and other elements in refractory metals, a fact undoubtedly of great technical importance.

a) The Technology of Boriding

Boriding of refractory metals and alloys as well as steels and cast irons is commonly carried out with liquid-phase and gaseous-phase (in powders) methods, but gaseous mixtures of boron halogenides or boron hydride with hydrogen can be used for this purpose. The most complete reviews on methods of boride coating on refractory metals have been given by POWELL (1970), SAMSONOV and EPIK (1973).

Dlectrolysis and liquid boriding is usually performed in melts based on borax and alkali metal halides with fluorine borates, boron carbide or amorphous boron as additions. Thus, MINKEVICH (1965) made a study of the process of liquid boriding in melts containing 60-70% borax and 30-40% boron carbide. ILUSCHENKO et al. (1969) studied the same process in a melt of 79% borax, 15% NaCl and 6% of boron. Electrolytic boriding of molybdenum in the eutectic melt KF + LiF with addition of 10% KBF_4 was studied by BONOMI et al. (1973). Boriding in this composition at 830°C ensures formation of the Mo_2B_5 layer with the rate of boron diffusion close to the diffusion rate of the gaseous-phase process. Vacuum boriding in powders of amorphous boron and boron carbide was investigated by MINKEVICH (1965), SAMSONOV and ZHUNKOVSKI (1970) and others. Analysis of the available data confirms that the saturation rate in all the methods and boriding compositions mentioned is relatively small. Moreover, these methods are inconvenient for practical applications.

To develop simple and more economical methods of boriding, SOSNOVSKI et al. (1972) investigated the process of saturation in boron carbide powder by the addition of haloid activators. All such activators were found to accelerate the saturation process to some extent but the most effective are those containing fluorine (e.g. KBF_4). Thus, on boriding at 1130°C for three hours in B_4C powder with addition of 1-1.5% KBF_4, a layer 54-57μ thick could be produced on molybdenum, and 55-55μ on niobium. At times, this result exceeds the thickness of boride layers in liquid and vacuum boriding, produced in the same temperature and time ranges. Moreover, the use of luorides as activators makes it possible to produce a good quality surface and prevents boron carvide particles from sticking to it.

KRZYMINSKI and KUNST (1973) investigated vacuum boriding of a number of metals and alloys including titanium, zirconium and their alloys, as well as hafnium, niobium and tantalum in amorphous boron powder. Boron was preheated in vacuum at 1000°C for three hours and cooled in argon environment. Investigation of the process kinetics and phase compositions of coatings showed that boron treated in such a way is highly active. The growth of boride layers (other conditions being equal) is maximum for metals of the VIA group, slightly lower for metals of the VA group, and minimum for metals of the IVA group.

Table 6 gives comparative data for growth kinetics of the boride layer on titanium, molybdenum and niobium borided in various saturating media at 1100°C.

Table 6. Kinetics of boriding Mo, Nb and Ti (temperature 1100°C)

No.	Composition of saturating medium	Metal	Total layer thickness, μ time of boriding hours 2	4	6
1.	60% $Na_2B_4O_7$ + 40% B_4C	Mo	120	160	195
		Nb	80	105	120
		Ti	4	7	9
2.	84% B_4C + 16% $Na_2B_4O_7$	Mo	30	42	50
		Nb	6	12	16
		Ti	3	5	7
3.	Amorphous boron, vacuum 5.10^{3} - 1.10^{3} torr	Mo	7	11	14
		Nb	8	12	15
		Ti	3	4	6
4.	Commercial B_4C	Mo	35	50	62
		Nb	30	45	56
		Ti	7	12	15
5.	Commercial B_4C + 2% KBF_4	Mo	45	62	76
		Nb	40	54	65
		Ti	10	15	20

b) Phase Composition, Structure and Properties of Coatings

Our investigations and analysis of reports by various authors show that phase composition and structure of boride coatings on refractory metals and alloys depend primarily on the activity of the saturating medium and the boriding temperature. At comparatively low temperatures (up to 1100°C) and high saturating ability of the boriding medium, higher borides of corresponding metals are formed. With the increase in saturation temperature, a tendency is observed to form phases with lower boron content. In other words, phase composition of coatings is determined by the ratio of boron delivery to the surface and its diffusion rate into metal depth. A definite influence is exerted also by the "metal-boron" phase equilibria and the thermodynamic stability of boride phases at saturating temperatures.

In most cases boride coating on metals of the IVA group are TiB_2, ZrB_2, HfB_2 diborides (at high saturating temperatures formation of monoborides on the metal boundary is possible). Similarly, boriding of VA group metals is accompanied by the formation of VB_2, NbB_2 and TaB_2 diborides, but lower borides are frequently formed underneath them. When VIA group metals are borided, coatings usually consist of at least two phases: CrB_2 and CrB, Mo_2B_5 and α-MoB, W_2B_5 and α-WB. However, in a number of cases Mo_2B boride (on the boundary with metal) and MeB_4 boride (on the external surface) can also be formed.

The structure of boride coatings is determined both by saturation conditions and conditions of subsequent cooling, involving essential differences in physical properties of metals and corresponding borides (modulus of elasticity, thermal expansion coefficient, etc.). In any case one should take into consideration that the properties of boride coatings (compactness, bond strength with the base, brittleness, residual stress, etc.) depend strongly on their phase composition and thickness. With metals of the VA and VIA groups, sufficiently high bond strength and compactness is shown in coatings whose thickness does not exceed 70-100μ and in metals of the IVA group 40-50μ. Typical structures of boride coatings on refractory metals are presented in Figure 2.

The hardness of corresponding phases of boride coatings decreases slightly with a decrease in boriding temperatures.

The properties of boride coating on refractory metals have been insufficiently investigated. According to the data of MINKEVICH (1965), wear resistance of borated molybdenum, tungsten and niobium under conditions of dry friction is much higher than that of steel U8 and cemented steel 30HT.

A study of abrasive wear resistance (tested on moving abrasive paper with SiC grains) of borided refractory metals showed that their wear resistance is ten times as high as that of starting metals and is 100 times greater in the case of borided niobium. Hardness of borated molybdenum and niobium in a temperature range of 100-900°C is on the level of high temperature hardness of standard metal ceramic alloys. Boride coatings, according to BURIKINA and GUBENKO (1973) are effective means of preventing the setting of alloys based on niobium, tantalum and molybdenum at temperatures up to 1500°C and loadings of up to 15 kg/mm^2 in a helium medium. SICHEV (1974) showed the potential of boride coatings (wear-resistant and antifrictional) on niobium, tantalum, molybdenum and tungsten while operating in vacuum, and in helium in a wide temperature range (up to 1200°C).

Scale and corrosion resistance of boride coatings on titanium, zirconium, niobium, molybdenum and upon oxidation in air at temperatures of 600-1000°C was reported by EPIK et al. (1965) and MINKEVICH (1965). In this temperature range boride coating considerably improve oxidation resistance of refractory metals. Thus, upon exposure for 24h at 950°C, scale resistance of borated titanium, zirconium and tantalum were proven to be higher by 14, 20 and 30 times respectively in comparison to untreated metals. In general, for all temperature ranges investigated, boride coatings can be compared favorably to the series of ZrB_2, TiB_2, TaB_2, NbB_2, W_2B_5 and Mo_2B_5 regarding their scale resistance. Our investigations have shown that boride coatings have high stability in solutions of inorganic acids and alkalies. Thus, corrosion resistance of borated titanium, zirconium, niobium, molybdenum and tungsten upon testing in concentrated and dilute (1:1) acids, such as: HCl, H_2SO_4, HNO_2, H_3PO_4 and also in 10% and 30% of solutions of KOH either cold for 20-30 days or boiling for 1-3 hours by far exceeded the stability of the corresponding pure metals. The data obtained agree well with the results in the papers investigating corrosion resistance of the corresponding boride powders.

EVTUSHOK and ZHUNKOVSKY (1973) showed that molybdenum and tungsten borides are "a barrier" to the carborization of these metals on contact with graphite and carbon in vacuum up to 1700-1750°C and 1800-1900°C respectively. In a hydrogen environment, corresponding "barrier" properties decrease up to 1650°C and 1750°C. The ability of boride coatings to sharply inhibit silicon diffusion and disilicide formation has been used by BYALOBZHESKY et al. (1973) to develop reliable heat-resistant coatings on molybdenum. Preliminary borating of molybdenum with subsequent deposition by diffusion of disilicide made it possible to attain a service life for coatings at 1300°C of up to 3000 h and at 1500°C of up to 1000 h in air.

EPIK et al. (1969) confirmed the high stability of boride coatings on titanium, niobium, molybdenum and tungsten in molten aluminum at 700 and 900°C for over five hours. Such coatings are used to develop stable current suppliers for determining thermodynamic properties of aluminum melts with emf method in 700-1100°C temperature range.

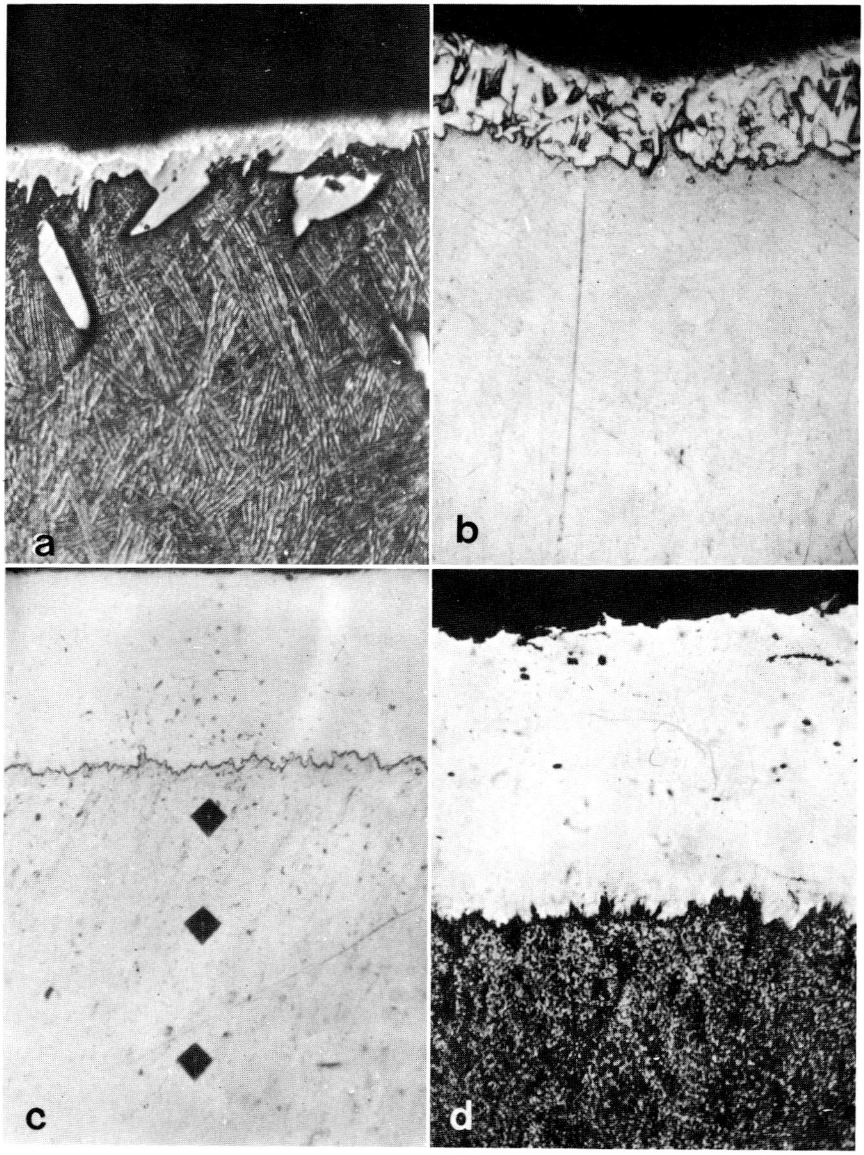

Fig. 2a-e. Typical structures of diffusion boride coatings on refractory metals. (a) Titanium (b) Zirconium (c) Niobium (d) Tantalum (e) Molybdenum

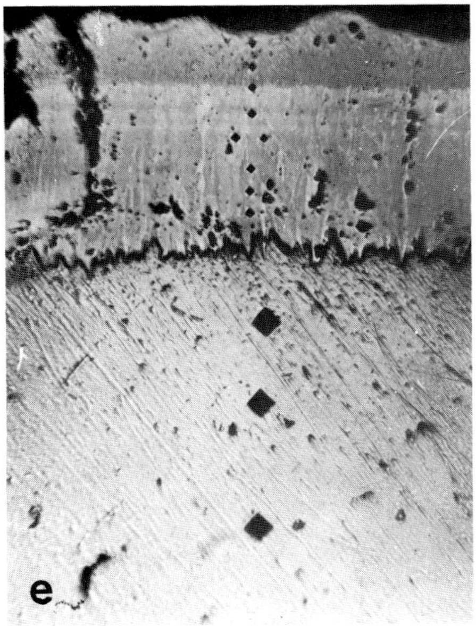

Fig. 2e

It should be mentioned that in the literature there are hardly any reports on practical applications of boride coatings on refractory metals and alloys. This can probably be explained by a lack of comparative studies of properties of boride coatings and considerations of alternatives. However, there is no doubt that the valuable technical properties of boride coatings on refractory metals will find wide applications in the near future.

References

BECKER, K.: Hochschmelzende Hartstoffe und ihre technische Anwendung, SS/34-42, 51-72, Berlin: Chemie 1933

BENTZ, G., PROVOST, G., URBAN, C.: Etude de la projection des berures et des nitrures en chalumeau à plasma. Bull. Soc. Ceram. 78, 5-17 (1968)

BONOMI, A., GIESS, H., CENTAZ, C.: Electrochemical boriding of molybdenum in molten salts. Electropos Surface Treat. 1, 419-427 (1973)

BURIKINA, A.L., GUBENKO, B.G.: Influence of diffusion boriding effect on setting of alloys, based on refractory metals. In: Friction and Wearability at High Temperatures. pp. 107-114, Moscow: Nauka 1973

BYALOBZHESKY, A., KRASILOV, B.I., TSIRLIN, M.S.: Modificated diffusion silicon countings for protectron of molibdenum from oxidation. In: Protective Coatings on Metals, Vol. VII, pp. 90-93, Kiev: Naukova Dumka 1973

CLEMENT, R.: Nouveau mode d'alimentation des chalumeau de projection d'oxydes ceramiques. Ind. Ceram. H 612, 859-861 (1968)

DERKACH, V.D., EPIK, A.P., KOTLYARENKO, L.A., SOSNOVSKY, L.A.: To the question of boron transfer mechanism to the saturating surface during boriding boron carbide powder of technical purity. In: Chemical-Heat Treatment of Metals and Alloys. pp. 15-19, Minsk (1971)

EPIK, A.P., BATALIN, G.I., BELOBRORDOV, E.A., KARPINOS, D.M., ZILBERBERG, V.G.: Investigation of refractory metals resistance with diffusion and plasma coatings in aluminium melts. In: Heat-Resistant and Heat-Stable Coatings. pp. 449-453, Leningrad: Nauka 1969

EPIK, A.P., BOVKUN, G.A., GOLUBCHIK, J.V., SINITSINA, L.P.: Some properties of diffusion carbide and boride coatings on refractory metals. In: Diffusion Coatings on Metals. pp. 127-142, Kiev: Naukova Dumka 1965

EPIK, A.P., DERKACH, V.D., KOTLYARENKO, L.A., SOSNOVSKY, L.A.: Boriding of machine elements and instruments in powder mixtures. Technol. Organ. Prod. 2, 69-73 (1970)

EVTUSHOK, T.M., ZHUNKOVSKY, G.L.: Reaction of boride coatings on molibdenum and tungsten with carbon. In: Protective coatings on Metals. VII, pp. 192-195, Kiev: Naukova Dumka 1973

GLUKHOV, V.P.: Boride Coatings on Iron and Steels. Kiev: Naukova Dumka 1970

ILUSCHENKO, N.G., ANFENOGENOV, A.I., BELYAEVA, G.I., PLOTNIKOVA, A.F., KORNILOV, N.I.: Diffusion coatings of metals in melting salts. In: Heat-Resistant and Heat-Stable Coatings. pp. 105-119, Leningrad: Nauka 1969

KARPENKO, G.V., POKHMURSKY, V.I., DALISOV, V.B., ZAMIKHOVSKY, V.S.: Influence of Diffusion Coatings on Strength of Steel Products. Kiev: Naukova Dumka 1971

KONOWROCKA, J., TACIKOWSKI, G.: Przeglad metod borowania dyfuzyjnego. Metaloznawstwo i obrob. diepl. 1, 26-32, (1973)

KRZYMINSKI, H., KUNST, H.: Borieren refraktarer Metalle. Harter Techn. Mitt. 28, 2, 100-109. Diskuss. 109-112 (1973)

LECRIVAIN, L., PROVOST, G.: Eigenschaften plasma-spritzenden Borides und Nitrides. Ber. Deut. Keram. Ges. 45, 347-351 (1968)

LYAKOVICH, L.S.: Boriding of Steel. Thesis, Minsk Physical-Technical Inst. Acad. Sci. USSR 1969

LYAKOVICH, L.S., TUROV, J.V., KRUKOVICH, L.G., HUSID, B.M., LEVITAN, S.P.: Liquid-phase boriding. In: Chemical Heat Treatment of Metals and Alloys. pp. 83-84, Minsk 1974

LYAKOVICH, L.S., VOROSHNIN, L.G.: Boriding of Steel. Moscow: Metallurgiya 1967

MATUSCHKA, A.: Borieren, Konstr. Elem.Meth. 10, 5, 54-58, 59 (1973)

MECHELBURG, E.: Überzüge nach Metall-Spritzverfahren. Draht 21, 941-951 (1970)

MINKEVICH, A.N.: Chemical Heat Treatment of Metals and Alloys. Moscwo: Mashinostroenie 1965

ORESHKIN, V.D., SVETLOPOLYANSKY, V.I., MASLENIKOV, V.D.: Investigation of high alloyed carbide-boride surface coatings. In: Surfacing Alloys Based on Refractory Compounds. pp. 62-73, Volgograd 1973

PANICH, G.G., LEPESHEVA, S.D.: About possibility of diffusion layer increasing by process of intensification by cyclical heat influences. In: Chemical-Heat Treatment of Metals and Alloys. pp. 51-53, Minsk 1974

PASECHNIK, S.Ya., KOROTKOV, V.D.: Electrolysis boriding of cast irons. In: Protective Coatings on Metals. pp. 40-45, Kiev: Naukova Dumka 1967

POWELL, K.: Chemical precipitation of non-metla materials, boron and borides. In: Precipitation from Gas Phase (K. Powell, J. Oxly, J. Blocher). Chapt. 10, pp. 276-344, Moscow: Atomizdat 1970

SAMSONOV, G.V., EPIK, A.P.: Coatings of High Temperature Materials. Moscow: Metallurgia 1973

SAMSONOV, G.V., ZHUNKOVSKY, G.L.: Powder Metal. 6, 44-51, (1970)

SICHEV, V.S.: Friction and wear of boride and chalcogenide coatings on refractory metals at high temperatures in vacuum and inert atmosphere. Candidate Thesis, Inst. Mat. Prob. Acad. Sci. USSR, Kiev 1974

SORKIN, L. M.: Strengthening of Details by Boriding. Moscow: Mashinostroenie 1972

SOSNOVSKY, L.A., EPIK, P.A., KRAPLYA, A.P.: Diffusion boriding of molibdenum and niobium in boron carbide powder. Powder Metall. 9, 75-78 (1972)

STEINECK, K.: Borieren mit borkarbidhaltigem Granulat. Schweiz. Maschinenmarkt 73, 70-74 (1973)

SVETLOPOLYANSKY, V.I., SEMENOV, F.G., ORESHKIN, V.D.: Increasing of shovel teeth wear-resistance strengthened by building-up materials, based on refractory compounds. In: Building up Alloys Based on Refractory Compounds. pp. 121-126, Volgograd (1972)

VII. Nuclear Applications of Boron and the Borides

D. E. MAHAGIN and R. E. DAHL

1. Introduction

One of the basic requirements associated with the development and operation of nuclear reactors is control and containment of the neutrons that sustain, and are in turn produced by, fission reactions. Materials that serve this function are often chosen because of their nuclear properties. However, physical and chemical characteristics and behavior may determine selection and operational success.

Boron is one of the very few elements to possess nuclear properties which warrant their consideration as neutron absorber materials. As a consequence, it has been used extensively in the nuclear industry. In many instances, however, the properties of boron and its compounds and their irradiation behavior impose limitations on the conditions under which they can be used. Therefore, in some situations, other materials are more suitable.

In this discussion, we review briefly the properties of boron materials which make them attractive as neutron absorbers. The characteristics which detract from their utility are also discussed. In order that all aspects of their application can be appreciated more fully, a summary of nuclear terminology and the selection criteria applied to neutron absorbers is presented. These discussions are concerned with control-system applications. Neutron-absorbing materials are also often used for shielding applications. The conditions and requirements are much less stringent than for control elements, but most of the considerations are similar.

Irradiation experience with boron and its compounds is reviewed and, finally, we compare the performance of these materials with the general trends in reactor evolution and speculate on their future in nuclear applications.

2. Nuclear Terminology and Criteria

Prior to considering the behavior of boron in nuclear applications, it seems prudent to discuss the manner in which it is employed and the reasons for its use. To do this, it will be necessary to review briefly certain aspects of reactor neutronics and operating principles. With this background, the utility of boron will become readily apparent.

a) Reactors and Their Control

Nuclear reactors operate by means of sustained chain reactions between neutrons and fissionable (fissile) isotopes. During these reactions, approximately two new neutrons are produced for every one that is absorbed. As long as one of them engages in another fission reaction, the process is self-sustaining. "Fission" neutrons are also lost from the system by leakage and parasitic captures in structural components of the core.

In order to describe quantitatively the neutron balance in a reactor, it is customary to use the infinite multiplication factor, k_∞. This dimensionless parameter is defined as the ratio of the number of fissions in any one generation to the number of fissions in the immediately preceding generation (LAMARSH, 1966) in a system of infinite size. Actual reactors are not infinite in size, of course, and the parameter used to define their criticality condition is the effective multiplication factor, k_{eff}. This is simply the infinite multiplication factor, reduced appropriately to account for the probability of neutron leakage during the time period between their birth in a fission reaction and their eventual capture. In order to sustain a chain reaction (the condition for reactor criticality), k_{eff} must be at least unity. When k_{eff} is exactly unity, the chain reaction proceeds at a constant rate. When it is greater than unity, the chain reaction increases (diverges). When it is less than unity, the chain reaction decreases, and eventually dies out. A reactor is considered to be critical when $k = 1$; supercritical when $k > 1$; and subcritical when $k < 1$.

To compensate for fuel burnup, varying demands, and account for uncertainties in temperature and other aspects of power operation, reactors are designed in such a manner that k_{eff} is slightly greater than unity. The concept of reactivity is introduced to describe the excess in the multiplication factor. To ensure that the reactor does not operate in the supercritical mode, even though k_{eff} is greater than unity, neutron absorbers are intentionally introduced into the reactor core as a part of the control system[1]. The effectiveness of these absorbers is measured in terms of their *worth*, which is simply the amount of reactivity that they can control. Reactivity is defined as

$$\text{Reactivity} = \rho = \frac{k_{eff}-1}{k_{eff}} = \frac{k_{excess}}{k_{eff}} \tag{1}$$

and is often represented as $\Delta k/k$ or $\delta k/k$ (or even δk).

Qualitatively, control elements which have a great deal of worth are called black, and those with lesser worth are often referred to as being gray. Both types are generally required in a reactor. A totally black rod would be one in which all neutrons incident upon its surface are absorbed (blackness = 1.0). Because the probability of a neutron's capture depends upon its energy, rods can be "black" to thermal[2] (or low energy) neutrons, and transparent to more energetic neutrons.

Control systems are necessary to vary the power in response to demands and, of course, to shut down the reactor. Several types of control elements are used in a given system to perform specific functions. These can be bundles of rods or plates of various shapes. They are generally inserted parallel to the vertical axis of the reactor so that they will fall upon release and provide "fail safe" protection (see Fig. 1). The nomenclature used to designate the various types of elements is: (1) safety elements, those which are kept above the core during operation and are dropped for emergency shutdown (i.e. scram); (2) regulating rods, which are partially inserted and are moved to adjust power; (3) and shim rods, which provide adjustment over long-term cycles (THOMPSON and BECKERLEY, 1969).

---

[1]Several methods of control have been used in operating reactors. Control by utilizing neutron absorbing materials is by far the most common, particularly in large plants.

[2]Neutrons are classified as "thermal" if their average energy is kT (∿0.025 eV at room temperature), which is equivalent to that of the atoms of the media due to thermal agitation. The energy of fast neutrons ranges from ∿0.1 to 10 MeV.

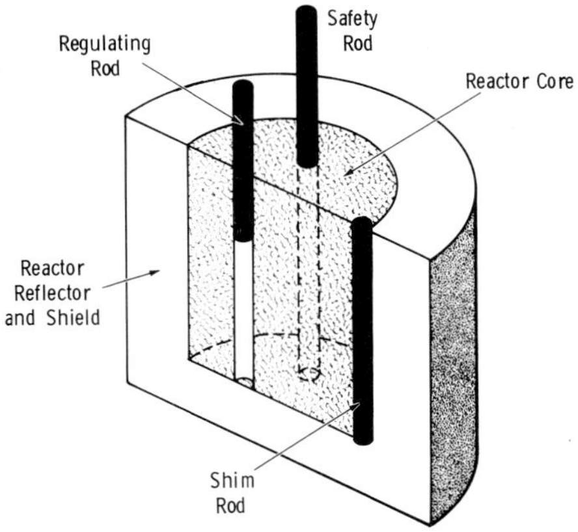

Fig. 1. Schematic view of reactor showing various types of control rods

Safety rods must be capable of moving into the reactor core very quickly and their combined worth (i.e. reactivity equivalent) must be significantly greater than the maximum reactivity of the reactor. These are the "blackest" in the system In thermal reactors, shim rods are often used for coarse control. As these rods are withdrawn, a reactor starts up and is brought close to the desired power level. These rods also have a fairly large reactivity equivalence (black) but are moved slowly by mechanical-drive mechanisms. As the desired power level of a reactor is approached, the position of the shim rods is such that the multiplication factor is almost unity and the task of bringing the reactor up to the operating power and maintaining the level is accomplished with regulating rods. The reactivity equivalence of regulating rods and the distance over which they can move is very limited, so they are unable to produce any excessive increases in reactor power or temperature (MURRAY, 1957). Typically, they contain "gray" neutron absorbers. In fast reactors, shims are often immovable and inserted to compensate for excess core reactivity.

An additional form of reactivity control is often incorporated in the overall system of large plants. This involves distribution of a burnable poison throughout the core with each loading of fresh fuel. Burnable poisons are neutron-absorbing materials which are consumed at approximately the same rate as the fuel. By employing them, it is possible to load extra fuel which compensates for burnup and the production of fission product poisons (i.e. fission products which capture neutrons parasitically) and yet avoids the use of high-worth regulating rods. Since the fuel and the absorber are depleted at essentially the same rate, utilization of burnable poisons also reduces or eliminates the need for a continuous withdrawal of the regulating rods in order to maintain the core reactivity constant during a power cycle[3].

There are general design requirements for all control systems (GLASSTONE, 1955). They should, for example, have sufficient neutron absorption capability to cover all contingencies and sufficient delicacy to permit fine adjustment. The control system should be simple and foolproof in action, thus assuring maximum reliability. It should be as economical as possible in the fabrication costs of the elements, and should conserve neutrons. Excessive neutron absorption increases the

---

[3]Water-cooled reactors often employ a soluble absorber - usually boric acid - in the coolant to provide this adjustment.

fuel loading, operating costs, and reduces power. In breeder reactors, it also reduces the breeding ratio, thus invoking a significant economic penalty.

b) Absorber Materials

Control systems such as those described in the preceding discussion depend upon neutron-absorbing materials. How, then, do these materials function and what properties must they possess? To answer the first of these questions, we define neutron absorbers as elements in which the nuclei capture neutrons without generating or emitting new neutrons. The two most common types of reactions involve neutron absorption by the target nucleus followed by the emission of gamma radiation (n,γ) or the ejection of an alpha particle (n,α). A neutron is not re-emitted or produced in either case.

The rate at which nuclear reactions proceed is determined largely by a quantity called the *nuclear cross section*, σ, which is a measure of the probability of the occurrence of that reaction. It is a property of the nucleus as defined for a particular reaction with incident particles, e.g. neutrons of a given energy. The dimensions of cross sections are cm^2. The absolute values are very small so the barn, 1×10^{-24} cm^2, is used in common terminology. Cross sections are greatest in the slow neutron energy range and, in general, decrease as neutron energy increases. Effective absorber materials have cross sections of at least 1000 barns in the neutron energy range below one electron volt (eV). In the fast neutron range, $E_n > 0.1$ MeV, the cross sections for all reactions are low; and an absorber having a cross section of $\sim$1 barn is very effective. Cross sections of absorber, fuel, and structural material isotopes are presented in Table 1 for comparison.

The cross section, σ, applies to a single nucleus and is called the microscopic cross section. Since the target material contains "N" nuclei/cm^3, the quantity $N\sigma$ is equivalent to the cross section of the material per cm^3 for a particular reaction. This is called macroscopic cross section and is the quantity used in evaluating absorber materials and calculating their effects. The macroscopic cross section is:

$$\Sigma = N\sigma \quad (cm^{-1}) \tag{2}$$

It has the dimensions of a reciprocal length; its inverse, λ, is the mean free path which a neutron would probably travel through a material before undergoing the particular reaction. Macroscopic cross sections (Σ) are calculated according to one of the following equations (MURRAY, 1957) which are applicable to elements, molecular compounds, and mixtures, respectively.

$$\text{Elements:} \quad \Sigma = \frac{\rho N_\alpha}{A} \tag{3}$$

$$\text{Molecular Compounds:} \quad \Sigma = \frac{\rho N_\alpha}{M}(\sigma_1\nu_1 + \sigma_2\nu_2 + \ldots \sigma_i\nu_i + \ldots) \tag{4}$$

$$\text{Mixtures:} \quad \Sigma = N_1\sigma_1 + N_2\sigma_2 + \ldots + N_i\sigma_i \ldots \tag{5}$$

where ρ is the density (gram/cm^3); N_α is Avogadro's number (6.023×10^{23}); A is the atomic weight; M is the molecular weight; N_i is the nuclear density of atoms of the kind "i"; and ν_i and σ_i are the fraction of atoms of the kind "i" and their microscopic cross section, respectively, in the molecule or mixture. The rate of reaction in a material is $R = \Sigma\phi$ per (cm^3 sec), where ϕ is the neutron flux, (neutrons/cm^2/sec). In evaluating the relative effectiveness of absorbers, macroscopic rather than microscopic cross sections should be compared.

Table 1. Absorption cross selections of reactor materials in typical thermal and fast reactors (DRAKE, 1970)

Isotope or element	Nominal density (g/cm^3)	Nuclear density atoms/cm^3 ($x\ 10^{-24}$)	$\bar{\sigma}_a$* ($cm^2 \times 10^{24}$)		$\bar{\Sigma}_a$ (cm^{-1})	
			LWR	LMFBR	LWR	LMFBR
Absorbers						
^{10}B	2.347	0.1411	3400	2.70	479.740	0.381
B natural	2.535	0.1411	670	0.54	94.537	0.076
Cd	8.650	0.0463	3300	0.15	152.790	0.007
Cladding						
Stainless Steel	7.850	0.8470	2.80	0,010	2.372	0.008
Zr	6.440	0.0425	0.16	0.015	0.007	0.001
Fuel						
U Natural	19.00	0.04833	--	--	--	--
^{235}U	18.70	0.04833	670	2,50	32.381	0.121
^{238}U	19.00	0.04833	2.90	0.40	0.137	0.019
^{239}Pu	19.60	0.04940	1300	2.20	64.220	0.109
Coolants						
H_2O	1.000	0.03343	0.664	--	0.0222	--
Na (liquid @ 400°C)	0.854	0.22408	0.470	1.3×10^{-3}	0.1053	0.0003

$*\bar{\sigma}_a = \frac{\int \phi(E)\ \sigma\ (E)\ dE}{\int \phi(E)\ dE}$ = spectral averaged cross section

To answer the second question concerning the requisite properties of neutron absorbers, it is necessary to consider in some detail the manner in which they are utilized. Some of this has been implied in the discussion of control systems, but will be repeated here for clarity.

First and foremost, a neutron absorber must have the capability for capturing neutrons efficiently to the degree required for reactivity control. In addition, the absorber material must retain this capability throughout the life of the core to assure that the reactor can be safely shut down at all times. The required absorption can be achieved by ensuring a sufficient concentration of absorber atoms in the core, but it is preferable to minimize the volume used by the control elements. This guarantees that maximum space is available for fuel which produces heat and energy. Thus, a high neutron-absorption cross section is essential for control materials. The energy dependence of the neutron-absorption cross section is also a very important consideration in selecting control elements for various types of reactors. There are some elements which absorb neutrons efficiently only in the thermal energy region ($\bar{E}_n \simeq 0.025$ eV). There are other materials which absorb neutrons at somewhat higher energies, but are relatively ineffectual in the fast neutron energy region ($E_n > 0.1$ MeV). Thus, a particular substance could be entirely satisfactory for one type of reactor and unacceptable for another.

Absorber materials are generally clad in a metallic sheath or dispersed in an inert matrix in order to maintain the structural integrity of the control rod. They should be chemically compatible with these materials for time periods in excess of their anticipated in-reactor residence. They should also exhibit at least short-term compatibility with the reactor coolant in order to preclude problems that might occur in the event of cladding rupture.

While in the reactor, an absorber material should exhibit predictable irradiation behavior under all its anticipated operating conditions, its performance should be such that it does not impose excessive lifetime or operational limitations on the control system.

Finally, the absorber material should be inexpensive to procure and fabricate in order to make the cost of controlling the reactor as economical as possible. The principal considerations in selecting burnable poisons are much the same as those for other absorbers, with the additional requirements that the poison be depleted at a rate comparable to the fuel so that the reactivity of the core will be nearly constant over each cycle.

c) Boron as a Neutron Absorber

The nuclear properties of boron are such that boron and its compounds are very effective as neutron absorbers in most types of reactors. The neutron-absorption cross section is sufficiently high in the low neutron energy range to make it an excellent candidate for use in thermal reactors. At higher energies, the cross section of most other isotopes becomes very small, often abruptly as in the case of cadmium, whereas boron-10 decreases monotonically with energy. Absolute values across the entire energy spectrum are of sufficient magnitude to make it very effective in intermediate and fast reactors. Figures 2 and 3 illustrate the boron-10 cross section in the low and high energy ranges, respectively (DRAKE, 1970). Cross sections of other absorbers, fuels, and structural materials are also included for comparison. The rectangular lines appearing in Figure 3 indicate the energy group structure used in some reactor physics analyses. Tabulated cross section data appearing

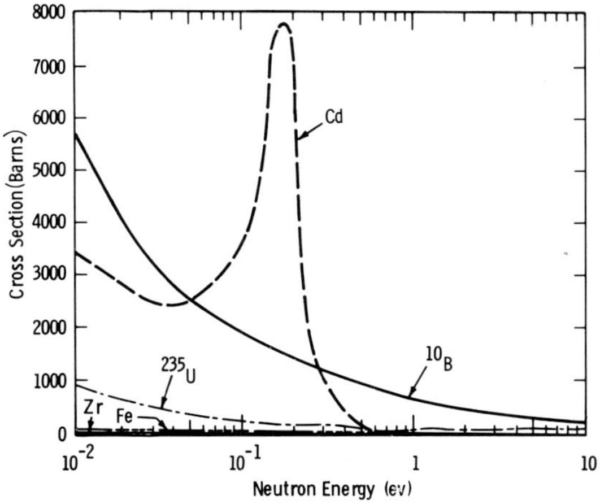

Fig. 2. Nuclear cross sections at low neutron energies - thermal reactors

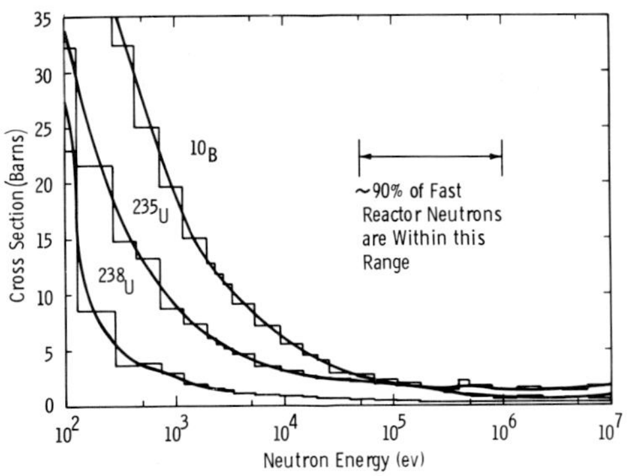

Fig. 3. Nuclear cross sections at high neutron energies - fast reactors

in Table 1 include nuclear densities and macroscopic cross sections. It is evident that the macroscopic cross section of boron carbide, and thus its worth per unit volume, compares very favorably with other candidate absorbers.

Boron has two principal isotopes, ^{10}B and ^{11}B, which are chemically identical but have quite different nuclear characteristics. The effectiveness of boron as a neutron absorber is due to the cross section of the ^{10}B isotope; ^{11}B is essentially non-reactive, (σ_a<<0.01 barns) from a nuclear standpoint. As boron occurs in nature, approximately 19.8% of the nuclei has a mass of ^{10}B, the remainder ^{11}B. This provides a great deal of flexibility in the use of boron and its compound in reactor systems, since this mixture can be adjusted. Boron-10 enrichments of up to 92% can be attained. Thus, the worth of an absorbing component can be increased by nearly a factor of 4.5 by substituting enriched for natural material without changing its dimensions. This is not possible with other candidate absorber materials.

In addition to a high absorption cross section, boron has another advantage over other potential control materials. This is the fact that the absorption reaction products - helium and lithium - are formed as stable, non-radioactive isotopes[4]. Inasmuch as they emit no nuclear radiation, decay heating problems during reactor shutdowns and transfer of depleted control rods are minimized[5].

The (n,α) reaction that produces stable helium and lithium, however, also leads to the most serious deterrent to the universal use of boron in nuclear control applications (ETHERINGTON, 1958). The formation of these products within a solid can lead to dimensional instability of the absorber. Helium is partially retained within solids, but the fraction which is released must be accommodated by some aspect of the control assembly design such as a plenum within an individual pin or a reliable pressure relief vent, since the volume of gas generated is appreciable. This is not unique to the design of control pins. Similar considerations are involved in fuel pin design in which gaseous fission products must be accommodated. A principal reason for selecting non-boron materials for nuclear control elements has been to avoid gas emission.

3. Irradiation Experience with Boron and the Borides

In the foregoing discussion, the unique properties that make boron a useful material in nuclear applications have been reviewed. In this section, the performance of boron and its compounds during irradiation is considered. The available data fall naturally into two broad classifications. The first of these includes irradiation experience in thermal reactors where the average neutron energies are relatively low and absorption cross sections are consequently high. In the second category are the results of fast reactor exposures where neutron energies are so high that very few elements other than boron can be considered for absorption applications.

a) Thermal Reactor Experience

The irradiation behavior of a wide variety of boron-bearing materials has been characterized in thermal reactor environments. The principal reason for this is the fact that boron is an extremely effective thermal neutron absorber. As such, it can be combined with a number of other substances and still yield compacts with macroscopic cross sections that meet control requirements. This means, in effect, that there are virtually no neutronic limitations on the selection of potential boride absorbers, and a considerable number have been investigated to select materials which offered clear-cut advantages from an irradiation performance standpoint.

α) Boron Compounds

For certain high-worth applications, boron must be used in either the elemental form or as a binary compound. In a typical thermal reactor, the operating temperature of the absorber is below 500 °C and its per-

---

[4]Another product generated by neutron reactions with boron is tritium (^3H), by the $^{10}_{5}B + ^{1}_{0}n \rightarrow 2^{4}_{2}He + ^{3}_{1}H$ process. The cross section for this reaction exists only above about 1 MeV and has a spectral average value of about 0.01b in fast spectra. Tritium from boron does not present a problem in thermal reactors since it is effectively gettered. In fast reactors, the amount of tritium generated by boron is small. It is partially retained in boron carbide and the remainder trapped.

[5]The $^{10}_{5}B + ^{1}_{0}n \rightarrow ^{7}_{3}Li + ^{4}_{2}He$ reaction is exoergic releasing 2.79 MeV (MAPLES et al., 1966) This causes internal heat generation within boron compounds. However, this ceases as the neutron flux is terminated.

formance is evaluated by monitoring gas release and swelling behavior, either of which may impose restrictions on potential control assembly designs.

In high-worth applications of this sort, a phenomenon known as self-shielding is often observed. It occurs because in the thermal neutron energy range absorption cross sections are very high and the mean free path is correspondingly low. Thus, neutrons do not penetrate a macroscopic body very deeply before they are absorbed, and this causes extreme gradients in irradiation damage. The inhomogeneity of the damage state can, in turn, lead to stress gradients within the compact, which ultimately result in its fracture. Much of the data concerning boride behavior in thermal reactor spectra must be interpreted by considering the effects of stress gradients due to self-shielding.

By far the most extensively studied boron compound for thermal reactor applications is boron carbide. It has been irradiated at temperatures as high as 650°C and exposure levels amounting to 100% depletion of the ^{10}B isotope (VALOVAGE, 1955; HAMMAN and SCHALL, 1960; HOYT and ZIMMERMAN, 1962; GRAY and LYNAM, 1963; MEFFORD and WILLIAMSON, 1968; PITNER and RUSSCHER, 1970). The effects of irradiation on the crystal structure have been reported (TUCKER and SENIO, 1955) and an attempt has been made to correlate performance parameters (i.e. gas release and swelling) with various material variables such as compact density, grain size, and compound chemistry.

A summary of the gas release from boron carbide is shown as a function of irradiation temperature in Figure 4. It is readily apparent that there is considerable scatter in the data, with similar exposure conditions resulting in gas release fractions that vary by factors of 4-6 in some cases. Much of this inconsistency can be attributed to stress gradients (caused by self-shielding) that result in significant amounts of cracking, spalling, and complete fracture of the solids. This deterioration enhances gas release, and since its occurrence cannot be correlated with exposure conditions, the gas-release data cannot be normalized to any baseline. In general, it appears that gas release fractions from powders are somewhat higher than those from hot-pressed shapes, but there is considerable overlap in the results. Other trends that have been observed are that gas release is inversely proportional to compact density and grain size (GRAY and LYNAM, 1963) and directly proportional to B/C ratio. This latter effect may result from unreacted boron in the carbide (COPELAND, et al., 1971).

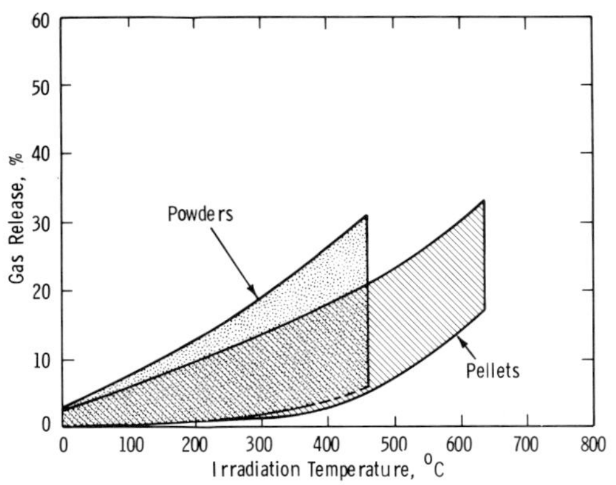

Fig. 4. Gas release from boron carbide during thermal reactor exposure

There is no definite relationship between gas release and ^{10}B burn-up. This is not surprising, inasmuch as self-shielding causes essentially total burnup within an ever-shrinking annulus in the solid. Hence, specimen geometry and neutron energy spectrum are quite possibly more important than burnup level for the control of gas release.

The swelling rate of boron carbide plates exhibits no temperature dependence below 300 °C and has been reported as 0.15% ΔV per 10^{20} captures/cm^3 to a depletion level of about 9 x 10^{21} captures/cm^3 (GRAY and LYNAM, 1963). Powders have been observed to swell about 0.5% ΔV per 10^{20} captures/cm^3 (HOYT and ZIMMERMAN, 1962). Both of these values should be treated with caution, however, because of the cracking and fragmentation that were referred to earlier. Both of these factors strongly influence measured swelling rates.

In Figures 5 and 6, the gas release and swelling behavior of a number of boron compounds as well as elemental boron are summarized (HOYT and ZIMMERMAN, 1962; CUNNINGHAM et al., 1962). All of the data shown in these graphs were obtained from tests at irradiation temperatures below 550 °C and average burnup levels below 2 x 10^{21} captures/cm^3. The diborides represented are HfB_2, TiB_2, VB_2, and ZrB_2. Tetraborides include YB_4 and DyB_4, and the hexaborides are DyB_6, EuB_6, SmB_6, and YB_6. It can be seen that, as a class, the diborides appear to release less gas and well more than the other compounds. At the other extreme, the hexaborides release more gas yet swell less. The tetraborides exhibit gas release and swelling behavior that is intermediate between the other classes. Boron, on the other hand, releases more gas and swells more than any of the compounds. Once again, caution in interpreting the gas release and swelling data is advised. These specimens were also subject to gross cracking and fragmentation and the experimental observations may or may not be indicative of their behavior during subsequent exposure under similar irradiation conditions.

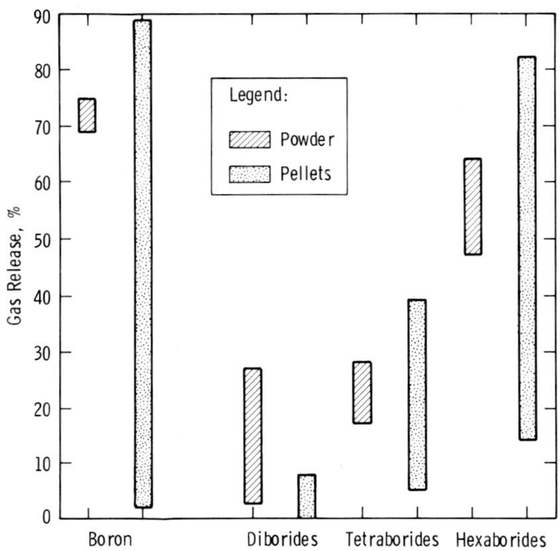

Fig. 5. Gas release from boron and selected borides during thermal reactor exposure

Boron nitride has also been investigated, primarily because it offers processing advantages over the other borides. It has been irradiated as both powder and hot-pressed shapes at temperatures from 20 to 1000 °C

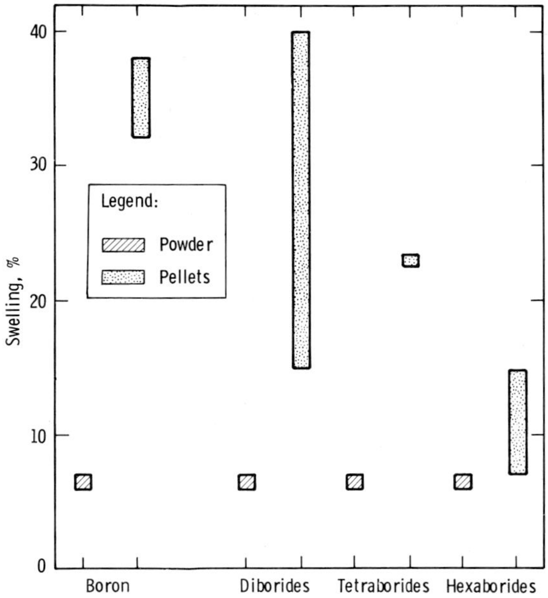

Fig. 6. Swelling of boron and selected borides during thermal reactor exposure

(HOYT and ZIMMERMAN, 1962; MORGAN et al., 1958) and burnup levels ranging from 4 to 99% ^{10}B depletion. Gas release data from this material are summarized in Table 2. In general, the helium release fractions are exceptionally high for nominal burnup. In addition, significant quantities of nitrogen have been detected in the boron nitride offgas. Accommodation of these large quantities of gas imposes severe restraints on potential control element designs and, as a consequence, boron nitride is generally conceded to be an undesirable absorber material for thermal reactor applications.

Table 2. Irradiation behavior of boron nitride

Specimen type	T_{irr} (^oC)	Burnup (% ^{10}B)	He release fraction (%)	Density change (%)
Hot-pressed	800	--	--	0.77
Hot-pressed	870	6.0	--	None
Hot-pressed	20	4.8	--	-1.3
Hot-pressed	20	4.8	--	-2.3
Hot-pressed	970	4.0	--	0.13
Hot-pressed	20	3.8	45.7	None
Hot-pressed	20	3.8	62.6	0.5
Hot-pressed	250	99.0	84.0	14.0
Powder	--	--	65.7	--
Powder	--	--	89.0	--
Powder	--	--	73.0	--

β) Mixtures Containing Borides

One method for alleviating the major problems associated with the use of boron and its compounds is the combination of relatively small amounts of these materials in an inert matrix material. Although this decreases the macroscopic neutron absorption cross section, it offers the advantage that the compact's irradiation behavior parallels that of the dominant phase and thus has the potential for superior irradiation performance compared to that of the boride. Low-worth control rods and burnable poison rods are commonly designed in this fashion.

In an absorber of this type, selection of the inert matrix is generally based on its compatibility with structural materials in the reactor core, its microscopic neutron absorption cross section, and its fabricability. For the most part, metal-base components are judged on their dimensional stability and mechanical properties after irradiation. Ceramic-base materials are evaluated by comparison of their gas release and swelling behavior with those of the pure boron compounds.

Boron has been alloyed with iron and ferritic steels (GRANGE et al., 1957), stainless steels (BERGER and KEENE, 1957), zirconium and its alloys (BEARD et al., 1957; MCKITTRICK and NEISZ, 1957), aluminum (THURBER et al., 1957), and several other metals (ANDERSON and KIESSLING, 1970; STEINITZ et al., 1952; BLUMENTHAL, 1952; BUDDERY and WELCH, 1951). Since boron solubility is limited in most cases, these alloys invariably contain a boride precipitate that may or may not be distributed homogeneously throughout the dominant phase. In general, the alloying process results in strength increases and ductility loss in the matrix metal. Irradiation damage in these alloys is characterized by considerable deterioration of the metal around the boride particles and swelling of the compact. Irradiation-induced dimensional instability of the alloys does not, in general, restrict potential designs, but the embrittlement associated with alloying and with the subsequent irradiation limits the capability of these alloys to serve as load-bearing members of the control element. Thus, they are not utilized to any great extent.

Dispersions of boron compounds in metals have also been studied extensively. The most common matrix materials are aluminum (TIPTON, 1960), zirconium and its alloys (BARNEY et al., 1958), and iron-base alloys (LEITTEN et al., 1959); but titanium (PRUS et al., 1959) and copper (LOSCO and SNYDER, 1958) have also been investigated. The use of aluminum-base material (boral) is restricted to low temperature applications because of the low melting point of the metal. Its irradiation stability appears to be quite satisfactory, however. Both boron and boron carbide have been irradiated in zirconium-base systems. The boron carbide compacts demonstrate better overall irradiation stability than boron dispersions, although both exhibit excellent gas retention and dimensional stability (BRAY, 1957). Both boron and boron carbide have been successfully incorporated in austenitic stainless steels, but examination of the compacts containing the carbide showed that excessive chemical reactions occurred during the process. Postirradiation examination of the boron dispersions indicated that they had excellent dimensional stability at a fluence of $10^{21} n/cm^2$, but that their mechanical properties (notably ductility) had deteriorated significantly (EICHENBERG, 1957). Improved performance has been achieved by introducing a boron concentration gradient into the steel to homogenize the irradiation damage (BEAVER and RIEHT, 1972).

Ceramic-base dispersions or solutions of boron compounds have also been studied. The principal matrix materials that have been employed are Al_2O_3, SiO_2, SiC, and graphite. Interest in the former three centers on their use as burnable poison elements in water-cooled reactors. For these applications, they must be clad in a suitable metallic sheath and their principal advantage over pure compounds is the fact that they can sustain longer in-core residence times without structural and dimensional degradation. Graphite-base dispersions are, for the most part, specified for gas-cooled reactor concepts (GULF GENERAL ATOMIC, INC., 1969). Since the coolant in these reactors is helium, they do not require cladding during operation, and their gas release characteristics are of minor significance. They should, however, exhibit swelling behavior that is compatible with the structural graphite in the core.

From the data summary in Table 3 (JACOBY et al., 1962; GRAY and LYNAM, 1963), it can readily be seen that the majority of the alumina-, silica-, and silicon carbide-base materials exhibit excellent gas retention capability and dimensional stability. The only notable exceptions are the borosilicate glasses and these can also be employed if the poison rod design is carefully established.

Table 3. Irradiation experience with ceramic-base boron dispersions

Material	Density (% TD)	Irradiation temperature (°C)	^{10}B Burnup (%)	He release (%)	Density change (%)
Al_2O_3 - 3.5v/o B_4C	65	230	100	1.7	--
"	84	230	99	0.8	--
"	78	170	93	1.9	--
"	777	170	93	2.0	--
"	82	170	93	1.4	--
"	85	170	93	1.6	--
Al_2O_3 - 6.9v/o B_4C	78	160	96	0.8	--
SiO_2 - 10.3v/o B_4C	--	365	94	--	-0.2
"	--	305	90	1.35	--
"	--	385	90	--	-0.7
SiO_2 - 27.3v/o B_2O_3	--	420	95	77	-0.7
"	--	305	90	78	--
"	--	380	90	71	-0.2
SiC - 30v/o B_4C[a]	99	250	41	0.5	--
SiC - 50v/o B_4C[a]	100	38	97	0.3	-30.4

[a]Average behavior of several samples

Boronated graphite mixtures containing between 5 and 43 wt% boron as boron carbide have been fabricated and tested (DAVIDSON and GATES, 1965; STANSFIELD and STANDEFUR, 1969; STANSFIELD, 1971, 1972). As mentioned earlier, gas release from these materials has virtually no effect on their utility. Hence, it has not been characterized. Irradiation-induced dimensional changes are important, however, and their response to several material and environmental variables has been studied extensively. The behavior of boronated graphite closely parallels that of non-boronated graphite, i.e. dimensional changes are anisotropic and can be related to the crystallographic basal plane orientation in a fabricated shape. At temperatures below 500°C and fast neutron ($E > 0.1$ MeV) fluences less than 10^{21}n/cm , extruded samples containing 5-7 wt% boron contract in a direction parallel to that of extrusion and expand in transverse directions (DAVIDSON and GATES, 1965). At higher temperatures and fluence levels $\sim 700°C$ and 6×10^{21} n/cm^2 ($E > 0.18$ MeV), respectively , expansions differing in degree have been observed in both directions (STANSFIELD, 1972). Dimensional changes are most readily correlated with fast neutron fluence and fabrication techniques, and no strong influence of boron content, boron carbide particle size, or ^{10}B burnup is apparent.

Very few of the boron-bearing materials that have been investigated for thermal reactor service have proven adequate for large-scale commercial utilization. For high-worth applications in water-cooled reactors, boron carbide is employed either as powder compacts or as low density ($\sim$75% TD) pellet columns clad in metallic sheaths. In gas-cooled reactors, boronated graphite containing $\sim$40 wt% boron as boron carbide meets the high-worth control rod requirements. Burnable poison elements in water reactors use clad pellet columns of Al_2O_3 - B_4C, boron dispersions in stainless steel, or borosilicate glass containing 10 to 15 wt% B_2O_3. Burnable poison elements in gas-cooled reactors are also boronated graphite ($\sim$5 wt% boron).

b) Fast Reactor Experience

In fast reactor spectra, the mean neutron energy is greater than 100 keV and the number of elements capable of serving as neutron absorbers in this range is very limited. Boron possesses clear-cut nuclear advantages over other substances for control applications and is being used in the control systems of virtually every fast reactor in the world.

Based on development efforts and favorable experience in thermal reactors, boron carbide has been the compound specified for fast reactor applications. Once again, performance criteria are based on gas release and swelling characteristics during exposure. Since fast reactor operating temperatures and neutron flux levels and energy distributions are significantly higher than those in thermal reactors, a considerable amount of development effort has been required to qualify boron carbide for fast reactor service.

PITNER (1971) has irradiated natural boron carbide powder compacts (80% TD) in the Experimental Breeder Reactor II (EBR-II) at temperatures ranging from 640 to 730°C. His gas release results (summarized in Table 4) indicated a strong temperature dependence and little, if any, effect of exposure time for burnup levels in the range 2.5 to 7.7×10^{20} captures/cm^3. These release fractions are reasonably consistent with previous experience at lower temperatures in thermal reactors. Similar to experience in thermal reactor spectra (MEFFORD and WILLIAMSON, 1968), a decrease in the average particle size in the compacts was observed and this break-up was verified by ceramographic examination. No direct measurement of compact density changes could be made, but cladding diameter increases of $\sim$0.5% were measured on the high exposure specimens and these were attributed to radial swelling of the boron carbide.

Experience with pellets irradiated in both the EBR-II (BASMAJIAN et al., 1972; and COPELAND and DONELLY, 1972) and Dounreay Fast Reactor (DFR) (MURGATROYD and BELL, 1973) has also been reported. A summary of the gas release data from these experiments is shown as a function of temperature in Figure 7. As is the case with thermal reactor irradiation, irradiation temperature appears to be the dominant

Table 4. Gas release from fast reactor-irradiated boron carbide powder

Sample	Irradiation temperature (^{o}C)	Burnup (10^{20} captures/cm^3)	Helium release fraction (%)
1	680	2.6	30.0
2	720	2.5	34.0
3	640	7.7	17.0
4	730	5.9	48.2

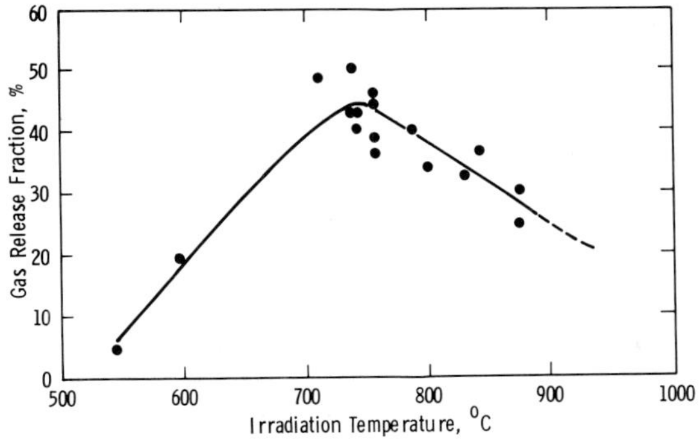

Fig. 7. Gas release from boron carbide during fast reactor exposure

variable governing gas release, rather than boron burnup[6]. Two aspects of Figure 7 are noteworthy. The first of these is the consistency in behavior (minimal data scatter) among the individual experiments in comparison with irradiation experience in thermal neutron spectra. This is attributed, in large extent, to the more uniform reaction profiles that prevail in fast neutron spectra. The absence of steep reaction rate gradients and their concomitant stress distributions in the pellets minimize cracking and general compact degradation and, hence, enhance behavioral consistency. The second significant feature is the occurrence of a gas release fraction maximum at an irradiation temperature of $\sim 750^{o}C$. Behavior of this sort is indicative of two competing mechanisms that control gas release, one of which must be trapping. This postulate is supported by transmission electron microscopy (CUMMINGS et al., 1972; and JOSTENS and DUBOSE, 1972) of the irradiated pellets, which demonstrates that a significant portion of the helium formed during irradiation is contained in flat, crack-like bubbles that lie on preferred lattice planes and strain the solid matrix severely (see Fig. 8).

Contrary to experience in thermal reactors, boron carbide swelling in fast reactors exhibits a definite temperature dependence. In the range $500\text{-}870^{o}C$, dimensional instability is greatest at low exposure temperatures and decreases as the irradiation temperature is raised. In this case, a boron burnup dependence is also noted with the swelling increasing monotonically up to exposures of $\sim 18 \times 10^{20}$ captures/cm^3.

[6]It has, in fact, been shown during an instrumented irradiation test of pellet samples that gas release rates actually decrease as boron burnup increases, eventually reaching a constant value that is dependent on irradiation temperature.

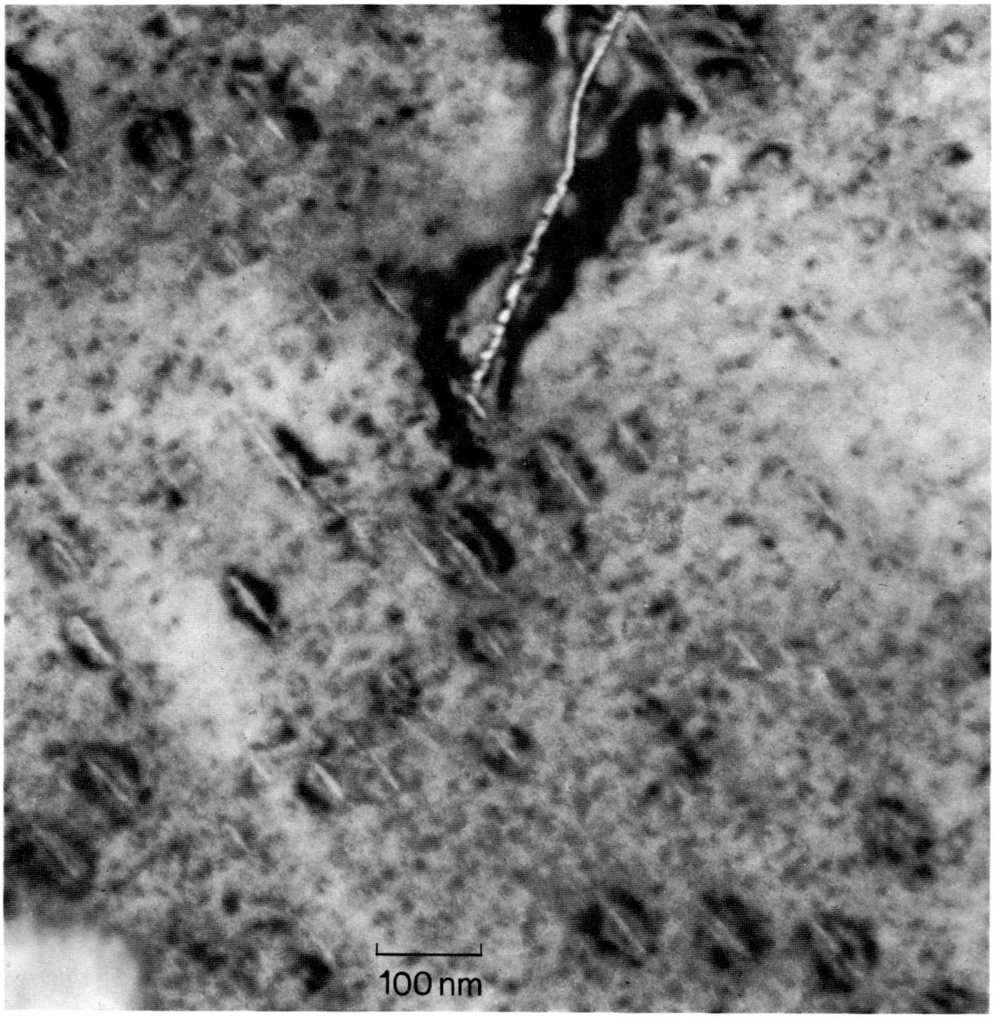

Fig. 8. Platelike helium bubbles formed in boron carbide during fast reactor exposure

The fast reactor swelling dependence (in terms of $\Delta D/D$) on both temperature and burnup is shown in Figure 9 as a surface that bounds all the available data. A least squares empirical equation describing this behavior is

$$\Delta D/D\ (\%) = C\left[-0.0211 + \frac{310.2}{9T + 160}\right] \quad (6)$$

where $\Delta D/D$ is given in percent, T is irradiation temperature (^{0}C), and C is boron burn-up (10^{20} captures/cm^3).

Also, in contrast to thermal reactor experience, gas release and swelling rates of boron carbide in fast reactor spectra are relatively insensitive to material variables such as pellet density, boron enrichment, and grain size. After a burnup level of $\sim 10 \times 10^{20}$ captures/cm^3

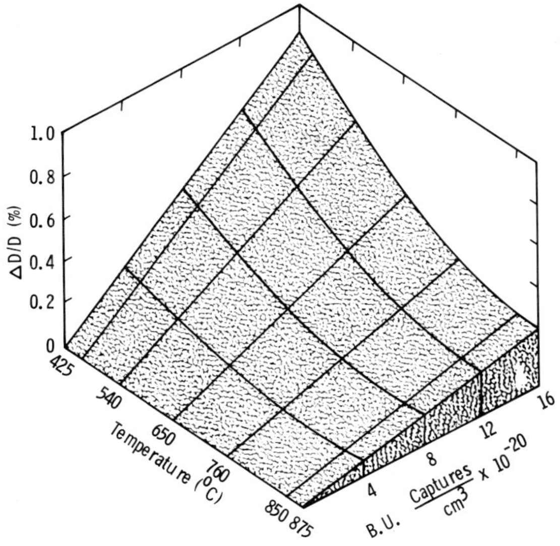

Fig. 9. Boron carbide swelling during fast reactor exposure

has been attained, it is virtually impossible to discern any influence of these parameters on irradiation behavior. Stoichiometry (or more correctly, B/C ratio) appears to affect the gas release rate in fast reactors. At burnup levels below 10 x 10^{20} captures/cm^3, samples with B/C ratios greater than 4.0 release less gas than B_4C. At higher burnup levels, B_4C releases significantly less helium than do specimens containing excess boron.

The adequacy of boron carbide as a fast reactor control material is demonstrated by its selection as a neutron absorber in all the fast reactors in the United States, Great Britain, France, and Russia. At the present time, it appears that most control rod designs in these plants have not been optimized, but present development efforts should soon permit this. With the advent of optimal control rod designs, borides (especially boron carbide) should serve as the control medium in several generations of fast breeder reactors.

4. Conclusions

We have, to this point, discussed the utility of boron in nuclear applications and reviewed its irradiation performance. It is now pertinent to summarize our findings.

It should first be noted that the outstanding nuclear properties of boron (absorption cross section, capacity for ^{10}B enrichment, etc.) make the borides leading contenders for use in most nuclear control systems. Unfortunately, their irradiation behavior is not totally satisfactory for some systems; and alternate materials have, on occasion, been used in their place.

The principal drawback to the use of the borides seems to be their inconsistent behavior in thermal neutron spectra. This difficulty can be overcome, however, by judicious material selection (low density compacts, cermets, ceramic mixtures, etc.) and the majority of large thermal reactors utilize boron – principally boron carbide – in at least one portion of their control systems.

In fast reactor spectra, boron is by far the best neutron absorber available. Fortunately, in this type of reactor, a much higher degree of behavioral consistency is observed and this alleviates many of the problems associated with its use. The combination of outstanding nuclear properties and predictable irradiation performance has made boron - once again as the carbide - the principal absorber material in the world's fast reactors.

Acknowledgments. The authors express their appreciation to several members of the Hanford Engineering Development Laboratory staff who assisted in the preparation of this report. Special thanks are extended to Dr. E.P. Lippincott for his analyses of nuclear cross sections, and to A.L. Pitner and J.A. Basmajian for their previously unpublished data pertaining to fast reactor irradiation experience with boron carbide, and to Dr. W.F. Sheely for a very comprehensive review of the text.

References

ANDERSON, L.H., KIESSLING, R.: Binary systems of boron with chromium, nickel, and thorium, including a discussion of the phase "TiB" in the titanium boron system. Acta. Chem. Scand. 4, 160 (1950)

BARNEY, W.K., SEHMEL, G.A., SEYMOUR, W.E.: The use of boron carbid for reactor control. Nucl. Sci. Eng. 4, 439 (1958)

BASMAJIAN, J.A., PITNER, A.L., MAHAGIN, D.E., RIPFEL, H.C.F., BAKER, D.E.: Irradiation effects in boron carbide pellets irradiated in fast neutron spectra. Nucl. Tech. 16, 238 (1972)

BEARD, A.P., HARRISON, J.W., CLARK, W.B.: Preparation of nuclear poison and control alloys: zirconium base-boron alloys. Schenectady, New York: Knolls Atomic Power Lab. KAPL-155 (1957)

BEAVER, R.J., RIEHT, A.E.: Boron concentration gradient for improved thermal reactor performance of boron-stainless steel control rods. Nucl. Tech. 16, 187 (1972)

BERGER, J.A., KEENE, W.L.: A new stainless steel for atomic energy. Steel 141, 104 (1957)

BLUMENTHAL, H.: Vanadium monoboride. J. Amer. Chem. Soc. 74, 2942 (1952)

BRAY, P.J.: Radiation damage studies in solids; nuclear resonance absorption technique. Providence, Rhode Island: Brown Univ. NYO-7624 (1957)

BUDDERY, J.H., WELCH, A.J.E.: Borides and silicides of the platinum metals. Nature 167, 362 (1951)

COPELAND, G.L., DONNELLY, R.G.: Evaluation of fast reactor-irradiated boron carbide pellets. Trans. ANS 15, No. 2 (1972)

COPELAND, G.L., DONNELLY, R.G., MARTIN, W.R.: Irradiation performance of boron carbide powders containing free boron. Trans. ANS 14, No. 2 (1971)

CUMMINGS, W.V., LAIDLER, J.J., MAHAGIN, D.E., MASTEL, B.: The microstructure of fast reactor-irradiated boron carbide. Trans. ANS 15, No. 2 (1972)

CUNNINGHAM, G.W., et al.: Boron and its compounds. In: Neutron Absorbers for Reactor Control (W.K. Anderson, J. S. Theilacker,eds.) USAEC, Washington, D.C., p.173 (1962)

DAVIDSON, J.M., GATES, L.O.: Boronated graphite irradiation studies final report of irradiation experiments conducted in a Hanford reactor. Richland, Washington: Battelle-Northwest, BNWL-96 (1965)

DRAKE, M.K., ed.: Data formats and procedures for the ENDF neutron cross section libraries. Upton, New York: Brookhaven Nat. Lab. BNL 50274, October 1970

EICHENBERG, J.D.: The effect of irradiation on the properties of boron stainless steel powder dispersion samples. Pittsburgh, Pennsylvania: Bettis Atomic Power Lab. WAPD-160 (1957)

ETHERINGTON, H., ed.: Nuclear Engineering Handbook, sec. 6, 7, and 10. New York, New York: McGraw-Hill 1958

GLASSTONE, S.: Principles of Nuclear Reactor Engineering, pp. 318-338. Princeton, New Jersey: D. Van Nostrand Co. Inc. 1955

GRANGE, R.A., SHORTSLEEVE, F.J., HILTY, D.C., BINDER, W.O., MCTOCK, G.T., OFFENHAUER, C.M.: Boron Calcium, Colubium and Zirconium in Iron and Steel. New York: John Wiley and Sons 1957

GRAY, R.G., LYNAM, L.R.: Irradiation behavior of bulk B_4C and B_4C + SiC burnable poison plates. Pittsburgh, Pennsylavania: Bettis Atomic Power Lab. WAPD-261 (1963)

GULF GENERAL ATOMIC, INC.: 40MW(E) prototype high-temperature gas-cooled reactor postconstruction research and development program: quarterly progress report for the period ending October 31, 1969. San Diego, California: Gulf Gen. Atomic, Inc., GA-9797 (1969)

HAMMAN, D.J., SCHALL, P.: Radiation effects on boron-containing compounds. Columbus, Ohio: Battelle Memorial Inst. BMI-1406 (1960)

HOYT, E.W., ZIMMERMAN, D.L.: Radiation effects in borides Part I - - helium release and swelling in irradiated borides. San Jose, California: Gen. Elec. GEAP-3743-Pt.I (1962)

JACOBY, W.R., MERTENS, P.R., THEILACKER, J.S.: Boron compounds dispersed in ceramic systems. In: Neutron Absorber Materials for Reactor Control (W.K. ANDERSON, J.S. THEILACKER, eds.) USAEC, Wash. D.C., p. 214 (1962)

JOSTENS, A., DUBOSE, C.K.H.: Microstructure of boron carbide after fast neutron irradiation. J. Nucl. Mat. 44, 91 (1972)

LAMARSH, J.R.: Introduction to Nuclear Reactor Theory, pp. 17-24, 109-132, 499-523. Reading, Massachusetts: Addison-Wesley Publ. Inc. 1966

LEITTEN, C.F., Jr., BEAVER, R.J., RIEHT, A.E.: Stainless steel clad dispersion of boron in iron for pressurized water reactors. J. Nucl. Mat. 1, 136 (1959)

LOSCO, E.F., SNYDER, H.J.: Development of boron-bearing materials for pressurized water reactor (PWR) control rods. Pittsburgh, Pennsylvania: Bettis Atomic Power Lab. WAPD-PWR-PMM-1806 (1958)

MAPLES, C., GOTH, G.W., CERNY, J.: Nuclear reaction Q-values (ERNEST O. LAWRENCE) Radiation Lab. Univ. California, UCRL 16964 (1966)

MCKITTRICK, G.F., NEISZ, W.A.: Development of zircalloy-clad discrete burnable poison elements for 53G/SAG. Schenectady, New York: Knolls Atomic Power Lab. KAPL-1726 (1957)

MEFFORD, C.R., WILLIAMSON, H.E.: The performance of boron-containing control rods in water-cooled power reactors. Nucl. App. 4, 425 (1968)

MORGAN, J.G., REAGAN, P.E., MORGAN, M.T.: Irradiation effects on boron-containing ceramics and cermets. Oak Ridge, Tennessee: Oak Ridge National Lab. ORNL-2522 (1958)

MURGATROYD, R.A., BELL, I.P.: Development of boron carbide absorber material for the U.K. fast reactor programme. Paper presented at the IAEA IWGFR Specialists Meeting, Dimitrovgrad, USSR, June 1973

MURRAY, R.L.: Nuclear Reactor Physics, Chaps I, II and VIII. Englewood Cliffs, New Jersey: Prentice-Hall, Inc. 1957

PITNER, A.L.: Fast reactor irradiation of boron carbide powder. Trans ANS 14 (1971)

PITNER, A.L., RUSSCHER, G.E.: Irradiation of boron carbide pellets and powders in Hanford thermal reactors. Richland, Washington: WADCO Corp. WHAN-FR-24 (1970)

PRUS, L.B., BYRON, E.S., VON PLINSKY, F.O. POREMBKA, S.W.: Enriched boron-titanium dispersions. Nucl. Sci and Eng. 6, 167 (1959)

STANSFIELD, O.M.: Irradiation-induced dimensional change in boronated graphite. Nucl. Tech. 16, 197 (1972)

STANSFIELD, O.M.: Neutron irradiation effects in boronated graphite, hafnated graphite, B_4C and HfC summary report on the BG-1 and BG-2 experiments. San Diego, California: Gulf Gen. Atomic, Inc. GA-10648 (1971)

STANSFIELD, O.M., STANDEFUR, N.L.: Irradiation-induced dimensional change in boronated graphite: results of the BG-1 (GA-19-1) experiment. San Diego, California: Gulf Gen. Atomic, Inc. GA-9276 (1969)

STEINITZ, R., BINDER, I., MOSKOWITZ, D.: System molybdenum-boron and some properties of the molybdenum borides. Trans AIME 194, 983 (1952)

THOMPSON, T.J., BECKERLEY, J.G., eds.: The Technology of Nuclear Reactor Safety, Vol. I, Chap. II. Cambridge, Massachusetts: M.I.T. Press 1964

THURBER, W.C., MILKO, J.A., BEAVER, R.J.: Boron-aluminum and boron-uranium-aluminum alloys for reactor application. Oak Ridge, Tennessee: Oak Ridge National Lab. ORNL-2149 (1957)
TIPTON, C.R., Jr., ed.: Reactor Handbook, 2nd Ed., Vol. 1, materials. p. 1057, New York: Weley Interscience 1960
TUCKER, C.W., Jr., SENIO, P.: X-ray scattering by neutron-irradiated single crystals of boron carbide I. Acta Cryst. 8, 371 (1955)
VALOVAGE, W.D.: Effect of irradiation on hot-pressed boron carbide. Schenectady, New York: Knolls Atomic Power Lab. KAPL-1403 (1955)

VIII. Use of Boron Compounds in Lightweight Armor

M. L. Wilkins

1. Introduction

Armor design to date has been achieved primarily with ballistic tests on a large variety of materials. A ballistic test consists of experimentally determining the maximum projectile velocity a given target can withstand without penetration. Correlating ballistic behavior and mechanical properties of materials has led to an empirical understanding of how armor functions. However, correlations have not been entirely satisfactory in determining which material properties are important. Ballistic tests have not yielded sufficient armor design information because there is more than one mechanism operating at the same time in the defeat of penetration. The separate effects cannot be readily sorted out by the normal ballistic test approach.

During a ballistic impact on a target, the load on the target varies from compression, to tension, to shear. No single material has a maximum strength for these combinations of loads. For example, a high compressive strength property of a material is usually accompanied by a low tensile strength. It would seem that producing a two-component armor combining the best of two different strength properties would provide an answer. But optimization of two-component armor has been difficult because of the interaction between the target and the projectile, and the interaction between the components of the armor.

Armor to protect aircraft, or to be worn by personnel, places a critical weight requirement on the choice of materials. The search for lightweight armor materials has led to the use of ceramics. One of the most effective lightweight armor materials has led to the use of ceramics. One of the most effective lightweight armor designs is a boron carbide faceplate backed by a lightweight material, such as aluminum or fiberglass. This composite armor is capable of stopping small arms fire at an areal density less than half that required by ballistic steel for the same ballistic threat.

Prerequisites to the development of better lightweight armors are the understanding of the physical processes occurring during penetration, and the identification of important material properties. Especially important is the need to identify conditions for failure of the materials.

With this information, material design criteria can be established. It may be possible that a better armor will result if some of the weak properties of armor materials are improved, even at the expense of a strong feature. Therefore, one objective of armor research should be to search for possible trade-offs in material properties.

The application of materials to light armor is unusual, because material properties are used in the region of material failure; that is, if the armor does not fail for a given ballistic threat, then it can be made lighter.

In the penetration of ceramic-faced armor, there are several competing material failure processes that lead to ultimate failure of the armor. The investigation presented here uses computer calculations to follow penetration processes step-by-step. The calculations are not used to replace experiments, but rather to check the understanding of experiments and to determine the important physical processes. A particular phenomenon can then be studied separately with calculations to determine the sensitivity of a process to physical parameters.

2. Theory

The theoretical analysis is provided by the elastic-plastic computer program, HEMP, which solves the equations of continuum mechanics by finite difference methods (WILKINS, 1969). Constitutive models were developed to describe the material behavior. The von Mises condition was used to describe the failure of brittle materials by fracture. Ultrasonic velocity measurements permitted the elastic constants to be calculated. Shock wave experiments were used to determine the yield, or failure strength in compression. Figure 1 shows results of shock wave experiments for boron carbide. Table 1 gives measured material parameters for several ceramics. These measurements provide data for the constitutive models (WILKINS, 1969) for use in computer simulations of the ballistic impact experiments. The objective of the calculations is to follow step-by-step the penetration process and identify the important material parameters.

3. Experiments

Penetration experiments were conducted by firing projectiles at targets. The experiments determined a ballistic limit velocity, V_{BL}, defined as the maximum projectile velocity that the target can sustain without penetration. The projectiles were made from a strong steel, with the general shape of a .30-caliber bullet (Fig. 2). The experimental method used flash radiographs to look into the target, and a streak camera to follow the projectile rear surface continuously during the penetration. This information, coupled with two-dimensional machine calculations that show the stress fields present, affords an effective way of studying impact phenomena.

Figure 3 shows the impact of a sharp, steel cylinder on a ceramic-aluminum target. The radiographs look through the edge of the target into the center of the impact region. The time sequence of events is as follows:

1. 0 to 9 μsec. The point of the projectile is destroyed. The projectile has slowed down by 20% of the original velocity. There is no target penetration; however, other experiments and calculations show that a fracture conoid has occurred in the ceramic.

2. 15.8 μsec. Partial penetration of the ceramic. First appearance of a fracture conoid, although the fracture occurred earlier.

3. 15.8 to 35.5 μsec. The aluminum back-up plate has yielded, the ceramic fracture conoid is pushed down, and the ceramic has been spread laterally. (The X-rays cannot penetrate the aluminum and the deflection of the ceramic-aluminum interface is obscured by the edges of the aluminum plate. Subsequent experiments use a 5-mil x 1/16-in gold foil, laid along the interface to provide contrast.

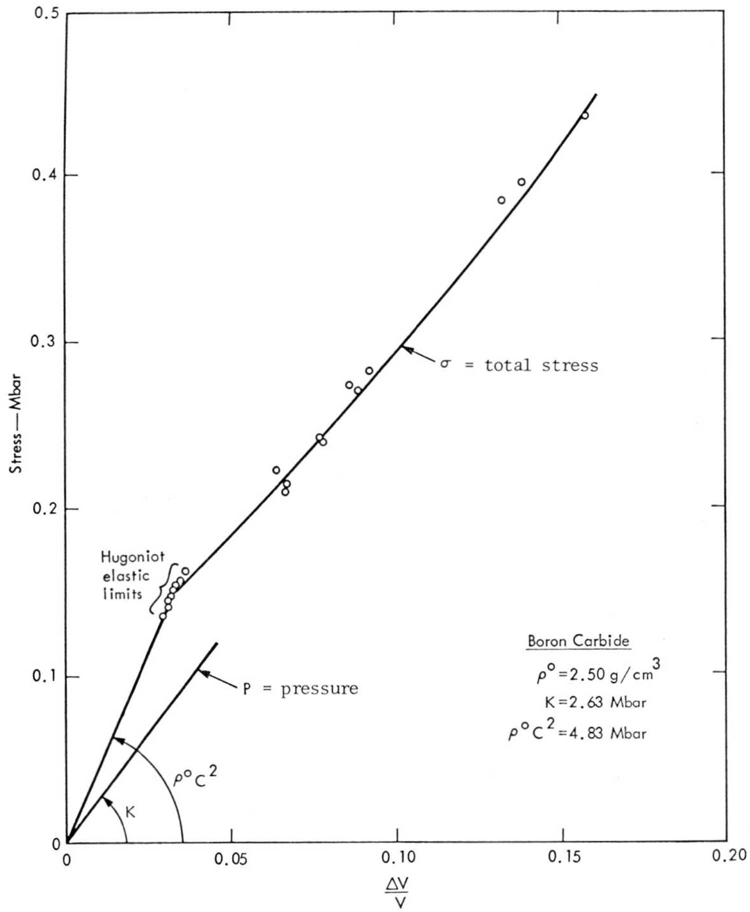

Fig. 1. Experimental data for boron carbide

4. Projectile Stopped. This experiment was conducted just below the ballistic limit, and the projectile was stopped by the target. If the projectile velocity is increased by 100 ft/sec, the portion of the original projectile shown in the last frame penetrates the target.

Figure 4 gives experimental results using a streak camera keyed to sequential flash radiographs, so that a continuous time-position record of the various interfaces can be made. A computer simulation of the experiment gave very close agreement with the results of Figure 4. It was thus possible to use the calculation to provide information on the details of the events occurring during the experiment. Figure 5 shows the time history of the projectile and target energies from the computer simulation. The following sequence of events corresponds to Figures 4 and 5:

1. From 0 to 9 μsec, the projectile tip is being destroyed. During this period, the back-up plate yields at the ceramic interface. Resulting tension in the ceramic as it follows the motion of the back-up plate initiates an axial crack.

Table 1. Elastic constants

Material	Density ρ^o (g/cm^3)	Bulk modulus K (Mbar)	Shear modulus μ (Mbar)	Sound speed C_ℓ (cm/μsec)	Hugoniot elastic limit (Mbar)	Yield strength Y^o (Mbar)	Elastic impedance $\rho^o C_\ell$
Sapphire (∿58° to "C" axis)	3.98	2.36	1.79	1.09	--	--	4.35
Hot-pressed alumina	3.92	2.50	2.49	1.07	0.140	0.093	4.19
Boron carbide	2.50	1.96	1.85	1.33	0.150	0.126	3.33
Beryllium oxide	2.84	1.85	1.43	1.15	0.085	0.064	3.27
Be_2B	2.03	1.26	1.23	1.20	0.067	0.056	2.40
Al B_{12}	2.58	2.00	1.82	1.31	0.096[a]	0.079[a]	3.38
B	2.36	2.09	2.01	1.42	--	--	3.35
Ti Be_{12}	2.28	1.36	1.27	1.16	0.054	0.045	2.64
Ti C (Ni bonded)	5.63	2.39	1.77	0.91	--	--	5.12
Ti B_2	4.52	2.60	2.41	1.13	--	--	5.11

[a]TAYLOR (1969)

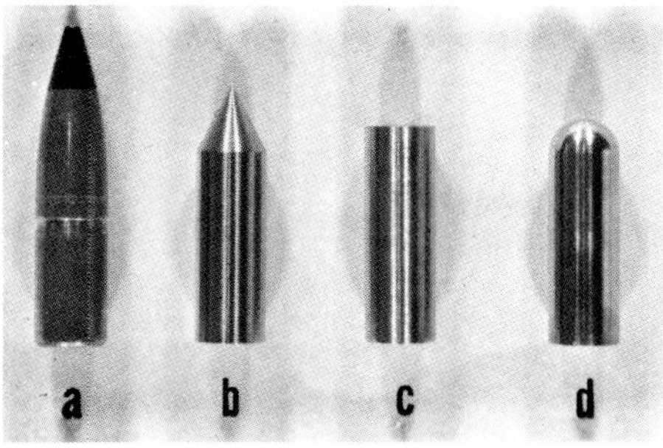

Fig. 2a-d. Projectiles: (a) 30-cal. AP Bullet. (b-d) LLL Bullet Simulator, steel hardness, Rc 54-56; length, 0.9 in.; diam., 0.3 in.; weight, 8.32 g

2. From 9 to 15 μsec, the projectile is eroded by the ceramic which has become a rubble from coalescing cracks within the fracture conoid. Approximately 40% of the projectile mass and 40% of the initial projectile energy is carried off by eroded projectile material. Erosion takes place because the stress level on the projectile is greater than the material strength of the projectile. Projectile material yields and flows perpendicular to the impact, and is then swept back as the projectile moves into the ceramic. The maximum decelerating force on the projectile is thus related to projectile yield strength. The erosion process continues until the prevailing stress level falls below the projectile's yield strength.

3. After 15 μsec, the erosion of the projectile stops. Figure 6 shows the calculation and a flash radiograph at time, 16 μsec. The initially sharp projectile is now a short, blunt cylinder. Both the calculation and the radiograph show that during the projectile erosion, material flows out in the radial direction of the projectile and is swept back by the ceramic. This material is subsequently stripped off the projectile by the ceramic. For later times, the projectile energy loss equals the target energy gained. Total energy in the target-projectile system is 60% of the original energy and is absorbed by the back-up plate.

The important energy-loss mechanism for a projectile that strikes a ceramic target is due to the loss of projectile mass during step (2) above. The rate of projectile energy loss shown in Figure 5 is such that an increase of 250 ft/sec in the ballistic limit would occur if the time duration of step (2) could be extended by 2 μsec. Because the duration of the projectile erosion phase is related to the time of break-up of the ceramic, it is important to maintain ceramic integrity for as long as possible. The calculations indicate that a fracture process starts along the axis of symmetry on the ceramic face opposite the impact, An improvement in target performance should result if a ceramic could be synthesized with improved tensile properties.

4. Important Parameters in Penetration Process

Experiments and calculations verify that the ballistic limit is very sensitive to the ceramic's tensile properties. The high-stress load transmitted through the ceramic to the back-up plate results in the back-up plate's yielding directly under the impact. The attendant motion of the ceramic as it follows the back-up plate is slight; but due to the high elastic modulus of the ceramic, a large tensile stress is set up in the ceramic. The axial crack that results is the beginning of the ceramic breakup under the projectile. A ceramic material property that would allow the ceramic to follow the motion

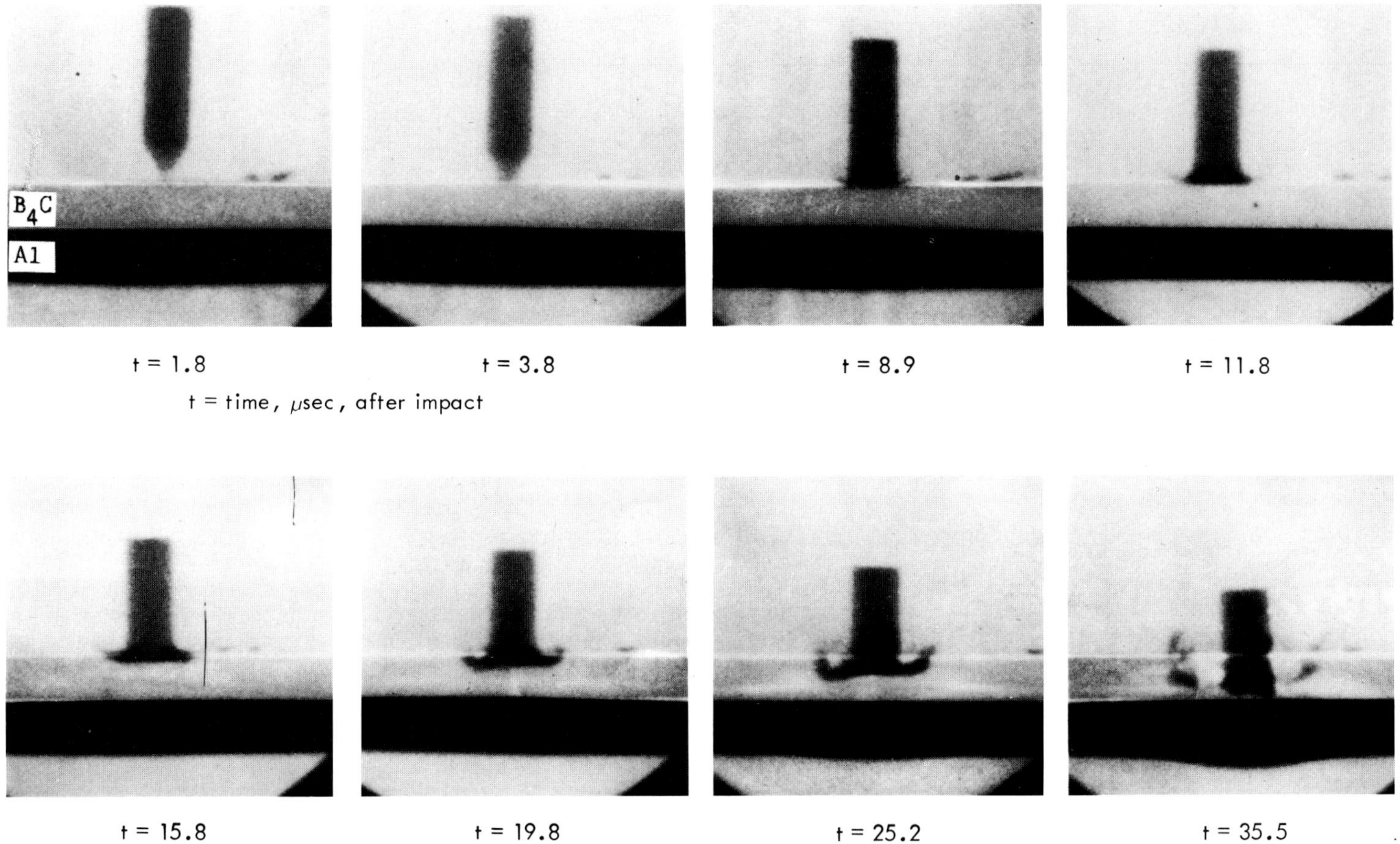

Fig. 3. Flash radiographs of a sharp steel projectile striking 0.285 in. boron carbide bonded to 0.25 in. 6061-T6 aluminum. Projectile velocity: 2300 ft/sec. (t = time, μsec, after impact)

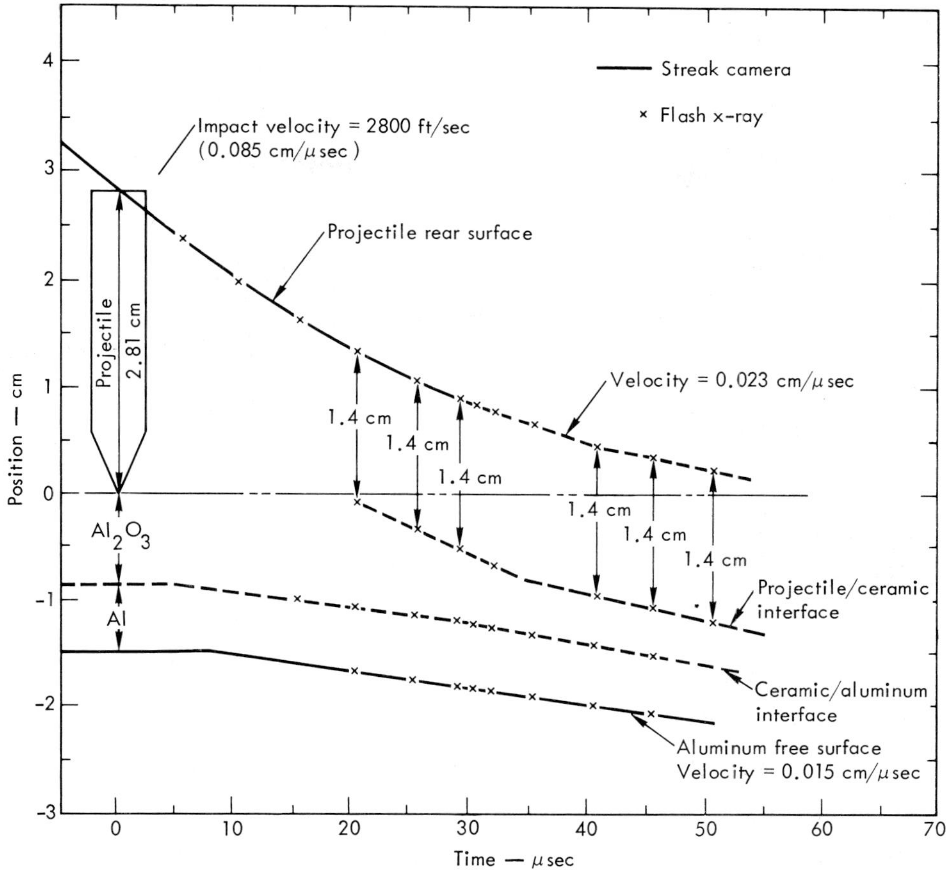

Fig. 4. Impact below ballistic limit for target of 0.34 in. Coors AD85 alumina backed by 0.25-in. 6061-T6 aluminum

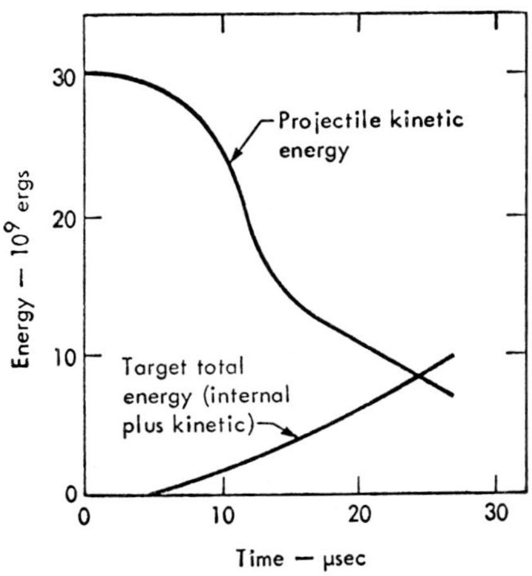

Fig. 5. Energy-versus-time for an impact at 2800 ft/sec of a sharp steel projectile with a target of 0.34-in. AD85 alumina backed by 0.25-in. aluminum

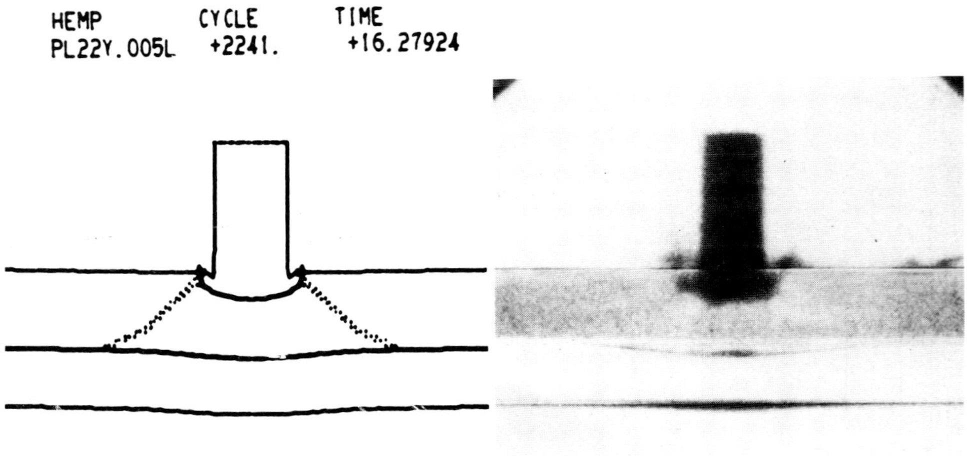

Fig. 6. Comparison of a calculation and an experiment at 16 sec after impact of a sharp steel projectile with 0.34-in. AD85 alumina bonded to 0.25-in. aluminum. Impact velocity = 2800 ft/sec. (The flash radiograph of the cylinder appears to have a smaller diameter than the actual diameter because of X-ray penetration of the edges)

of the back-up plate further before failure is ductility. Two computer calculations were made (Fig. 7) to translate an improvement in ceramic tensile properties into a time delay before fracture. These calculations show fracture patterns for different times. The calculated fractured portions of the ceramic (alumina) are indicated by the darkened section. A fracture conoid can be seen emanating from the impact surface and an axial crack starting from the surface opposite impact. In one calculation, the ceramic is considered to fracture when the tension reaches 3 kb; and in the second calculation, when the tension reaches 8 kb. The 3-kb tension corresponds to about 0.1% strain, and the 8-kb tension corresponds to about 0.3% strain. Therefore the calculations compare the fracture pattern as a function of time when one ceramic is capable of a greater elongation before fracture. It should be emphasized that it does not matter if the material properties are such, that a high tensile stress can be reached before fracture occurs, or if the material deforms plastically at a lower stress magnitude, but fractures at the same value of strain. The results will be the same because the important parameter is the magnitude of the strain when fracture occurs. Thus, the expression, "improved tensile properties," is used in the sense of an increase in strain before fracture from a tensile load.

Figure 7 shows that the accumulated damage to the ceramic is about the same at times, t = 4.15 μsec and t = 5.87 μsec, for the calculations with fracture at 3-kb tension and fracture at 8-kb tension, respectively. Thus, for these calculations, a 0.2% increase in strain before fracture results in a 1.7 μsec increase in the breakup time of the ceramics. When the ductile property of the ceramic is increased so that the time of breakup has been increased by 2 μsec, the calculated increase in the ballistic limit is 250 ft/sec.

There is not just one, but several important characteristics which a target needs to defeat penetration. A ballistic impact on a ceramic creates large compressive stresses, and subsequent tensile stresses, which can lead to fracture. Adding ductility in a ceramic to improve the tensile properties, obviously, would bring about an improvement in ballistic performance if other parameters remained unchanged. But

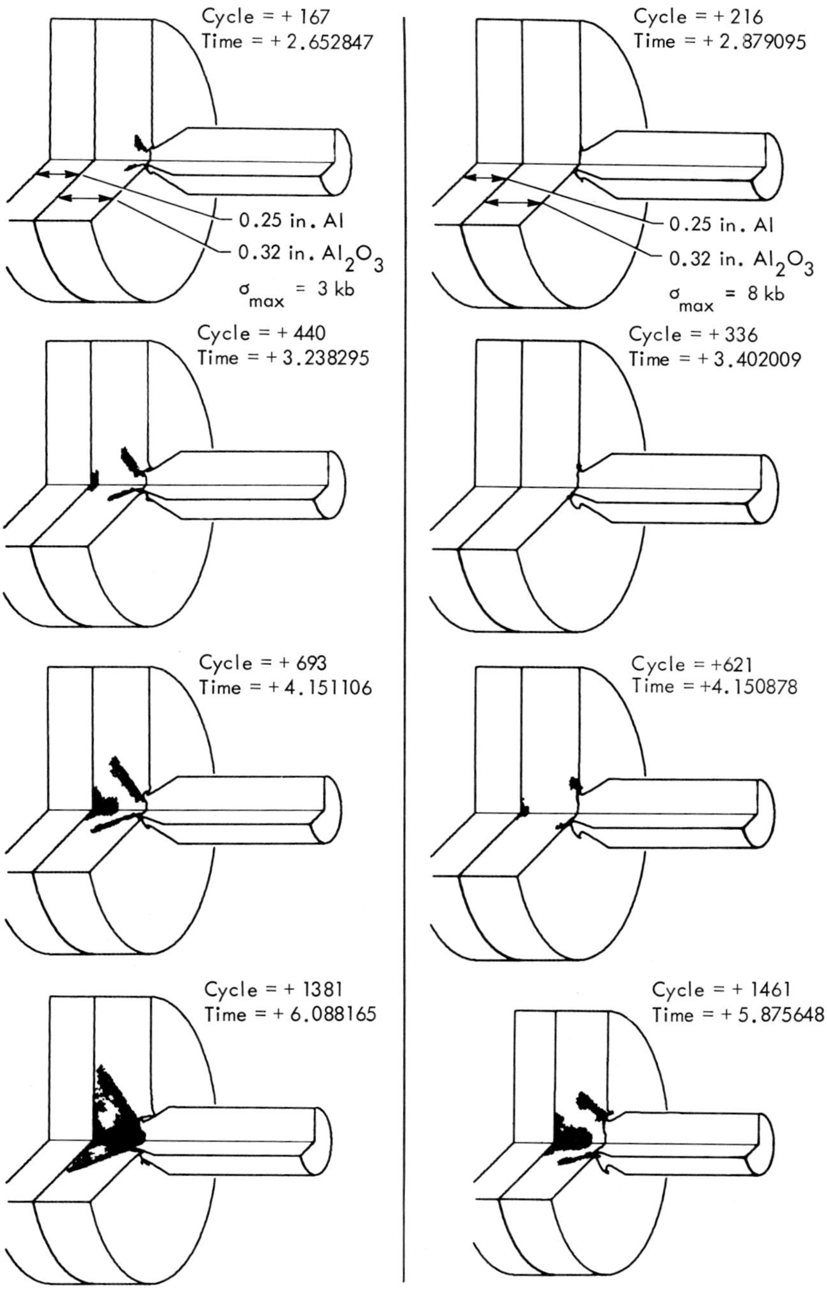

Fig. 7. Effect of fracture strength on time for fracture (left column: fracture strength = 3-kb tension; right column: fracture strength = 8-kb tension); time in μsec

most ceramics have essentially no ductility. However, they are effective in armor because they present a high resistance to deformation from compressive loads. We can identify several parameters associated with this characteristic with the aid of Figure 8. This figure shows Hugoniot curves for several ceramics and the reflected Hugoniot for steel. The intersection of the reflected Hugoniot curve for steel with a ceramic curve gives the stress state for a one-dimensional strain impact. The penetration process, of course, involves more than one space dimension. However, some of the important aspects of material properties can be identified with reference to a one-dimensional strain impact. The slope of the curves, up to the Hugoniot elastic limit, corresponds to the elastic impedance, $\rho^{o}C_{\ell}$. For a given projectile impact velocity, a higher elastic impedance implies a higher impact stress. The sharp point of a projectile is vulnerable to a high stress impact.

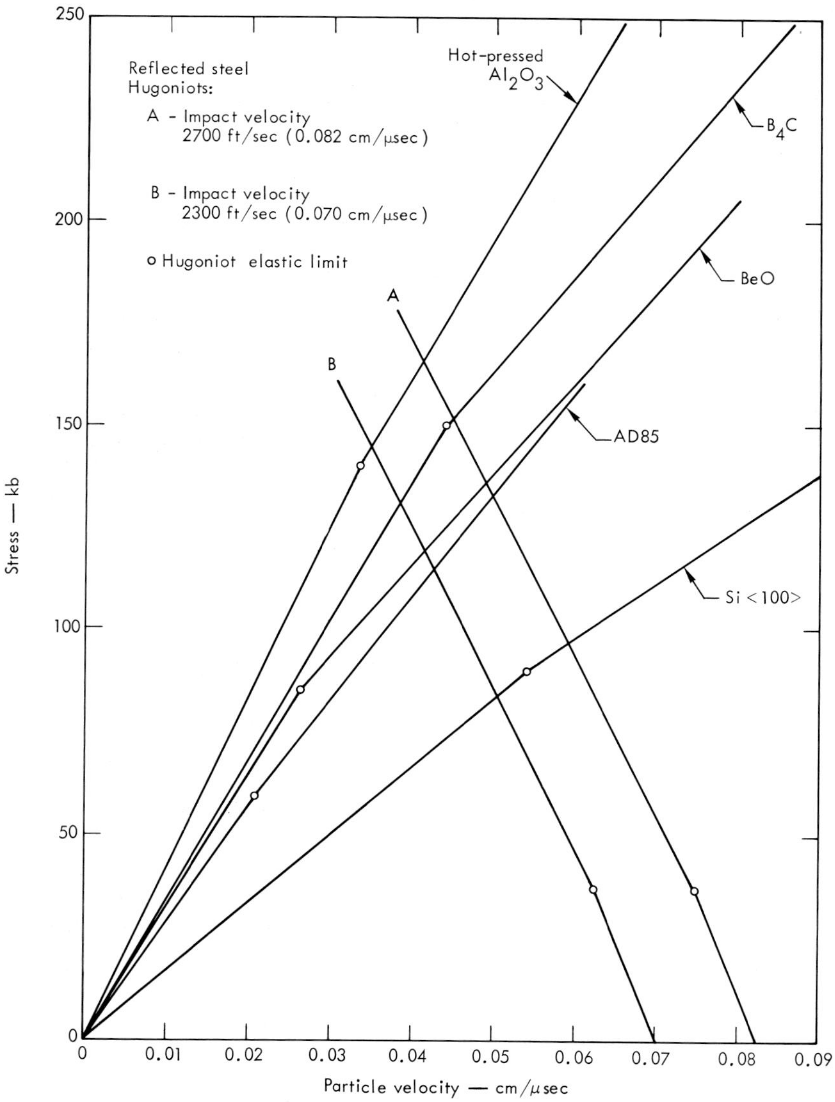

Fig. 8. One-dimensional strain impacts of steel on ceramics

Another parameter implicit in these curves is the compressive yield strength, Y^o. The elastic impedance is a measure of the resistance to distortion of the material, while the yield strength is a measure of the amount of distortion that may occur before an elastic failure. It is not the elastic impedance at zero stress level that is important, but the impedance or resistance to distortion at the impact stress level. The compressive yield strength, Y^o, is the material parameter that determines the stress level where the impedance of the material changes value. A high compressive strength, Y^o, when taken alone does not necessarily imply a good candidate for armor material. This point can be appreciated by referring to the curve for Si in Figure 8. The yield strength for Si is $Y^o = 0.09$ Mbar. This is higher than the yield strength for AD 85, which is $Y^o = 0.038$ Mbar. However, the ballistic limit for a target with Si is much lower than for a target with AD 85 The low ballistic limit for Si can be understood by noting that the impact stress level is much lower for Si than it is for AD 85 (Fig. 8). Actually, high impedance and high yield strength are components of a lumped parameter which is incompressibility at the impact stress.

From this discussion, it is obvious that a high elastic impedance and a high yield strength are important parameters when they exist together in the material. For these characteristics to be maintained as long as possible during the impact process, the tensile properties of the ceramic become important; i.e. ultimately a tensile load is placed on the material and a tensile failure becomes the weakest link.

Thus, the optimum ceramic should have a balance between elastic impedance, compressive yield strength, and the ability to withstand tensile stresses bevore fracture. These parameters are importent in addition to low density. For a given target density, a low ceramic density permits a thicker ceramic facing, which in turn reduces the stress level on the back-up plate. The stress at the back-up plate is a critical factor, since it governs the motion and, hence, tensile stress set up in the ceramic at the ceramic back-up plate interface. Thus, low ceramic density for lightweight armor is a non-trivial requirement.

5. Ductility in a Ceramic

It is natural to examine ballistic limit experiments with various ceramics with the idea that perhaps ductility had contributed to the ballistic limit. Of the experiments available, all the ballistic limits, except BeO, correlated with the material properties of incompressibility at the impact stress and low density. When tested at equal ceramic thickness, the ballistic limit for BeO is greater than the value for B_4C and is almost equal to the value for hot-pressed Al_2O_3.

Figure 8 shows that the Hugoniot curve for BeO lies below the curves for B_4C and hot-pressed Al_2O_3. It was suspected that BeO might possess better tensile properties than the other materials; this would compensate for the lower elastic impedance and compressive strength of BeO. Experiments that measure the shear strength of materials as a function of confining pressures showed that B_4C and hot-pressed Al_2O_3 failed in a brittle manner at confining pressures up to 7.5 kb. However, BeO yielded plastically and reached strains as high as 5% when the confining pressure was about 3 kb, and greater. The conclusion is that BeO has the desirable physical property of elongation before fracture, as long as a confining pressure greater than 3 kb is present. This is just the calculated stress range that is operating at the ceramic back-up-plate interface during the first 20 μsec after the impact.

Optical examinations of the BeO taken from high-pressure shear tests indicated that the material had indeed deformed in a plastic manner. Additional evidence of plastic flow was obtained from BeO specimens subjected to ballistic tests and examined by microscopy.

Figure 9 compares B_4C, BeO, and AD85 targets struck by a sharp steel projectile. For the BeO target, the load on the back-up plate has been distributed over a larger area and more projectile erosion has occurred when compared to the other two. A series of experiments was performed using shadowgraph X-rays to measure the residual mass and residual velocity for a sharp projectile penetrating AD85 and BeO. Results showed the most projectile energy is lost for an impact on a BeO target.

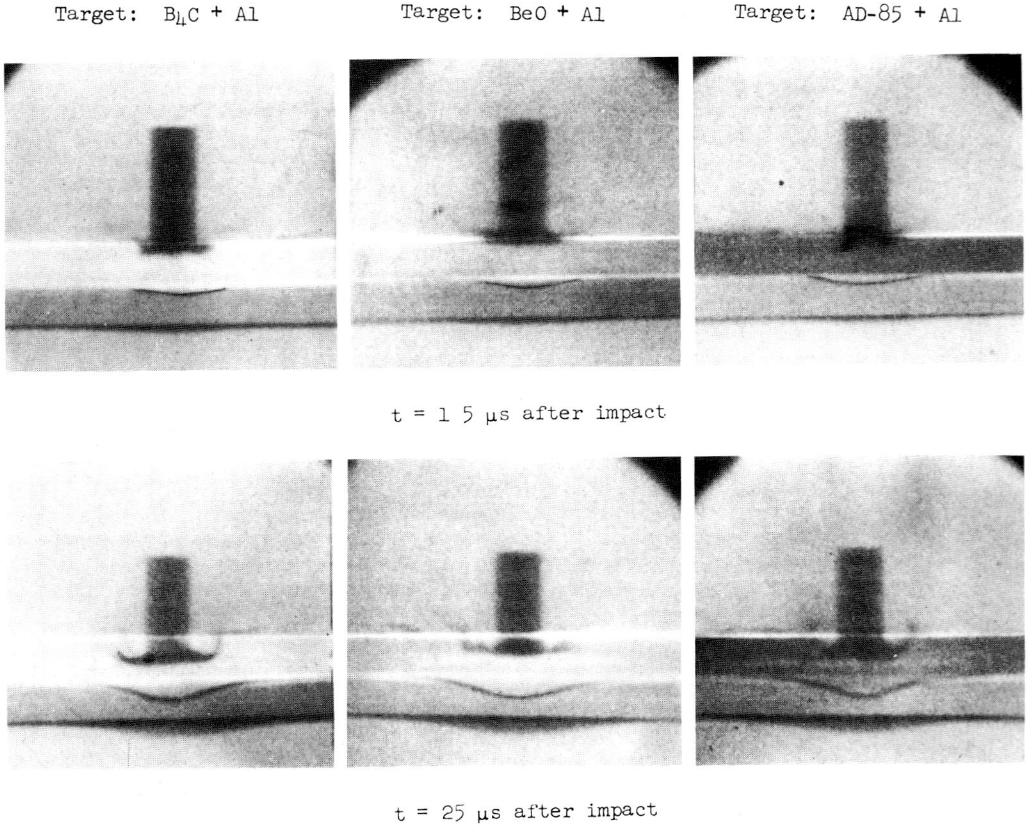

Fig. 9. Flash radiographs of a sharp steel projectile striking targets of 0.25-in. B_4C, BeO and AD-85, each bonded to 0.25" 6061-T6 aluminum. Projectile impact velocity = 2400 ft/sec

These results mean that the capability for plastic flow, which delays the onset of tensile stresses, permits BeO to remain integral for a longer period of time. Thus, BeO is able to present an incompressible barrier to the projectile for a longer time than other ceramics that fail sooner when a tensile load occurs. These results are thought to substantiate the results of the computer calculation that ductility can be a key parameter.

6. Cermets

The indication that improved ductility might lead to a higher ballistic performance for ceramics leads to the class of materials known as "cermets," which are combinations of ceramic and metals. The cracks which eventually compromise the ceramic armor facing originate from the interface between the ceramic and the armor backing. The computer calculations indicated the armor would be improved by introducing some ductility at this interface, thereby inhibiting crack formation.

Any improvement in the ductility of a material would be at the expense of the compression strength which is a critical property for any armor material. However, since the high compressive strength property is required at the impact surface, and the ductility property required at the opposite surface, the possibility exists for satisfying both requirements. The trade-off in material properties, strength for ductility, can be made at the ceramic surface opposite the impact surface. This change in material properties could be accomplished by a cermet where the metal-to-ceramic proportions change through the thickness dimension of the material. Specifically, a cermet is required where the metal-to-ceramic ratio is low on the impact surface, and increases smoothly to some higher ratio on the opposite surface. This material can be termed a "graded cermet."

In the synthesis of a metal-ceramic system there are two fundamental requirements: a metal must be found that (1) wets the ceramic phase, and (2) forms a strong bond with the ceramic. In addition, for the requirement at hand, both components must possess low density. Boron carbide is currently the optimum low-density armor material and would be a natural starting point to fabricate a cermet. Preliminary work showed it would be difficult to find a metal phase that satisfied all of the above requirements.

7. Compounds of Boron and Beryllium

For the ultimate in lightweight material, boron and beryllium compounds are especially interesting for possible use as ceramic armor. The borides and the beryllides combine low density, high hardness, and high sound speeds. Properties of the beryllium-boron compounds are summarized in Table 2; the elastic properties of several beryllium compounds as measured at LLL are given in Table 3. Two interesting materials for possible application to lightweight armor are diberyllium boride (Be_2B) and tetraberyllium boride (Be_4B). These materials have unusually low densities and a hardness that is more than adequate for applications to armor. Hardness has been found to correlate well with compressive strength as determined by Hugoniot elastic limit experiments. The experimental compressive strength of Be_4B, Y^0 = 0.058 Mb, given in Table 3, is high enough to destroy the point of steel armor-piercing projectiles. Previous work showed that a ceramic with a yield strength, Y^0 = 0.04 Mb, is effective in the defeat of steel projectiles. The adequate compressive strength coupled with the high elastic modulii, Table 3, indicate that the material is a possible light-weight armor candidate.

Figure 10 shows experimental results for several lightweight ceramics. It is seen that Be_4B is superior to B_4C. Experiments with Be_2B gave results similar to those of Be_4B.

Table 2. Properties of beryllium boride compounds

Compound	Theoretical density ρ^0 (g/cm^3)	Crystal structure	Melting point (°C)	Hardness (HK_{100})
Be_4B	1.94	Tetragonal	~1160	1370
Be_2B	1.89	Cubic	~1500	~1300
BeB_2	2.42	Hexagonal	>1970	3180
BeB_6	2.35	Tetragonal	2020-2120	2580
BeB_{12}	2.42	Hexagonal	--	--

Table 3. Measured elastic properties of beryllium compounds

Compound	Density ρ^0 (g/cm^3)	Longitudinal sound speed C_ℓ (cm/μsec)	Shear wave velocity C_s (cm/μsec)	Elastic impedance $\rho^0 C$	Bulk Modulus K (Mbar)	Shear modulus μ (Mbar)	Yield strength Y^0 (Mbar)
BeO[a]	2.84	1.15	0.703	3.27	1.85	1.45	0.064
$TiBe_{12}$	2.28	1.14	0.750	2.60	1.23	1.28	0.045
Be_4B	1.94[b]	1.275	0.838	2.45	1.32	1.35	0.058
Be_2B	2.03[b]	1.27	0.800	2.58	1.52	1.30	0.057
BeB_2	2.25[c]	1.28	0.828	2.88	1.63	1.54	--

[a]Brush-thermalox grade BeO
[b]The fabricated density includes contributions from BeO and BeB_2 impurities and from some porosity
[c]Density is lower than the theoretical density (ρ^0 = 2.42) because of a small amount of Be_2B

8. Graded Cermet

The beryllium-boron system appears to offer promise for developing a cermet where beryllium would be the metal phase that must both wet and bond with the ceramic phase. Research on Be_4B demonstrated that the ratio of boron-to-beryllium could be shifted to yield a material with a higher Be content. Thus, it is possible to make a cermet with Be_4B using Be as the metal phase. Microstructure studies of hot-pressed specimens show only a slight increase in grain size as the Be content is increased. An increase in grain size is thought to be detrimental to the fracture properties of a ceramic. The result that the microstructure did not change substantially made it appear possible to make a graded armor material using Be_4B.

Discs used for ballistic test were made with a dispersion gradient of Be metal in Be_4B. The half of the material from the impact surface to the center contained only Be_4B. The Be powder was introduced at the center of the material thickness and increased to 10 vol% at the side opposite the impact. This ensured that the high B content was at the impact surface of the material to increase the material hardness, and the high Be content was at the back-up plate interface to improve the ductility and reduce the density. The specimens were resistance-heated and hot-pressed.

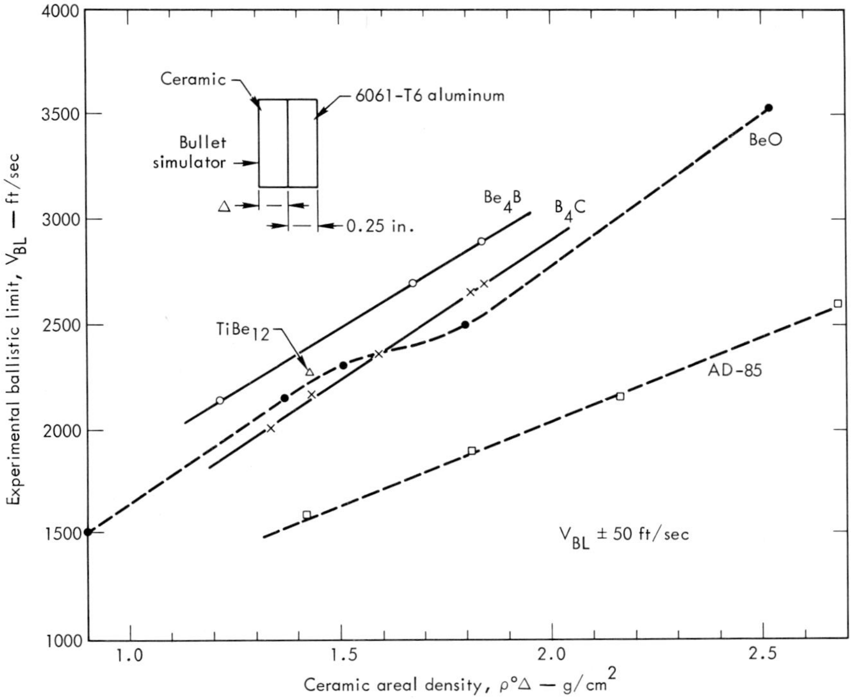

Fig. 10. Experimental ballistic limit versus ceramic areal density for ceramic targets backed by aluminum

Ballistic test of the discs were made with .30-cal. AP M2 rounds. The ballistic test data in Figure 11 compare Be-graded Be_4B discs to Be_4B discs without added beryllium. The data are given as a figure of merit defined as the weight of B_4C armor divided by the weight of a given armor required to defeat the same ballistic threat. The results of these tests indicate a weight-saving of 14 percent can be obtained with Be_4B graded armor compared to Be_4B armor for a .30-cal. AP M2 round at an impact velocity of 2500 ft/sec. Post-test X-ray diffraction measurements near the back-up plate interface showed that, actually, there was little free Be present. This indicates that the addition of Be probably shifted the stoichiometry of the Be_4B to a higher Be content ($Be_{4.5}B$). Because only a small amount of free Be metal was present at the back-up plate interface where it was intended to introduce approximately 10% free metal, it is assumed that the material tested was not an optimum material.

An alternative approach to gradient armor is to introduce Be honeycomb into the back half of the discs. Preliminary experiments indicated that close control of the hot-pressing time and temperature was required to densify the Be_4B and to bond it with the Be honeycomb without causing an excessive reaction between the two materials. An increase of only 30°C can cause a wide reaction zone to occur. The reaction zone is apparent in the microstructure and in the microhardness. The ballistic performance of Be honeycomb discs (Fig. 11) indicates a 15% weight-saving compared to B_4C armor for protection from a .30-cal. AP M2 round, striking at 1900 ft/sec. Although the Be metal was introduced into the ceramic in the form of a honeycomb, it could have been in the form of thin, corrugated or dimpled sheets.

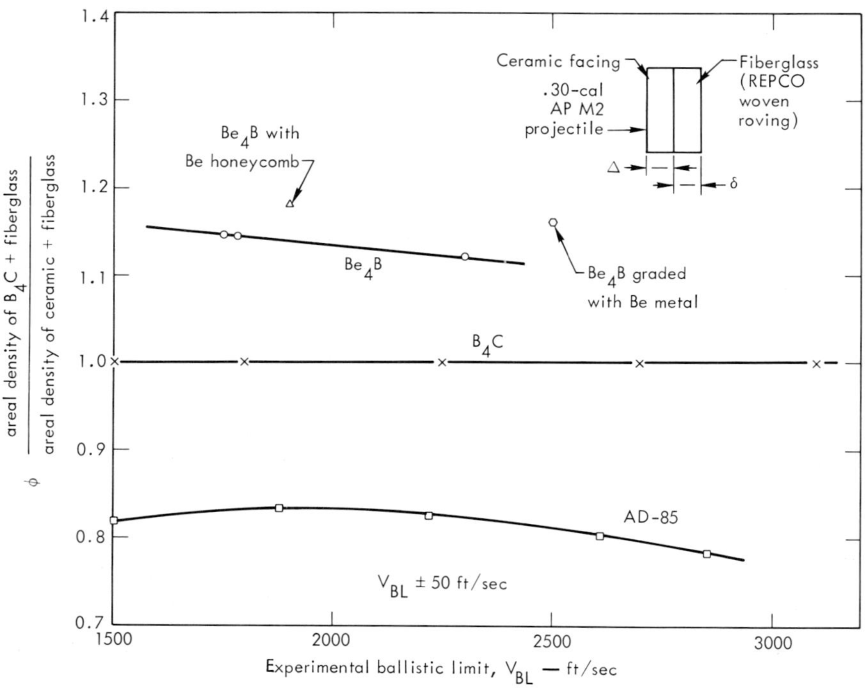

Fig. 11. Experimental ballistic limit of targets with various ceramic impact surfaces compared to targets with B_4C impact surfaces

In conclusion, the results of these tests and investigations demonstrate that ceramic armor can be improved when ductility is introduced into the side opposite the impact. Additional improvement can be expected by optimizing the amount and location of the metal in the ceramic. Compounds of boron and beryllium offer the ultimate material system for developing lightweight armor.

Acknowledgments. This work was part of a project sponsored by the Defence Advanced Research Projects Agency under the auspices of the U.S. Atomic Energy Commission, to study lightweight armor materials. I would like to acknowledge the contributions of the Lawrence Livermore Laboratory scientific personnel. William Gust and Hugh Heard performed the physical properties of materials measurements. Charles Honodel conducted the balistic experiments. E. Dick Giroux programmed the constitutive models for the HEMP code. Carl Cline and Richard Landingham conducted the research program to develop new compounds with boron.

References

WILKINS, M.L.: Calculation of Elastic-Plastic Flow. Lawrence Livermore. Livermore Rep. UCRL-7322, Rev. I (1969)

TAYLOR, J.W.: Los Alamos Sci. Lab. Los Alamos Report LA-4113 MS, 2 (1969)

Subject Index